Particle Strengths

Particle Strengths

Extreme Value Distributions in Fracture

Robert F. Cook

The American Ceramic Society

WILEY

Published by John Wiley & Sons, Inc., Hoboken, New Jersey.

Published simultaneously in Canada.

Library of Congress Cataloging-in-Publication Data:

Names: Cook, Robert F. (Independent scientist), author.
Title: Particle strength : extreme value distributions in fracture / Robert F. Cook.
Description: Hoboken, New Jersey : Wiley ; American Ceramic Society, [2023] | Includes bibliographical references and index.
Identifiers: LCCN 2022057645 (print) | LCCN 2022057646 (ebook) | ISBN 9781119850939 (hardback) | ISBN 9781119850946 (adobe pdf) | ISBN 9781119850953 (epub) | ISBN 9781119850960 (ebook)
Subjects: LCSH: Strength of materials. | Particle dynamics. | Materials science.
Classification: LCC TA405 .C844 2023 (print) | LCC TA405 (ebook) | DDC 620/.43–dc23/eng/20230105
LC record available at https://lccn.loc.gov/2022057645
LC ebook record available at https://lccn.loc.gov/2022057646

Cover Design: Wiley
Cover Image: Courtesy of Robert F. Cook

Set in 9.5/12.5pt STIXTwoText by Integra Software Services Pvt. Ltd, Pondicherry, India

To Michelle,
who always said this should be done,
and was never-endingly encouraging.

Contents

Preface

This book is concerned with strengths and strength distributions of particles. As such, it is a much-altered and expanded version of a review of the subjects published on-line several years earlier. The goal now, as then, is to present a comprehensive, unified, and objective view of particle strength measurements and the analyses required to interpret and apply the results of such measurements. The book draws from the fields of materials science and engineering, mechanics of materials, and fracture mechanics.

Particles are considered here as entities—both natural and man-made, and nearly always solid—that are of comparable extent in all three dimensions. The extents are usually, but not exclusively, small. Examples include grains of sand and crystals of table salt, but also aspirin tablets, fertilizer pellets, glass marbles, and river rocks. Attention is restricted to brittle particles, those that fail mechanically by fracture with negligible associated plastic deformation. The particle strengths of interest here are thus brittle fracture strengths, characterized by the unstable and unchecked propagation of a crack from a flaw in a component under the influence of a tensile stress. An intriguing aspect of particle failure is that cracks propagate under tensile stress—and yet the predominant form of particle loading is compression. Strictly, "diametral compression," in which the particle is squeezed across a diameter. Diametral compression applied to the exterior of a particle leads to development of a zone within the particle of tensile stress perpendicular to the compression direction—a phenomenon quantified by Hertz in the 1880s.

Particles encounter diametral compression in a variety of ways: between displaced platens in a laboratory setting, the most common method of particle strength measurement; by impact against, or by, a hard surface; squeezed between rollers; or by the weight of other overlying particles in a packed particle array. The diversity of particle loading methods is matched by the diversity of particle materials. The majority of particles are inorganic materials—glasses and crystalline ceramics, rocks, and minerals—although there is a significant minority of organic materials—foods, pharmaceuticals, and hydrogels, and those that contain both materials—seashell fragments, coal, and coral. The variety of loading methods and diversity of materials have lead to a plethora of presentation schemes for reporting particle strengths and related failure metrics, significantly impeding scientific and engineering application of particle strength information. Thus, in order to advance applications of particle strength information, a major feature of the book is the presentation in a single, unbiased format of a large number of measured strength distributions of particles. The format enables identification of critical features of particle strength distributions and thence identification of patterns of behavior and underlying causes.

In particular, a major point made throughout the book is that it is the population of strength-controlling flaws that is the fundamental physical property that determines measured strength distributions of particles. The flaws are characterized by size and spatial distributions that are material properties. The measurements are characterized by the number and size of the particles tested. As the test parameters change, the strength distributions change. Hence, a second important feature of the book is the clear and quantitative development and application of analysis that relates fundamental properties of the invariant flaw population to observed characteristics of resultant strength distributions. Two important attributes of the strength distribution presentation method and analysis are used throughout the book. First, presentation of particle strength distributions in unbiased coordinates permits simple identification of predominant size effects. Second, fits to measured particle strength distributions are used simply as intermediate smoothing steps to infer underlying flaw populations. The commonly used linearized presentation scheme and fit to strength data in transformed coordinates are shown to be obscuring and misleading.

The idea and motivation for this book started some years ago during consideration of strength predictions for small microelectromechanical systems (MEMS) components. Progress in the MEMS area required the disentangling of much strength distribution folklore, especially associated with linearized distributions, leading to the development of a coherent (and simple) probability framework for strength and flaw distributions. Application of the developed framework to ceramic materials lead to analysis of refractory particle strengths and the idea that particles might be the end state of size effects: particles might contain only *one* strength-controlling flaw and thereby exhibit no variation in strength distribution from that set by the flaw population. Motivation for this book is then in part a pursuit of answers to the related questions of "What do the strength distributions of particles look like?" and "How did they get that way?" My research career prior to this had deliberately avoided such questions through experimental use of controlled-flaw strength techniques; deliberately introducing the one strength-controlling flaw by indentation. I was, however, extremely familiar with strength testing of many, small, brittle specimens under controlled conditions, and the subsequent analysis to extract underlying material properties.

The book chapters can be viewed in groups that address the above questions, although not in sequence. Chapter 1 describes the major particle types considered here, their shapes and sizes, their applications, and their qualitative failure mechanisms. Chapter 2 describes the mechanics of particle loading and diametral compression and quantitative measures of particle strength. The introductory concepts of Chapters 1 and 2 are used in a detailed outline of the book at the end of Chapter 2.

Chapters 3 and 4 provide the analytical framework for much of the book. Clear physical and mathematical descriptions are given for flaw populations, strength distributions, and how they are related. Chapter 3 describes forward analysis from flaws to strengths and Chapter 4 describes reverse analysis from strengths to flaws. Many of the subsequent chapters rely on analysis developed in Chapter 4. The analyses in these chapters are general and apply to strength distributions of all brittle components, not just particles.

Chapters 5 and 6 provide surveys of the measured strength distributions of "conventional" extended components and particles, respectively. The surveys show clear differences in the strength distribution shapes and relative widths of the two component types, and the different ways size effects are manifested in strength distributions. Both chapters make extensive use of Chapter 4 analyses to deconvolute strength behavior into underlying flaw distributions.

Chapters 7, 8, and 9 include the majority of the particle strength distributions presented, analyzed, and discussed here. As above, all three chapters make extensive use of Chapter 4 analyses to deconvolute strength behavior into underlying flaw distributions. Similarities and differences in particle strength and flaw distributions within and between material groups are emphasized, along with the different ways particle size influences strength behavior.

Chapters 10 and 11 describe and discuss the strength behavior of agglomerate and hydrogel particles, respectively; particles very different from the "hard," inorganic particles considered in the earlier chapters. The majority of agglomerate and hydrogel particles are "soft" and largely organic. Although strength distributions are presented in both chapters, the emphasis in Chapter 10 is on fabrication effects on strength and the emphasis in Chapter 11 is on pre-failure deformation.

Chapters 12 and 13 are the most analytical in the book with regard to fracture, addressing basic fracture mechanics of particles and applications of fracture mechanics in engineering contexts, respectively. Chapter 12 makes clear the basic physics of particle strengths, in which external work generates internal elastic energy changes that are transformed into surface energy changes on fracture, and how that physics is applied in flaw and strength analyses. Chapter 13 extends the fracture ideas to the use and manipulation of strength distributions in engineering applications, relying on fundamental scaling principles.

Overall, the book emphasizes the development and application of descriptive analyses. The analyses enable extraction of the information contained in particle strength and strength distribution measurements (or, at least, most of the information). A self-imposed test of the veracity of a particular analysis, used throughout the book, was whether simulated data could pass for experimental measurements. The reader can judge. The book contains very little in the way of theory, except perhaps the assumptions that flaws act independently (although that is relaxed in Chapter 9) and that particle materials are linear elastic (and that is relaxed in Chapter 11). In addition, the book contains almost no statistics, except the occasional provision of means and standard deviations. Fit parameters are not given; plots are intended to convey all the information. There is, however, probability analysis required to translate strength distribution characteristics into flaw distribution characteristics and vice versa. More refined next-step probability ideas such as Bayesian (conditional) probability and the related hazard concept, clearly applicable to particle reliability, are not addressed.

My thinking on the subjects in this book has developed and benefited over the years from many discussions and interactions with colleagues, including: S.J. Bennison, S.J. Burns, F.W. DelRio, E.R. Fuller Jr, B.R. Lawn, D.B. Marshall, M.L. Oyen, G.M. Pharr IV, and M.V. Swain.

Robert F. Cook
Winterville NC
August 2022

Abbreviations and Symbols

cdf	cumulative distribution function
ccdf	complementary cumulative distribution function
edf	empirical distribution function
pdf	probability density function
SIF	stress-intensity factor
a	contact radius, impression semi-diagonal
b	component dimension, deformation zone dimension
c	crack length
d	component dimension
f	population pdf
h	sample pdf, component dimension
k	number of flaws in component (size of component), component stiffness
l	component dimension
m	particle mass
t	time
u	displacement, contact displacement
v	displacement, particle or mass velocity
w	displacement, particle diametral displacement under load, work
A	crack area
B	proportionality factor relating strength and crack length
$\mathrm{B}(\alpha, \beta)$	beta function
D	diameter of particle (size of particle)
E	Young's modulus, particle failure energy
F	population cdf, force applied to component, particle failure force
$\bar{F}$	population ccdf
$\mathcal{G}$	mechanical energy release rate
H	sample cdf, hardness
$\hat{H}$	estimate of sample cdf
$\bar{H}$	sample ccdf
K	stress-intensity factor
L	length of component
M	elastic modulus
N	number of components in sample (size of sample)
P	force applied to component, porosity
$Pr(\sigma)$	sample strength edf
R	radius of particle, exterior radius of convex tablet, fracture resistance
T	toughness
U	energy

V	particle or component volume
W	work
Y	yield stress
α	beta function exponent
β	beta function exponent
γ	surface energy, elastic-plastic geometry term
δ	contact displacement under load, microstructural traction zone length
ε	strain
η	relative width of strength distribution
λ	component compliance, flaw spatial density, grain size
μ	relative strength, microstructural geometry term
ν	relative crack length, Poisson's ratio
ξ	relative crack length
ρ	material mass density
σ	strength, stress
τ	viscous deformation time constant
χ	residual stress geometry term
ψ	applied stress geometry term
ΔV	fundamental volume element
Ω	population volume
$\sim$	"varies as"
$\approx$	"approximately equal to"

Subscripts, superscripts, indices, dummy variables, and less-frequent uses of these and other symbols are defined in context.

1

Introduction to Particles and Particle Loading

The importance of particles as structural components in a range of engineering applications is described, providing an introduction and motivations for the work to follow. The geometrical aspects of particles in the load-bearing configuration of diametral compression are defined and used in a description of the observed failure modes of loaded particles. Images of particles commonly and not-so-commonly encountered in everyday life are presented and discussed. The geometry of particle loading is followed by an analysis of particle shape. A discrete Fourier analysis method is used to describe a range of particle shapes—multi-lobed, rough, angular, or variable—and the effects of particle shape on estimates of particle size are considered. The similarities of the analyses of particle shape and the following analyses of particle strength are discussed.

1.1 Particle Failure and Human Activity

Particles occur in nature and are produced by human industry. In many cases, particles are placed under load. This section introduces particles, how they are loaded, and how they may fail to support loads in application.

1.1.1 Particles as Structural Components

The force required to break or crush a particle is critical in many areas of human activity. Particles—objects that are limited in extent in three dimensions and often small—are usually not thought of as *structural components*, for which the ability to support a mechanical force is a distinguishing property. The familiar load-bearing structural components, columns in compression, rods or bars in tension, beams in bending, and shafts in torsion are all extended in one dimension to form the component axes. For columns and bars in structural applications, the directions of the forces associated with applied loading are parallel to the component axes; inward for columns, outward for bars. For beams and shafts in structural applications, moments are associated with applied loading but differ in orientation with respect to the component axes; perpendicular for beams, parallel for shafts. The cross-sections perpendicular to the axes of all four of these simple components are usually uniform, greatly simplifying structural analyses (Mase 1970; Timoshenko and Goodier 1970; Sadd 2009). Such structural components are usually large, and their load-bearing function obvious, for example, a pillar or column supporting a building. Figure 1.1 shows schematic diagrams of loaded simple structural components: (a) column, (b) bar, (c) beam, and (d) shaft. In Figure 1.1 applied forces are shown as dark shaded arrows; unless otherwise stated, this scheme is used throughout the book.

Particles *are*, however, usually load-bearing in their many applications and are best thought of as short columns with nonuniform cross-sections. The column-like loading configuration, in which a diameter of a particle is placed in compression, is known as *diametral compression*. (Terms in italics are *definitions* to be used throughout—some are specific to this book.) Figure 1.2 shows schematic diagrams of particles in diametral compression, (a) a spherical particle and (b) a cylindrical particle. Both particles are sitting on flat platens and the sphere and cylinder are loaded symmetrically by a point force and a line force, respectively (the cylinder axis is parallel to the platen). The particle dimensions parallel to the loading axes are the diameter of the sphere and the diameter of the cylindrical face. The maximum cross-sectional dimensions perpendicular to the loading axes are the diameter, in the case of the sphere, and the diameter or the thickness, for the cylinder. (Consistent with the idea that particles are distinguished by limited extent in all dimensions, the thickness of the cylinder

Particle Strengths: Extreme Value Distributions in Fracture, First Edition. Robert F. Cook.

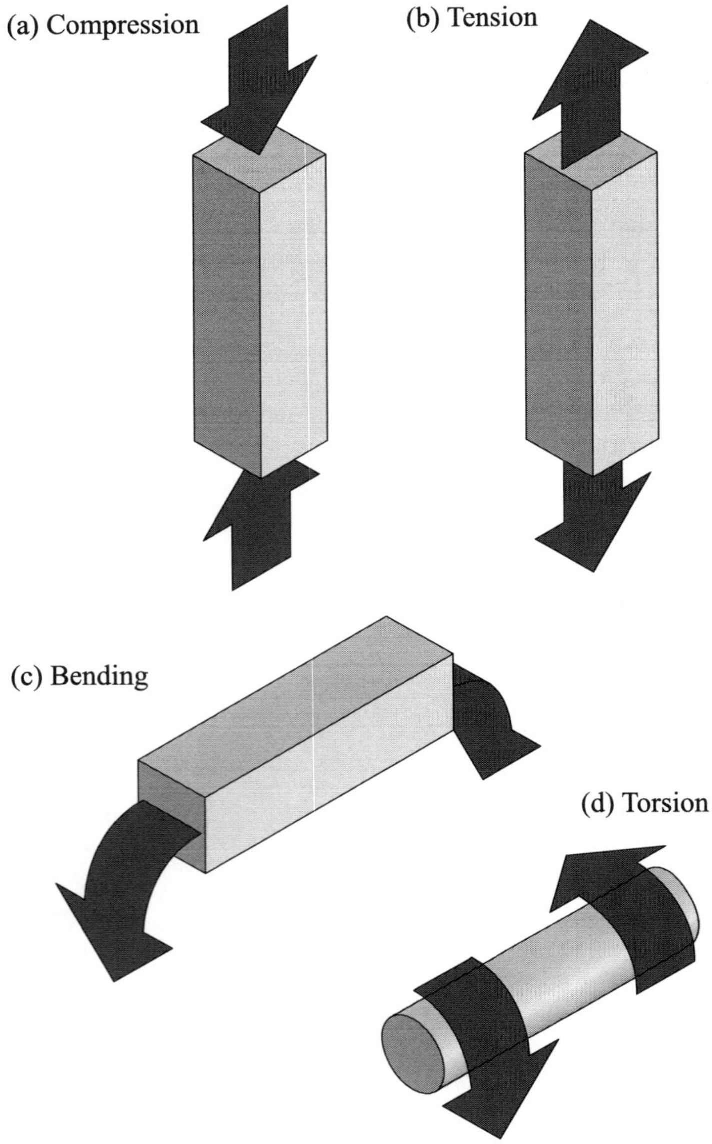

Figure 1.1 Schematic diagrams of loading geometry for the four major forms of structural components: (a) compression of a column; (b) tension of a bar; (c) bending of a beam; (d) torsion of a shaft. Applied forces in (a) and (b) and applied moments in (c) and (d) shown in dark shading.

is comparable to the face diameter). The particle cross-sections perpendicular to the loading axes are nonuniform—the cross sectional areas of the particles are larger at the center and smaller at the extremities where the forces are applied. The load-bearing capacity of particles in diametral compression is a major focus of this book.

Examples of the particles to be considered here are shown in the images of Figure 1.3. The Figure exemplifies the diversity of particles in load-bearing applications, natural, industrial, and domestic, and also provides examples of the fact that some particles are commonly encountered and some are not. The particles vary in material (organic and inorganic), size (tens of micrometers to hundreds of millimeters), and shape (from smooth and spherical to rough and angular), and include: (a) Cr_2O_3 particles dispersed in a cordierite ($2MgO.2Al_2O_3.5SiO_2$) ceramic matrix (the particles are bright, the matrix is gray, and pores are dark). (b) Sand (largely quartz, crystalline silica, SiO_2). (c) Fertilizer granules. (d) Salt (NaCl) crystals. (e) Peppercorns. (f) Aspirin tablets. (g) Ceramic pie weights. (h) Taconite (iron ore, Fe_3O_4 and Fe_2O_3) pellets. (i) Glass (largely amorphous silica and other oxides) spheres. (j) Plaster (gypsum, $CaSO_4$) spheres. (k) Tumbled river rocks (largely quartz). (l) Jagged railway ballast (largely granite, a mixture of quartz and other minerals). Throughout the book, as various aspects of particle mechanical behavior are considered, details regarding specific particle applications, materials, sizes, and shapes will be given in context. However, independent of particle size and shape, an overall description of particle loading geometry in diametral compression can be developed. In the next Section, 1.1.2, this geometrical description is outlined and then applied in an overview of observations of how particles may fail in use as structural components.

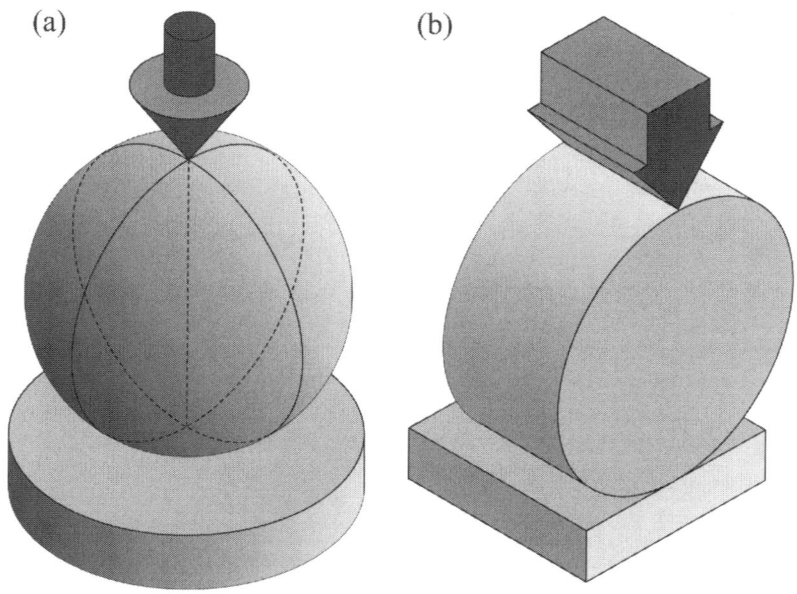

Figure 1.2 Schematic diagrams of the diametral compression loading geometry for two forms of particles: (a) sphere loaded by a point force; (b) cylinder loaded by a line force. Applied forces shown in dark shading.

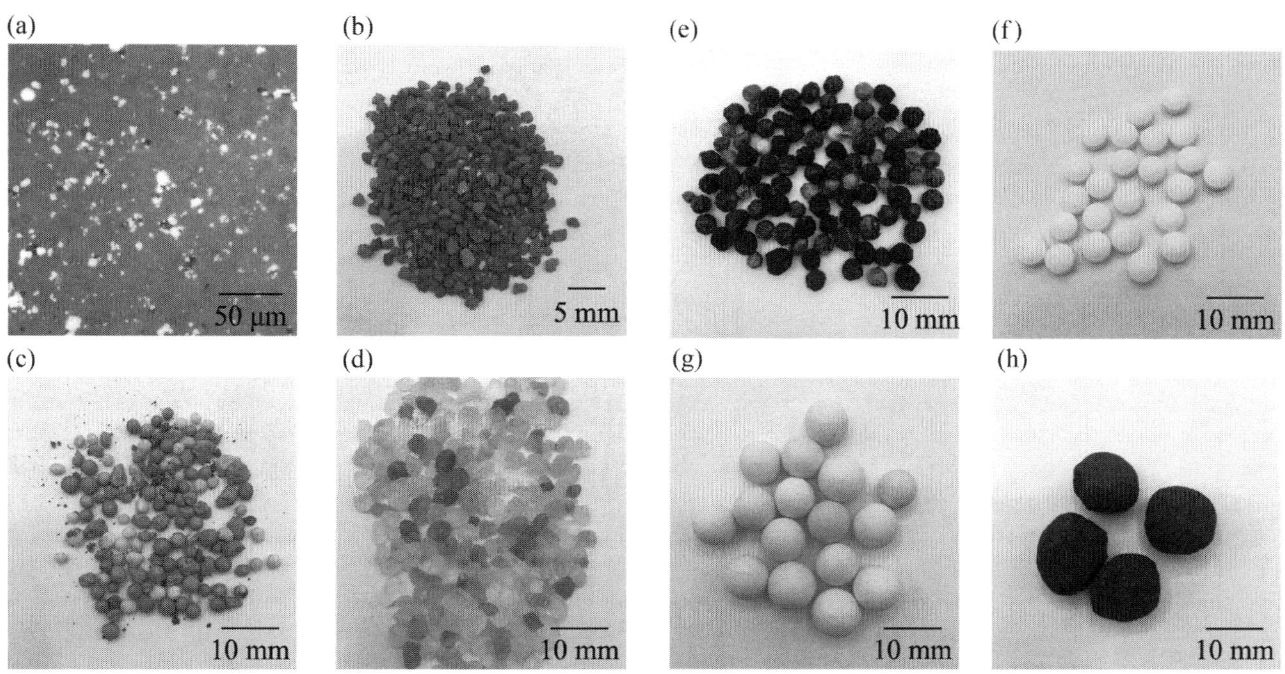

Figure 1.3 Images of groups of particles, demonstrating the wide variety of particles in natural, industrial, and domestic applications: (a) Cr_2O_3 particles in a cordierite ceramic matrix composite (polished section); (b) Sand grains; (c) Fertilizer granules; (d) Salt grains; (e) Peppercorns; (f) Aspirin tablets; (g) Ceramic pie weights; (h) Taconite (iron ore) pellets;

(Continued)

Figure 1.3 (Cont'd) (i) Glass marbles; (j) Plaster (gypsum) moulded spheres; (k) River rocks; (l) Railway track ballast (wooden sleeper and metal rail visible in foreground). All images optical. Note variations in shape and size between and within groups. *Source:* Robert F. Cook.

1.1.2 Particle Loading

Detailed consideration of loading of particles in diametral compression requires a clear description of the loading geometry. Particles are distinct from columns in that locations and orientations within a loaded particle are specified relative to the axis of loading rather than an axis of the particle. A near equidimensional particle may take any orientation relative to that of an applied force whereas the axis of a column is nearly always colinear with the direction of the force (Figures 1.1 and 1.2). Figure 1.4 shows schematic cross sections of the particles loaded in diametral compression from Figure 1.2, illustrating important locations and directions. Both are axial cross sections (containing the loading axis) and both are discs. For such particles, the points of force application (top of particle) or displacement application (bottom of particle) are the poles (Figure 1.4a). The line joining the poles is oriented along the loading axis, in the axial direction (Figure 1.4b). Midway along the line joining the poles is the center of the particle (Figure 1.4a). Directions perpendicular to the axial direction in the axial plane are transverse directions (Figure 1.4b). The loaded spherical particle system of Figure 1.2a exhibits rotational symmetry, such that all angularly separated axial cross sections, and therefore transverse directions within those sections, are equivalent. The loaded cylindrical particle system of Figure 1.2b exhibits translational symmetry, such that all spatially separated cross sections parallel to the cylinder face ("through the thickness") are equivalent. The terms pole, center, axial, and transverse in the cylindrical case refer to points and directions in such face sections.

Figure 1.5 shows schematic cross sections perpendicular to the loading axis through the centers of the particles loaded in diametral compression, illustrating important lines and planes. These are transverse sections, containing transverse directions indicated by arrows. The perimeter of the transverse cross section through the spherical particle is a circle and that for the cylinder is a rectangle. For the spherical particle, this perimeter is the equator; see Figure 1.5a (shown in bold). For the cylindrical particle, lines on the perimeter perpendicular to the axial and transverse directions are equatorial lines (Figure 1.5b) (bold). For both sphere and cylinder, the sections containing the equators are the labeled equatorial planes. Sections perpendicular to transverse directions that contain the axis are meridional planes, shown and labeled in the schematic diagrams of Figure 1.6. The perimeter of a meridional plane in the spherical particle is a circle and that in the cylinder is a rectangle. For the spherical particle, this perimeter is the meridian (Figure 1.6a) (bold), and spherical meridional planes are identical in shape to axial planes. For the cylindrical particle there is only

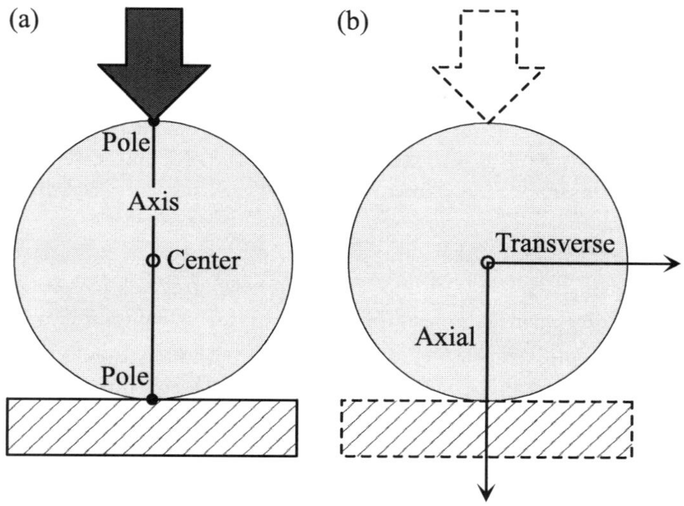

Figure 1.4 Schematic diagrams of the axial sections of (a) spherical and (b) cylindrical particles loaded in diametral compression, illustrating locations and directions in the axial plane. Particle orientations as in Figure 1.2; both sections are discs.

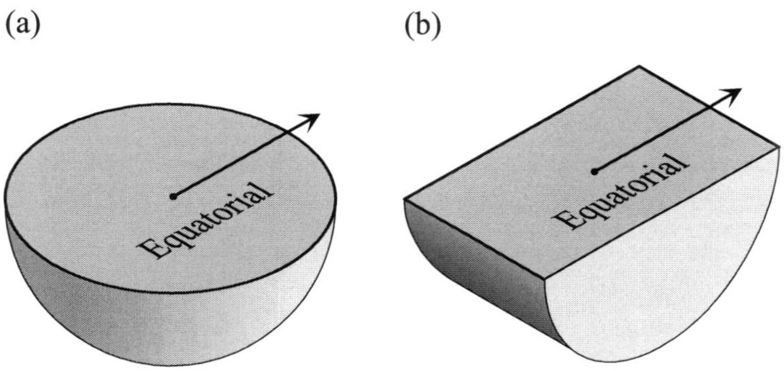

Figure 1.5 Schematic diagrams of the transverse sections of (a) spherical and (b) cylindrical particles loaded in diametral compression, illustrating locations and directions in the equatorial plane. Particle orientations as in Figure 1.2. Particle equators are indicated as bold lines and are (a) a circle and (b) a rectangle for the spherical and cylindrical particle, respectively.

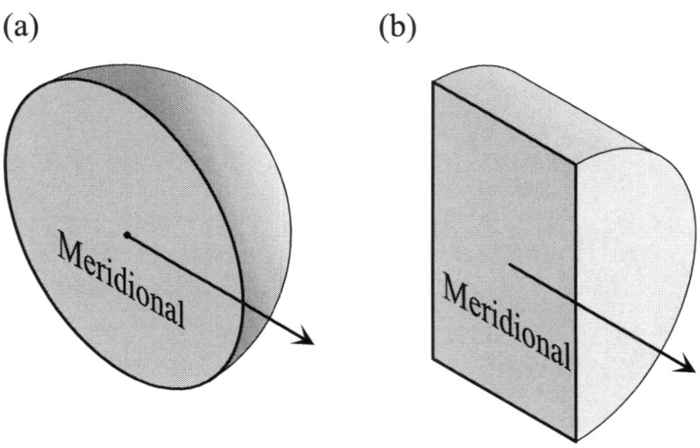

Figure 1.6 Schematic diagrams of the meridional sections of (a) spherical and (b) cylindrical particles loaded in diametral compression, illustrating locations and directions in the meridional plane. Particle orientations as in Figure 1.2. Particle meridians are indicated as bold lines and are (a) a circle and (b) a rectangle for the spherical and cylindrical particle, respectively.

one meridional plane, perpendicular to the cylinder face passing through the axis (Figure 1.6b). Meridional directions are indicated by arrows.

These geometrical specifications, Figures 1.4–1.6, are of two-fold importance: (1) Diametral compression is the most common form of loading for particles in testing and application. Individual tests and applications vary in detail, but the overall diametral compression form is maintained. A clear geometric description thus enables comparison of particle behavior in different tests and applications. (2) Particle failure in diametral compression is characterized by reversible deformation (Timoshenko and Goodier 1970; Sadd 2009), irreversible deformation (Mase 1979), and fracture (the formation of new surfaces, Lawn 1993). The deformation modes and fracture exhibit different geometrical characteristics. A clear geometric description of particle loading thus enables identification and description of the material constitutive laws governing particle failure. Points (1) and (2), test variations and particle failure behavior, respectively, are discussed in the following, drawing on the geometrical considerations.

Figure 1.7 shows schematic cross-sections of a range of particle mechanical tests, illustrating variations on the diametral compression form for spherical and cylindrical particles. Particles are shown in light shading, test fixture elements are shown hatched, and velocities are shown as solid lines with arrow heads; unless otherwise stated, this scheme is used throughout the book. The loading variation implemented most frequently is shown in Figure 1.7a, in which an upper platen is displaced by a test machine toward a particle resting on a fixed lower platen. The test machine generates the requisite force to maintain an imposed displacement rate and compress the particle. Typically, the imposed displacement rate is slow, such that the test is regarded as quasi static. Examples of such tests extend back many decades and continue into recent times (Kapur and Furstenau, 1967; Hooper 1971; Wynnyckyj 1985; Vallet and Charmet 1995; Antonyuk et al. 2005; Wang et al. 2019).

Other test variations include: (b) concentrated loading (Hiramatsu and Oka 1966), (c) passing particles between rotating rollers (Brecker 1974; Huang et al. 1993, 1995), (d) instrumented ball drop (King and Bourgeois 1993; Tavares and

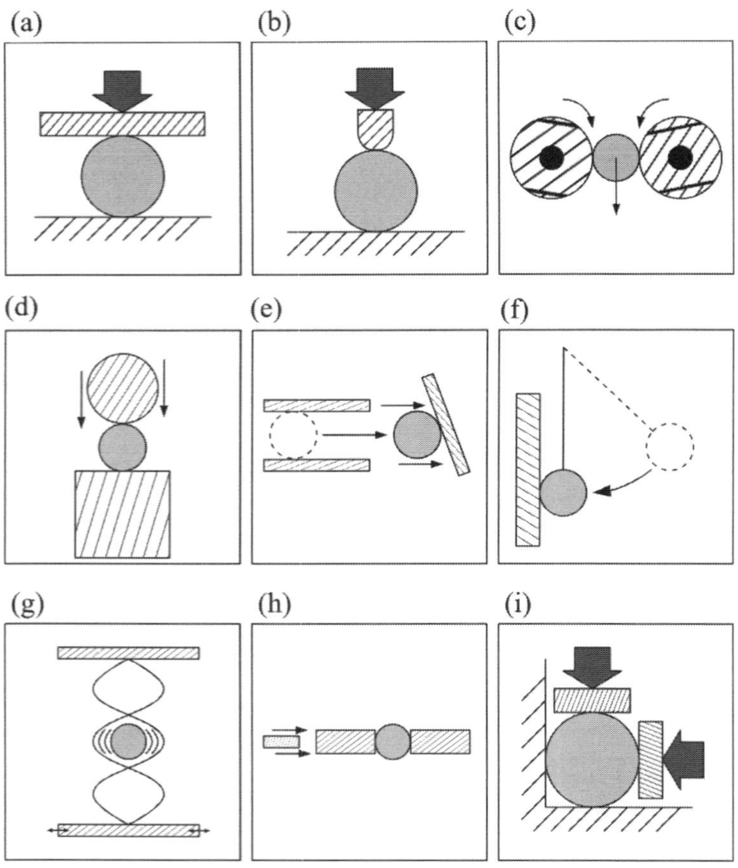

Figure 1.7 Schematic cross-section diagrams of commonly implemented mechanical tests of particles leading to diametral compression: (a) quasi-static loading between displaced platens—the most common; (b) quasi-static loading between a displaced probe and a platen; (c) crushing between two counter-rotating rollers; (d) impact by a dropped spherical weight onto a fixed particle; (e) impact against a platen of a moving particle impelled by a gas gun; (f) impact against a platen of a moving particle swung by a pendulum; (g) vibration by acoustic waves generated in a fluid; (h) impact by a striker impelled by a gas gun of a fixed particle; (i) quasi-static loading between two perpendicular sets of displaced platens leading to biaxial loading.

King 1998; Tavares 1999; Tavares et al. 2018), (e) impact on hard surfaces of moving particles impelled by a gas gun (Salman et al. 2002, 2003) or (f) by gravity controlled by a pendulum (Kantak et al. 2005), (g) ultrasonic fragmentation (Knoop et al. 2016), (h) ballistic impact of moving surfaces onto particles (Gustafsson et al. 2017; Xiao et al. 2019b), and (i) biaxial loading (Satone et al. 2017). Test configurations Figures 1.7a, 1.7b, and 1.7i are quasi-static and displacement controlled; configuration Figure 1.7c is rapid and displacement controlled; configurations Figures 1.7d, 1.7e, 1.7f, and 1.7h are regarded as impact-based and are rapid—d and h utilize deceleration of moving test fixture masses to generate compressive forces on stationary particles and e and f utilize deceleration of moving particles by stationary platens to generate compressive forces; configuration Figure 1.7g applies cyclic compressive forces to particles via acoustic waves transmitted through fluids. Variations on some of these configurations include: particles freely dropped onto a platen (Kantak and Davis 2004; Kantak et al. 2005) rather than impelled; contact is made with another single particle, rather than an element of the test fixture, leading to particle-particle collisions (Foerster et al. 1994; Chandramohan and Powell 2005); and contact is made between large probes or spheres and a bed of fine particles, leading to multi-particle contacts (Studman and Field 1984; Huang et al. 1998). A brief compilation of test configurations is given by Antonyuk et al. (2010).

The diversity of test configurations in Figure 1.7 reflects a diversity of mechanical measurement goals: Figures 1.7a and 1.7b enable careful observation of deformation and fracture during loading and failure of large scale particle components; Figures 1.7c, 1.7d, 1.7e, and 1.7g enable high-throughput testing of large numbers of small particle components; Figures 1.7e, 1.7f and 1.7h enable measurement of rate effects in deformation and fracture; and Figures 1.7e and 1.7j enable measurement of mixed-mode loading effects. The various test configurations, however, all generate a diametral compression loading geometry, whether the loading axis is vertical (e.g. Figure 1.7a), horizontal (e.g. Figure 1.7h), or inclined (e.g. Figure 1.7e). In addition, the various test configurations all seek to replicate specific loading cycles imposed on particles as structural components in application: By overlying weights, including those of other particles, by crushing jaws or rollers, or by colliding surfaces, including those from other particles.

Here and throughout, the terms *load* and *loading* are used as broad descriptors to refer to any configuration that generates applied forces on components, here in particular on particles. Loading may arise from direct application of a force, as in an overlying mass giving rise to a gravitational weight (a "load"). Loading may arise from an imposed displacement, in which case the force arises indirectly as a reaction to the displacement, as in platen, jaw, or roller configurations. Loading may also arise from an imposed change in displacement rate, in which case the force arises indirectly as a consequence of deceleration of a moving mass, as in impact configurations. A *loading cycle* is thus composed of *load-unload* segments, separated by the *peak force* generated in the cycle.

The force required to cause a structural component to fail in its load-bearing function is the component *failure force* (sometimes referred to as the supportable force, Ashby 1999). Thus, from performance and reliability perspectives, loading cycles applied to structural components can be characterized qualitatively by simple comparison of the peak force in a cycle to the failure force of a component. In this scheme, peak forces are conveniently classified into three domains: (i) small peak forces that do not exceed the component failure force; (ii) moderate peak forces that may exceed the component failure force; and (iii) large peak forces that definitely exceed the component failure force. Obviously, component design seeks to avoid domain (iii), often by identifying material responses in domains (i) and (ii) that lead to component failure. Observations of fracture and deformation during particle loading, also extending back many years, map onto this three domain classification: (i) At small peak forces, localized regions of irreversible deformation form and particles remain intact. (ii) At moderate peak forces, the regions of irreversible deformation expand and one or two fracture surfaces form. The fracture surfaces do not extend through particles and particles are thus partially fractured or cracked. (iii) At large peak forces, the regions of irreversible deformation continue to expand, significantly altering the shape of particles, and multiple fracture surfaces form. The fracture surfaces extend through particles and particles are thus completely fractured or cracked. These observations are now discussed in more detail, using the three domains of peak force as a framework. Attention is focused initially on the symmetric platen-particle-platen quasi-static loading configuration (Figure 1.7a), as this is the most common. The symmetric configuration provides a basis for consideration of the less common asymmetric particle-platen impact loading configuration (Figure 1.7e).

(i) Following a load-unload cycle to a small peak force, flat areas are observed at particle poles, indicative of localized irreversible deformation. The *deformation flats* are parallel to the equatorial plane, centered on particle axes, and, for spherical and cylindrical particles respectively, are circular or rectangular in outline. Inspection of the flats suggests that they

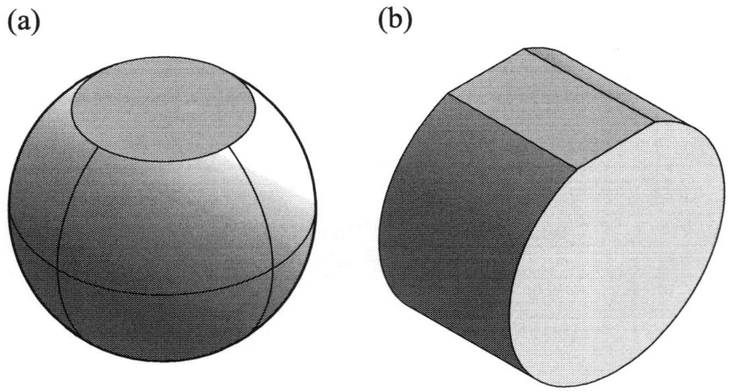

Figure 1.8 Schematic diagrams of the irreversible deformation flats formed on the poles of particles loaded between displaced platens in diametral compression as in Figure 1.7a. (a) Spherical and (b) cylindrical particle.

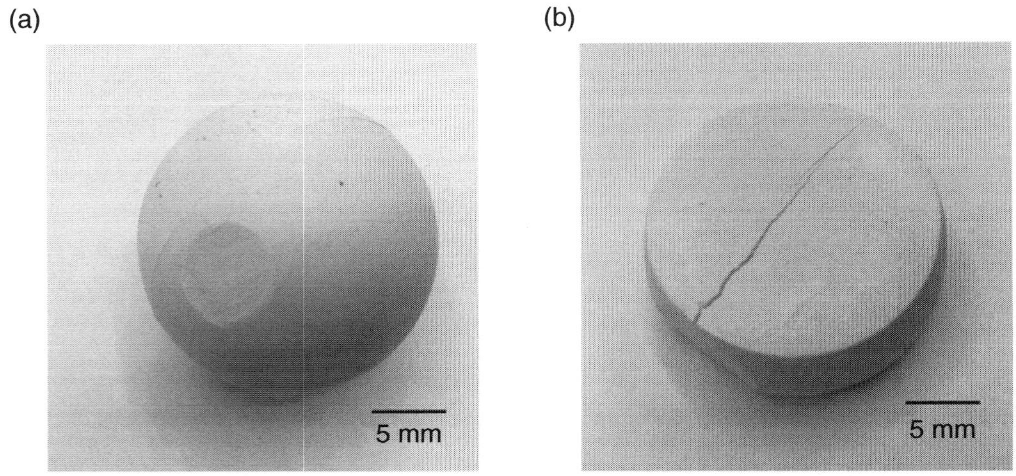

Figure 1.9 (a) Image of irreversible deformation flat formed on the pole of a spherical plaster particle loaded between displaced platens in diametral compression. (b) Image of irreversible deformation flat and pre-failure meridional cracks formed in a cylindrical plaster particle loaded between displaced platens in diametral compression. *Source:* Robert F. Cook.

are the surface faces of sub-surface regions of compacted material. Schematic diagrams of the flats formed at the poles of spherical and cylindrical particles are shown in Figure 1.8. An optical image of the flat formed at the pole of a spherical plaster particle is shown in Figure 1.9a. A similar flat is shown in an optical image of a sodium benzoate (food additive) particle (Antonyuk et al. 2005).

(ii) Deformation and fracture features formed during or following a load-unload cycle to a moderate peak force have been observed frequently. In situ observations have shown flats forming and expanding during loading at the poles of a variety of particles: epoxy cylinders and irregular plates (Hiramatsu and Oka 1966), plaster spheres and cylinders (Tsoungui et al. 1999), spherical fertilizer granules (Salman et al. 2003), plaster and cellulose cylinders (Procopio 2003), a sodium benzoate sphere (Pitchumani et al. 2004), a zeolite catalyst sphere (Antonyuk et al. 2005), and plaster cylinders (Wong and Jong 2014). Many observations following load-unload cycles to moderate peak forces also show flats: at the poles of a spherical limestone pellet (Kapur and Furstenau 1967), a plaster sphere (Tsongui et al. 1999), a polystyrene sphere (Schönert 2004), plaster spheres (Wu et al. 2004), metal coated polymer spheres (Zhang et al. 2007), concrete spheres (Khanal et al. 2008), cellulose cylinders and rounded tablets (Shang et al. 2013a, 2013b), silica spheres (Paul et al. 2014, 2015), and ceramic spheres (Pejchal et al. 2018). An obvious feature in these moderate force observations is the appearance of cracks such that the particles are partially fractured. The in situ observations show that the cracks form and extend during loading (Tsongui et al. 1999; Procopio 2003; Salman et al. 2003; Pitchumani et al. 2004; Antonyuk et al. 2005; Wong and Jong 2013). The cracks are visible as surface traces of crack opening displacement along particle meridians, frequently as traces from one polar flat through the particle equator to the other polar flat. The particles are only partially fractured as these meridional cracks do not extend completely through the particle. Schematic diagrams of meridional cracks and polar flats in spherical and cylindrical particles, similar to an earlier diagram (Kapur and Furstenau 1967), are shown in Figure 1.10. An optical image of the pre-failure flat and cracks formed in a cylindrical plaster particle is shown in Figure 1.9b.

(iii) By definition, loading a structural component to a large force leads to component failure and the loading cycle is truncated at the failure force of the component. For example, rods or bars in tension often fail by fracture, columns in compression often fail by buckling. The predominant feature of loading a particle to a large peak force is failure of the particle by complete fracture. In a large number of particle failure observations, fracture on meridional planes extends to the edges of a particle in the absence of observable irreversible deformation: irregular rock pieces (Hiramatsu and Oka 1966), marble cylinders (Jaeger 1967), iron ore spheres (Wynnyckyj 1985), glass and mineral spheres (Ryu and Saito 1991), glass and sapphire spheres (Shipway and Hutchings 1993), ceramic cylinders with milled flats (Fahad 1996), glass cylinders (Schönert 2004), ceramic spheres (Luscher et al. 2007), mannitol (a drug) spheres (Adi et al. 2011), rock cylinders (Erarslan and Williams 2012; Erarslan et al. 2012), iron ore pellets (Gustafsson et al. 2013), irregular rock particles (Wang and Coop 2016), ceramic spheres (Satone et al. 2017), and glass spheres and irregular rock pieces (Silva et al. 2019). In many of these cases, fracture is on a single meridional plane and a single crack propagates through the particle to generate two separated hemispherical fragments from spherical particles (e.g. Ryu and Saito 1991; Luscher et al. 2007; Silva et al. 2019), or two semicircular fragments from cylindrical particles (Fahad 1996). Figure 1.11 shows schematic diagrams of separated fragments from particles that failed in this simple manner. (*Fragments* are pieces generated by fracture—they can be large or

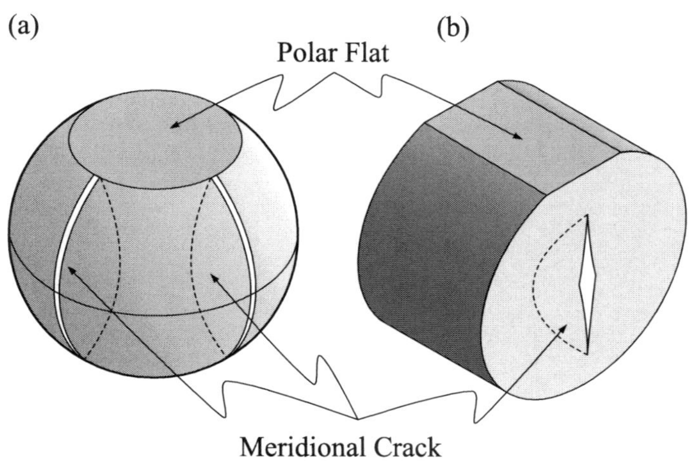

Figure 1.10 Schematic diagrams of irreversible deformation flats and meridional cracks generated in (a) a spherical and (b) a cylindrical particle loaded in diametral compression. The surface traces of the cracks are shown in white and the internal crack fronts are shown as dashed lines. Note that as the cracks do not extend through the particles, the particles are only partially fractured.

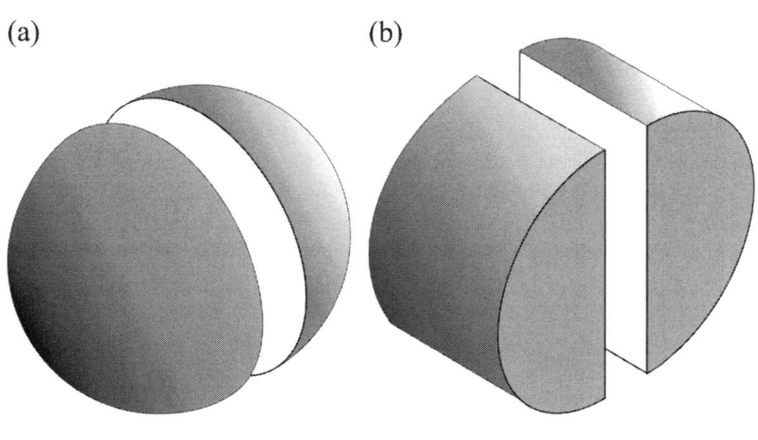

Figure 1.11 Schematic diagrams of complete fracture by a single meridional crack in (a) a spherical and (b) a cylindrical particle loaded in diametral compression. The cracks faces or fracture surfaces are shown in white. The cracks extend through particles leading to the generation of two fragments.

small fractions of components, e.g. Cook 2021.) Similar failure by simple meridional fracture was observed in early experiments on short, dried, plaster columns (Fairhurst 1960). In some cases, particle fragments remain attached to each other after particle failure (e.g. Wynnyckyj 1985; Gustafsson et al. 2013) (sphere) or (Jaeger 1967; Erarslan and Williams 2012) (cylinder) such that the particle retains its initial shape. In some cases, simple fragmentation is accompanied by minor irreversible deformation at the poles (e.g. Wynnyckyj 1985; Ryu and Saito 1991; Wu et al. 2004). An example is shown in Figure 1.9b.

In the more striking cases of particle failure, fracture is on multiple meridional planes, and multiple cracks propagate through the particle to generate many separated fragments (e.g. Ryu and Saito 1991; Zhang et al. 2007). This failure mode is restricted to spherical particles. Failure is preceded by significant irreversible deformation such that large flats are formed at the poles and the incipient failure geometry is thus a barrel. The fragments generated on failure are of three typical forms. The largest are wedge shaped: near spherical wedges bound by meridional planes that terminate at the perimeter of the flat, typically three to six in number. The second largest are cone shaped: axially oriented deformed spherical sectors that terminate on the faces of the flats, typically two. The smallest are very small randomly shaped and oriented fragments that are extremely numerous. The largest fragments originate from the exterior of the particle, the smallest fragments originate from the interior of the particle. Figure 1.12 shows a schematic diagram of the separated fragments of a spherical particle that failed in this multiple fracture manner. Similar patterns are shown in images of a fragmented glass particle (Ryu and Saito 1991) and a fragmented sand particle broken in constrained impact loading, Figure 1.7h, (Xiao et al. 2019a). X-Ray tomography images of failed particles in an aggregate array suggest that both single and multiple fracture modes can be operative simultaneously (Zhao et al. 2020).

Optical images of the large force failure sequence of a spherical plaster particle are shown in Figure 1.13. In this case, loading by platen displacement was halted and reversed immediately after the formation of visible surface traces of meridional cracks (Figure 1.13a). With some care the particle could be reconstructed into a spherical shape (Figure 1.13b), as

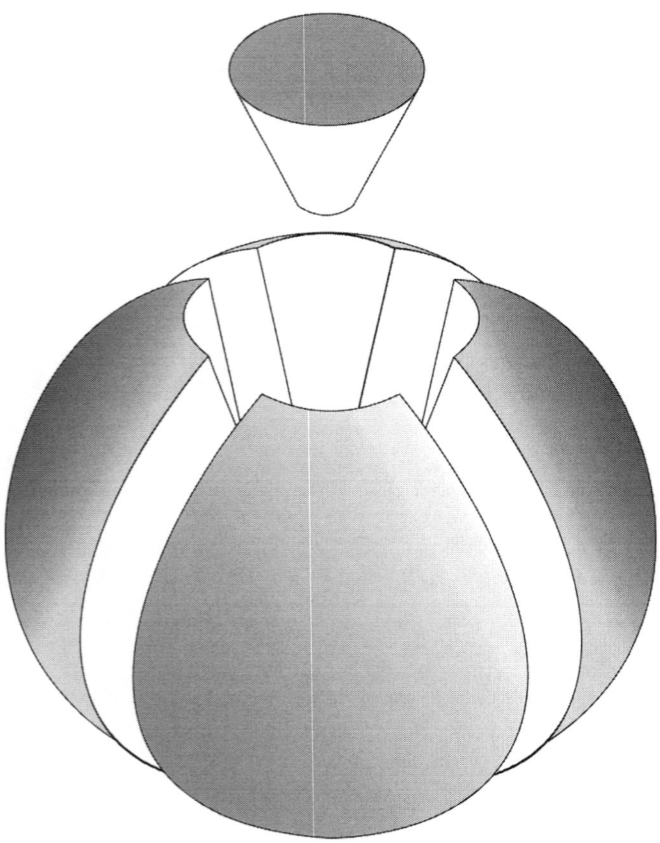

Figure 1.12 Schematic exploded view diagram of complete fracture by multiple meridional cracks in a spherical particle loaded in diametral compression. The cracks faces or fracture surfaces are shown in white. The cracks extend through the particle leading to the generation of many fragments, including wedges and truncated cones topped by irreversible deformation flats (lower cone not visible).

in Figure 1.12. Far more common in large force particle failure is the case in which loading is applied well after the initial formation of meridional cracks. In these cases, deformation of the particle proceeds by expansion of the irreversible deformation compaction zones at the polar contact regions. The fractured bulk of the particle, now with greatly increased compliance, deforms elastically. On unloading, the bulk of the particle forms wedges as discussed, see Figure 1.13c. The compacted regions typically disintegrate into fine fragments and powder and are removed on unloading. Optical images of the very large force failure patterns in cylindrical and spherical plaster particles are shown in Figure 1.14. The meridional cracks formed on loading are labelled C and and the unrecoverable powdered zones of compacted material that were removed on unloading are labeled R. Small chips or flakes of material, parallel to the cylinder face, often form on loading from the polar contacts in cylindrical geometries and are removed; an example here is labelled F.

Observations of particles after failure in free-particle impact experiments, Figure 1.7e, exhibit some similarities and two clear differences to those just discussed for quasi-static diametral compression tests, Figure 1.7a. The similarities are that particle failure is caused by both single and multiple meridional fracture and that the fragmentation morphology is the same, hemispheres for single fracture (alumina spheres, Salman et al. 1988, 1995; glass and PMMA spheres, Lecoq et al. 2003) and wedges and cones for multiple fracture (glass spheres, Salman and Gorham 2000; Salman et al. 2002, 2003, 2004; concrete spheres, Khanal et al. 2008). The first difference is obvious from Figure 1.7: impact tests are asymmetric and thus

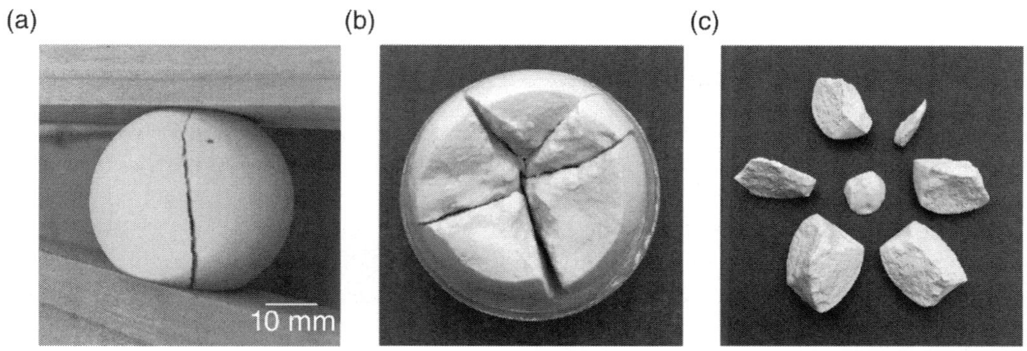

Figure 1.13 Images of the failure sequence of a spherical plaster particle loaded in diametral compression (vertical in image). (a) At peak load there is extensive deformation at the poles of the particle and distortion from the initial spherical shape. Complete fracture by a single crack is shown. (b) Reconstruction of the particle after unload shows that the particle has returned to a spherical shape and that multiple meridional cracks extend through the particle leading to the generation of many fragments, including wedges and a truncated cone topped by an irreversible deformation flat (removed). (c) Deconstruction of the particle after unload shows the multiple wedge fragments and the single cone fragment. *Source:* Robert F. Cook.

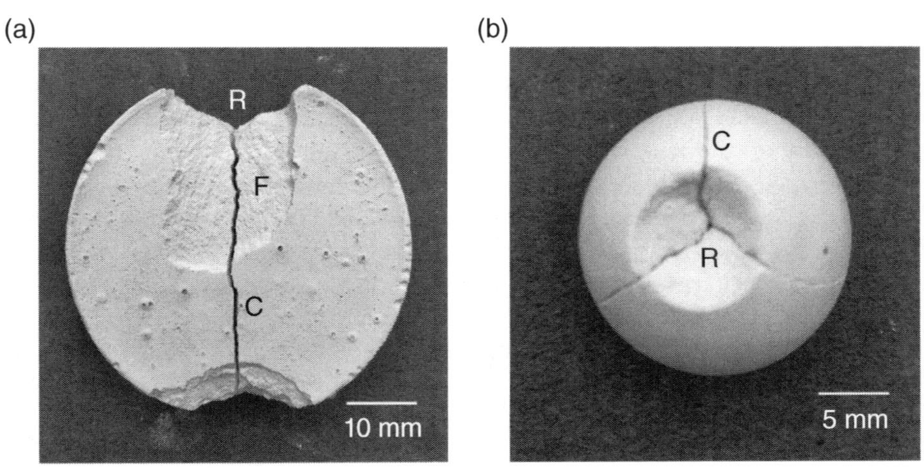

Figure 1.14 Images of failed plaster particles loaded excessively in diametral compression. (a) Cylinder and (b) Sphere. The meridional cracks are labelled C, the removed compaction zones are labelled R, and the removed flake of material in the cylinder geometry is labelled F. The compaction zones are a dominant feature of the failed components and are (a) prismatic triangular and (b) conical. *Source:* Robert F. Cook.

an irreversible deformation flat is formed on only one side of a tested particle. The second difference is that a deformation flat is observed on nearly every failed particle, even when the failure mode is of the simple, single meridional fracture type. In some cases, the cone of material including the deformed flat is retrievable after failure (e.g. Salman et al. 2004), but in some cases, the cone is absent, pulverized, and mixed in with the other small fragments (e.g. Khanal et al. 2008).

The diversity of particle measurement configurations, Figure 1.7, and identification of the diversity of particle failure mechanisms, Figures 1.10–1.14, provide information that enables optimization of particles as structural components and avoidance of particle failure. The next Section, 1.1.3, considers the origins of forces leading to particle failure in various applications and the associated engineering goals in these applications.

1.1.3 Particles in Application

The fracture processes illustrated in Figures 1.10–1.14 are common to all particle applications. In some applications, fracture is the desired outcome. In these cases, the engineering goal is usually to minimize the work required from the applied loading system in generating new, smaller particles (fragments of the original larger, broken, particles) with attendant new surfaces and associated surface energy. An example in this category is the crushing of rock in mining operations so as to enable transport of rock fragments and refinement of rock into metal ore. Reducing the size, comminution, of rock particles in the processing of ore is by far the most energy intensive activity in mining and accounts for ≈ 1 % of *all* energy usage in industrialized countries with large mining sectors (Tromans 2008). However, the efficiency of rock comminution processes (the ratio of work performed:surface energy created) is estimated as 1–2 %, considerably less than the maximum limiting efficiency of the process, estimated as 5–10 %. These values suggest that considerable improvement in energy efficiency in rock and ore processing is possible, leading to a much smaller effect of mining on the environment. Clearly, a greater understanding of the fracture processes shown in Figures 1.10–1.14 for the comminution of a single particle, in combination with industrial field tests (Unland and Szczelina 2004; Unland and Al-Khasawneh 2009), will contribute to improving the energy efficiency and minimizing the environmental burden of rock mining.

Another application in which particle fracture is the desired outcome, and minimization of energy consumption is an engineering goal, is cement and concrete production. Cement is formed by grinding particles of limestone ($CaCO_3$), clay (primarily oxides of Si and Al), and perhaps a few other minerals, to a powder that is then heated and tumbled. The process leads to a chemical and physical reaction that forms cement clinkers. Clinkers are small, round particles of cement suitable for transport and clinker production takes place in dedicated fabrication facilities worldwide. Concrete is formed by grinding cement clinkers to a powder, mixing with water and stony aggregate and allowing the resulting mixture to set. Comminution of raw materials and clinkers is the most (electrical) energy intensive activity in cement and concrete production (Afkhami et al. 2015) and accounts for over 1 % of all energy usage in most countries, industrialized and developing (Schneider et al. 2011). As with rocks and mining, a greater understanding of the fracture process for single particles of cement will contribute to improving energy efficiency in construction.

In most particle applications, fracture is an undesired outcome. In these cases, the engineering goal is usually to maximize the force required for particle breakage. As loading in these application is either deliberate, or incidental but inescapable, focus thus switches from loading efficiency to the failure force of the particle. The nature of the loading is important, however, in optimizing the overall performance of particles in specific applications. In the rock and cement crushing examples, loading was applied by a driven mechanical crushing or cracking device. In many applications in which particle failure is to be avoided, loading is applied by quasi-static or passive means. Perhaps the most common particle application that involves deliberate quasi-static loading is that of railway ballast, large rocky particles that bind together in a layer to fix railway tracks in place and support the weight of passing trains. Improvement in ballast particle supportable force would lead to fewer railway track repairs and greater allowed train weights. Most passive loading is also gravitational, the weight of particles themselves. When held in a container or otherwise confined, particles in upper layers compress those in lower layers, leading to passive static loading. Such loading occurs during bulk transport of particles, e.g. in rail cars, ship holds, and in applications involving fixed packs of particles through which fluids flow, e.g. in static catalyst arrays or on beaches and in river beds (Matthews 1983; Greene 2002; Novák-Szabó et al. 2018). A modern-day example of an application that requires quasi-static loading of a packed particle array that must also permit fluid flow is that of a proppant bed. Proppants are small ceramic particles that are pumped into underground fissures after hydraulic rock fracture, "prop" open the crack walls, and enable hydrocarbons to be extracted (the entire process is often termed "fracking") (Liang et al. 2016; Wang and Elsworth 2018; Feng et al. 2021). Improvements in particle failure force brought about by greater understanding of the diametral compression fracture process will lead to more efficient transportation of particles, greater catalytic efficiency, better environmental remediation, and more efficient hydrocarbon extraction. The combination of particle properties that maximizes particle performance, however, differs with application. For example, railway ballast performance is increased by maximizing particle failure force. Catalyst array and proppant bed performance are increased by maximizing the particle failure force/mass ratio. Particle performance is considered in Chapter 13.

Another form of deliberate or inescapable loading is impact. Particle weight can be converted into particle kinetic energy by dropping particles from heights. If the resulting motion is subsequently rapidly stopped by a barrier, the large deceleration of the particle leads to impact loading. Such loading occurs when particles are poured, e.g. into a ship hold or rail car, and the barrier is the bottom or sides of the container or hopper or a previously poured layer of particles. Both passive loading by particle weight and impact loading by particle dropping occur in transportation of cement clinkers and iron ore pellets. These operations lead to the generation of cement or iron ore fines that have unwanted environmental effects, lead to material losses on transport, and reduce the efficiency of the particles in application. Degradation of particles by impact is a similar issue, with perhaps graver consequences, in handling of fuel particles in advanced nuclear reactors (Hong et al. 2007). Impact loading also occurs when particles are moved by fluid entrainment, e.g. in pipes or fluidized catalyst beds during chemical manufacturing, and collisions with other particles or pipe walls occur. Many industrial processes involving particle impact in transport and handling could thus be greatly improved if particle failure force were better understood and could be enhanced without degrading other properties such as chemical reactivity. Natural world impact loading of extremely large volumes and masses of particles leading to particle failure, comminution, and transport occurs in rock avalanches (Dufresne and Dunning 2017). Many natural processes involving particle impact could thus also be better understood and guarded against if particle failure were better understood.

In contrast to free particles loaded by impact, there are many applications in which bound particles are imbedded in solids and loaded by deformation of the surrounding matrix material. The materials are commonly encountered: porcelain (domestic applications and electrical insulators), concrete (construction), steel (domestic and structural applications), and metal matrix composites (grinding, drilling, and structural applications). In porcelain, alumina (Al_2O_3) particles are dispersed in a glassy matrix. The deformation and fracture properties of the resulting composite depends on the particles retaining their load-bearing ability (Austin et al. 1946; Davidge and Green 1968). In concrete, considered earlier, aggregate particles are dispersed in a cement matrix. The overall deformation and fracture properties of concrete materials depend on the load-bearing ability of the aggregate particles. In steel, iron carbide (Fe_3C) particles are formed by heat treatment in an iron matrix. The deformation and fracture properties of steels depend on the load-bearing ability of the carbide particles. Observations of deformation in steel show that bound carbide particles exhibit simple meridional fracture on axial compression and simple equatorial fracture on axial tension (Gurland, 1972), consistent with the unbound particle observations describe in Section 1.2. In metal matrix composites, deformation- and fracture-resistant ceramic particles are mixed into metals leading to dispersion-based enhancements in mechanical properties. The failure forces of the resulting components depend on the particles retaining their load-bearing ability so as to maximally impede irreversible deformation in the metal matrices (Wallin et al. 1987; Hall et al. 1994; Nan and Clarke 1996). In all cases, it is clear that improved composite materials would result from a better understanding of the load-bearing ability of the incorporated particles and the effects of particle size (e.g. Hall et al. 1994) and shape (e.g. Gurland 1972).

Applications utilizing both loose and bound particles, and that involve deliberate particle contact with a surface, are grinding and drilling. In these applications, the goal is the efficient removal of material from a surface by the action in shear of particles, or a matrix containing particles. The geometry of contact between particle and surface is critical: Particles with sharp, angular exteriors are best for removing material from surfaces under the imposed loading combination of diametral compression and transverse displacement. Metal matrix composite grinding wheels and drill faces utilize imbedded sharp ceramic particles as the abrasive elements to remove material. Metal, ceramic, glass, and fabric-coated wheels in combination with loose sharp particles, often suspended in fluids, compressed between the wheels and surfaces are also used as abrasive elements. In both cases, particle size and shape are critical to the efficiency of the grinding and drilling processes. Smaller, broken, particles of abrasive media are often completely ineffective at material removal and such particles are often generated by the diametral compressive loading that is a necessary feature of grinding. Hence, the failure forces of grinding media particles are critical to the effectiveness and longevity of grinding wheels and drills. A better understanding of particle failure force would lead to more efficient grinding and drilling processes.

Finally, in some applications designed particle fracture is a desired outcome. In these cases, the engineering goal is to manufacture particles such that they remain unbroken by packaging and transport and then broken at the appropriate force in application. An example of this type of particle application has already been encountered in cement clinkers. Ideally, clinkers are strong enough to be handled and transported with little breakage and creation of dust or fines and then weak enough to be crushed using little energy, prior to mixing with water and aggregate to form concrete. Two more common examples of this type are pharmaceutical tablets and food. Pharmaceutical tablets are formed by first mixing powders of pharmaceutically active ingredients, inactive ingredients such as flavorants, and binders and fillers. The mixed powder is then formed into tablets by compaction in a mold and the tablets may then be coated. The completed tablets must then be strong enough to withstand handling, packaging, and transport prior to consumer acquisition (Muzzio et al. 2002). For tablets that are to be divided in use, the tablets must then be weak enough to be broken, e.g. halved, fragmented back into powder, by consumers. There are, of course, other factors such as pharmaceutical dose that must be considered in designing for this "middle ground" of failure force. Some food products are directly analogous to pharmaceuticals in this regard, sugar-coated chocolates and nuts, for example: They must withstand packaging and transport and then be breakable by human jaws and teeth to be edible. The overall chewability of food particles directly affects nutrition—some nuts and seeds are so strong that they cannot be cracked open by human jaws and thus cannot be effectively digested. Thus, the supportable force of food particles is of great concern to manufacturers designing food products, and also to anthropologists and zoologists determining early human and animal diets from fossil teeth and bones (Constantino et al. 2010; Lee et al. 2011).

The emphasis in this Section, 1.1, has been definition of major geometrical aspects of the particle diametral loading configuration (e.g. pole position, axial direction, equatorial plane) and application of the definitions in descriptions of experimental observations of particle failure (e.g. irreversible deformation on transverse sections, fracture on meridional planes). In defining and applying these geometrical ideas, consideration of ideal particle shapes, sphere and cylinder, were sufficient as attention was focused on general particle behavior relative to the loading orientation. However, the example particle images in Figure 1.3 make clear that many particles are not spherical or cylindrical, but have irregular geometrical shapes and outlines. However, Figure 1.3 also makes clear that the irregularity of shape within a set of particles is constrained: particles within a set "look the same." The next Section, 1.2, thus moves attention from the overall geometry of particle loading and failure to the more specific consideration of particle shape. In particular, attention is given to the exterior forms of particles, and how those forms can be characterized quantitatively and used to distinguish one set of particles from another. In addition, quantification of particle shape can be used as a basis for quantification of particle size, a critical particle characteristic throughout the book in the consideration of particle failure.

1.2 Particle Shapes and Sizes

The particles of different materials in Figure 1.3 exhibit very different shapes and sizes. For example, the sand grains, pie weights, and railway ballast all appear very different. Qualitative and semi-quantitative description and classification of particles by size and shape are thus easily performed by visual inspection. In terms of size, visual comparison of the largest dimensions of the example particles enables a size classification in which the sand grains are smaller than the pie weights, which in turn are smaller than the ballast rocks. In terms of shape, visual assessment and comparison of the ratios of the largest:smallest dimensions enables a shape classification in which the sand grains and ballast rocks have greater aspect ratios than the pie weights. Conversely, comparison of the ratios of the largest:smallest radii of the exterior profiles of the particles enables a shape classification in which the pie weights are more round than the sand grains or the ballast rocks. Clearly, size classification can be extended to be made quantitative by direct measurements of particles or their images. Projections of particle outlines to form two dimensional (2-D) images such as Figure 1.3 enable simple measurements of particle dimensions or areas to establish ranking by size. Shape classification can also be made quantitative, typically by measuring particle 2-D image parameters more complicated than length or area, for example, exterior local curvature. Analysis of such parameters can lead to scalar indices characterizing "elongation" (approximately the inverse of aspect ratio), "roundness," "circularity," "regularity," "angularity," "sphericity," and so on. Reviews, definitions, and descriptive examples of these terms are given by Pourghahramani and Forrsberg (2005a, 2005b), Blott and Pye (2008), and Mollon and Zhao (2012). Measurements in three dimensions (3-D) using optical topography (Ferellec and McDowell 2008) or X-ray tomography (Garboczi 2002) enable measurements of particle volume and hence ranking of particles by size. More typically, however, such measurements are used to classify particles by shape, and are discussed below.

As noted by Garboczi (2002), there are three major reasons for developing quantitative descriptions of particle shape: (i) to distinguish samples of particles, perhaps formed of different materials, perhaps obtained from different sources, or perhaps fabricated by different processes; (ii) to relate particle properties to particle structural features; and (iii) to enable mathematical descriptions of particles suitable for numerical simulation. As the scalar measures noted above are directed toward characterization rather than prediction, they provide the means to address points (i) and (ii) but are unable to address point (iii). In addition, the scalar measures lack precision as they reflect the summation or integration of many aspects of a particle surface or outline in order to generate a single quantity. Here, a brief overview is given of a quantitative description of particle shape that addresses all three points above. The method is based on discrete Fourier transform analysis of 2-D projections of particle shape and characterizes shapes by spectra of values, leading to greater precision than characterized by a single quantity. A major goal of this shape overview is to address Garboczi reason (ii) above and establish quantitative terms for shape characteristics that influence particle mechanical properties. Such characteristics provide information on how shape influences measures of particle size and how parameters such as roundness and roughness influence the load-bearing ability of a particle.

In order to classify sediments and rocks, measurements of particle or "grain" shapes have long been studied by geologists, in both qualitative and quantitative terms (Ehrlich and Weinberg 1970; Barclay and Buckingham 2009). Similarly, in order to characterize and simulate the mechanical and transport properties of the cement+aggregate composite of concrete, measurements of the shapes of aggregate particles have been performed (Wittmann et al. 1984; Garboczi 2002). The measurements of shape have been based on Fourier analysis of the two dimensional outlines of particles. Interpretation of such outlines has been extended via complex analysis to specification of the outline centroid, enclosed area, and moments of inertia (Kiryati and Maydan 1989), although attention here will be restricted to real analysis of outline shape and enclosed area.

Analysis begins by representing the projected outline of a particle in 2-D polar coordinates, $r(\theta)$, where r is the distance of a point on the outline measured from the center of mass or centroid of the particle and θ is an angle measured from a reference direction, Figure 1.15. The complete discrete Fourier series representation of the particle outline is

$$r(\theta) = r_0 \sum_{j=0}^{M} d_j \cos(j\theta + \phi_j), \tag{1.1}$$

where the index j represents an individual Fourier spatial frequency component, r_0 is the average distance from the centroid and a characteristic length scale of the outline, d_j represents a component relative amplitude, ϕ_j represents a component phase, and M is the integer number of components. The $j = 0$ component represents a circle, $r = r_0$, that is the basis of the outline. The $j = 1$ component represents a rigid translation of this circle that does not affect outline shape and hence $d_1 = 0$ is set. The effective series representation of the particle outline is thus

$$r(\theta) = r_0 + r_0 \sum_{j=2}^{M} d_j \cos(j\theta + \phi_j). \tag{1.2}$$

For a large group of particles described by Eq. (1.2), randomly oriented (such that $0 \leq \phi_j \leq 2\pi$ forms a uniform distribution with the mean value of $\phi_j = \pi$), the series term vanishes when averaged over all orientations $0 \leq \theta \leq 2\pi$. The average "size" of the particles in the group, given by the diameter D_{circle}, is then related to the base circle radius by

$$D_{\text{circle}} = 2r_0. \tag{1.3}$$

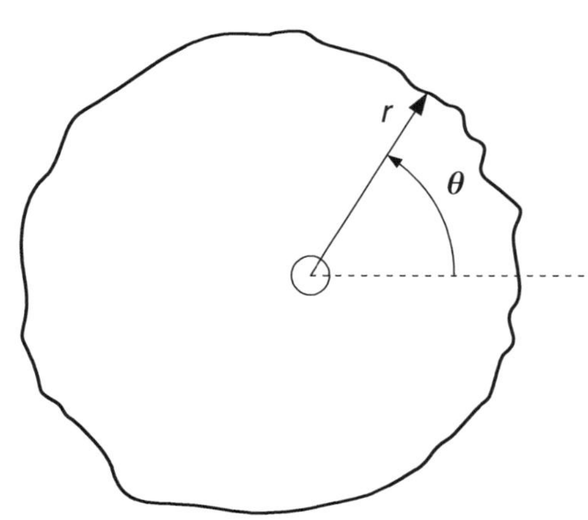

Figure 1.15 Diagram of a particle 2-D section or projection illustrating the polar (r, θ) coordinate system used to describe particle shape.

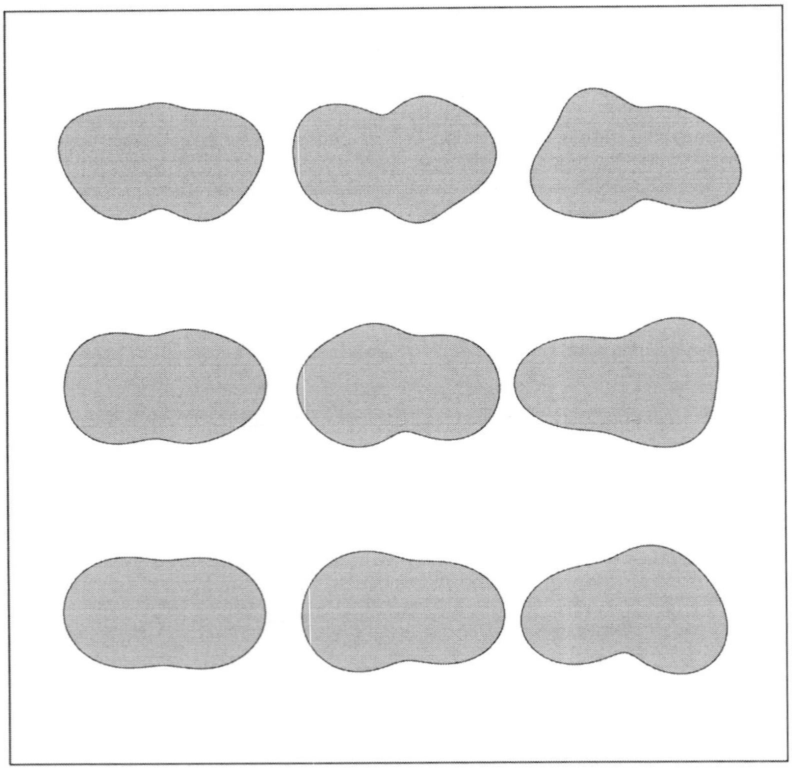

Figure 1.16 An array of 2-D particle shapes simulated using a Fourier-based analysis and varying $j = 3$ and 5 spatial frequency components superposed on a base $j = 2$ shape (lower left). The array illustrates the evolution of smooth multi-lobed particle shapes.

Figure 1.16 shows an array of particle outlines generated using only the $j = 2$, 3, and 5 components of the series given by Eq. (1.2) in addition to the r_0 component. The amplitude terms d_j were determined by the empirical decreasing relation observed for sand grains (Ehrlich and Weinberg 1970; Barclay and Buckingham 2009)

$$d_j = d_2(2/j)^{5/3}, \quad j \geq 3, \tag{1.4}$$

where $d_2 = 0.15$ was used here. The phase terms ϕ_j were determined by random selection from a uniform distribution between 0 and 2π. To enable easy comparison, the longest axes of the generated outlines were set horizontal. The outline in the bottom left of the array in Figure 1.16 was generated using the d_2 component alone and appears as a symmetric lozenge shape. The outline in the bottom right of the array was generated using d_2 and d_3 components and appears as a three-lobed figure. The lack of symmetry is a consequence of the different random values selected for ϕ_2 and ϕ_3. The outline in the center of the bottom row of the array is intermediate between the left and right outlines and results from a weak d_3 perturbation of the d_2 outline. The outline in the upper left of the array was generated using d_2 and d_5 components and appears as a slightly asymmetric five-lobed figure. The outline in the center of the left column of the array is intermediate between the lower and upper outlines and results from a weak d_5 perturbation of the d_2 outline. The outline in the upper right of the array was generated using the full d_2, d_3, and d_5 components. In progressing from the lower left to the upper right of the array, the two-fold symmetry of the outlines, a consequence of the greater d_2 value, Eq. (1.4), is progressively weakened such that the upper right outline displays almost no symmetry. Physically, the outlines in the lower row of the diagram represent the typical particle forms of river rocks, e.g., Figure 1.3k, shaped by the strong erosive power of water (Matthews 1983; Novák-Szabó et al. 2018). The other outlines represent the typical particle forms of desert rocks shaped by the weaker abrasive power of wind-borne grit (e.g. Lancaster 1984).

Using Eq. (1.2), the area A enclosed by the outlines in Figure 1.16 is given by (Kiryati and Maydan 1989; Garboczi 2002),

$$A = \frac{1}{2}\int_0^{2\pi} |r(\theta)|^2 d\theta = \pi r_0^2 + \frac{\pi r_0^2}{2}\sum_{j=2}^{M} d_j^2. \tag{1.5}$$

Noting that the area A enclosed by a circle is related to the diameter D by $A = \pi D^2/4$, the area enclosed by a particle outline given by Eq. (1.5) can be used to define particle size as an equivalent circular diameter D_{outline} by

$$D_{\text{outline}} = 2r_0\left[1 + (1/2)\sum_{j=2}^{M} d_j^2\right]^{1/2}. \tag{1.6}$$

This value is larger than that given by the radius of the base circle Eq. (1.3), $D_{\text{outline}} \geq D_{\text{circle}}$. Estimates of particle size based on area are larger than those based on radius and increase as more and greater Fourier components are used to describe particle shape. As a consequence, as perceptions of area are greater than those of linear dimension, the outlines in Figure 1.16 appear to increase slightly in size progressing from lower left to upper right.

Figure 1.17 shows an array of particle outlines generated using the $j = 2, 3$, and 7–32 components of the series given by Eq. (1.2) in addition to the r_0 component. The amplitude terms d_j, phase terms ϕ_j, and orientation of the outline long axes were set as described above. The bottom row of the array in Figure 1.17 was formed in the same manner as that in Figure 1.16 and shows a progression left to right from smooth two-lobed outline to smooth three-lobed outline. The outline in the upper left of the array in Figure 1.17 was generated using d_2 and d_7–d_{32} components and appears as a roughened two-lobed figure.The outline in the center of the left column of the array is intermediate between the lower and upper outlines and results from a weak d_7–d_{32} perturbation of the d_2 outline. The outline in the upper right of the array was generated using the full d_2, d_3, and d_7–d_{32} components. In progressing from the lower left to the upper right of the array, the outline becomes increasingly three-fold symmetric on a scale of the outline and increasingly rougher on a much smaller scale. Mathematically, small values of j correspond to small spatial frequencies and thus long wavelength components that perturb the overall shape of the particle. Large values of j correspond to large spatial frequencies and small wavelength components that appear as roughness on particles. Physically, the "rough" outline in the upper right of Figure 1.17 represents a common form of a volcanic rock particle that has not yet been weathered (Riley et al. 2003; Liu et al. 2015) and the center row represents the crushed sand particles of Figure 1.3b.

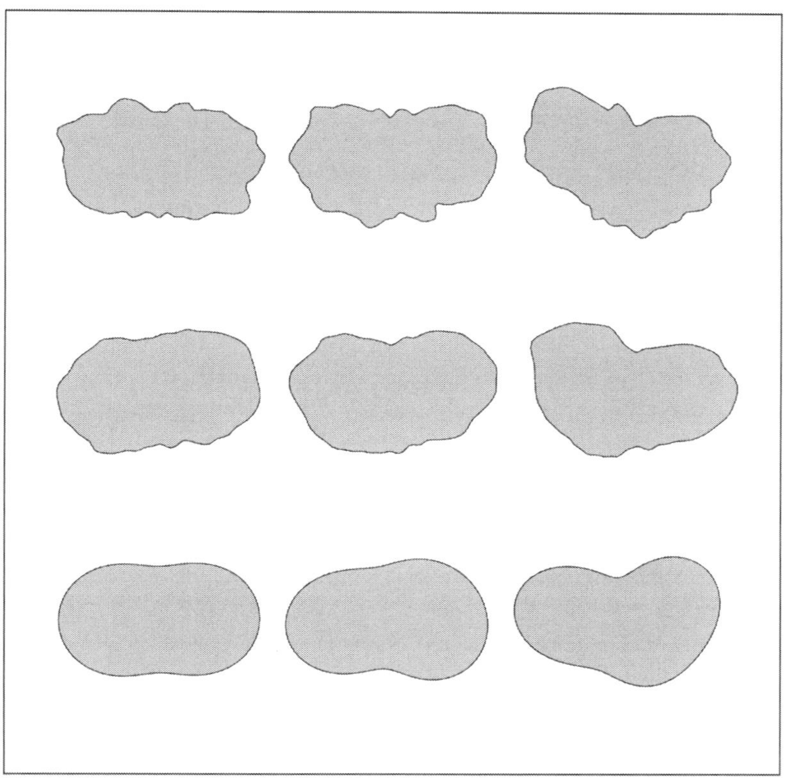

Figure 1.17 An array of 2-D particle shapes simulated using a Fourier-based analysis and varying $j = 3$ and 7–32 spatial frequency components superposed on a base $j = 2$ shape (lower left). The array illustrates the evolution of rough particle shapes.

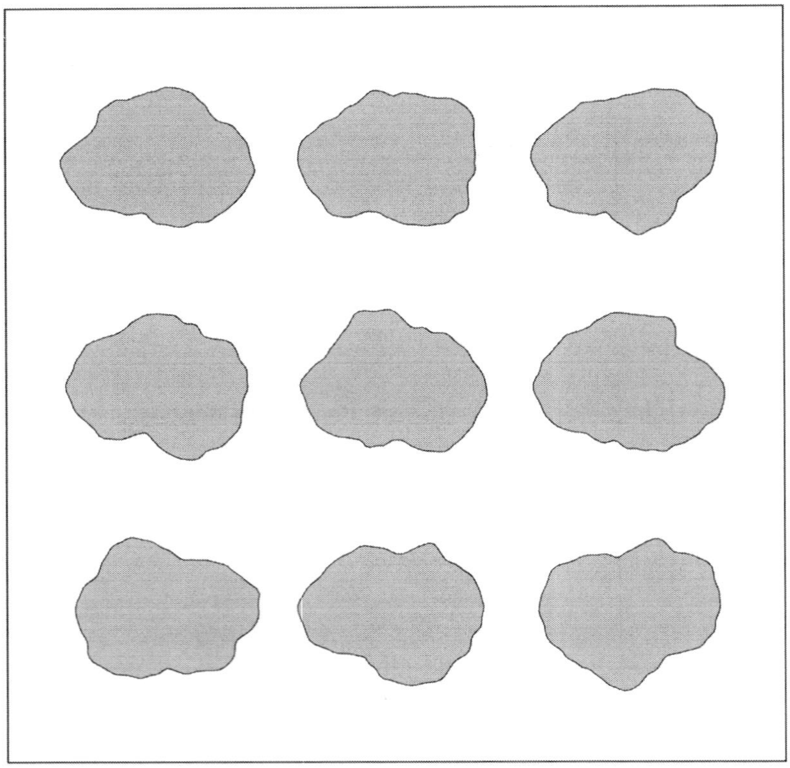

Figure 1.18 An array of 2-D particle shapes simulated using a Fourier-based analysis and randomly selected phases of $j = 2-32$ spatial frequency components. The array illustrates the variation of particle shapes created by a stochastic process.

Figure 1.18 shows an array of particle outlines generated using the full complement of $j = 2$–32 components of the series given by Eq. (1.2) in addition to the r_0 component. The amplitude terms d_j, phase terms ϕ_j, and orientation of the outline long axes were set as described above. All outlines in Figure 1.18 were generated similarly. In particular, an identical set of amplitude terms d_j set by Eq. (1.4) was used for each outline, leading to an array of outlines of similar size and aspect ratio. However, different sets of phases ϕ_j selected randomly from the uniform domain were used for each outline, leading to the variability in shapes within the array. The synthetically generated full-spectrum particle outlines of Figure 1.18 are more equidimensional and slightly larger than those of the split spectrum shapes of Figure 1.17, a consequence of filling in by the intermediate spatial frequency components. The similarity of outlines in Figure 1.18 to many of the groups of real particles in the images of Figure 1.3, particularly the salt of Figure 1.3d, suggest that the Fourier-based algorithm reflects major features of particle shape generation. In particular, the similarity of the outlines suggests that many real particles are formed by a spectrum of processes across a domain of spatial frequencies that decrease in amplitude effect with increasing spatial frequency. That is, amplitude decreases as wavelength decreases, in this case given by the spectrum of Eq. (1.4). Although the spectrum is fixed for a population or sample of particles, the phases of the effects are randomly distributed, suggesting that the effects at different scales are uncorrelated. In addition, the synthetic outlines in Figure 1.17 are similar to the groups of real particles in Figure 1.3 that would be regarded as possessing "rough" surfaces. The similarity of the outlines in these cases suggests that rough particles are formed by a bi-modal spectrum of processes consisting of a band of large amplitude, small spatial frequency components determining particle shape and a separate band of small amplitude, large spatial frequency components determining roughness. Arrays of 2-D outlines similar to those in Figures 1.16–1.18, generated using a similar algorithm and arriving at similar conclusions, have been presented in geological and technological particle contexts (Wittmann et al. 1984; Garboczi 2002; Barclay and Buckingham 2009; Mollon and Zhao 2012, 2014).

Successful as the discrete Fourier component analysis is for describing many aspects of particle shape, the compact form, Eq. (1.2), and realistic outlines, Figures 1.16–1.18 and in earlier works, come at the expense of assuming that the amplitude d_j and phase ϕ_j of a component are uncorrelated. As a consequence, as noted previously (Wittmann et al. 1984; Mollon and Zhao 2013), the analysis is not able to handle particles with facets in a simple way (straight-line segments in an outline require a large number of correlated components for each segment). The railway ballast rocks of Figure 1.3i are an example of faceted particle shapes that are not well handled by the uncorrelated spectra used in Figure 1.18. However, a simple extension of the idea above, that rough particles are described by a bi-modal spectrum of components, can generate perturbed, faceted outlines. The extension starts with a geometrically defined faceted outline and then superposes an uncorrelated Fourier component to arrive at the final outline. An example is shown in Figure 1.19 of an array of outlines generated in this way. The lower left outline in the array shows a pentagonal shape with rounded vertices, formed by interpolating circular arcs between the facets of a pentagon. The left column of the array shows increasing distortion of the rounded pentagon from bottom left to top left by decreasing the radii of the interpolated arcs and stretching the resulting shape horizontally. The right column of the array shows the outlines resulting from superposing a single d_{11} Fourier component on the distorted faceted shapes. The center column is intermediate between the left and right columns. In progressing from the lower left to the upper right of the array, the outline changes from a rounded, slightly faceted, symmetric shape, to a rough, faceted, asymmetric shape resembling a neolithic arrow head. In progressing from the upper left to the lower right of the array, the outline changes from a faceted shape resembling a cleaved crystal fragment to a rough, almost round, outline resembling a pollen grain. It is clear from Figure 1.19 that the sequence of rounded polygon + distortion + Fourier component is capable of generating realistic faceted outlines.

The above considerations of particle shape are an example of "forward" analysis. Through application of mathematical models that use assumed initial parameters, such analyses generate predictions as final results. In this case, sets of particle outlines, the results, were predicted by forward analysis using a Fourier model and assuming various initial parameters for component amplitudes and phases. Details of the initial assumptions in this case included spectra of component amplitudes and random selection of component phases from an assumed distribution. Particle shape is also amenable to "reverse" analysis. Such analysis generates interpretations as final results. By applying mathematical models that use a given, usually empirical, observation as the initial condition or constraint, descriptive parameters of the observation are obtained. In this case, a particle outline, the observation, can be interpreted by reverse analysis as a Fourier series and parameters obtained characterizing components of the series.

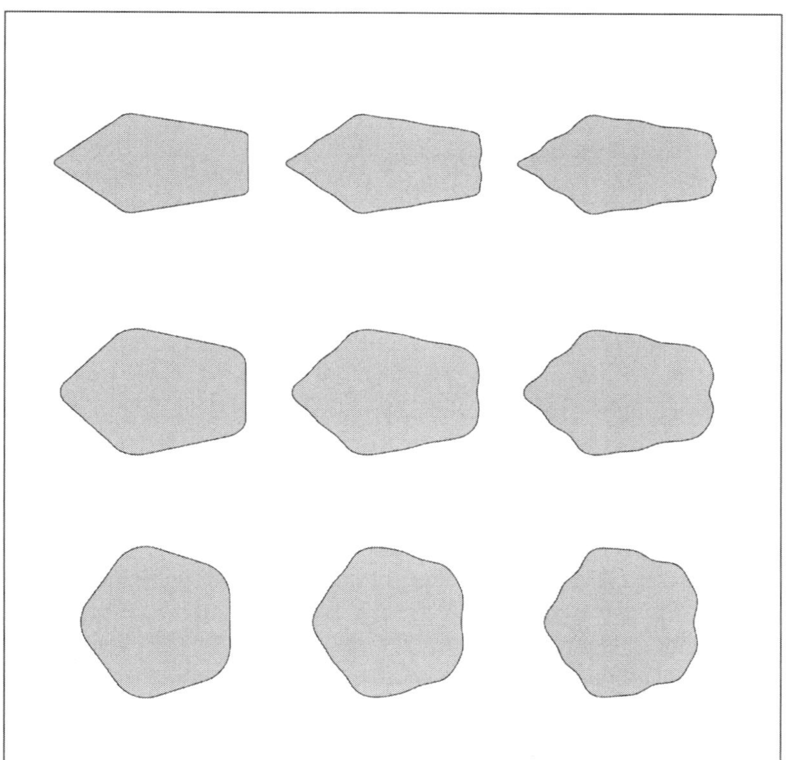

Figure 1.19 An array of 2-D particle shapes simulated using a base rounded pentagon and varying superposed elongations and Fourier-based roughness (j = 7–32 spatial frequency components). The array illustrates the evolution of angular particle shapes.

Reverse analysis here begins with an observed particle outline represented as a discrete set of N radius-angle pairs, $r_i(\theta_i)$. The index i identifies a point on the observed outline within domain 1 to N and r_i is measured from the centroid of the particle. The average radius of the particle, r_0, is given by

$$r_0 = \frac{1}{N}\sum_{i=1}^{N} r_i. \tag{1.7}$$

The observed discrete set of points $r_i(\theta_i)$ can be described by a continuous outline $r(\theta)$ expressed as a sum of M Fourier components

$$r(\theta) = r_0 + r_0\sum_{j=2}^{M}[a_j\cos(j\theta) + b_j\sin(j\theta)], \tag{1.8}$$

where the M amplitude coefficients of the Fourier components are given by sums over the N observations

$$a_j = \frac{1}{Nr_0}\sum_{i=1}^{N}[r_i\cos(j\theta_i)] \tag{1.9}$$

$$b_j = \frac{1}{Nr_0}\sum_{i=1}^{N}[r_i\sin(j\theta_i)] \tag{1.10}$$

and $a_1 = b_1 = 0$ is set as $j = 1$ terms correspond to rigid shifts. Resolution is limited by $M < N$, and usually adequate descriptions within observational scatter are provided for $M < N/10$. The (a_j, b_j) pair is related to the (d_j, ϕ_j) pair described earlier by

$$d_j^2 = a_j^2 + b_j^2 \tag{1.11}$$

and

$$\phi_j = \tan^{-1}\left(\frac{b_j}{a_j}\right). \tag{1.12}$$

An example of reverse analysis of a 2-D particle outline is shown in Figure 1.20. The open symbols represent an outline generated using forward analysis as in Figure 1.18 and then sampled as in an experimental observation: $N = 100$ points were selected on the outline at equal angular increments and $\pm$ 5 % white noise superposed on the radii to simulate measurement scatter. The solid line is a fit to the simulated observation using the above reverse analysis and $M = 8$. Most details of the observation are described by the fitted line. Similar analyses and decompositions of observed outlines of particle have been presented in earlier works considering particle shape (Ehrlich and Weiberg 1970; Garboczi 2002; Mollon and Zhao 2012, 2013, 2014).

The outlines in Figures 1.16–1.20 can be interpreted as either 2-D planar cross-sections or 2-D planar projections of 3-D particles. Extensions of the ideas above from these 2-D representations to full representations of the 3-D surfaces and volumes of 3-D particles have proceeded in several ways. The simplest extension in conceptual terms is to represent the surface of a particle as a weighted sum of spherical harmonics, in a 3-D analogy to the 2-D sum of Eq. (1.2). This approach was developed and implemented by Garboczi (2002) and used X-ray tomography information from concrete aggregate as the input. Another simple extension in conceptual terms is to generate the 3-D surface of a particle by interpolation between three perpendicular 2-D projections, each described by Fourier series. This approach was developed and implemented by Mollon and Zhao (2013, 2014) and used scanning electron microscopy (SEM) images of sand grains as the input. An extension to 3-D representation of particles that is somewhat different in not using Fourier-related techniques is to model the volume and surface of a particle by a clump of overlapping spheres. This approach was developed and implemented by Lu and McDowell (2007, 2010) and Ferellec and McDowell (2008, 2010) using stereo optical images of railway ballast as input and considerably refined computationally by Zhou et al. (2019) in considerations of roadway asphalt composites. A motivation for both the 2-D interpolation and overlapping sphere methods of generating 3-D numerical representations of particles was to enable mechanical simulation of 3-D multi-particle ensembles, point (iii) above noted by Garboczi (2002). In this book, the focus is on experimental mechanical behavior of single particles. Considerations of 2-D characterization can provide sufficient insight into the effects of particle shape on single particle measurements. In addition, the differences in 2-D characterization reflect the different processes by which particles were generated. In the example cases above,

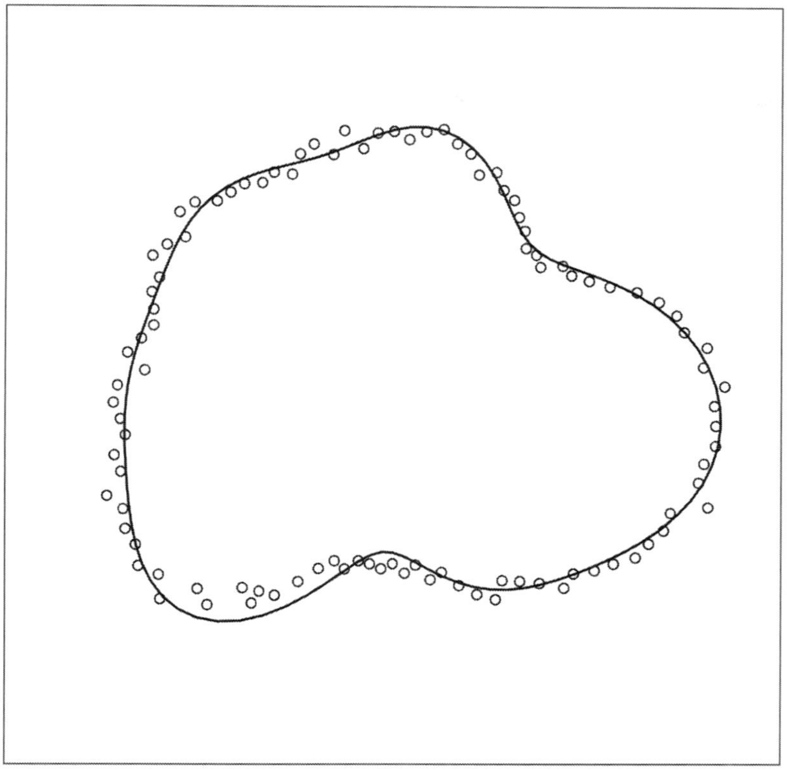

Figure 1.20 A 2-D digitized particle outline (symbols) best fit by a Fourier-based shape description analysis (solid line).

the processes included natural abrasion of sand grains by wind and water, engineered mixing and firing to form ceramic weights, and industrial crushing of rocks to produce ballast. Thus the 2-D characterization methods address points (i) and (ii) explained earlier.

An important factor in many measurements of particle mechanical properties is particle size, typically specified as a linear dimension D that characterizes particle diameter. Three methods of particle size determination are considered here: caliper measurement, sieve selection, and area equivalence. The first (caliper) method measures a length between two points on the surface of a particle. Typically, the points lie on opposite sides of the particle centroid. In 2-D, this method is the direct measurement of a chord length between points on the particle perimeter. The second (sieve) method estimates an upper bound to the smallest distance separating points on opposing surfaces of a particle. In 2-D, this method is a measurement of the upper bound of the least spanning chord between points on the perimeter of a particle outline. The third (area) method provides a length that describes a section or projection of the particle in terms of the area of an equivalent disc. The method is inherently 2-D and has been described in Eq. (1.6). The three methods are shown in Figure 1.21, illustrating the different lengths sensed by each for an identical particle in 2-D. The dashed circles in Figures 1.21a and 1.21b represent the base circle of the particle, with diameter $D_{\text{circle}} = 2r_0$. The dashed circle in Figure 1.21c represents the equivalent area circle for the particle, with diameter D_{outline}.

A framework for estimating differences in particle size is provided by the Fourier method used above to describe particle shape. The example particle used to illustrate the differences in Figure 1.21 is elongated with a dominant Fourier shape component $d_2 = 0.15$. Using this particle as an example, estimation of size differences begins by noting typical implementation techniques for the methods in Figure 1.21. In caliper-type particle size measurements, the particle is typically held with the long axis of the particle perpendicular to the caliper jaws, such that the caliper size measurement D_{caliper} is an estimate of the dimension of the shortest axis. Comparison of Figure 1.21 with the shape examples of Figure 1.16 suggests that $D_{\text{caliper}} \approx 2r_0(1 - 2d_2)$. In sieve-type particle size measurements, the particle is more likely to pass through a sieve with the long axis perpendicular to the sieve opening. The sieve size measurement D_{sieve} is then an estimate of the largest dimension perpendicular to the longest axis (the shoulders not the waist of the particle). Comparison of Figure 1.21 and Figure 1.16 suggests that $D_{\text{sieve}} \approx 2r_0(1 - d_2)$. For the elongated particle considered here, the area-based size measurement, Eq. (1.6), is approximated by $D_{\text{outline}} \approx 2r_0[1 + (d_2/2)^2]$. The maximum disparity between the mechanical

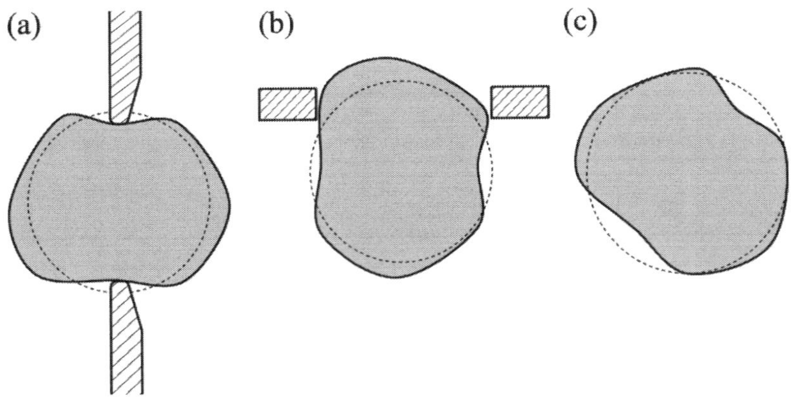

Figure 1.21 Schematic 2-D diagrams illustrating three different size measurement techniques for particles: (a) physical caliper; (b) physical sieve; (c) imaged area. Note that the physical techniques measure smaller dimensions than the image technique.

measurements and the image-based area measurement is estimated by combining the caliper approximation and the area approximation to give $D_{\text{outline}}/D_{\text{caliper}} \approx [1 + (d_2/2)^2]/[1 - 2d_2]$. Taking into account the difference between the caliper and sieve measurement estimations and accumulated second order effects associated with additional Fourier components in the area measurement, the difference between mechanical and image-based size measurements, to leading order in d_2, is $D_{\text{outline}}/D_{\text{caliper}} \approx (1 + 2d_2) \pm d_2$. For elongated particles described by $d_2 \approx 0.15$, projected area measurements can thus overestimate D determined by the more common mechanical measurements by a factor of 1.3 ± 0.15. This factor is small compared with the relative ranges in size of 10 or greater often observed for particle populations (Brown and Wohletz 1995; Altuhafi and Coop 2011). However, the disparity between image-based and mechanical measurements may well influence the relative numbers of particles classified by size within a population. The most common method of performing such a classification is by application of a sequence of sieves (Brecker 1974; Bertrand et al. 1988; Sikong et al. 1990; Huang et al. 1993; Tavares and King 1998; Aman et al. 2010). The sieve openings in such sequences are typically separated by factors of 2 in area such that the linear dimensions of the sieve openings are separated by factors of $2^{1/2} \approx 1.4$. Relative particle sizes in classified samples of particles within a population are thus often in the sequence ... 0.7, 1.0, 1.4, 2.0, 2.8, 4.0, ... The ratio of linear dimensions between sieved samples (≈ 1.4) is only slightly greater than the effect of shape (≈ 1.3) on perceived size, suggesting that shape considerations may be important in interpreting size measurements and size effects.

In particular, shape should be considered in interpreting particle size effects in mechanical measurements. The simplest physical effect is that in the common diametral compression geometries, Figures 1.7a, 1.7b, and 1.7d, elongated particles placed on platens tend to "lie down," such that long axes are parallel to the platens and short axes are parallel to the applied forces. This effect biases both mechanical responses and particle sizes estimated from the tests (Figures 1.7a and 1.7b resemble blunt caliper measurements), although Chapter 2 will show that shape bias in mechanical response is probably weak. A more subtle analytical effect is that mechanical measurements tend to underestimate the volume of elongated particles. For equi-axed particles, the three perpendicular 2-D outlines are described by identical representations of Eq. (1.2), and the true volume V of the particle well approximated by $V \approx V_{\text{outline}} = (\pi/6)D^3_{\text{outline}}$. If the particle is prolate such that one axis is relatively elongated, the volume given by caliper or sieve measurement is $V/(1 + 2d_2)^2 < V$; if oblate, such that one axis is relatively shortened, the volume is $V/(1 + 2d_2) < V$. Three key points to note are that: (i) as d_2 is usually small, these errors introduced by mechanical size measurements are also small; (ii) if a single technique is used throughout a study and the studied particle shape does not vary with scale, experimental assessments of the *scaling* of mechanical properties with particle volume are only weakly affected by such errors, and (iii) both points also apply to other, greater spatial frequency, Fourier components describing particle roughness.

The discussion of shape suggests that a major focus of this book—the scaling of particle failure force distributions with particle size—is not significantly affected by particle shape variations. As a consequence, for consistency and simplicity, values of D will be presented here as reported in published works. Usually, but not always, the values of D will have been determined by sieving. D will be considered a characteristic particle diameter and will be taken as the particle size throughout. Further size details will be considered in cases in which the measurement technique or the particle shape is an influence on interpretation of size effects on mechanical properties.

1.3 Summary: Particle Loading and Shape

This chapter has introduced the concept of particles as structural components and described the common load-bearing geometry of a particle in diametral compression (Figure 1.2). The diversity of loading configurations, in both particle testing and application, leading to diametral compression has been summarized (Figure 1.7), while emphasizing the importance and common occurrence of symmetric, platen based configurations. In particular, quasi-static loading implemented via displacement controlled platen-particle-platen configurations is frequently implemented in particle mechanical testing (Figure 1.7a), enabling observation of particle behavior during failure, as well as assessment of particle load-bearing ability. The observed irreversible deformation, fracture, and failure behavior of loaded particles has been described in geometrical terms specified with reference to the diametral loading geometry. Applications of particles in which fracture is important were shown to extend across a wide range of human activities, including mining (crushing and transport of rock, recovery of oil by fracking); construction and fabrication (preparation and storage of cement, grinding and drilling of parts); transportation (selecting railway track ballast); chemical engineering (stability of catalysts and catalyst beds); environmental remediation (selection of beach and river sand and pebbles); materials engineering (dispersion enhancement of mechanical properties of composites); healthcare (fabrication of pharmaceutical tablets); and nutrition (chewability of food). The importance of particle fracture in these applications was divided into three broad categories: those in which fracture is desired and energy minimization is a goal (e.g. rock and cement crushing); those in which fracture is to be avoided and maximization of the load-bearing force for a particle is a goal (e.g. railway track ballast, catalyst beds); and, those in which fracture is to be designed and specification of the load-bearing force for a particle is a goal (e.g. pharmaceutical tablets, food).

Images of a variety of particles were shown, Figure 1.3, emphasizing the everyday encounter with some particles. The diversity of size and shape of particles in the natural, domestic, and industrial worlds was noted. A Fourier-based analysis method for particle shapes in 2-D, either sections or projections, was described. The method was demonstrated in simulating the shapes of particles with increasing elongation and numbers of lobes, increasing roughness, diversity within samples, and the effects of angularity. The method was also demonstrated in describing the shape of a particle from existing data. The potential effects of particle shape on measurements of particle size by various techniques, including the frequently implemented sieving method, were discussed. Although sizing by sieving probably leads to slight underestimation of the volumes of elongated particles, shape effects on experimental measurements of scaling particle mechanical properties with size are likely small.

The observations of particle failure catalogued in Section 1.1 provide the basis for most of the physical phenomena considered in this book. It is clear that the failure of a particle as a structural component is almost completely controlled by global fracture of the particle material, with possible minor influence of localized material deformation. The linkage of failure with fracture raises two fundamental questions: (1) As fracture is driven by tension in a material, what is the nature of tension in a particle loaded in diametral compression? (2) What limits the maximum tension that a particle material can support, which in turn limits the load-bearing ability of a particle? Answers to the first question are considered in Chapter 2, which describes results from key mechanics analyses addressing the tension and compression arising in loaded particles. Answers to the second question are considered in Chapters 3 and 4, which describe the construction and use of a probability-based analytical framework that addresses the maximum sustainable tension of materials and thus the failure forces of loaded components formed of those materials. More detailed outlines of Chapters 3–13 are provided in Chapter 2 after the concepts, terms, and results of mechanics analyses of loaded particles are introduced. The descriptions of fracture and deformation in this Chapter 1 are largely qualitative. The descriptions in the following Chapters 2–13 become progressively more quantitative.

The considerations of particle shape, Section 1.2, in addition to providing a quantitative basis for descriptions of particle forms, included several important thematic aspects of analysis that will re-appear in considerations of particle failure:

- The first aspect in common between shape and failure considerations is the importance and usefulness of 2-D analyses. Such analyses often contain the essential physics of a topic using mathematics that is more simply expressed and analytically tractable, and which thus leads to results that are more easily visualized, than their 3-D counterparts. 2-D loading analyses (Chapter 2) will be seen to be very informative in interpreting failure measurements of 3-D particles.
- The second aspect in common between shape and failure considerations is the clear distinction between, and application of, forward and reverse analyses. Such analyses are built on a common mathematical framework but differ in several key aspects. These aspects are (i) the direction of analysis, forward vs reverse, (ii) the nature of the starting point, assumption vs observation, and (iii) the intent of the analysis and the nature of the end point, prediction vs interpretation.

Here, forward analysis prediction of failure distributions from tension-limiting material flaw populations (Chapter 3) and reverse analysis interpretation of failure distributions in terms of flaws (Chapter 4) will be seen to be two aspects of the same framework. A related point is that in both shape and failure considerations, forward analysis is continuum-based (e.g. Figure 1.16) and reverse analysis is based on discrete, noisy, data (e.g. Figure 1.20).

- The third aspect in common between shape and failure considerations is that the analytic descriptions of both sets of observations include a stochastic element. A *stochastic* process is one in which a random selection is made from a from a defined domain of values according to a fixed probability distribution (Van Kampen 2007). Stochastic selection of Fourier component phases was required to model realistic appearing particle outlines (the distribution in this case was uniform with domain 0 to 2π). Stochastic selection will be required to enable prediction of realistic distributions of sampled failure forces. Chapter 2 touches on all of these points by examining the application of 2-D loading analyses in particle failure, the formation of failure distributions as the starting point for reverse analysis of failure forces, and the inherent stochastic nature of the observations.

References

Adi, S., Adi, H., Chan, H.-K., Finlay, W.H., Tong, Z., Yang, R., and Yu, A. (2011). Agglomerate strength and dispersion of pharmaceutical powders. *Journal of Aerosol Science* 42: 285–294.

Afkhami, B., Akbarian, B., Beheshti, N., Kakaee, A.H. and Shabani, B. (2015). Energy consumption assessment in a cement production plant. *Sustainable Energy Technologies and Assessments* 10: 84–89.

Altuhafi, F.N. and Coop, M.R. (2011). Changes to particle characteristics associated with the compression of sands. *Géotechnique* 61: 459–471.

Aman, S., Tomas, J., and Kalman, H. (2010). Breakage probability of irregularly shaped particles. *Chemical Engineering Science* 65: 1503–1512.

Antonyuk, S., Tomas, J., Heinrich, S., and Mörl, L. (2005). Breakage behaviour of spherical granulates by compression. *Chemical Engineering Science* 60: 4031–4044.

Antonyuk, S., Heinrich, S., Tomas, J., Deen, N.G., van Buijtenen, M.S., and Kuipers, J.A.M. (2010). Energy absorption during compression and impact of dry elastic-plastic spherical granules. *Granular Matter* 12: 15–47.

Ashby, M.F. (1999). *Materials Selection in Mechanical Design*. Butterworth-Heinemann.

Austin, C.R., Schofield, H.Z., and Haldy, N.L. (1946). Alumina in whiteware. *Journal of the American Ceramic Society* 29: 341–354.

Barclay, D.R. and Buckingham, M.J. (2009). On the shapes of natural sand grains. *Journal of Geophysical Research* 114: B02209.

Bertrand, P.T., Laurich-McIntyre, S.E., and Bradt, R.C. (1988). Strengths of fused and tabular alumina refractory grains. *American Ceramic Society Bulletin* 67: 1217–1221.

Blott, S.J. and Pye, K. (2008). Particle shape: A review and new methods of characterization and classification. *Sedimentology* 55: 31–63.

Brecker, J.N. (1974). The fracture strength of abrasive grains. *Journal of Engineering for Industry* 96: 1253–1257.

Brown, W.K. and Wohletz, K.H. (1995). Derivation of the Weibull distribution based on physical principles and its connection to the Rosin-Rammler and lognormal distributions. *Journal of Applied Physics* 78: 2758–2763.

Chandramohan, R. and Powell, M.S. (2005). Measurement of particle interaction properties for incorporation in the discrete element method simulation. *Minerals Engineering* 18: 1142–1151.

Constantino, P.J., Lee, J.J.-W., Chai, H., Zipfel, B., Ziscovici, C., Lawn, B.R., and Lucas, P.W. (2010). Tooth chipping can reveal the diet and bite forces of fossil hominins. *Biology Letters* 6: 826–829.

Davidge, R.W. and Green, T.J. (1968). The strength of two-phase ceramic/glass materials. *Journal of Materials Science* 3: 629–634.

Dufresne, A. and Dunning, S.A. (2017). Process dependence of grain size distributions in rock avalanche deposits. *Landslides* 14: 1555–1563.

Ehrlich, R. and Weinberg, B. (1970). An exact method for characterization of grain shape. *Journal of Sedimentary Research*, 40: 205–212.

Erarslan, N. and Williams, D.J. (2012). Experimental, numerical and analytical studies on tensile strength of rocks. *International Journal of Rock Mechanics and Mining Sciences* 49: 21–30.

Erarslan, N., Liang, Z.Z., and Williams, D.J. (2012). Experimental and numerical studies on determination of indirect tensile strength of rocks. *Rock Mechanics and Rock Engineering* 45: 739–751.

Fahad, M.K. (1996). Stresses and failure in the diametral compression test. *Journal of Materials Science* 31: 3723–3729.

Fairhurst, C.W. (1960). Compressive properties of dental gypsum. *Journal of Dental Research* 39: 812–824.

Feng, Y.C., Ma, C.Y., Deng, J.G., Li, X.R., Chu, M.M., Hui, C., and Luo, Y.Y. (2021). A comprehensive review of ultralow-weight proppant technology. *Petroleum Science* 18: 807–826.

Ferellec, J.F. and McDowell, G.R. (2008). A simple method to create complex particle shapes for DEM. *Geomechanics and Geoengineering: An International Journal* 3: 211–216.

Ferellec, J.F. and McDowell, G.R. (2010). A method to model realistic particle shape and inertia in DEM. *Granular Matter* 12: 459–467.

Garboczi, E.J. (2002). Three-dimensional mathematical analysis of particle shape using X-ray tomography and spherical harmonics: Application to aggregates used in concrete. *Cement and Concrete Research* 32: 1621–1638.

Greene, K. (2002). *Beach nourishment: A review of the biological and physical impacts.* Atlantic States Marine Fisheries Commission, ASMFC Habitat Management Series # 7.

Gurland, J. (1972). Observations on the fracture of cementite particles in a spheroidized 1.05 % C steel deformed at room temperature. *Acta Metallurgica* 20: 735–741.

Gustafsson, G., Häggblad, H.-Å., and Jonsén, P. (2013). Characterization modelling and validation of a two-point loaded iron ore pellet. *Powder Technology* 235: 126–135.

Gustafsson, G., Häggblad, H.-Å., Nishida, M., Larsson, S., and Jonsén, P. (2017). Fracture probability modelling of impact-loaded iron ore pellets. *International Journal of Impact Engineering* 102: 180–186.

Hall, J.N., Jones, J.W., and Sachdev, A.K. (1994). Particle size, volume fraction and matrix strength effects on fatigue behavior and particle fracture in 2124 aluminum-SiCp composites. *Materials Science and Engineering A* 183: 69–80.

Hiramatsu, Y. and Oka, Y. (1966). Determination of the tensile strength of rock by a compression test of an irregular test piece. *International Journal of Rock Mechanics and Mining Sciences & Geomechanics Abstracts* 3: 89–99.

Hong, S.G., Byun, T.S., Lowden, R.A., Snead, L.L., and Katoh, Y. (2007). Evaluation of the fracture strength for silicon carbide layers in the tri-isotropic-coated fuel particle. *Journal of the American Ceramic Society* 90: 184–191.

Hooper, J.A. (1971). The failure of glass cylinders in diametral compression. *Journal of the Mechanics and Physics of Solids* 19: 179–188.

Huang, H., Huang, Q.K., Zhu, X.H., and Hu, X.Z. (1993). An experimental investigation of the strengths of individual brown corundum abrasive grains. *Scripta Metallurgica* 29: 299–304.

Huang, H., Zhu, X.H., Huang, Q.K., and Hu, X.Z. (1995). Weibull strength distributions and fracture characteristics of abrasive materials. *Engineering Fracture Mechanics* 52: 15–24.

Huang, H., Dallimore, M.P., Pan, J., and McCormick, P.G. (1998). An investigation of the effect of powder on the impact characteristics between a ball and a plate using free falling experiments. *Materials Science and Engineering: A*, 241: 38–47.

Jaeger, J.C. (1967). Failure of rocks under tensile conditions. *International Journal of Rock Mechanics and Mining Sciences* 4: 219–227.

Kantak, A.A. and Davis, R.H. (2004). Oblique collisions and rebound of spheres from a wetted surface. *Journal of Fluid mechanics* 509: 63–81.

Kantak, A.A., Galvin, J.E., Wildemuth, D.J., and Davis, R.H. (2005). Low-velocity collisions of particles with a dry or wet wall. *Microgravity-Science and Technology* 17: 18–25.

Kapur, P.C. and Fuerstenau, D.W. (1967). Dry strength of pelletized spheres. *Journal of the American Ceramic Society* 50: 14–18.

Khanal, M., Schubert, W., and Tomas, J. (2008). Compression and impact loading experiments of high strength spherical composites. *International Journal of Mineral Processing* 86: 104–113.

King, R.P. and Bourgeois, F. (1993). Measurement of fracture energy during single-particle fracture. *Minerals Engineering* 6: 353–367.

Kiryati, N. and Maydan, D. (1989). Calculating geometric properties from Fourier representation. *Pattern Recognition* 22: 469–475.

Knoop, C., Todorova, Z., Tomas, J., and Fritsching, U. (2016). Agglomerate fragmentation in high-intensity acoustic standing wave fields. *Powder Technology* 291: 214–222.

Lancaster, N. (1984). Characteristics and occurrence of wind erosion features in the Namib Desert. *Earth Surface Processes and Landforms* 9: 469–478.

Lawn, B.R. (1993). *Fracture of Brittle Solids, 2nd ed.* Cambridge.

Lee, J.J.-W., Constantino, P.J., Lucas, P.W., and Lawn, B.R. (2011). Fracture in teeth—a diagnostic for inferring bite force and tooth function. *Biological Reviews* 86: 959–974.

Lecoq, O., Chouteau, N., Mebtoul, M., Large, J.F., and Guigon, P. (2003). Fragmentation by high velocity impact on a target: A material grindability test. *Powder Technology* 133: 113–124.

Liang, F., Sayed, M., Al-Muntasheri, G.A., Chang, F.F., and Li, L. (2016). A comprehensive review on proppant technologies. *Petroleum* 2: 26–39.

Liu, E.J., Cashman, K.V., and Rust, A.C. (2015). Optimising shape analysis to quantify volcanic ash morphology. *GeoResJ* 8: 14–30.

Lu, G. and McDowell, G.R. (2007). The importance of modelling ballast particle shape in the discrete element method. *Granular Matter* 9: 69–80.

Lu, M. and McDowell, G.R. (2010). Discrete element modelling of railway ballast under monotonic and cyclic triaxial loading. *Géotechnique* 60: 459–467.

Luscher, W.G., Hellmann, J.R., Segall, A.E., Shelleman, D.L., and Scheetz, B.E. (2007). A critical review of the diametral compression method for determining the tensile strength of spherical aggregates. *Journal of Testing and Evaluation* 35: 624–629.

Mase, G.E. (1970). *Continuum Mechanics*. McGraw-Hill.

Matthews, E.R. (1983). Measurements of beach pebble attrition in Palliser Bay, southern North Island, New Zealand. *Sedimentology* 30: 781–799.

Mollon, G. and Zhao, J. (2012). Fourier-Voronoi-based generation of realistic samples for discrete modelling of granular materials. *Granular matter* 14: 621–638.

Mollon, G. and Zhao, J. (2013). Generating realistic 3D sand particles using Fourier descriptors. *Granular Matter* 15: 95–108.

Mollon, G. and Zhao, J. (2014). 3D generation of realistic granular samples based on random fields theory and Fourier shape descriptors. *Computer Methods in Applied Mechanics and Engineering* 279: 46–65.

Muzzio, F.J., Shinbrot, T., and Glasser, B.J. (2002). Powder technology in the pharmaceutical industry: The need to catch up fast. *Powder Technology* 124: 1–7.

Nan, C.-W. and Clarke, D.R. (1996). The influence of particle size and particle fracture on the elastic/plastic deformation of metal matrix composites. *Acta Materialia* 44: 3801–3811.

Novák-Szabó, T., Árpád Sipos, A., Shaw, S., Bertoni, D., Pozzebon, A., Grottoli, E., Sarti, G., Ciavola, P., Domokos, G., and Jerolmack, D.J. (2018). Universal characteristics of particle shape evolution by bed-load chipping. *Science Advances* 4: eaao4946.

Paul, J., Romeis, S., Tomas, J., and Peukert, W. (2014). A review of models for single particle compression and their application to silica microspheres. *Advanced Powder Technology* 25: 136–153.

Paul, J., Romeis, S., Mačković, M., Marthala, V.R.R., Herre, P., Przybilla, T., Hartmann, M., Spiecker, E., Schmidt, J., and Peukert, W. (2015). In situ cracking of silica beads in the SEM and TEM – Effect of particle size on structure-property correlations. *Powder Technology* 270: 337–347.

Pejchal, V., Fornabaio, M., Žagar, G., Riesen, G., Martin, R.G., Medřický, J., Chráska, T., and Mortensen, A. (2018). Meridian crack test strength of plasma-sprayed amorphous and nanocrystalline ceramic microparticles. *Acta Materialia* 145: 278–289 .

Pitchumani, R., Zhupanska, O., Meesters, G.M.H., and Scarlett, B. (2004). Measurement and characterization of particle strength using a new robotic compression tester. *Powder Technology* 143–144: 56–64.

Pourghahramani, P. and Forssberg, E. (2005a). Review of applied particle shape descriptors and produced particle shapes in grinding environments. Part I: Particle shape descriptors. *Mineral Processing & Extractive Metallurgy Review* 26: 145–166.

Pourghahramani, P. and Forssberg, E. (2005b). Review of applied particle shape descriptors and produced particle shapes in grinding environments. Part II: The influence of comminution on the particle shape. *Mineral Processing & Extractive Metallurgy Review* 26: 167–186.

Procopio, A.T., Zavaliangos, A., and Cunningham, J.C. (2003). Analysis of the diametrical compression test and the applicability to plastically deforming materials. *Journal of Materials Science* 28: 3629–3639.

Riley, C.M., Rose, W.I., and Bluth, G.J. (2003). Quantitative shape measurements of distal volcanic ash. *Journal of Geophysical Research* 108(B10): 2504.

Ryu, H.J. and Saito, F. (1991) Single particle crushing of nonmetallic inorganic brittle materials. *Solid State Ionics* 47: 35–50.

Sadd, M.H. (2009) *Elasticity.* Elsevier.

Salman, A.D., Szabo, M., Angyal, I., and Verba, A. (1988). Particle degradation. *Periodica Polytechnica Mechanical Engineering* 32: 233–251.

Salman, A.D., Gorham, D.A., and Verba, A. (1995). A study of solid particle failure under normal and oblique impact. *Wear* 186: 92–98.

Salman, A.D. and Gorham, D.A. (2000). The fracture of glass spheres. *Powder Technology* 107: 179–185.

Salman, A.D., Biggs, C.A., Fu, J., Angyal, I., Szabó, M., and Hounslow, M.J. (2002). An experimental investigation of particle fragmentation using single particle impact studies. *Powder Technology* 128: 36–46.

Salman, A.D., Fu, J., Gorham, D.A., and Hounslow, M.J. (2003). Impact breakage of fertiliser granules. *Powder Technology* 130: 359–366.

Salman, A.D., Reynolds, G.K., Fu, J.S., Cheong, Y.S., Biggs, C.A., Adams, M.J., Gorham, D.A., Lukenics, J., and Hounslow, M.J. (2004). Descriptive classification of the impact failure modes of spherical particles. *Powder Technology* 143: 19–30.

Satone, M., Iimura, K., Teraoka, T., Hanafusa, T., Hisatani, S., Nishiwaki, M., and Suzuki, M. (2017). Analysis of granule fracture under biaxial compression. *Ceramics International* 43: 16835–16842.

Schneider, M., Romer, M., Tschudin, M., and Bolio, H. (2011). Sustainable cement production–present and future. *Cement and Concrete Research* 41: 642–650.

Schönert, K. (2004). Breakage of spheres and circular discs. *Powder Technology* 143: 2–18.

Shang, C., Sinka, I. C., Jayaraman, B., and Pan, J. (2013a). Break force and tensile strength relationships for curved faced tablets subject to diametrical compression. *International Journal of Pharmaceutics* 442: 57–64.

Shang, C., Sinka, I C., and Pan, J. (2013b). Modelling of the break force of tablets under diametrical compression. *International Journal of Pharmaceutics* 445: 99–107.

Shipway, P.H. and Hutchings, I.M. (1993). Fracture of brittle spheres under compression and impact loading II. Results for lead-glass and sapphire spheres. *Philosophical Magazine A* 67: 1405–1421.

Sikong, L., Hashimoto, H., and Yashima, S. (1990). Breakage behavior of fine particles of brittle minerals and coals. *Powder Technology* 61: 51–57.

Silva, N.V., Angulo, S.C., da Silva Ramos Barboza, A., Lange, D.A., and Tavares, L.M., (2019). Improved method to measure the strength and elastic modulus of single aggregate particles. *Materials and Structures* 52: 1–15.

Studman, C.J. and Field, J.E. (1984). The influence of brittle particles on the contact between rigid surfaces. *Journal of Physics D: Applied Physics* 17: 1631–1646.

Tavares, L.M. (1999). Energy absorbed in breakage of single particles in drop weight testing. *Minerals Engineering* 12: 43–50.

Tavares, L.M., Cavalcanti, P.P., de Carvalho, R.M., da Silveira, M.W., Bianchi, M., and Otaviano, M. (2018). Fracture probability and fragment size distribution of fired iron ore pellets by impact. *Powder Technology* 336: 546–554.

Tavares, L.M. and King, R.P. (1998). Single-particle fracture under impact loading. *International Journal of Mineral Processing* 54: 1–28.

Timoshenko, S.P. and Goodier, J.N. (1970). *Theory of Elasticity.* McGraw-Hill.

Tromans, D. (2008). Mineral comminution: Energy efficiency considerations. *Minerals Engineering* 21: 613–620.

Tsoungui, O., Vallet, D., Charmet, J.-C., and Roux, S. (1999). Size effects in single grain fragmentation. *Granular Matter* 2: 19–27.

Unland, G. and Szczelina, P. (2004). Coarse crushing of brittle rocks by compression. *International Journal of Mineral Processing* 74: S209–S217.

Unland, G. and Al-Khasawneh, Y. (2009). The influence of particle shape on parameters of impact crushing. *Minerals Engineering* 22: 220–228.

Vallet, D. and Charmet, J.C. (1995). Mechanical behaviour of brittle cement grains. *Journal of Materials Science* 30: 2962–2967.

Wallin, K., Saario, T., and Törrönen, K. (1987). Fracture of brittle particles in a ductile matrix. *International Journal of Fracture* 32: 201–209.

Wang, C., Cheng, Y., He, X., Yi, M., and Wang, Z. (2019). Size effect on uniaxial compressive strength of single coal particle under different failure conditions. *Powder Technology* 345: 169–181.

Wang, W. and Coop, M.R. (2016). An investigation of breakage behaviour of single sand particles using a high-speed microscope camera. *Géotechnique* 66: 984–998.

Wang, J. and Elsworth, D. (2018). Role of proppant distribution on the evolution of hydraulic fracture conductivity. *Journal of Petroleum Science and Engineering* 166: 249–262.

Wittmann, F.H., Roelfstra, P.E., and Sadouki, H. (1985). Simulation and analysis of composite structures. *Materials Science and Engineering* 68: 239–248.

Wong, L.N.Y. and Jong, M.C. (2014). Water saturation effects on the Brazilian tensile strength of gypsum and assessment of cracking processes using high-speed video. *Rock Mechanics and Rock Engineering* 47: 1103–1115.

Wu, S.Z., Chau, K.T., and Yu, T.X. (2004). Crushing and fragmentation of brittle spheres under double impact test. *Powder Technology* 143–144: 41–55.

Wynnyckyj, J.R. (1985). The correlation between the strength factor and the true tensile strength of agglomerate spheres. *The Canadian Journal of Chemical Engineering* 63: 591–597.

Xiao, Y., Wang, L., Jiang, X., Evans, T.M., Stuedlein, A.W., and Liu, H. (2019a). Acoustic emission and force drop in grain crushing of carbonate sands. *Journal of Geotechnical and Geoenvironmental Engineering* 145: 04019057.

Xiao, Y., Yuan, Z., Chu, J., Liu, H., Huang, J., Luo, S.N., Wang, S., and Lin, J. (2019b). Particle breakage and energy dissipation of carbonate sands under quasi-static and dynamic compression. *Acta Geotechnica* 14: 1741–1755.

Zhang, Z.L., Kristiansen, H., and Liu, J. (2007). A method for determining elastic properties of micron-sized polymer particles by using flat punch test. *Computational Materials Science* 39: 305–314.

Zhao, B., Wang, J., Andò, E., Viggiani, G., and Coop, M.R. (2020). Investigation of particle breakage under one-dimensional compression of sand using X-ray microtomography. *Canadian Geotechnical Journal* 57: 754–762.

Zhou, C., Yue, H., Li, Y., Zhang, M., Liu, J., and Aijaz, S. (2019). A sphere filling algorithm for irregular aggregate particle generation based on nonlinear optimization method. *KSCE Journal of Civil Engineering* 23: 120–129.

2

Particles in Diametral Compression

The predominant method of particle strength measurement—diametral compression—is considered in some detail. Relevant concepts and terms regarding extensive and intensive mechanical properties of particles are described and defined, followed by consideration of observations and development of models describing quasi-static diametral compression force-displacement behavior. The results of stress analyses of cylindrical and spherical particles are described, enabling the definition of particle strength. Comparisons of experimental strength measurements from a range of quasi-static and impact particle loading configurations support the analyses. Brief overviews of particle size effects on strength and the strength empirical distribution function for describing strength variations introduce key themes of the book. The subjects to be treated in detail in subsequent chapters are outlined.

2.1 Extensive and Intensive Mechanical Properties

The mechanical performance characteristics of structural components are *extensive quantities*: measures of component performance that depend on the size or "extent" of the component. Examples include the maximum force supportable by a component, the relation between applied force and resulting displacements of component parts, and the relation between applied force and resulting lifetime of a component. Other quantities characterizing component performance, such as mass and electrical resistance, are similarly extensive. Two related features of extensive quantities are that a quantity is measured at a component exterior or external boundary, and, as a consequence, a quantity is measurable without regard to the details of component geometry or the material forming the component. An illustration of these features is provided by the loaded column in Figure 1.1a: The maximum force supportable by the column is measurable if the column is behind a wall. In this case, only the column ends are accessible for measurement and it cannot be known if the column is of large or small cross section or formed of steel or concrete.

In contrast, mechanical properties of materials are *intensive quantities*: such quantities depend only on the internal details of a material and not on the size or extent of the material element considered. The extensive performance characteristics of a component are related to the intensive properties of materials forming the component by geometrical characteristics of the component. A simple example is particle mass m, which is determined by the volume V of the particle and the (assumed constant) density ρ of the particle material and given by $m = \rho V$. This relation makes clear that mass m is an extensive quantity as it depends on the extent V of the component and can be measured (e.g. by a balance or scale) independently of ρ or V. Similarly, the relation makes clear that material density ρ is an intensive material quantity as it depends only on the choice of particle material, and can be measured independently of the absolute values of m or V (e.g. by the Archimedes method). An illustration is provided by consideration of a large body, volume Ω, of uniform material density ρ. If samples, volume V, are taken from Ω and the sample density determined, all measurements will return a value of ρ, independent of V.

In the context of load-bearing ability of structural components, relations similar to the component mass equation can be generated to describe mechanical phenomena. In particular, the extensive mechanical characteristics of particles, e.g. as listed above, the failure force or the force-displacement relation, can be expressed in terms of intensive mechanical properties of particle materials and particle geometry factors. In Chapter 1, the extensive mechanical characteristic of the failure force of a particle was discussed in the context of qualitative descriptions of particle failure mechanisms. Specifically, at small applied forces irreversible deformation was observed in the form of flats generated at particle poles. At large applied forces, fracture was observed in the form of cracks generated on particle meridians. Development of measurable quantitative criteria for the formation of flats and cracks requires relations between the force applied to the exterior of a particle, the

Particle Strengths: Extreme Value Distributions in Fracture, First Edition. Robert F. Cook.

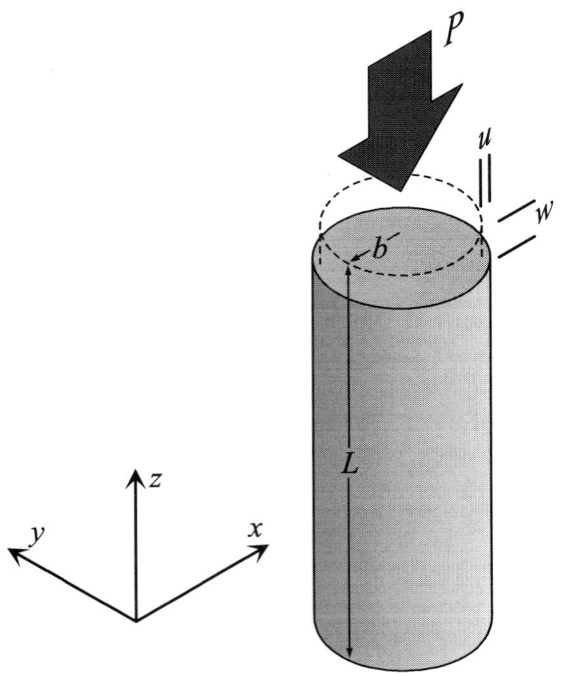

Figure 2.1 A schematic diagram of a cylindrical column, initial dimensions L and b indicated. Under the action of an applied force P, the column contracts axially by w and expands transversely by u.

mechanical deformation and fracture properties of the particle material, and the particle geometry. Such relations have been established in detailed analyses including materials properties and component geometry for many mechanical performance measures of common structural components (Ashby 1999). Here, the terms and concepts required to establish the relations between extensive and intensive mechanical behavior of particles in diametral compression are developed. Analysis begins by consideration of a simple compressive column.

Figure 2.1 shows a uniform circular section column acted on by a compressive applied force P that is distributed uniformly over the disc face of the column. Both the axis of the column and the direction of the force are oriented vertically, along the z direction of the xyz coordinate system shown. The initial length of the column is L, indicated by the dashed outline, and the lower end of the column is fixed at $z = 0$. The initial cross-sectional radius and area of the column are b and πb^2, respectively, also shown dashed. Under the action of the force, the upper end has contracted by displacement w, indicated by the bold outline. The displacement w is taken to be *elastic*, such that the $w(P)$ response is reversible and represents a set of equilibrium configurations. Further, the displacement behavior is taken to be linear, such that the displacement response is specified as $w(P) = \lambda P$, where λ is invariant. Such linear elastic behavior for most components usually requires $w/L \ll 1$. The constant λ is termed the *compliance* of the column. The linear elastic behavior can also be expressed as $P = kw$, where invariant k is termed the *stiffness* of the column and $k = 1/\lambda$. P, w, k, and λ are all extensive quantities. In this example, P and w are negative. k and λ are positive.

In order to compare columns of different dimensions, it is convenient to define the *engineering stress* σ acting on the column by $\sigma = P/\pi b^2$ and the *engineering strain* of the column ε by $\varepsilon = w/L$. The linear elastic behavior of the column expressed in terms of these quantities is thus $\varepsilon = (\lambda \pi b^2/L)\sigma$ or $\sigma = (kL/\pi b^2)\varepsilon$. The usual condition for observation of linear elasticity can thus be restated as one in which the strain is very small, $\varepsilon \ll 1$. Experiments show that these stress-strain expressions describe deformation of columns formed of the same material but with differing dimensions using an invariant combination of extensive parameters $kL/\pi b^2$. The implications are that σ and ε are intensive quantities describing the loading on the column and the deformation of the column, respectively, and that the invariant extensive combination can be expressed as $k = E\pi b^2/L$, where E is an invariant intensive material property. If the column is of moderate aspect ratio, $2 < L/b < 10$, the quantity E is termed the material *Young's modulus*. The expression for the intensive linear elastic stress-strain response of the material forming the column is thus $\sigma = E\varepsilon$. In this example, σ and ε are negative. E is positive. (If L/b is outside the range given, other modes of deformation can occur beyond simple compression, Ashby 1999; Sadd 2009).

In addition to the axial contraction w generated by the force P, internal connectivity of the column material leads to associated transverse expansions u and v in the x and y directions, respectively. For the column here, unconstrained transversely along its length, symmetry leads to identical displacements $u = v$ at all points on the exterior circular radius b of the column ($x^2 + y^2 = b^2$). For a linear elastic column, observations of the transverse displacements show that $(u/b) = -\nu(w/L)$. ν is a positive, invariant material property termed the *Poisson's ratio*. If the transverse strain is written as $\varepsilon_T = u/b$, the relation between transverse and axial strain is $\varepsilon_T = -\nu\varepsilon$. ν is intensive. An elastically isotropic material is one in which the stress-strain response of the material is the same in all directions and for which E and ν completely describe material deformation. Such a material is also elastically homogeneous, in which the deformation response is the same at all points. If the uniform column of Figure 2.1 is composed of an elastically isotropic material and is in equilibrium with the applied force, stress and strain are distributed homogeneously throughout the column. In terms of the density illustration above, consider the uniform density material of the large body Ω to be also elastically isotropic. If sample columns are taken from Ω and the sample elastic moduli determined, all measurements will return values of (E, ν), independent of b or L.

The work W performed in deformation of the linear elastic column by the applied force is expressed in extensive terms as $W = Pw/2 = \lambda P^2/2 = kw^2/2$. Substituting the expressions for stress and strain into the extensive work expressions gives $W = (\sigma\varepsilon/2)(\pi b^2 L) = (\sigma^2/2E)(\pi b^2 L) = (E\varepsilon^2/2)(\pi b^2 L)$. These expressions embody the first law of thermodynamics and make clear that the work performed by the applied force generates a change in the internal energy of the column. The change can be expressed as $W = U_E$, where the E subscript indicates *internal elastic energy* and $U_E = 0$ is set at $w = 0$. For example, internal elastic energy can thus be expressed as $U_E = (\sigma^2/2E)(\pi b^2 L)$; a product of a combination of intensive parameters within the first set of parentheses and a combination of extensive parameters in the second set. The other combinations of intensive parameters are similarly expressed. The combination of extensive parameters is recognized as the volume of the column, $V = \pi b^2 L$. The implication is that the combinations of parameters within the first sets of parentheses must represent the mean *elastic energy density* (energy/volume) of the column. Denoting the elastic energy density as $\mathcal{U}_E$ gives $U_E = \mathcal{U}_E V$, where, for example, $\mathcal{U}_E = \sigma^2/2E$ and the direct analogy with the mass equation is noted. U_E is extensive and $\mathcal{U}_E$ is intensive, both are positive. As stress and strain are distributed homogeneously throughout the column, elastic strain energy is also distributed homogeneously and $\mathcal{U}_E$ describes the elastic energy density at all points in the column.

Generalizations of the above concepts will be required in consideration of mechanical behavior of particles. The generalizations include non-linear force-displacement responses arising from both geometrical and material non-linearities, irreversible responses arising from inelastic material deformation, and inhomogeneous and anisotropic stress distributions arising from non-uniform component sections. Detailed considerations of stress, strain, elasticity, and work-energy relations that describe these generalizations are given elsewhere (Mase 1970). In brief, the extensive quantities of compliance and stiffness are generalized as derivatives, $\lambda = \mathrm{d}w/\mathrm{d}P$ and $k = \mathrm{d}P/\mathrm{d}w$, respectively. For linear elastic components, these quantities devolve to invariant values, otherwise λ and k may be functions of w, P, or time, and may even be negative. Examples are discussed in Section 2.2.1. Similarly, the material constitutive behavior, expressed as a relation between the intensive quantities of stress and strain is generalized to a derivative, $\mathrm{d}\sigma/\mathrm{d}\varepsilon = f(\varepsilon)$, where $f(\varepsilon)$ is a function of strain. For isotropic linear elastic materials, the function devolves to invariant combinations of E and ν, otherwise the function may be increasing or decreasing, describing "strain hardening" or "strain softening," respectively, and may also be a function of time. Examples are discussed in Section 2.2.2 and Chapter 11.

Extension to other stress states is made most simply by considering a uniform bar under tensile loading (Figure 1.1b), and reversing the directions and signs of the force, displacement, stress, and strain from those describing the column under compression discussed earlier. Schematic cross-section diagrams of the stress states in these components is shown in Figure 2.2, using arrows to represent the tractions (Mase 1970) acting on faces of square elements of material. The arrows facing inward to the elements uniformly along the length of the column indicate the uniform compressive stress state. The arrows facing outward similarly indicate uniform tension in the bar. An overview of particle stress behavior, to be discussed in detail in Section 2.3, is shown in Figure 2.3 using the same pictorial notation. The Figure shows a schematic cross section of the stress state in a cylindrical particle loaded in diametral compression. The stress state is inhomogeneous and anisotropic. The predominant stress component in the loaded particle is axial compression, consistent with the close geometrical relation to the column. As discussed in Chapter 1, particles have non-uniform sections. Hence, at the center of the particle, the axial compression is a minimum, reflecting the largest transverse diameter of the particle at that point and hence the largest cross-sectional area. At locations on the axis of the particle midway between the particle center and the poles, the axial compressive stress increases, as the transverse cross-sectional area decreases. At locations on the axis near poles, the axial stress is very large and depends less on the overall shape of the particle and more on the details of the contact. The secondary stress component in the loaded particle is transverse tension. The non-uniform sections under compressive stress give rise to non-uniform Poisson expansion effects that lead to tensile stress. At the center of the particle, the transverse

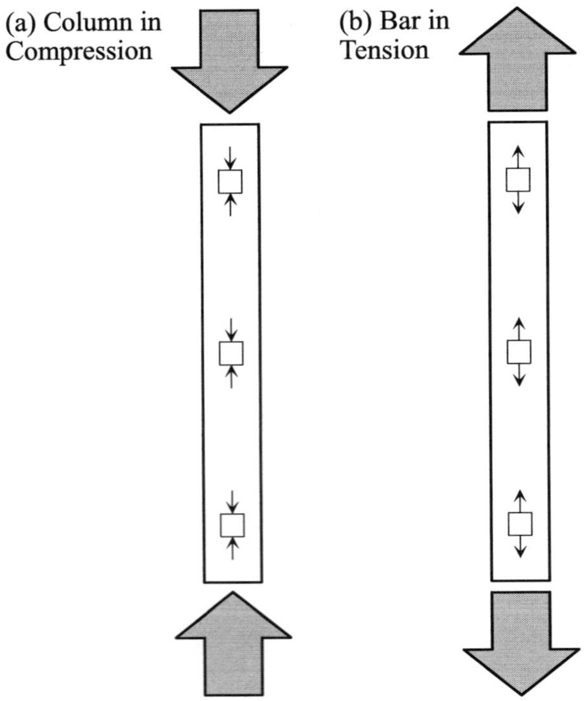

Figure 2.2 Schematic diagrams of uniform compressive and tensile stress states in a loaded column and a loaded bar. The large arrows indicate the directions of the applied forces. The small arrows indicate the directions of the resulting tractions acting on elements of material in each structural component.

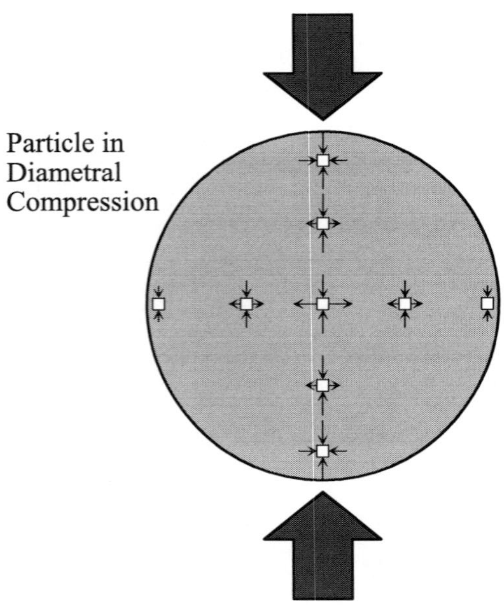

Figure 2.3 Schematic diagram of the inhomogeneous stress state generated in a cylindrical particle loaded in diametral compression. The axial compressive force is indicated by the large arrows. The small arrows indicate the directions of the resulting tractions acting on elements of material within the particle. The axial stress state is compressive and the transverse stress state is tensile; both sets of stress components decrease in magnitude near the particle periphery.

tension is a maximum, reflecting the combined effects of the entire strained particle. At locations on the equatorial plane midway between the particle center and the equator, the transverse tension decreases as Poisson effects decrease away from the axis. At locations on the equatorial plane near the equator, the transverse stress tends to zero as the traction-free surface is approached. The stress states of both cylindrical and spherical loaded particles are discussed in Section 2.3.

The following Section 2.2 builds on the concepts and terms introduced here regarding extensive and intensive quantities to describe observations and analyses of mechanical properties of particles. In particular, experimental observations described in extensive force-displacement terms are used as the foundation for developing models of material deformation described in intensive stress-strain terms. In particular, the brief summary of stress developed in a loaded particle given above, expressed in scalar terms, is sufficient for this purpose.

2.2 Particle Behavior in Diametral Compression

2.2.1 Force-Displacement Observations

Figure 2.4 shows in cross-section a typical configuration for a quasi-static single particle mechanical test. The particle is placed between two stiff, yield- and fracture-resistant, parallel platens that are displaced toward each other: contacting, deforming, and eventually breaking the particle. The relative platen displacement w is usually controlled by a universal mechanical testing machine that generates force P to maintain the imposed displacement and $P(w)$ is monitored as the particle is deformed. As in most contact problems, the force P and relative platen displacement w are positive and compressive when directed into the contact and across the particle diameter (Johnson 1985). This is opposite to the convention used to describe loading of most structural components, e.g. the column and bar above. The configuration of Figure 2.4 is one of several diametral loading or diametral compression configurations applied to particles (Figure 1.7). Unless stated otherwise, the *contact coordinate* direction convention will be used throughout this book in descriptions of force and displacement in all diametral compression configurations for particles. Stress and strain terms and definitions are unchanged from convention.

Many of the important phenomena associated with particles in diametral compression were established in an early study by Kapur and Fuerstenau (1967) on pelletized limestone ($CaCO_3$). Limestone powder, consisting of sub-particles approximately 30 μm in size, was rotated in a ball drum with water, about 48 % by volume fraction, to generate near-spherical agglomerate particles or pellets, with relative density of about 70 % and sizes ranging from $D = 8$ mm to 20 mm. The mechanical behavior of the limestone agglomerate particles was measured in diametral compression under constant displacement rate loading. A typical force-displacement $P(w)$ response, derived from the published work, is shown in Figure 2.5 (for brevity, here and throughout, data in Figures are identified by the first author and publication year of the source; citations to data sources are included in Figure captions). Distinctive features of the response are the linearity throughout loading until peak force P_{max} is reached, followed by an abrupt decrease in the force to near zero, both characteristics of *brittle* failure. The post-failure decrease in force exhibited a slope of much greater magnitude than the pre-failure increase in force. Inspection of the particles after testing showed that the tops and bottoms of the particles, the poles, formerly in contact with the platens of the test machine, were deformed into flat, horizontal, circular surfaces (orientation as in Figure 2.4). In addition, the particles had split into two hemispheres by fracture along a vertical plane containing opposing meridians. An example of the fracture and deformation pattern in a gypsum ($CaSO_4$, plaster) particle is shown in Figure 1.9a. Kapur and Fuerstenau also noted a weaker than expected increasing trend in particle failure force with increasing particle size, along with a considerable dispersion about the trend. Trends in failure force with particle size and dispersion in force will both be considered in much greater detail in Section 2.5, Section 2.6, and throughout; the section here will focus on pre-failure force-displacement behavior. Figure 2.5 also shows linear, brittle, force-displacement responses for two sets of porous alumina (Al_2O_3) particles: catalyst particles $D = 1.7$ mm to 2.0 mm, with data derived from the published work of Subero-Couroyer et al. (2003); and particles $D \approx 2$ mm formed by a sol-gel method, with data derived from the published work of Satone et al. (2017) (using the symbol $\approx$ to mean "approximately equal to"). In these cases, the post-failure unloading slopes were extremely steep. The failure of the alumina particles (Satone et al. 2017) was characterized by meridian fracture similar to Figure 1.9b. Other particle systems that exhibited similar failure characteristics are discussed in Section 1.2.

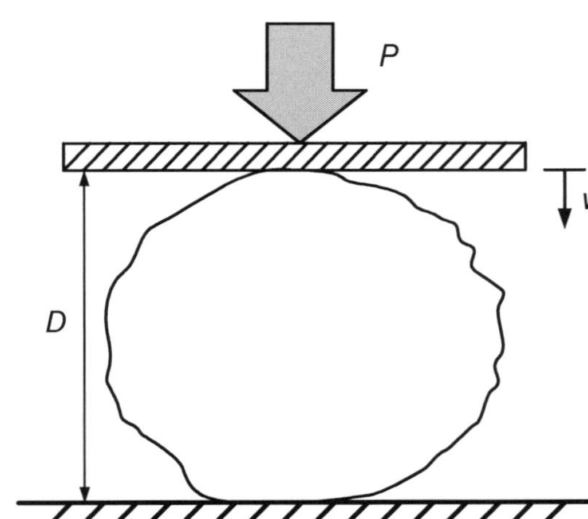

Figure 2.4 Schematic diagram of a particle loaded in diametral compression using a conventional platen apparatus. The particle, size characterized by diameter D, is deformed between two platens by imposing a displacement, w, and monitoring the required force, P. Often, contact is set by $P = w = 0$ when particle stiffness dP/dw exceeds a given small value.

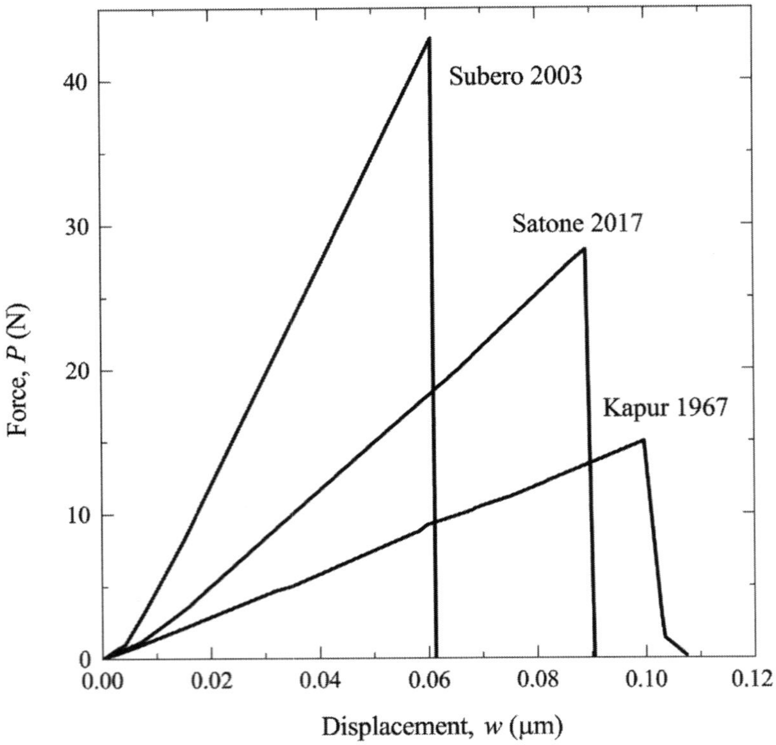

Figure 2.5 Linear force-displacement, *P-w*, responses for brittle particles tested in diametral compression. Responses for a $D = 6$ mm cement clinker (Adapted from Vallet, D et al., 1995), a $D = 1.7–2.0$ mm porous alumina catalyst particle (Adapted from Subero-Couroyer, C et al., 2003), and a $D = 2$ mm dense alumina particle (Adapted from Satone, M et al. 2017).

The simple linear, brittle force-displacement behavior of Figure 2.5 and the localized fracture and deformation morphology of Figure 1.9 provide a foundation for consideration of more complicated particle diametral compression responses. In particular, the most commonly observed force-displacement response, including nearly all materials and sizes of particles, is a small, but significant, variation on Figure 2.5: Most particle force-displacement responses include a short, curved, convex (upward curving) segment on initial loading followed by a predominant linear segment prior to peak load and brittle failure. Examples of this behavior are shown in Figure 2.6, arranged in order of increasing failure load P_{max}, with data derived from the published works: (a) Iron ore agglomerate, $D \approx 20$ mm, $P_{max} \approx 12$ N (Wynnyckyj 1985); (b) Polycrystalline alumina, $D = 1.9$ mm to 2.6 mm, $P_{max} \approx 80$ N (Portnikov and Kalman 2018; (c) Glass, $D = 1.0$ mm–1.25 mm, $P_{max} \approx 250$ N (Aman et al. 2010); (d) Marble-based ($CaCO_3$, $MgCO_3$, SiO_2) rock, $D \approx 25$ mm, $P_{max} \approx 4500$ N (Wang et al. 2015). A clear feature of Figure 2.6 is the similar shaped convex + linear force-displacement responses observed for ranges of particle material microstructure (agglomerate, polycrystalline, amorphous, multi-phase), particle size (1 mm to 25 mm), and particle failure load (12 N to 4500 N). The dashed lines are straight lines as guides to the eye. The similarity extends to many other particle systems, including materials both engineered and natural: cement (Vallet and Charmet 1995; Tavares and Cerqueira 2006) and concrete (Khanal et al. 2008); ceramics, including zirconia (ZrO_2) (Portnikov and Kalman 2014), blast furnace slag (Ribas et al. 2014), titania (TiO_2) (Herre et al. 2017), and a lithia (Li_2O)-based battery material (Dang et al. 2019); coal (largely carbon, C) (Dong et al. 2018; Wang et al. 2019); coral (largely limestone) (Ma et al. 2019); food and food additives, including sodium benzoate (a preservative) (Pitchumani et al. 2004; Antonyuk et al. 2005, 2010) and peppercorns (Singh et al. 2016); water ice (Kuehn et al. 1993); the polymer polypropylene (Portnikov and Kalman 2018); salt (NaCl) (Aman et al. 2010; Portnikov et al. 2013; Portnikov and Kalman 2014, 2018, 2019; Liburkin et al. 2015); and many silica (SiO_2)-based systems, including quartz and rice husk ash (Ribas et al. 2014), and siliceous and calcareous sands (Yasufuku and Kwag 1999; Kwag et al. 1999; McDowell and Amon 2000; Nakata et al. 2001b; Wang and Coop 2016; Liu and Wang 2018; Jarrar et al. 2020). The similarity of many particle force-displacement responses to those shown in Figure 2.6 is further noted by recognizing that the above list includes a broad diversity of particle sizes. The sizes extend from the very small, $D \approx 600$ nm TiO_2 particle tested in a scanning electron microscope (SEM) (Herre et al. 2017) and $D \approx 5$ μμm–20 μm Li_2O-based particles tested in a nanoindenter (Dang et al. 2019), through the more usual mm-scale, e.g. $D \approx 1.1$ mm to 1.8 mm sodium benzoate particles tested in a custom robotic instrument (Pitchumani et al. 2004), to the very large, $D \approx 150$ mm concrete spheres tested in a universal testing machine and exhibiting failure loads $P_{max} \approx 80$ kN (Khanal et al. 2008).

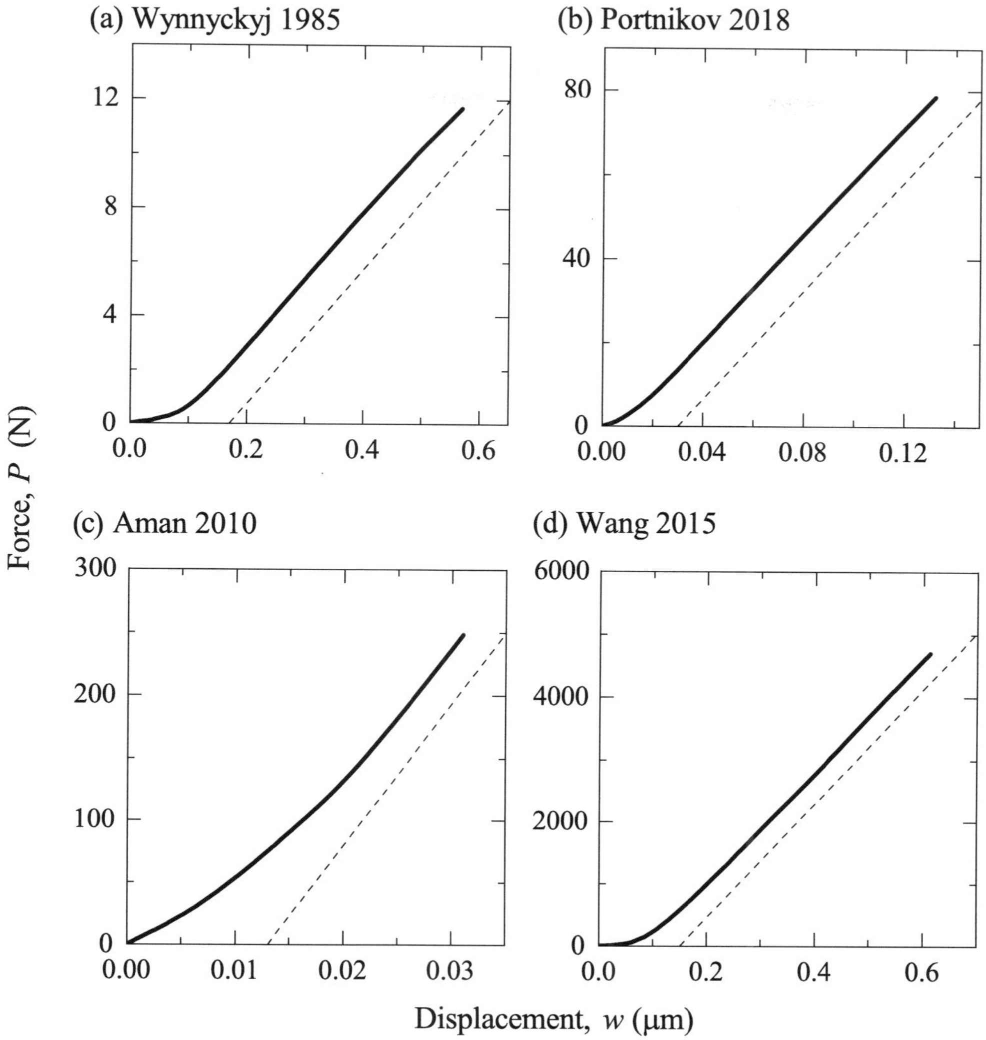

Figure 2.6 Convex + linear force-displacement, *P*-*w*, responses for particles tested in diametral compression. (a) Iron ore agglomerate, $D \approx 20$ mm, (Adapted from Wynnyckyj, J.R 1985); (b) polycrystalline alumina, $D = 1.9–2.6$ mm, (Adapted from Portnikov, D et al., 2018); (c) glass, $D = 1.0–1.25$ mm, (Adapted from Aman, S et al. 2010); and (d) marble-based ($CaCO_3$, $MgCO_3$, SiO_2) rock, $D \approx$ 25 mm, (Adapted from Wang, Y et al. 2015). Dashed lines are a guide to the eye.

A frequently observed variation on the smooth $P(w)$ loading responses of Figure 2.6 is the superimposition of small-amplitude erratic force and displacement behavior. An example, observed on loading of a coral particle (Ma et al. 2019) is shown in Figure 2.7a, using data derived from the published work. In mechanical terms, the erratic behavior is described by intermittent compliance, $\mathrm{d}w/\mathrm{d}P$, increases superimposed on the smooth compliance behavior of the responses of Figure 2.6 (stiffness, $\mathrm{d}P/\mathrm{d}w = 1/\text{compliance}$). If the mechanical test system is displacement controlled, intermittent compliance increases lead to force or "load" drops (e.g. Herre et al. 2017). If the system is force or load controlled, intermittent compliance increases lead to displacement excursions (e.g. Dang et al. 2019). Much more usual, however, is the intermediate case in which the test system is compliance controlled. In this case load and displacement are imposed on the particle through an effective spring represented by "machine compliance" λ_{machine}. Such compliance characterizes the combined effects of flexure of the loading fixture, flexure of the test instrument, and control of the test protocol. In compliance controlled loading, intermittent compliance increases can generate events in the observed mechanical response characterized by simultaneous load drops and displacement excursions. The event slopes in $P(w)$ graphs are given by $-1/\lambda_{\text{machine}}$. An example of many such events is observed in the loading of the coral particle in Figure 2.7a, leading to a somewhat erratic force-displacement response, although, as expected, most events have similar descending slopes. The overall linear loading behavior is clear as is a compliant initial loading segment. The large number of compliance events in addition to the initial compliance segment are somewhat unusual, and are shown here to illustrate the effects. More usual is one or two compliance events, typified by observations on sand (Yasufuku and Kwag 1999; Kwag et al. 1999; McDowell and Amon

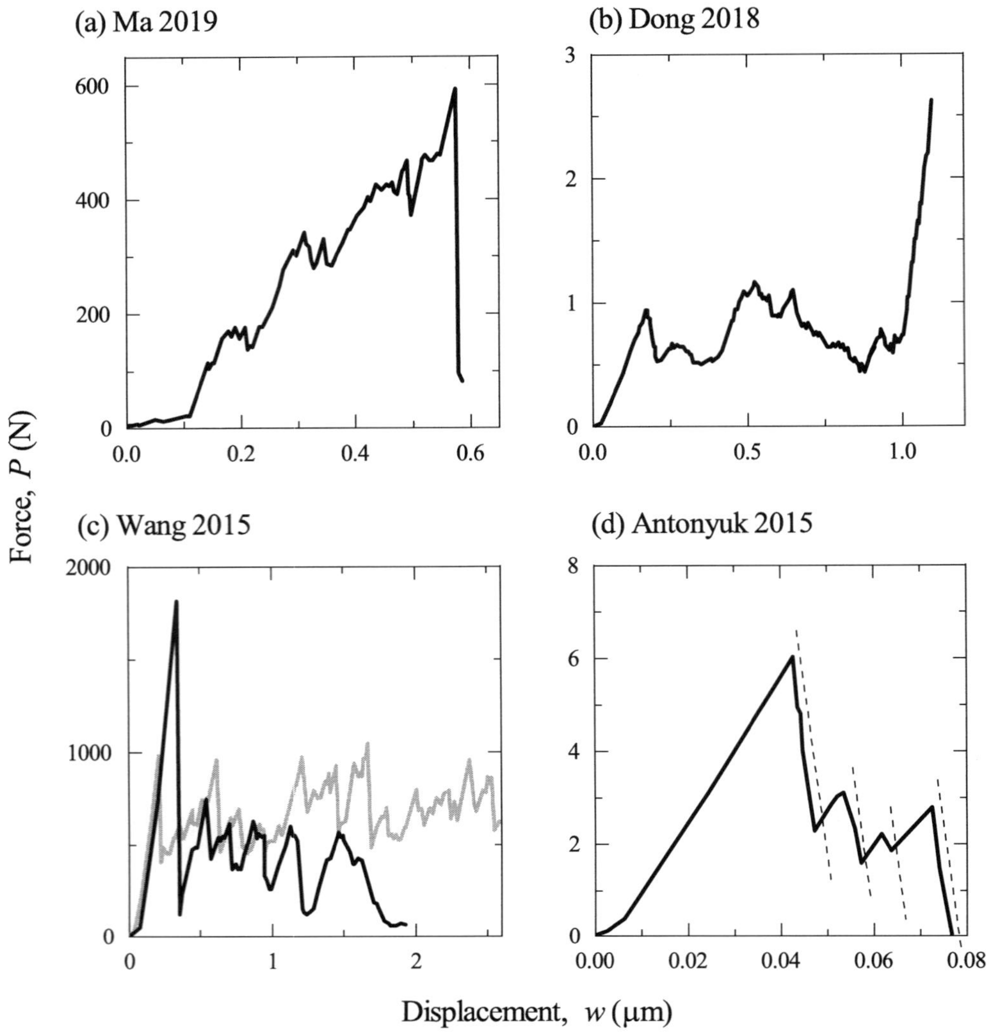

Figure 2.7 Variations on smooth $P(w)$ loading responses for particles in diametral compression: (a) Superimposition of small-amplitude erratic force and displacement behavior in a coral particle (Adapted from Ma, L et al. 2019); (b) intermittent large-amplitude compliance increases superimposed on almost invariant mean force in a coal particle (Adapted from Dong, J et al. 2018); (c) intermittent large-amplitude compliance increases following peak load with slowly decreasing mean load in rock particles (Adapted from Wang, Y et al. 2015); and (d) erratic force decreases, all of similar negative slope, until zero force is reached following peak load in a zeolite particle (Adapted from Antonyuk, S et al. 2005). Dashed lines are a guide to the eye.

2000; Nakata et al. 2001b; Jarrar et al. 2020). Other particle systems that exhibit intermittent compliance effects on loading, similar to the materials illustrated in Figure 2.7a, are: alumina (Verspui et al. 1997; D. Wu et al. 2007; Ogiso et al. 2007; Antonyuk et al. 2005, 2010; Calvié et al. 2012, 2014; Portnikov and Kalman 2018); agglomerates and porous media (Kendall and Wiehs 1992; Müller et al. 2013), including silica gel, zeolites, and fertilizer pellets (Antonyuk et al. 2005, 2010; Portnikov and Kalman 2014, 2018; Portnikov et al. 2019); iron ore pellets (Cavalcanti and Tavares 2018, Tavares et al. 2018); glass spheres (Portnikov and Kalman 2014; Paul et al. 2014, 2015; J. Huang et al. 2014; Parab et al. 2017; Pejchal et al. 2017; Shan et al. 2018; Silva et al. 2019), and rocks, including granite, limestone, and shale (Ribas et al. 2014; Nad and Saramak 2018; Silva et al. 2019).

A series of intermittent large-amplitude compliance increases on particle loading prior to P_{max} is observed less frequently. The increases lead to erratic $P(w)$ behavior superimposed on almost invariant mean force with increasing displacement. An example, observed on loading of a coal particle (Dong et al. 2018) is shown in Figure 2.7b, using data derived from the published work. In contrast to this example, behavior that is observed extremely frequently is a series of intermittent large-amplitude compliance increases *after* peak load, leading to erratic $P(w)$ behavior with slowly decreasing mean load. Two examples observed on loading of rock particles by Wang et al. (2015) are shown in Figure 2.7c, using data derived from the

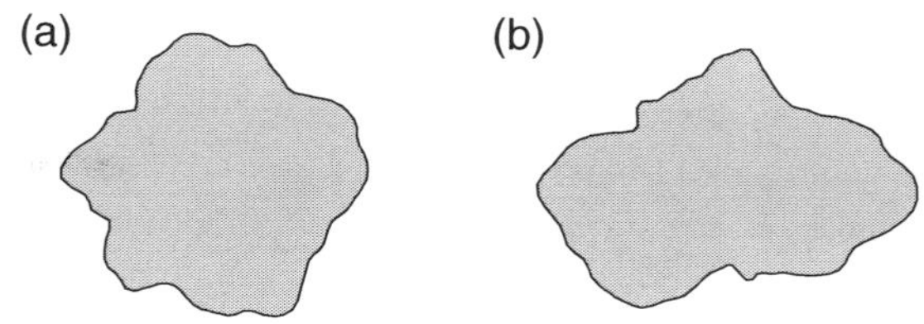

Figure 2.8 Simulated shapes of rock particles using the Fourier-based analysis of Section 1.2 for (a) an equiaxed particle and (b) a prolate particle, similar to those studied by and adapted from Wang, Y et al. 2015. Equiaxed particles exhibited less fragmentation on failure in diametral compression, implying a smaller strength.

published work. The overall $P(w)$ behavior of the rock particles is similar. Loading is smooth and exhibits a similar slope for both particles (suggesting that the particles have similar compliance). Following P_{max}, the decrease in force is abrupt (suggesting that the machine compliance is small) and followed by a segment of erratic behavior at near constant force for both particles. In detail, P_{max} differs between the particles by a factor of approximately two and the particle with the smaller P_{max} value (grayed line) exhibits an erratic region that is extended in displacement and has greater mean force. The P_{max} observations of Wang et al. are in agreement with the particle size effects noted by Kapur and Furstenau (see Section 2.5); of interest here are the effects of particle shape on the erratic $P(w)$ behavior following P_{max}. Wang et al. (2015) noted that equiaxed rock particles exhibited $P(w)$ behavior with the greater P_{max} value and reduced erratic segments (black line). An example of such a particle, simulated using the Fourier-based shape analysis of Section 1.2 and setting $d_2 = d_3 = 0$, is shown in Figure 2.8a. Conversely, rock particles that were prolate, with aspect ratios of 2 or greater exhibited behavior with the lesser P_{max} value and extended erratic segment. An example of such a particle, simulated using $d_2 = 0.25$ and using Eq. (1.3), is shown in Figure 2.8b, setting the long axis of the particle horizontal. Following initial failure, equiaxed particles exhibited few fragments separated by well-defined fracture planes containing the particle meridians. Prolate particles exhibited many fragments with many, intersecting, fracture planes. The likely explanation, following many observations of component fragmentation on brittle fracture (Cook 2021), is that both equiaxed and prolate particles failed in the meridional crack manner described in Section 1.1.2, generating fragments on failure that were crushed after peak force. As noted by Wang et al., the long axes of the prolate particles tended to lie horizontally (Figure 2.8b) on the test platen such that the fragments generated at failure differed from those of the equiaxed particles. It is unlikely that more asperities contacted the platen during loading of the prolate particles leading to greater fragmentation on failure. The elastic properties of the particle material would preclude such multi-asperity contact and such fragmentation has not been observed elsewhere. In addition, as will be discussed in detail below, the stress component of relevance for particle failure is not greatly perturbed by surface asperity effects.

An additional observation regarding erratic behavior following P_{max} is illustrated in Figure 2.7d, which shows the $P(w)$ behavior of a zeolite particle (Antonyuk et al. 2005), using data derived from the published work. The loading to P_{max} is smooth as in Figure 2.6 and is followed by a number of erratic force drops until zero force is reached. The force drops represent the machine compliance and are all of similar negative slope, $-1/\lambda_{machine}$. Figure 2.7d is a particularly clear example of this phenomenon and the dashed lines in the figure are a guide to the eye and highlight the slope similarity. These sloped decreases, observed after peak load, are exactly analogous to those observed prior to peak load in Figure 2.7a. Figure 2.7c and 2.7d make clear that the force exerted by the loading system to maintain extended platen displacement following peak load is often significant. As a consequence, the work performed by the loading system (given by the area under the $P(w)$ response) following peak load is often much greater than the work performed prior to peak load. This observation is of great importance in applications in which the energy required to crush particles (e.g. concrete production, rock mining) is a key factor. Work and energy in particle failure are discussed in Chapter 13. It should be noted that all the phenomena discussed here regarding force-displacement behavior were reported as force-*time* behavior from the earliest studies of rapid particle loading: Linear loading, initial convex or compliant segments, and force drops and erratic behavior before and after P_{max} have been reported as functions of time for alumina abrasive particles (Brecker 1974; Huang et al. 1993), aluminosilicate agglomerates (Wong et al. 1987), quartz and other mineral particles (King and Bourgeois 1993; Tavares and King 1998), pharmaceutical agglomerates (Adi et al. 2011), and iron ore pellets (Gustafsson et al. 2017). In most of these cases, loading was applied by impact or counter-rotating rollers (and is discussed further below) and in many of these cases, the work performed or energy expended by the loading system during particle failure was of primary interest (see Chapter 13).

The next section focuses on particle behavior prior to peak force and develops models for force-displacement responses consistent with the observations in this section (Figures 2.5 and 2.6) and other observations of pre-failure mechanical responses of particles.

2.2.2 Force-Displacement Models

Development of a model for mechanical behavior of particles in diametral compression begins by noting that the force-displacement trajectories of such particles are similar to those for spherical indentation of planar surfaces. Figure 2.9 highlights the similarity by displaying force-displacement measurements from load-unload cycles in the two tests configurations: diametral compression of near-spherical particles, left column, and spherical indentation of flat surfaces, right column. At the head of each column is a schematic cross-section of the loading configuration: components are shown shaded, testing fixtures are shown hatched. The similarity of the component-fixture contact geometries is clear. The components for diametral compression were 1 mm to 10 mm in diameter and measurements were performed in universal testing machines operated in displacement control. The probes for spherical indentation were $\approx 10\ \mu m$ in diameter and measurements were performed in nanoindenters in force control.

Figure 2.9a (top left) shows the diametral compression $P(w)$ behavior of an alumina particle during load-unload cycles (Antonyuk et al. 2010), using data derived from the published work. The arrows indicate the directions of loading, consistent with Figure 2.4 and the schematic diagram. The force-displacement cycles for two different P_{max} levels are shown and are hysteretic. The increasing force response on loading is common to both levels and consists of predominantly linear behavior, similar to the loading responses shown in Figure 2.5. The solid straight lines in Figure 2.9 are guides to the eye. The decreasing force responses are also predominantly linear and of steeper slope. Figure 2.9b (top right) shows the $P(\delta)$ behavior of a flat steel block indented in a load-unload cycle with a spherical probe (Field and Swain 1993), using data derived from the published work. Displacement δ for indentation is characterized relative to the indenter-surface contact point at which $P = \delta = 0$ is set. The overall force-displacement behavior is hysteretic, the loading response is linear, and the unloading response is predominantly linear of steeper slope. A similar linear spherical indentation response was observed in aluminum, copper, and tungsten (Field and Swain 1995; Ni et al. 2004). Figure 2.9c (bottom left) shows the cyclic $P(w)$ behavior of an iron ore particle in diametral loading (Gustafsson et al. 2013a), using data derived from the published work. In this case, the increasing force response consists of the commonly observed combination of an initial convex segment followed by a linear segment, similar to the loading responses shown in Figure 2.6. The decreasing force responses are initially linear and become convex at small forces. Similar convex + linear hysteretic cyclic responses were observed for a zeolite particle in diametral loading (Antonyuk et al. 2010). Figure 2.9d (bottom right) shows the $P(\delta)$ behavior of soda-lime silicate glass indented in a load-unload cycle with a spherical probe (Swain 1999), using data derived from the published work. The overall force-displacement behavior is hysteretic and the loading and unloading responses are convex+linear. Similar responses have been observed for cyclic spherical indentation of gold (Bell et al. 1991) and hardened steel (Field and Swain 1993). The similarities of the shapes of the force-displacement responses in Figure 2.9a and 2.9b (predominantely linear) and in Figure 2.9c and 2.9d (convex + linear) are clear. As might be anticipated, the similar responses reflect the similar contact geometries: diametral compression—deformable curved surfaces loaded by stiff flats; indentation—deformable flat surface loaded by a stiff sphere. Hence, the mechanics of particles in diametral compression closely follows spherical contact mechanics.

Elastic contact between a sphere and a flat is a well analyzed problem in mechanics (Gladwell 1980; Johnson 1985; Hills et al. 1993; Maugis 2000; Ling et al. 2002). Analysis of the problem lies in a broader set of analyses considering elastic contacts between different materials (characterized by different elastic moduli) comprising components of different geometries (characterized by different curvatures). The case considered most often is that of a rigid (zero compliance) spherical indenter contacting a flat (zero curvature) surface of an elastic material at normal incidence, shown in cross-section in Figure 2.10a. As shown, it is convenient to use indentation coordinates to describe such contacts: the indentation force, P, and displacement, δ, applied by the sphere to the material are zero at the point of contact and positive when directed into the surface (as used in Figure 2.4). If P or δ are controlled such that the resulting contact radius, a, is small relative to the sphere radius, R, ($a \ll R$), the contact is referred to as Herztian, following original analysis of this contact problem by Hertz (1896) (originally published 1881, see also Johnson 1985). Simple dimensional scaling arguments provide the essential Hertzian results. Using the symbol $\sim$ to mean "varies as," the characteristic indentation stress generated in the material during contact is given by $\sigma \sim P/a^2$ and the characteristic indentation strain is $\varepsilon \sim a/R$. As the contact is small, the strains are small and the material behaves linearly elastically, $\sigma = M\varepsilon$. M is an indentation elastic modulus that takes into account geometrical details such as the exact shape of the indenter and material anisotropy. For smooth contacts by rigid spheres with elastically isotropic materials, $M = E/(1 - \nu^2)$, where E and ν are the Young's modulus and Poisson's ratio of the material, respectively (and M in this case is often referred to as the plane strain modulus). Combining the stress and strain relations gives the variation of the contact radius with indentation load as $a^3 \sim RP/M$. Further, for such

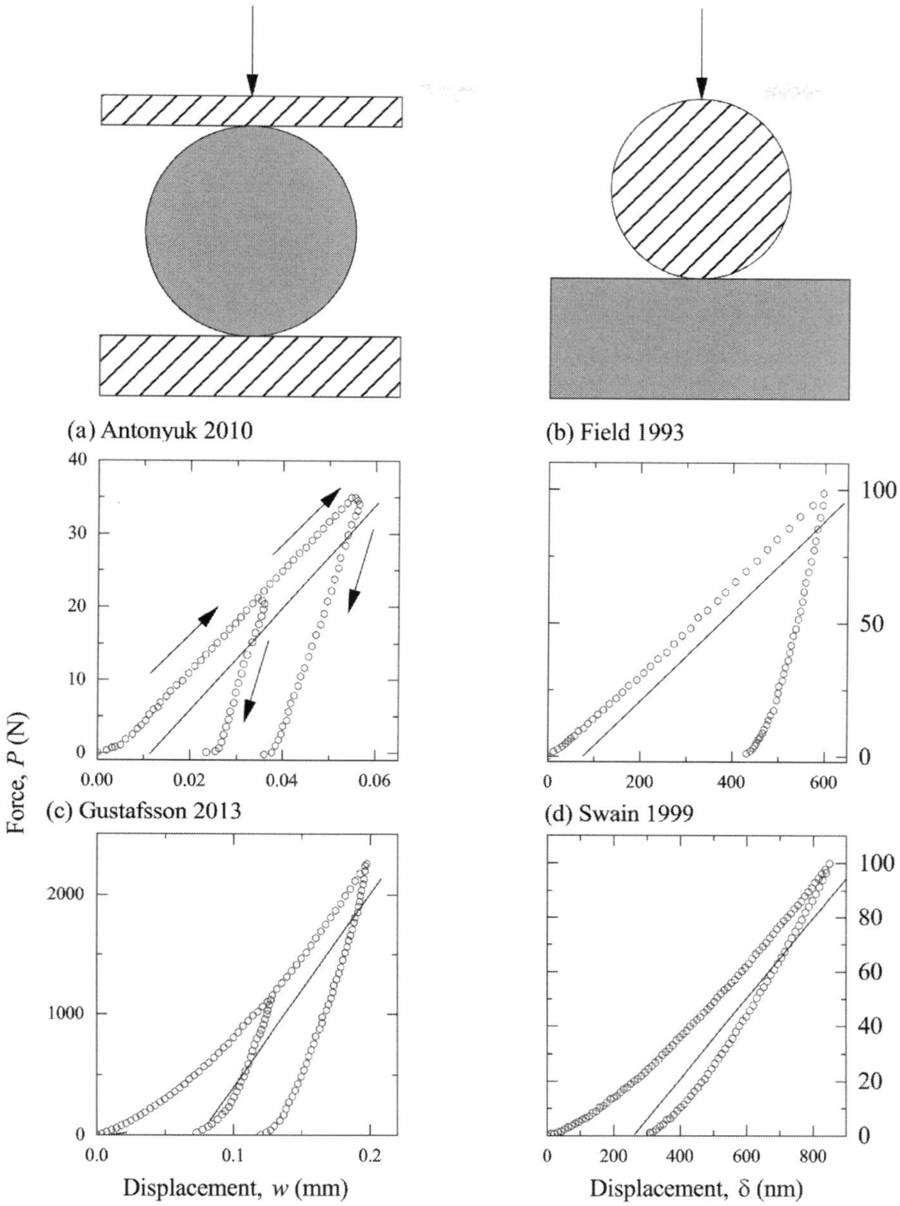

Figure 2.9 Plots of force-displacement trajectories showing similar responses for particles in diametral compression, left column, and spherical indentation of planar surfaces, right column. (a) Diametral compression $P(w)$ behavior of an alumina particle during load-unload cycles (Adapted from Antonyuk, S et al. 2005); (b) spherical probe indentation $P(\delta)$ behavior of a flat steel block indented in a load-unload cycle (Adapted from Field, J.S et al. 1993); (c) Diametral compression $P(w)$ behavior of an iron ore particle during load-unload cycles (Adapted from Gustafsson, G et al. 2013a); and (d) spherical probe indentation $P(\delta)$ behavior of soda-lime silicate glass indented in a load-unload cycle (Adapted from Swain, M.V 1999). Solid lines are a guide to the eye.

small contacts, the indentation contact radius and displacement are related by a parabolic approximation, $\delta \sim a^2/R$, such that the Hertzian force-displacement relation is thus $P \sim MR^{1/2}\delta^{3/2}$ (for $\delta/R \ll 1$). Elastic (reversible) Hertzian force-displacement behavior has been observed at small indentations in glass (Bell et al. 1991; Ni et al. 2004; Oyen 2011), steel (Field and Swain 1993), and silicon (Swain and Menčík 1994). Further analytical details, including specification of the proportionality constants in the above expressions and the distributions of stress and strain in the inhomogeneous indentation deformation field, can be found in the extensive earlier works (Gladwell 1980; Johnson 1985; Hills et al. 1993; Maugis 2000; Ling et al. 2002).

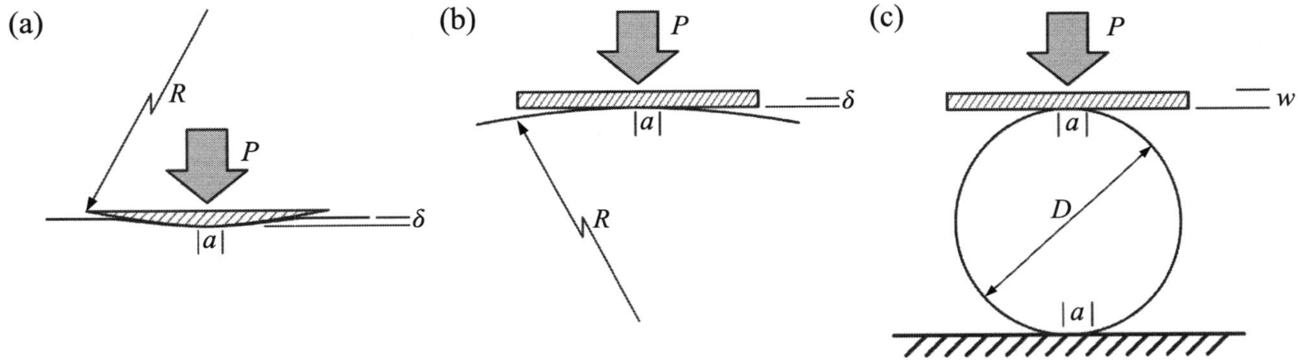

Figure 2.10 Schematic cross-section diagrams illustrating spherical-flat contacts. (a) Rigid sphere, radius R, indenting a flat material surface with force P, leading to contact displacement δ and contact radius a. This is the most common configuration of elastic Hertzian contact. (b) Rigid flat compressing a curved material surface, radius R, with force P, leading to contact displacement δ and contact radius a. This configuration may also be elastic Hertzian contact and represents particle impact. (c) Flat platens compressing a particle, diameter D, with force P, leading to relative platen displacement w and contact radii a. Each contact may be Hertzian and this configuration represents the common particle test of diametral compression.

The detailed analyses make clear that the Hertzian contact is symmetric—the same relations hold if the geometry of contact is reversed such that a rigid flat indents an elastic spherical surface. In both cases, non-linear force-displacement responses arise from geometrical non-linearity of systems composed of linear elastic materials. Distinct from the column considered in Section 2.1, in which the diameter $2b$ supporting the force applied to the column is invariant, Figure 2.1, the diameters $2a$ of the spherical-flat contacts in Figure 2.10a and 2.10b change with contact force, leading to geometrical non-linearity. The geometry of the system of interest here, a sphere compressed between two rigid platens, Figure 2.10c, can be examined similarly to the spherical-flat contacts. In this case, the measured displacement w in symmetric diametral compression loading is given by $w \approx 2\delta$. For the very small displacements imposed during initial sphere compression, the elastic Hertzian analysis above pertains, $P \sim \delta^{3/2}$, as has been demonstrated in axial force-displacement measurements in many particle systems, e.g. (Kendall and Weihs 1992; Antonyuk et al. 2005; Satone et al. 2017).

Transverse force-displacement measurements, in particular of the contact radius a, have also been demonstrated for particles, consistent with Hertzian analysis. Chen et al. (2006, 2007) used an in situ optical technique to measure pressure variation and infer contact radius for spherical ruby particles loaded in diametral compression. The particles, radius $R = 78$ μm, were compressed between two sapphire platens in a displacement controlled custom instrument. A schematic diagram of the test configuration is shown in Figure 2.11a. Ruby and sapphire are both single crystal Al_2O_3 (corundum) and differ only in doping, usually Cr in ruby and Ti in sapphire. In this case, the indentation modulus is given by the "reduced" modulus value $M = E/2(1 - \nu^2) \approx 214$ GPa, where $E \approx 390$ GPa and $\nu \approx 0.3$ are the Young's modulus and Poisson's ratio of Al_2O_3 and the factor of 2 accounts for joint deformation in the particle and the platen (Johnson 1985). The applied force in the contact experiments was $P = 4.3 \pm 0.3$ N, where the uncertainty represents instrument calibration effects; particle failure occurred at $P = 7$ N. Pressure variation was measured by spectroscopic determinations of the shift in the ruby fluorescence lines using a scanning confocal microscope arrangement. Recent reviews and presentations of analysis methods in application and calibration of stress measurement in Al_2O_3 using ruby fluorescence are given in Cook and Michaels (2017, 2019). The analysis method is applied here to the particle loading data of Chen et al. in determination of the "mean stress," reported as a pressure. The focus in this section is on pressure variation at a contact. Section 2.3 briefly discusses global pressure variations throughout a particle.

Figure 2.11b shows the variation in pressure $p(r)$ obtained from spectroscopic scans performed adjacent to loaded particle poles, using data derived from the published work. The scans were located along a transverse direction, indicated by the filled symbols in Figure 2.11a; scan location r is measured relative to the loading axis. The results from four separate particle scans are shown as the symbols and lines in Figure 2.11b. The variation in pressure beneath a Hertzian contact is given by $p = p_0(1-r^2/a^2)^{1/2}$, where a is the contact radius seen earlier and p_0 is the peak pressure generated in the material beneath the probe on the load axis ($r = 0$) of the contact (Johnson 1985). For ideal Hertzian behavior, the contact radius is given by $a = (3RP/2M)^{1/3}$ and the peak pressure is given by $p_0 = 3P/2\pi a^2$. The contact configuration parameters provided above

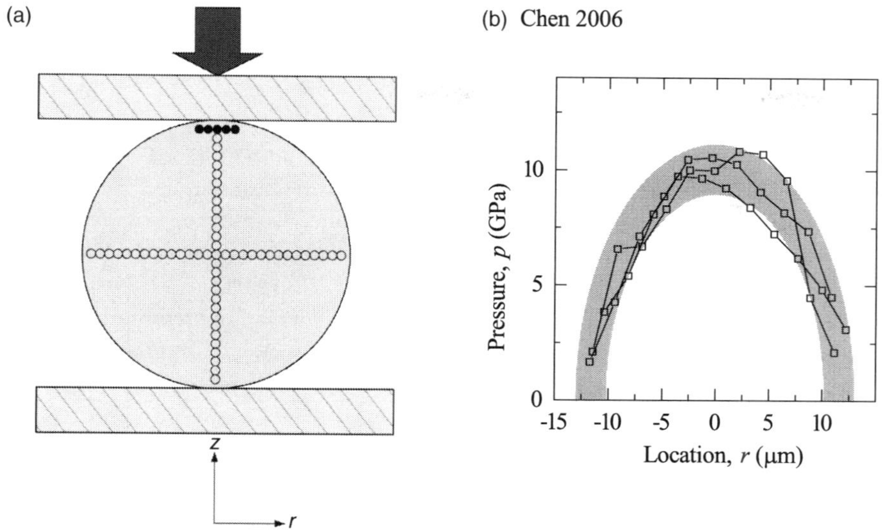

Figure 2.11 (a) Schematic cross-section diagram of a ruby sphere compressed between two sapphire flat platens (sphere and platens are both single crystal Al_2O_3). The filled and open symbols indicate the location of fluorescence scans used to determine pressure variation in situ in a loaded particle. (b) Plot of transverse pressure variations measured at the filled symbol locations adjacent to a particle pole (Chen et al. 2006, 2007). Symbols represent separate scans, shaded band represents best-fit elastic Hertzian contact behavior.

give predicted values of $a = (13.3 \pm 0.4)$ μm and thus $p_0 = (11.6 \pm 1.0)$ GPa, where the uncertainties reflect the uncertainty in the applied force. The shaded band in Figure 2.11b shows a visual best fit to the scan data using the parabolic Hertzian pressure variation expression, yielding fit parameters of $a = (11.7 \pm 1.3)$ μm and $p_0 = (10.0 \pm 1.1)$ GPa, where the values and uncertainties indicate the bounds of the shaded band and fit. The sapphire-ruby flat-on-sphere contacts at this applied force are well described by ideal Hertzian expressions, both qualitatively (the parabolic variation) and quantitatively (the fit parameters are consistent with predictions). The values of a and R imply the characteristic indentation strain in these tests was $\varepsilon \approx 0.15$ and that the related platen displacement was $\delta \approx 1.8$ μm. Although the indentation strain was large and probably near the upper limit of applicability of ideal Hertzian elastic analysis, no permanent deformation flats similar to those described in Chapter 1 were observed, although residual circular contact markings were visible following contact at either the measurement force or the failure force. This behavior is consistent with observations of the permanent contact deformation pressure ("hardness") of sapphire of about 20 GPa (Cook 2019), about twice the pressure generated in these experiments.

For subsequent, larger displacements, in these particle systems and others discussed previously, e.g. Figures 2.5, 2.6, and 2.9, significant deviations from the Hertzian response are observed, requiring modification of the above analysis. Consideration of the system suggests that the required modification is to the material constitutive law: The dynamics of the system require the characteristic contact stress σ to scale as $\sigma \sim P/a^2$ whatever the contact radius, and in fact $\sigma = P/(\pi a^2)$ is the magnitude of the mean contact stress or pressure. The energetics of the system then require the conjugate characteristic strain magnitude ε to scale as $\varepsilon \sim a/R$ (Mase 1970), where the particles are assumed to be sufficiently regular such that $D \approx 2R$. It is recognized that local values of stress and strain within the inhomogeneous contact field may vary considerably from σ and ε, respectively, particularly close to the contact. The much smaller ratios δ/R or w/D are characteristic strains for an entire particle and are useful for scaling diametral compression data from different size particles (e.g. Gustafsson et al. 2013a) but are not conjugate to the contact stress σ. Such ratios are considered below. Similarly, the kinematics of typical contacts suggest that the parabolic approximation $\delta \sim a^2/R$ is usually appropriate and that the full expression that makes circular corrections of order $(a/R)^2$ is not required. ("Rough" particles, for which local contact values of R may be much less than $D/2$ and at which variations in deformation and stress are significant are considered briefly in Chapter 12.)

Spherical indentation behavior as in Figure 2.9 is well described by material constitutive laws that include reversible and irreversible components. For most materials, the reversible component is linear elastic and characterized by an invariant value M. (For elastomers and gels, the reversible component is non-linear. Elastic variations in M are considered in detail in Chapter 11.) For metals, the irreversible component is associated with dislocation-mediated plasticity (Field and Swain 1993). For ceramics and rocks, the irreversible component is associated with shear-fault mediated damage or quasi-plasticity (Hagan and Swain 1978; Hagan 1979; Rhee et al. 2001; Momber 2004). The similarity of the hysteretic $P(w)$ and $P(\delta)$ responses in Figure 2.9 implies that particle material constitutive laws in diametral compression also consist of reversible and irreversible components. For the brittle particles of interest here, it is likely that the irreversible component consists of sub-surface pore closure, in addition to shear faulting and sliding, along with surface asperity crushing (pores and asperities are not factors in indentation of most dense and smooth ceramic and rock components, Hagan 1979; Rhee et al. 2001; Momber 2004). The onset of irreversible deformation is marked by a slight discontinuity in derivative of $P(w)$ loading responses and subsequent suppressed P values, Figure 2.9, and the effects of irreversible deformation are observed in the permanent flats formed at the contacts, Figure 1.9. Irreversible deformation associated with particles in diametral compression will be referred to here as "compaction," with recognition that the term encompasses densification, shear, and crushing.

Modification to the material constitutive law to include compaction begins by noting that for the elastic Hertzian system, $d\sigma/d\varepsilon = M$. At a critical strain on loading, compaction effects decrease the stress increment required to generate a strain increment for a material element, such that

$$d\sigma/d\varepsilon = M\,[1 - (x+1)(\varepsilon - \varepsilon_C)^x], \quad \varepsilon > \varepsilon_C, \tag{2.1}$$

where ε_C is a critical strain for the onset of compaction and $x \geq 0$ is an exponent that controls the form of the stress decrease. More insight is provided by integrating the above equations and re-arranging to give,

$$\frac{\sigma}{M} = \varepsilon, \quad \varepsilon \leq \varepsilon_C, \tag{2.2}$$

and

$$\frac{\sigma}{M} = \varepsilon - (\varepsilon - \varepsilon_C)^{x+1}, \quad \varepsilon > \varepsilon_C. \tag{2.3}$$

If the completion of a loading half-cycle is specified by peak strain of ε_{peak}, the conjugate stress σ_{peak} is thus given by

$$\frac{\sigma_{peak}}{M} = \varepsilon_{peak} - (\varepsilon_{peak} - \varepsilon_C)^{x+1}. \tag{2.4}$$

The unloading half-cycle from this state is elastic and is given by

$$\frac{\sigma}{M} = \frac{\sigma_{peak}}{M} + (\varepsilon - \varepsilon_{peak}), \quad \varepsilon \leq \varepsilon_{peak}. \tag{2.5}$$

Figure 2.12 illustrates Eqs. (2.2)–(2.5) for two combinations of material parameters, (a) $M = 100$ GPa, $\varepsilon_C = 0.01$, $x = 0.003$ and (b) $M = 50$ GPa, $\varepsilon_C = 0.01$, $x = 0.1$, for two values of the test cycle parameter, $\varepsilon_{peak} = 0.06$ and 0.045. In both (a) and (b) the stress-strain cycles are hysteretic, consisting of elastic loading followed by the onset of compaction and significant strain and, finally, elastic unloading. In both cases, the peak stress reached is similar, although in (b) the significant strain hardening leads to an approximate doubling of the stress on compaction, whereas in (a) the almost complete absence of strain hardening leads to almost no stress increase. Elastic recovery on unloading is greater for (b) than (a) due to the lesser elastic modulus. The limit $x = 0$ corresponds to the equivalent of elastic-perfectly plastic behavior with no strain hardening and leads to significant hysteresis for a material element undergoing the complete load-unload strain cycle. Conversely, $x \approx 1$ corresponds to the equivalent of extreme strain hardening and little hysteresis. As noted, the analysis here is similar to that for metals, in which irreversible contact strain is usually associated with dislocation plasticity for strains greater than a yield strain and the term "work hardening" is used to describe non-linear plastic deformation, or "working" beyond this strain. For both metals and particles here, the term "hardening" refers to increases in stress after the onset of yield or compaction.

Intensive material behavior is expressed in stress-strain coordinates by Eqs. (2.2)–(2.5). The behavior is converted to extensive component contact behavior expressed in force-displacement coordinates using the expressions for characteristic stress and strain, $P = \sigma a^2$ and $a = \varepsilon R$, and the kinematic relation $\bar{\delta} = a^2/R$. The overbar emphasizes that displacement here, $\bar{\delta}$, characterizes contact deformation alone. Multiplying Eq. (2.2) by ε^2 and using these expressions returns Hertzian

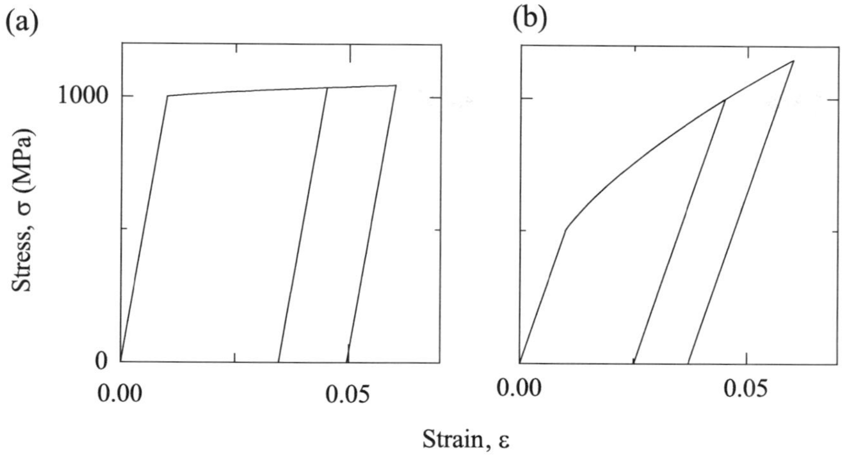

Figure 2.12 Plots of hysteretic stress-strain σ-ε responses for material elements exhibiting reversible elastic deformation and irreversible compaction at contacts during particle diametral compression. There is initial elastic loading followed by elastic + compaction deformation and final elastic unloading. Two peak strain levels are shown. (a) A material exhibiting little work hardening after the onset of compaction. (b) A material exhibiting moderate work hardening.

behavior as above,

$$\frac{P}{MR^2} = \left(\frac{\bar{\delta}}{R}\right)^{3/2}. \tag{2.6}$$

However, multiplying Eq. (2.3) by ε^2 and using the dynamic and kinematic expressions gives

$$\frac{P}{MR^2} = \left(\frac{\bar{\delta}}{R}\right)^{3/2} - \left(\frac{\bar{\delta}}{R}\right)\left[\left(\frac{\bar{\delta}}{R}\right)^{1/2} - \varepsilon_C\right]^{x+1}, \quad \left(\frac{\bar{\delta}}{R}\right) \geq \varepsilon_C^2, \tag{2.7}$$

which is an expression for the loading response of a particle exhibiting compaction in diametral compression. The expression includes the experimental variables $P, \bar{\delta}$, material parameters M, ε_C, x, and particle geometry term R. Limits includes Hertzian behavior in the first term on the right side and linear force-displacement behavior obtained by setting $x = 0$ in the second term on the right side. However, it should be noted, just as in elastic-plastic indentation testing (Cheng and Cheng 2004), that there are many non-unique combinations of ε_C and x that can give rise to linear or other functional dependencies over selected $\bar{\delta}$ domains for $P(\bar{\delta})$ loading behavior. Multiplying Eq. (2.5) by ε^2 and using the dynamic and kinematic expressions and Eq. (2.4) gives

$$\frac{P}{MR^2} = \left(\frac{\bar{\delta}}{R}\right)^{3/2} - \left(\frac{\bar{\delta}}{R}\right)\left(\varepsilon_{\text{peak}} - \varepsilon_C\right)^{x+1}, \quad \left(\frac{\bar{\delta}}{R}\right) \leq \varepsilon_{\text{peak}}^2, \tag{2.8}$$

which is an expression for the unloading response of a particle that exhibited compaction during diametral compression.

Eq. (2.6)–(2.8) describe the contact behavior of a flat on curved surface contact geometry of Figure 2.10b, which is one component of the full sphere between two platens diametral compression geometry of Figure 2.10c. The contact deformation field within a sphere in diametral compression decays over a distance of approximately $3a$, typically much smaller than the size of the sphere, characterized by R. The magnitude of the mean axial compressive stress at the center of the sphere σ_{mean} will be a useful concept throughout and is given in this case by $\sigma_{\text{mean}} = P/\pi R^2$. The related strain is thus $\sim \sigma_{\text{mean}}/M$. The axial compressive strain throughout the sphere increases only weakly from this value (inversely as sphere cross sectional area) for locations remote from the contacts at the poles. The characteristic bulk strain in the sphere is thus $\varepsilon_{\text{bulk}} \approx P/MR^2$ where the subscript indicates a bulk value, remote from a contact. The displacement associated with the bulk deformation of the sphere is thus $\delta_{\text{bulk}} \approx \varepsilon_{\text{bulk}} R \approx P/MR$. The displacement at a platen δ is the sum of the contact and bulk displacements

$$\delta(P) = \bar{\delta}(P) + \delta_{\text{bulk}}(P) \tag{2.9}$$

where $\bar{\delta}(P)$ is an implicit inversion function of Eqs. (2.6)–(2.8) and $\delta_{bulk}(P)$ is the simple linear dependence given above. Figure 2.13 illustrates Eq. (2.9) using the parameters from Figure 2.12. In this case, and most others, $\delta_{bulk}(P)/\bar{\delta}(P) \approx 0.1$. The open symbols represent data generated numerically and then sampled as in an experimental observation: 50 points were selected in each cycle segment in approximately equal displacement increments and ± 2 % white noise superposed on the forces to simulate measurement scatter. The simulations of Figure 2.13 clearly replicate the shapes of the force-displacement load-unload cycles for diametral compression of particles shown in Figures 2.9a and 2.9c. The simulations also replicate the values of the scaled quantities, P/R^2 and w/R, observed in both load-unload experiments (e.g. Gustafsson et al. 2013a) and the loading behavior of Figures 2.5 and 2.6. Linear diametral compression behavior is clearly associated with material compaction on loading with little strain hardening. Convex + linear behavior is associated with material compaction and considerable strain hardening. Bulk elastic deformation contributes little to the overall linear response.

The simulations also replicate the spherical indentation data of Figures 2.9c and 2.9d. A similar analysis using a slightly different constitutive law was shown to provide very good fits to data in the original elastic-plastic spherical indentation studies (Field and Swain 1993). Setting $x = 0$ in Eq. (2.7) and $\delta_{bulk} = 0$ in Eq. (2.9) gives the force-displacement relation of $P/\delta = MR\varepsilon_C$, where $MR\varepsilon_C$ is an invariant apparent stiffness. This expression is in agreement with a diametral compression analysis and observations of particle behavior in the large-displacement asymptotic limit (Antonyuk et al. 2010). The analysis assumes a non-strain hardening elastic-plastic material response and identifies $\sigma_C = M\varepsilon_C$ as a yield stress. Physical insight is gained from this equation by taking advantage of the kinematic compression relation to write $P = \sigma_C a^2$, making clear that the linearly increasing force is due to a linearly increasing area of compacted material $\sim a^2$ supporting constant stress σ_C. A different insight is gained by using the kinematic relation to gain $P = M[a^2/(\delta/\varepsilon_C)]\delta$, where the term in square brackets represents the aspect ratio of a column of area a^2 and length δ/ε_C. The linearity of the force-displacement behavior implies that once compaction in the material has spread across the contact plane, a steady-state configuration pertains on continued loading such that the aspect ratio of a column of material deformed by the contact remains invariant.

Another facet of force-displacement behavior in particle diametral compression is time-dependence. Figure 2.14a shows the diametral compression $P(w)$ behavior of a sodium benzoate particle during a load-unload cycle (Antonyuk et al. 2010), using data derived from the published work. The force-displacement cycles for two different P_{max} levels are shown and are hysteretic. The increasing force response on loading is common to both levels and consists of convex + linear behavior, similar to the loading responses shown in Figures 2.9a and 2.9b. The decreasing force responses, however, are completely different from those presented earlier. On unloading, the displacement clearly continues to increase before decreasing, leading to a bowed response or "nose" and negligible displacement recovery with decreasing force. The response is similar for both values of P_{max}. The vertical line is a guide to the eye and emphasizes the nose. This response is typical of that observed for many polymeric materials exhibiting time-dependent behavior on indentation (Oyen and Cook 2003; Cook and Oyen 2007; Cook 2018). Figure 2.14b shows the $P(\delta)$ behavior of a flat poly(methyl methacrylate) (PMMA) surface

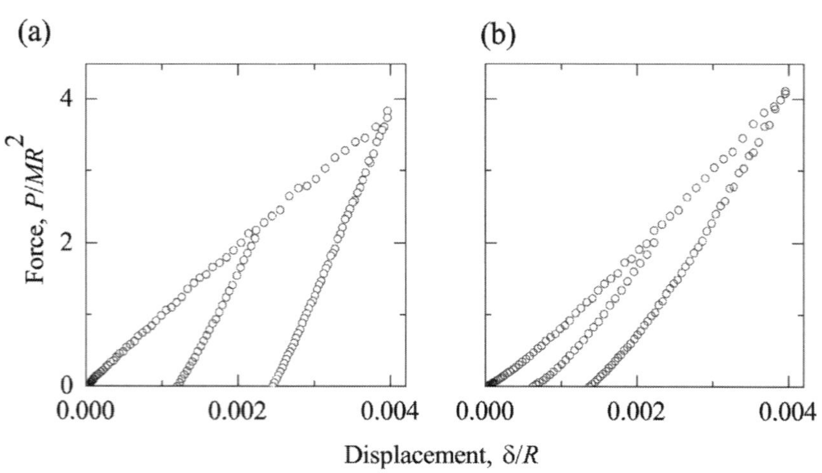

Figure 2.13 Plots of simulated hysteretic force-displacement P-δ responses for particle contacts exhibiting reversible elastic deformation and irreversible compaction during particle diametral compression. R is particle radius, M is particle elastic contact modulus, hysteresis parameters are those used in Figure 2.12. (a) Little work hardening on contact, (b) moderate work hardening on contact. Compare with Figure 2.9.

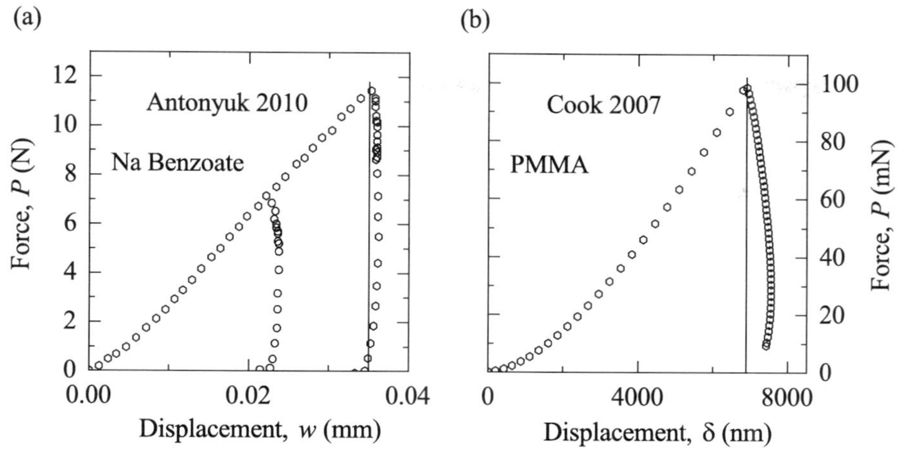

Figure 2.14 Plots of force-displacement trajectories showing similar responses for particles in diametral compression, left, and indentation of planar surfaces, right. (a) Diametral compression $P(w)$ behavior of a sodium benzoate particle during load-unload cycles (Adapted from Antonyuk, S et al. 2010). (b) Berkovich indentation $P(\delta)$ behavior of a flat poly(methyl methacrylate) (PMMA) surface indented in a load-unload cycle (Adapted from Cook, R.F et al. 2007). Note the advancing "nose" in each case, indicating time-dependent viscous deformation. Solid lines are a guide to the eye.

indented in a load-unload cycle with a Berkovich pyramidal probe (Cook and Oyen 2007), using data derived from the published work. The vertical line is a guide to the eye. The similarity in the shapes of the force-displacement responses in Figures 2.14a and 2.14b, especially the forward nose on unloading, is clear, implying similar time-dependent effects in indentation and particle deformation.

Time-dependent effects in indentation have been interpreted as the superimposition of viscous deformation on elastic-plastic deformation. The viscous deformation is regarded as exhibiting time dependence associated with molecular and microstructural rearrangement and the elastic-plastic deformation is regarded as time-independent (as above). Quasi-phenomenological models, based on non-linear viscous, elastic, and plastic mechanical elements framed in extensive coordinates, have been shown to describe and predict time-dependent viscous-elastic-plastic (VEP) indentation phenomena for a variety of materials and contact geometries (Oyen and Cook 2003; Oyen et al. 2004; Cook and Oyen 2007; Gayle and Cook 2016; Cook 2018). A simple VEP model will thus be implemented here to describe time-dependent effects in particle compression. Two phenomena will be examined: the forward-advancing nose on unloading in single cycle tests (Figure 2.14a) and the related peak-force tip rounding (Figures 2.9a and 2.9b) and cyclic hysteresis (see the following paragraphs).

A starting point in the development a VEP model for particle deformation in diametral compression is to recognize that the total deformation is described by independent deformation mechanisms responding to a common supported force. In mechanical terms, this corresponds to elements in series, such that displacements add. Here

$$\delta(P) = \delta_V(P) + \bar{\delta}(P), \tag{2.10}$$

where the first term on the right side represents the displacement arising from time-dependent viscous processes and the second term represents elastic-plastic displacement as above (for simplicity, the minor bulk elastic deformation is neglected here). Following indentation VEP analysis (Cook and Oyen 2007), the displacement rate in the viscous element is related to the supported force in direct analogy to the relationship between displacement and force in the elastic component. Here, from the Hertzian relation, Eq. (2.6),

$$P = MR^{1/2}\tau^{3/2}\left(\frac{d\delta_V}{dt}\right)^{3/2}, \tag{2.11}$$

where t is time and τ is a time constant for viscous deformation. Inverting this non-linear differential equation and re-arranging gives

$$\left(\frac{\delta_V(t)}{R}\right) = \frac{1}{\tau}\int_0^t \left(\frac{P(u)}{MR^2}\right)^{2/3} du, \tag{2.12}$$

which is an integral expressing the viscous displacement in the scaled variables of Eqs. (2.6)–(2.8), where u is a dummy variable for time. Previous VEP analyses have been oriented toward descriptions of force-controlled instrumented indentation experiments in which $P(t)$ is known and thus the integral of Eq. (2.12) can be closed to give $\delta_V(t)$. The experiments here are displacement-controlled and thus, in general, Eq. (2.12) embodies a set of implicit equations that require iterative solution. However, for moderate viscous perturbation of the elastic-plastic displacement, as here, the nose is not too great, the integral can be closed by approximating the elastic-plastic displacement by the imposed displacement. Hence, for constant macroscopic strain rate loading

$$\frac{\bar{\delta}}{R} \approx \frac{\delta}{R} = \varkappa t \tag{2.13}$$

where $\varkappa$ is the imposed strain rate. Combining Eqs. (2.12) and (2.13) gives

$$\left(\frac{\delta_V}{R}\right) = \left(\frac{1}{\varkappa\tau}\right) \int_0^{\bar{\delta}/R} \left(\frac{P(\upsilon)}{MR^2}\right)^{2/3} d\upsilon, \tag{2.14}$$

where υ is a dummy variable for $\bar{\delta}/R$. The integral of Eq. (2.14) can thus be closed as the integrand is known explicitly from Eqs. (2.6)–(2.8) as illustrated in Figure 2.13. The full VEP-based displacement is then given by Eq. (2.10). The familiar product of a rate and a time constant, $(1/\varkappa\tau)$, then acts a tuning factor to characterize more or less viscous deformation.

Figure 2.15 illustrates Eq. (2.10), implementing Eq. (2.14) using the parameters $M = 1$ GPa, $\varepsilon_C = 0.2$, $x = 0.01$, $\varepsilon_{peak} = 0.5$, and $\varkappa\tau = 0.07$. (As $\varkappa\tau$ is a dimensionless term that scales the shape of the force-displacement response, the data in Figure 2.15 are displayed in arbitrary units.) The parameters characterize a compliant particle that requires considerable deformation prior to the onset of compaction at near constant stress and does so with only a moderate amount of time-dependent deformation. As in Figure 2.13, the open symbols represent data generated numerically and then sampled as in experimental observations (for simplicity, noise was omitted here). The simulation of Figure 2.15 clearly replicates the shape of the force-displacement behavior of Figure 2.14a. As alluded to above, other choices of parameters may be used to generate similarly shaped force-displacement trajectories, but it is clear that the forward-advancing displacement nose is due to time-dependent deformation. If the strain rate is increased, the nose is diminished, consistent with experimental observations of spherical and Berkovich indentation of polymer surfaces (Oyen 2006).

Materials that exhibit a single-cycle indentation nose also exhibit indentation creep, in which the displacement increases at constant force, and indentation relaxation, in which the force decreases at constant displacement (Cook and Oyen 2006; Oyen 2006). A related phenomenon is simultaneous creep and relaxation, in which a "rounding" effect appears at peak force. The effect arises from the test system and thus applies to both indentation and particle testing of all materials. Two, common, test system features are required: (i) the loading of an indenter or platen is by an externally imposed displacement through a time-dependent compliant element; and (ii) there is a pause in the imposed displacement between the loading and unloading half cycles of a test sequence. As the constraint is neither constant force nor constant displacement, simultaneous creep and relaxation result. An example of time-dependent compliance behavior in particle diametral compression is visible in Figure 2.9a, in which there is a clear rounding effect at peak force. In this case, the average slope of the rounded creep + relaxation regions is $-1/\lambda_{machine}$, but the time in the region unknown from the Figure. Time-dependent test system effects are usually distinguished from material behavior by peak rounding rather than an unloading nose and may give rise to cyclic hysteresis. An example of this behavior is shown in Figure 2.16, which shows the behavior of a zeolite particle in cyclic diametral compression (Antonyuk et al. 2010), using data derived from the published work. The first cycle is shown as the solid symbols. This cycle exhibits hysteresis, indicative of irreversible compaction as seen earlier, and peak force rounding, indicative of time-dependent system effects. The second cycle is shown as open symbols that exhibit hysteresis and bracket the unloading response of the first cycle, also a clear indication of time-dependent effects. Subsequent cycles are shifted in displacement, with similar shape to the second cycle. The 20th cycle is shown, also as open symbols. It is clear that the majority, but not all, of the hysteresis in the second and subsequent cycles arises from peak rounding, suggesting that both system and material time-dependant effects are active. Similar behavior was exhibited by sodium benzoate and other zeolite particles (Antonyuk et al. 2010; Russell et al. 2014). VEP analyses applied to indentation systems also describe such behavior (Cook and Oyen 2007; Cook 2018).

In terms of the earlier considerations of particle shape it is important to note that particle size is often measured prior to mechanical tests by first sieving a large population of particles into smaller sub-populations classified by diameter (Brecker 1974; Bertrand et al. 1988; Sikong et al. 1990; Huang et al. 1993; Tavares and King 1998; Aman et al. 2010). Separate samples

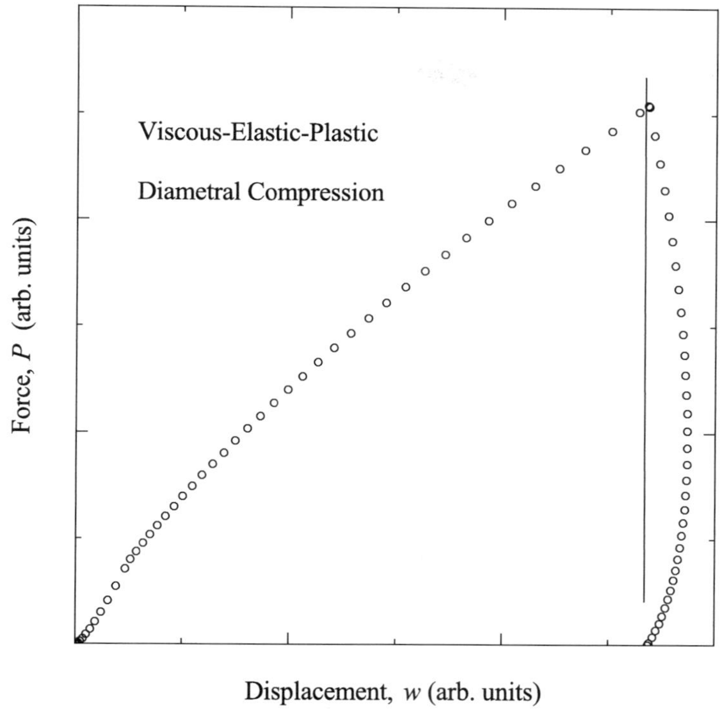

Figure 2.15 Plot of simulated hysteretic force-displacement *P-w* response for a particle in diametral compression in which the material exhibits viscous-elastic-plastic (VEP) deformation. Compare with Figure 2.14. Solid line is a guide to the eye.

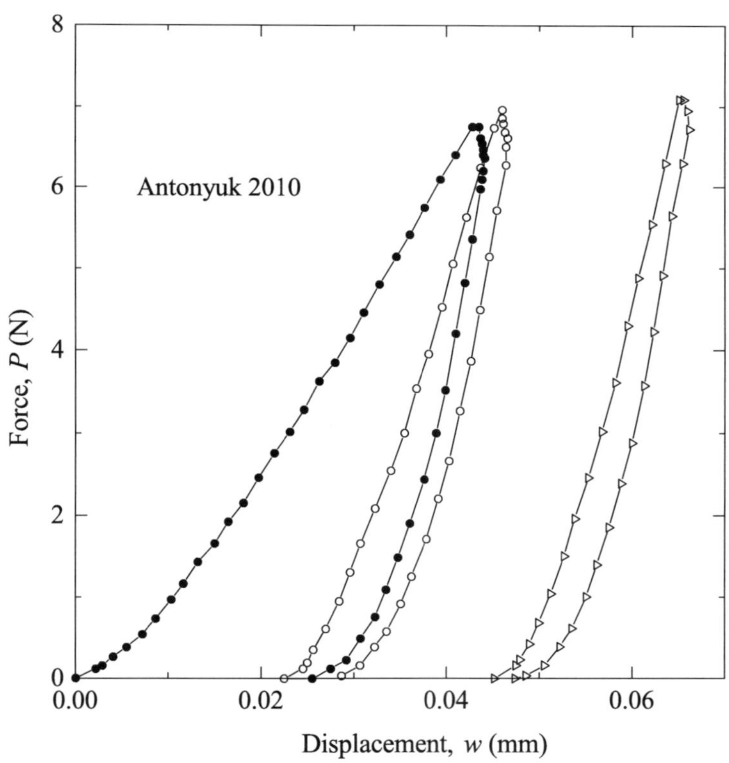

Figure 2.16 Plot of force-displacement *P-w* trajectories observed during loading of a zeolite particle in cyclic diametral compression (Adapted from Antonyuk, S et al. 2010), in which the loading system exhibits time-dependent deformation. Cycles 1, 2, and 20 shown. Note the small advancing displacement segment localized to peak force and responsible for hysteresis after cycle 1.

are then taken from the sub-populations for diameter and mechanical measurements. An important factor in pre-failure behavior is that the effective particle diameter in a mechanical test, D, is often complicated by the fact that particles frequently exhibit exterior angular facets and are not ideally spherical, as discussed in Section 1.2. In addition, particles often display exterior surface layers that are more compliant and softer than the interior core, indicative of surface asperities and surface porosity, as discussed by Vallet and Charmet (1995). Sometimes this issue is overcome by taking the effective particle diameter as the instantaneous separation of the platens when the contact stiffness, dP/dw, exceeds a specified threshold during a mechanical test, setting $P = 0$ and $w = 0$ at this condition to provide a "self-consistent" diameter measurement (Luscher et al. 2007; Gustafsson et al. 2013a). In terms of the above considerations, surface asperities can be characterized by significant contributions to particle shape by small spatial frequency components (e.g. $d_7, d_8, d_9, \ldots$) and soft surface layers over hard interiors leading to compaction variation can be characterized by a significant strain hardening rate (i.e. $x \gg 0$). Non-spherical shapes and angular exteriors can be characterized by significant large spatial frequency components (e.g. d_2, d_3), such that the effective contact radius R is given by a local shape parameter, e.g. $R \approx d_2$, rather than a global size parameter, $R \approx D/2$. However, although self-consistent diameter measurements performed during mechanical tests can overcome surface layer effects, bias toward smaller values, as in caliper measurements of non-spherical shapes, can still occur. Similarly, concentrated loading or weight drop tests of a particle sitting on a platen or tests made by passing a particle between rollers will also bias the peak force measurement toward the response along a smaller dimension of a particle. As discussed above, shape effects are only likely to cause small biases in particle size measurement relative to sampled size dispersion. Similarly, shape effects are only likely to cause small biases in particle failure force measurements (e.g. Figure 2.8) relative to sampled force dispersion—the dispersions are large and, as shown below, the stress developed in a loaded particle is only weakly influenced by shape. Nevertheless, attention in this book is focused on the variability of failure force values and the underlying distribution of particle strengths (see Section 2.3) and hence potential shape effects leading to additional variability and bias should be borne in mind.

This Section (2.2.2) and the previous Section (2.2.1) have considered in some detail the force-displacement behavior of particles in diametral compression. The most prevalent experimental phenomena, convex + linear force-displacement trajectories, pre- and post-peak force erratic behavior, and advancing displacement and hysteresis on cyclic loading, have been surveyed and shown to have clear analogs in indentation behavior. Analytical descriptions have been developed for force-displacement trajectories, drawing on stress-strain analyses of indentations. The descriptions are based on reversible + irreversible stress-strain formulations, here elastic Hertzian + compaction, using the concepts of characteristic contact stress and strain to connect intensive material behavior with extensive component behavior. In addition, analytical descriptions have been developed for time-dependent effects in loading cycles, in this case drawing on indentation analyses framed directly in extensive terms. A common thread through these considerations has been the focus on compressive force and displacement and compressive stress and strain as these are the important factors in determining the observed *axial* deformation behavior, particularly at the particle poles. The next section switches the focus to the accompanying *transverse* tensile stress and strain as these are the important factors in determining fracture behavior at the particle center.

2.3 Stress Analyses of Diametral Compression

Figure 2.17 shows the force-displacement diametral compression behavior for several particles of (a) cement clinkers (Vallet and Charmet 1995) and (b) iron ore pellets (Cavalcanti and Tavares 2018), using data derived from the published works. In (a) diameters were measured for individual particles to generate a mean $D = 6$ mm. In (b) particles were sieved from a large population into a smaller sub-population classified by diameter range as $D = 9$ mm–12.5 mm. For both systems, the $P(w)$ loading responses exhibited the features discussed above: initial compliance, small positive dP/dw, and load drops, negative dP/dw, before exhibiting stiffer, linear, constant dP/dw, regions prior to failure of the particles at peak force, P_{max}. In both cases, the linear pre-failure regions for individual particles exhibited similar slopes, as discussed in analytical terms above ($MR\varepsilon_C$ is invariant) and in experimental terms by Vallet and Charmet (1995). However, the peak force values exhibited considerable variability. These combined observations suggest that the particle and contact geometries were relatively invariant just prior to failure but that the condition for instability was variable.

Similar variability in peak force at failure is observed in other geometries used for single particle mechanical testing, illustrated in Figure 1.7. Variation from the overall form of the $P(w)$ responses shown in Figure 2.17, and sometimes observed in the other test geometries of Figure 1.7, arises from specific test and particle configurations. The variations include: the

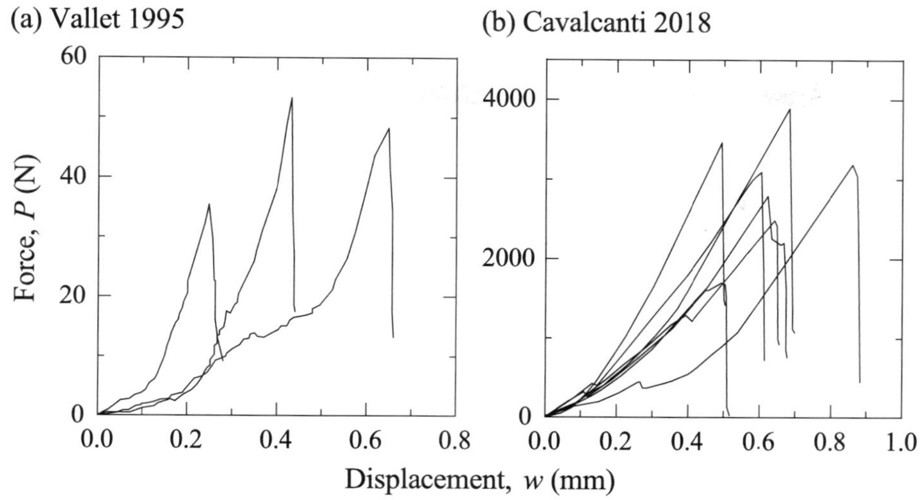

Figure 2.17 Plots of the force-displacement *P*-*w* diametral compression behavior for several particles of (a) cement clinkers, mean D = 6 mm (Adapted from Vallet, D et al. 1995) and (b) iron ore pellets, range D = 9 mm–12.5 mm (Adapted from Cavalcanti, P.P et al. 2018). Note the similar *P*-*w* behavior for each set of particles prior to peak load and the variability in the peak load.

commonly observed absence of the initial compliant region if the particles are "hard"; greater load drops if the testing system itself is compliant; and, a small remnant supported force and small load drops after peak force, indicative of particle crushing, if platen or roller displacement is not halted by the test system. In all cases, force-displacement behavior during particle initial loading is regarded as separate from the behavior approaching peak force: Force-displacement behavior during loading is variable and associated with irreversible deformation at the poles of a particle. Force-displacement behavior near peak force is invariant, except for the exact truncation point at peak force, and associated with fracture at the center of the particle. The implication is that, independent of the details of the loading method (e.g. by platens, impact, or rollers), particle behavior under load is characterized by axial compression concentrated at the particle poles and transverse tension concentrated at the particle center. Hence, distinct from Section 2.2, the focus here is the tensile stress distribution in loaded particles. Knowledge of such stress distributions enables peak force values characterizing particle failure to be interpreted in terms of particle strengths.

The starting point for analysis of stress distributions in loaded particles is consideration of a cylinder, radius $R = D/2$, thickness t, diametrally loaded by two line forces P/t, as shown in Figure 2.18a, along with the Cartesian xyz coordinate system to be used. The cylinder is loaded axially in the z direction and in this case the magnitude of the mean compressive stress on the central equatorial xy plane of the cylinder is $\sigma_{\text{mean}} = P/(tD)$. *Line forces* are forces/length and as the thickness of the cylinder is t, each force is of magnitude P distributed along the thickness dimension in the y direction, perpendicular to the plane of the diagram. If t is small relative to R, the cylinder is in a state of plane stress in xz. If t is large relative to R, the cylinder represents an axial cross-section of a rod and is in a state of plane strain in the xz plane; the two deformation states are simply related (Timoshenko and Goodier 1970; Sadd 2009) by a Poisson's ratio factor. The stress distribution throughout the cylinder in the xz plane was first determined by Hertz (1896), who recognized that the superposition of three stress fields provided a radial traction-free condition for the boundary of the cylinder. The three stress fields consist of two arising from each of two opposing line forces acting on the surfaces of semi-infinite half spaces. Together, these generate a state of equibiaxial compressive stress on a circle between the line forces. The third stress field is a superposed uniform equibiaxial tensile stress that compensates the field from the line forces on the circle. The three superposed fields leave the circle radially and circumferentially traction free, thus meeting the required boundary condition for a cylinder and providing the internal stress distribution; the procedure is described using more recent notation by Timoshenko and Goodier (1970), Hu et al. (2001), and Sadd (2009). The stress components are (Fahad 1996):

$$\sigma_{xx} = \frac{-P}{\pi R t}\left[\frac{2Rx^2(R-z)}{\beta_1^4} + \frac{2Rx^2(R+z)}{\beta_2^4} - 1\right] \tag{2.15}$$

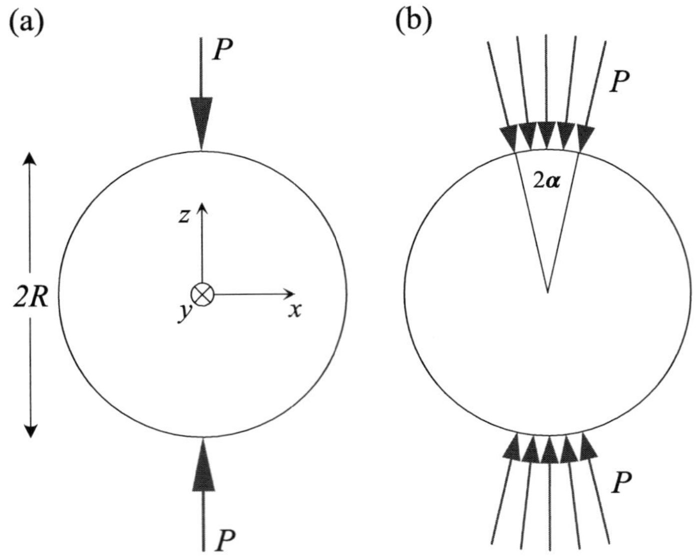

Figure 2.18 Schematic cross-section diagrams of cylinders loaded in diametral compression. cylinder diameter $2R$ in xz plane, thickness t (not shown) in yz plane. (a) Line loading by localized force P. (b) Pressure loading by force P distributed over included angle 2α.

$$\sigma_{zz} = \frac{-P}{\pi R t}\left[\frac{2R(R-z)^3}{\beta_1^4} + \frac{2R(R+z)^3}{\beta_2^4} - 1\right] \tag{2.16}$$

$$\sigma_{xz} = \frac{P}{\pi R t}\left[\frac{2Rx(R-z)^2}{\beta_1^4} + \frac{2Rx(R+z)^2}{\beta_2^4}\right], \tag{2.17}$$

where, following the notation of Procopio et al. (2003), $\beta_1^2 = (R-z)^2 + x^2$ and $\beta_2^2 = (R+z)^2 + x^2$ define radii in the cylinder coordinate system. The first two terms in each expression derive from the two opposed line forces and the third term, where present, derives from the superposed equibiaxial field. Note the overall proportionality to the line force P/t.

The axial stress in the cylinder is σ_{zz}, responsible for the compressive deformation considered in the previous section. On the equatorial plane, setting $z = 0$ in Eq. (2.16),

$$\sigma_{zz}(x,0) = \frac{-P}{\pi R t}\left[\frac{4R^4}{(R^2+x^2)^2} - 1\right] \tag{2.18}$$

and as $x \le R$, $\sigma_{zz}(x,0) \le 0$. At the free surface of the cylinder, $x = R$ on this plane and the stress $\sigma_{zz}(R,0) = 0$, consistent with the traction free surface model. At the center of the cylinder, $x = 0$ on this plane,

$$\sigma_{zz}(0,0) = \frac{-P}{\pi R t}[3], \tag{2.19}$$

noting that $\sigma_{zz}(0,0) \approx -2\sigma_{\text{mean}}$. Of more interest here is the transverse stress in the cylinder, σ_{xx}. On the equatorial plane, setting $z = 0$ in Eq. (2.15),

$$\sigma_{xx}(x,0) = \frac{P}{\pi R t}\left[\frac{(R^2-x^2)^2}{(R^2+x^2)^2}\right]. \tag{2.20}$$

and thus $\sigma_{xx}(x,0) \ge 0$ over the entire plane. At the free surface of the cylinder, $\sigma_{xx}(R,0) = 0$, consistent with the stringent radial-traction-free surface condition. On the axial yz plane, setting $x = 0$ in Eq. (2.15),

$$\sigma_{xx}(0,z) = \frac{P}{\pi R t}[1], \tag{2.21}$$

noting that the stress is tensile and independent of z. In particular, at the center of the cylinder $\sigma_{xx}(0,0) = -\sigma_{zz}(0,0)/3 \approx 0.67\sigma_{\text{mean}}$: The transverse tensile stress at the center of the cylinder is one third of the axial compressive stress. Hence, the configuration of Figure 2.18a is ideal for measuring the tensile deformation and fracture properties of brittle materials without the complications of component gripping. The configuration is also known as the "Brazilian" or "indirect" tensile test, first considered in detail by Jaeger and Hoskins (1966) and Jaeger (1967) in the context of rock measurements and more latterly reviewed by Darvell (1990) and Li and Wong (2013).

The maximum stress supportable by a material is the material *strength*, characterizing failure of the material. If the material fails in tension, by crack propagation and fracture, the strength is the fracture strength. The maximum tensile stress generated in a cylinder loaded elastically in diametral compression is given by Eq. (2.21). If P_{max} measured in a diametral compression experiment on a cylinder of diameter D is substituted for P in Eq. (2.21), the fracture strength of the cylinder material σ is related to the failure force by the widely used expression,

$$\sigma = \frac{2P_{max}}{\pi D t} \quad \text{(Brazil)}. \tag{2.22}$$

Strengths measured in this way are known as *indirect* tensile strengths, as the measured failure quantity P_{max} is compressive. A *direct* tensile test would use the bar configuration of Figure 2.2b, in which a measured failure force is tensile. Tensile strengths of particle materials measured using Eq. (2.22) or similar (see following paragraphs) are often referred to as "particle strengths," and this term will be used throughout to refer to the tensile strengths of materials tested in particle form by diametral compression (recognizing that extensive and intensive terms are mixed).

Figure 2.19 shows filled contour maps of (a) σ_{xx} and (b) σ_{zz} using a fixed contour set and Eqs. (2.15) and (2.16). The most notable features of these maps are the central columns of compression or tension in each, extending axially and transversely in the cylinders. In particular, the existence of an extended region of near uniform transverse tensile stress in the center of the cylinder provides support for the application of the Brazilian strength test: Small variations in the location of strength limiting features near the center of cylinder components of rocks or other materials will not significantly alter the strengths measured using Eq. (2.22) as the stress variations are also small. Minor features in the maps are that the predominantly tensile transverse stress component σ_{xx} is compressive close to the poles of the cylinder and, not very evident but clear from Eq. (2.16), is that the compressive axial stress component σ_{zz} is divergent at the poles. These minor features reflect the proximity of the (infinitesimally) localized line force contact at the poles and the associated rapidly varying stress field.

A clear extension of the above analysis is to use the stress solutions Eqs. (2.15)–(2.17) themselves in superposition. In particular, the solutions for line-force loading of a cylinder can be used as the basis for pressure loading of cylinder, Figure 2.18b. If a compressive force P is uniformly distributed on the perimeter of a cylinder over an arc of included angle 2α, in addition to extension along the thickness dimension t, the pressure over the loaded region is $P/(2\alpha R t)$. Such a pressure can be described by the angular distribution of a series of diametral line forces imposed on the exterior of a cylinder and the stress distribution interior to such a cylinder given by the superposition of stresses arising from the series of line forces. The superposition calculation was originally performed by Hondros (1959) using a cylindrical polar coordinate system and Fourier series expansions to describe the exterior pressure distribution and the resulting interior stress field. The full-field cylinder stress fields were presented in the form of compact combined radial and trigonometric series, along with expressions for the axial and equatorial stresses. Recent analyses by Hung and Ma (Hung and Ma 2003; Ma and Hung 2008) have extended the work of Hondros and derived the stress and displacement fields in closed, although extended, functional form, permitting comparison and agreement with photoelastic measurements. The results of Hondros simplify significantly on the axial and equatorial planes and these important limiting variations are often cited (Fahad 1996; Hung and Ma 2003; Procopio et al.

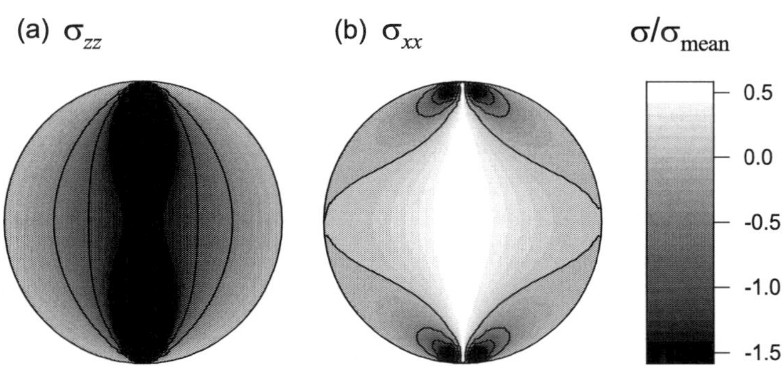

Figure 2.19 Filled contour maps of stress component amplitudes in a line loaded cylinder (Figure 2.18a). (a) Axial stress σ_{zz} exhibits a column-like zone of compression. (b) Transverse stress σ_{xx} exhibits a central zone of tension. Common contours and fill shading.

2003; Ma and Hung 2008; Li and Wong 2013). Noting that the magnitude of the mean compressive stress on the equatorial plane is still $\sigma_{mean} = P/(tD)$, the variations of the resultant transverse and axial stress components in Cartesian coordinates are, on the axial plane,

$$\sigma_{xx}(x=0) = \frac{P}{\pi\alpha Rt}\left[\frac{(R^4 - R^2z^2)\sin 2\alpha}{R^4 - 2R^2z^2\cos 2\alpha + z^4} - \tan^{-1}\left(\frac{R^2+z^2}{R^2-z^2}\tan\alpha\right)\right] \tag{2.23}$$

$$\sigma_{zz}(x=0) = \frac{-P}{\pi\alpha Rt}\left[\frac{(R^4 - R^2z^2)\sin 2\alpha}{R^4 - 2R^2z^2\cos 2\alpha + z^4} + \tan^{-1}\left(\frac{R^2+z^2}{R^2-z^2}\tan\alpha\right)\right] \tag{2.24}$$

and on the equatorial plane,

$$\sigma_{xx}(z=0) = \frac{P}{\pi\alpha Rt}\left[\frac{(R^4 - R^2x^2)\sin 2\alpha}{R^4 + 2R^2x^2\cos 2\alpha + x^4} - \tan^{-1}\left(\frac{R^2-x^2}{R^2+x^2}\tan\alpha\right)\right] \tag{2.25}$$

$$\sigma_{zz}(z=0) = \frac{-P}{\pi\alpha Rt}\left[\frac{(R^4 - R^2x^2)\sin 2\alpha}{R^4 + 2R^2x^2\cos 2\alpha + x^4} + \tan^{-1}\left(\frac{R^2-x^2}{R^2+x^2}\tan\alpha\right)\right]. \tag{2.26}$$

Note the weak sensitivity to the included angle of loading 2α and the overall proportionality to the pressure $P/(\alpha Rt)$. As the basis stress field solutions, Eqs. (2.15)–(2.17), met the boundary conditions so do the solutions arising from superposition, Eqs. (2.23)–(2.26): Setting $x = R$ in the solutions for stress on the equatorial plane, Eqs. (2.25) and (2.26), leads to the traction free surface condition, $\sigma_{xx} = \sigma_{zz} = 0$. Similarly, setting $z = R$ in the solutions for stress on the axial plane, Eqs. (2.23) and (2.24), leads to divergence of σ_{xx} and σ_{zz} at the poles. At the center of the cylinder, $x = z = 0$, both axial and equatorial equations give

$$\sigma_{zz}(0,0) = \frac{-P}{\pi Rt}\left(\frac{\sin 2\alpha}{\alpha} + 1\right) \tag{2.27}$$

and

$$\sigma_{xx}(0,0) = \frac{P}{\pi Rt}\left(\frac{\sin 2\alpha}{\alpha} - 1\right). \tag{2.28}$$

For very small angular distribution of pressure, $2\alpha \ll 1$, Eqs. (2.27) and (2.28) converge to Eqs. (2.19) and (2.21), respectively. For non-zero 2α, the magnitudes of these stress values are less than the line-force loading analogs with the limit of uniform compression $\sigma_{yy} = \sigma_{xx} = -P/(\pi Rt)$ for $2\alpha = \pi$.

Figure 2.20 shows the variation of the axial and transverse stresses on the equatorial and axial planes determined using the Hertzian line force loading analysis, Eqs. (2.15)–(2.17), and the Hondros pressure loading analysis, Eqs. (2.23)–(2.26), setting $\alpha = 0.2$ for the latter. All quantities are scaled by σ_{mean}. The increase in the magnitude of the stress components on the equatorial plane, from zero at the cylinder surface to maxima in the cylinder center is clear in Figure 2.20a. The decrease in the stress components on the axial plane from maxima at the cylinder center to divergence at the cylinder poles is also clear in Figure 2.20b. Although distributed pressure loading uniformly diminishes the stress variation, in both planes there is very little difference between the Hertz line force behavior and Hondros distributed pressure behavior, even for the relatively large subtended angle here of $2\alpha \approx 22°$, a result supported by the finite element analysis of Erarslan and Williams (2012). The implication is that for cylinder-shaped particles (e.g. pharmaceutical tablets) tested in the Brazil component geometry the contact details of loading (line force or distributed) will not greatly influence the indirect tensile strength measurements determined using Eq. (2.22) (although there are contrary opinions, Mazel et al. 2016). (Strength measurements of cylinder components containing central notches or holes (Shetty et al. 1985; Chen and Hsu 2001; Zhu et al. 2004) are considered in Chapter 12).

The superposition and compensation method implemented by Hertz in determination of the two dimensional stress field in a diametrally loaded cylinder is not completely applicable to the three-dimensional stress field in a diametrally loaded sphere. As discussed by Hu et al. (2001), superposition of the three-dimensional stress fields arising from two opposed and separated point forces applied to semi-infinite half spaces, Boussinesq fields (Johnson 1985), does not lead to an easily formed traction free surface. In particular, on the three-dimensional spherical surface intersecting the two separated points, the stress field includes varying normal and shear components that are not as easily compensated as those of the analogous

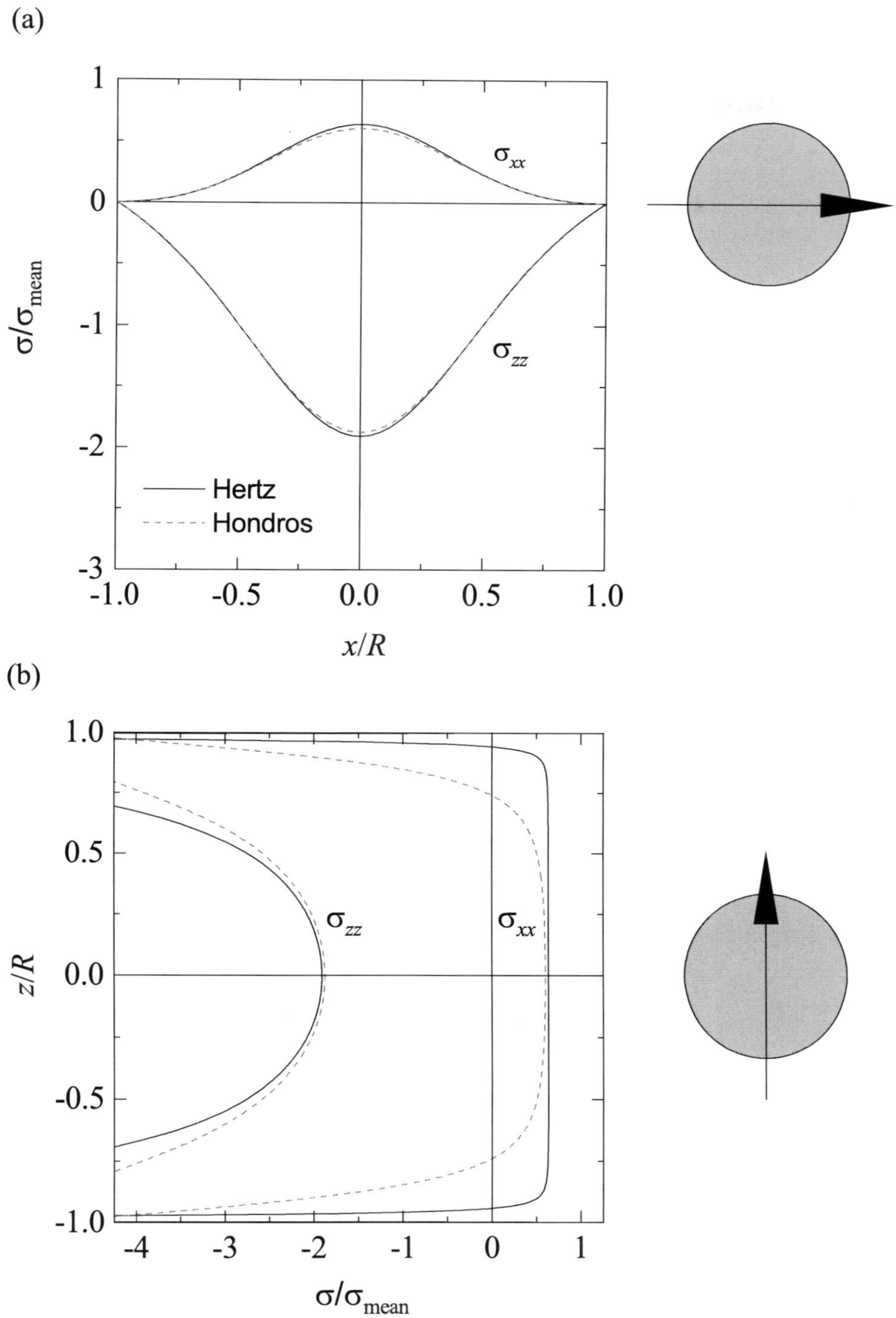

Figure 2.20 Plots of the variations in axial σ_{zz} and transverse σ_{xx} stresses in a cylinder loaded in diametral compression by a line force (Hertz solutions) and a distributed pressure (Hondros solutions), Figure 2.18a and Figure 2.18b. (a) Equatorial ($z = 0$) plane. (b) Axial ($x = 0$) plane. Schematic diagrams of plot directions shown on right.

two-dimensional circular line. However, as noted by Hu et al. (2001), symmetry and principal stress considerations dictate that shear stress vanish on the equatorial plane. Thus, on that plane, normal stress compensation can be superposed on the double Boussinesq field to leave the equatorial perimeter of the spherical surface almost traction free.

A sphere, radius $R = D/2$, diametrally loaded by two point forces P, is shown in cross section in Figure 2.21 along with the cylindrical $r\theta z$ coordinate system to be used. The cross section is described by rz. The magnitude of the mean compressive

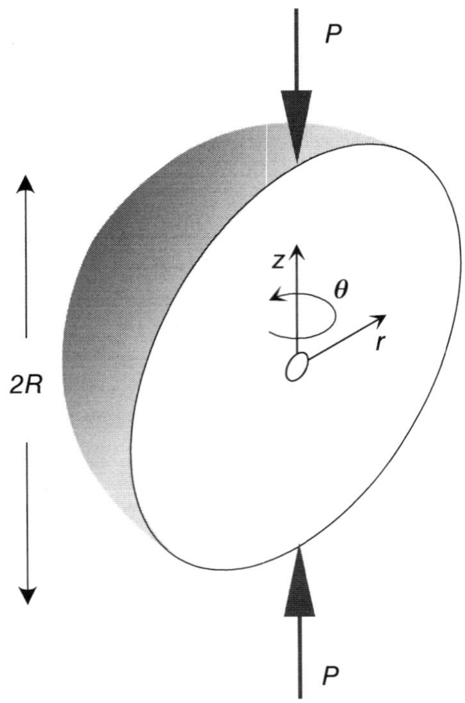

Figure 2.21 Schematic cross-section diagram of a sphere, diameter $2R$, loaded in diametral compression by a point force, P. Cylindrical coordinate system $r\theta z$ shown.

stress on the central equatorial plane of the sphere is $\sigma_{\text{mean}} = P/\pi R^2$. As the loaded sphere system has cylindrical symmetry, all rz sections are identical on rotation by angle θ about the z axis in Figure 2.21 (just as all xz sections were identical on translation along y in Figure 2.18). Hence, the three normal stress components, σ_{rr}, $\sigma_{\theta\theta}$, and σ_{zz} that define the stress state on the equatorial plane, $z = 0$, are functions of r alone. The three stress components are (Hu et al. 2001)

$$\sigma_{rr} = \frac{P}{\pi R^2}\left[\frac{R^2}{\beta^2}\left(-\frac{3r^2R}{\beta^3} + \frac{(1-2\nu)\beta}{\beta+R}\right) - \frac{1}{2}\left(\frac{(1-2\nu)\sqrt{2}}{\sqrt{2}+1} - \frac{3}{2\sqrt{2}}\right)\right] \quad (2.29)$$

$$\sigma_{\theta\theta} = \frac{P}{\pi R^2}\left[\frac{R^2(1-2\nu)}{\beta^2}\left(\frac{R}{\beta} - \frac{\beta}{\beta+R}\right) - \frac{1}{2}\left(\frac{(1-2\nu)\sqrt{2}}{\sqrt{2}+1} - \frac{3}{2\sqrt{2}}\right)\right] \quad (2.30)$$

$$\sigma_{zz} = \frac{-P}{\pi R^2}\left[\frac{3R^5}{\beta^5} - \frac{3\sqrt{2}}{8}\right], \quad (2.31)$$

where $\beta^2 = R^2 + r^2$ once again defines a radius in the coordinate system of the sphere cross section and ν is the Poisson's ratio of the sphere material. Note the overall proportionality to the magnitude of the mean stress $P/\pi R^2$ and the weak dependence on ν. The mathematical structure of the stress components, Eqs. (2.29)–(2.31), is similar to that of Eqs. (2.15)–(2.17). The components consist of terms that are coordinate functions, representing the opposed forces, plus invariant terms. For σ_{rr} and σ_{zz}, the invariant terms represent the imposed traction free condition, such that the axial and radial stresses vanish on the perimeter of the equatorial plane, $\sigma_{zz} = \sigma_{rr} = 0$ at $r = R$. The circumferential stress $\sigma_{\theta\theta} \neq 0$ at $r = R$ and is tensile. For $\nu = 0.25$, Eq. (2.30) simplifies considerably and

$$\sigma_{\theta\theta}(r = R) = \frac{P}{\pi R^2}\left[\sqrt{2} - 1\right] \approx 0.41\sigma_{\text{mean}}. \quad (2.32)$$

At the center of the cylinder, $r = 0$, the magnitudes of all three stress components are maxima and the distinction between σ_{rr} and $\sigma_{\theta\theta}$ vanishes. For $\nu = 0.25$, the three stress components are

$$\sigma_{rr}(r=0) = \sigma_{\theta\theta}(r=0) = \frac{P}{\pi R^2}\left[\frac{5-\sqrt{2}}{4\sqrt{2}}\right] \approx 0.63\sigma_{\text{mean}}. \tag{2.33}$$

$$\sigma_{zz}(r=0) = \frac{-P}{\pi R^2}\left[3 - \frac{3\sqrt{2}}{8}\right] \approx -2.47\sigma_{\text{mean}}. \tag{2.34}$$

Relative to the characteristic mean stress, these values for the transverse and axial stresses at the center of a sphere are comparable to those determined above at the center of a cylinder. In two dimensions (cylinder-like configurations) the transverse tensile stress is uniaxial whereas in three dimensions (sphere-like configurations) the transverse tensile stress is biaxial. An important limit, discussed below, is the value of the transverse stress at the center of a sphere for $\nu = 0$, $\sigma_{rr}(r=0) = \sigma_{\theta\theta}(r=0) \approx 0.74\sigma_{\text{mean}}$.

Figure 2.22a shows the variation of the normal stresses on the equatorial plane determined using the Hu et al. (2001) analysis, Eqs. (2.29)–(2.31), setting $\nu = 0.25$. All quantities are scaled by σ_{mean}. The magnitudes of all three stress components increase from minima at the sphere surface to maxima at the sphere center. In the cases of σ_{rr} and σ_{zz} the minima are zero and the behavior of these stress components is strikingly similar to that of the analogous quantities describing stress variation on the equatorial plane for a cylinder, Figure 2.20a. Stress variation for the sphere is somewhat more localized at the center than that for the cylinder. The stress $\sigma_{\theta\theta}$ is not zero at the sphere surface and $\sigma_{\theta\theta}$ variation on the equatorial plane is limited. Comparisons between stress variations, and their underlying analyses in cylinders and spheres, highlighted both qualitatively and quantitatively in Figures 2.20a and 2.22a, is made possible through the use of σ_{mean}. Other, similar, comparisons, including results derived from explicitly three-dimensional analyses are discussed shortly.

Using the restricted two-dimensional superposition and compensation scheme, Hu et al. (2001) also determined the transverse tensile stress σ_{rr} at the center of a sphere diametrally loaded by pressure distributions. For both uniform and parabolic (Hertzian) pressure distributions, the values above were not significantly altered. The implication is that for spherical particles tested in diametral compression the contact details of loading (point force or distributed) will not greatly influence strength measurements. Hu et al. (2001) also noted that the axial compressive stresses adjacent to three different loading conditions, point loading, uniform pressure loading, and parabolic pressure loading, differed only near the surface contact, suggesting that for particles tested in diametral compression the exact shape of the particle will not greatly influence bulk strength measurements.

The full three-dimensional stress field throughout a diametrally loaded sphere, including the axial variation in stress, was considered in the works of Hiramatsu et al. (1965a) (in Japanese) and, most notably and most cited, Hiramatsu and Oka (1966). Earlier, Hiramatsu et al. had investigated the strength of rocks in direct tension, indirect tension in the Brazil geometry, and bending on impact (Hiramatsu et al. 1954, 1955). In addition, Hiramatsu and Oka had investigated stress fields in rocks adjacent to holes and tunnels, through analyses and photoelastic and strain gauge measurements (Hiramatsu and Oka 1956, 1962, 1964, 1965b). The 1965 and 1966 works that considered stress fields in spheres under diametral compression and strengths of irregularly shaped rocks should be probably be viewed as parts of an ongoing program by Hiramatsu and Oka that considered rock stress and strength measurement broadly. Later analyses considered ring, cylinder, and indentation test geometries (Hiramatsu et al. 1969) and the 1965 and 1966 analysis was repeated in a study considering rock comminution (Oka and Majima 1970). Using spherical polar coordinates, Hiramatsu and Oka (1966) expressed the full displacement field within a sphere loaded by point forces as a set of series of radial terms weighted by angular coefficients expressed as Legendre polynomials. Conventional elasticity derivative relations were then used to express the stress components in similar series involving the Lamé elastic constants. The mathematical techniques used by Hiramatsu and Oka were very similar to those developed earlier in the extensive treatise of Lur'e (1964) considering three-dimensional problems in elasticity (originally published in Russian in 1955). In particular, both Lur'e and Hiramatsu and Oka considered pressure loading by a distribution of radially inward-directed point forces centered on a region bounded by a circle at the poles of the sphere. This configuration is the three-dimensional analog of the two-dimensional case considered by Hondros (1959), in which point forces were distributed over an arc at the poles of a cylinder. Lur'e and Hiramatsu and Oka specified the numerical values of the Legendre polynomial coefficients in terms of the angle subtended by the circle enclosing the loaded region and the loading pressure by boundary condition matching to the distributed loading configuration.

A major result of the Lur'e (1964) work was specification of the principal stress state at the center of a sphere loaded by vanishingly small regions of pressure, approximating diametral point loading. In current notation, the radial tensile stress given by Lur'e at the center of the sphere was

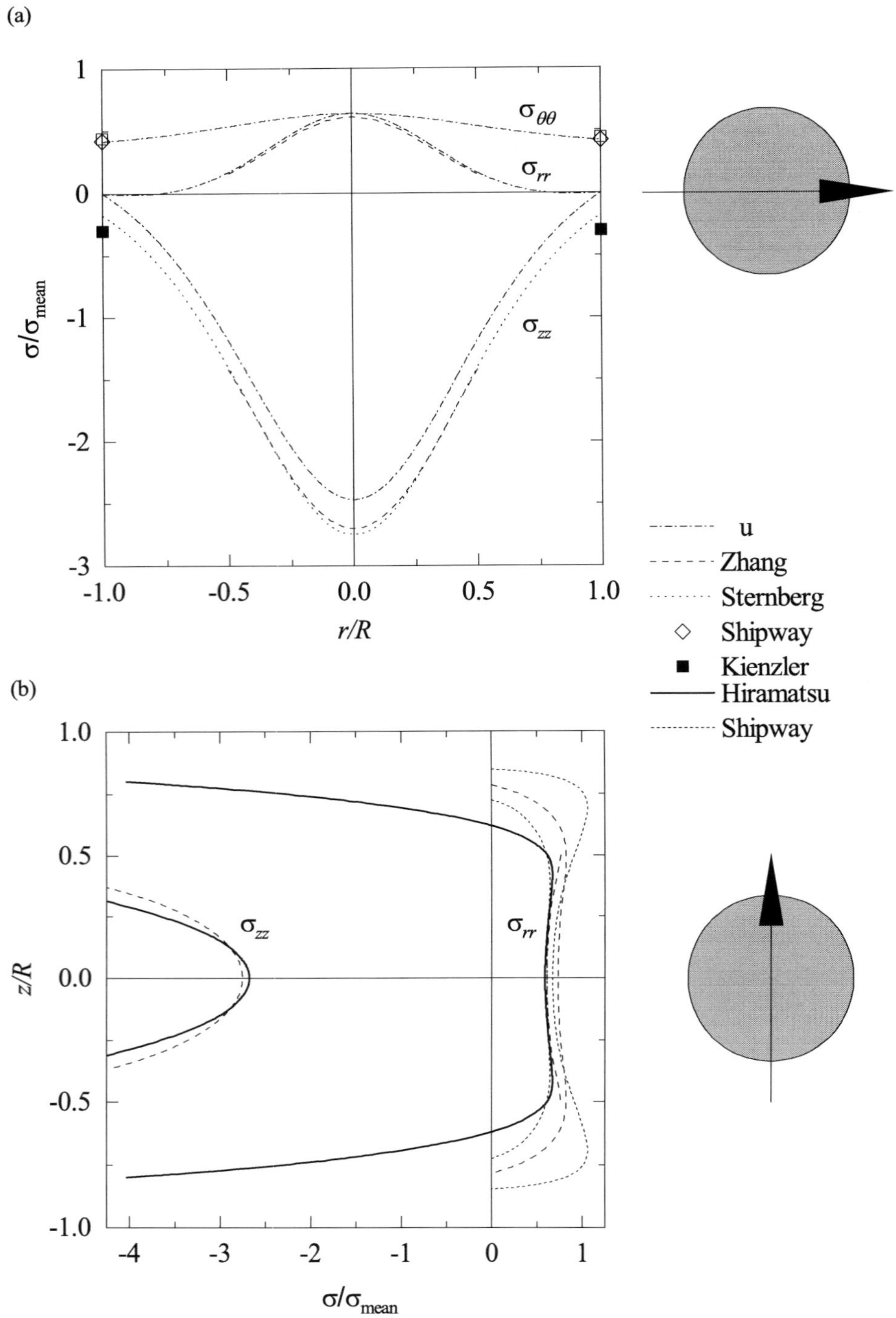

Figure 2.22 Plots of the variations in axial σ_{zz} and transverse σ_{rr} stresses in a sphere loaded in diametral compression by a point force, Figure 2.21a (Hu et al. 2001, Zhang 1991, Sternberg and Rosenthal 1952, and Kienzler and Schmitt 1990 solutions) and a distributed pressure (Hiramatsu and Oka 1966, Shipway and Hutchings 1993a solutions). (a) Equatorial ($z = 0$) plane. (b) Axial ($r = 0$) line. Schematic diagrams of plot directions shown on right.

$$\sigma_{rr}(0,0) = \sigma_{\theta\theta}(0,0) = \frac{21}{4(7+5\nu)}\sigma_{\text{mean}} \approx 0.636\,\sigma_{\text{mean}}, \tag{2.35}$$

and the axial compressive stress at the center of the sphere was

$$\sigma_{zz}(0,0) = \frac{-3(14+5\nu)}{2(7+5\nu)}\sigma_{\text{mean}} \approx -2.77\,\sigma_{\text{mean}}. \tag{2.36}$$

In both cases, the numerical values are specified using using $\nu = 0.25$. Relative to the mean stress, the radial stress at the center of the sphere, Eq. (2.35), is comparable to that at the center of a cylinder, Eq. (2.21), and to that estimated using the Hu et al. (2001) equatorial analysis of a sphere, Eq. (2.33). In similar terms, the magnitude of the axial stress at the center of the sphere, Eq. (2.36), is greater than that at the center of a cylinder, Eq. (2.19), and slightly greater than that estimated from equatorial analysis, Eq. (2.34).

A major result of the work of Hiramatsu and Oka (1966), obtained by numerical calculation using the series analysis outlined, was also the state of principal stress at the center of a sphere. For pressure loading on the sphere, by perimeter domains of subtended angle $\leq 10°$ and for material Poisson's ratio $0.15 \leq \nu \leq 0.33$, Hiramatsu and Oka found, in present cylindrical coordinate notation,

$$\sigma_{rr}(r = 0) = \sigma_{\theta\theta}(r = 0) \approx 0.64\,\sigma_{\text{mean}} \tag{2.37}$$

$$\sigma_{zz}(r = 0) \approx -2.5\,\sigma_{\text{mean}}, \tag{2.38}$$

with little variation due to subtended angle or Poisson's ratio effects—the coefficient for the transverse tensile stress varied from 0.61 to 0.68. The values are very close to those determined by Lur'e, Eq. (2.35) and Eq. (2.36), and those estimated using the Hu et al. (2001) equatorial analysis, Eqs. (2.33) and (2.34). The axial variation of $\sigma_{rr}(z)$ and $\sigma_{zz}(z)$ (current notation) determined by Hiramatsu and Oka for subtended angles $\approx 5°$ and material Poisson's ratio ≈ 0.3 is shown in Figure 2.22b, using data derived from the published work. The overall decrease of the stress components along the axis from maxima in the center of the sphere to negative divergence at the poles is clear. Also clear is that a zone of tension occupies about one half of the center of the sphere and that there is a slight, ≈ 10 %, increase in σ_{rr} within this zone; both effects noted by Hiramatsu and Oka. As above, the axial behavior of these stress components is strikingly similar to that of the analogous quantities describing stress variation on the axial plane for a cylinder, Figure 2.20b. The axial stress variation for the sphere is more clearly localized toward the center than for the cylinder.

The three-dimensional stress analysis results of Lur'e (1964) and Hiramatsu and Oka (1966) were to some extent anticipated by the earlier work of Sternberg and Rosenthal (1952). In a highly mathematical work, Sternberg and Rosenthal applied the Boussinesq stress function method in both spherical and dipolar coordinates to arrive at a full three-dimensional description of the stress field in an elastic sphere diametrally loaded by point forces. The description was expressed as a series of Legendre polynomials weighted by Poisson's ratio terms. Sternberg and Rosenthal took great care to ensure that in the limit of a very large sphere the derived description included stress divergence at the poles of the sphere identical to that for a point loaded surface plane of a half space (Boussinesq loading). In this way, along with matching other boundary conditions, Sternberg and Rosenthal ensured that their solution was valid. Sternberg and Rosenthal presented an exact expression for the axial compressive stress at the center of the sphere, identical to that later presented by Lur'e (1964), Eq. (2.36). In addition, Sternberg and Rosenthal presented an expression for the axial compressive stress variation on the equatorial plane,

$$\sigma_{zz}(z = 0) = -\left(\frac{1-p}{2}\right)^{1/2} \left[3p^2/4 - (2-\nu)p + (1-2\nu)^2\right] \sigma_{\text{mean}} \tag{2.39}$$

where

$$p = \frac{(r/R)^2 - 1}{(r/R)^2 + 1} \tag{2.40}$$

is a relative radial coordinate. Eq. (2.39) is a near complete summation of the series solution and agrees with Eq. (2.36) at $r = 0$ to within 0.1 %. The complete variation of the axial compressive stress on the equatorial plane of the sphere is shown in Figure 2.22a using Eq. (2.39). The functional form of the variation is in agreement with that from the equatorial analysis, Eq. (2.31), but uniformly decreased by about 10 %. The most obvious difference is that the full analysis leads to non-zero axial stress, $\sigma_{zz} \neq 0$ at the sphere perimeter, representing an in-plane traction on the sphere periphery, similar to that for $\sigma_{\theta\theta} \neq 0$, Eq. (2.32). These non-zero tractions do not violate the out of plane free surface condition and instead suggest that the equatorial compensation imposed by Hu et al. (2001) leading to Eq. (2.31) was a little too constraining. Although Sternberg and Rosenthal showed that their axial stress result Eq. (2.39) agreed with photoelastic measurements, no other easily implemented analytical results were presented. (A similarly mathematical follow-on work regarding the displacement field in a loaded sphere was presented by Titovich and Norris (2012), but again, no easily implemented analytical results were presented.)

The three-dimensional stress analysis results of Sternberg and Rosenthal (1952), Lur'e (1964, and earlier), and Hiramatsu and Oka (1966) have been confirmed and refined in subsequent works. Wijk (1978) showed that the calculated axial stress was sensitive to the number of terms used in the Legendre polynomial series solutions. In particular, Wijk varied the number of terms and showed that the range and average of the axial variation of the tensile stress in the central zone were likely greater than those estimated by Hiramatsu and Oka (1966). The values of all stress components at the center of the zone and the compressive stress variation were unaffected. These findings had been recognized by Hiramatsu and Oka (1967) almost immediately after the original 1966 publication. Zhang (1991) considered the equatorial and axial variation of the stress components in loaded particles of various shapes (sphere, ellipsoid, prism) using a numerical finite difference method. For a point loaded sphere, the radial variations of the equatorial values of σ_{rr} and σ_{zz} determined by Zhang are shown in Figure 2.22a and the axial variations of σ_{rr} and σ_{zz} are shown in Figure 2.22b. In the axial case, the values are in agreement with the calculations of Hiramatsu and Oka. In the equatorial case, the σ_{rr} values are in agreement with Hu et al. but the σ_{zz} values are in agreement with Sternberg and Rosenthal, supporting the over-constraint suggestion above. Zhang also showed agreement with photoelastic measurements and that more prolate particles (characterized by large d_2 values; see Section 1.2) exhibited greater stress variations. Chau, Wei and colleagues (Wei et al. 1999; Chau et al. 2000; Wei and Chau 2013; Wei et al. 2019) have extended the Hiramatsu and Oka analysis to include parabolic (Hertzian) distributed pressure diametral loading of spheres and cylinders. In the first case, presenting the solutions as series of Legendre polynomials and in the second case, presenting the solutions as series of Bessel functions. In terms of characteristic mean stress, the results are comparable to those of Hiramatsu and Oka, (Figure 2.22b). Chau et al. (2000) and Wei et al. (2019) place their calculations in context by also providing detailed chronologies of prior stress field analyses. (Such chronologies are also provided by Luscher et al. (2007) and Li and Wong (2013)).

Based on superposition of the stresses developed in a sphere decelerated by impact, Shipway and Hutchings (1993a) developed expressions for the full stress distribution in a diametrally loaded sphere. The impact related stress expressions were derived earlier by Dean et al. (1952) in the context of ballistics, and were also expressed as a series of Legendre polynomials. Shipway and Hutchings noted that expressions for stress in a diametrally loaded sphere derived from the superposed impact solutions were somewhat simpler than those derived by Hiramatsu and Oka (1966) but were in numerical agreement. In addition, Shipway and Hutchings considered the detailed variation of the circumferential stress component, noting its dominance on the equatorial plane. The open symbols in Figure 2.22a show the surface values of $\sigma_{\theta\theta}$ calculated by Shipway and Hutchings using the superposed Dean et al. solutions and by the finite element analysis (FEA) of Kienzler and Schmitt (1990); the values are in agreement with the Hu et al. calculation. (Kienzler and Schmitt also noted the agreement of their FEA work with the analyses of Lur'e.) A focus of the Shipway and Hutchings work was the effects of contact radius a on stress variation. Figure 2.22a shows the axial variation of σ_{rr} calculated for $a/R = 0.1$ and 0.2. The values and variation are in overall agreement with those of Hiramatsu and Oka (1966) and Zhang (1991). In detail, the stress range and spatial extent are somewhat greater, particularly for the smaller contact radius, reflecting the series effect noted by Wijk (1978) and an effect that also appears in the Zhang (1991) and Chau et al. (2000) calculations.

The above discussions make clear that stress distributions in particles loaded in diametral compression are well described and understood in terms of a variety of analyses. The analyses are validated by analytical relation to other contact systems (e.g. Hertz, Boussinesq) and verified by close agreement between numerical results. A predominant motivation for the analyses was to provide a quantitative connection between the extensive parameter characterizing particle failure, the maximum sustainable force P_{max}, and the intensive parameter characterizing the constituent material failure, the maximum supportable stress, or strength σ. Connections between extensive and intensive quantities are made by geometrical factors and here an obvious term to include in such a factor is the particle radius R or diameter D. For the brittle particles considered here, the strength of interest is the fracture strength and thus the stress of relevance is tensile. For the majority of particles, those of near circular cross section, the analyses predict tensile stress oriented transverse to the axis of particle loading in a region extending about the particle center of approximately half the particle diameter, Figures 2.20 and 2.22. For cylindrical components, the peak force-strength relation reflecting these findings is Eq. (2.22) describing the Brazil or indirect tensile strength test.

For spherical components, the analogous peak force-strength relation is that first proposed by Hiramatsu and Oka (1966),

$$\sigma = 0.9\frac{P_{max}}{D^2} \qquad \text{(HO)}. \tag{2.41}$$

This equation is used in the majority of works considering particle strength and is the basis of analysis for most of the strengths reported here. Equation (2.41) will be identified as HO. The origin of the HO equation is Eq. (2.37), modified by

Hiramatsu and Oka to enable experimental comparisons: R was replaced with D in σ_{mean}, the $\approx \pm 0.04$ variation in the numerical coefficient 0.64 characterizing tensile stress at the center of a particle was recognized, and the $\approx 10\,\%$ increase of the tensile stress within the central zone was recognized. The 0.9 coefficient in HO should thus be interpreted with $\approx 12\,\%$ relative uncertainty, reflecting the effects of summarizing lengthy analyses. In analytical terms, the 0.9 coefficient is in agreement, well within uncertainty, with the calculations of Wijk (1978), Kienzler and Schmitt (1990), Zhang (1991), Shipway and Hutchings (1993a), Chau and Wei (2000), and Hu et al. (2001) (who empahsized the $\nu = 0$ agreement with Eq. (2.33)). In experimental terms, the 0.9 coefficient was well supported in the original Hiramatsu and Oka work on rock strength. The following considers these and related experiments in detail. (Extensions to HO in cases in which P_{max} and D are not reported directly are considered in the Section 2.4.)

Figure 2.23 is a logarithmic plot of the relationship between experimental fracture strengths measured by three sets of two different techniques for a range of materials, using data derived from the published works cited. The strong correlation in the plot provides significant support for the HO equation and the relation between particle strengths and "conventional" strength measurements, a correlation that has been disputed (Darvell 1990) in the absence of data such as shown in Figure 2.23. The two different experimental strengths in each set are labeled "Control" and "Test" in Figure 2.23 and the techniques specified below. The starting point in consideration of the plot is comparison of the strengths of rocks and minerals generated from the compilations of Li and Wong (2013) and Perras and Diederichs (2014) for measurements by direct tensile tests and indirect Brazil tests (lower open symbols). The correlation between these measurements provides support for the application of indirect tensile strength measurements. The next point of consideration is comparison of the strengths of rocks and iron ore agglomerates determined by Hiramatsu and Oka (1966) (upper solid symbols) and Wynnyckyj (1985) (lower solid symbols), respectively, through measurements by two-dimensional Brazil cylinder tests and three-dimensional particle tests. The correlation between these measurements provides support for application of the HO equation in particle strength measurements. The upper solid symbols represent the original HO work, in which irregularly shaped test pieces were used in the three-dimensional measurements. The final point of consideration is comparison of the

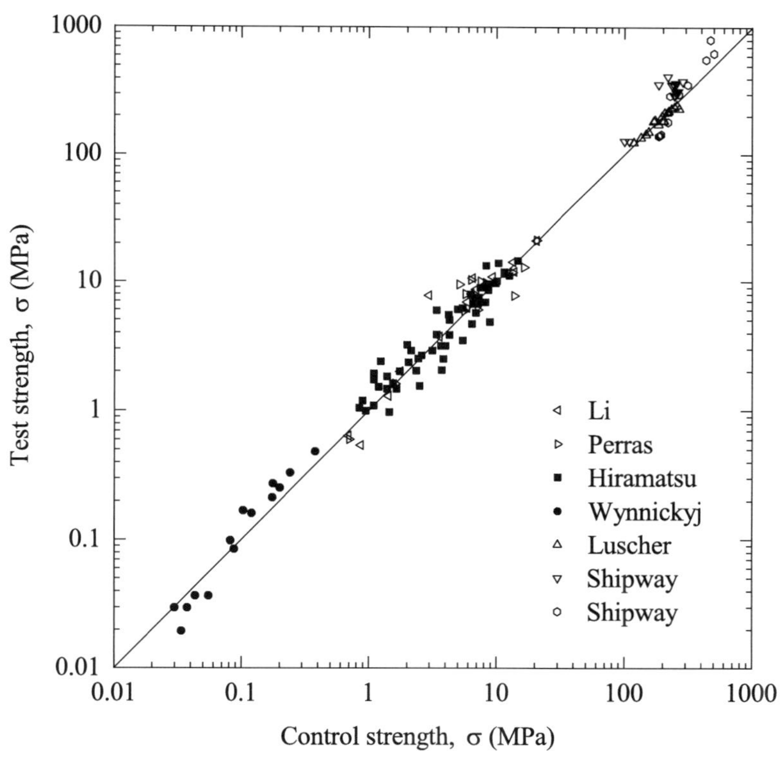

Figure 2.23 Plot demonstrating correlation of strength measurements relevant to particle strengths determined by the HO equation, Eq. (2.41): Direct tensile tests and indirect Brazil tests (Li and Wong; Perras and Diederichs 2014); two-dimensional Brazil cylinder tests and three-dimensional particle tests (Hiramatsu and Oka 1966; Wynnyckyj 1985); and comparison of the HO equation with those of Wijk (1978) (Luscher et al. (2007) and Shipway and Hutchings (1993b, 1993c). The upper solid symbols (1 MPa–10 MPa) represent the original HO work, in which irregularly shaped test pieces were used in three-dimensional measurements.

strengths of lead glass and sapphire spheres and kaolinite and bauxite agglomerates (upper open symbols) determined by Shipway and Hutchings (1993b, 1993c) and Luscher et al. (2007) through particle tests analyzed by HO and the Shipway and Hutchings analysis (Shipway and Hutchings 1993a). The correlation between these measurements provides support for application of the HO equation in terms of encompassing the variations in stress arising from implementation of series solutions as noted by Wijk (1978), and subtended angle of loading and materials Poisson's ratio effects as noted by Hiramatsu and Oka (1966), Chau and Wei (2000), and others cited above. In detail on this last point, Luscher et al. (2007) used internally consistent measurements of particle diameter (discussed earlier), peak load, and contact area, and showed that although the HO equation underestimated the internally-consistent strength calculation by a relative factor of about 5 %, the relative experimental strength uncertainty varied from 4 % to 9 % at the 90 % confidence level. Comparison of Eqs. (2.32) and (2.33) suggests a smaller coefficient of ≈ 0.6 that may pertain for particles known to fail from equatorial surfaces (Shipway and Hutchings 1993a) (this effect is considered in Chapter 13). The scatter in the experimental data of Figure 2.23 encompasses this variation. Hence, the overall conclusion from Figure 2.23, representing over four orders of magnitude of strength measurements and most materials classes, is that the HO equation is valid for determining the tensile strength of particles and the HO equation will be used here throughout.

In some particle failure studies, the failure characteristics of particles were reported as P_{max} vs D, e.g. (Capes, 1971), such that strength σ could be evaluated directly from HO. In some works, failure characteristics were reported as P_{max} vs m, (e.g. Kapur and Fuerstenau 1967), where m is the mass of the particle. In those cases, the diameter of the particle was first calculated from $D = (6m/\pi\rho)^{1/3}$, where ρ is the effective density of the particle (taking porosity into account) before application in HO.

2.4 Impact Loading

Two experimental configurations leading to diametral compression of particles can be regarded as impact loading (Figure 1.7). In these configurations, the velocity, and thus kinetic energy, of the platen or the particle is significant on initial platen-particle contact and force on the particle arises from deceleration of the platen or particle during the ensuing contact process. In conventional quasi-static diametral compression tests, Figure 1.7a, the particle is stationary and the platens have negligible velocity and kinetic energy. In counter-rotating roller tests, Figure 1.7c although the particle translates and the rollers rotate, the changes in particle and roller kinetic energy are usually negligible relative to the work of compression. In both cases, the deformation is displacement or compliance controlled. In *drop weight* impact tests, Figure 1.7d, the particle is stationary and one platen has kinetic energy E on initial contact. In *free particle* impact tests, Figure 1.7e, the platen is stationary and the particle has velocity v on initial contact. In both impact cases, the particle deformation and failure is energy controlled.

In most drop weight impact works, failure characteristics are reported as E_m, e.g. (Tavares and King 1998), where $E_m = E/m$ is the specific energy, energy/particle mass, required for particle failure. If the kinetic energy of the platen E (the dropped weight) is completely converted to work performed at the contacts

$$E = 2\int_0^{\delta_{max}} P(\delta)\,d\delta, \tag{2.42}$$

using the notation above. A quasi-static contact approximation is valid as the contact velocities are usually of order 1–10 m s^{-1} leading to contact durations of 0.1–10 ms (King and Bourgeois 1993; Tavares and King 1998; Tavares 1999). Such velocities are slow relative to the speed of acoustic waves in solids, ≈ 10^3 m s^{-1}, and the durations long relative to atomic vibration periods, ≈ 10^{-10} s (Ashcroft and Mermin 1976). Hence, the often-observed linear quasi-static force-displacement behavior noted earlier, $P(\delta) = MR\varepsilon_C\delta$, gives

$$E = (2MR\varepsilon_C)\frac{{\delta_{max}}^2}{2} = \frac{2{P_{max}}^2}{MD\varepsilon_C}. \tag{2.43}$$

The mass of the particle is $m = \rho\pi D^3/6$ and thus the specific failure energy is given by

$$E_m = \frac{E}{m} \sim \left(\frac{P_{max}}{D^2}\right)^2 \frac{1}{\rho M\varepsilon_C}, \tag{2.44}$$

recognizing that the first term is particle specific and the second term involves only material properties. Using the HO equation gives a relation between particle strength and specific failure energy

$$\sigma \sim E_m^{1/2}(\rho M \varepsilon_C)^{1/2}. \tag{2.45}$$

Figure 2.24 is a logarithmic plot of σ vs E_m for the drop weight impact failure of particles. The solid lines are predictions of behavior following Eq. (2.45) using measurements from the extensive work of Tavares and King (1998), in which σ and E_m were measured separately for a range of materials. A single force-displacement response was analyzed for each particle, and σ determined from the peak and E_m determined from the area enclosed. The measurements are thus quasi-independent as the form of $P(w)$ and thus $P(\delta)$ was not assumed. The upper and lower lines in Figure 2.24 represent the extremes of the observed material behavior for D = 2.0 mm–2.8 mm particles: (upper) glass and alumina and (lower) gilsonite, a tar like substance. The center line represents the behavior of the majority of materials examined: rocks, ores, and minerals. The symbols in Figure 2.24 represent (σ, E_m) measurement pairs of individual particles of (from least energy to greatest): limestone (Yashima et al. 1987); marble, feldspar, quartz, borosilicate glass (Kanda et al. 1985; Ryu and Saito 1991); and, soda lime silicate glass (Tavares and King 1998), using data derived from the published works cited. The overall trend in the data is well described by Eq. (2.45) over four orders of magnitude of observed specific failure energies. There are geometry dependencies with regard to the observed energy ranges (not all particles

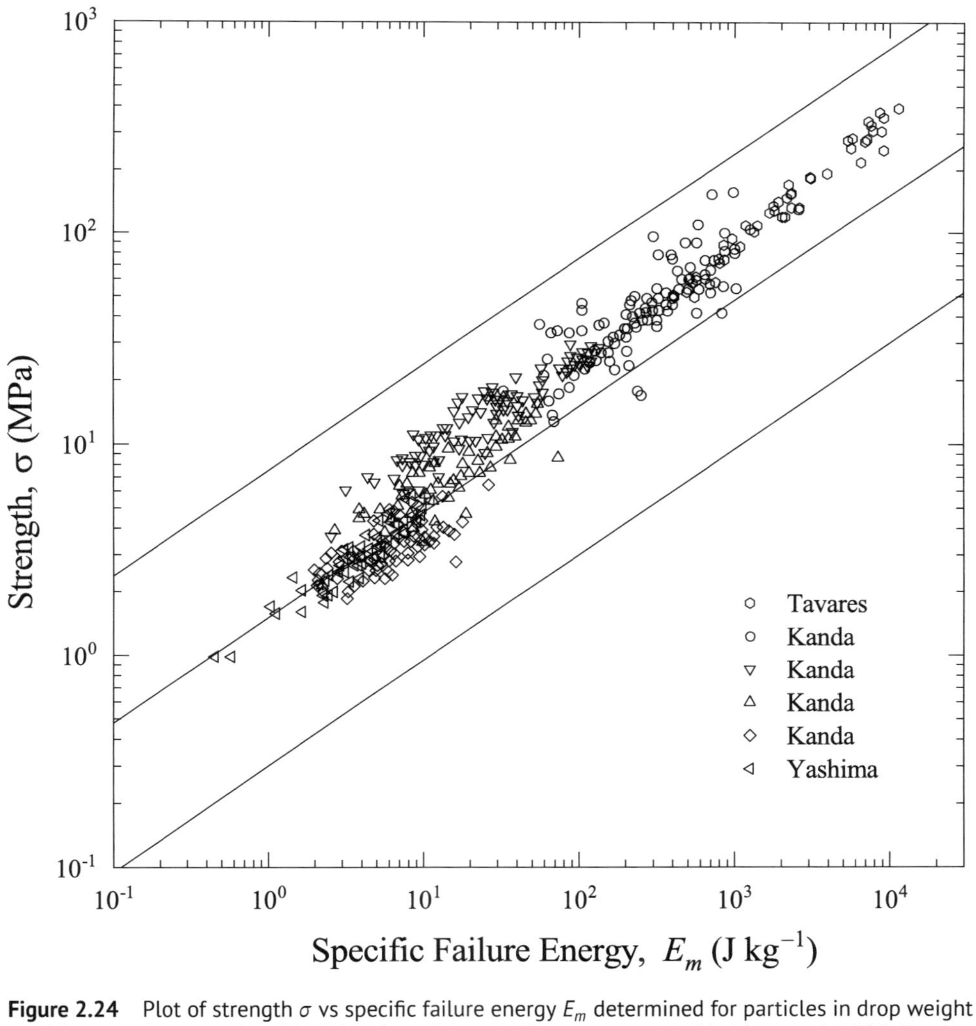

Figure 2.24 Plot of strength σ vs specific failure energy E_m determined for particles in drop weight impact tests. Symbols represent (σ, E_m) measurement pairs of individual particles of different materials (Yashima et al. 1987; Kanda et al. 1985; Ryu and Saito 1991; Tavares and King 1998). Lines are predictions from independent measurements of the central response and bounds of behavior (Tavares and King 1998).

were the same size and shape) and strengths (not all particles had the same deformation properties). However, the central trend of $\sigma = 1.5$ MPa/(J kg$^{-1/2}$) $E_m^{1/2}$ is a good quantitative description for most particles and will be used here throughout.

Alternative $P(\delta)$ laws can be used in Eq.(2.42) and thus modifications to Eqs. (2.43)–(2.45) follow. The most commonly analyzed is the elastic Hertzian response, $P(\delta) = MR^{1/2}\delta^{3/2}$, appropriate at small displacements and thus for weak particles. At perhaps the other extreme, with a focus on plasticity, is the elastic-plastic response, $P(\delta) = H\delta^2\tan^2\psi/(1 + \gamma H/M)$, appropriate for geometrically similar sharp contacts. In this relation, H is the hardness, characterizing material resistance to localized contact plastic deformation, 2ψ is the included contact angle, and γ is a dimensionless geometry term taking into account elastic deformation on contact (Marshall et al. 1983; Cook 2019). Note the central role of H here. For very soft (small H) materials, the plastic response is dominant. This relation describes a rough or faceted platen contacting a particle and thus also probably pertains at small displacements. The Hertzian response has been analyzed in many of the works cited above, leading to $\sigma \sim E_m^{3/5}$ in Eq. (2.45). This relation is consistent, within experimental scatter, with the reported observations. However, as definitive evidence of Hertzian force-displacement behavior is not given and linear $P(w)$ behavior *is* predominantly observed (discussed previously), the simpler relationship is preferred. For particles with deliberately milled end flats (Brecker 1974; Fahad 1996; Mazel et al. 2016) or naturally occurring flat facets at the poles, deformation is likely to be linearly elastic with no change in contact area during loading. In such cases, $E_m = \mathcal{U}_\sigma/\rho$, where $\mathcal{U}_\sigma$ is the strain energy density, energy/volume, at failure, expressed in terms of particle strength as $\mathcal{U}_\sigma \sim \sigma^2/2M$. In this case, the relationship between strength and specific energy is given by $\sigma \sim (\rho M)^{1/2}E_m^{1/2}$ and the scaling of Eq. (2.45) also pertains in this case.

In most free particle impact works, characteristic velocities v are reported for particle failure. In some cases, v is an estimate of the median velocity (Shipway and Hutchings 1993b, 1993c), in some cases v is an estimate of the lower bound velocity (Cheong et al. 2003), and in some cases the full distribution of velocities is reported (Salman et al. 1988, 1995, 2002, 2003; Rozenblat et al. 2013). In such experiments, particles are usually accelerated to a terminal velocity by an air or gas gun and the particle velocity v prior to impact with a stationary target recorded by optical timing devices. If the kinetic energy associated with the particle velocity is completely converted to work performed at the contact

$$\frac{1}{2}mv^2 = \int_0^{\delta_{max}} P(\delta)\,d\delta, \tag{2.46}$$

as in Eq.(2.42), where m is the particle mass. Integrating and rearranging as in Eqs.(2.43) and (2.44) gives

$$\sigma \sim v(\rho M \varepsilon_C)^{1/2}, \tag{2.47}$$

noting that Eqs. (2.47) and (2.45) have identical materials properties dependence. (The full differential equation of motion of the particle is $mu\,du/d\delta = P(\delta)$ where u is the instantaneous velocity, for which Eq. (2.47) is one solution.)

Figure 2.25 is a logarithmic plot of σ vs v for impact failure of free particles. The solid lines are predictions of bounds of behavior following Eq. (2.47). The predictions use the drop weight behavior of Figure 2.24 to set the width of the bounds and measurements of the impact of soda-lime silicate glass spheres against a steel platen from the work of Cheong et al. (2013) to fix the upper bound position (the upper bound in Figure 2.24 also represents steel on glass). The symbols in Figure 2.25 represent measurements from several materials, each composed of two samples of particles. Each was sampled separately from a large population and tested separately in quasi-static diametral compression to measure failure strength and in free particle impact to measure failure velocity. The measurements (from least strength to greatest) are of potash, salt, and fertilizer (Han et al. 2006; Rozenblat et al. 2013), soda lime silicate glass (Cheong et al. 2003), and lead silicate glass (Shipway and Hutchings 1993b, 1993c). For the lead silicate glass the particle diameter was fixed and the platen material varied, for all other particles the platen was fixed and the particle diameter varied. The bounds encompass the observations over about three orders of magnitude of impact velocity and from porous potash to dense glass materials. Individual materials are very well described in some cases by the predicted linear variation of strength with velocity. Some materials, most notably the salt, display clear deviations from the linear variation and marked particle diameter effects (some smaller particles are stronger than the predominant impact velocity trend). Overall, however, Eq. (2.47) describes material and velocity effects in free particle impact failure for most particles and will be used here throughout.

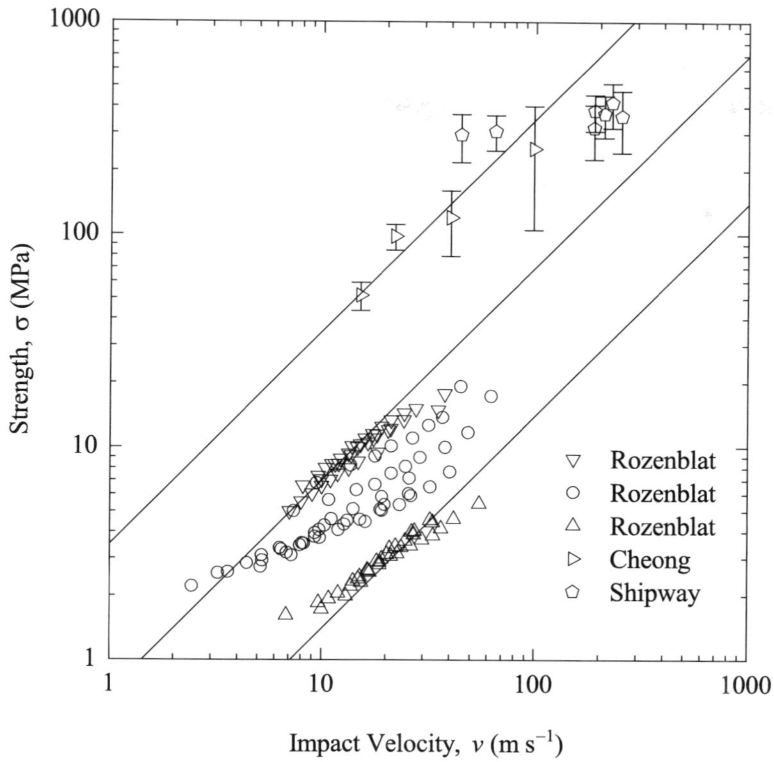

Figure 2.25 Plot of strength σ vs velocity v determined for particles in free impact tests. Symbols represent (σ, v) measurement pairs of samples of particles tested separately for different materials (Rozenblat et al. 2013; Cheong et al. 2003; Shipway and Hutchings 1993b, 1993c). Lines represent bounds determined from Figure 2.24 and the measurements of Cheong et al. (2003).

2.5 Strength Observations

Figure 2.26 is a set of logarithmic plots of particle strength, σ, as a function of particle diameter, D, for a range of particle materials. The strengths were determined using the HO equation, using data derived from the published works cited. The materials are limestone pellets (Kapur and Furstenau 1967), sand agglomerates (Capes 1971), cement clinkers (Vallet and Charmet 1995), coral sand (Shen et al. 2020), coal (Wang et al. 2019), and quartz sand (Nakata et al. 2001b). The symbols represent strength measurements of individual particles. The total number of strength measurements of individual particles derived from a published work and presented in the Figure and used in analysis, N_{tot}, is provided in the Figure Caption. (This information, where relevant, is provided similarly throughout.) The plots in the figure are drawn to the same scale (vertical × horizontal = 100 × 30) in logarithmic coordinates. The solid lines are power-law guides to the eye of slopes −0.5 and −1.0. The Figure includes a diversity of particle types and sizes, from porous agglomerates of smaller substituents probably predominantly bound by hydrogen and van der Waals bonds, the limestone pellets and sand agglomerates, Figures 2.26a and 2.26b, porous materials bound by primary bonds, the cement clinkers and coral, Figures 2.26c and 2.26d, to dense, single-phase particles of coal and sand, Figures 2.26e and 2.26f. The collection of particles includes a range of sizes from approximately $D = 20$ mm, Figure 2.26a to $D = 0.1$ mm, Figure 2.26f. A typical individual particle study extended over about a factor of 10 in particle diameter. The collection of particle types exhibited strengths ranging from less than 0.1 MPa, Figure 2.26a, to 500 MPa, Figure 2.26f. A typical individual particle study exhibited about a factor of 10 in strength range.

Figure 2.26 exemplifies many of the characteristics observed in the particle strength measurements considered in this book:

The predominant characteristic of Figure 2.26 is that all particle types exhibited decreases in strength with particle diameter; the solid lines of negative slope encompass the trends. Previous decreasing trends in strength with particle diameter for these and other observations have been characterized by empirical power-law slopes, including -0.35 ± 0.2 (Kapur and Furstenau 1967), -0.4 ± 0.3 (Capes 1971), -0.5 ± 0.2 (Studman and Field 1984), -0.5 ± 0.4 (Vallet and Charmet 1995),

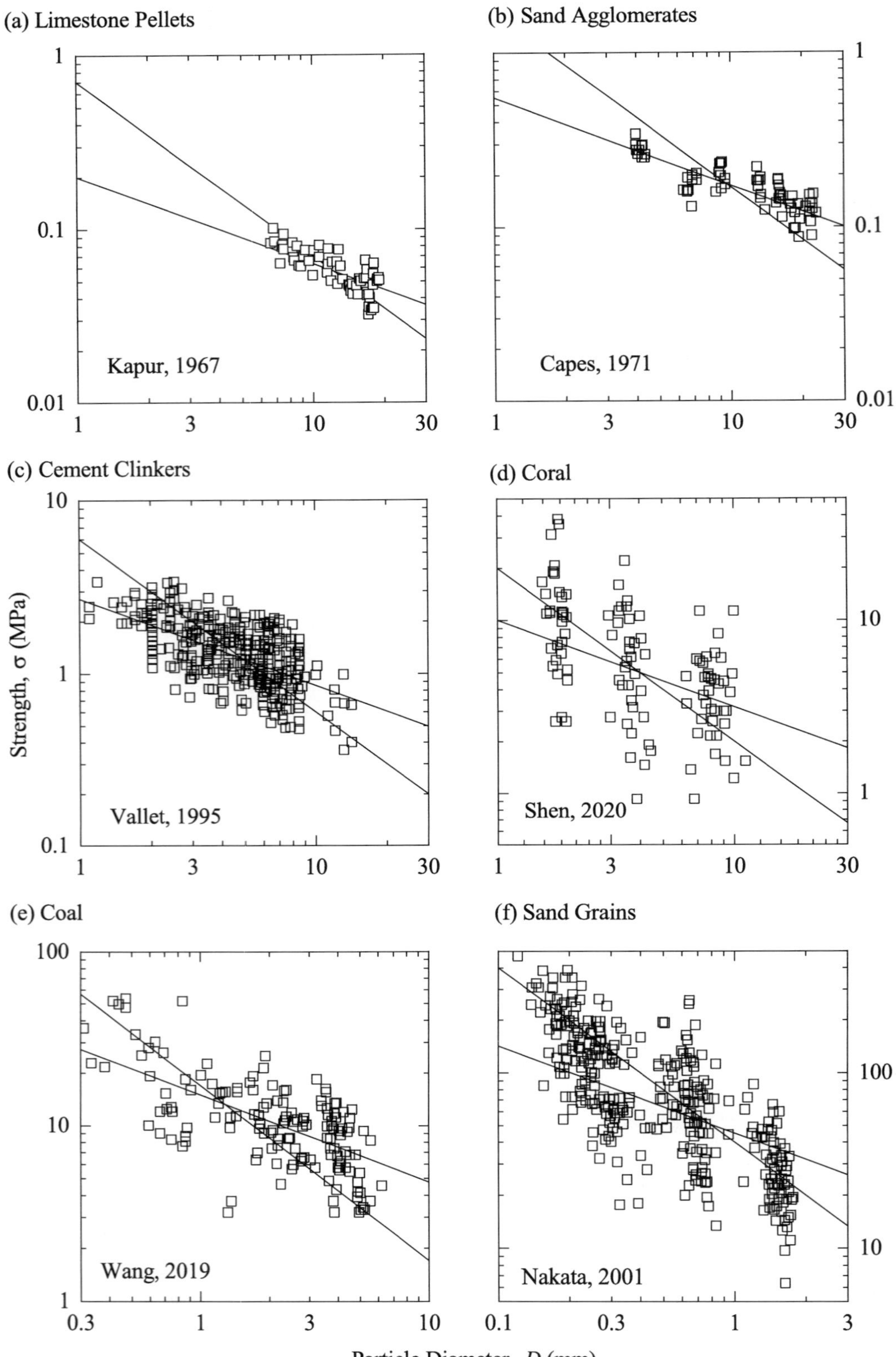

Figure 2.26 Logarithmic plots of strength σ vs diameter D for six particle systems, showing measurements for individual particles, illustrating decreasing strength trends and dispersions. (a) Limestone pellets, number of components, N_{tot} = 62 (Adapted from Kapur, P.C et al. 1967). (b) Sand agglomerates, N_{tot} = 67 (Adapted from Capes, C.E 1971). (c) Cement clinkers, N_{tot} = 309 (Adapted from Vallet, D et al. 1995). (d) Coral grains, N_{tot} = 107, (Adapted from Shen, J et al. 2020). (e) Coal, N_{tot} = 145 (Adapted from Wang, C et al. 2019). (f) Sand grains, N_{tot} = 368 (Adapted from Nakata, Y et al. 2001b).

−0.9 ± 0.3 (Yasufuku and Kwag 1999), −0.34 to −0.42, as cited by McDowell and Amon (2000), −0.79 ± 0.3 (Nakata et al. 2001b), −0.75, as cited by Nakata et al. (2001b), −0.7 ± 0.5 and −1.1 ± 0.2 (Lobo-Guerrero et al. 2006b), −0.52 ± 0.25 (Wang et al. 2015), and −0.42 ± 0.3 (Xu et al. 2016), where the uncertainties are estimated from the published data. (A compilation of empirical trends is provided by Xu et al. in the context of fragmentation, but provides no estimates of uncertainties.) The decreasing trend extends out to very large sizes, as slopes of −0.26 to −0.43 were determined for 3 mm to 1600 mm "particles" of coal (Moomivand 1999), a slope of −0.5 was determined for 30 mm to 500 mm crushed rocks (Unland and Szczelina 2004), and a decreasing trend was also observed in large plaster particles, up to 60 mm in size (Tsoungui et al. 1999, Chau et al. 2000). As will be seen, the negative values and large relative uncertainties well describe particle behavior but are probably not estimates of a universal dependence.

The second clear characteristic of Figure 2.26 is that the strength measurements exhibit considerable dispersion or variation. For a selected particle diameter, the observed strengths can exhibit a factor of 10 range from the least to the greatest. This behavior is well illustrated in the observations of Figure 2.26d, in which the coral particles were sorted into samples of similar diameters prior to strength testing, and Figure 2.26f, in which the sand particles were similarly sorted into samples. In each case, each sample exhibited a range of 10 in strength.

This last point highlights a third characteristic of Figure 2.26. The *relative* strength dispersion was invariant with diameter for all particles implying that the *absolute* strength dispersion decreased with increasing diameter. An alternative description, summarizing the above, is that although there are clear material dependencies, large particle strengths are weaker than small particle strengths and large particle strength distributions appear narrower than small particle distributions. Representative particle diameters are a few millimeters and representative particle strengths are a few tens of megapascals.

A fourth and final characteristic in Figure 2.26, considering the range of materials as a whole, is that there is a clear tendency for strengths to increase as particle size decreases, independent of material. This point will be considered in greater detail in Chapter 13, and further evidence of this tendency will be provided throughout the book in consideration of agglomerates and in consideration of very small particles. Very few studies of the latter category of particles include systematic studies of size effects, but the limited observations are consistent with the noted trend. Examples include mineral particles of diameter a few μm that exhibited strengths of nearly 1 GPa (Sikong et al. 1990), glass spheres of diameter approximately 100 μm that exhibited strengths of 2 GPa (Studman and Field 1984), and diamond particles of diameter 50 μm that exhibited strengths of 8 GPa (List et al. 2006).

The central questions suggested by Figure 2.26 to be addressed in this book are then: what is the strength distribution for a given particle diameter? And, how does the particle strength distribution depend on diameter and material?

2.6 Strength Empirical Distribution Function

The predominant vehicle used for presentation and comparison of strength measurements of samples of components is the *empirical distribution function*, the edf. The components may be "large" components with strengths given by appropriate compression or tension formulae, as in Section 2.1, or "small" particles with strengths given by the HO formula of Eq. (2.41) or similar. As the name implies, the edf characterizes *empirical* observations or experimental measurements of strengths. A follow-on implication of the term empirical is that the number of strength measurements used to create the edf, N, is finite, as the number of experimental measurements is also necessarily finite. As will be seen, as each component exhibits only one strength this also implies that the number of components in the sample under consideration is also N. In probabilistic and statistical terms, the group of N components is usually considered a sample from a much larger population of components, and hence a sample of strengths from a population of strengths (Spiegel 1961; Walpole and Meyers 1972). As is the case for most such sampling procedures, the statistical goal is to use a sample to provide information about the population. As is also usually the case, probabilistic analyses are used to generate such information from quantitative attributes of the sample measurements. Here the strength edf of a sample is used to generate information regarding the distribution of strengths in the population.

The edf generated from sample measurements differs from other statistical properties of a sample such as the sample mean or the sample standard deviation. In particular, a strength edf describes the *distribution* of strengths within the domain of the N measurements constituting the sample. Thus, the strength edf provides more information than the single values of the mean strength of the sample or the standard deviation of the strengths, and hence is specified as a *function* of the N strength values. The strength edf is a discrete function constructed as follows. The N strength measurements σ of the sample are ranked from least to greatest and identified with an index i from 1 to N as σ_i, such that σ_1 is the least strength in the sample and σ_N is the greatest strength. The quantity $Pr(i) = (i - 0.5)/N$ is then generated for each value of i. The sequences σ_i and $Pr(i)$ then form a parametric set with i as parameter. Elimination of i leads to the discrete function $Pr(\sigma)$ that is the strength edf. $Pr(\sigma)$ is the probability that a randomly selected member of the set i has a strength less than σ_i (hence the commonly used Pr notation). More importantly, $Pr(\sigma)$ has the property that it provides an estimate of the probability that a randomly selected member of the population has a strength less than σ. Alternative forms for $Pr(i)$ expressed as $(i + a)/(N + b)$, where a and b are small values, are discussed briefly by Castillo (1988), and by Forbes et al. (2011) and elsewhere (Wikipedia 2022a). For $N \gg 1$, all forms of $Pr(i)$ lead to convergence of $Pr(\sigma)$ to the distribution function describing the population (this is discussed in detail in Chapters 3 and 4). The form here, $Pr(i) = (i-0.5)/N$ (using $a = -0.5, b = 0$), is preferred as the variables i and N are unambiguously related to the finite, countable, sample strength indices and the limiting experimental values $Pr(\sigma_1) = 0.5/N$ and $Pr(\sigma_N) = (N - 0.5)/N = 1 - Pr(\sigma_1)$ are distinguished from the mathematical asymptotic values of 0 and 1 and are symmetric with respect to i as $Pr(i)$ uniformly fills the range. $Pr(\sigma)$ thus extends over the domain σ_1 to σ_N and range ≈ 0 to ≈ 1.

Figure 2.27 is a plot of the bursting strength of glass bottles measured as part of commercial production. A sequence of 200 consecutive bottle strengths is shown, sampled from a production run of 1000 bottles. Figure 2.27 reflects the measurements of Hunter as supplied to Preston (1937), using data derived from the published work. The bursting strength is given as a pressure in pounds/square inch (psi, 1 psi $\approx$ 6.895 kPa) and appears quite erratic. The mean $\pm$ standard deviation of the bottle strengths for the entire production run of 1000 was (461 ± 74) psi, shown as the gray band in Figure 2.27. (The possible weak upward "drift" of the sample of 200 relative to the mean and standard deviation of the population of 1000 was noted by Preston; such sequence effects, although important in manufacturing, will not be considered here.) Figure 2.28 shows the strength edf constructed as described above, using the data of Figure 2.27. The strength domain was bounded by $\sigma_1 \approx$ 300 psi and $\sigma_{200} \approx$ 750 psi. As will be discussed in Chapter 5, the edf variation between these bounds is typical for large components and is *sigmoidal* ("S" shaped). Sigmoidal edf curves are monotonic increasing and exhibit a single inflection point within the domain, at which the second derivative $d^2Pr(\sigma)/d\sigma^2$ passes through zero and changes sign from positive

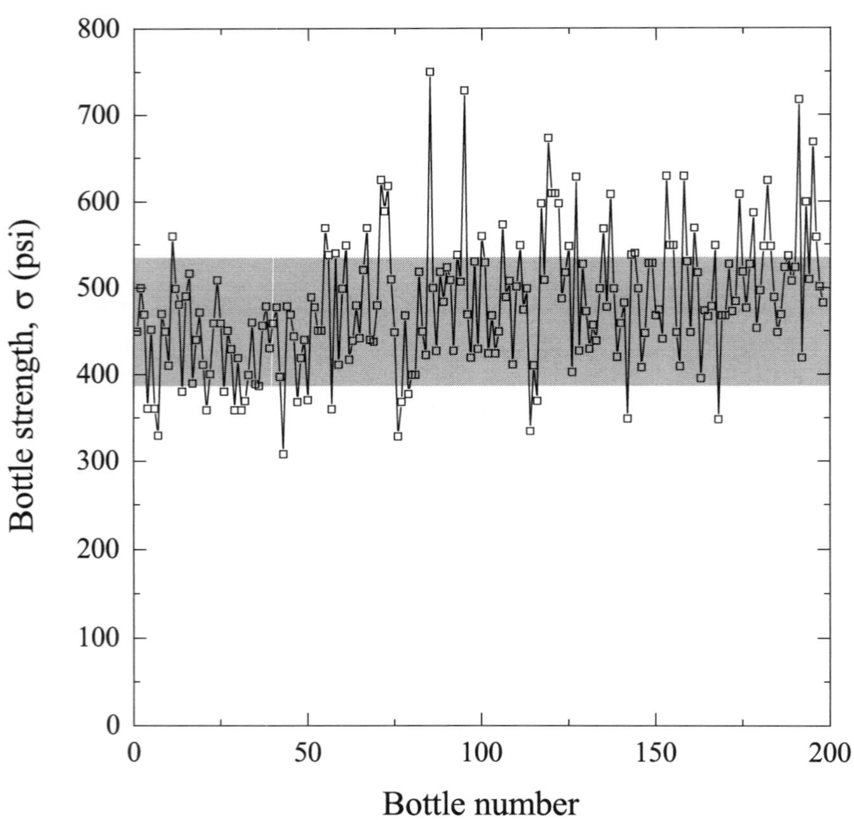

Figure 2.27 Plot of strength σ vs number in production sequence for burst pressure testing of 200 glass bottles (Adapted from Preston, F.W 1937). Symbols represent individual strength tests. Shaded band represents the mean strength and standard deviation bounds of the entire production run of 1000 bottles.

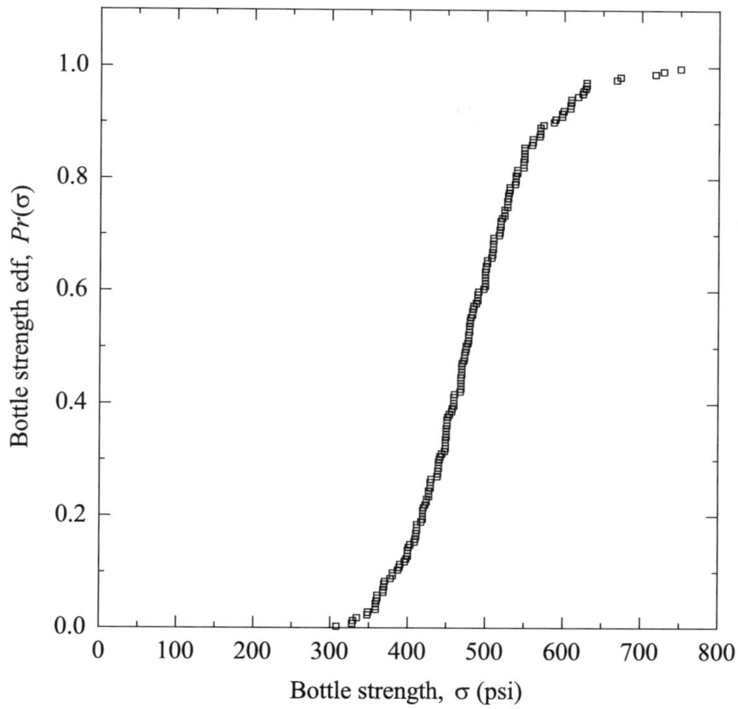

Figure 2.28 Strength edf $Pr(\sigma)$ for the burst strength of glass bottles taken from Figure 2.27. Note the clear lower bound strength of the edf and the sigmoidal shape extending over a factor of two in strength, typical of large components.

to negative as σ increases. The second derivative may be near zero with no change in sign in limited regions at the bounds of the domain, leading to straight sections of near constant derivative. These sections may be "flat" and have near-zero first derivative, $dPr(\sigma)/d\sigma$, describing the classic sigmoid. More commonly, the sections are "sloped" and have clear non-zero positive first derivative, especially at the lower bound. The definition of the edf guarantees the first monotonic property. The simultaneous application of the monotonic and derivative properties requires a large derivative at the center of the domain. As will also be discussed in Chapter 5, the relative width of the domain, $\sigma_N/\sigma_1 \approx 2$, is also typical for large components, that is, the strongest component in the sample was about twice the strength of the weakest component.

Figure 2.29 shows strength edf plots of (a) coral particles and (b) sand particles, using the data of Shen et al. and Nakata et al. from and 2.26f, respectively. The strength data have been sorted by size, $D \approx 2$ mm, 4 mm, and 8 mm for coral and $D \approx 0.1$ mm, 0.4 mm, and 1 mm for sand, to generate three edf responses for each material. The forms of the edf responses and the effects of particle size are similar in each case.

Figure 2.29 exemplifies many of the characteristics observed in the particle strength measurements considered in this book:

The most noticeable feature of Figure 2.29, especially in comparison with Figure 2.28, is that the edf variations are *concave*. Concave edf curves are monotonic increasing by definition. However, distinct from sigmoidal curves, concave edf curves have negative second derivatives, $d^2Pr(\sigma)/d\sigma^2$, or downward curvature, throughout the domain and no change in sign. Similar to sigmoidal curves, the second derivative may be near zero in regions at the bounds of the domain, and this is usually the case, leading to straight sections of near constant derivative. The derivatives $dPr(\sigma)/d\sigma$ are large at the lower bounds and decrease monotonically, notwithstanding some scatter, to much smaller values at the upper bounds. Although there is a weak tendency to small derivatives at the lower bounds in Figure 2.29a, the overall edf variations in Figure 2.29 are not sigmoidal.

A second noticeable feature of Figure 2.29, again especially in comparison with Figure 2.28, is that the relative width of the domains, $\sigma_N/\sigma_1 \approx 20$, or greater. Such extended domains are typical for particle components, in which the strongest component in a sample is many times the strength of the weakest component.

A third noticeable feature of Figure 2.29, inferred previously in consideration of Figure 2.26, is that the domains of the edf responses contract in absolute terms as the particle size D increases. A related observation is that the contraction appears to occur toward an invariant lower bound. That is, σ_1 in separate edf responses for a given material does not appear to depend on D: all responses converge to the same stress.

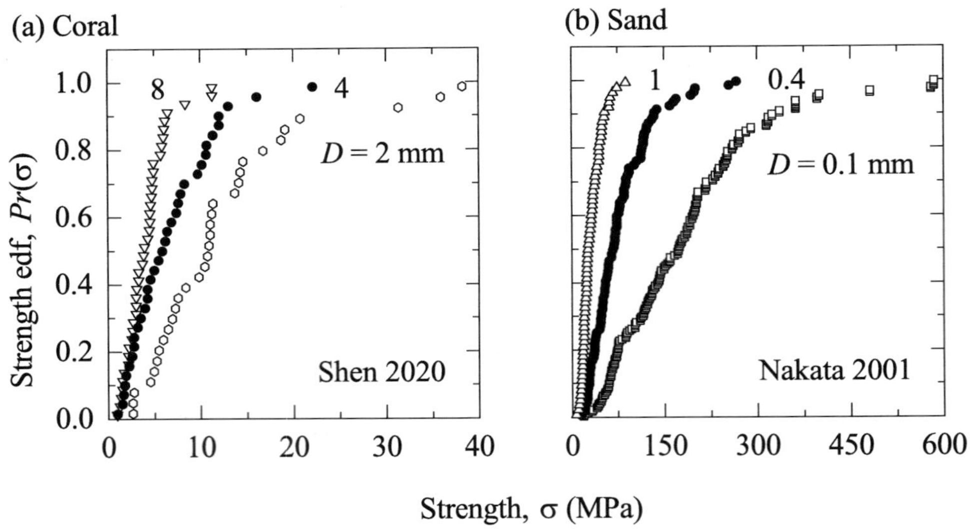

Figure 2.29 Strength edf $Pr(\sigma)$ plots for the diametral compression strength of particles sorted by size from Figure 2.26: (a) Coral particles, D = 2 mm, 4 mm, and 8 mm (Adapted from Shen, J et al. 2020). (b) Sand particles D = 0.1 mm, 0.4 mm, and 1 mm (Adapted from Nakata, Y et al. 2001b). Note the extremely small lower bound strength of the edfs and the concave shape extending over more than a factor of 10 in strength, typical of particle components.

Figures 2.28 and 2.29 raise many questions that are the basis for presentation of results, analysis, and discussion in this book. The questions include:

- What gives rise to the variability in strength quantified by the edf responses of Figures 2.28 and 2.29?
- Why are the strength edf responses of Figure 2.28 (sigmoidal) and Figure 2.29 (concave) different?
- Why are the relative strength domains of large components and particles so different?
- Why do the strength domains of particles contract toward a common small strength as particle size increases?
- How representative are the edf responses for large components, Figure 2.28, and particles, Figure 2.29?

The following section outlines the structure of this book in addressing these questions, and how the answers provide a basis for discussing many other aspects of particle strength.

2.7 Outline of *Particle Strengths*

2.7.1 Individual Topics

This Chapter 2, and the previous chapter, Chapter 1, have introduced many of the concepts required to discuss particle strengths: examples of the common particle materials, descriptions of particle shapes, measurements and analysis of force-displacement behavior of particles in diametral compression, and stress analysis of loaded particles. These concepts provided a basis for the introduction of the main focus of this book, Section 2.6: the strength distributions of particles. Such distributions are at the center of materials science and engineering considerations of particles. In a materials science sense, particle strengths stand at the properties end of the materials science structure-properties linkage. For brittle particles, the linkage is one described by fracture and thus strength measurements provide insight into the structural element of the strength-controlling flaw (Lawn 1993). Strength distribution measurements thus provide insight into flaw distributions. Flaws themselves stand at one end of the processing-structure linkage of materials engineering. Particle processing determines particle structure, including particle constituent phases and porosity, as well as flaws and fracture resistance, and thus strength distribution measurements provide guidance to particle manufacturing engineers in optimizing particle materials. Strength also stands at one end of a materials engineering linkage, that of properties-performance (Kingery et al. 1975). The material property of strength is a key factor in determining component performance measures, such as the load-bearing capability of a particle or the energy required to crush a particle (Ashby 1999). Hence, strength distribution measurements enable engineers working with particles to optimize component designs, optimize component efficiency, and make component reliability predictions.

The following two chapters, Chapters 3 and 4, address the first four questions put forth earlier by providing a clear analytical foundation for discussing strength variability of particles. In particular, Chapter 3 develops a "forward analysis," in which a population of flaws in a material is assumed and distributions of strengths in samples of the material are predicted. The major goal of the analysis is to predict the strength distributions of samples of material in the form of particles, but the analysis will be completely general and applicable to all components and component forms (e.g. tensile bars). Chapter 4 develops a "reverse analysis," in which a distribution of strengths of components is given, as in an experiment, and the underlying population of flaws inferred. Again, although the analysis will have as a goal the understanding of particle strength distributions, the analysis will be generally applicable to strength distributions of all component forms. Throughout Chapters 3 and 4 it will become clear that strength distributions are variable "extreme value" distributions characterizing samples from an invariant distribution describing a more numerous population. Experimental measurements of extreme value distributions depend critically on the sample "size" relative to the population. For strength distributions, the physical size of the strength components tested and the numerical size of the test sample (number of components tested) are both important. Chapters 3 and 4 make clear connection with extreme value distribution analysis and investigate the effects on strength distributions of both component physical size and sample numerical size. Chapters 3 and 4 also make clear that measurements of strength distributions of particles provide the opportunity to examine the physics of extreme value behavior in the limits of small physical component size and large numbers of components.

The next two chapters, Chapters 5 and 6, address the fifth and last question by surveying and analyzing experimental measurements of strength distributions of large "extended" components and particles. Here, an extended component is taken to be one that contains multiple flaws that experience significant stress as the component is loaded. Such components include traditional tensile bars, compression blocks, and bend beams. Chapter 5 will present strength distributions for a wide range of materials in all these component geometries and apply the reverse analysis of Chapter 4 to infer underlying flaw distributions. The survey of extended components in Chapter 5 will provide a reference point for the examination of particles. A particle here will be assumed to be "small" and to contain one or a few flaws that experience significant stress as the particle is loaded. The distinction is made for two reasons. First, particles are physically small. If flaw densities (number of flaws/volume) are reasonably invariant for a material, then a small particle will simply contain fewer flaws than a large tensile bar. Second, stress distributions in loaded particles are localized. If the transverse tensile stress in a particle of diameter D is restricted to a central region of characteristic size $D/2$, as discussed earlier, only 1/8 of the flaws in a particle will experience tension on loading. This is not so in a tensile bar; all flaws in the bar experience tension on loading. Chapter 6 will present strength distributions for a wide range of particles, including a variety of materials and sizes. The reverse analysis of Chapter 4 will be used to infer underlying flaw distributions in these particles and comparisons made with the flaw distributions inferred for the large components of Chapter 4. The experimental observations of Chapters 5 and 6 highlight the effects of component size on extreme value distributions.

Chapters 7–9 consider particle strength distributions in which particle size displays an increasingly direct effect. The limits of behavior are discussed in Chapters 7 and 9. In Chapter 7, the flaw population that controls particle strength distributions is considered invariant and independent of particle size. The observed strength distributions are described as resulting from "stochastic" extreme value size effects, as particle size simply alters the probability that a particular flaw will appear in the ensemble of flaws within a particle. Stochastic size effects will be shown to give rise to the convergence of strengths to a common small value as particle size increases, as in Figure 2.29, thereby addressing the fourth question stated earlier. In Chapter 9, the flaw population that controls particle strength distributions is considered to be dependent on particle size. The observed strength distributions are described as resulting from "deterministic" extreme value size effects, as particle size determines the flaw population. Deterministic size effects will be shown to give rise to the majority of strength distributions of particles, thereby addressing the fifth question. In Chapter 8, an intermediate example is considered, in which particle size influences only one aspect of the flaw population, the maximum flaw size. The materials considered here are ceramic particles in the green, unfired porous state and the fired dense state. The particles are those used in proppant beds in oil "fracking" operations and this chapter constitutes a case study in materials science application of particle strength distributions.

Chapters 10 and 11 consider the strength distributions of particles formed from materials much weaker and much more compliant than the majority of those considered in earlier chapters (they were mostly rocks, minerals, and ceramics). Chapter 10 considers the strength distributions of agglomerate particles, those formed from the binding together of many smaller sub-particles, such as the sand agglomerates of Figure 2.26b. Included in this chapter are considerations of industrially important particles such as catalysts, and domestically important particles such as food and pharmaceutical tablets. As the binding in these particles is usually controlled by localized secondary bonds between adjacent sub-particles, these particles are weak. Chapter 11 considers the strength distributions of hydrogel particles, those formed from sparse organic networks with very large volume fractions ($\approx 99\ \%$) of included water. As hydrogel materials are composed almost entirely of liquids, they are ideal for uses as diverse as medical devices, agricultural products, and micro-fluidic applications. As the mechanical properties of these particles are controlled by conformational changes of the network and flow in the weakly constrained fluid, the particles are very compliant. As a consequence, the force-displacement behavior of hydrogel particles, and that of the related elastomer particles, at large deformations prior to failure, will be considered. This chapter also constitutes a case study in materials science application of particle strength distributions.

Chapter 12 considers fracture mechanics analyses of particle failure with the materials science goal of enabling greater quantitative interpretation of particle strengths. At this point in the book attention had been focused on the nature of the distributions within a measured strength domain. Interpretation of the distributions in term of flaw size had been constrained only weakly by particle size. In this chapter, the flaw-strength linkage is considered in some detail in order to enable interpretations of the numerical values of particle strengths in terms of materials parameters, and thus controllable by particle processing. The materials parameters include elastic modulus, hardness, and toughness, and microstructural factors such the sizes of grains, incorporated pores, or inclusions. Flaw behavior under the action of applied stress, usually crack extension, is considered in a fracture mechanics framework developed mostly in stress terms. Fracture systems considered include homogeneous and inhomogeneous stiff particles, elastic-plastic particles, and porous particles. This chapter contains much analytical material.

Chapter 13 considers design analyses that include particle failure with the materials engineering goal of enabling quantitative application of particle strength information. The performance of particles in mechanical applications in terms of stiffness, load-bearing capability, reliability and lifetime, and energy consumption in production and comminution are analyzed. The engineering aspects of particle failure controlled by energy, force, mass, and impact velocity are considered. Chapter 13 also emphasizes the need for unbiased edf presentation schemes.

2.7.2 Overall Themes

The goal of this book is to provide a single view of particle failure extending across all (brittle) particle materials and particle sizes. The lens of the view is the single failure metric of characteristic particle strength, given by the HO equation, Eq. (2.41), and unbiased presentation and analysis of particle strength distributions, given by the edf, as in Figures 2.28 and 2.29. The ensuing estimation of strength-determining flaw sizes is the first and major step in determining particle failure mechanisms. The book will: (i) place particle strengths in the broad context of material strengths; (ii) apply modern analysis techniques to extract maximum information from particle strength tests; and (iii) enable design with particles in applications in which mechanical properties are important in manufacturing and handling.

The philosophy here is very much that of an experimentalist. The field of particle strength behavior is first surveyed and analyzed to identify the most common and typical behaviors and then descriptions and models developed as needed to describe those behaviors. This approach is different from that in many textbooks, in which mathematical models and analyses are developed and then published works used selectively to provide examples that illustrate the models. The structure of the present book regarding strength distributions might look like the latter. Chapters 3 and 4 are analytical and Chapters 5–11 are illustrative, but the reader should be assured that the intellectual development order was reversed from this and that the illustrative data are representative and supported by extensive references. Chapter 12 is of necessity more analytical and the derivations contained therein could be omitted at a first reading. Very little solid mechanics, fracture mechanics, or probability and statistics mathematics is required to understand the strength and flaw distribution analyses, simply a basic knowledge of functions and calculus. A familiarity with early chapters of the elasticity texts of Timoshenko and Goodier (1970) or Sadd (2009), the fracture text of Lawn (1993), and the probability and statistics texts of Lipschutz (1965) and Spiegel (1961), or similar, might be helpful.

Brief engineering applications for each particle system are described at the beginning of most sections. After Chapter 4, readers can pick and choose topic areas as interested. Where possible, chapters and sections have been divided into consideration of small, medium, and large particles, following a scheme used in an earlier review (Cook 2020). Small particles, diameter $D < 1$ mm, provide the best chance of containing a single strength-containing flaw and enabling direct assessment of a flaw population. Medium particles, 1 mm $< D <$ 10 mm, are the most common, contain a few flaws each, and often exhibit stochastic size effects. Large particles, $D >$ 10 mm, contain many flaws each, usually exhibit stochastic size effects, and often exhibit superposed deterministic size effects. The distinctions in size and behavior are not absolute and most particles in these divisions are natural materials such as salt, sand, quartz, soil, coal, limestone, rocks and minerals, and railway ballast. An important class of particles that are predominantly of medium size, are the engineered particles, including alumina, glass, iron ore pellets, cement, catalysts, ice, and food. In all, over 270 experimental strength distributions including over 13,000 individual particle measurements in more than 140 materials systems, are examined.

It is important to note here that, historically, application of particle strength measurements to advance many of the science and engineering areas discussed above has been impeded by presentation of most experimental strength data in a biased format. That format is the linearized transform of the powered exponential function, commonly referred to as the "Weibull" distribution (Stoyan 2013). There is much confusion regarding the fundamental nature of the powered exponential description of strength distributions. It is not mathematically fundamental or necessary: It is one of many widely known arithmetic descriptions of a sigmoidal trend (Wikipedia 2022b) and the linearized transform is thus also one of many (Castillo 1988). It is not physically based or unique: If strength-controlling flaws are independent (also known as "weakest link" behavior), all flaw populations, sigmoidal or otherwise, lead to the same, stochastic, inverse size effect on component strength (Zok 2017). The current book reveals many new phenomena by presenting strength distributions in an unbiassed way and applying new analysis methods. In many cases, published data were digitized and back-transformed to an unbiassed state prior to analysis. Chapters 3 and 4 discuss the powered exponential description in detail.

The emphasis throughout the book is on the quantitative phenomenology of strength distributions and presentation and analysis of numerical values of strength. Many groups have published extensive measurements in this area, including:

- The early work of the Bradt group (1986–1988) on ceramic-related materials (Kschinka et al. 1986; Wong et al. 1987; Bertrand et al. 1988).
- The Salman group (1988–2005) that focused on free particle impact velocity effects (Salman et al. 1988, 1995; Salman and Gorham 2000; Salman et al. 2002, 2003, 2004; Cheong et al. 2003; Gorham and Salman 2005).
- The work of the King and Tavares groups (1993–2021) that established the initial specific energy measurement methods and drop weight tests, mostly applied to rocks and minerals (King and Bourgeois, 1993; Tavares and King 1998; Tavares 1999, 2004, 2021; Tavares and Cerqueria 2006; Tavares and das Neves 2008; Barrios et al. 2011; Ribas et al. 2014; Cavalcanti and Tavares 2018; Tavares et al. 2018; Silva et al. 2019).
- The McDowell group (1996-2020) that mostly performed work on soils, but more lately on railway ballast (McDowell et al. 1996; McDowell and Amon 2000; McDowell 2001, 2002; McDowell and Humphreys 2002; Lim et al. 2004; Lim and McDowell 2005; Lu and McDowell 2007, 2010; Ferellec and McDowell 2008, 2010; de Bono and McDowell 2018, 2020; deBono et al. 2020).
- The Kalman, Peukert, Tomas, and Portnikov groups that did much to advance quasi-static testing intrumentation and observations (2000–2019) Kalman 2000; Vogel and Peukert 2002; Petukhov and Kalman 2004; Han et al. 2006; Antonyuk et al. 2005, 2010; Khanal et al. 2008; Aman et al. 2010; Rozenblat et al. 2011, 2013; Müller et al. 2013; Portnikov et al. 2013; Russell et al. 2014; Portnikov and Kalman 2014, 2018, 2019; Paul et al. 2014, 2015; Liburkin et al. 2015; Knoop et al. 2016; Herre et al. 2017).
- The Gustafsson group that performed work on iron ore pellets (2009–2017) (Gustafsson et al. 2009, 2013a, 2013b, 2013c, 2017).

Many other aspects of particle behavior are important in interpreting such values, particularly observations of particle failure mechanisms (Shipway and Hutchings 1993b, 1993c; Tsoungui et al. 1999; Salman and Gorham 2000; Nakata et al. 2001a; Procopio et al. 2003; Chaudhri 2004; Wu et al. 2004; Gorham and Salman 2005; Khanal et al. 2008; Paul et al. 2015; Wang and Coop 2016; Norazirah et al. 2016; Satone et al. 2017; Parab et al. 2017; Gustafsson et al. 2017; Wang et al. 2019). Interpretations of particle failure are also aided by numerical calculations, especially discrete element modeling (DEM) of single particle failure, e.g. (Carmona et al. 2008; Hanley , et al. 2015; Nguyen et al. 2015), multiple particle failure, e.g. (Lobo-Guerrero and Vallejo 2006a; de Bono and McDowell 2018, 2020), finite element modeling (Žagar et al. 2018), and peridynamic modeling (Zhu and Zhao 2019). Characterization of particle size distributions, including the effects of attrition and bulk comminution on such distributions, and the fluid-based handling and transport of of bulk particles, or powders, provides chemical engineering context for the work here (Rumpf 1990; Salman et al. 2007; Rhodes 2008). The surface-dominated behavior of superfine particles provides a chemistry and electrical and magnetic properties context (Ichinose et al. 1992). These aspects will not be addressed in this book.

References

Adi, S., Adi, H., Chan, H.-K., Finlay, W.H., Tong, Z., Yang, R., and Yu, A. (2011). Agglomerate strength and dispersion of pharmaceutical powders. *Journal of Aerosol Science* 42: 285–294.

Aman, S., Tomas, J., and Kalman, H. (2010). Breakage probability of irregularly shaped particles. *Chemical Engineering Science* 65: 1503–1512.

Antonyuk, S., Tomas, J., Heinrich, S., and Mörl, L. (2005). Breakage behaviour of spherical granulates by compression. *Chemical Engineering Science* 60: 4031–4044.

Antonyuk, S., Heinrich, S., Tomas, J., Deen, N.G., van Buijtenen, M.S., and Kuipers, J.A.M. (2010). Energy absorption during compression and impact of dry elastic-plastic spherical granules. *Granular Matter* 12: 15–47.

Ashby, M.F. (1999). *Materials Selection in Mechanical Design*. Butterworth-Heinemann.

Ashcroft, N.W. and Mermin, N.D. (1976). *Solid State Physics*. Holt, Rinehart and Winston.

Bell, T.J., Bendeli, A., Field, J.S., Swain, M.V. and Thwaite, E.G. (1991). The determination of surface plastic and elastic properties by ultra micro-indentation. *Metrologia* 28: 463–469.

Bertrand, P.T., Laurich-McIntyre, S.E., and Bradt, R.C. (1988). Strengths of fused and tabular alumina refractory grains. *American Ceramic Society Bulletin* 67: 1217–1222.

Brecker, J.N. (1974). The fracture strength of abrasive grains. *Journal of Engineering for Industry* 96: 1253–1257.

Calvié, E., Joly-Pottuz, L., Esnouf, C., Clément, P., Garnier, V., Chevalier, J., Jorand, Y., Malchère, A., Epicier, T., and Masenelli-Varlot, K. (2012). Real time TEM observation of alumina ceramic nano-particles during compression. *Journal of the European Ceramic Society* 32: 2067–2071.

Calvié, E., Réthoré, J., Joly-Pottuz, L., Meille, S., Chevalier, J., Garnier, V., Jorand, Y., Esnouf, C., Epicier, T., Quirk, J.B., and Masenelli-Varlot, K. (2014). Mechanical behavior law of ceramic nanoparticles from transmission electron microscopy in situ nano-compression tests. *Materials Letters* 119: 107–110.

Capes, C.E. (1971). The correlation of agglomerate strength with size. *Powder Technology* 5: 119–125.

Castillo, E. (1988). *Extreme Value Theory in Engineering*. Academic Press.

Cavalcanti, P.P. and Tavares, L.M. (2018). Statistical analysis of fracture characteristics of industrial iron ore pellets. *Powder Technology* 325: 659–668.

Chau, K.T., Wei, X.X., Wong, R.H.C., and Yu, T.X. (2000). Fragmentation of brittle spheres under static and dynamic compressions: Experiments and analyses. *Mechanics of Materials* 32: 543–554.

Chen, Y., Best, A., Butt, H.J., Boehler, R., Haschke, T., and Wiechert, W. (2006). Pressure distribution in a mechanical microcontact. *Applied Physics Letters* 88: 234101.

Chen, Y., Best, A., Haschke, T., Wiechert, W., and Butt, H.J. (2007). Stress and failure at mechanical contacts of microspheres under uniaxial compression. *Journal of Applied Physics* 101: 084908.

Chen, C.S. and Hsu, S.C. (2001). Measurement of indirect tensile strength of anisotropic rocks by the ring test. *Rock Mechanics and Rock Engineering* 34: 293–321.

Cheng, Y.T. and Cheng, C.M. (2004). Scaling, dimensional analysis, and indentation measurements. *Materials Science and Engineering: R: Reports* 44: 91–149.

Cheong, Y.S., Salman, A.D., and Hounslow, M.J. (2003). Effect of impact angle and velocity on the fragment size distribution of glass spheres. *Powder Technology* 138: 189–200.

Cook, R.F. (2018). A flexible model for instrumented indentation of viscoelastic-plastic materials. *MRS Communications* 8: 586–590.

Cook, R.F. (2019). Fracture sequences during elastic–plastic indentation of brittle materials. *Journal of Materials Research* 34: 1633–1644.

Cook, R.F. (2020). Single particle strength distributions: Heavy tails and extreme values. http://doi.org/10.5281/zenodo.4024618 (accessed July 4, 2021).

Cook R.F. (2021). Dynamic fragmentation patterns in flexed ceramic discs. http://doi.org/10.5281/zenodo.4439463 (accessed December 4, 2021).

Cook, R.F. and Michaels, C.A. (2017). Review: Coefficients for stress, temperature, and composition effects in fluorescence measurements of alumina. *Journal of Research of the National Institute of Standards and Technology* 122: 43. https://doi.org/10.6028/jres.122.043. (accessed December 5, 2021).

Cook, R.F. and Michaels, C.A. (2019). Stress measurements in alumina by optical fluorescence: Revisited. *Journal of Research of the National Institute of Standards and Technology* 124: 124020. https://doi.org/10.6028/jres.124.020 (accessed December 5, 2021).

Cook, R.F. and Oyen, M.L. (2007). Nanoindentation behavior and mechanical properties measurement of polymeric materials. *International Journal of Materials Research* 98: 370–378.

Dang, D., Wang, Y., and Cheng, Y.-T. (2019). Fracture behavior of single $LiNi_{0.33}Mn_{0.33}Co_{0.33}O_2$ particles studied by flat punch indentation. *Journal of the Electrochemical Society* 166: A2749–A2751.

Darvell, B.W. (1990). Review Uniaxial compression tests and the validity of indirect tensile strength. *Journal of Materials Science* 25: 757–780.

Dean W.R., Sneddon I.M., and Parsons, H.W. (1952). *Selected Government Research Reports: Strength and Testing of Materials: Part II: Testing Methods and Test Results* London: H.M.S.O. 212–234.

de Bono, J.P. and McDowell, G.R. (2018). On the micro mechanics of yielding and hardening of crushable granular soils. *Computers and Geotechnics* 97: 167–188.

de Bono, J.P. and McDowell, G.R. (2020). On the packing and crushing of granular materials. *International Journal of Solids and Structures* 187: 133–140.

Dong, J., Cheng, Y., Hu, B., Hao, C., Tu, Q., and Liu, Z. (2018). Experimental study of the mechanical properties of intact and tectonic coal via compression of a single particle. *Powder Technology* 325: 412–419.

Erarslan, N. and Williams, D.J. (2012). Experimental, numerical and analytical studies on tensile strength of rocks. *International Journal of Rock Mechanics and Mining Sciences* 49: 21–30.

Fahad, M.K. (1996). Stresses and failure in the diametral compression test. *Journal of Materials Science* 31: 3723–3729.

Ferellec, J.F. and McDowell, G.R. (2008). A simple method to create complex particle shapes for DEM. *Geomechanics and Geoengineering: An International Journal* 3: 211–216.

Ferellec, J.F. and McDowell, G.R. (2010). A method to model realistic particle shape and inertia in DEM. *Granular Matter* 12: 459–467.

Field, J.S. and Swain, M.V. (1993). A simple predictive model for spherical indentation. *Journal of Materials Research* 8: 297–306.

Field, J.S. and Swain, M.V. (1995). Determining the mechanical properties of small volumes of material from submicrometer spherical indentations. *Journal of Materials Research* 10: 101–112.

Forbes, C., Evans, M., Hastings, N., and Peacock, B. (2011). *Statistical Distributions, 4th ed.* Wiley.

Gayle, A.J. and Cook, R.F. (2016). Mapping viscoelastic and plastic properties of polymers and polymer-nanotube composites using instrumented indentation. *Journal of Materials Research* 31: 2347–2360.

Gladwell, G.M.L. (1980). *Contact Problems in the Classical Theory of Elasticity.* Sijthoff & Noordhoff.

Gorham, D.A. and Salman, A.D. (2005). The failure of spherical particles under impact. *Wear* 258: 580–587.

Gustafsson, G., Häggblad, H.-Å., and Knutsson, S. (2009). Experimental characterization of constitutive data of iron ore pellets. *Powder Technology* 194: 67–74.

Gustafsson, G., Häggblad, H.-Å., and Jonsén, P. (2013a). Characterization modelling and validation of a two-point loaded iron ore pellet. *Powder Technology* 235: 126–135.

Gustafsson, G., Häggblad, H.-Å., and Jonsén, P. (2013b). Multi-particle finite element modelling of the compression of iron ore pellets with statistically distributed geometric and material data. *Powder Technology* 239: 231–238.

Gustafsson, G., Häggblad, H.-Å., and Jonsén, P., and Markland, P. (2013c). Determination of bulk properties and fracture data for iron ore pellets using instrumented confined compression experiments. *Powder Technology* 241: 19–27.

Gustafsson, G., Häggblad, H.-Å., Nishida, M., Larsson, S., and Jonsén, P. (2017). Fracture probability modelling of impact-loaded iron ore pellets. *International Journal of Impact Engineering* 102: 180–186.

Hagan, J.T. (1979). Cone cracks around Vickers indentations in fused silica glass. *Journal of Materials Science* 14: 462–466.

Hagan, J.T. and Swain, M.V. (1978). The origin of median and lateral cracks around plastic indents in brittle materials. *Journal of Physics D: Applied Physics* 11: 2091–2102.

Han, T., Petukhov, Y., Levy, A., and Kalman, H. (2006). Theoretical and experimental study of multi-impact breakage of particles. *Advanced Powder Technology* 17: 135–157.

Herre, P., Romeis, S., Mačković, M., Przybilla, T., Paul, J., Schwenger, J., Torun, B., Grundmeier, G., Spiecker, E., and Peukert, W. (2017). Deformation behavior of nanocrystalline titania particles accessed by complementary in situ electron microscopy techniques. *Journal of the American Ceramic Society* 100: 5709–5722.

Hertz, H. (1896) *Miscellaneous Papers*. Macmillan & Co.

Hills, D.A., Nowell, D., and Sackfield, A. (1993). *Mechanics of Elastic Contacts.* Butterworth-Heinemann.

Hiramatsu, Y., Nishihara, M., and Oka, Y. (1954). A discussion on the methods of tension test of rock. *Journal of the Mining and Metallurgy Institute of Japan* 70: 285–289.

Hiramatsu, Y., Nishihara, M., Oka, Y., and Niihori, Y. (1955). Impact flexure tests of rocks. *Zairyō shiken* 4: 204–206.

Hiramatsu, Y. and Oka, Y. (1956). On the failure appearing in the rock around an underground opening. *Nippon Kōgyō Kaishi* 72: 439–444.

Hiramatsu, Y. and Oka, Y. (1962). Stress conditions for fracture of rock around underground openings. *Nippon Kōgyō Kaishi* 78: 803–808.

Hiramatsu, Y. and Oka, Y. (1964). Measurements of the variation in stress in rocks. *Nippon Kōgyō Kaishi* 80: 356–361.

Hiramatsu, Y., Oka, Y., and Kiyama, H. (1965a). Rapid determination of the tensile strength of rocks with irregular test pieces. *Nippon Kōgyō Kaishi* 81: 1024–1030.

Hiramatsu, Y. and Oka, Y. (1965b). Stress in the rock around underground openings. *Zairyō* 14: 480–485.

Hiramatsu, Y. and Oka, Y. (1966). Determination of the tensile strength of rock by a compression test of an irregular test piece. *International Journal of Rock Mechanics and Mining Sciences & Geomechanics Abstracts* 3: 89–99.

Hiramatsu, Y. and Oka, Y. (1967). Authors' reply to discussion by J. Salençon. Determination of the tensile strength of rock by a compression test of an irregular test piece. *International Journal of Rock Mechanics and Mining Sciences & Geomechanics Abstracts* 4: 363–365.

Hiramatsu, Y., Oka, Y., and Kiyama, H. (1969). Investigations on the disc test, ring test and indentation test for rocks. *Nippon Kōgyō Kaishi* 85: 8–14.

Hondros, G. (1959). The evaluation of Poisson's ratio and the modulus of materials of low tensile resistance by the Brazilian (indirect tensile) test with particular reference to concrete. *Australian Journal of Applied Science* 10: 243–268.

Hu, G., Otaki, H., and Lin, M. (2001). An index of the tensile strength of brittle particles. *Minerals Engineering* 14: 1199–1211.

Huang, J., Xu, S., Yi, H., and Hu, S. (2014). Size effect on the compression breakage strengths of glass particles. *Powder Technology* 268: 86–94.

Huang, H., Huang, Q.K., Zhu, X.H., and Hu, X.Z. (1993). An experimental investigation of the strengths of individual brown corundum abrasive grains. *Scripta Metallurgica* 29: 299–304.

Hung, K.M. and Ma, C.C. (2003). Theoretical analysis and digital photoelastic measurement of circular disks subjected to partially distributed compressions. *Experimental Mechanics* 43: 216–224.

Ichinose, N., Ozaki, Y., and Kashu, S. (1992). *Superfine Particle Technology.* Springer-Verlag.

Jaeger, J.C. and Hoskins, E.R. (1966). Rock failure under the confined Brazilian test. *Journal of Geophysical Research* 71: 2651–2659.

Jaeger, J.C. (1967). Failure of rocks under tensile conditions. *International Journal of Rock Mechanics and Mining Sciences* 4: 219–227.

Jarrar, Z.A., Alshibli, K.A., and Al-Raoush, R.I. (2020). Three-dimensional evaluation of sand particle fracture using discrete-element method and synchrotron microcomputed tomography images. *Journal of Geotechnical and Geoenvironmental Engineering* 146: 06020007.

Johnson, K.L. (1985). *Contact Mechanics.* Cambridge.

Kalman, H. (2000). Particle breakage and attrition. *KONA Powder and Particle Journal* 18: 108–120.

Kanda, Y., Sano, S., Saito, F., and Yashima, S. (1985). Relationships between particle size and fracture energy for single particle crushing [translated]. *KONA Powder and Particle Journal* 3: 26–31.

Kapur, P.C. and Fuerstenau, D.W. (1967). Dry strength of pelletized spheres. *Journal of the American Ceramic Society* 50: 14–18.

Kendall, K. and Weihs, T.P. (1992). Adhesion of nanoparticles within spray dried agglomerates. *Journal of Physics D: Applied Physics* 25: A3–A8.

Khanal, M., Schubert, W., and Tomas, J. (2008). Compression and impact loading experiments of high strength spherical composites. *International Journal of Mineral Processing* 86: 104–113.

Kienzler, R. and Schmitt, W. (1990). On single-particle comminution; numerical analysis of compressed spheres. *Powder technology* 61: 29–38.

King, R.P. and Bourgeois, F. (1993). Measurement of fracture energy during single-particle fracture. *Minerals Engineering* 6: 353–367.

Kingery, W.D., Bowen, H.K., and Uhlmann, D.R. (1975). *Introduction to Ceramics.* Wiley.

Knoop, C., Todorova, Z., Tomas, J., and Fritsching, U. (2016). Agglomerate fragmentation in high-intensity acoustic standing wave fields. *Powder Technology* 291: 214–222.

Kschinka, B.A., Perrella, S., Nguyen, H., and Bradt, R.C. (1986). Strengths of glass spheres in compression. *Journal of the American Ceramic Society* 69: 467–472.

Kuehn, G.A., Schulson, E.M., Jones, D.E., and Zhang, J. (1993). The compressive strength of ice cubes of different sizes. *Journal of Offshore Mechanics and Arctic Engineering* 115: 142–148.

Kwag, J.M., Ochiai, H., and Yasufuku, N. (1999). Yielding stress characteristics of carbonate sand in relation to individual particle fragmentation strength. In: *Engineering for Calcareous Sediments* (ed. A.K. Al-Shafei), 79–86. Balkema.

Lawn, B.R. (1993). *Fracture of Brittle Solids, 2nd ed.* Cambridge.

Li, D. and Wong, L.N.Y. (2013). The Brazilian disc test for rock mechanics applications: Review and new insights. *Rock Mechanics and Rock Engineering* 46: 269–287.

Liburkin, R., Portnikov, D., and Kalman, H. (2015). Comparing particle breakage in an uniaxial confined compression test to single particle crush tests–model and experimental results. *Powder Technology* 284: 344–354.

Lim, W.L. and McDowell, G.R. (2005). Discrete element modelling of railway ballast. *Granular Matter* 7: 19–29.

Lim, W.L., McDowell, G.R., and Collop, A.C. (2004). The application of Weibull statistics to the strength of railway ballast. *Granular Matter* 6: 229–237.

Ling, F.F., Lai, W.M., and Lucca, D.A. (2002). *Fundamentals of Surface Mechanics.* Springer-Verlag.

Lipschutz, S. (1965). *Probability.* McGraw-Hill, Inc.

List, E., Frenzel, J., and Vollstaedt, H. (2006). A new system for single particle strength testing of grinding powders. *Industrial Diamond Review* 1: 42–54.

Liu, S. and Wang, J. (2018). Static fatigue of sand particles. *Canadian Geotechnical Journal* 55: 1682–1687.

Lobo-Guerrero, S. and Vallejo, L.E. (2006a). Discrete element method analysis of railtrack ballast degradation during cyclic loading. *Granular Matter* 8: 195–204.

Lobo-Guerrero, S. and Vallejo, L.E. (2006b). Application of Weibull statistics to the tensile strength of rock aggregates. *Journal of Geotechnical and Geoenvironmental Engineering* 132: 786–790.

Lu, G. and McDowell, G.R. (2007). The importance of modelling ballast particle shape in the discrete element method. *Granular Matter* 9: 69–80.

Lu, M. and McDowell, G.R. (2010). Discrete element modelling of railway ballast under monotonic and cyclic triaxial loading. *Géotechnique* 60: 459–467.

Lur'e, A.I. (1964). *Three-Dimensional Problems of the Theory of Elasticity.* Wiley.

Luscher, W.G., Hellmann, J.R., Segall, A.E., Shelleman, D.L., and Scheetz, B.E. (2007). A critical review of the diametral compression method for determining the tensile strength of spherical aggregates. *Journal of Testing and Evaluation* 35: 624–629.

Ma, L., Li, Z., Wang, M., Wei, H., and Fan, P. (2019). Effects of size and loading rate on the mechanical properties of single coral particles. *Powder Technology* 342: 961–971.

Ma, C.C. and Hung, K.M. (2008). Exact full-field analysis of strain and displacement for circular disks subjected to partially distributed compressions. *International Journal of Mechanical Sciences* 50: 275–292.

Marshall, D.B., Evans, A.G., and Nisenholz, Z. (1983). Measurement of dynamic hardness by controlled sharp-projectile impact. *Journal of the American Ceramic Society* 66: 580–585.

Mase, G.E. (1970). *Continuum Mechanics.* McGraw-Hill.

Maugis, D. (2000). *Contact, Adhesion and Rupture of Elastic Solids.* Springer-Verlag.

Mazel, V., Guerard, S., Croquelois, B., Kopp, J.B., Girardot, J., Diarra, H., Busignies, V., and Tchoreloff, P. (2016). Reevaluation of the diametral compression test for tablets using the flattened disc geometry. *International Journal of Pharmaceutics* 513: 669–677.

McDowell, G.R., Bolton, M.D., and Robertson, D. (1996). The fractal crushing of granular materials. *Journal of Mechanics and Physics of Solids* 44: 2079–2102.

McDowell, G.R. and Amon, A. (2000). The application of Weibull statistics to the fracture of soil particles. *Soils and Foundations* 40: 133–141.

McDowell, G.R. (2001). Statistics of soil particle strength. *Géotechnique* 51: 897–900.

McDowell, G.R. (2002). On the yielding and plastic compression of sand. *Soils and Foundations* 42: 139–145.

McDowell, G.R. and Humphreys, A. (2002). Yielding of granular materials. *Granular Matter* 4: 1–8.

Momber, A.W. (2004). Deformation and fracture of rocks loaded with spherical indenters. *International Journal of Fracture* 125: 263–279.

Moomivand, H. (1999). Effect of size on the compressive strength of coal. In: *99 International Symposium on Mining Science and Technology* (ed. X. Heping and T.S. Golosinski), 399–404. Beijing, China. 29–31 August 1999.

Müller, P., Seeger, M., and Tomas, J. (2013). Compression and breakage behavior of γ-Al_2O_3 granules. *Powder Technology* 237: 125–133.

Nad, A. and Saramak, D. (2018). Comparative analysis of the strength distribution for irregular particles of carbonates, shale, and sandstone ore. *Minerals* 8: 37.

Nakata, Y., Hyodo, M., Hyde, A. F. L., Kato, Y., and Murata, H. (2001a). Microscopic particle crushing of sand subjected to high pressure one-dimensional compression. Soils and Foundations 41: 69–82.

Nakata, Y., Kato, Y., Hyodo, M., Hyde, A.F.L., and Murata, H. (2001b). One-dimensional compression behavior of uniformly graded sand related to single particle crushing strength. *Soils and Foundations* 41: 39–51.

Ni, W., Cheng, Y.T., Cheng, C.M., and Grummon, D.S. (2004). An energy-based method for analyzing instrumented spherical indentation experiments. *Journal of Materials Research* 19: 149-157.

Ogiso, H., Yoshida, M., Nakano, S., and Akedo, J. (2007). Effects of Al ion implantation on the strength of Al2O3 particles. *Surface and Coatings Technology* 201: 8180–8184.

Oka, Y. and Majima, H. (1970). A theory of size reduction involving fracture mechanics. *Canadian Metallurgical Quarterly* 9: 429–439.

Oyen, M.L. (2006). Analytical techniques for indentation of viscoelastic materials. *Philosophical Magazine*, 86: 5625–5641.

Oyen, M.L. (2011). Mechanics of indentation. In: *Handbook of Nanoindentation with Biological Applications* (ed. M.L. Oyen), 123–151. Pan Stanford Publications.

Oyen, M.L. and Cook, R.F. (2003). Load-displacement behavior during sharp indentation of viscous-elastic-plastic materials. *Journal of Materials Research* 18: 139–150.

Oyen, M.L., Cook, R.F., Emerson, J.A., and Moody, N.R. (2004). Indentation responses of time-dependent films on stiff substrates. *Journal of Materials Research* 19: 2487–2497.

Parab, N., Guo, Z., Hudspeth, M. C., Claus, B. J., Fezzaa, K., Sun, T., and Chen, W. W. (2017). Fracture mechanisms of glass particles under dynamic compression. *International Journal of Impact Engineering* 106: 146–154.

Paul, J., Romeis, S., Tomas, J., and Peukert, W. (2014). A review of models for single particle compression and their application to silica microspheres. *Advanced Powder Technology* 25: 136–153.

Paul, J., Romeis, S., Mačković, M., Marthala, V.R.R., Herre, P., Przybilla, T., Hartmann, M., Spiecker, E., Schmidt, J., and Peukert, W. (2015). In situ cracking of silica beads in the SEM and TEM – Effect of particle size on structure–property correlations. *Powder Technology* 270: 337–347.

Pejchal, V., Žagar, G., Charvet, R., Dénéréaz, C., and Mortensen, A. (2017). Compression testing spherical particles for strength: Theory of the meridian crack test and implementation for microscopic fused quartz. *Journal of Mechanics and Physics of Solids* 99: 70–92.

Perras, M.A. and Diederichs, M.S. (2014). A review of the tensile strength of rock: Concepts and testing. *Geotechnical and Geological Engineering* 32: 525–546.

Petukhov, Y. and Kalman, H. (2004). Empirical breakage ratio of particles due to impact. *Powder technology* 143: 160–169.

Pitchumani, R., Zhupanska, O., Meesters, G.M.H., and Scarlett, B. (2004). Measurement and characterization of particle strength using a new robotic compression tester. *Powder Technology* 143–144: 56–64.

Portnikov, D., Kalman, H., Aman, S., and Tomas, J. (2013). Investigating the testing procedure limits for measuring particle strength distribution. *Powder Technology* 237: 489–496.

Portnikov, D. and Kalman, H. (2014). Determination of elastic properties of particles using single particle compression test. *Powder Technology* 268: 244–252.

Portnikov, D. and Kalman, H. (2018). The effect of temperature on the mechanical characteristics of individual particles. *Powder Technology* 336: 393–405.

Portnikov, D. and Kalman, H. (2019). Material comminution functions of wet particles. *Powder Technology* 343: 29–39.

Preston, F.W. (1937). Concerning the strength of the weakest bottles Applicability of the "normal" curve of errors to statistical analyses of strength tests of glassware. *Journal of the American Ceramic Society* 20: 329–336.

Procopio, A.T., Zavaliangos, A., and Cunningham, J.C. (2003). Analysis of the diametrical compression test and the applicability to plastically deforming materials. *Journal of Materials Science* 28: 3629–3639.

Rhee, Y.W., Kim, H.W., Deng, Y., and Lawn, B.R. (2001). Brittle fracture versus quasi plasticity in ceramics: a simple predictive index. *Journal of the American Ceramic Society* 84: 561–565.

Rhodes, M. (2008). *Introduction to Particle Technology, 2nd ed.* Wiley.

Ribas, L., Cordeiro, G.C., Toledo Filho, R.D., and Tavares, L.M. (2014). Measuring the strength of irregularly-shaped fine particles in a microcompression tester. *Minerals Engineering* 65: 149–155.

Rozenblat, Y., Portnikov, D., Levy, A., Kalman, H., Aman, S., and Tomas, J. (2011). Strength distribution of particles under compression. *Powder Technology* 208: 215–224.

Rozenblat, Y., Levy, A., Kalman, H., and Tomas, J. (2013). Impact velocity and compression force relationship – Equivalence function. *Powder technology* 235: 756–763.

Russell, A., Müller, P., and Tomas, J. (2014). Quasi-static diametrical compression of characteristic elastic–plastic granules: Energetic aspects at contact. *Chemical Engineering Science* 114: 70–84.

Rumpf, H. (1990). *Particle Technology.* Springer.

Ryu, H.J. and Saito, F. (1991). Single particle crushing of nonmetallic inorganic brittle materials. *Solid State Ionics* 47: 35–50.

Sadd, M.H. (2009). *Elasticity.* Elsevier.

Salman, A.D., Szabo, M., Angyal, I., and Verba, A. (1988). Particle degradation. *Periodica Polytechnica Mechanical Engineering* 32: 233–251.

Salman, A.D., Gorham, D.A., and Verba, A. (1995). A study of solid particle failure under normal and oblique impact. *Wear* 186: 92–98.

Salman, A.D., Ghadiri, M., and Hounslow, M. (2007). *Particle Breakage.* Elsevier.

Salman, A.D. and Gorham, D.A. (2000). The fracture of glass spheres. *Powder Technology* 107: 179–185.

Salman, A.D., Biggs, C.A., Fu, J., Angyal, I., Szabó, M., and Hounslow, M.J. (2002). An experimental investigation of particle fragmentation using single particle impact studies. *Powder Technology* 128: 36–46.

Salman, A.D., Fu, J., Gorham, D.A., and Hounslow, M.J. (2003). Impact breakage of fertiliser granules. *Powder Technology* 130: 359–366.

Salman, A.D., Reynolds, G.K., Fu, J.S., Cheong, Y.S., Biggs, C.A., Adams, M.J., Gorham, D.A., Lukenics, J., and Hounslow, M.J. (2004). Descriptive classification of the impact failure modes of spherical particles. *Powder Technology* 143: 19–30.

Satone, M., Iimura, K., Teraoka, T., Hanafusa, T., Hisatani, S., Nishiwaki, M., and Suzuki, M. (2017). Analysis of granule fracture under biaxial compression. *Ceramics International* 43: 16835–16842.

Shan, J., Xu, S., Liu, Y., Zhou, L., and Wang, P. (2018). Dynamic breakage of glass sphere subjected to impact loading. *Powder Technology* 330: 317–329.

Shen, J., Dongsheng, X.U., Liu, Z., and Wei, H. (2020). Effect of particle characteristics stress on the mechanical properties of cement mortar with coral sand. *Construction and Building Materials* 260: 119836.

Shetty, D.K., Rosenfield, A.R., and Duckworth, W.H. (1985). Fracture toughness of ceramics measured by a chevron-notch diametral-compression test. *Journal of the American Ceramic Society* 68: C-325–C-327.

Shipway, P.H. and Hutchings, I.M. (1993a). Fracture of brittle spheres under compression and impact loading. I. Elastic stress distributions. *Philosophical Magazine A* 67: 1389–1404.

Shipway, P.H. and Hutchings, I.M. (1993b). Fracture of brittle spheres under compression and impact loading II. Results for lead-glass and sapphire spheres. *Philosophical Magazine A* 67: 1405–1421.

Shipway, P.H. and Hutchings, I.M. (1993c). Attrition of brittle spheres by fracture under compression and impact loading. *Powder Technology* 76: 23–30.

Sikong, L., Hashimoto, H., and Yashima, S. (1990). Breakage behavior of fine particles of brittle minerals and coals. *Powder Technology* 61: 51–57.

Silva, N.V., Angulo, S.C., da Silva Ramos Barboza, A., Lange, D.A., and Tavares, L.M. (2019). Improved method to measure the strength and elastic modulus of single aggregate particles. *Materials and Structures* 52: 1–15.

Singh, D., McGlinchey, D., and Crapper, M. (2016). Breakage functions of particles of four different materials subjected to uniaxial compression. *Particulate Science and Technology* 34: 494–501.

Spiegel, M.R. (1961) *Statistics.* McGraw-Hill.

Sternberg, E. and Rosenthal, F. (1952). The elastic sphere under concentrated loads. *Journal of Applied Mechanics* 23: 413–421.

Stoyan, D. (2013). Weibull, RRSB or extreme-value theorists? *Metrika* 76: 153–159.

Studman, C.J. and Field, J.E. (1984). The influence of brittle particles on the contact between rigid surfaces. *Journal of Physics D: Applied Physics* 17: 1631–1646.

Subero-Couroyer, C., Ghadir, M., Brunard, N., and Kolenda, F. (2003). Weibull analysis of quasi-static crushing strength of catalyst particles. *Chemical Engineering Research and Design* 81: 953–962.

Swain, M.V. (1999). Plastic deformation of brittle materials. In: *Key Engineering Materials* (ed. E. Yasuda, F. Wakai, L.M. Manocha, and Y. Tanabe), 166: 41–46. Trans Tech Publications.

Swain, M.V. and Menčík, J. (1994). Mechanical property characterization of thin films using spherical tipped indenters. *Thin Solid Films* 253: 204–211.

Tavares, L.M. (1999). Energy absorbed in breakage of single particles in drop weight testing. *Minerals Engineering* 12: 43–50.

Tavares, L.M. (2004). Optimum routes for particle breakage by impact. *Powder Technology* 142: 81–91.

Tavares, L.M. (2021) Review and further validation of a practical single-particle breakage model. *KONA* online 1–22.

Tavares, L.M., Cavalcanti, P.P., de Carvalho, R.M., da Silveira, M.W., Bianchi, M., and Otaviano, M. (2018). Fracture probability and fragment size distribution of fired iron ore pellets by impact. *Powder Technology* 336: 546–554.

Tavares, L.M. and Cerqueria, M.C. (2006). Statistical analysis of impact-fracture characteristics and microstructure of industrial Portland cement clinkers. *Cement and Concrete Research* 36: 409–415.

Tavares, L.M. and das Neves, P.B. (2008). Microstructure of quarry rocks and relationships to particle breakage and crushing. *International Journal of Mineral Processing* 87: 28–41.

Tavares, L.M. and King, R.P. (1998). Single-particle fracture under impact loading. *International Journal of Mineral Processing* 54: 1–28.

Timoshenko, S.P. and Goodier, J.N. (1970). *Theory of Elasticity.* McGraw-Hill.

Titovich, A.S. and Norris, A.N. (2012). Green's function for symmetric loading of an elastic sphere with application to contact problems. *Journal of Mechanics of Materials and Structures* 7: 701–719.

Tsoungui, O., Vallet, D., Charmet, J.-C., and Roux, S. (1999). Size effects in single grain fragmentation. *Granular Matter* 2: 19–27.

Unland, G. and Szczelina, P. (2004). Coarse crushing of brittle rocks by compression. *International Journal of Mineral Processing* 74: S209–S217.

Vallet, D. and Charmet, J.C. (1995). Mechanical behaviour of brittle cement grains. *Journal of Materials Science* 30: 2962–2967.

Verspui, M.A., de With, G., and Dekkers, E.C.A. (1997). A crusher for single particle testing. *Reviews of Scientific Instruments* 68: 1553–1556 .

Vogel, L. and Peukert, W. (2002). Characterisation of grinding-relevant particle properties by inverting a population balance model. *Particles & Particle System Characterisation* 19: 149–157.

Walpole, R.E. and Myers, R.H. (1972). *Probability and Statistics for Engineers and Scientists.* Macmillan.

Wang, C., Cheng, Y., He, X., Yi, M., and Wang, Z. (2019). Size effect on uniaxial compressive strength of single coal particle under different failure conditions. *Powder Technology* 345: 169–181.

Wang, W. and Coop, M.R. (2016). An investigation of breakage behaviour of single sand particles using a high-speed microscope camera. *Géotechnique* 66: 984–998.

Wang, Y., Dan, W., Xu, Y., and Xi, Y. (2015). Fractal and morphological characteristics of single marble particle crushing in uniaxial compression tests. *Advances in Materials Science and Engineering* 2015: 537692.

Wei, X.X., Chau, K.T., and Wong, R.H.C. (1999). Analytic solution for axial point load strength test on solid circular cylinders. *Journal of Engineering Mechanics* 125: 1349–1357.

Wei, X.X. and Chau, K.T. (2013). Three dimensional analytical solution for finite circular cylinders subjected to indirect tensile test. *International Journal of Solids and Structures* 50: 2395–2406.

Wei, X.X., Chau, K.T. and Wong, R.H.C. (2019). Theoretical and experimental validation of point load strength test for irregular lumps. *Journal of Engineering Mechanics* 145: 04019065.

Wijk, G. (1978). Some new theoretical aspects of indirect measurements of the tensile strength of rocks. *International Journal of Rock Mechanics and Mining Sciences and Geomechanics Abstracts* 15: 149–160.

Wikipedia (2022a). *Empirical distribution function.* en.wikipedia.org/wiki/Empirical_distribution_function. (accessed August 14, 2022).

Wikipedia (2022b). *Sigmoid function.* en.wikipedia.org/wiki/Sigmoid_function. (accessed May 14, 2022).

Wong, J.Y., Laurich-McIntyre, S.E., Khaund, A.K., and Bradt, R.C. (1987). Strengths of green and fired spherical aluminosilicate aggregates. *Journal of the American Ceramic Society* 70: 785–791.

Wu, D., Zhou, J., and Li, Y., (2007). Mechanical strength of solid catalysts: Recent developments and future prospects. *AIChE Journal* 53: 2618–2629.

Wynnyckyj, J.R. (1985). The correlation between the strength factor and the true tensile strength of agglomerate spheres. *The Canadian Journal of Chemical Engineering* 63: 591–597.

Xu, Y., Song, D., and Chu, F. (2016). Approach to the Weibull modulus based on fractal fragmentation of particles. *Powder Technology* 292: 99–107.

Yashima, S., Kanda, Y., and Sano, S. (1987). Relationships between particle size and fracture energy or impact velocity required to fracture as estimated from single particle crushing. *Powder technology* 51: 277–282.

Yasufuku, N. and Kwag, J.M. (1999). Significance of soil particle fragmentation strength related to soil crushability. In: *Soil Mechanics and Geotechnical Engineering*, (ed. S.S. Chung, S.W. Hong, C.C. Ihm, H.T. Kim, M.M. Kim, B.S. Lee, S.R. Lee, J.B. Park, and G.S. Yang). CRC.

Zhang, L.C. (1991). Numerical analysis of the shape effect on the tensile strength evaluation of axisymmetric specimens. *International Journal for Numerical and Analytical Methods in Geomechanics* 15: 241–250.

Zhu, W.C., Chau, K.T. and Tang, C.A. (2004). Numerical simulation on failure patterns of rock discs and rings subject to diametral line loads. In *Key Engineering Materials* 261: 1517–1522 Trans Tech.

Zhu, F. and Zhao, J. (2019). A peridynamic investigation on crushing of sand particles. *Géotechnique* 69: 526–540.

Zok, F.W. (2017). On weakest link theory and Weibull statistics. *Journal of the American Ceramic Society* 100: 1265–1268.

3

Flaw Populations

This chapter defines and describes in clear, simple mathematical terms the physical basis of how the characteristics of a population of flaws (which is usually not observed) determines the characteristics of a distribution of strengths (which is what is measured) through the mediating influences of the sample and component sizes (which are what is controlled). This is forward analysis and is focused on answering the question "what could we expect?" This chapter is quite pedagogical and the examples ideal as they will all be determined by analysis. The analysis is based on stochastic extreme value size effects on strength, in which a strength-controlling flaw in a component is drawn from an invariant population and the flaw probability is determined solely by component size. The analysis is applicable to all brittle components.

3.1 Flaw Sizes and Strengths

The mechanical behavior of particles in diametral compression, including force-displacement behavior on loading and the nature of particle failure at the maximum supported force, was described in some detail in Chapters 1 and 2. Particle failure and fragmentation in diametral compression were marked by fracture on particle meridional planes, consistent with elastic stress analyses. In many cases, irreversible compaction localized to the poles of particles significantly modified diametral compression contact geometries and lead to predominantly linear force-displacement responses. However, the local effects of axial compression at the loading points and global transverse tension on the meridional planes were physically well separated and no observable irreversible deformation was associated with fracture. The materials forming the particles in Chapters 1 and 2 (and throughout this book) can thus be classified as brittle: those that respond to tensile stress by elastic deformation followed by fracture, with negligible associated plastic deformation. The greatest supportable tensile stress—or strength—of such materials is thus the *brittle fracture strength*.

Fracture in brittle materials, encompassing many of the particle materials considered in Chapters 1 and 2—rocks, minerals, ceramics, glasses, salts—has been well studied and analyzed (Lawn 1993; Maugis 2000). In brief, brittle fracture is a process of sequential interatomic bond rupture. On tensile loading, the vast majority of a component formed from a brittle material is deformed elastically. Fracture is restricted to an expanding plane of ruptured bonds—a *crack*—delimited by a surrounding linear boundary region, localized to a few bond spacings, separating intact from ruptured bonds—the crack front, often referred to as the crack tip. A *flaw* in a material—typically including a pre-existing crack—concentrates stress on loading such that the resulting deformation local to the flaw is accommodated by bond rupture. If the flaw does not include a pre-existing crack, this process is known as crack initiation. If the flaw does include a crack, this process is known as crack propagation. If the loaded fracture system is unstable, the crack will propagate through the material, separating the component into two or more pieces by rupturing bonds one at a time. The maximum stress at which this process occurs is the strength. The fracture process can be viewed from an energy perspective: The work performed by the applied loading external to the component generates an internal energy change within the component associated with elastic strain energy. On fracture, an internal energy conversion takes place: the elastic strain energy, distributed throughout the component, is transformed by crack propagation into surface energy, localized to the newly formed fracture surfaces within the component. The final state is one in which the component pieces, the fragments, are in their initial undeformed configuration but possess new peripheral surfaces formed by broken bonds.

Flaws in brittle materials that limit material strengths, and thus limit the maximum supportable forces of components formed from the materials, are extremely varied in form. By definition, all concentrate stress, leading to fracture. Examples

Particle Strengths: Extreme Value Distributions in Fracture, First Edition. Robert F. Cook.

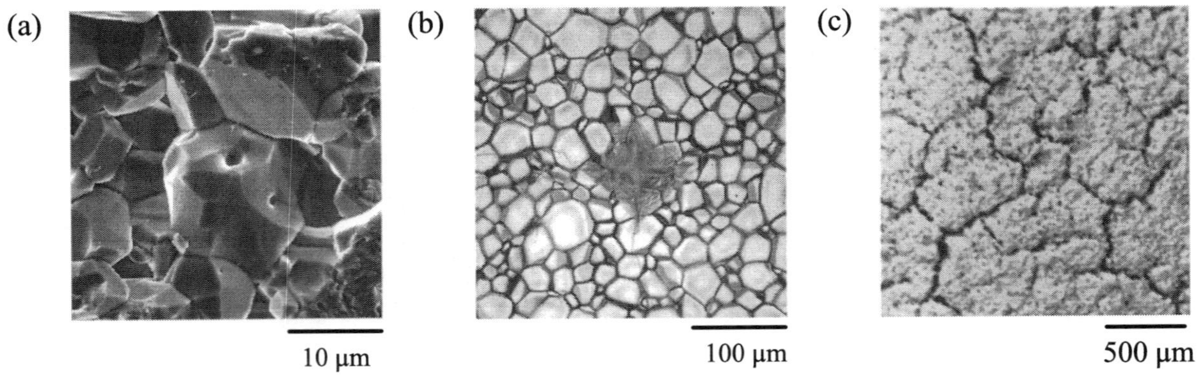

Figure 3.1 Images of flaws in materials that lead to fracture of brittle components. (a) Scanning electron micrograph of the fracture surface of a polycrystalline refractory alumina (Al_2O_3) material. The angular grains (different Al_2O_3 crystal orientations) and cracked grain boundaries are visible, along with smaller, near-spherical grain boundary pores. (b) Optical micrograph of the surface of a polycrystalline alumina material containing an indentation flaw. The grains and grain boundaries in the material are visible, along with the square residual contact impression of the indentation and radial cracks emanating from the impression corners. (c) Optical micrograph of the surface of a dried powder compact of cordierite ($2MgO.2Al_2O_3.5SiO_2$) material. The network of dessication cracks is visible. *Source:* Robert F. Cook.

of flaws relevant to particle failure are shown in Figure 3.1, in order of increasing scale: (a) the fracture surface of polycrystalline alumina refractory; (b) the fired surface of a polycrystalline alumina containing an indentation; and (c) the surface of a dried ceramic powder compact. Some flaws exhibit no obvious associated crack: pores in fired materials that are almost spherical (Figure 3.1a), grain boundaries in polycrystalline materials with associated decreased bond density (Figure 3.1b). Some flaws appear as simple cracks: cracked grain boundary facets arising from thermal expansion mismatch effects (Figure 3.1a), and some are associated with localized contact (Figure 3.1b). Some materials have distributed and interconnected flaws (Figure 3.1c). Material microstructure features, grains and powder particles, clearly affect fracture and crack propagation. The indentation flaw of Figure 3.1b is both a good model for natural flaws and a good test vehicle for studying strength. The contact impression, generated by a square Vickers diamond pyramid in Figure 3.1b, is a region of localized compression and thus mimics natural localized surface contacts in brittle materials, including the formation of compacted shear faults (visible in Figure 3.1b). The contact impression also mimics localized inclusion and strain effects as in composite and polycrystalline materials with inhomogeneous (e.g. Figure 1.3a) or anisotropic thermal expansion behavior. Indentations flaws are ideal for studying brittle material strengths as the location and size of the indentation can be controlled.

Figure 3.2 shows examples of the effects of flaw size, characterized as crack lengths c, on brittle material strengths σ, in a series of logarithmic plots. The plots are drawn to the same scale (vertical × horizontal = 10 × 100) and include a diversity of materials and flaw types. Some of the earliest work in this area, and the most influential, was that of Griffith (1921), in which the strengths of glass bulb components were measured by pressure loading. Prior to strength testing, scratches and cracks of various lengths were formed in the bulbs and annealed. The results are shown in Figure 3.2a, using data derived from the published work. Symbols represent individual strength measurements at the crack lengths indicated, in the units used by Griffith (psi and in, 1 psi = 6,895 kPa, 1 in = 25.4 mm). The solid line represents the strength behavior predicted by Griffith, based on the energy perspective given above and the stress analysis of Inglis (1913). The behavior is encapsulated in the Griffith equation

$$\sigma = Bc^{-1/2} \qquad (\text{Griffith}), \tag{3.1}$$

in which B is a material- and flaw geometry-dependent quantity. Griffith estimated B from measurements of the surface energy of glass as a function of temperature. The agreement between measurements and prediction, both quantitatively and qualitatively is clear. Eq. (3.1) will be used extensively throughout this book.

Figure 3.2b shows more recent strength measurements of glass, biaxially flexed square plates containing annealed Vickers indentation flaws from the work of Glaesemann et al. (1987), using data derived from the published work. Symbols represent individual strength measurements. The solid line represents a best fit to the data consistent with Eq. (3.1). The fit well describes the data and, as noted by Glaesemann et al., is consistent with other indented glass measurements. The similarity between the Griffith (1921) and Glaesemann et al. (1987) measurements is apparent. Figure 3.2c shows strength measurements of notched bend beams of titania (TiO_2) agglomerates from the work of Kendall (1988), using data derived from the published work. Symbols represent individual strength measurements. The solid line represents a best fit to the data

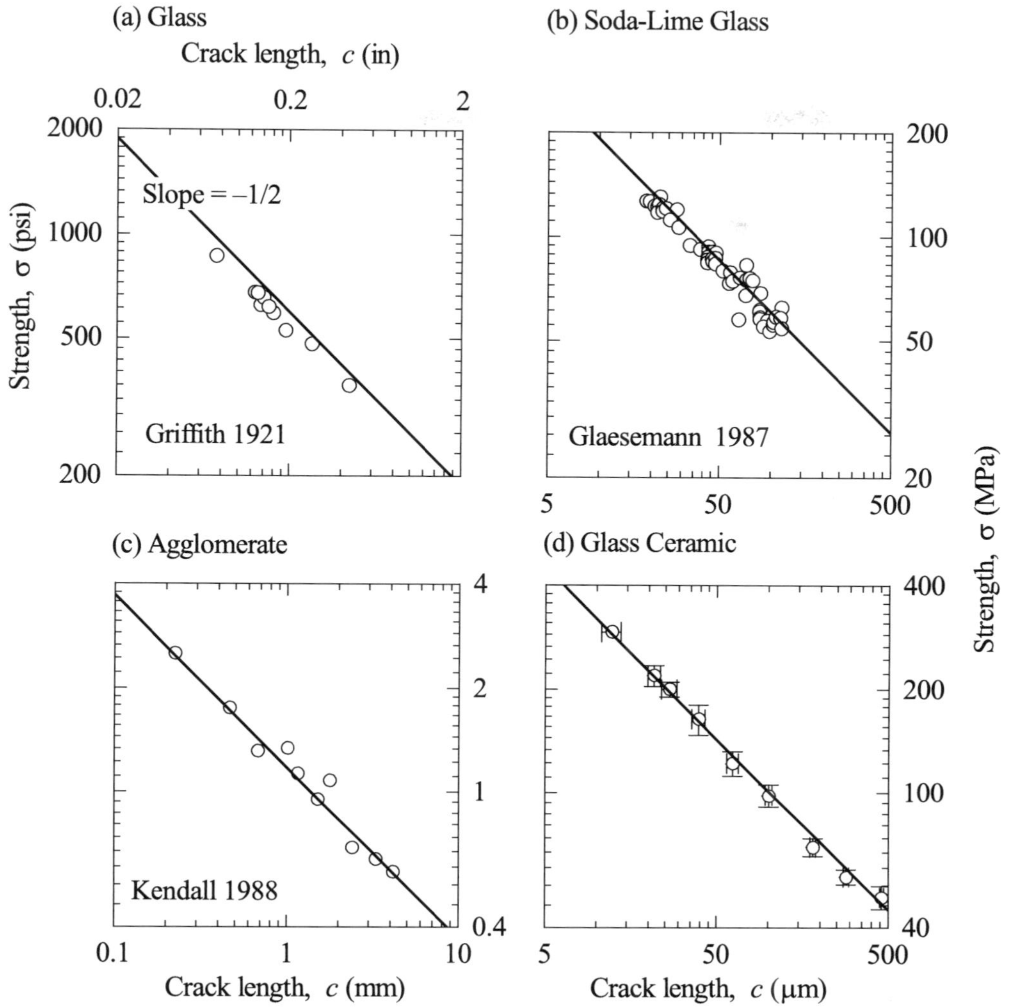

Figure 3.2 A series of strength σ vs crack length c plots in logarithmic coordinates for brittle materials, illustrating the Griffith equation, Eq. (3.1). (a) Glass bulbs containing annealed scratches Adapted from Griffith, A.A 1921. (b) Glass plates containing annealed indentations (Adapted from Glaesemann, G.S et al. 1987). (c) Titania (TiO_2) powder compact agglomerates containing notches Adapted from Kendall, K (1988). (d) Cordierite ($2MgO.2Al_2O_3.5SiO_2$) glass ceramic cylinders containing indentations. The Griffith equation has been well verified by experiment for over 100 years.

consistent with Eq. (3.1) and again, the fit well describes the data. At the other end of the strength range, Figure 3.2d shows strength measurements of biaxially flexed circular cordierite plates containing Vickers indentation flaws. Symbols and bars represent means and standard deviations of multiple crack length and strength measurements at different indentation forces. The solid line represents a best fit to the data consistent with Eq. (3.1) and again, the fit well describes the data.

The Griffith equation, Eq. (3.1), is a result of application of the first law of thermodynamics, which, in this case, describes work by applied loading to create elastic energy that is then converted into surface energy by the process of fracture. The successful application of the equation is exemplified in Figure 3.2. The fundamental basis of the equation is observed in the expression for B expressed in modern notation (Lawn 1993; Maugis 2000) $B = (ER)^{1/2}/\psi$, where E is material Young's modulus as discussed earlier, R is the material fracture resistance, and ψ is a dimensionless term characterizing the geometry of the fracture system. R can be a function of crack length, $R = R(c)$, but the simplest form and interpretation is that $R = 2\gamma$ where γ is an invariant, intensive, material property characterizing the surface energy areal density (energy/area) required to create an equilibrium fracture surface. The term "areal density" is usually omitted and implied, the *surface energy* is an "excess" energy, measured relative to the bulk material bound state, and the factor of two in $R = 2\gamma$ accounts for the two surfaces created on fracture. 2γ is a lower bound to R. R may be increased in some materials by microstructural effects leading to irreversible deformation and energy loss (plasticity in metals and polymers are the prime examples). B thus contains information regarding the work performed, via the crack geometry term, ψ, the elastic energy, via E, and the surface energy, via R. The material property combination $(ER)^{1/2}$ is termed the material *toughness* $T = (ER)^{1/2}$ and takes an invariant lower bound $T_0 = (2E\gamma)^{1/2}$. B may thus be written compactly as $B = T/\psi$. Unless otherwise stated, B will be considered an invariant, intensive, material property throughout this book. A characteristic value for brittle materials is $B = 1$ MPa m$^{1/2}$ (Lawn, 1993) such that a crack length of $c = 100$ μm corresponds to a strength of $\sigma = 100$ MPa. Chapter 12 discusses the fracture mechanics of particle failure in detail, including a broad consideration of flaws such as cracks associated with pores, inclusions, and indentations, and a broad consideration of surface energy, such as the effects of material microstructure on fracture resistance. These effects are quantified by variations in ψ and R. The Griffith equation makes clear that strength is a "composite" variable as the right side of Eq. (3.1) is the product of intensive and extensive quantities.

Section 3.2, discusses how the Griffith equation can be incorporated into intensive descriptions of flaws and strengths. Although the ultimate goal here is description of the strength behavior of particles, the analysis in this chapter, and Chapter 4 is general and applies to the strengths of all brittle materials and components.

3.2 Populations of Flaws and Strengths

3.2.1 Population Definitions

The strengths of the brittle components shown in Chapter 2, glass bottles and coral and sand particles, quantified by the empirical distribution function (edf), exhibited significant variability. Although strength variability is common in brittle components (see Chapters 5 and 6 for more examples), the nature of the strength variability exhibited by the coral and sand particles appeared distinct from that of the larger, extended, brittle components, the glass bottles. Eq. (3.1) and Figure 3.2 make clear that the underlying cause of the strength variability was variability in the strength controlling flaw size in each particle or bottle. Hence, the differences in the forms of the edf behavior for the particles and the bottles reflected differences in the flaw size distributions. Hence, analysis of the fundamentals of strength distributions of brittle components begins with the distributions of strength-controlling flaws that occur in the components. Such distributions are drawn from populations of flaws in the materials forming the components, and flaw and strength populations are thus the starting point of analysis here. The analysis is similar to that presented in recent works considering strength distributions in ceramics and silicon (Si) components (Cook and DelRio 2019a, 2019b; Cook et al. 2019, 2021; DelRio et al. 2020). The analysis in this chapter is *forward analysis* and begins with a fundamental intensive material property, the flaw population, and proceeds to predict an extensive component performance measure, the strength distribution. Chapter 4 considers reverse analysis.

A large body of material, volume Ω, Figure 3.3a, can be characterized by two flaw population attributes. These attributes characterize (1) the spatial distribution of flaws in the material and (2) the size distribution of these flaws. The spatial distribution of flaws is most easily specified by the *flaw spatial density*, λ, the number of strength-controlling flaws/volume. The reciprocal $1/\lambda = \Delta V$ is the average volume occupied by a single flaw and is thus termed a *fundamental volume element*, Figure 3.3b. The total number, or population, of flaws in the body of material is thus $\Omega\lambda$. Here, the size of the flaw in each fundamental element is specified by a crack length c, Figure 3.3b (other specifications are considered in Chapter 12). The crack length may be different for each fundamental element and locally determines the strength, such that the strength of each fundamental volume element is σ, Figure 3.3c. If the population of elements and thus flaws, $\Omega\lambda$, is large, the crack lengths may be treated as a continuum with a continuous *probability density function* (pdf), $f(c)$ (Lipschutz 1965; Walpole and Myers 1972). The crack length pdf, $f(c)$ describes the frequency of crack lengths over the region of support or domain of f, $c_{min} \leq c \leq c_{max}$, where c_{min} and c_{max} are the minimum and maximum crack lengths of the population, respectively. $f(c) = 0$ for $c < c_{min}$ and $c > c_{max}$. λ and $f(c)$ (and thus c_{min} and c_{max}) are fundamental attributes of the flaw population. Both λ and $f(c)$ are invariant and intensive. If a component volume V is sampled from Ω, the component contains $k = V\lambda$ strength-controlling flaws, and both k and V are extensive. The similarity to the mass density equation of Chapter 2 is apparent. In Figure 3.3c $k = 1$. Section 3.3 considers the effects of variations in λ and k on strength distributions of samples. The rest of this section considers the effects of variations in $f(c)$ on strength distributions of populations (i.e. all of Ω tested as in Figure 3.3c).

The continuous *cumulative distribution function* (cdf), $F(c)$, of the crack length population is given by integration of the pdf, $f(c)$,

$$F(c) = \int_0^c f(u)du, \tag{3.2}$$

where u is a dummy crack length variable. $F(c)$ gives the proportion of elements in the population with cracks smaller than c and is therefore the probability that an element selected at random from the population will contain a crack of length smaller than c. Normalization of the pdf requires

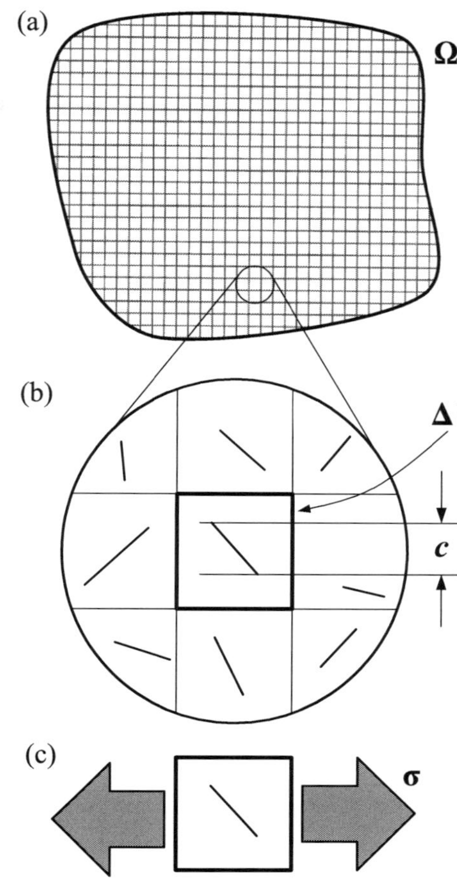

Figure 3.3 (a) A schematic diagram of a large body of material, volume Ω, containing a population of fundamental volume elements. (b) Each volume element, volume ΔV, contains one strength-controlling crack, length c, belonging to a material population of flaw sizes. (c) The brittle fracture strength of each volume element is σ, belonging to a material population of strengths.

$$\int_{c_{min}}^{c_{max}} f(c)\mathrm{d}c = 1, \tag{3.3}$$

(as all crack lengths lie between c_{min} and c_{max}), such that the range of $F(c)$ is 0 to 1 over the domain $c_{min} \leq c \leq c_{max}$. More useful here, as discussed in Chapter 4, is that the pdf is expressed as the derivative of the cdf:

$$f(c) = \mathrm{d}F(c)/\mathrm{d}c, \tag{3.4}$$

such that $f(c)$ can be estimated from $F(c)$. Eqs. (3.2) and (3.4) make clear that the crack length cdf $F(c)$ is equally fundamental as the crack length pdf $f(c)$.

Similarly, given the transformation relation of the Griffith equation, Eq. (3.1), the strength cdf $F(\sigma)$ and strength pdf $f(\sigma)$ are also fundamental descriptions of the population of elements. The region of support or domain for $F(\sigma)$ and $f(\sigma)$ is conjugate to the crack length domain and bounded by

$$\sigma_u = Bc_{min}^{-1/2}, \tag{3.5}$$

where the upper bound to the element strengths, σ_u, corresponds to the minimum crack length, and

$$\sigma_{th} = Bc_{max}^{-1/2}, \tag{3.6}$$

where the lower bound to the element strengths, σ_{th}, corresponds to the maximum crack length. The lower bound to the strength domain, σ_{th}, is known as the *threshold strength* and is critical in engineering design of mechanical loading safety factors. (σ_{th} is the threshold stress for material failure—no failures occur for applied stresses less than the threshold.) $f(\sigma)$ is defined in relation to $F(\sigma)$ similar to Eqs. (3.2) and (3.4)

$$F(\sigma) = \int_0^{\sigma} f(v)\mathrm{d}v, \tag{3.7}$$

where v is a dummy strength variable, and

$$f(\sigma) = \mathrm{d}F(\sigma)/\mathrm{d}\sigma, \tag{3.8}$$

and normalization requires

$$\int_{\sigma_{th}}^{\sigma_u} f(\sigma)\mathrm{d}\sigma = 1, \tag{3.9}$$

such that the range of $F(\sigma)$ is 0 to 1 over the domain $\sigma_{th} \leq \sigma \leq \sigma_u$. The cdf $F(\sigma)$ gives the proportion of elements in the population with strengths less than σ and is the probability that an element selected at random from the population will exhibit a strength less than σ.

The inverse relationship between strength and crack length, Eq. (3.1), makes it useful to define the strength *complementary cumulative distribution function* (ccdf), $\bar{F}(\sigma)$, by

$$\bar{F}(\sigma) = 1 - F(\sigma). \tag{3.10}$$

$\bar{F}(\sigma)$ gives the proportion of elements in the population with strengths greater than σ and is thus the probability that a randomly selected element will exhibit a strength greater than σ. The domain and range of $\bar{F}(\sigma)$ are identical to those of $F(\sigma)$. The crack length ccdf, $\bar{F}(c)$, is similarly defined

$$\bar{F}(c) = 1 - F(c). \tag{3.11}$$

with similar properties. The power of the ccdf $\bar{F}(\sigma)$ is that on substitution of c for σ using the transformation of variables given by Eq. (3.1), noting that large strengths imply small cracks and vice versa, the relationships between cdf responses for crack length and strength are (Walpole and Myers 1972)

$$\bar{F}(\sigma) = F(c) \tag{3.12}$$

and

$$\bar{F}(c) = F(\sigma). \tag{3.13}$$

Eqs. (3.12) and (3.13) pertain for any plausible strength-crack length relation for which $\mathrm{d}\sigma/\mathrm{d}c < 0$ (Walpole and Myers 1972) (e.g. $\sigma \sim c^{-x}$ for cracks at pores as discussed in Chapter 12). Combining Eq. (3.2), Eq. (3.11), and Eq. (3.13) gives

$$F(\sigma) = 1 - \int_0^c f(u)\mathrm{d}u \tag{3.14}$$

where u is a dummy crack length variable, enabling determination of the population strength cdf $F(\sigma)$ from the population crack length pdf $f(c)$, where σ and c are conjugate. The functions $f(x)$, $F(x)$, and $\bar{F}(x)$ will be used throughout to denote characteristics of the population, where the argument x will define the function in context. Usually, x will be c or σ or a dummy variable for these used in analyses. The dimension of $f(x)$ is x^{-1}. F and $\bar{F}$ are dimensionless.

3.2.2 Population Examples

Within the framework of materials science and engineering, Eq. (3.14) represents the materials science connection between a structural measure (the population of crack lengths) and a property measure (the distribution of fracture strengths). From a materials engineering perspective, the pdf $f(c)$ is determined by materials processing and the cdf $F(\sigma)$ is a key factor in determining component performance. As a consequence, flaw populations are more often considered in terms of the pdf $f(c)$ of crack lengths and load-bearing properties more often considered in terms of the cdf $F(\sigma)$ of strengths. The following sections thus consider the relation between the forms of these two (conjugate) intensive material characteristics and the effects of flaw populations on extensive strength distributions of samples from the population. In particular, the population cdf $F(\sigma)$ provides an upper bound to the sample edf $Pr(\sigma)$ (e.g. for the particles in Chapter 2)—a point discussed in detail in Section 3.3 and Chapter 4. Characterizing the forms of strength populations is often the first step in identifying the nature of strength-controlling flaws.

In engineering considerations, the absolute values of c and σ are important as these determine component capability, e.g. for the tensile bar in Chapter 2, $P_{max} \sim b^2Bc^{-1/2}$. In material science considerations, only relative measures of c and σ

are important as these determine the forms of materials behavior. The relative measures of σ and c used here are μ and ν, given by

$$\mu = \frac{\sigma - \sigma_{\mathrm{th}}}{\sigma_{\mathrm{u}} - \sigma_{\mathrm{th}}} \tag{3.15}$$

and

$$\nu = \frac{c - c_{\min}}{c_{\max} - c_{\min}}, \tag{3.16}$$

such that μ and ν are bounded by $0 \leq \mu, \nu \leq 1$. The scaled domain of the strength population, η, is defined by

$$\frac{\sigma_{\mathrm{u}}}{\sigma_{\mathrm{th}}} = \eta > 1, \tag{3.17}$$

a concept encountered in the strength edf discussions of Chapter 2 as the experimental ratio σ_N/σ_1. Using Eq. (3.1), the scaled domain of the crack length population is thus

$$\frac{c_{\max}}{c_{\min}} = \eta^2. \tag{3.18}$$

Using Eqs. (3.5) and (3.6), Eqs. (3.15) and (3.16) can then be re-written as

$$\mu = \frac{(\sigma/\sigma_{\mathrm{th}}) - 1}{\eta - 1} \tag{3.19}$$

and

$$\nu = \frac{\eta^2(\sigma/\sigma_{\mathrm{th}})^{-2} - 1}{\eta^2 - 1}. \tag{3.20}$$

Eliminating $(\sigma/\sigma_{\mathrm{th}})$ between Eqs. (3.19) and (3.20) gives an expression relating the relative strength and relative crack length μ and ν as

$$\nu = \frac{\eta^2[\mu(\eta - 1) + 1]^{-2} - 1}{\eta^2 - 1}. \tag{3.21}$$

Eq. (3.21) has (μ, ν) solutions of (1,0) and (0,1) and for $\eta \gg 1$ Griffith behavior is recovered, $\mu \rightarrow (\eta^2\nu)^{-1/2}$. On substitution of μ for σ and ν for c using the transformations of variables given by Eqs. (3.15) and (3.16), the pdf and cdf relationships of Eqs. (3.2)–(3.14) are identically expressed in absolute and relative coordinates. In particular, the relative form of Eq. (3.14) is

$$F(\mu) = 1 - F(\nu) = 1 - \int_0^{\nu} f(u)\mathrm{d}u \tag{3.22}$$

where u here is a dummy relative crack length variable. Eq. (3.22) enables determination of the population relative strength cdf $F(\mu)$ from the population relative crack length pdf $f(\nu)$, where μ and ν are conjugate.

The simplest form of crack length pdf is the uniform density, $f(\nu) = 1$, such that $F(\nu) = \nu$ and $F(\mu) = 1 - \nu$ in Eq. (3.22). Substitution of Eq. (3.21) into this equation gives

$$F(\mu) = \frac{1 - [\mu(\eta - 1) + 1]^{-2}}{1 - \eta^{-2}} \tag{3.23}$$

which returns $F(0) = 0$ and $F(1) = 1$ as required, but otherwise Eq. (3.23) is not readily interpreted. Figure 3.4 shows a plot of (a) the uniform relative crack length pdf and (b) the conjugate relative strength cdf given by Eq. (3.23) using $\eta = 2$. The domain and range of both functions is [0, 1] and the abscissa of the plots is extended about the domains to emphasize that the crack lengths and strengths are bounded. To guide the eye, the domain of the cdf resulting from the pdf is shown shaded. The uniform crack length density gives rise to a concave strength distribution—the probability of a strength value in the population, $F(\mu)$, exhibits the greatest increase at small strengths and very small increases at large strengths.

The next simplest forms of crack length pdf are the linearly increasing and decreasing densities, $f(\nu) = 2\nu$ and $f(\nu) = 2(1 - \nu)$, such that $F(\nu) = \nu^2$ and $F(\nu) = 2\nu - \nu^2$ and thus $F(\mu) = 1 - \nu^2$ and $F(\mu) = (1 - \nu)^2$, respectively. An obvious extension of this form is a power-law decreasing density, such that there are many more small flaws than large flaws,

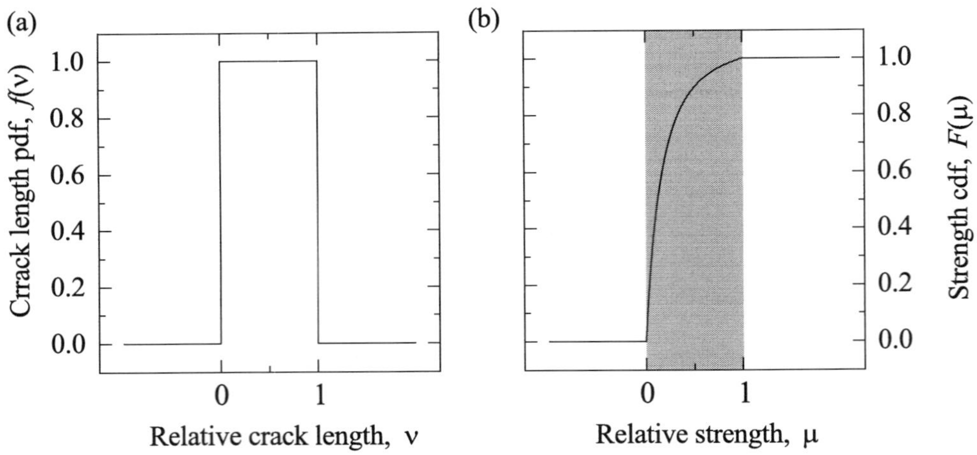

Figure 3.4 (a) Plot of the probability density function $f(\nu)$ as a function of relative crack length ν for a population of crack lengths of uniform density (a "top hat" function). (b) Plot of the cumulative distribution function $F(\mu)$ as a function of relative strength μ conjugate to the population in (a). Although crack lengths are uniformly distributed, strengths are not.

$f(\nu) = (x+1)(1-\nu)^x$, and thus $F(\nu) = 1-(1-\nu)^{x+1}$ and $F(\mu) = (1-\nu)^{x+1}$. Substitution of Eq. (3.21) into these equations for $F(\mu)$ gives the correct limits, but again the resulting closed-form expressions are similar to Eq. (3.23) and are not readily interpreted. However, the general form of Eq. (3.22) is simply evaluated using any selected form of $f(\nu)$ and Eq. (3.21) to generate $F(\mu)$ by numerical integration, and this procedure was followed throughout the book. Figure 3.5 shows plots of the linear and power-law pdf variations (left column) and the conjugate cdf curves (right column). The format is similar to Figure 3.4 and the conjugate pairs are (a)-(b), (c)-(d), and (e)-(f). In order to permit easy comparisons, the pdf variations, $f(\nu)$, are shown scaled by the maximum values, f_{max}, such that $f(\nu)/f_{max}$ occupy the range [0, 1]. The connections and trends in Figure 3.5 are representative of the reciprocal effects between crack length probability densities and strength probability distributions, to be observed throughout this book. First, crack length pdf variations skewed to large cracks as in Figure 3.5a lead to strength cdf variations with large increases at small strengths as in Figure 3.5b. As the crack length pdf variation is skewed to smaller crack lengths in the sequence (a) → (c) → (e), the strength probability increase shifts to greater strengths as in the sequence (b) → (d) → (f). Second, linear pdf variations are associated with non-linear cdf variations and vice verse, as in Figures 3.5a-b and 3.5e-f.

A more natural and flexible form of crack length pdf is related to the *beta function*, $B(\alpha, \beta)$, given by

$$B(\alpha, \beta) = \int_0^1 u^{\alpha-1}(1-u)^{\beta-1}du, \tag{3.24}$$

where u is a dummy variable and the exponents $\alpha, \beta > 0$. Extensive mathematical details regarding $B(\alpha, \beta)$ are given elsewhere (Gupta and Nadarajah 2004) and applications are discussed in (Krishnamoorthy 2016). Here, exponents $\alpha, \beta > 1$ will be used to describe $f(\nu)$ variations, as follows. The relative crack length cdf, $F(\nu)$, is given by the incomplete beta function

$$F(\nu) = \frac{1}{B(\alpha, \beta)} \int_0^\nu u^{\alpha-1}(1-u)^{\beta-1}du \tag{3.25}$$

and thus the relative crack length pdf, $f(\nu)$, is given by

$$f(\nu) = \frac{\nu^{\alpha-1}(1-\nu)^{\beta-1}}{B(\alpha, \beta)}, \tag{3.26}$$

where it is clear that $B(\alpha, \beta)$ acts as a normalization term such that Eq. (3.3) is fulfilled. It is more intuitive to re-parameterize Eq. (3.26), defining the *concentration* parameter as the sum of the exponents $\varkappa = \alpha + \beta$. The *mode* of the pdf, ω, the most probable relative crack length (the peak in the probability density), is then given by $\omega = (\alpha - 1)/(\varkappa - 2)$. The mean and variance (μ, var) of the pdf are similarly simply expressed, and, although these quantities are not used here, it is noted that the physical meaning of $\varkappa$ is as an inverse measure of the relative variance (the width of the peak), $\varkappa \sim \mu(1-\mu)/\text{var}$ (Gupta and Nadarajah 2004). The exponents in Eq. (3.26) are simply expressed in these more intuitive terms as

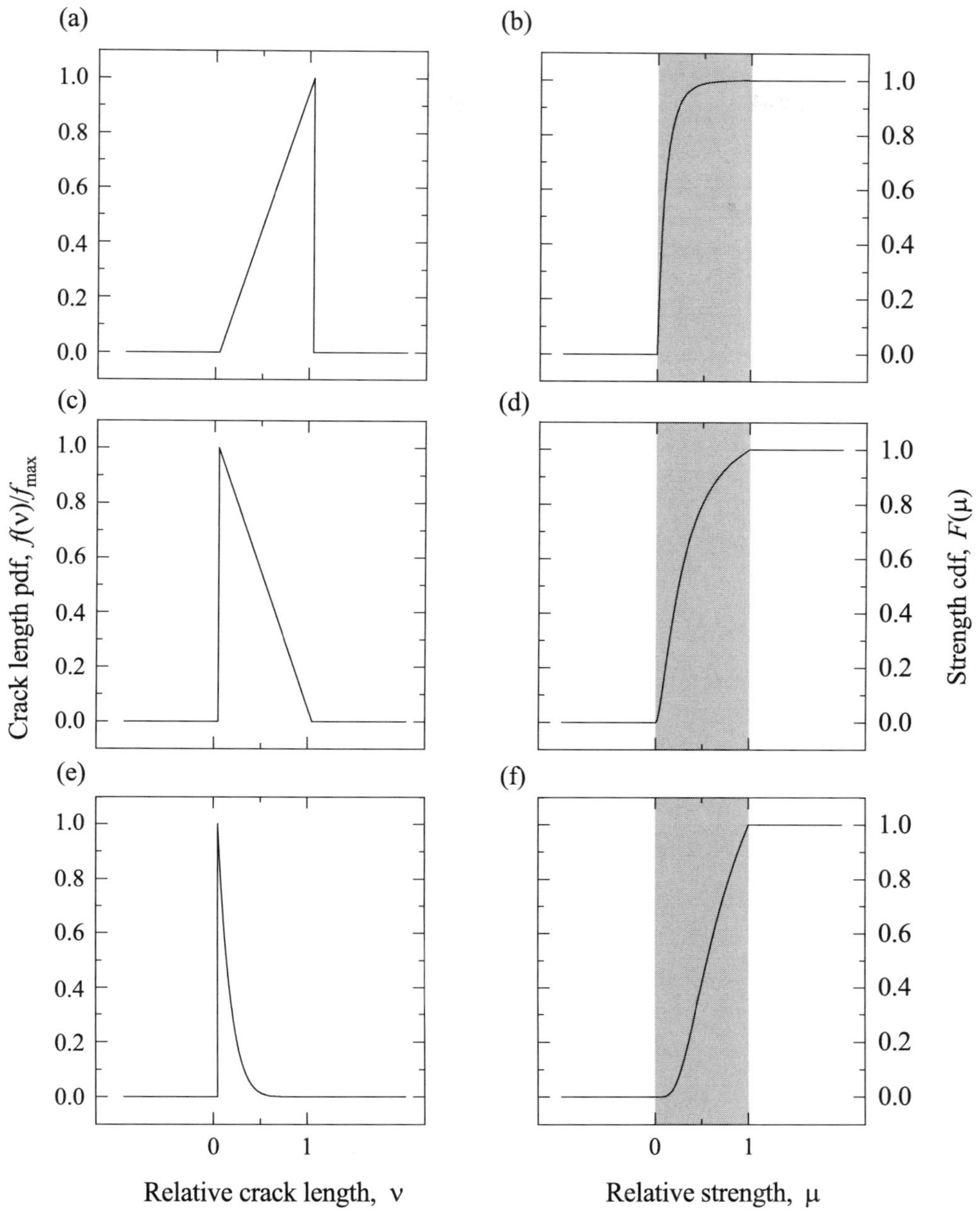

Figure 3.5 Plots of probability density functions $f(\nu)$ as a function of relative crack length ν and conjugate cumulative distribution functions $F(\mu)$ as a function of relative strength μ for non-uniform populations of crack lengths. Conjugate pairs are (ab), (cd), and (ef). As the crack length density becomes "more peaked" at small cracks, the strength distribution becomes less uniform.

$$\alpha = 1 + (\kappa - 2)\omega \tag{3.27}$$

and

$$\beta = 1 + (\kappa - 2)(1 - \omega). \tag{3.28}$$

Figure 3.6 shows plots of beta function pdf variations, left column, Eq. (3.26), and the resulting cdf variations, right column, determined by numerical integration of Eq. (3.22) and application of Eq. (3.21). The format is similar to Figure 3.5 with the exceptions of changes in scales and removal of shading. The conjugate pairs are (a)-(b), (c)-(d), and (e)-(f) and the pdf variations are shown scaled by the maximum values, $f(\nu)/f_{max}$. The resulting strength cdf variations were obtained using $\eta = 2$. Figure 3.6a shows a symmetric "bell-shaped" crack length pdf, obtained using $\omega = 0.5$ and $\kappa = 6$. The conjugate strength cdf is shown in Figure 3.6b and is an asymmetric sigmoid, heavily skewed to small strengths. Figure 3.6c shows an asymmetric crack length pdf skewed to small crack lengths, obtained using $\omega = 0.25$ and $\kappa = 25$. The resulting strength cdf is shown in Figure 3.6d and is a symmetric sigmoid. Figure 3.6e also shows an asymmetric crack length pdf, in this case skewed to large crack lengths, obtained using $\omega = 0.75$ and $\kappa = 4$. The resulting strength cdf is shown in Figure 3.6f and is concave. The connections and trends in Figure 3.6 reinforce the observations made above that there are reciprocal effects between crack length probability densities and strength probability distributions. The asymmetric crack length pdf skewed to small cracks, Figure 3.6c, lead to a symmetric sigmoidal strength cdf variation, Figure 3.6d. The asymmetric crack length pdf skewed to large cracks, Figure 3.6e, led to an asymmetric concave strength cdf skewed to small strengths, Figure 3.6f. The non-linear beta function pdf variation, Eq. (3.26), thus leads to some forms that are easily recognized (bell, sigmoid, and concave), although comparison of the left and right columns of Figure 3.6 reinforces the point that the detailed form of a strength cdf is often not readily apparent from a crack length pdf.

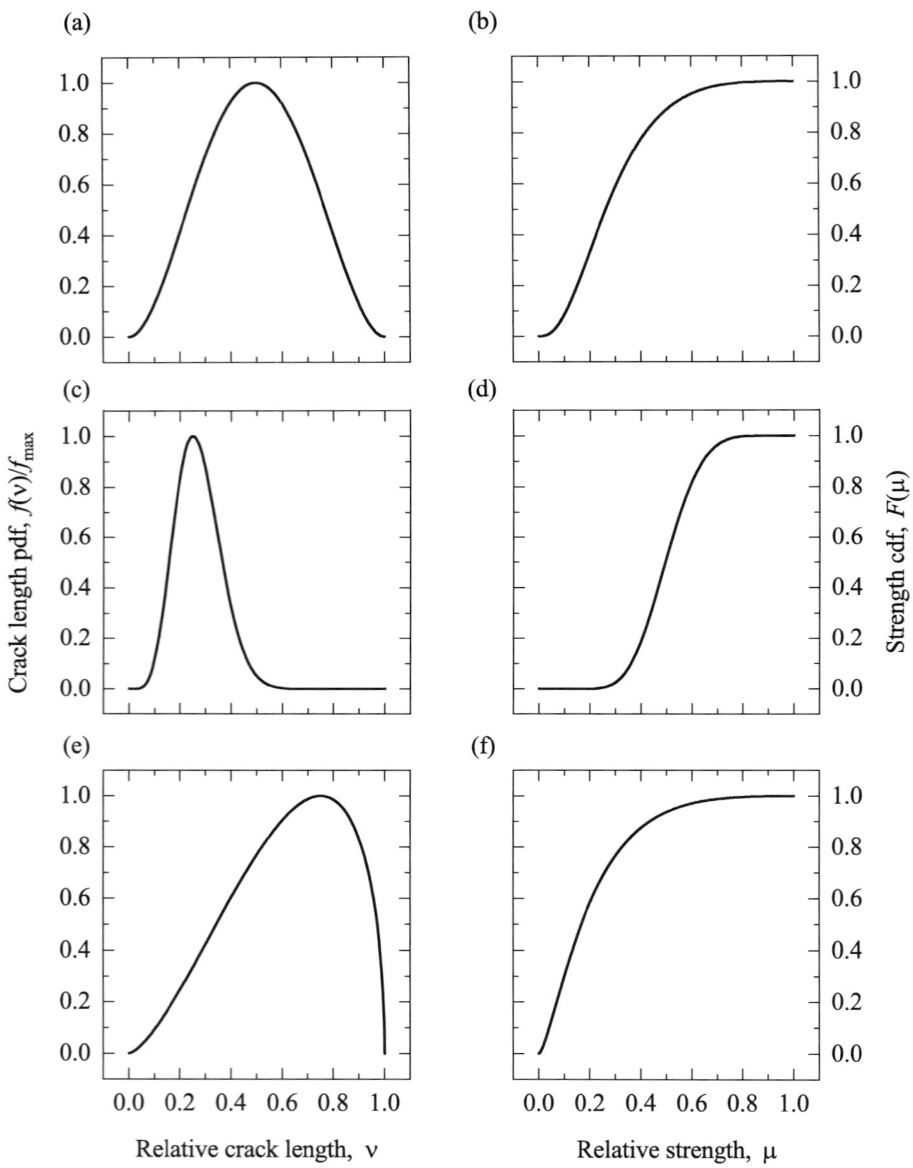

Figure 3.6 Plots of probability density functions $f(\nu)$ as a function of relative crack length ν and conjugate cumulative distribution functions $F(\mu)$ as a function of relative strength μ for smoothly varying populations of crack lengths described by the Beta function, Eq. (3.26). Conjugate pairs are (ab), (cd), and (ef). Asymmetric crack length populations, (c) and (e) give rises to commonly observed strength distribution shapes, (d) sigmoidal and (f) concave. (c) is described as "light tailed."

Important but less familiar pdf forms are generated by adding linear perturbations to the beta function formulation, as shown (using the same format as Figure 3.6) in the examples of Figure 3.7. The simplest addition is a constant value, shown as the dot-dash line in Figure 3.7a at $\approx 0.4f_{\max}$, with a beta function superposed using $\omega = 0.3$ and $\kappa = 4$, and the total shown as the solid line. (Providing 3.3 is fulfilled, there is no requirement for $f(0) = f(1) = 0$ at the domain limits of $f(\nu)$.) The resultant strength cdf is shown in Figure 3.7b and is a clear concave form with linear variation at small strengths ($\mu \to 0$). The next simplest addition is linear, shown as the descending dot-dash line in Figure 3.7c at $\approx 0.3f_{\max}$, with a beta function superposed using $\omega = 0.07$ and $\kappa = 12$, and the total shown as the solid line. The resultant strength cdf is shown in Figure 3.7d and is approximately linear throughout the entire strength domain. Figures 3.6 and 3.7 demonstrate the flexibility of the beta function and its perturbations to describe a wide range of crack length and strength distribution forms.

The central strength regions of the cdf variations in Figures 3.6 and 3.7, the variations about the medians, $F(\mu) \approx 0.5$, are predominantly linear, $\mathrm{d}F(\mu)/\mathrm{d}\mu \approx$ constant, independent of the detailed shape of the crack length pdf $f(\nu)$. The technologically important small strength regions, near the small μ thresholds, $F(\mu) \approx 0$, are determined by the large crack regions of the crack length pdf, $f(\nu)$ for $\nu \to 1$. Two common threshold strength behaviors were shown here: sigmoidal behavior, in which $\mathrm{d}F(\mu)/\mathrm{d}\mu \to 0$, implying $\mathrm{d}^2F(\mu)/\mathrm{d}\mu^2 > 0$ as $\mu \to 0$, and concave behavior, in which $\mathrm{d}F(\mu)/\mathrm{d}\mu \to$ constant, implying $\mathrm{d}^2F(\mu)/\mathrm{d}\mu^2 \to 0$ as $\mu \to 0$. These behaviors are typified by Figures 3.6d and 3.7b. Consideration of the large crack regions of the conjugate crack length populations, Figures 3.6c and 3.7a, respectively, shows that sigmoidal strength behavior is associated with a population that contains relatively few large cracks and concave behavior is associated with a population that contains relatively many large cracks. A broad distinction can be thus made between crack length pdf variations based on the "weight" of the crack length population tail at large crack lengths. A population similar to that in Figure 3.6c will be referred to as representing a light-tailed distribution and a population similar to that in Figure 3.7a will be referred to as representing a heavy tailed distribution. As these terms refer to the relative numbers of large cracks and hence small

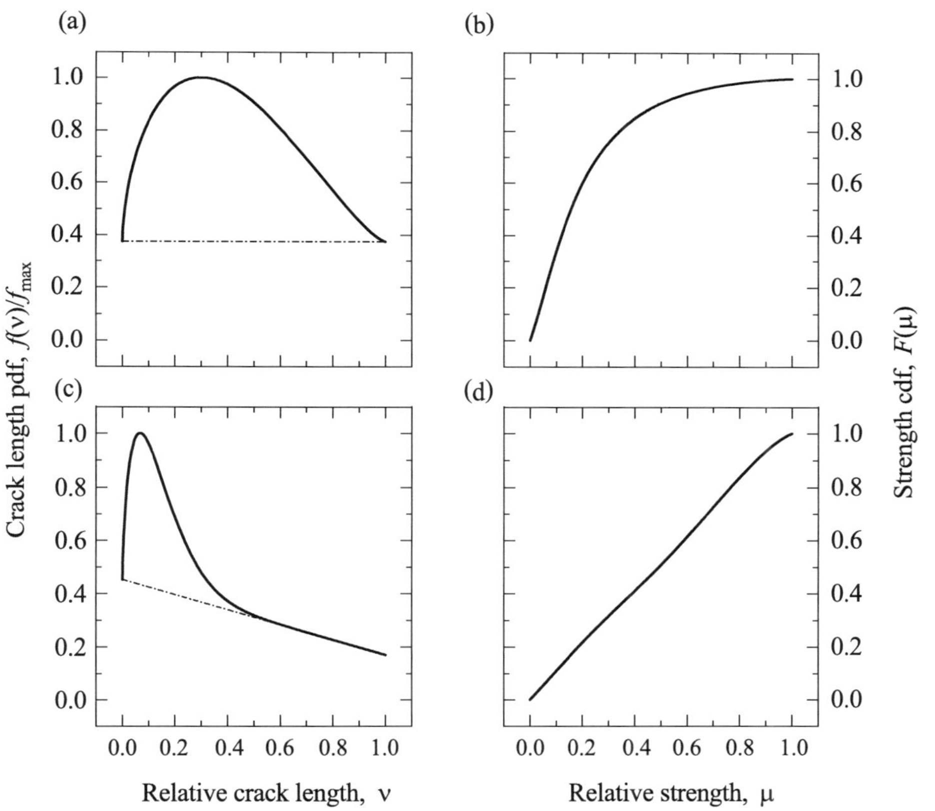

Figure 3.7 Plots of probability density functions $f(\nu)$ as a function of relative crack length ν and conjugate cumulative distribution functions $F(\mu)$ as a function of relative strength μ for populations of crack lengths described by perturbed (shifted) Beta functions. Conjugate pairs are (ab) and (cd). The perturbed populations give rise to commonly observed strength distribution shapes, (b) ideal concave and (d) linear. Both (a) and (c) are described as "heavy tailed." Combination (ab) is most prevalent for particles.

strengths in the material population, the categorisation is important in selection of materials and estimating reliability of structural components. Section 3.4 discusses these descriptive terms in mathematical detail.

This section discussed one of the two defining features of a brittle fracture flaw population: the probability density of crack lengths, $f(c)$ and, via the Griffith equation, the conjugate cumulative distribution of strengths, $F(\sigma)$, for the fundamental volume elements of a material. $f(c)$ and $F(\sigma)$ are intensive material properties, as is the second defining feature of a flaw population, the spatial density of volume elements and flaws, λ, the number of flaws/volume. The following section discusses the application of these intensive population characteristics in the prediction of an extensive characteristic: the strength distribution of components of material sampled from the population. As in the considerations of mass and elastic energy in Chapter 2, the volume of the sample will be shown to be a key factor in connecting an intensive quantity to an extensive quantity.

3.3 Samples of Flaws and Strengths

3.3.1 Sample Definitions

Prediction of component strength distributions begins with consideration of k fundamental volume elements sampled from the population of elements $\Omega\lambda$. The mechanical properties of the population are described by the element number density λ and the crack length pdf $f(c)$, or equivalently by the related cdf $F(c)$, strength pdf $f(\sigma)$ or cdf $F(\sigma)$, or crack length or strength ccdf $\bar{F}(c)$ or $\bar{F}(\sigma)$. The elements are randomly sampled and $k \ll \Omega\lambda$ such that the elements are independent and identically distributed. If $k = 1$, the probability that this single sampled element contains a crack of length less than c is $F(c)$. As the elements are independent (Lipschutz 1965) for $k = 2$ the probability that both elements in the sample contain cracks of lengths less than c is $F(c)^2$. For k elements selected, the probability $H_U(c)$ that all elements contain cracks of lengths less than c is $H_U(c) = F(c)^k$. $H_U(c)$ is the probability that an element selected at random from the sample will have length less than c. Thus, $H_U(c)$ is also the probability that c is the upper bound or greatest length of all cracks in the sample. Conversely, for $k = 1$, $\bar{F}(c)$ is the probability that a single selected element contains a crack of length greater than c. The probability, $\bar{H}_L(c)$, that all elements in the sample contain cracks of lengths greater than c is $\bar{H}_L(c) = \bar{F}(c)^k$, where sample crack length ccdf is defined identically to that for the population, $\bar{H}(c) = 1 - H(c)$. $\bar{H}_L(c)$ is the probability that an element selected from the sample will have length greater than c and is thus the probability that c is the lower bound or smallest length of all cracks in the sample. Using the definitions of $\bar{F}(c)$ and $\bar{H}(c)$, it is straightforward to show that $H_L(c) = 1 - [1 - F(c)]^k$.

The sample crack length cdf variations, $H_U(c)$ and $H_L(c)$ have the same region of support as $F(c)$ and in relative crack length terms $H_U(\nu) = F(\nu)^k$ and $H_L(\nu) = 1 - [1 - F(\nu)]^k$. Figure 3.8 shows the variations of $H_U(\nu)$, $H_L(\nu)$, and $F(\nu)$ for $k = 1, 2$, and 5, for the simple uniform distribution case of $f(\nu) = 1$ and $F(\nu) = \nu$. For $\nu = 0$ and $\nu = 1$ all distributions converge as the domains and ranges are identical. For $k = 1$, the sample probability bounds are identical to the population variation. For greater k values the bounds split and become increasingly separated. There are two ways to interpret Figure 3.8, using $\nu = 0.5$ as an example, for which $F(\nu) = 0.5$ here. At fixed crack length, increasing k leads to increasing probability that the smallest crack in the sample is smaller than $\nu = 0.5$ and decreasing probability that the largest crack in the sample is smaller than $\nu = 0.5$. At fixed probability, increasing k leads to a crack length smaller than $\nu = 0.5$ that is probably the smallest in the sample and a crack length larger than $\nu = 0.5$ that is probably the largest in the sample.

The sample strength upper and lower probability bounds, $H_U(\sigma)$ and $H_L(\sigma)$, can be related to the strength cdf of the population $F(\sigma)$ in the same way as described above for the crack length bounds. $H_U(\sigma)$ and $H_L(\sigma)$ thus have the same region of support as $F(\sigma)$. However, greater physical insight is gained by beginning strength analysis with the crack length bounds. The sample strength ccdf is defined by $\bar{H}(\sigma) = 1 - H(\sigma)$ and thus in relative terms, the crack and strength cdf and ccdf are related by $\bar{H}(\mu) = H(\nu)$ and $\bar{H}(\nu) = H(\mu)$, in direct analogy to 3.10 and 3.11. The sample upper bound crack length cdf $H_U(\nu)$ is converted to a sample lower bound strength cdf, $H_L(\mu)$, by

$$
\begin{aligned}
H_U(\nu) &= F(\nu)^k \\
\bar{H}_L(\mu) &= \bar{F}(\mu)^k \\
H_L(\mu) &= 1 - [1 - F(\mu)]^k.
\end{aligned}
\tag{3.29}
$$

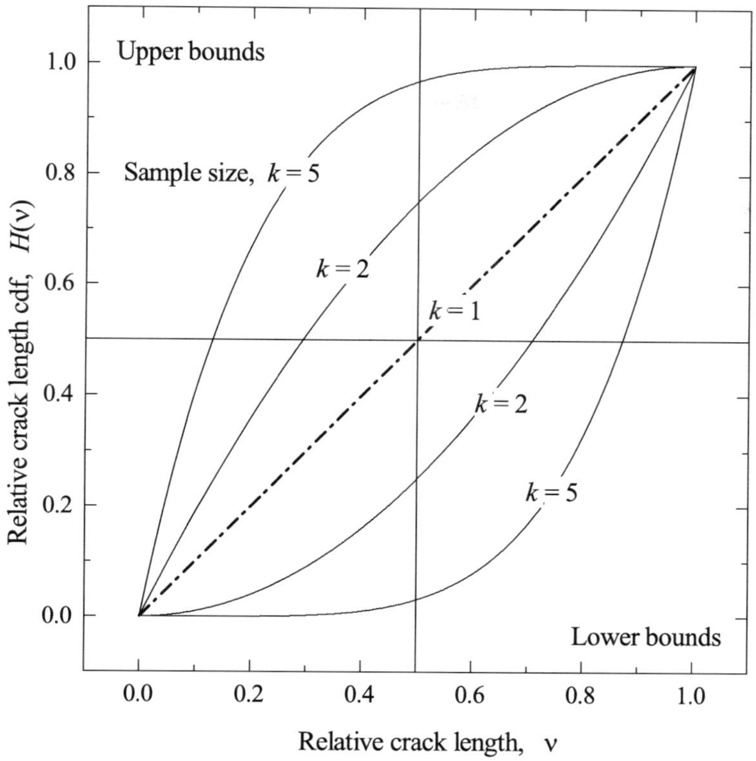

Figure 3.8 Plot of the cumulative distribution functions $H(\nu)$ as a function of relative crack length ν for a uniform population of cracks. $H(\nu)$ gives the probability for the upper and lower bounds on crack lengths in samples of 1, 2, and 5 elements drawn from the population. As sample size increases, the probability of a large crack increases.

Similarly, the sample lower bound crack length cdf $H_L(\nu)$ is converted to a sample upper bound strength cdf, $H_U(\mu)$, by

$$\begin{aligned} H_L(\nu) &= 1 - [1 - F(\nu)]^k \\ \bar{H}_U(\mu) &= 1 - F(\mu)^k \\ H_U(\mu) &= F(\mu)^k. \end{aligned} \tag{3.30}$$

The inverse relationships between σ and c imbedded in the Griffith relation, Eq. (3.1), and μ and ν in the relative form, Eq. (3.21), lead to reciprocal relationships between the upper and lower probability bounds for strength and crack length.

Figure 3.9 shows the variations of $H_U(\mu)$, $H_L(\mu)$, and $F(\mu)$ for $k = 1$, 2, and 5, for the simple uniform crack length distribution case of $f(\nu) = 1$ and $F(\nu) = \nu$ considered above. The $F(\mu)$ variation is identical to that shown in Figure 3.4b. Apart from the concave nature of this base response, the interpretation of Figure 3.9 is identical to that of Figure 3.8. For $\mu = 0$ and $\mu = 1$ all distributions are identical. For $k = 1$, the sample probability bounds are identical to the population variation. For greater k values the bounds split and become increasingly separated. There are two ways to interpret Figure 3.9, considering an initial $\mu = 0.5$ point. At fixed strength, increasing k leads to increasing probability that the smallest strength in the sample is less than $\mu = 0.5$ and decreasing probability that the largest strength in the sample is less than $\mu = 0.5$. At fixed probability, increasing k leads to a strength less than $\mu = 0.5$ that is probably the weakest in the sample and a strength greater than $\mu = 0.5$ that is probably the strongest in the sample.

H_U and H_L are *extreme value distributions*—they characterize probability distributions of the extreme values, the maxima or minima, of samples from the population. The above discussion considered crack length and strength probability for an element selected from a sample of k elements. The interpretation of H_U and H_L as extreme value distributions enables consideration of the probability of the extreme crack length or strength for a sample selected from N such samples. If N samples, each consisting of multiple elements, are taken from a population, $H_U(c)$ gives the probability that a

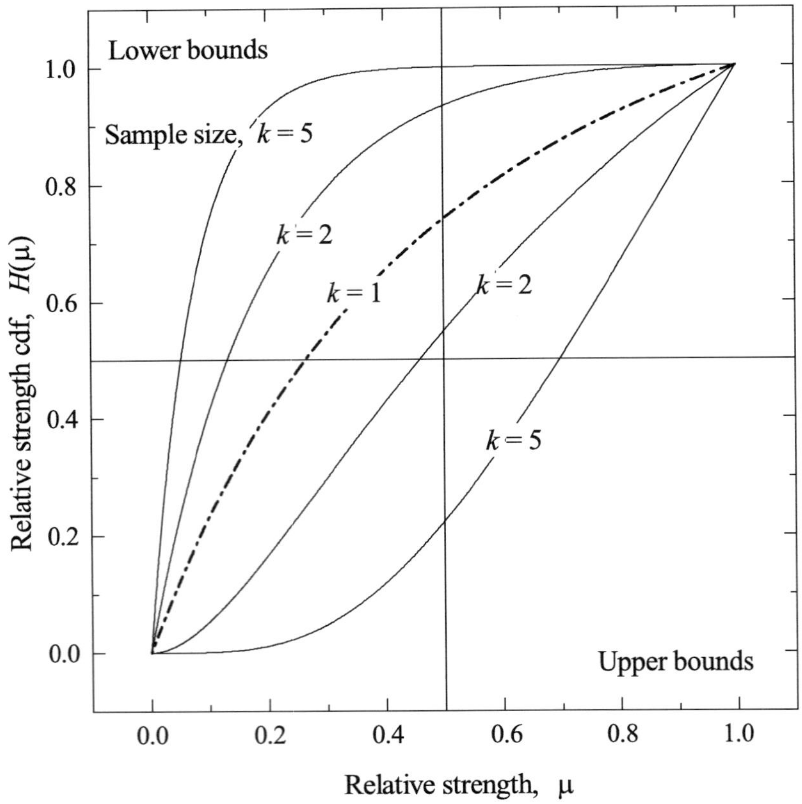

Figure 3.9 Plot of the cumulative distribution functions $H(\mu)$ as a function of relative strength μ for a uniform population of cracks. $H(\mu)$ gives the probability for the upper and lower bounds on strengths in samples of 1, 2, and 5 elements drawn from the population. As sample size increases, the probability of a small strength increases.

sample selected at random from N will contain a largest crack of length c. Similarly, if N samples are taken from a population, $H_L(\sigma)$ gives the probability that a sample selected at random from N will contain an element of least strength σ. In relative terms, $H_U(\nu)$ gives the probability that a randomly selected sample contains a largest crack ν or $H_L(\mu)$ gives the probability that a randomly selected sample contains a least strength μ. If the number of elements in the samples taken from a population is large, $k \gg 1$, the asymptotic behavior of the extreme value distribution at the maximum (as e.g. $F(\nu) \rightarrow 1$) is $H_U(\nu) = F(\nu)^k \rightarrow \exp[-k(1-F(\nu))]$ and the asymptotic behavior at the minimum (as e.g. $F(\mu) \rightarrow 0$) is $H_L(\mu) = 1-[1-F(\mu)]^k \rightarrow 1-\exp[-kF(\mu)]$. (Epstein 1948a). An example application of maximum extreme value analysis using the asymptotic approximation is shown in a study of pore sizes sampled from cross-sections of silicon nitride, for which the sample size was $k \approx 130$ from a population of $\Omega\lambda \approx 4200$ and the pore number density was $\lambda \approx 8.1 \times 10^4$ mm^{-3} (Chao and Shetty 1992). The focus here is on application of minimum extreme value analysis in sampled strengths of brittle components.

The interpretation of strength extreme value distributions (Figure 3.9), is refined if the fundamental elements in a sample are joined to form a component. The simplest case is a component, volume $V = k/\lambda$, in which the elements act independently (one crack does not influence another), loaded such that stress is applied homogeneously to all k elements (and k cracks), and component failure occurs if one element fails (e.g. a tensile bar). In this case, if the applied stress exceeds the strength of the weakest element, the component fails and further increases in stress beyond this smallest strength are not possible. The strengths of all elements except the weakest are thus not sensed in a component strength test. Hence, only the strength lower bound probability $H_L(\mu)$ in Figure 3.9 is of physical relevance, noting that this is related to the crack length upper bound, Eq. (3.29). A non-destructive evaluation technique could determine the ensemble of crack sizes within a component, such that both upper and lower crack length bounds in Figure 3.8 are of relevance. However, only the largest or most extreme crack in a component is sensed in a component strength measurement.

The guiding equation for strength distributions of components, each consisting of k independent elements, is thus

$$H(\sigma) = 1 - [1 - F(\sigma)]^k, \tag{3.31}$$

expressed in absolute terms. It is implicit in writing Eq. (3.31) that the components are homogeneously stressed. As the elements are randomly selected and independent so are the components, and thus an entire ensemble of (N) components will be referred to hereafter as a *sample*. An alternative mathematical statement of the independence and identical distribution of the elements and components is that the population from which the sample is drawn is invariant with respect to the constituent component size k, i.e. $\mathrm{d}F/\mathrm{d}k = 0$. Eq. (3.2) and Eqs. (3.10)–(3.13) enable Eq. (3.31) to be written as

$$H(\sigma) = 1 - \left[\int_0^c f(u)\mathrm{d}u\right]^k, \tag{3.32}$$

where u is a dummy crack length variable, enabling determination of the sample strength cdf $H(\sigma)$ from the population crack length pdf $f(c)$ (and is a generalization of Eq. (3.14)), where σ and c are conjugate. Using the relations $H(c) = \bar{H}(\sigma)$ and $h(c) = \mathrm{d}H(c)/\mathrm{d}c$ gives

$$h(c) = k\left[\int_0^c f(u)\mathrm{d}u\right]^{k-1} f(c) \tag{3.33}$$

where $h(c)$ is the sample crack length pdf. It is implicit in writing Eqs. (3.31)–(3.33) that H and h represent the extreme value probability distribution and density of a sample and hence for simplicity, in reference to components, the "L" and "U" subscripts are here and hereafter omitted. The term "size" is used in several ways throughout and the meaning made clear in context: *Component size* refers to k, *sample size* refers to N. Notably, Eqs. (3.31)–(3.33) are independent of N. The functions $h(x)$, $H(x)$, and $\bar{H}(x)$ will be used throughout to denote extreme value characteristics of a sample, where the argument x will define the function in context. Usually, x will be c or σ or a dummy variable for these used in analyses. The dimension of $h(x)$ is x^{-1}. H and $\bar{H}$ are dimensionless. It is important to recognize that $h(x)$, $H(x)$, and $\bar{H}(x)$ are asymptotic continuum responses for very large sample sizes, $N \gg 1$.

Eqs. (3.31)–(3.33) provide several insights. The first is that the sample strength distribution and conjugate crack length density, $H(\sigma)$ and $h(c)$, are extensive—they depend on the extent of a sample as quantified by the number of fundamental volume elements comprising a component, $k = \lambda V$. There are N such components in the sample. For $k = 1$, the sample distribution and density revert to the intensive population distribution and density, and extreme value effects vanish. Thus, a major motivation for consideration of $H(\sigma)$ and $h(c)$ for small components (e.g. MEMS elements) and small particles is the possibility that such small components may contain one or a few strength-controlling cracks and enable direct insight into material flaw populations as $k \to 1$ and $H(\sigma) \to F(\sigma)$ and $h(c) \to f(c)$.

In this regard, it is important to note that although the number of cracks k is general, the spatial distribution of cracks or flaws may be restricted. If cracks are limited to a layer of thickness d in a component, the number of relevant fundamental volume elements and thus cracks may be written as $k = (\lambda d)(V/d)$. The product λd is thus an intensive areal density of cracks (number/area) and V/d is the extent of the component in terms of area. Examples include surface layers, in which cracks are associated with surface roughness or contacts. If cracks are limited to a layer of cross-sectional area d^2 in a component, the number of relevant fundamental volume elements and thus cracks may be written as $k = (\lambda d^2)(V/d^2)$. The product λd^2 is thus an intensive linear density of cracks (number/length) and V/d^2 is the extent of the component in terms of length. Examples include cracks in fibers or along the sidewalls of MEMS elements. Finally, if cracks are limited to a volume of d^3 in a component, the number of relevant fundamental volume elements and thus cracks may be written as $k = (\lambda d^3)(V/d^3)$. The product λd^3 is thus a number of cracks and V/d^3 is the extent of the component in terms of the number of crack containing regions. Examples include cracks at stress concentrations in shaped, notched, or filleted components (DelRio et al. 2020; Cook et al. 2021) or at controlled indentation flaws (e.g. Figure 3.2). A limit of this behavior is $V = d^3 = 1/\lambda$ such that $k = 1$. An example in this case might be small particles for which particle diameter $D \approx d$.

A second insight provided by Eqs. (3.31)–(3.33) is a clear prediction of the relation between strength distributions of samples containing flaws from the same population but composed of different sized components. For two separate samples containing components of two different sizes k_1 and k_2, the sample strength distributions are

$$H_1(\sigma) = 1 - [1 - F(\sigma)]^{k_1} \tag{3.34}$$

and

$$H_2(\sigma) = 1 - [1 - F(\sigma)]^{k_2}. \tag{3.35}$$

Eliminating $F(\sigma)$ between these two equations gives

$$H_2(\sigma) = 1 - [1 - H_1(\sigma)]^{k_2/k_1}, \tag{3.36}$$

noting that the exponent k_2/k_1 is the ratio of the numbers of flaws in the different sized components. Eq. (3.36) shows that the strength distributions of such components, containing independent flaws drawn from the same population, are related by this ratio and that for $k_2 > k_1$, $H_2 > H_1$. A similar result was noted earlier by Shih (1980) in an extensive analytical work considering failure probability and by Wong et al. (1987) for an assumed volumetric distribution of flaws. Eq. (3.36) is a generalization of Eq. (3.31) and in both cases it is important to note that the domain of σ is unaffected by component size and k—in particular the threshold is unaffected: $H(\sigma_{th}) = F(\sigma_{th}) = 0$ and $H_1(\sigma_{th}) = H_2(\sigma_{th}) = 0$.

Although explicit absolute information (distribution, density) regarding the flaw population is lost, systems that obey Eq. (3.36) imply the existence of a single underlying population of independent strength-controlling flaws. Component size thence influences a sample strength distribution only through the probability that a component contains an element exhibiting an extreme strength, selected from the invariant population, Eq. (3.31). Systems with this strength behavior will thus be referred to as exhibiting a *stochastic extreme value size effect*. Conversely, systems that do not obey Eq. (3.36) imply that a single underlying population of independent strength-controlling flaws does not exist. Physically, the implementation of different fabrication and flaw generation methods for components of different sizes, e.g. different heat treatment, surface finishing, or sieving techniques, or interaction between fundamental volume elements so as to remove flaw independence, e.g. by coalescence or shielding, lead to the same mathematical modification of Eq. (3.31): A size dependence is incorporated into the strength cdf such that $F = F(\sigma, k)$ and the simple multiplicative use of k indicative of independence is removed. In this case, the strength distribution of a sample is determined by component size through both the population of strengths and the probability that a component contains a particular extreme strength from the now variable population. Systems with this strength behavior will thus be referred to as exhibiting a *deterministic extreme value size effect*. A third insight of Eq. (3.31) is then the clear distinction between stochastic and deterministic size effects. A key feature of systems exhibiting stochastic effects in strength distributions is that σ_{th} is invariant, i.e. the strength of the weakest component in a sample, the threshold strength, is independent of component size. Conversely, component size-dependent threshold strengths are indicative of deterministic size effects.

The final step in forward analysis is to predict the strength edf $Pr(\sigma)$ for a sample size N of components size k. This step is considered in the next section using the population pdf $f(c)$ as a basis and stochastic size effects as discussed in this section. The two pdf $f(c)$ variations considered as examples are the light tail (Figure 3.6c) and heavy tail (Figure 3.7)a crack densities, giving rise to the commonly observed (see Chapter 2) sigmoidal and concave strength cdf variations, $F(\sigma)$, Figures 3.6d and 3.7b respectively. Deterministic size effects and strength distributions are considered in detail in Chapter 9.

3.3.2 Sample Examples

Stochastic extreme value size effects in samples of brittle fracture components are summarized in the relative forms of Eqs. (3.32) and (3.33), both based on the population relative crack length pdf $f(\nu)$. For cracks, the sample relative crack length upper bound pdf, $h(\nu)$, is given by

$$h(\nu) = k\left[\int_0^{\nu} f(u)\mathrm{d}u\right]^{k-1} f(\nu), \tag{3.37}$$

where u is a dummy relative crack length variable. As a consequence, for strengths the sample relative strength lower bound cdf, $H(\mu)$, is given by the parametric set

$$H(\mu;\nu) = 1 - \left[\int_0^{\nu} f(u)\mathrm{d}u\right]^{k}, \tag{3.38}$$

where $\mu(\nu)$ is the relative form of the Griffith equation given by inversion of Eq. (3.21),

$$\mu(\nu) = \frac{\eta[\nu(\eta^2 - 1) + 1]^{-1/2} - 1}{\eta - 1}. \tag{3.39}$$

The size of the components is quantified by the number of cracks contained, k, which is proportional to the volume of the components and the spatial density of cracks in the population, $k = \lambda V$. The relative strength domain is quantified by the ratio η of the strongest/weakest volume elements in the population. The size of a sample is quantified by the number of components measured, N.

Simulation or prediction of a strength edf, $Pr(\sigma)$, begins with a first step of selection of intensive parameters, $f(\nu)$ and η, characterizing the relative shape of the population crack length pdf and the relative extent of the population strength domain, respectively. The second step is the selection of the extensive parameter k characterizing the size of the components. At this stage, Eqs. (3.38) and (3.39) can be evaluated, providing an expression for the sample relative strength cdf $H(\mu)$. As might be anticipated from Figure 3.9, $H(\mu)$ converges to the population continuum asymptote, $H(\mu) \rightarrow F(\mu)$ as components become small, $k \rightarrow 1$, and diverges from the population response, $H(\mu) > F(\mu)$ as components become large, $k > 1$. The fidelity with which $H(\mu)$ is represented by a sample increases with increasing N. For arbitrary sized components, $H(\mu)$ is the asymptotic distribution for large samples, $N \gg 1$. Discrete values of component relative strength, μ_i, can be selected from a discretized population of components by random sampling according to the probability distribution $H(\mu)$ (Walpole and Myers 1972). The samples can be ranked and relative strength edf plots, $Pr(\mu)$, can then be constructed as described in Chapter 2. For the examples here, the population of components was 1000, with an implied fundamental element population of $\Omega\lambda = 1000k$, and samples were taken from the populations without replacement. The prediction examples were evaluated numerically.

The simplest sample is of N components consisting of single elements ($k = 1$, $V = \Delta V$) drawn from the population and strength tested individually (Figure 3.10), giving rise to a sequence of strength values $\sigma_1, \sigma_2, \sigma_3, \ldots \sigma_N$. As N is finite, the strengths are analyzed as discrete values by an edf, as opposed to the population-based continuum cdf calculations above, e.g. Figure 3.9. Such a sample enables assessments of the effects of N on empirical descriptions of different populations. Figure 3.11 shows a sequence of nine different $Pr(\mu)$ variations, describing samples from the above sigmoidal strength distribution using $k = 1$ and $N = 30$. The solid lines show the asymptotic response, in this case of the population $F(\mu)$. Symbols represent the strengths of individual components, in this case of individual fundamental elements. The $Pr(\mu)$ variations are shown Figure 3.11a–i in approximate order of increasing median strength and apparent shape, concave–linear–convex. There are two predominant, related, features of the samples in Figure 3.11. First, the sample strengths are restricted to a limited domain of μ in which the variation of the cdf $H(\mu)$ is large, noting that this corresponds to the limited domain in which the related pdf $h(\mu) = \mathrm{d}H(\mu)/\mathrm{d}\mu$ is large, in this case $0.2 \leq \mu \leq 0.8$. Second, although the median strengths of the $Pr(\mu)$ variations well approximate the median strength of the population, the restricted sample domain, truncated at both small and large strengths, obscures the true shape of population cdf as implied by $Pr(\mu)$. Although there are exceptions, Figure 3.11d exhibits a very small strength and Figures 3.11f and 3.11h exhibit weak sigmoidal tendencies; an unbiassed assessment of the average forms of the $Pr(\mu)$ variations in Figure 3.11 would be linear. Figure 3.12 shows a similar sequence of nine different $Pr(\mu)$ variations describing samples from the concave strength distribution, also using $k = 1$ and $N = 30$. The solid lines show the asymptotic response of $F(\mu)$ and symbols represent the strengths of individual fundamental element components. The $Pr(\mu)$ variations are shown Figure 3.12a–i in approximate order of increasing median strength. Relative

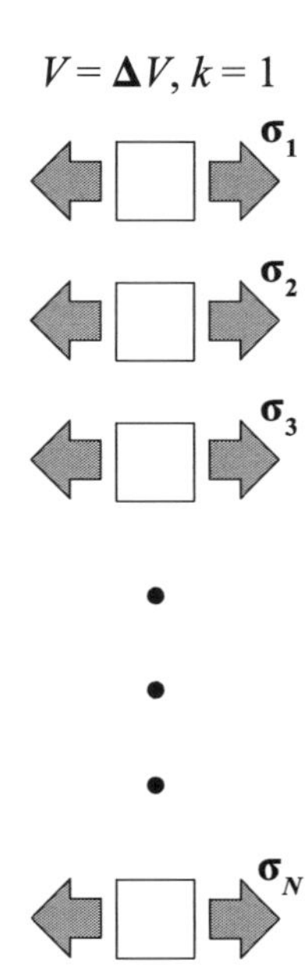

Figure 3.10 Schematic diagram of a strength test sample consisting of N components, each consisting of a ($k = 1$) single element from the population. The test generates a sequence of strength values $\sigma_1, \sigma_2, \sigma_3, \ldots \sigma_N$.

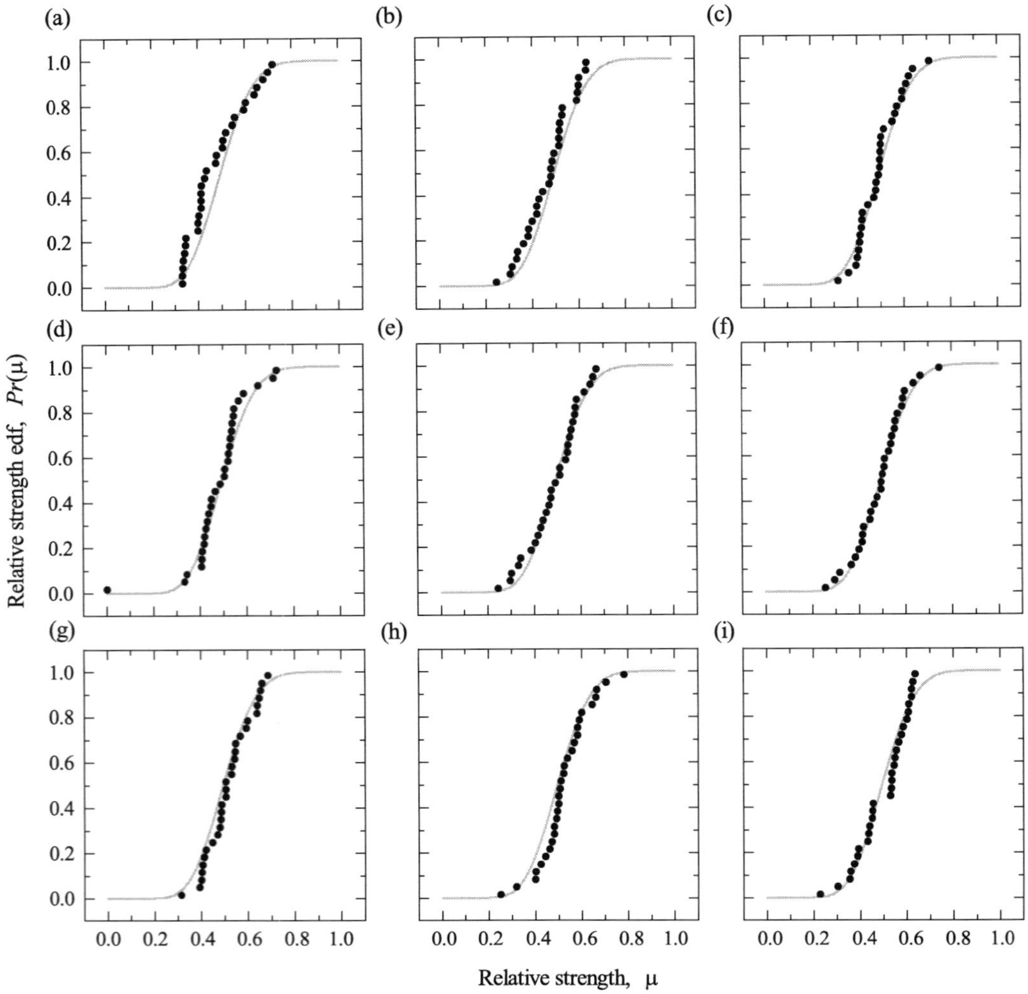

Figure 3.11 Plots of empirical distribution functions $Pr(\mu)$ as a function of relative strength μ for repeated samples of $N = 30$ components drawn from a sigmoidal population of strengths. The components were single elements as in Figure 3.10.

to Figure 3.11 there are two predominant features of the samples in Figure 3.12. First, the sample strengths are distributed over the full domain of μ, although similar to Figure 3.11 there are fewer at large μ in which the variation of the cdf $H(\mu)$ is small. Second, the median strengths and shapes of the $Pr(\mu)$ variations well approximate those of the population. An unbiassed assessment of the average forms of the $Pr(\mu)$ variations in Figure 3.12 would be concave.

Just as the underlying component or population cdf $H(\mu)$ or $F(\mu)$ affects the form of the sample edf $Pr(\mu)$, so does the sample size, N. Figure 3.13 shows two sequences of three $Pr(\mu)$ variations for the sigmoidal and concave strength distributions using $k = 1$. The sample sizes increase in order $N = 10$, 30, 100 for Figures 3.13a, c, e and 3.13b, d, f, respectively. The variations for $N = 30$ are identical to those in Figures 3.11 and 3.12. For $N = 10$, no real information regarding either distribution can be gained except that the medians are well approximated. For $N = 100$ (corresponding to 10 % of the population), the sigmoidal response is still restricted to a limited domain and the shape barely discernible. The concave response now completely covers the domain. The conclusion to be drawn from Figure 3.13 is that increasing N, the size of the sample, increases the precision with which the underlying distribution can be described over regions of the domain in which the distribution varies significantly, but does not increase the accuracy with which the distribution can be specified over the entire domain. In practical terms, thresholds and upper bounds of strength for sigmoidal strength distributions are not well assessed, even by large sample sizes.

The component cdf, $H(\mu)$, affects the form of the sample edf $Pr(\mu)$, through systematic variation of $F(\mu)$ by the component size, k. A simple extension of Figure 3.10 is a sample of N components, each consisting of multiple elements ($V = \Delta V$), and strength tested to give rise to a sequence of component strength values $\sigma_1, \sigma_2, \sigma_3, \ldots \sigma_N$, Figure 3.14.

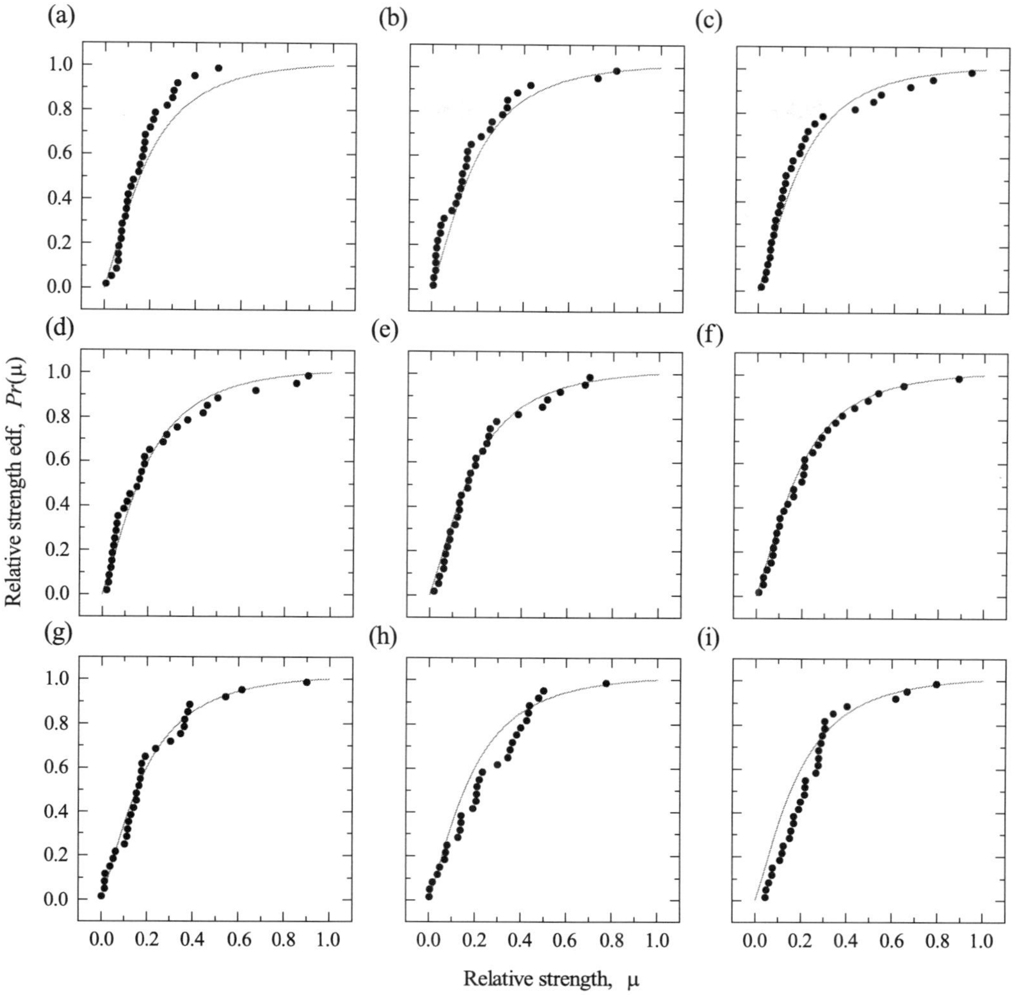

Figure 3.12 Plots of empirical distribution functions $Pr(\mu)$ as a function of relative strength μ for repeated samples of $N = 30$ components drawn from a concave population of strengths. The components were single elements as in Figure 3.10.

The strengths are analyzed as discrete values by an edf and such a sample enables assessments of the effects of k on empirical descriptions of different-sized components. Figure 3.15 shows two sequences of three $Pr(\mu)$ variations for the sigmoidal and concave strength distributions using fixed sample size $N = 30$. The component sizes increase in order $k = 3, 10, 30$ for Figures 3.15a, c, e and 3.15b, d, f, respectively. The solid lines show the invariant population $F(\mu)$ behavior as seen earlier. The dashed lines show the k-dependent asymptotic $H(\mu)$ responses. Symbols represent the strengths of individual components, in this case of increasing size, top to bottom. For components drawn from both populations, the component asymptotic $H(\mu)$ responses are approximately the same shape as that of the population $F(\mu)$. As k increases, the $H(\mu)$ responses occupy increasingly limited domains in which $H(\mu)$ varies. As the range of the functions is invariant $[0, 1]$, this effect leads to an increase in the slope of the responses about an invariant threshold. In the case of the concave strength population, the threshold is in common with that of the population, $\mu = 0$. In the case of the sigmoidal strength population, the effective threshold is shifted from that of the population to about $\mu \approx 0.2$, as noted above for samples of individual elements from the population (Figure 3.11). For both populations, the discrete samples well approximate the median values and shapes of the asymptotic $H(\mu)$ component responses, although sample information is, as above, restricted to the domain in which $H(\mu)$ varies significantly. As the minimum strengths are relatively invariant, the increasing domain limitation as k increases arises from increasing truncation of the upper bound maximum strength values.

The physical phenomenon represented by Figure 3.15 is that larger components contain a greater number flaws k and that the distribution of flaws in a component approaches that of the population as $k \gg 1$. Larger components thus exhibit a

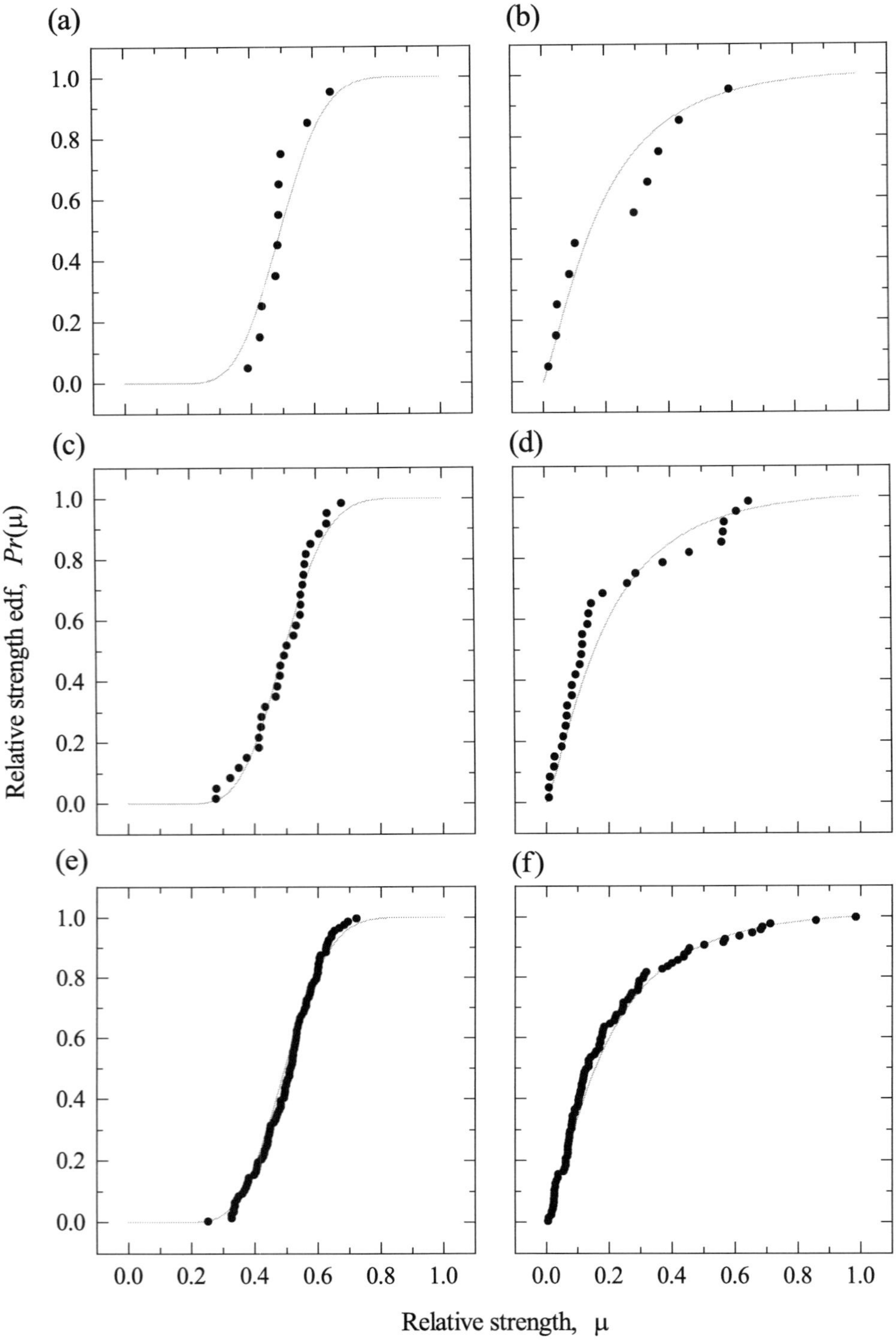

Figure 3.13 Plots of empirical distribution functions $Pr(\mu)$ as a function of relative strength μ for samples of (ab) $N = 10$, (cd) $N = 30$, and (ef) $N = 100$ components drawn from (ace) sigmoidal and (bdf) concave population of strengths. The components were single elements as in Figure 3.10.

greater probability that the largest, most extreme, flaw in the ensemble of flaws in a component approaches the largest flaw in the population. As a consequence, larger components exhibit a greater probability that component strength approaches the population lower bound threshold strength. This is the stochastic size effect on component strength. Figure 3.15 shows this effect very clearly for the concave strength population. The effect is less clearly shown for the sigmoidal strength population as strengths of larger components converge on an apparent threshold associated with a restricted domain

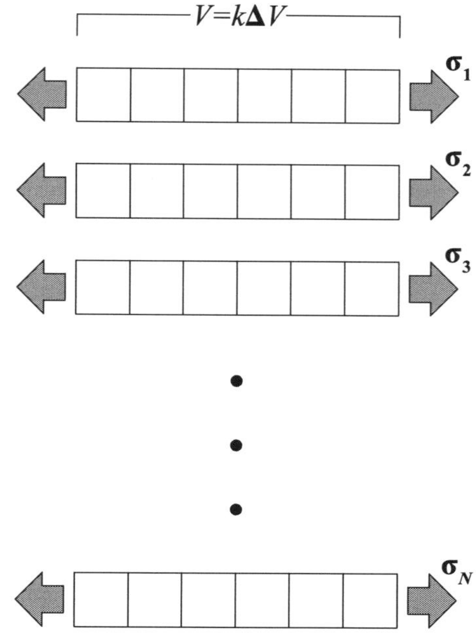

Figure 3.14 Schematic diagram of a strength test sample consisting of N components, each consisting of k multiple elements drawn from the population ($k = 6$ is shown as an example). The test generates a sequence of strength values $\sigma_1, \sigma_2, \sigma_3, \ldots \sigma_N$.

as discussed above. Two obvious overall limits are that a single component the size of the population must contain the largest flaw and therefore must exhibit the threshold strength, and that a sample of the entire population of single elements must contain the largest and smallest flaws and therefore must exhibit the complete domain of strengths. In the first case, component size is large, $k = \Omega\lambda$ and $N = 1$, and in the second case, sample size is large, $k = 1$ and $N = \Omega\lambda$.

Stochastic size effects are shown in greater detail in a plot of sample strength as a function of component size, Figure 3.16. The symbols show individual strength evaluations, including data from Figure 3.15; in order to better gauge variation, strength values are shown relative to the threshold strength, $\sigma/\sigma_{\text{th}}$. For components of individual elements ($k = 1$), the population $N = 1000$ was used. For larger components ($k > 1$), the sample $N = 30$ was used, similar to Figures 3.11–3.15. The relative strength was obtained from Eqs. (3.15) and (3.17):

$$\sigma/\sigma_{\text{th}} = 1 + \mu(\eta - 1), \tag{3.40}$$

which has bounds of $(1, \eta)$. The results of size effect predictions from the sigmoidal strength population are shown in Figure 3.16a, using $\eta = 2$ as above, and the results from the concave strength population are shown in Figure 3.16b, using $\eta = 4$. The bounds are shown as shaded bands. The overall behavior for both strength populations is similar. For the smallest components, the individual elements ($k = 1$), the strengths fill the $\sigma/\sigma_{\text{th}}$ range $(1, \eta)$. For larger components, the strength range is significantly truncated, at both large and small strengths for the sigmoidal population and at large strengths for the concave population. As examples, for the sigmoidal population, Figure 3.16a, at $k = 3$, the strength range $\sigma/\sigma_{\text{th}}$ is (1.2, 1.6) about a median of $\sigma/\sigma_{\text{th}} \approx 1.4$, and for the concave population, Figure 3.16a, at $k = 3$, the strength range $\sigma/\sigma_{\text{th}}$ is (1.0, 1.7) about a median of $\sigma/\sigma_{\text{th}} \approx 1.3$. In both cases, for even larger components, both populations approach invariant relative strength values, approximately 1.2 for the sigmoidal population and 1.0 for the concave population.

It is clear from Figure 3.16 that the greatest change to component strength distributions, e.g. as characterized by medians and ranges, brought about by stochastic size effects occurs for small components. For both strength populations in Figure 3.16, increasing the size of the components from a single fundamental volume element to three elements brought about the greatest reduction in strength, approximately a factor of two, with very small reductions, less than a factor of 0.5, on further size increases. The greatest changes in strengths occurred for small components drawn from the concave strength population with a greater domain (η). The implication is that experimental observation of such stochastic size effects would usually be very difficult with large components: the changes in median strengths are comparable to the strength ranges and the changes in component size required would exceed most experimental capabilities. However, MEMS components and, in particular, particles provide viable experimental strength test vehicles for observing stochastic size effects. From an extensive perspective, particles are small, such that the number of included cracks, k may be small. Particles are numerous, such that the number of measurable components, N, may be large. From an intensive perspective, particles often exhibit broad strength distributions, such that η is large, and that are concave, such that $H(\mu)$ is not truncated at the bounds. These aspects are considered in detail in Chapters 5, 6, and 7.

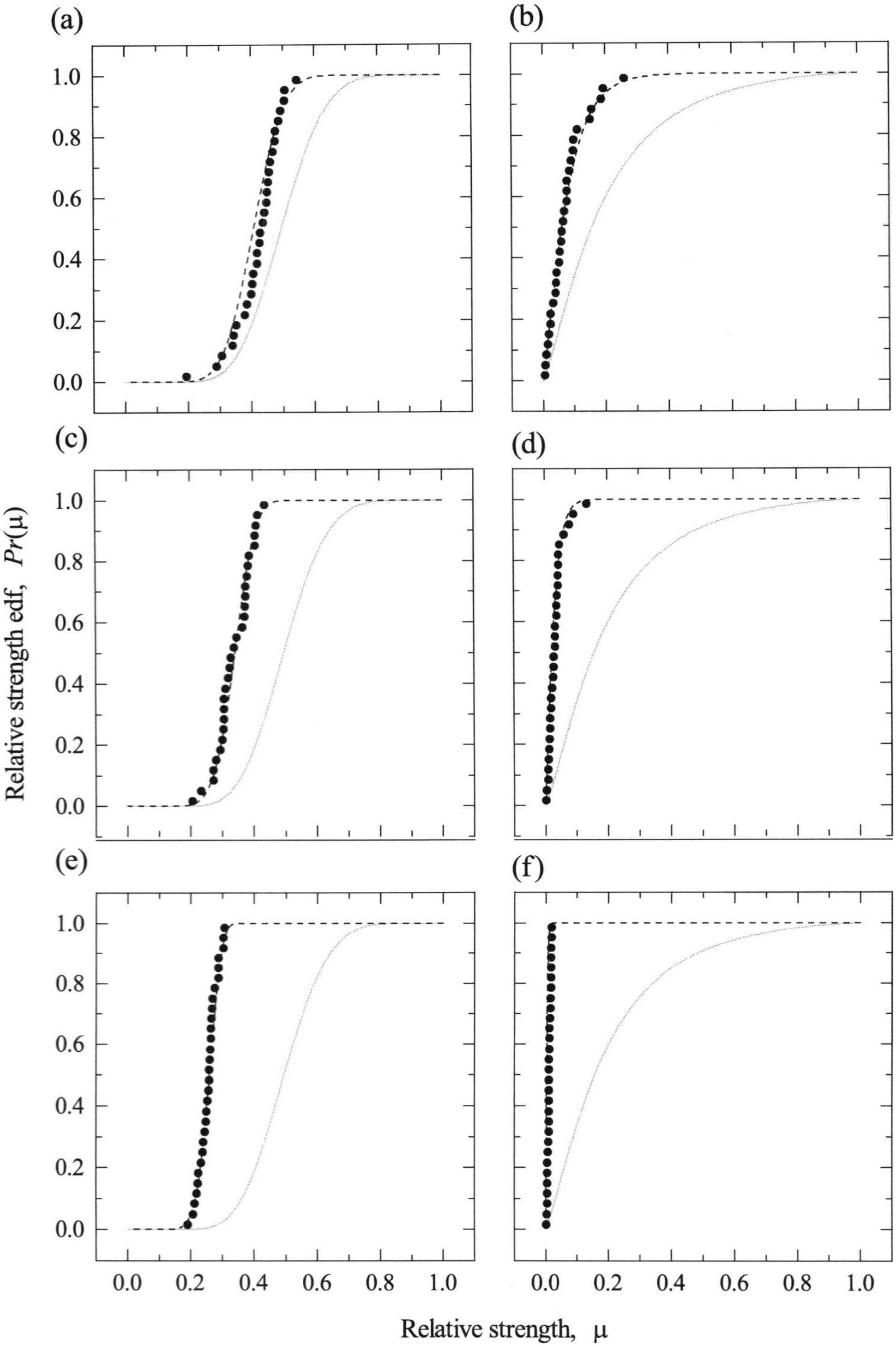

Figure 3.15 Plots of empirical distribution functions $Pr(\mu)$ as a function of relative strength μ for samples of $N = 30$ components drawn from (ace) sigmoidal and (bdf) concave populations of strengths. The components were multiple elements as in Figure 3.14 consisting of (ab) $k = 3$, (cd) $k = 10$, and (ef) $k = 30$ elements.

The predictive capabilities of the probabilistic strength framework developed in this chapter are shown in the example simulated strength edf $Pr(\sigma)$ responses of Figure 3.17. These responses mimic the experimental responses shown in Chapter 2 for glass bottles and 2 mm coral sand particles. The experimental pdf variations were simulated by forward analysis from an intensive crack population $f(\nu)$ modified by extensive component and sample size parameters, k and N. In both cases, the crack population was described by a perturbed beta distribution, as in Figure 3.7. For the glass bottles of Figure 3.17a,

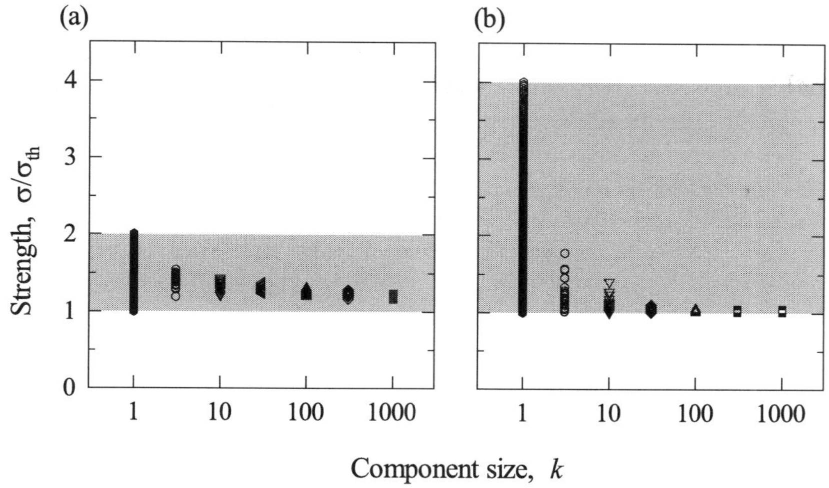

Figure 3.16 Plot of relative strength σ/σ_{th} as a function of component size k for components drawn from (a) sigmoidal and (b) concave populations of strengths. A significant size effect on strength is only observed for small components.

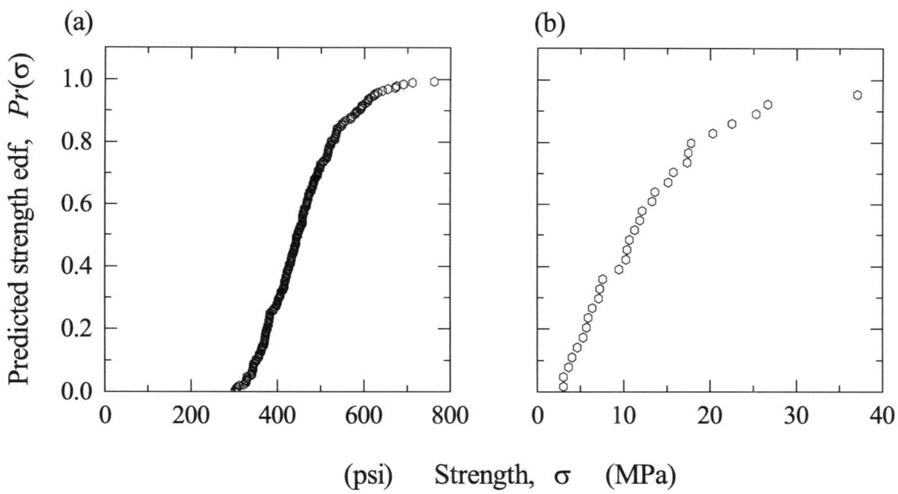

Figure 3.17 Simulated empirical distribution functions $Pr(\sigma)$ for (a) glass bottles and (b) coral particles matching the experimental observations presented in Chapter 2. In each case, the strength-controlling crack population is described by a perturbed beta function and the strength values determined by stochastic size effects.

the intensive parameters were $\omega = 0.20$, $\kappa = 6.0$, linear shifts of 0.040–0.035, and a relative strength domain of $\eta = 10$. The extensive component size was $k = 3$ and the results were matched to the data of sample size $N = 200$ and threshold strength $\sigma_{th} = 300$ psi. For the coral particles of Figure 3.17b, the intensive parameters were $\omega = 0.02$, $\kappa = 40.0$, linear shifts of 0.10–0.01, and a relative strength domain of $\eta = 40$. The extensive component size was $k = 1$ and the results were matched to the data of sample size $N = 32$ and threshold strength $\sigma_{th} = 2.6$ MPa. The intensive parameters and extensive component size are not unique; other combinations of values can also mimic the data. The similarity of the simulations to the experimental observations (Figures 2.28 and 2.29a)—qualitatively and quantitatively—is striking. The implication is that the physical basis of the probabilistic strength framework is sound.

3.4 Heavy-Tailed and Light-Tailed Populations

The behavior of the large-crack tail of a material flaw population and the implications for the form of the conjugate strength distribution were discussed briefly in qualitative terms in Section 3.2. Here, the terms are discussed quantitatively, using the strength simulations of Figure 3.17 as illustrative examples.

In systems described by "expected" probability behavior, large events, in the tail of the probability distribution well away from a central characteristic, occur only rarely and thus have very small probability. Such events (e.g. a 100-year flood, a stock-market crash, a total bridge collapse) are usually disastrous. Hence, there is great design interest in assuring that the tails of distributions that describe such events (e.g. flood heights) are consistent with conventional expectations. Such expectations are that the probability of a large event decreases exponentially with the size of the event. Large, disastrous events that occur more frequently than conventional expectations can be described by probability distributions with heavy tails. For an event x, the ccdf $\bar{F}(x)$ gives the probability that an event selected at random from the population of events, $f(x)$, will have a value greater than x. Formally, a distribution for which $\bar{F}(x)$ approaches 0 less rapidly than exponentially as x increases is termed a *heavy-tailed distribution* (Bryson 1974; Nair et al. 2013). Conversely, a distribution for which $\bar{F}(x)$ approaches 0 more rapidly than exponentially as x increases is termed a *light-tailed distribution*. These ideas and terms are usually applied to distributions on unbounded regions of support, such that, mathematically,

$$\lim_{x\to\infty} \frac{\bar{F}(x)}{\exp(-x)} = \infty. \quad \text{(Heavy-tailed distribution)} \tag{3.41}$$

The crack length population considered here, $f(c)$, exists on a bounded region of support, $c_{\min} \leq c \leq c_{\max}$. The tail behavior of interest is at the large crack bound, $c \to c_{\max}$ of the probability distribution. $\bar{F}(c)$ represents the probability that an element selected at random from the population of elements will contain a crack length larger than c. Hence, a population of elements described by a heavy-tailed crack length distribution will contain more large cracks and thence more elements of small strength than an exponential distribution. A mathematical definition of a heavy tail for a distribution on a bounded domain, is more limited than Eq. (3.41):

$$c_{\text{med}} \leq c < c_{\max}, \; \frac{\bar{F}(c)}{\exp(-c/l)} \geq 1, \quad \text{(Heavy-tailed bounded distribution)} \tag{3.42}$$

where c_{med} is the median crack length in the distribution and l is an exponential scaling factor, such that $F(c_{\text{med}}) = \bar{F}(c_{\text{med}}) = \exp(-c_{\text{med}}/l) = 0.5$. Similar to the strength considerations of Eq. (3.40), variation here is better assessed using crack lengths relative to the minimum crack length, obtained from Eqs. (3.16) and (3.17):

$$c/c_{\min} = 1 + \nu(\eta^2 - 1), \tag{3.43}$$

which has bounds of $[1, \eta^2]$.

Figure 3.18 shows plots of the crack length populations $f(c/c_{\min})$ underlying the simulated strength edf behaviors of Figure 3.17. In the forward analysis scheme considered here, these were the starting points of analysis, for which the plots of Figure 3.17 were the end points. Figure 3.18a, describing the glass bottles of Chapter 2, is a broad-lobed function, similar to Figure 3.6c. Figure 3.18b, describing the coral particles of Chapter 2, is narrow peaked function with a significant large crack tail, similar to Figure 3.5e. It is very difficult to assess from plots such as Figure 3.18 whether the distributions are

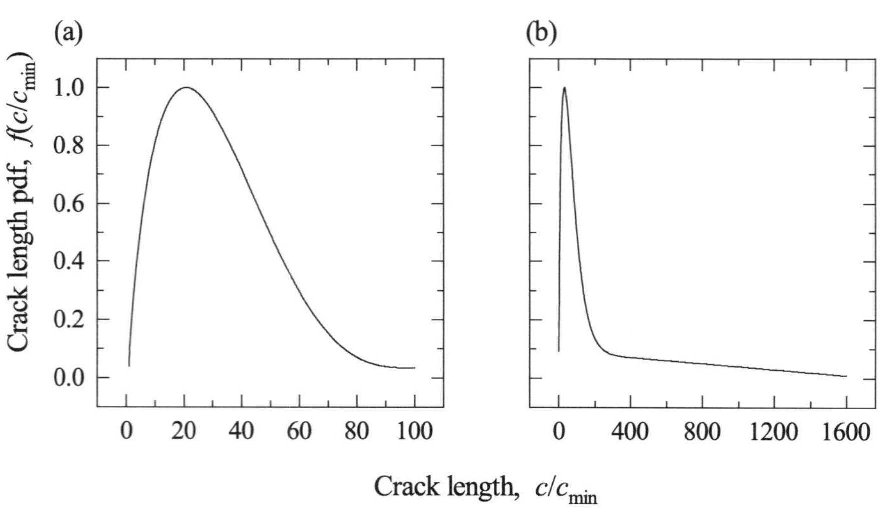

Figure 3.18 Plots of probability density functions $f(c/c_{\min})$ as a function of crack length $c/c_{\min}$ for the two simulated strength responses of Figure 3.17. (a) Glass bottles. (b) Coral particles. Note the clear tail in b indicating many large cracks far from the peak of the distribution.

light tailed or heavy tailed, although it is clear that Figure 3.18b is a candidate for heavy-tailed classification. In addition, heavy- and light-tailed behavior can also be quantified by evaluation of the pdf $f(c)$ kurtosis, which quantifies the relative number of elements far from the mean (Spiegel 1961; Krishnamoorthy 2016); again Figure 3.18b is a candidate for heavy-tailed classification. Figure 3.19 show plots of the ccdf variations $\bar{F}(c/c_{min})$ for the populations in Figure 3.18. Semi-logarithmic coordinates are used to accommodate the large spans of the crack length populations. The $\bar{F}$ variations are shown as the solid bold lines; the exponential functions matched to the $\bar{F}$ medians are shown as the fine dashed lines, and intersect the distributions at $c/c_{min} = 0.5$ (the coordinates distort the curves). The lower set of lines represents the crack length population of Figure 3.18a and illustrates a light-tailed distribution; the $\bar{F}$ variation is less than the exponential variation from the median to the maximum crack length. The upper set of lines represents the crack length population of Figure 3.18b and illustrates a heavy-tailed distribution; the $\bar{F}$ variation is greater than the exponential variation from the median to the maximum crack length.

The relationships between the forms of the crack length pdf $f(c)$ and the strength cdf $F(\sigma)$ can be assessed using domain based relative coordinates ν and μ, Figures 3.4–3.7. Assessments of heavy-tailed vs light-tailed distributions are best performed in coordinates related to the extensive crack length, Figures 3.18 and 3.19. Nevertheless, clear similarities are apparent in comparisons of the decreasing power-law crack population of Figure 3.5e and the beta distribution based population of Figure 3.18b, and the consequent, predominantly linear, strength distributions of Figure 3.5f and Figure 3.17b. Power-law based distributions, typified by the *Pareto distribution*, $\bar{F}(x) \sim x^{-\alpha}$, where $\alpha > 1$ is a characteristic exponent (Forbes et al. 2011; Krishnamoorthy 2016), are commonly studied examples of heavy-tailed distributions (Nair et al. 2013; Krishnamoorthy 2016). A feature of Pareto distributions is scale invariance,

$$\bar{F}(xy) = g(y)\bar{F}(x), \quad x \to \infty \tag{3.44}$$

where y is an arbitrary multiplicative ratio and $g(y)$ is a function of y only. A *regularly varying distribution* is given by $\bar{F}(x) = x^{-\rho}L(x)$, where ρ is another characteristic exponent and $L(x)$ is a slowly-varying function. If a heavy-tailed crack length distribution is regularly varying, $\bar{F}(c) = c^{-\rho}L(c)$, then

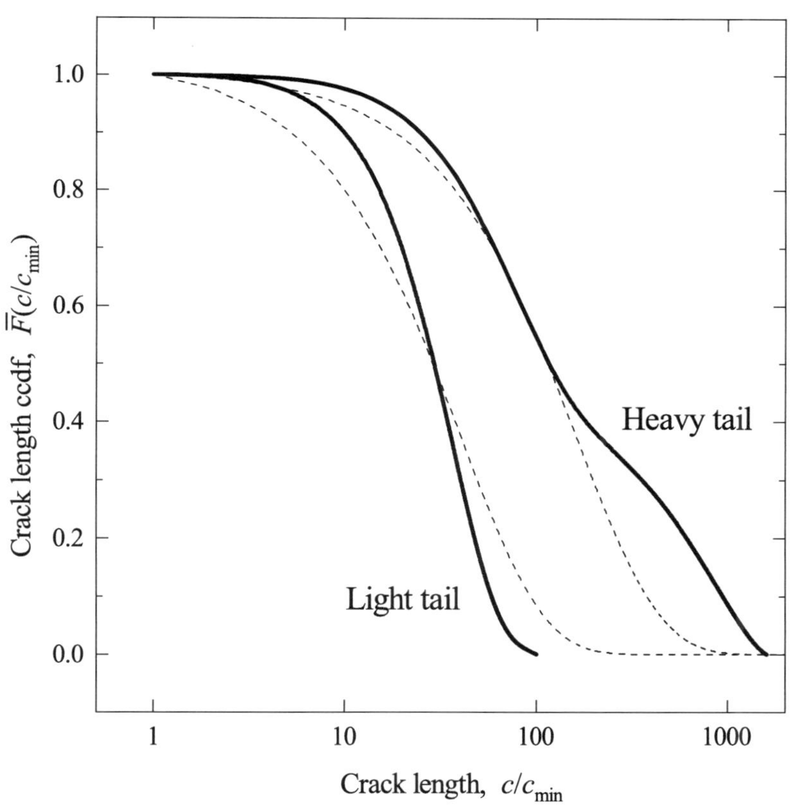

Figure 3.19 Plot in semi-logarithmic coordinates of complementary cumulative distribution functions $\bar{F}(c/c_{min})$ as a function of crack length c/c_{min} for the populations of Figure 3.18. Bold solid lines show $\bar{F}$. Fine dashed lines show exponential comparisons matched to median values. The upper solid line indicates a heavy-tailed distribution and the lower solid line indicates a light-tailed distribution.

$$\bar{F}(yc) = y^{-\rho}\frac{L(yc)}{L(c)}\bar{F}(c) \approx y^{-\rho}\bar{F}(c), \quad c \to c_{\max} \tag{3.45}$$

Hence, the large crack tail behavior of a regularly varying crack length distribution behaves similarly to a Pareto distribution. Much of the mathematics of Pareto distributions, which are simply expressed, can thus be applied to heavy-tailed crack length distributions, as Eq. (3.45) is consistent with Eq. (3.44). In particular, the property of scale invariance suggests crack length, and hence component strength, distributions may well be scale invariant. This feature is discussed in detail in Chapter 7.

A common feature of light-tailed distributions is that they are often also long-tailed. Whereas the tail weight of a probability distribution was classified by comparison to an expectation of exponential decrease, the tail length is classified by comparison to an expectation of domain width. In systems described by expected probability behavior, very large events, well away from the central characteristic and peak in the probability density do not occur. The expectation is that a very large event is outside the domain of events. Very large events that occur infrequently, but at all, contrary to conventional expectations can be described by probability distributions with long tails. Formally, a distribution for which $\bar{F}(x)$ approaches a fixed value as x increases is termed a *long-tailed distribution* (Nair et al. 2013). Again, these ideas and terms are usually applied to distributions on unbounded regions of support, such that, mathematically,

$$\lim_{x\to\infty}\frac{\bar{F}(x+\Delta x)}{\bar{F}(x)} = 1, \text{ for all } \Delta x \quad \text{(Long-tailed distribution)} \tag{3.46}$$

where Δx is an arbitrary increment in the distribution quantile. As discussed above, the crack length population considered here, $f(c)$, exists on a bounded region of support, $c_{\min} \le c \le c_{\max}$, and the tail behavior of interest is at the large crack bound, $c \to c_{\max}$. Hence, a population of elements described by a long-tailed crack length distribution will contain very large cracks and thence elements of very small strength considerably weaker than the most frequently observed strengths. A mathematical definition of a long-tailed crack length distribution on a bounded domain is more limited than Eq. (3.46):

$$c_{\text{med}} \le c < c_{\max}, \frac{\bar{F}(c+\Delta c)}{\bar{F}(c)} \to 1, \ \Delta c < c_{\max} - c \quad \text{(Long-tailed bounded distribution)} \tag{3.47}$$

Neither of the distributions shown in Figure 3.19 would be considered long tailed. Examples of long-tailed distributions will be considered in Chapters 4 and 5.

3.5 Discussion and Summary

The analytical framework developed and explored here in Chapter 3 enabled a clear and explicit connection between an intensive characteristic of material structure—the material flaw population—and an extensive measure of component mechanical performance—a component strength distribution. The framework closely followed that developed and applied in strength studies of micro-scale silicon (Si) MEMS components containing nano-scale flaws (Cook and DelRio 2019a, Cook et al. 2019, 2021; DelRio et al. 2020) and in strength and fracture studies of meso-scale ceramic components containing micro-scale flaws (Cook and DelRio 2019b). The framework characterizes flaws in the material by spatial density and a continuum population of crack lengths, and characterizes strengths of components by discrete distributions derived from samples. Analysis using the framework is possible in two directions: In forward analysis, suited to theoretical considerations, the material flaw population is assumed a priori, and strength distributions are predicted based on component size and sample size, reflecting components formed from the material and sampled in groups. In reverse analysis, the practical focus of this book, strength distributions of components (here particles) are specified by experimental measurements and the underlying material flaw populations are determined by deconvolution techniques. The framework is developed most intuitively by forward analysis beginning with an assumed flaw population and this was the approach in this Chapter. In reverse analysis, the developed steps are implemented in reverse and this is discussed in Chapter 4, including identification of additional steps required in application of the framework to experimental data.

Specific elements of the analysis developed here are shown in example schematic density and distribution function plots of Figure 3.20; the forward path from material characteristic to predicted component behavior is indicated by the diagonal gray arrow. All functions are drawn in relative coordinates with domains and ranges of [0, 1]. Intermediate or unused plots are shaded gray. The analysis is based on stochastic extreme value size effects. Analysis begins with a continuum description

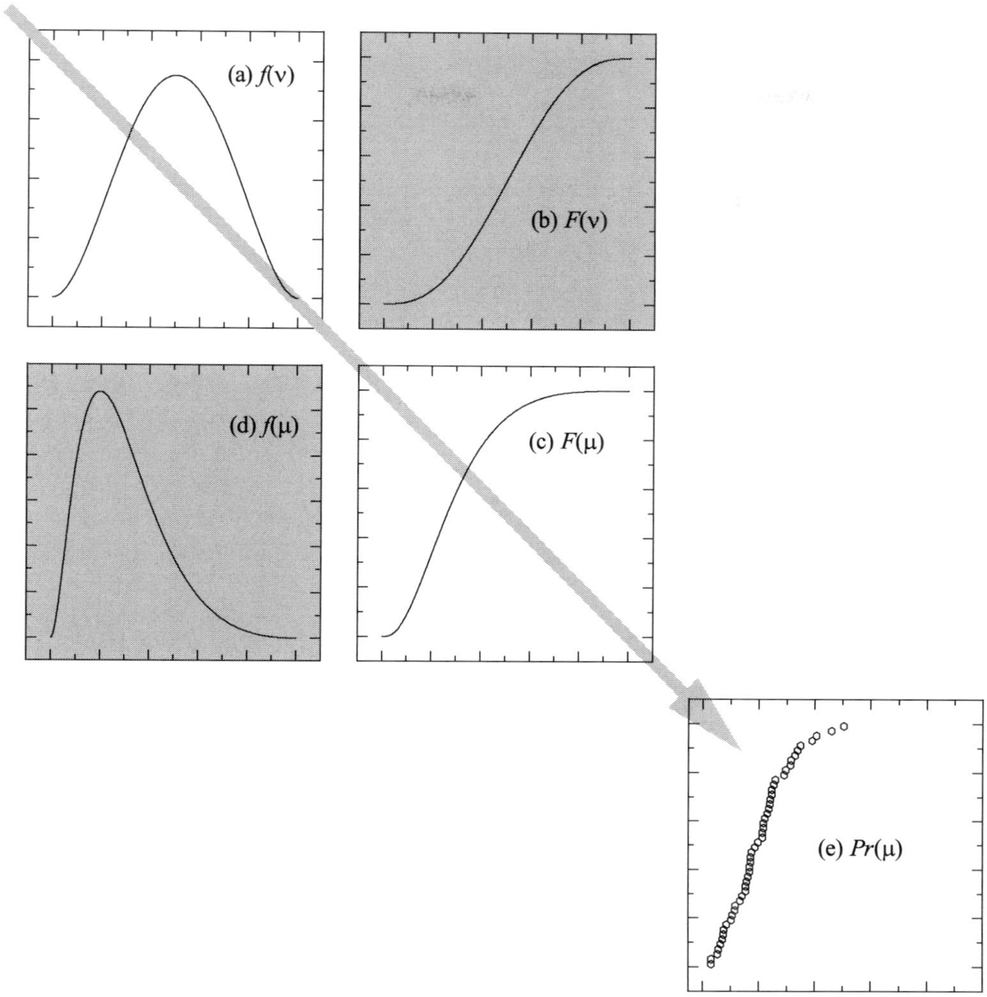

Figure 3.20 Schematic diagram illustrating forward analysis within a probabilistic framework to predict component strength distributions from material flaw populations. The arrow indicates the direction of analysis. (a) Relative crack length probability density function, pdf, $f(\nu)$. (b) Relative crack length cumulative distribution function, cdf, $F(\nu)$. (c) Relative strength cumulative distribution function, cdf, $F(\mu)$. (d) Relative strength probability density function, pdf, $f(\mu)$. (e) Relative strength empirical distribution function, edf, $Pr(\mu)$.

of the crack length population pdf $f(\nu)$, top left, Figure 3.20a, using as an example here a symmetric bell shaped density function. The bounded beta density, Eq. (3.26), was shown to be an extremely flexible and effective description for crack populations. Integration of $f(\nu)$ leads to the intermediate population crack length cdf, $F(\nu)$, top center, Figure 3.20b. Two substitutions, crack lengths by strengths and probabilities by their complements, lead to the population strength cdf, $F(\mu)$, center (Figure 3.20c). Differentiation of $F(\mu)$ leads to the population strength pdf, $f(\mu)$, left center (Figure 3.20d). Although unused, comparison of Figure 3.20d with Figure 3.20a emphasizes that symmetric crack length populations do not give rise to symmetric strength populations. Selection of the size of the components in terms of the number of included cracks k leads to determination of the asymptotic sample strength cdf, $H(\mu)$, from $F(\mu)$. $H(\mu)$ is the lower bound strength probability distribution for samples of components. Selection of the size of the sample N and selection of discrete strengths according to the sample strength probability distribution leads to prediction of the sample strength edf, $Pr(\mu)$, Figure 3.20e, bottom right.

There are several points to note regarding the analysis schematised in Figure 3.20. Specification of the relationship between fracture strength and crack length, $\sigma(c)$, here the Griffith equation, Eq. (3.1), enables development of an expression connecting relative strength and crack length, $\mu(\nu)$, Eq. (3.21), on bounded domains. Probability analysis then enables fracture behavior described by any one of Figures 3.20a, 3.20b, 3.20c, or 3.20d to generate the remaining three. Hence, the four plots in the top left square of Figure 3.20 are equivalent in defining the population of cracks and strengths. As relative coordinates are used, the parameter B characterizing the Griffith relationship between absolute strength and crack length is not used here. Specification of a component size, here by the number of single-crack containing volume elements k, enables any one of Figures 3.20a, 3.20b, 3.20c, or 3.20d to generate upper and lower probability bounds for crack length or strength in a component. Hence, an additional eight plots (not shown) are also equivalent in defining the asymptotic continuum behavior of a component and the population. As the number of cracks is used, the parameter λ characterizing the absolute crack density and thus the absolute component size is not used here. Specification of a sample size, here by the number of components N, enables selection of discrete strength values according to the component strength probability lower bound. The selected strength values can then be ranked and an edf, Chapter 2, constructed (Figure 3.20e, bottom right). As a stochastic method was used to generate the edf, the information in Figure 3.20e cannot be used unambiguously to assess the population, and hence the (one way) arrow in Figure 3.20 reflects the direction of strength prediction by forward analysis.

The loss of information in Figure 3.20e relative to Figure 3.20abcd takes two forms. The first is that most components are composed of more than one fundamental volume element and thus contain more than one crack. Hence, if the elements are independent, component strengths are given by the lower bound sample probability. As a consequence, component strengths are described by *contracted strength distributions* dependent on component size effects (Figure 3.15). (Formally, the domain over which $dH(\mu)/d\mu > 0$ is contracted.) Second, stochastic selection is biassed toward strengths of greater probability and hence component strengths are preferentially selected from these probability ranges. As a consequence, component strengths are described by *truncated strength distributions* dependent on sample size effects (Figure 3.13). (Formally, the domain over which $H(\mu) \neq 0$ is truncated.) Contraction and truncation effects are visible in comparison of Figures 3.20e and 3.20c, but are made more clear in Figure 3.21, which combines the concave strength population and sample data of Figure 3.15 to illustrate component and sample size effects. The solid lines in Figure 3.21 represent the population strength cdf $F(\mu)$. The different symbols represent sample strength edf $Pr(\mu)$ behavior for components of increasing size, indicated by the arrow in the expanded view. Figure 3.21 clearly illustrates increasing contraction and truncation effects as component size increases. For all component sizes, the largest values of the discrete strength data lie well below the upper bound of the underlying population. The strength distributions of components related by stochastic size effects appear as "fanned out" responses originating at a common small strength threshold, whether the population is light tailed or heavy tailed.

It is important to note that the parameter characterizing component size and domain restriction, k, need not be an integer. The flaw spatial density λ is an average over the entire population of flaws existing within the large body Ω of material considered. The body is envisaged as a sandy beach, a coral reef, a container ship filled with ore pellets, a billet of ceramic, a production run of glass bottles, or a monolithic mountain of rock, for example. The average number of cracks in a component

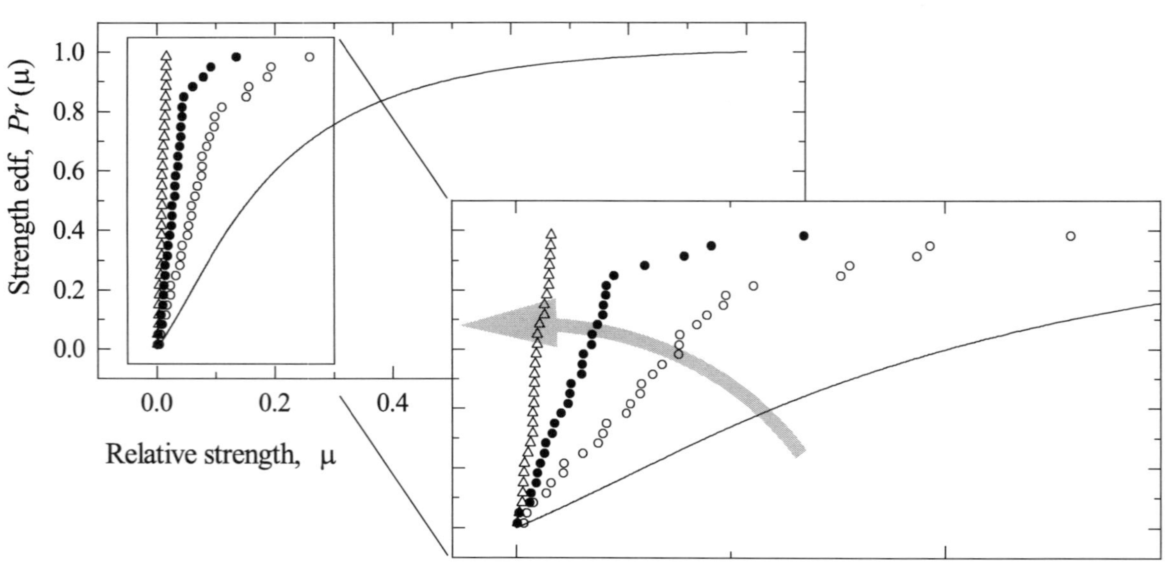

Figure 3.21 Relative strength empirical distribution functions, edf, $Pr(\mu)$ for components of increasing size, indicated by arrow. The solid line represents the behavior of the strength population. Larger components exhibit significantly restricted strength domains.

of volume V is then defined by $k = V\lambda$. As noted above, the greatest changes in strength occur when $k \approx 1$ (Figure 3.16), suggesting that size effects on strength will be greatest if the factor relating V and $1/\lambda$ is small, e.g., $V = 1.1/\lambda$, $V = 1.5/\lambda$, and so on. It is also important to re-iterate that V could be limited in one or more physical dimensions, such that k reflects an areal or lineal density, for example. A priori it is not known how variations in the physical dimensions of a component affect variations in k and such scaling is addressed in Chapters 7 and 9.

The strength distribution prediction framework developed here has emphasized stochastic phenomena: the volume elements and the contained strength-controlling cracks are independent and component size influences crack probability. This behavior is commonly referred to as "weakest link behavior" and a loaded chain often used as an example. The framework here is also able to address deterministic phenomena that is not simple weakest link: the volume elements and the contained strength-controlling cracks are not independent and component size influences the crack population. This behavior is commonly exhibited by particles and is addressed in Chapter 9. The framework is also able to address multimodal effects, such as concurrent flaw populations, concurrent component sizes, and variable flaw spacing effects (the last two are equivalent). These effects are discussed briefly in Chapter 4. The framework can also be applied to more advanced probabilistic concepts such as the hazard rate and Bayesian estimation methods.

Finally, it is important to place the development here in physical and historical context. Stochastic effects in mechanical behavior of small-scale components are observed in all modes of deformation. Patterned dielectric lines on compressive substrates exhibit stochastic variation in elastic buckling of the lines arising from stochastic variation of line stress and dimensions (Friedman et al. 2016). Blunt indentation of metals exhibits stochastic variation in initiation of plastic deformation arising from stochastic variation of slip center effectiveness (Morris et al. 2011). Microelectromechanical systems (MEMS) devices exhibit stochastic variations in tensile fracture strengths arising from stochastic variation in strength-controlling flaw size (Cook et al. 2019a). It is the third of these phenomena, fracture, specifically the fracture strength distributions of brittle single particles, that is the subject of this book. Small particles provide insight into the physics of fracture by enabling the possibility that each particle contains only one strength-controlling flaw and strength distributions are thus a direct measure of material flaw populations, as illustrated for very small MEMS components (Cook et al. 2021). Physics of fracture is extended by the study of larger particles that contain multiple flaws and exhibit strengths drawn from distributions reflecting stochastic and deterministic extreme value size effects. The book applies analyses developed in the framework here to an extensive survey of strength distributions of particles and other components dating back over 60 years and generated from newly digitized measurements.

It has long been known and simply explained that there are stochastic effects in the failure of brittle components. Based on the work of Inglis (1913) considering stress concentrations at notches, Griffith (1921) used energy balance methods to show that the strengths of components formed from brittle materials were controlled by small (sub millimeter) cracks: the larger the crack, the smaller the strength (Figure 3.2). The exact relationship is determined by the fracture resistance of the material and the strain energy change generated by the crack in the component. Soon after the work of Griffith, it was recognized by Weibull (1939) and Epstein (1948a, 1948b) that (in current terms) the strength distribution of sampled components containing independent flaws was an extreme value distribution determined by the material flaw population and the number of flaws in each component, and hence, by implication, the physical size of the components. Many of the concepts developed here in Section 3.2 and Section 3.3 are represented in these earlier works. Examples of the "terms" and symbols used by Weibull (1939), almost exclusively in the context of mechanical strength, included:

Here $\leftarrow$ Weibull (1939)
independent elements $\leftarrow$ "elementary laws of probability"
fundamental volume element ΔV $\leftarrow$ "small volume element" dv
population $\leftarrow$ "material function" $n(\sigma)$
population cumulative distribution function F $\leftarrow$ "distribution curve for unit volume" S_0
bounded population σ_{th}, σ_u $\leftarrow$ "range of distribution may be limited" σ_2, σ_1
restricted dimension d $\leftarrow$ "surface layer of small thickness" h
sample cumulative distribution function H $\leftarrow$ "distribution curve" S
size of component k $\leftarrow$ "volume of stressed system" V
size of sample N $\leftarrow$ "total number of measurements" N

Many of the concepts used by Weibull were used by Epstein (1948a). Examples of the "terms" and symbols used by Epstein, motivated in part by considerations of electrical breakdown, included:

Here ← Epstein (1948a)
extreme values ← "extreme values in samples"
sample strength lower bound cdf H_L ← "smallest value in samples d.f." H
population spatial density λ ← "flaws per unit length" Δ

The distributions described qualitatively by Weibull (1939) and quantitatively by Epstein (1948a, 1948b) thus represented a stochastic extreme value size effect. Epstein provides a brief historical summary of extreme value size effects in strength known to that time. Although it is clear that both Weibull and Epstein were aware that strength values were a consequence of flaw sizes, the focus in both of these early works was on descriptions of the distributions of strengths per se. Weibull introduced a power law expression for $\bar{F}(\sigma)$ (current term) as a vehicle to discuss probability behavior, although later (Weibull 1951) this expression was applied empirically. Epstein showed that this and other distributions (Laplace (exponential), Gauss) were all candidates for describing strength distributions. The similarity of Figure 3.17 to experimental observations makes clear that strength distributions are a consequence of flaw populations, via forward analysis within the framework developed here. That framework, embodying many of the concepts articulated by Weibull and Epstein, is used in reverse analysis in Chapter 4 to interpret strengths in terms of fundamental crack length populations.

References

Bryson, M.C. (1974). Heavy-tailed distributions: Properties and tests. *Technometrics* 16: 61–68.

Cook, R.F., Boyce, B.L., Friedman, L.H., and DelRio, F.W. (2021). High-throughput bend-strengths of ultra-small polysilicon MEMS components. *Applied Physics Letters* 118: 201601.

Cook, R.F. and DelRio, F.W. (2019a). Material flaw populations and component strength distributions in the context of the Weibull function. *Experimental Mechanics* 59: 279–293.

Cook, R.F. and DelRio, F.W. (2019b). Determination of ceramic flaw populations from component strengths. *Journal of the American Ceramic Society* 102: 4794–4808. (typographical error in Eq. (9)).

Cook, R.F., DelRio, F.W., and Boyce, B.L. (2019). Predicting strength distributions of MEMS structures using flaw size and spatial density. *Microsystems & Nanoengineering* 5: 1–12.

DelRio, F.W., Boyce, B.L., Benzing, J.T., Friedman, L.H., and Cook, R.F. (2020). Shoulder fillet effects in strength distributions of microelectromechanical system components. *Journal of Micromechanics and Microengineering* 30: 125013.

Epstein, B. (1948a). Application of the theory of extreme values in fracture problems. *Journal of the American Statistical Association* 43: 403–412.

Epstein, B. (1948b). Statistical aspects of fracture problems. *Journal of Applied Physics* 19: 140–147.

Forbes, C., Evans, M., Hastings, N., and Peacock, B. (2011). *Statistical Distributions, 4th ed.* Wiley.

Friedman, L.H., Levin, I., and Cook, R.F. (2016). Stochastic behavior of nanoscale dielectric wall buckling. *Journal of Applied Physics* 119: 114305.

Glaesemann, G.S., Jakus, K. and Ritter Jr, J.E. (1987). Strength variability of indented soda-lime glass. *Journal of the American Ceramic Society* 70: 441–444.

Griffith, A.A. (1921). The phenomena of rupture and flow in solids. *Philosophical Transactions of the Royal Society London A* 221: 163–198.

Gupta, A.K. and Nadarajah, S. (eds) (2004). *Handbook of Beta Function and its Applications.* Taylor and Francis.

Inglis, C.E. (1913). Stress in a plate due to the presence of cracks and sharp corners. *Transactions of the Institution of Naval Architects* 55: 219–241.

Kendall, K. (1988). Agglomerate strength. *Powder Metallurgy* 31: 28–31.

Krishnamoorthy, K. (2016). *Handbook of Statistical Distributions with Applications, 2nd ed.* Taylor and Francis.

Lawn, B.R. (1993). *Fracture of Brittle Solids, 2nd ed.* Cambridge.

Lipschutz, S. (1965). *Probability.* McGraw-Hill, Inc.

Maugis, D. (2000). *Contact, Adhesion and Rupture of Elastic Solids.* Springer-Verlag.

Morris, J.R., Bei, H., Pharr, G.M., and George, E.P. (2011). Size effects and stochastic behavior of nanoindentation pop in. *Physical Review Letters* 106: 165502.

Nair, J., Wierman, A., and Zwart, B. (2013). The fundamentals of heavy tails. https://adamwierman.com/wp-content/uploads/2020/09/2013-SIGMETRICS-heavytails.pdf (accessed January 28, 2022).

Spiegel, M.R. (1961). *Statistics.* McGraw-Hill.

Todinov, M.T. (2010). The cumulative stress hazard density as an alternative to the Weibull model. *International Journal of Solids and Structures* 47: 3286–3296.

Walpole, R.E. and Myers, R.H. (1972). *Probability and Statistics for Engineers and Scientists.* Macmillan.

Weibull, W. (1939). A statistical theory of the strength of materials. *Proceedings of the Royal Swedish Institute of Engineering Research* 151: 1–45.

Weibull, W. (1951). A statistical distribution function of wide applicability. *Journal of Applied Mechanics* 18: 293–297.

4

Strength Distributions

This chapter describes in clear, simple operational terms the numerical basis of how the characteristics of strength distributions (which are what is measured) can be used to infer the characteristics of populations of strength-controlling flaws (which are usually not observed) through the choice of sample and component sizes (which are what can be controlled). This is reverse analysis and is focused on the answering the question "what can we learn?" This chapter is practical and the examples near-ideal as they will be selected from published works to illustrate specific points. The terms and inter-relationships defined in Chapter 3 will be used extensively. The analysis is applicable to all brittle components.

4.1 Brittle Fracture Strengths

4.1.1 Samples of Components

The strength distributions of components sampled from a large body of a brittle material were considered in some detail in Chapter 3. Brittle materials exhibit elastic deformation prior to failure by fracture at a peak stress, the brittle fracture strength. Fracture strengths are determined by the stress concentrating effects of flaws in a material. Based on the work of Inglis (1913) considering stress concentrations at notches, Griffith (1921) used energy balance methods to show that the strengths of components formed from brittle materials were controlled by small (sub millimeter) cracks: planar regions of broken bonds. The larger the crack, the smaller the strength (Figure 3.2), with the exact relationship determined by the surface energy or fracture resistance of the material and the strain energy change generated by the crack in the component. If an intended structural component contains a single crack or flaw, determination of component strength as a function of flaw geometry and residual stress state, material microstructure, component environment, and applied loading rate is now a well-understood fracture mechanics problem (Lawn 1993). Attention in the previous chapter was focused on the collective strength behavior of many components containing many flaws.

If a group of components is assembled, each containing a single, but different, flaw and the strength of each component subsequently measured, the resulting strength distribution reflects the distribution of flaw characteristics within the group, often dominated by crack length. Usually, however, each component contains a multitude of flaws, usually all different. The behavior of a component in a strength test is then still easily determined if the flaws act independently, as attention can then focus on the behavior of the largest, most potent, or strength-controlling flaw that exhibits the least resistance to failure or smallest strength. (Other more resistant flaws will respond to the applied stress but will not cause component failure.) If a group of such components containing multiple flaws is then assembled and tested, the resulting strength distribution reflects the distribution of the smallest, or minimally extreme, strengths in the group of components and is conjugate to the distribution of the largest, or maximally extreme, flaws. The strength distribution of a group of components, each containing multiple flaws, is then an extreme value distribution. Usually the flaws in a group of components are a subset of a much larger set constituting the entire population of flaws in the bulk material from which the test components were sampled. Quantification of these phenomena in Chapter 3 employed forward analysis to begin with a fundamental intensive material property, the flaw population, and proceeded to predict an extensive component performance measure, a strength distribution. The strength distribution resulting from a sample of components was shown to be a function not only of the material flaw population but the experimental variables of component size and sample size. The ability of the analytical

Particle Strengths: Extreme Value Distributions in Fracture, First Edition. Robert F. Cook.

framework developed in Chapter 3 to describe the material flaw-component strength linkage is illustrated in comparison of simulated strength distributions, Figure 3.19, with experimental examples, Figures 2.28 and 2.29. This chapter considers *reverse analysis*, beginning with a component strength distribution and applying the framework developed in Chapter 3 to infer the underlying material flaw population.

Section 4.1.2 develops a description for sample strength distributions that is amenable to reverse analysis—the description and reverse analysis procedure are used throughout the book. In the following Section 4.2, the description and analysis procedure are verified and some examples considered. Verification consists of comparison of two flaw probability densities, the first inferred from sample strength distributions and the second describing the known population used to (numerically) generate the strengths. Uncertainty in the crack length probability density inferred from a given strength distribution is considered, along with the effects of experimental variability in strength measurements. The overall reverse analysis method is demonstrated by application to two contrasting example sets of experimental data: (i) sigmoidal strength distributions describing silicon nitride beams tested in bending and (ii) a concave strength distribution describing calcareous sand (coral) particles tested in diametral compression. A major philosophical element of these sections is that descriptions of sample strength distributions by arithmetic functions are regarded as necessary, but intermediate, steps in larger analyses. Such descriptions are merely smoothing functions used as "stepping stones" to enable analysis of strength behavior in terms of fundamental material flaw populations. In terms of the functions discussed in Chapter 3, an arbiter of analytic success here is not whether $Pr(\sigma)$ is well described, but whether $f(c)$ is well estimated.

4.1.2 Analysis of Sample Strength Distributions

To summarize some concepts from Chapter 3 required for reverse analysis: A continuous cdf $H(\sigma)$ describes the extreme value probability distribution of component strengths sampled from a population of strengths, $F(\sigma)$. The strength population derives from an invariant population of flaws. Flaw size is characterized by crack length c with conjugate strengths σ. The flaw spatial density (number of flaws/volume) is given by λ and the population of flaw sizes described by a pdf $f(c)$. The relationship between strength and crack length is assumed to be the Griffith equation, $\sigma = Bc^{-1/2}$, where B is related to material toughness. λ, $f(c)$, and B are all intensive quantities characterizing the material crack length and strength populations. Probabilistic considerations enable the population to be characterized equivalently by other pdf and cdf variations, including the population strength cdf $F(\sigma)$. Component size is characterized by a dimensionless term $k = \lambda V$, where V is the volume of the component. Sample size is given by N, the number of components tested in strength measurements giving rise to $Pr(\sigma)$. The experimental strength edf $Pr(\sigma)$ can be used as a discrete estimator of the continuous cdf $H(\sigma)$ and thence as an estimator of the sample extreme value crack length pdf $h(c)$. Estimates of $H(\sigma)$ and $h(c)$ improve if N is large. The connection between $H(\sigma)$ and $F(\sigma)$ or between $h(c)$ and $f(c)$ can only be made if the nature of the extreme value size effect is known. If the size effect is purely stochastic, the connection between $H(\sigma)$ and $F(\sigma)$, and $h(c)$ and $f(c)$ depends on the dimensionless size of the component, k.

A plausible method for estimating $h(c)$ from $Pr(\sigma)$ data is to select a B value and use the Griffith equation to construct $Pr(c)$ from the discrete sequence σ_1–σ_N of measured strength values. $Pr(c)$ is a discrete estimator of the continuous extreme value cdf $H(c)$. In principle, an estimate of the underlying pdf, $h(c) = \mathrm{d}H(c)/\mathrm{d}c$, could thus be obtained by discrete differentiation of $Pr(c)$: $h(c) \approx \Delta Pr(c)/\Delta c$, where Δ indicates the changes between discrete values. In practice, even with many of the extensive data sets to be considered here, such discrete differentiation leads to results with excessive noise or variation that are not easily interpreted or applied in comparisons or predictions (Cook and DelRio 2019a). Hence, a *smoothing function* is required prior to numerical analysis. The simplest smoothing operation is perhaps the moving average, which generates a new data set with usually the same step size as the original (Spiegel 1961). Although sometimes useful for revealing trends, the moving average method does not usually smooth enough to enable low-noise discrete differentiation of strength-derived data. A better method is to fit a smoothly varying function to the entire data set, such that numerical differentiation of the fitting function is also smoothly varying. The simplest and most common fitting functions result from regression of the data to linear or polynomial functional forms, which generate new continuum data sets (Walpole and Myers 1972). Such data sets are smooth and smoothly varying on numerical differentiation. The beta function related pdf and cdf of Chapter 3, Eqs. (3.24)–(3.26), are such polynomial functions, and were demonstrated as effective descriptors of flaw distributions. Although analytically tractable, there are several issues with this approach. First, and perhaps most important, the fitting procedure is now removed from the experimental strength data. This has the consequence that intuition regarding fitting parameters is almost lost and that there is no simply specified strength distribution function. Second, to describe many crack length edf data sets, in particular commonly observed heavy tailed data sets, a perturbed beta distribution is required, increasing the number of inter-related fitting parameters. Third, selection of the $\sigma(c)$ relation (usually the Griffith equation) is required prior to fitting, constraining the interpretation of strength measurements.

A better method of analyzing $Pr(\sigma)$ data is to estimate $H(\sigma)$ by fitting a smoothly varying function $\hat{H}(\sigma)$ to the $Pr(\sigma)$ data, use the Griffith equation and the probability relations to thence generate a smooth estimate of $\hat{H}(c)$, and finally numerically differentiate this function to obtain an estimate of the sample extreme value pdf $\hat{h}(c) = \mathrm{d}\hat{H}(c)/\mathrm{d}c$. This method obviates many of the issues identified above leaving only a decision regarding the form of the function $\hat{H}(\sigma)$. A guiding principle of the analysis here was that the form of $\hat{H}(\sigma)$ should enable the best estimate of $h(c)$, rather than be the best fit to $Pr(\sigma)$ of a pre-selected form (e.g. error function).

Major attributes of the required fitting function are a domain bounded by non-zero values and a predominant linear variation. The domain of the fitting function is given by $\sigma_L \le \sigma \le \sigma_U$, where σ_L and σ_U are empirical lower and upper strength bounds, respectively. It is anticipated that strength fitting bounds will be close to the experimentally observed bounds, $\sigma_L \approx \sigma_1$ and $\sigma_U \approx \sigma_N$. As in Chapter 3 it is convenient to express the fitting function by a relative measure of strength. In this case the relative strength measure $\hat{\mu}$ is given by

$$\hat{\mu} = \frac{\sigma - \sigma_L}{\sigma_U - \sigma_L} \tag{4.1}$$

with domain $0 \le \hat{\mu} \le 1$. The fitting function selected here is composed primarily of three linear variations of the form $y_i = a_i\hat{\mu} + b_i$, where $i = 1, 2, 3$ and

$$\begin{aligned} y_1 &= a_1\hat{\mu} \\ y_2 &= a_2\hat{\mu} + (0.5 - a_2\mu_m) \\ y_3 &= a_3\hat{\mu} + (1 - a_3). \end{aligned} \tag{4.2}$$

The functions y_1 and y_3 pertain at the bounds as $y_1(0) = 0$ and $y_3(1) = 1$. The function y_2 pertains at the center of the domain, as $y_2(\mu_m) = 0.5$ and μ_m is recognized as the median of the quantile y_2. The parameters a_1, a_2, a_3 are all empirical asymptotic slopes of the full fitting function at the lower bound, center, and upper bound of the relative strength domain, respectively. $a_1, a_2 \ge 0$ and $0 \le a_3 \le 1$. Smooth variation between these asymptotes is obtained by forming greater or lesser *p-norm* interpolation functions. For y_1 and y_2 the *p*-norm is

$$y_{12}^{p_1} = y_1^{p_1} + y_2^{p_1}, \tag{4.3}$$

where $p_1 > 0$ is a fitting exponent such that y_{12} approaches the greater of y_1 and y_2 as asymptotes and generates a smooth transition between y_1 and y_2. Extending this idea, for y_{12} and y_3 the *p*-norm is

$$y_{123}^{p_2} = y_{12}^{p_2} + y_3^{p_2}, \tag{4.4}$$

where $p_2 < 0$ is a fitting exponent such that y_{123} approaches the lesser of y_{12} and y_3 as asymptotes and generates a smooth variation from $y_1 \to y_2 \to y_3$. Simple linear transformation adjusts the range of y_{123} to $[0, 1]$ over the domain $[0, 1]$ to form $\hat{H}(\hat{\mu})$:

$$\hat{H}(\hat{\mu}) = \frac{y_{123} - \min(y_{123})}{\max(y_{123}) - \min(y_{123})}. \tag{4.5}$$

Eqs. (4.1)–(4.5) form a parametric set to describe the required smoothly varying *tri-linear fitting function*.

Figure 4.1 shows examples of the sample fitting function $\hat{H}(\hat{\mu})$ and the y_i components. The full functions are shown as bold lines and the asymptotic components are shown as fine lines. Figure 4.1a shows the creation of a sigmoidal response using the asymptotic parameters $a_1 = 0.1$, $a_2 = 3$, $\mu_m = 0.55$, $a_3 = 0.2$, the interpolation parameters $p_1 = 6$ and $p_2 = -8$, and the domain parameters $\sigma_L = 100$ MPa and $\sigma_U = 300$ MPa. The full $\hat{H}(\hat{\mu})$ response passes from y_1 through the predominant y_2 center asymptote to y_3. Figure 4.1b shows the creation of a concave response using the asymptotic parameters $a_1 = 5$, $a_2 = \mu_m = 0$, $a_3 = 0.1$, the interpolation parameters $p_1 = 1$ and $p_2 = -6$, and the domain parameters as before. The full *bi-linear* $\hat{H}(\hat{\mu})$ response has y_2 suppressed and passes from the predominant y_1 asymptote directly to y_3. The examples of Figure 4.1 are clearly able to replicate the strength edf responses shown in Chapter 2. The next section, 4.2, implements the

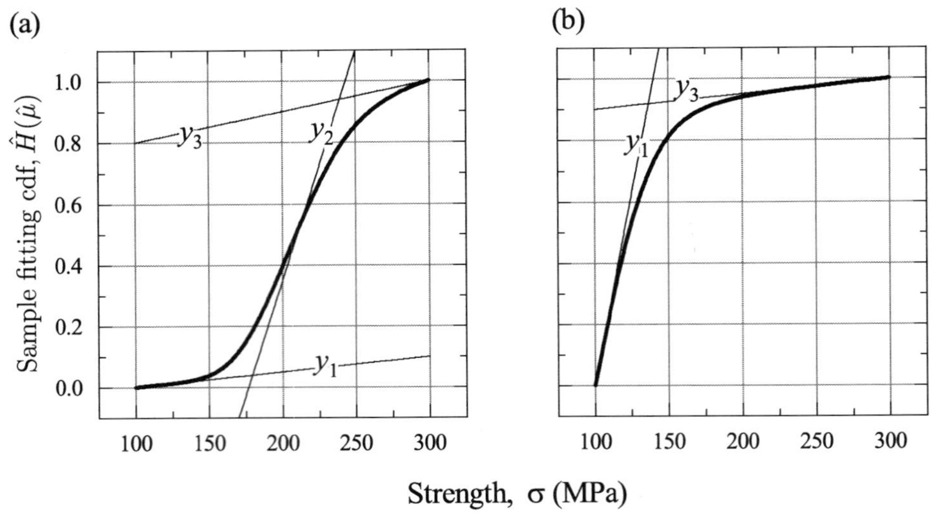

Figure 4.1 Smoothly varying fitting function $\hat{H}(\hat{\mu})$ used as continuous estimators in analysis of discrete experimental strength measurements. Full functions are shown as bold lines, asymptotic components are shown as fine lines (a) Sigmoidal response (b) Concave response.

full methodology described above for estimating the extreme value crack length pdf $h(c)$ from simulated and experimental strength edf $Pr(\sigma)$ data, utilizing $\hat{H}(\hat{\mu})$ as fitting functions.

4.2 Sample Strength Distributions

4.2.1 Sample Analysis Verification

A necessary first step in establishing sample strength analysis is *verification* of the methodology used to deconvolute strength distributions in terms of crack length probability density functions. That is, address the question "does the methodology perform correctly?" In this case, agreement between analyses of strength data with known underlying crack length data provides such verification. Forward analysis techniques, applied within the probabilistic framework of Chapter 3, will be used in a first step. Two contrasting example crack populations characterized by two different crack length pdf $f(c)$ variations will be used and cracks randomly sampled from each. The resulting conjugate strength samples will be used to create two simulated strength edf $Pr(\sigma)$ variations. In the second step, reverse analysis will be used by fitting $\hat{H}(\sigma)$ functions to each set of strengths and thence estimating the underlying crack length cdf and pdf variations, $\hat{H}(c)$ and $\hat{h}(c)$. As the cracks are selected singly from populations, the component size $k = 1$ in each case and $\hat{h}(c)$ can be compared directly with $f(c)$. In both cases the sample size is $N = 100$.

The first example crack population is characterized by crack length bounds $c_{min} = 25$ μm and $c_{max} = 100$ μm and the asymmetric peaked pdf shape shown in Figure 3.6c. Setting $B = 1$ MPa m$^{1/2}$ gives the strength bounds $\sigma_{th} = 100$ MPa and $\sigma_u = 200$ MPa and strength cdf shape of the symmetric sigmoid shown in Figure 3.6d. Figure 4.2 shows the strength edf plot $Pr(\sigma)$ resulting from sample selection from the population. Open symbols indicate individual strengths. As in Figure 3.15e, the sampled strengths occupy a limited domain within the population, in this case approximately 120–180 MPa. The fine solid line in Figure 4.2 represents a visual best fit to the strength data of the smooth tri-linear form $\hat{H}(\sigma)$ given by Eqs. (4.1)–(4.5). The shaded band represents the range of upper and lower bounds of fits to the data, maintaining the fit domain bounds. The formulation describes the data well. Figure 4.3 shows the estimated crack length pdf plot $\hat{h}(c)$ resulting from the best fit to the strength data. As the estimation was part of a verification process, the known value of $B = 1$ MPa m$^{1/2}$ was used. The solid line is the best estimate of the crack length pdf and the shaded band indicates the upper and lower uncertainty bounds, all derived from the analogous strength fits. The dashed line in Figure 4.3 represents the known $f(c)$ variation of the crack length population over domain 25 μm–100 μm. Over the domain of the crack length probability density prediction, there is agreement with that of the initial population, particularly with regard to the location and width of the peak. The domain of the prediction is limited to approximately 30–70 μm, smaller than that of the population, a consequence of the restricted domain of the strength observations relative to the strength population. The prediction does not extend to the light tail at large crack lengths.

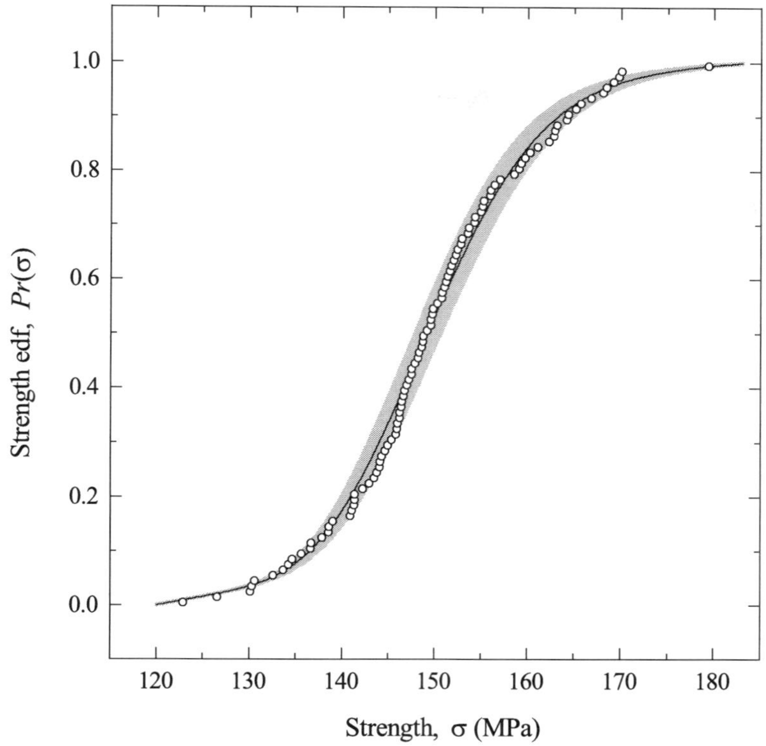

Figure 4.2 Sigmoidal strength edf plot $Pr(\sigma)$ resulting from numerical sampling from a light-tailed crack population. Open symbols indicate individual strengths. Line and shaded band represent mean and bounds of smooth fits to the data.

The second example crack population is characterized by crack length bounds $c_{\text{min}} = 0.1$ mm and $c_{\text{max}} = 1.6$ mm and the elevated, rounded pdf shape shown in Figure 3.7a. Setting $B = 0.4$ MPa m$^{1/2}$ gives the strength bounds $\sigma_{\text{th}} = 10$ MPa and $\sigma_{\text{u}} = 40$ MPa and concave strength cdf shape shown in Figure 3.7b. The weaker variation in $f(c)$ and weaker departure of $F(\sigma)$ from linearity provide a stricter test of the methodology than the anlogous peak and sigmoid of Figure 4.2 and Figure 4.3. Figure 4.4 shows the strength edf plot $Pr(\sigma)$ resulting from sample selection from the population. Open symbols indicate individual strengths. As in Figure 3.15f, the sampled strengths occupy the entire domain within the population, in this case approximately 10–40 MPa. The fine solid line in Figure 4.4 represents a visual best fit to the strength data of the smooth form $\hat{H}(\sigma)$ given by Eqs. (4.1)–(4.5), suppressed here to bi-linear form. The shaded band represents the range of upper and lower bounds of fits to the data. The formulation describes the data well. Figure 4.5 shows the estimated crack length pdf plot $\hat{h}(c)$ resulting from the best fit to the strength data. As the estimation was part of a verification process, the known value of $B = 0.4$ MPa m$^{1/2}$ was used. The solid line is the best estimate of the crack length pdf and the shaded band indicates the upper and lower uncertainty bounds, all derived from the analogous strength fits. The dashed line in Figure 4.5 represents the known $f(c)$ variation of the crack length population over domain 0.1–1.6 mm. The domain of the prediction is nearly identical, a consequence of similar domains of the strength observations relative to the strength population. The relative uncertainty of the crack length pdf prediction is greater than that above, a consequence of the smaller relative underlying population variation. Even considering uncertainty, there is weaker agreement between the crack length probability density prediction and that of the initial population, although the locations and widths of the peaks are similar. The prediction increases slightly rather than decreasing at small crack lengths.

Overall, the results of Figures 4.3 and 4.5 verify that the sample strength analysis methodology of Section 4.1 performs correctly. Most important, for both examples of crack and strength populations considered, the predicted crack length pdf peak position and width were in agreement with the known generating pdf. Over the entire generating domain, however, in both cases the predicted pdf was not totally in agreement with the generating pdf, although the disagreements were small. As perhaps anticipated, strength domain restriction in the sigmoidal case restricted the domain of predicted crack lengths. Less initially obvious, near linearity in the concave case depressed the range of predicted crack length probability densities relative to the generating pdf.

The uncertainty bounds indicated by the shaded bands in Figure 4.2 and Figure 4.4 represent scatter in probability values introduced by the discretization process used to generate $Pr(\sigma)$ on sampling from $F(\sigma)$ (an N effect). The strength values

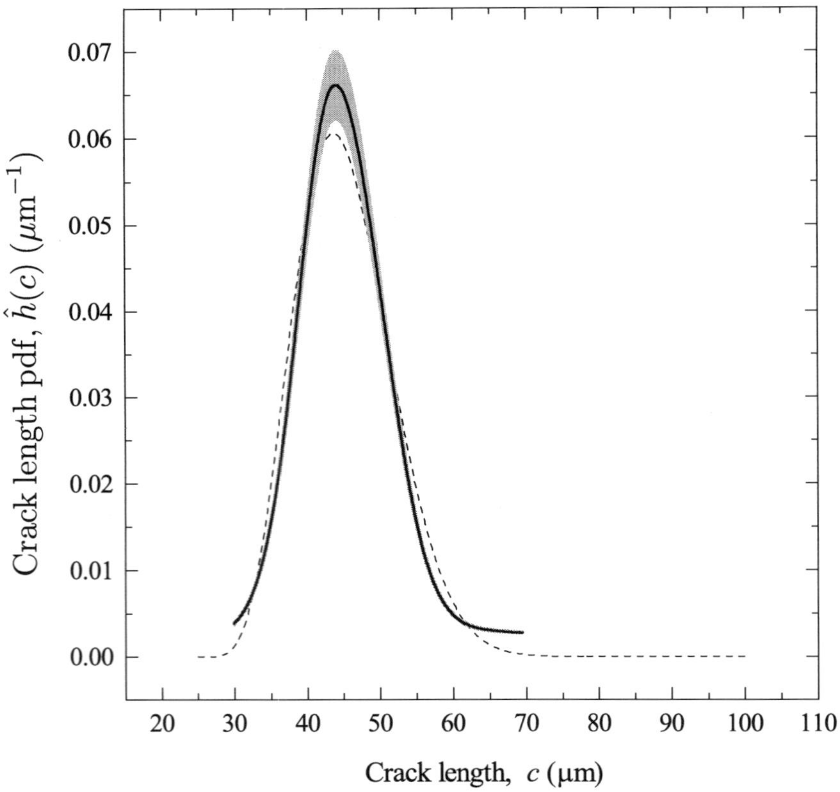

Figure 4.3 Estimated crack length pdf plot $\hat{h}(c)$ resulting from best fit to the strength data of Figure 4.2. Solid line is best estimate and shaded band indicates uncertainty bounds. Dashed line represents the $f(c)$ variation of the known crack length population. There is agreement over the domain of the prediction.

corresponding to the probability values were generated from $F(c)$ directly. In a strength experiment, scatter in strength values can be introduced by local variations in the fracture criterion, such that a sampled crack length can give rise to a range of strengths (a B effect). Scatter in experimental strength values can also be introduced by variations in testing procedures (component dimension, alignment, or failure force measurement effects) (Quinn et al. 2009). Figure 4.6a shows the effect of scatter in strength on the edf of Figure 4.2 if the underlying crack length ranking is unchanged and the conjugate strengths varied. The variation was generated by a random relative factor of up to $\pm$ 0.03, typical of relative uncertainty in strength tests. The strength ranking is now incorrect and many strength measurements fall outside the uncertainty bounds established in Figure 4.2. However, in an experiment, crack lengths are unknown and $Pr(\sigma)$ is generated solely from measured strength values. Hence, Figure 4.6b shows the data of Figure 4.6a sorted by strength as in an experiment. The strength ranking is now correct and nearly all strength measurements fall inside the established uncertainty bounds. An implication of Figure 4.6 is that variations in strength arising from material inhomogeneity or experimental factors may only weakly perturb resultant $Pr(\sigma)$ variations. The reason for this is that the ranking process obscures strength variation. As an example, if adjacent ranked strength values, σ_i and σ_{i+1} separated by $\Delta\sigma$ within $Pr(\sigma)$ in Figure 4.2, are varied with opposing errors to give $\sigma_i + \Delta\sigma$ and $\sigma_{i+1} - \Delta\sigma$ as in Figure 4.6a, re-sorting simply reverses the ranking such that $Pr(\sigma)$ is unchanged, similar to Figure 4.6b. The predicted crack length pdf is then also unchanged.

An additional step in establishing the sample strength analysis is *validation* of the methodology. In this case, the question to be addressed is not whether the methodology correctly calculates crack lengths, but "does the methodology calculate the correct crack length?" In this case, agreement between predicted crack lengths and independently measured crack lengths provides such validation. Verification of a methodology usually relies more on correct implementation of mathematical and algorithmic principles whereas validation usually relies more on the correct embodiment of physical principles. Verification may be completely intensive, but validation must be extensive. In this case, validation requires use of the correct component dimensions and selection of the correct failure criterion (for the Griffith criterion, the correct value of B). Validation of methodologies similar to that described in Section 4.1 were shown by Chao and Shetty (1992) in consideration of strength-controlling pores in silicon nitride, Cook (2017) in consideration of scratches on alumina, Cook and DelRio (2019b) in consideration of cracked grain boundary facets in alumina particles, and, particularly, Cook et al. (2019, 2021) in

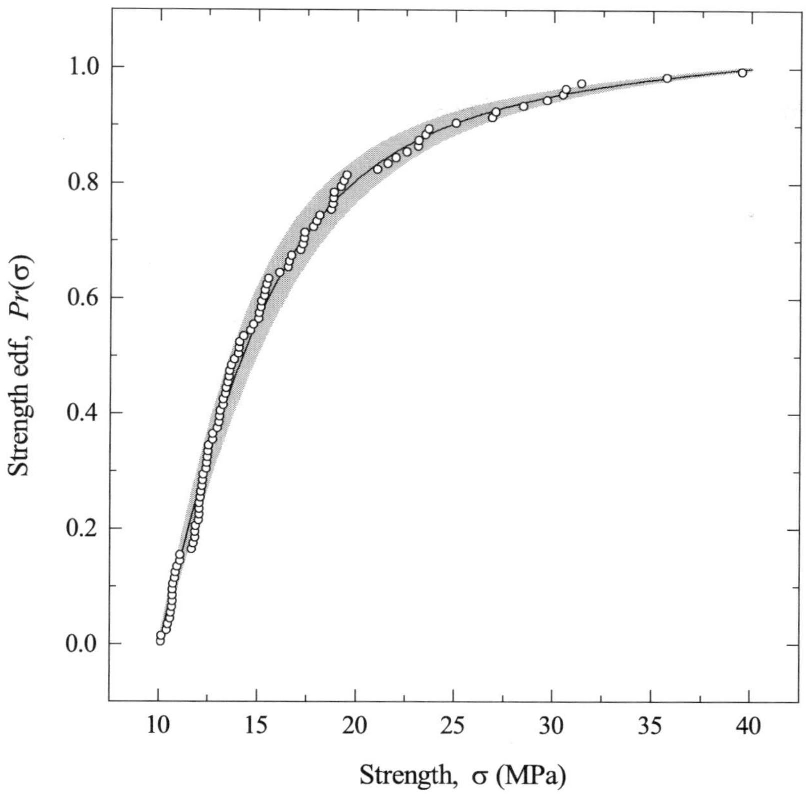

Figure 4.4 Concave strength edf plot $Pr(\sigma)$ resulting from numerical sampling from a heavy-tailed crack population. Open symbols indicate individual strengths. Line and shaded band represent mean and bounds of smooth fits to the data.

consideration of grain boundary grooves on the side-walls of MEMS components. The strength variability effects considered in Figure 4.6 are partially validation issues. Many more validation examples will be considered throughout the book.

An important physical consideration related to validation is that the maximum inferred crack length must be bounded by the finite energy used to form the component or flaw. The maximum inferred crack length, c_U, is conjugate to the minimum inferred strength, $c_U = (B/\sigma_L)^2$ and the minimum inferred crack length, c_L, is conjugate to the maximum inferred strength, $c_L = (B/\sigma_U)^2$. Analytically, the energy consideration implies that the maximum inferred crack length must be smaller than the component size, for particles $c_U < D$ (otherwise the particle is already broken). The minimum inferred crack length must be non-zero, $c_L > 0$ (otherwise the crack is not strength-controlling).

Interpretations of flaw distributions, characterized by $h(c)$ and $f(c)$ deconvoluted from strength data, should include the above caveats regarding methodology verification and validation. This section makes clear that the deconvolution process is enabled by a verified method that provides anticipated uncertainties and domain limitations. Method uncertainties and domain limitations can be somewhat minimized by increased sample size N. The following section applies the deconvolution methodology to two sets of experimental strength measurements, similar to those analyzed above in consideration of verification. The selection of material fracture criterion B and component size k are shown to be important factors in method validation.

4.2.2 Sample Examples

This section applies the reverse analysis methodology described above to experimental strength measurements. Two contrasting example sets of strength data, derived from two different materials tested in two different geometries will be examined. The first material is a dense ceramic, Si_3N_4, tested in bending after machining bend beams to the same size, volume V, from two different batches of bulk material. The data are publicly available as part of a larger statistical analysis study (NIST 1996) and include $N = 240$ tests/batch, extending over a strength domain of approximately 300 MPa–800 MPa. The second material is a carbonaceous sand (largely coral, $CaCO_3$) tested as particles with $D \approx 0.8$ mm–1.0 mm (Xiao et al. 2019), using data derived from the published work. The data include $N = 40$ tests, extending over a strength domain of

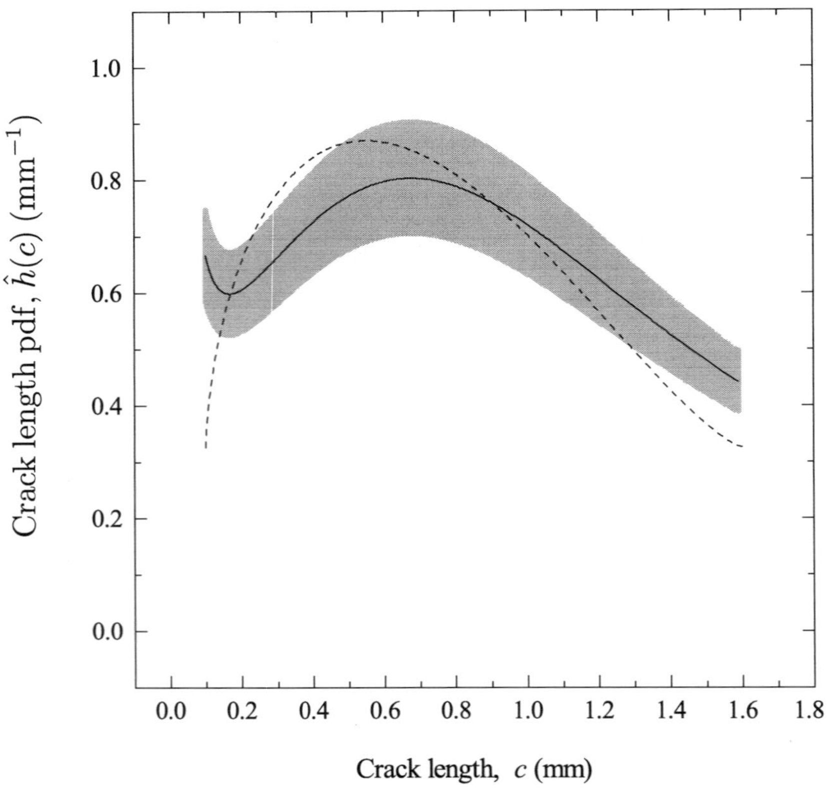

Figure 4.5 Estimated crack length pdf plot $\hat{h}(c)$ resulting from best fit to the strength data of Figure 4.4. Solid line is best estimate and shaded band indicates uncertainty bounds. Dashed line represents the $f(c)$ variation of the known crack length population. There is agreement over the domain of the prediction.

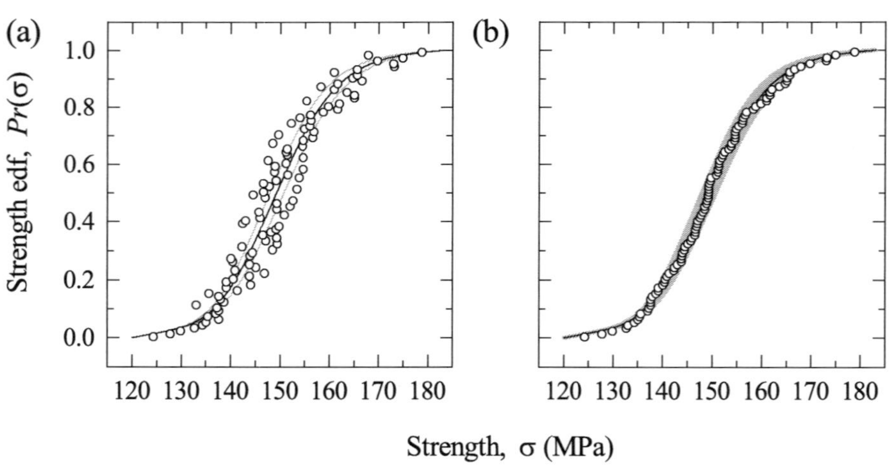

Figure 4.6 Sigmoidal strength edf plot $Pr(\sigma)$ resulting from numerical sampling from the light-tailed crack population of Figure 4.2 with superposed $\pm$ 0.03 strength scatter. Open symbols indicate individual strengths. Line and shaded band represent mean and bounds of smooth fits to the original data (a) Unsorted (b) Scatter removed by re-sorting.

approximately 3 MPa–65 MPa. Analysis of these data will outline many of the factors considered throughout the book in interpreting strength data in terms of flaw sizes and arriving at valid flaw size estimates.

Figure 4.7 shows strength edf $Pr(\sigma)$ plots for the Si_3N_4 bend beams machined from the two different batches of material. The two sets of strengths are clearly not the same, and establishing this difference via a range of statistical tests—none implementing $Pr(\sigma)$ or assessing the physical reason for the difference—was the focus of the original study led by Jahanmir (NIST 1996). The focus here is to interpret the observed sigmoidal strength distribution behavior in terms of underlying

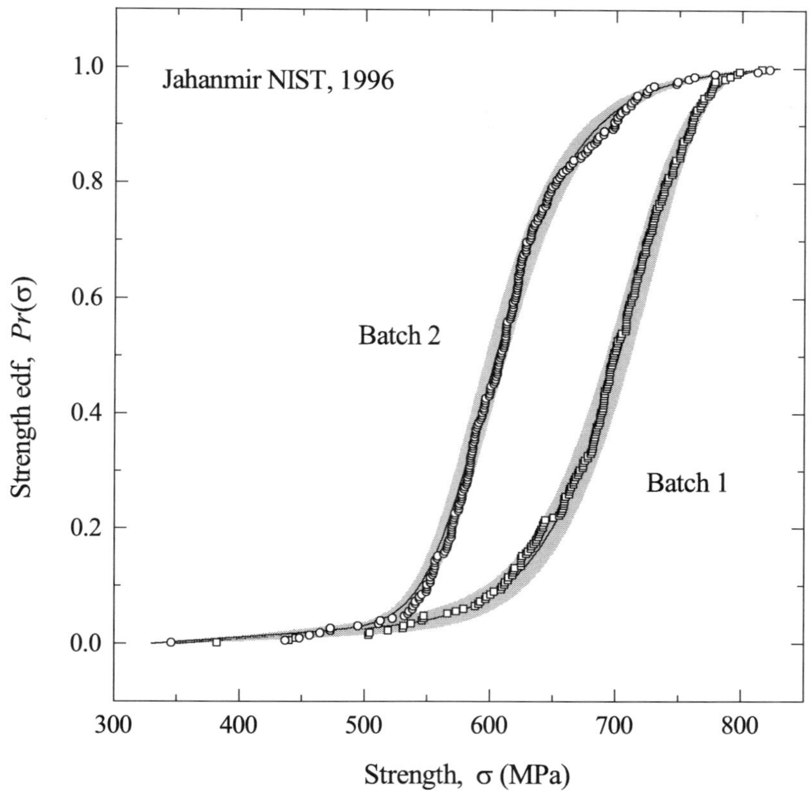

Figure 4.7 Strength edf $Pr(\sigma)$ plots for Si_3N_4 bend beams machined from two different batches of material. Open symbols represent individual strength measurements. Solid lines represent best fits to the strength data. Shaded bands represent fit bounds. Data adapted from NIST study led by Jahanmir (1996), N_{tot} = 480.

strength-limiting flaws. The open symbols in Figure 4.7 represent individual strength measurements sorted by batch. The fine solid lines represent visual best fits to the strength data, $\hat{H}(\sigma)$ using Eqs. (4.1)–(4.5). The shaded bands represent ranges of the upper and lower bounds of fits to the data, maintaining the fit domain bounds. The fit domain bounds were approximately those of the experimental observations, $\sigma_L \approx \sigma_1$ and $\sigma_U \approx \sigma_N$, noting that the experimental domain width $\sigma_N/\sigma_1 \approx$ 2.7 here for both batches. The $\hat{H}(\sigma)$ formulation describes both sets of data well.

There are three possible interpretations of the strength difference between batch 1 (stronger) and batch 2 (weaker) shown in Figure 4.7. In physical terms, the interpretations are (i) that batch 1 has smaller flaws, (ii) that batch 1 is more fracture resistant, or (iii) that batch 1 has fewer flaws. In the analytical terms of the framework above, these three interpretations involve variation between the batches by one of (i) the flaw size probability density, $f(c)$; (ii) the criterion for fracture, B; or (iii) the spatial density of flaws, λ. Using the subscripts 1 and 2 to indicate the batches, perhaps the simplest interpretation is that batches 1 and 2 exhibited the same criterion for fracture, $B_1 = B_2$, and the same flaw spatial density, $\lambda_1 = \lambda_2$, but that batch 2 contained a preponderance of larger cracks through a variation in (i), the flaw size probability, $f_1(c) \neq f_2(c)$. A single value of $B = 4$ MPa m$^{1/2}$, approximating the toughness of dense bulk Si_3N_4 (Chao and Shetty 1992), was used to estimate the crack length pdf $\hat{h}(c)$ from best fits to the strength data for both batches. (As a single value λ was also assumed, comparison of $h(c)$ is the same as comparison of $f(c)$.) The best fit estimates are shown as the solid lines in Figure 4.8 and the shaded bands indicate the upper and lower uncertainty bounds, all derived from the analogous strength fits. Both estimated flaw populations extend over the same crack length domain, about 20–140 μm, typical of advanced ceramics. Both flaw populations are peaked at small crack lengths adjacent to the minimum and have long tails extending to the maximum. The mode crack length (quantile of greatest probability density, the peak) and peak width are smaller for batch 1 than batch 2.

The second interpretation is that batches 1 and 2 exhibited the same flaw population, $f_1(c) = f_2(c)$ and $\lambda_1 = \lambda_2$, but that batch 1 was more fracture resistant ("tougher") than batch 2, through a variation in (ii), the fracture criterion, $B_1 > B_2$. A revised estimate of the crack length pdf $\hat{h}(c)$ from the best fit to the strength data for batch 1 is shown in Figure 4.9. The

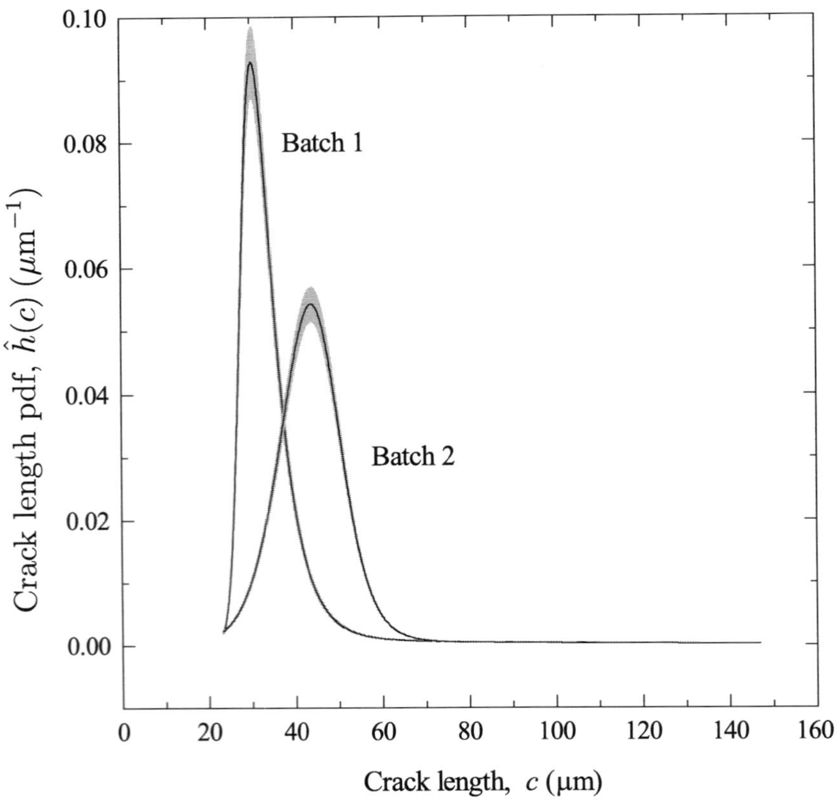

Figure 4.8 Estimated crack length pdf plots $\hat{h}(c)$ resulting from best fit to the strength data of Figure 4.7. Solid lines are best estimates and shaded bands indicate uncertainty bounds. Although both flaw populations extend over the same crack length domain and are peaked at small crack lengths with long tails extending to large crack lengths, the two populations are not the same.

estimate for batch 2 is unchanged from Figure 4.8 and is shown as the shaded band and fine line. An estimate for batch 1, using an increased value of $B = 4.8$ MPa m$^{1/2}$, not unreasonable for Si$_3$N4, was used to determine a revised $\hat{h}(c)$ variation such that the inferred flaw populations were similar. This revised estimate is shown as the bold solid line in Figure 4.9 indicating similar peak position, peak width, and extended tail for the two populations.

The third interpretation is that batches 1 and 2 contained the same distribution of flaw sizes and resistance to fracture, $f_1(c) = f_2(c)$ and $B_1 = B_2$, but that components of batch 2 contained more flaws, and therefore a greater number of large flaws, through variation in (iii), the flaw spatial density, $\lambda_2 > \lambda_1$. Recognizing that the number of flaws in a component is given by $k = \lambda V$, and that component size V was the same for both batches, enables Eq. (3.36) to be rewritten as

$$H_2(\sigma) = 1 - [1 - H_1(\sigma)]^{\lambda_2/\lambda_1}. \tag{4.6}$$

Equation 4.6 assumes that the strengths are related by a stochastic size effect, in this case involving a flaw density change rather than a physical size change. A predicted estimate of the strength edf $Pr(\sigma)$ for batch 2 from the best fit to the strength data for batch 1, using Eq. (4.6), is shown in Figure 4.10. The strength data for batch 2 are unchanged from Figure 4.7 and are shown as the open symbols. The best fit to the strength data for batch 1 is unchanged from Figure 4.7 and is shown as the dashed line. A strength estimate for batch 2, using a flaw spatial density ratio of $\lambda_2/\lambda_1 = 7.5$ was used to determine a revised $Pr(\sigma)$ variation such that the inferred median strengths were similar. This revised estimate is shown as the bold solid line in Figure 4.10. The estimate does not fit the data very well, underestimating the strengths (or overestimating the probability) in most cases.

The above analyses of the sampled Si$_3$N$_4$ component strengths make clear that batches 1 and 2, regarded as nominally identical, exhibited different fracture characteristics. It is unlikely that the third interpretation considered—that batch 2 material contained many more flaws than batch 1 material—solely explained the difference, as the strength predictions do not match very well, and, as the original researchers were surprised by the batch 1 and 2 strength differences, it is unlikely that such a material difference, a factor of >7 in the number of flaws, would have gone unnoticed. Similarly, it is also unlikely that the second interpretation considered—that batch 1 material was more fracture resistant than batch 2 material—solely explained the strength difference. Although such a difference can explain a strength difference based on a

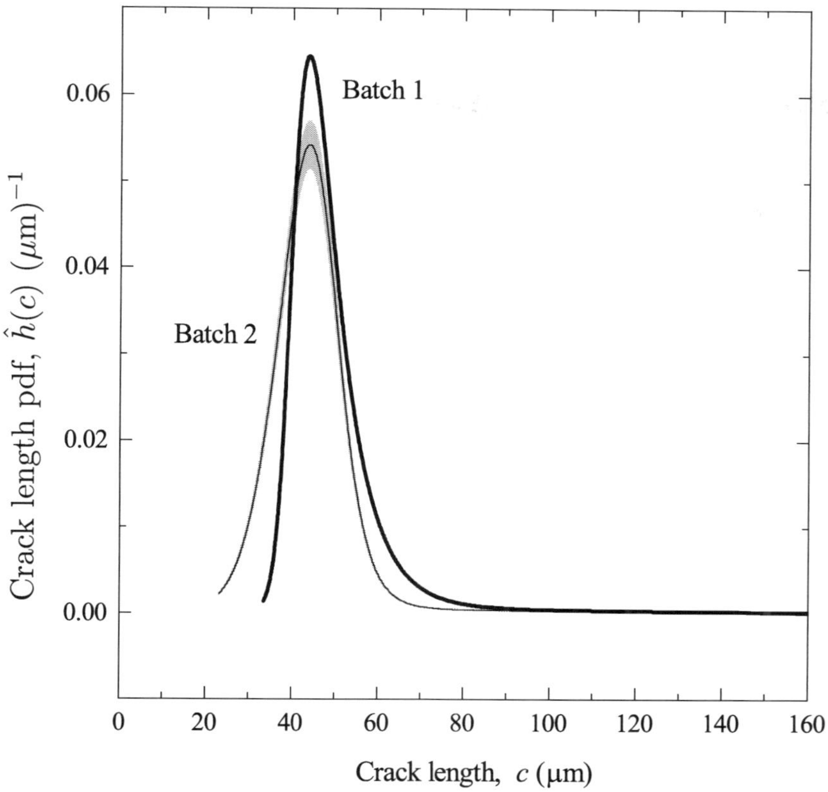

Figure 4.9 Estimated crack length pdf plots $\hat{h}(c)$ resulting from best fit to the strength data of Figure 4.7. Fine solid line and shaded band is batch 2 response from Figure 4.8. Bold solid line indicates batch 1 response in which a similar $\hat{h}(c)$ response was generated by interpreting greater strength as reflecting greater fracture resistance.

similar flaw population, it is again unlikely that that such a material difference, a factor of approximately 1.2 in toughness, would have gone unnoticed. Hence, it is likely that the strength difference is explained primarily by the first interpretation considered—the flaw distribution in the batch 1 material contained fewer large cracks than the distribution in the batch 2 material, although both distributions extended over the same domain. Such a flaw distribution change would have pass unnoticed, and could have been generated by a very small change in materials processing (such that associated flaw density and toughness changes would have also been generated as secondary effects).

Based on the means and standard deviations of strength samples from the two batches, (1) $\sigma = 689 \pm 66$ MPa and (2) $\sigma = 611 \pm 62$ MPa, the original study concluded from statistical analyses, that batches 1 and 2 exhibited different strengths, although also noting that such sample-level assessments ignored "nuisance factors" as batch 2 sample "outliers" and "skewness." The analysis here makes clear that such factors do not need to be invoked and that neither batch exhibited more outliers nor was particularly more skewed relative to the other. The analysis provides clear physical explanations in quantitative terms for the observed strength differences. The analysis also illustrated the advantages of the methodology described in Section 4.1. Based on initial functional fits to strength data, subsequent explanations in terms of crack length, crack density, or material toughness were generated simply.

Figure 4.11 shows a strength edf $Pr(\sigma)$ plot for the carbonate sand particles. The plot includes particles that exhibited substantial compliance increases both before and after peak force failure and investigating this difference in behavior via acoustic emission was the focus of the original study (Xiao et al. 2019). The focus here is to interpret the observed concave strength distribution behavior in terms of underlying strength-limiting flaws. The open symbols in Figure 4.11 represent individual strength measurements. The fine solid line represents a visual best fit to the strength data, $\hat{H}(\sigma)$ using Eqs. (4.1)–(4.5) (suppressing y_2). The shaded band represents the range of the upper and lower bound fits to the data, maintaining the fit domain bounds. The fit domain bounds were approximately those of the experimental observations, $\sigma_L \approx \sigma_1$ and $\sigma_U \approx \sigma_N$, noting that the experimental domain width $\sigma_N/\sigma_1 \approx 25$ here. The $\hat{H}(\sigma)$ formulation describes the data well and the response is clearly different from Figure 4.7.

Figure 4.12 shows as solid lines the estimated crack length pdf $\hat{h}(c)$ determined from the best fit to the strength data, using a value of $B = 0.1$ MPa m$^{1/2}$. The inset shows the behavior using an expanded ordinate; the shaded band indicates

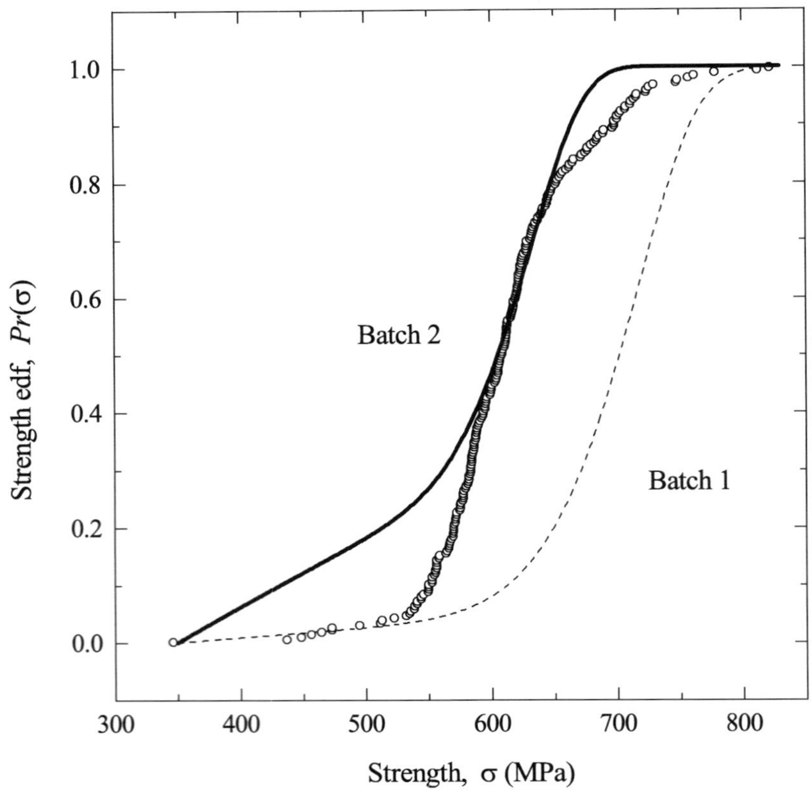

Figure 4.10 Strength edf $Pr(\sigma)$ plots for Si_3N_4 bend beams machined from two different batches of material. Dashed line represents best fit mean response for batch 1 from Figure 4.7. Open symbols represent individual strength measurements from batch 2. Solid line indicates batch 2 $Pr(\sigma)$ response generated from batch 1 (matched medians) by interpreting lesser strength as reflecting greater flaw density.

the upper and lower uncertainty bounds derived from the strength fits. The crack length pdf is very different from that of Figure 4.8. The crack length domain extends from 0.002–0.605 mm, the very wide domain reflects the very wide strength domain. The pdf is dominated by a very large value at the domain minimum that decreases sharply to a very long tail that extends to the domain maximum. The inset shows that the tail does not reach zero across the entire domain. The value of B used was selected such that the largest estimated crack, corresponding to the smallest strength, did not exceed the particle size. For comparison, the toughness of glass in moist air is approximately 0.5 MPa $m^{1/2}$ suggesting that carbonate sand particles, in this case predominantly polycrystalline coral, act as weakly bonded, perhaps porous, glass, consistent with experience. The choice of B does not affect the inferred shape of the crack length pdf, simply altering the scale. Selection of B so as to be consistent with particle size and material considerations will be seen to be a common factor throughout the book.

A final consideration in interpretation of these experimental data is classification of the tail behavior (see Section 3.4). Figures 4.13 and 4.14 show the variations of crack length ccdf $\hat{H}(c)$ for the Si_3N_4 bend beams and coral particles, respectively. The solid lines in Figure 4.13 reflect the crack length behavior of Figure 4.8. The dashed lines are exponential decreases matched to the experimental median quantiles. The solid line in Figure 4.14 reflects the crack length behavior of Figure 4.12 and Figure 4.14 uses semi-logarithmic coordinates to accommodate the wide crack length domain. The dashed line is an exponential decrease matched to the experimental median; the semi-logarithmic coordinates distort the curve. The experimental responses in Figure 4.13 are both less than the exponential behaviors for crack lengths larger than the median, classifying the Si_3N_4 flaw populations as light-tailed. The experimental response in Figure 4.14 is greater than the exponential behavior for crack lengths larger than the median, classifying the coral flaw population as heavy-tailed. For crack lengths larger than the medians in Figure 4.13, the experimental responses are significantly extended, approaching (very small) invariant values as the maximum crack length is approached. Such behavior classifies the Si_3N_4 flaw populations as long-tailed. As multi-flaw stochastic effects increase the tail weight (see Chapter 3), the responses in Figure 4.13 representing the Si_3N_4 beams are strictly lower bounds to the population response. As single-flaw behavior has been incorporated

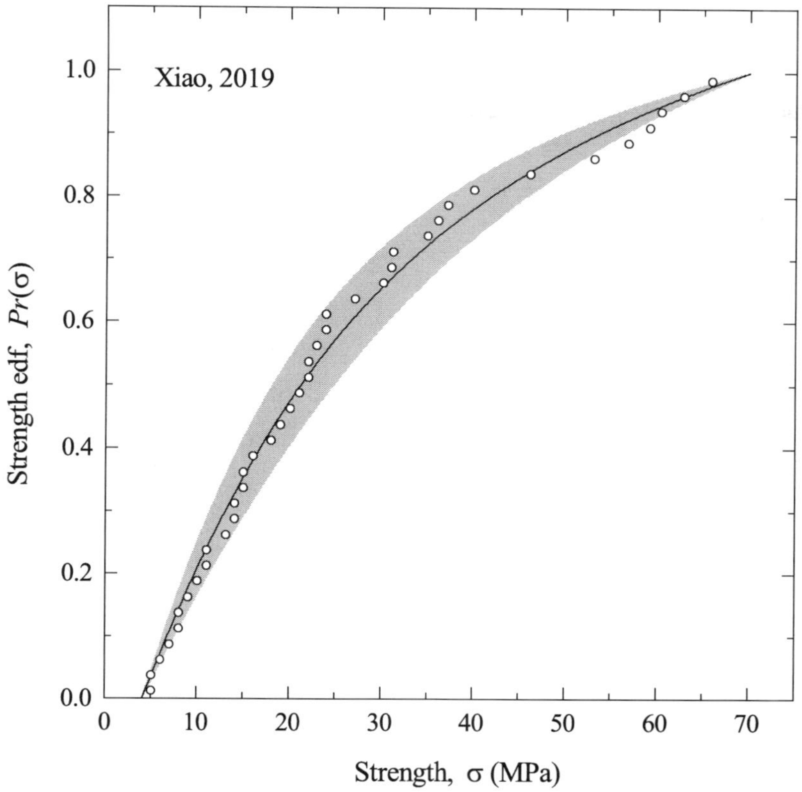

Figure 4.11 Strength edf $Pr(\sigma)$ plots for carbonate sand (coral) particles. Open symbols represent individual strength measurements. Solid line represents best fit to the strength data. Shaded band represents fit bounds. Data adapted from Xiao et al. (2019), $N_{\text{tot}} = 40$.

into the coral response through the selection of B, the response in Figure 4.14 is probably that for the population. In any event, light- and long-tailed crack length distributions are associated with sigmoidal strength behavior and heavy tailed crack length distributions are associated with concave strength behavior.

4.3 Discussion and Summary

The analysis developed and applied here in Chapter 4 enables a clear and explicit connection between an extensive measure of component mechanical performance—a component strength distribution—and an intensive characteristic of material structure—the material flaw population. The analysis was constructed within the probabilistic framework linking crack lengths of flaws and strengths of sampled components developed in Chapter 3. In contrast to Chapter 3, the framework was utilized here in reverse analysis, rather than forward analysis: estimation of underlying crack lengths from experimental strength measurements, rather than prediction of strengths from assumed flaw populations. A key step in reverse analysis is differentiation of a function characterizing the discrete experimental data. In order to reduce noise and scatter introduced by discrete differentiation, an extra step is added in reverse analysis that fits a smooth function to the experimental data prior to differentiation.

Specific elements of the analysis developed here are shown in example schematic probability distribution and density function plots of Figure 4.15; the reverse analysis path, from observed component behavior to estimated material characteristic is indicated by the diagonal gray arrow. All functions are drawn in absolute coordinates extending over the experimental domains. A supplementary plot, unused in direct analysis, is shown shaded gray. Analysis begins with a discrete description of the component strength sample $Pr(\sigma)$, bottom right (Figure 4.15a), using as an example here edf data from the above Si_3N_4 batch 2. Fitting of $Pr(\sigma)$, here by the flexible interpolated tri-linear function, Eqs. (4.1)–(4.5), leads to mean and upper and lower bound continuum estimates of the data, $\hat{H}(\sigma)$, center, Figure 4.15b. Two substitutions, strengths by crack lengths and probabilities by their complements, lead to the crack length cdf, $\hat{H}(c)$, not shown. Differentiation of $\hat{H}(c)$ leads to the sought sample crack length pdf, $\hat{h}(c)$, top left, Figure 4.15c. Extension of the sample analysis to estimate the population $f(c)$

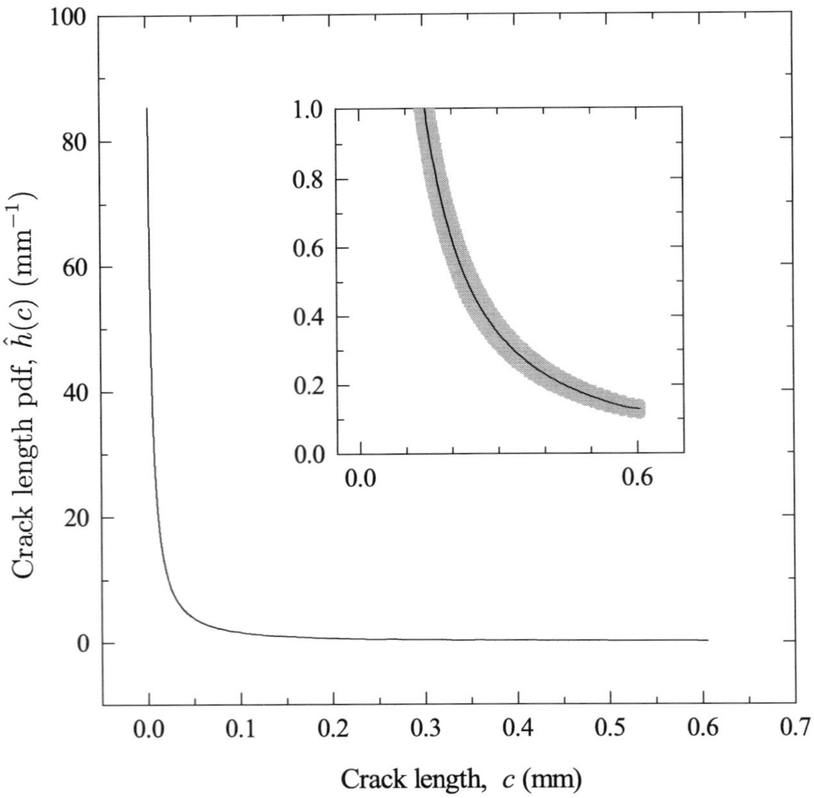

Figure 4.12 Estimated crack length pdf plots $\hat{h}(c)$ resulting from best fit to the strength data of Figure 4.11. Solid lines are best estimates and shaded band indicates uncertainty bounds. The inset indicates that the tail behavior does not reach zero over the extent of the crack length domain.

from $\hat{h}(c)$ requires knowledge or assumption of the nature of size effects determining the extreme flaws in components and the component size, expressed as a multiple k of the fundamental material volume element. An ancillary plot, the crack length ccdf $\bar{H}(c)$, top center, Figure 4.15d, is easily determined from $\hat{H}(c)$ and enables assessment and classification of the tail behavior of the sample, and potentially the population, crack length distribution. The example in Figure 4.15 is a long- and light-tailed crack length distribution, shown here to lead to sigmoidal strength distributions.

A critical factor in establishing the reverse analysis outlined here was method verification—checking that the method was implemented correctly. The method verification procedure used consisted of two sequential steps. First, as in Chapter 3, a forward analysis was carried out using an input population crack length pdf $f(c)$ to generate an output sample strength edf $Pr(\sigma)$. Second, a reverse analysis using this edf as input was then carried out to generate as output an estimated crack length pdf $\hat{h}(c)$. The verification procedure was thus implementation of Figure 3.20 followed implementation of Figure 4.15. Positive agreement and verification was shown between $f(c)$ and $\hat{h}(c)$ for two different crack length populations, light-tailed and heavy-tailed, that generated intermediate sigmoidal and concave $Pr(\sigma)$ sample strength distributions, respectively. In both cases, locations and widths of peaks in the input crack length probability densities were reproduced by the means and bounds of the output estimates. Thus the analysis method outlined in Section 4.1 was verified.

Another important factor in establishing applications of analyses is method validation—checking that a method generates the correct output. In this case, method validation consists of implementing reverse analysis on experimental strength measurements and generating an estimate of the underlying crack length distribution consistent with experimental observations. The analysis step in method validation was demonstrated here. Contrasting experimental strength measurements, of Si_3N_4 beams and coral particles, were analysed and crack length distributions estimated. As expected from the verification process, application of the methodology to experimental strength data was straightforward, independent of the shape of the experimental strength distribution. Estimates typical for these systems of the material fracture parameter B and component size k generated values of c consistent with expectations: tens of μm for Si_3N_4 and sub mm for coral. More complete validation requires independent experimental measurements of $Pr(\sigma)$, $Pr(c)$, B, and k, in addition to application of analysis. The complete set of experimental measurements is usually not available. The two major studies that demonstrated overall

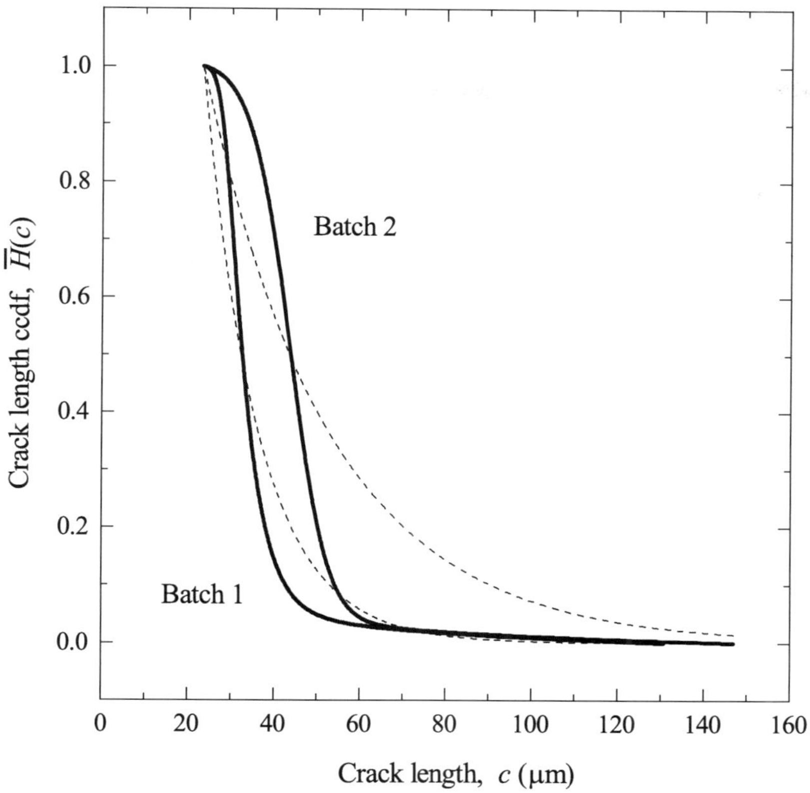

Figure 4.13 Variations of crack length ccdf $\hat{H}(c)$ for Si_3N_4 bend beams. The solid lines reflect the crack length behavior of Figure 4.8. The dashed lines are exponential decreases matched to the experimental medians. The small values of $\hat{H}(c)$ relative to the exponential trends at large crack lengths indicate that the crack populations is light-tailed.

validation were involved and extensive, considering cracked pores in Si_3N_4 (Chao and Shetty 1992) and grain boundary grooves in polycrystalline Si (Cook et al. 2019). In both cases, despite careful observations and measurements, the authors identified accurate evaluation of the flaw distribution factors $Pr(c)$ and k as the most uncertain step in overall validation and, as here, this is usually the case. However, as demonstrated here, bounds on crack lengths expressed as $Pr(c)$ can be established via the explicit dependencies of B and k incorporated into the analytical framework.

It is notable that here and in the earlier works (Chao and Shetty 1992; Cook et al. 2019), smoothing fit functions were constructed and implemented based on ability to describe experimental data, rather than based on expression as accepted forms. For example, the expression used by Chao and Shetty (1992) to describe the pore size distribution $Pr(R)$, in Si_3N_4 was (current notation)

$$\hat{H}(R) = 1 - \exp\left[-\exp\left(b_1 + b_2 \ln R + b_3(\ln R)^2 + b_4(\ln R)^3\right)\right], \tag{4.7}$$

where b_i are fitting parameters. In the words of the authors, "The function did give a better fit to the pore-size distribution .... This was particularly true in the asymptotic large-pore-size regime of the distribution ... ." The domain of R was unbounded $[0, \infty]$. Similarly, the expression used to describe the relative strength distribution in Si and other brittle materials (Cook and DelRio 2019b; Cook et al. 2019, 2021) was

$$\hat{H}(\hat{\mu}) = 30\left(\hat{\mu}^{3p}/3 - \hat{\mu}^{4p}/2 + \hat{\mu}^{5p}/5\right), \tag{4.8}$$

where p is a fitting parameter. The domain of $\hat{\mu}$ was bounded [0,1]. In its simplest form, the concave (y_2 suppressed) expression for the interpolated relative strength distribution used above is

$$\hat{H}(\hat{\mu}) = \left[(a_1\hat{\mu})^p + \left(a_3\hat{\mu} + (1 - a_3)\right)^p\right]^{1/p}, \tag{4.9}$$

where a_i and p are fitting parameters. The domain of $\hat{\mu}$ is bounded [0,1]. Although all three functions in Eqs. (4.7)–(4.9) involve polynomial-based non-linearity, and might be regarded as "uncommon," once fit to experimental data all three

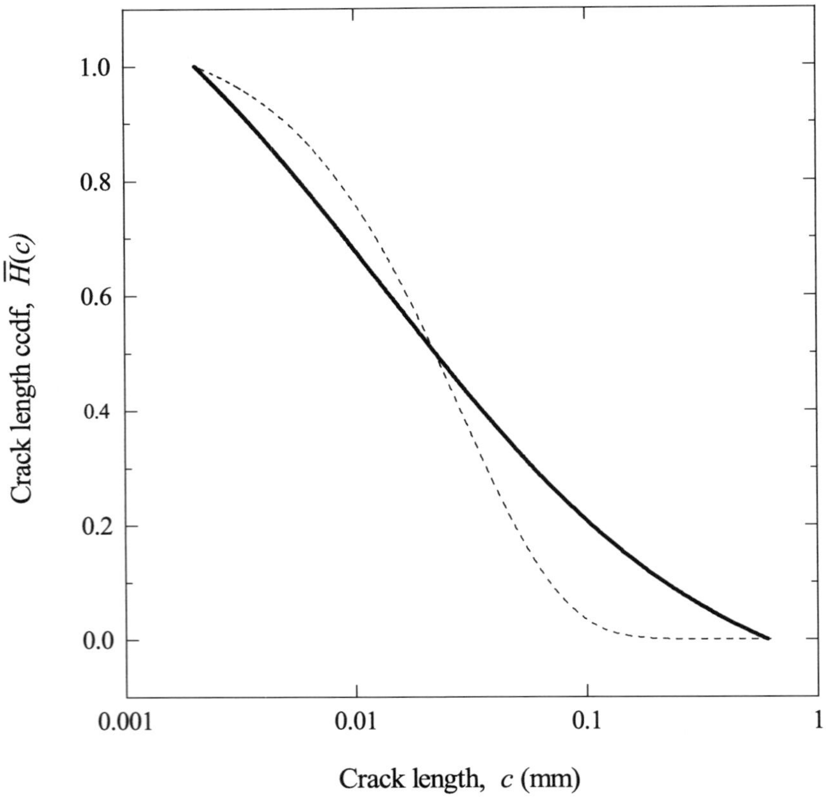

Figure 4.14 Variations of crack length ccdf $\hat{H}(c)$ for coral particles. The solid line reflects the crack length behavior of Figure 4.12. The dashed line is an exponential decrease matched to the experimental median. The large values of $\hat{H}(c)$ relative to the exponential trend at large crack lengths indicate that the crack population is heavy-tailed.

are easily differentiated and integrated numerically in data analysis. Each has advantages relative to describing observed behavior: Eq. (4.7) is designed to describe wide domains of observed pore sizes and hence the expansion in $\ln R$; Eq. (4.8) is designed to describe asymmetric sigmoids, and has the feature, independent of symmetry, that $\mathrm{d}\hat{H}/\mathrm{d}\hat{\mu} = 0$ at the domain bounds; and Eq. (4.9) and the related sigmoidal form are designed to describe potentially heavy-tailed distributions in which $\mathrm{d}\hat{H}/\mathrm{d}\hat{\mu}$ values are specified at the domain bounds (see also Cook (2021) for a quadratic variation). It is also notable that although some fit parameters (e.g. b_i, p) were given in these works, very little attention was given to the values of the parameters or the functions in which they were used. Attention in these works was focused on the fidelity of a fit to the considered data and subsequent application in method validation.

In contrast, simply expressed functions, that are usually easily linearized, are commonly fit to strength edf $Pr(\sigma)$ data and fit parameters provided as empirical descriptors. Most such functions are sigmoidal and unbounded, and, in simplest form, include

$$
\begin{aligned}
&H(x) = 1 - \exp(-x/b) && \text{(a) (Exponential)}\\
&H(x) = 1 - \exp(-x^2/b) && \text{(b) (Rayleigh)}\\
&H(x) = 1 - \exp(-x^m/b) && \text{(c) (Weibull)}\\
&H(x) = \exp(-x^{-m}/b) && \text{(d) (Fréchet)}\\
&H(x) = 1 - \exp(-\exp(x/b)) && \text{(e) (Gumbel)}\\
&H(x) = \frac{1}{1 + \exp(-x/b)} && \text{(f) (Logistic)}\\
&H(x) = \mathrm{erf}(x/b) && \text{(g) (Error)}\\
&H(x) = 3x^2 - 2x^3 && \text{(h) (Bounded cubic)}
\end{aligned}
\qquad (4.10)
$$

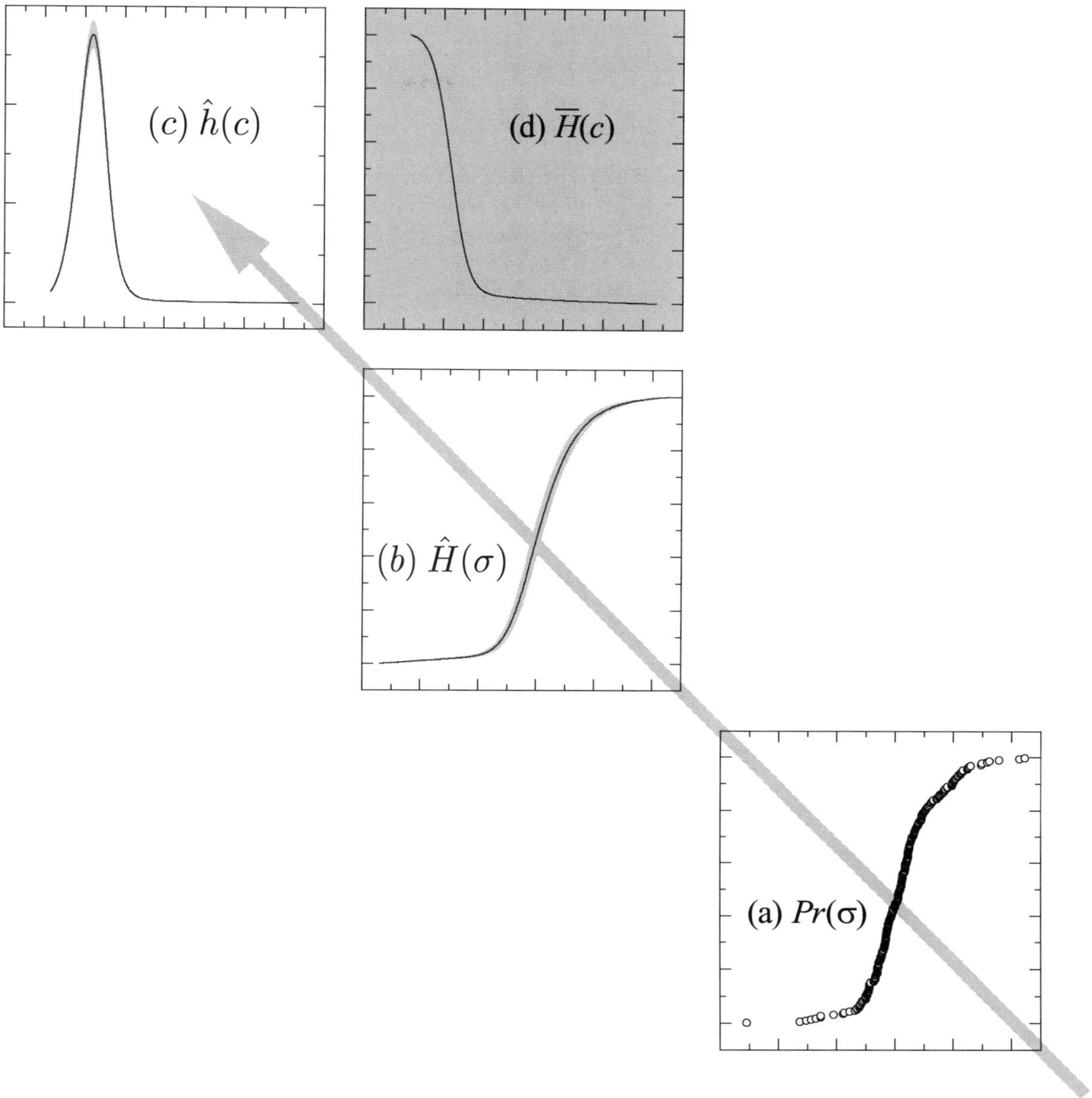

Figure 4.15 Schematic diagram illustrating reverse analysis within a probabilistic framework to estimate material flaw populations from component strength distributions. The arrow indicates the direction of analysis (a) Strength empirical distribution function, edf, $Pr(\sigma)$ (b) Fitted smooth strength cumulative distribution function, cdf, $\hat{H}(\sigma)$ (c) Estimated crack length probability density function, pdf, $\hat{h}(c)$ (d) Complementary crack length cumulative distribution function, ccdf, $\bar{H}(c)$.

The argument x is usually strength and all can be shifted to describe a finite lower bound by $x \to x - a$ and transformed to describe non-linear behavior by $x \to x^m$ or $x \to \ln x$. There are also trigonometric descriptions. More details can be found elsewhere (Johnson et al. 1994; Forbes et al. 2011; Krishnamoorthy 2016; Wikipedia 2022). Examination of Eq. (4.10) shows many interrelationships: the Rayleigh and Weibull distributions are nonlinear extensions of the Exponential distribution; the Fréchet and Weibull distributions are inverses; the Rayleigh, Weibull, and Gumbel distributions are all simply stretched or powered exponentials; and Eqs. (4.7) and (4.8) are non-linear extensions of the Gumbel and Bounded Cubic distributions, respectively. In works considering the related particle and fragment size distributions Wohletz et al. (1989) and Brown and Wohletz (1995) provide brief historical reviews of distribution development and note that the "Weibull" distribution (Weibull 1939, 1951) was preceded by some years in the work of Rosin and Rammler that used a similar distribution (to describe particles, 1933). A detailed historical review of distribution development and naming post Rosin and Rammler is provided by Stoyan (2013), who notes that a neutral and historically fairer term for Eq. (4.10c) might be that suggested by Rammler, the "powered exponential distribution." Such a distribution had been the subject of earlier work by Fréchet (on extreme values, 1927). Subsequent work has shown that the Fréchet, Weibull, and Gumbel distributions are the three limiting asymptotic forms (sample size $N \rightarrow \infty$) of extreme value distributions (Castillo 1988; Hansen 2020). Equation (4.10c) is also the form of the "Avrami" equation developed in 1937 and 1939 by a number of researchers to describe phase transformations (Cantor 2020).

Despite many recent comparative studies (e.g. Rozenblat et al. 2011; R'Mili et al. 2012; Bertalan 2014; Cavalcanti and Tavares 2018), no real definitive choice can be made between the sigmoidal forms in Eq. (4.10). To paraphrase Vesilind (1980), from the context of the sigmoid in particle size distributions, "None of these functions [in Eq. (4.10)], it should be emphasized, is 'correct'. Some simply describe the shape of [a strength] distribution curve better than others." The diversity of sigmoidal responses presented to this point (Chapters 2–4), supports this view, as only the predominant and necessary central increase is in common. To quote Smith (1982), from the context of the sigmoid in materials structure, "Both the beginning and the end depend on highly localized conditions and are unpredictable in detail." From the practical point of view of this Chapter, the truth of both of these statements leads to the lack of suitability of any of the expressions in Eq. (4.10) as smoothing fits to $Pr(\sigma)$ strength data. The fixed functional forms are too constraining to provide broadly applicable fits, all require shifts to accommodate non-zero strength thresholds, all do not handle non-zero slopes well at domain bounds, and the lack of upper bounds impedes fitting. The flexible tri-linear interpolated form of Eqs. (4.1)–(4.5) overcomes these limitations and hence will be used throughout.

Nevertheless, in the intervening decades since the contributions of Weibull (1939) and Epstein (1948a, 1948b) to considerations of probability in strength distributions, failure analyses have been overshadowed by the myriad applications of the powered exponential function popularized by Weibull (1951). The powered exponential function provides for simple arithmetic description of distributions arising in many phenomena, including groups of component strengths, and has been incorrectly associated with size effects and flaw independence, often termed weakest link behavior, in strength measurements. The two-parameter powered exponential function of Eq. (4.10) is easily linearized, leading to a common method of transforming measured strength data for presentation, in most cases severely obscuring the underlying strength distribution and providing the completely misleading impression that the function has a physical basis. In addition, the linearized fits are usually poor. However, the majority, but not all, of the particle strength distribution data included in this book were published in powered exponential function transformed coordinates, although other transformed coordinates were sometimes used, e.g. log normal used by the King and Tavares groups (King and Bourgeois 1993; Tavares and King 1998). The transformations rendered comparisons between distributions, comparisons with analyses, and assessments of stochastic or other size effects almost impossible. To enable the book and allow for direct comparisons, the data here are presented in uniform formats of unbiassed edf plots of observed strengths. In all cases, the data were obtained by extensive manual digitization of published figures and post-digitization analyses, including "back" transformation, conversion from particle failure force or failure energy units, and conversion from particle mass units. In most of the particle research considered here, strength distributions were presented as the object of study and the effects of particle size were noted. In very few studies were the widths and forms of the strength distributions discussed. In almost no case was there quantitative analysis of the strength distributions and the underlying strength-controlling flaw populations.

The analysis and methodology developed in this chapter will be applied throughout the book to strength distributions of particles, Chapters 6–12. A primary goal in application of the analysis is determination of the distributions of cracks and flaws that determine the strengths of particles loaded in diametral compression (Figure 1.2), and determination of the effects of particle material and size on such distributions. In order to provide foundation and context for the particle analyses, the following chapter applies the analysis and methodology in a survey of strength distributions of components in the common structural loading configurations of tension, compression, and, particularly bending, Figure 1.1. Such components are often large and tensile stress generated by loading usually extends throughout the component.

References

Bertalan, Z., Shekhawat, A., Sethna, J.P., and Zapperi, S. (2014). Fracture strength: Stress concentration, extreme value statistics, and the fate of the Weibull distribution. *Physical Review Applied* 2: 034008.

Brown, W.K. and Wohletz, K.H. (1995). Derivation of the Weibull distribution based on physical principles and its connection to the Rosin-Rammler and lognormal distributions. *Journal of Applied Physics* 78: 2758–2763.

Cantor, B. (2020). *The Equations of Materials.* Oxford.

Castillo, E. (1988). *Extreme Value Theory in Engineering.* Academic Press.

Cavalcanti, P.P. and Tavares, L.M. (2018). Statistical analysis of fracture characteristics of industrial iron ore pellets. *Powder Technology* 325: 659–668.

Chao, L.-K. and Shetty, D.K. (1992). Extreme-value statistics analysis of fracture strengths of a sintered silicon nitride failing from pores. *Journal of the American Ceramic Society* 75: 2116–2124.

Cook, R.F. (2017). Fracture mechanics of sharp scratch strength of polycrystalline alumina. *Journal of the American Ceramic Society* 100: 1146–1160.

Cook, R.F., Boyce, B.L., Friedman, L.H., and DelRio, F.W. (2021). High-throughput bend-strengths of ultra-small polysilicon MEMS components. *Applied Physics Letters* 118: 201601.

Cook, R.F. and DelRio, F.W.(2019a). Material flaw populations and component strength distributions in the context of the Weibull function. *Experimental Mechanics* 59: 279–293.

Cook, R.F. and DelRio, F.W. (2019b). Determination of ceramic flaw populations from component strengths. *Journal of the American Ceramic Society* 102: 4794–4808. (typographical error in Eq. (9)).

Cook, R.F., DelRio, F.W., and Boyce, B. L. (2019). Predicting strength distributions of MEMS structures using flaw size and spatial density. *Microsystems & Nanoengineering* 5: 1–12.

Epstein, B. (1948a). Statistical aspects of fracture problems. *Journal of Applied Physics* 19: 140–147.

Epstein, B. (1948b). Application of the theory of extreme values in fracture problems. *Journal of the American Statistical Association* 43: 403–412.

Forbes, C., Evans, M., Hastings, N., and Peacock, B. (2011). *Statistical Distributions, 4th ed.* Wiley.

Griffith, A.A. (1921). The phenomena of rupture and flow in solids. *Philosophical Transactions of the Royal Society London A* 221: 163–198.

Hansen, A. (2020). The three extreme value distributions: An introductory review. *Frontiers in Physics* https://doi.org/10.3389/fphy.2020.604053 (accessed February 4, 2022).

Inglis, C.E. (1913). Stress in a plate due to the presence of cracks and sharp corners. *Transactions of the Institution of Naval Architects* 55: 219–241.

Johnson N.L., Kotz S., and Balakrishnan N. (1994). *Continuous Univariate Distributions volume 1, 2nd ed.* Wiley.

King, R.P. and Bourgeois, F. (1993). Measurement of fracture energy during single-particle fracture. *Minerals Engineering* 6: 353–367.

Krishnamoorthy, K. (2016). *Handbook of Statistical Distributions with Applications, 2nd ed.* Taylor and Francis.

Lawn, B.R. (1993). *Fracture of Brittle Solids, 2nd ed.* Cambridge.

NIST (1996). *Ceramic Strength*. https://www.itl.nist.gov/div898/handbook/eda/section4/eda42a.htm (accessed January 29, 2022).

Quinn, G.D., Sparenberg, B.T., Koshy, P., Ives, L.K., Jahanmir, S., and Arola, D.D. (2009). Flexural strength of ceramic and glass rods. *Journal of Testing and Evaluation* 37: 1–23.

R'Mili, M., Godin, N., and Lamon, J. (2012). Flaw strength distributions and statistical parameters for ceramic fibers: The normal distribution. *Physical Review E* 85: 051106.

Rozenblat, Y., Portnikov, D., Levy, A., Kalman, H., Aman, S., and Tomas, J. (2011). Strength distribution of particles under compression. *Powder Technology* 208: 215–224.

Smith, C.S. (1982). *A Search for Structure*. MIT Press.

Spiegel, M.R. (1961). *Statistics.* McGraw-Hill.

Stoyan, D. (2013). Weibull, RRSB or extreme-value theorists? *Metrika* 76: 153–159.

Tavares, L.M. and King, R.P. (1998). Single-particle fracture under impact loading. *International Journal of Mineral Processing* 54: 1–28.

Vesilind, P.A. (1980). The Rosin-Rammler particle size distribution. *Resource Recovery and Conservation* 5: 275–277.

Walpole, R.E. and Myers, R.H. (1972). *Probability and Statistics for Engineers and Scientists.* Macmillan.

Wikipedia (2022). Sigmoid function. https://en.wikipedia.org/wiki/Sigmoid_function (accessed February 4, 2022).

Wohletz, K.H., Sheridan, M.F., and Brown, W.K. (1989). Particle size distributions and the sequential fragmentation/transport theory applied to volcanic ash. *Journal of Geophysical Research: Solid Earth* 94: 15703–15721.

Xiao, Y., Wang, L., Jiang, X., Evans, T.M., Stuedlein, A.W., and Liu, H. (2019). Acoustic emission and force drop in grain crushing of carbonate sands. *Journal of Geotechnical and Geoenvironmental Engineering* 145: 04019057.

5

Survey of Extended Component Strength Distributions

This chapter uses the analysis of Chapter 4 in a survey of the strength distributions observed for extended components, including variations in component material, geometry, and size, and interprets the distributions in terms of underlying flaw distributions. The chapter is a background to the survey of strength distributions observed for particles in the following Chapter 6. Extended components include loaded tensile bars, fibers, and films, compressive columns, bend beams and flexed plates, and pressurized containers. The materials forming these components are predominantly structural ceramics and glasses. Emphasized concepts include the common sigmoidal strength response and strength threshold and stochastic and deterministic size effects.

5.1 Introduction

Development and application of analyses for determination of crack length distributions underlying strength distributions of brittle components were considered in detail in Chapter 4. The analyses were applied to two sources of strength data: computationally generated data, in verification of the methodology outlined, and experimentally measured data, in demonstration of the capabilities of the methodology. In both cases, contrasting sets of strength distribution data were considered: a sigmoidal distribution, typical of "conventional" structural components in extended form and a concave distribution, typical of the subject of this book, structural components in particle form. In this chapter, attention is focused on the brittle fracture strength and crack length behavior of *extended components*. Extended components are those that contain many strength-controlling flaws and associated cracks that are exposed to significant stress on component loading. Typically, such components are physically large, such as beams and columns in buildings, plates and bottles for domestic use, "chips" and their substrates in microelectronic devices, wear resistant components for high temperature operation in engines, and long lengths of reinforcing fibers in composites. As such components are large, the number of fundamental volume elements, and thus cracks, contained in each is large. However, if the volume elements, and cracks, are small, physically small components can also be "extended." Examples include micromachined MEMS components and the transverse dimensions of fibers. Physically limited volume elements can also characterize extended components, as in flaws/area on substrate surfaces and flaws/length in fibers. The major implication of the term extended in consideration of strength distributions of extended components is that the strength distributions are likely to be controlled by stochastic extreme value size effects.

This chapter surveys the experimental observations of strength, and thus crack length, distributions of extended components—observations that extend back almost 100 years. The observations here are presented in order of approximate increasing complexity. The next section considers the behavior of typical brittle materials, primarily loaded in bending, the most common loading geometry for brittle materials: glasses (borosilicate and soda-lime silicate), ceramics (alumina, silicon nitride, and porcelain), and semiconductors (single crystal and polycrystalline silicon). Brittle materials loaded in the less common configurations of tension, compression, and flexure are also considered. This is followed by two sections considering size effects on strength distributions of extended brittle components. First, stochastic size effects, the most common, and second, less commonly observed, deterministic size effects. All $Pr(\sigma)$ strength edf data will be fit by the tri-linear interpolated continuum form described by Eqs. (4.1)–(4.5). The strength fits will be deconvoluted into crack length pdf variations as described in Chapter 4. The crack length pdf conjugate to a strength fit will be notated simply as $h(c)$, omitting

Particle Strengths: Extreme Value Distributions in Fracture, First Edition. Robert F. Cook.

the caret notation ($\hat{x}$), as here and following all crack length pdf variations are to be understood as best fit estimates generated from experimental data. All fitting and deconvolution steps were performed computationally. Unless required for discussion, parameters associated with intermediate strength fitting steps and bounds associated with strength and crack length probability uncertainties are not given.

5.2 Materials and Loading Survey

5.2.1 Glass, Bending and Pressure Loading

Figure 5.1 shows strength edf $Pr(\sigma)$ plots for two sets of glass strength measurements. Figure 5.1a shows the behavior of the pressure loaded (probably soda-lime silicate) glass bottles from the work of Preston (1937), Figure 2.28, with failure pressure converted to wall strength using a bottle radius/wall thickness ratio of 10. Figure 5.1b shows the behavior of borosilicate glass rods measured in bending from the work of Phani and De (1987), using data derived from the published work. (For ease of visualization here and following, author details and details of component and sample sizes, dimensions and numbers, respectively, are provided in the figure caption; note that N_{tot} and N are often different.) To assess and compare $Pr(\sigma)$ for each material easily and clearly the plots are provided in an unbiassed format, to be used throughout: The data are presented in linear coordinates, the abscissa extends from approximately the minimum observed strength, σ_1 to approximately the maximum observed strength, σ_N, and the ordinate extends from −0.1 to 1.1. The abscissa plotting scheme, starting at strength σ_1, enables consideration of trends and fits of experimental data for a single system; the plotting scheme in Figure 2.28, starting at strength 0, enables consideration of relative strength domains in analysis of multiple systems (Cook 2020). Symbols represent individual strength measurements. Lines represent visual best fits. (Here and throughout, Eqs. (4.1)–(4.5) are used in fitting. Eq. (4.8), used in earlier works, provides almost the same information, but is slightly less flexible.) In Figure 5.1, both sets of data are well described by sigmoidal behavior, although there is a greater tendency to nonzero $\mathrm{d}Pr(\sigma)/\mathrm{d}\sigma$ derivative behavior at the data bounds in Figure 5.1b compared with Figure 5.1a. Both sets of data exhibit strength domain widths σ_N/σ_1 of approximately 2.5. Detailed information regarding glass compositions can be found elsewhere (Kingery et al. 1975).

Figure 5.2 shows the crack length pdf $h(c)$ variations underlying the glass strength behavior, determined using $B = 0.5$ MPa $\mathrm{m}^{1/2}$, typical for silicate glasses in moist air. The $h(c)$ variations are shown in two ways. In Figure 5.2a, linear coordinates are used, making clear that the glass bottles of Figure 5.1a contained strength-controlling cracks much larger than the glass rods of Figure 5.1b. The curves are labeled (i) and (ii), respectively, and the area under each curve is 1. Consideration of Figure 5.1 shows that the strength distributions do not overlap. As a common B value was used for both sets of components, the crack length populations also do not overlap. Consideration of Figure 5.1 also shows that the strength distributions are of similar shape, implying that the crack length populations should also be of similar shape, a finding difficult to assess in

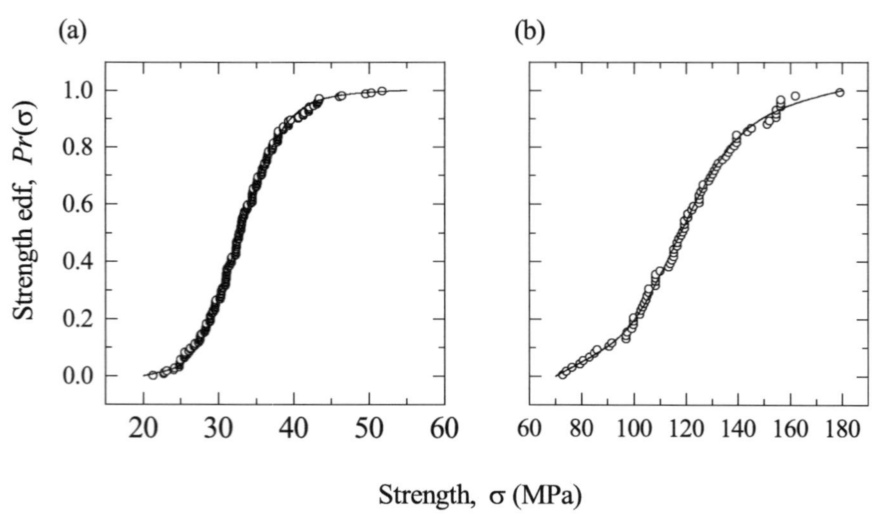

Figure 5.1 Plots of strength edf behavior, $Pr(\sigma)$, for glass components. (a) Commercially manufactured bottles, burst test under pressure loading, number of components $N = 200$ (Adapted from Preston, F.W 1937), tests by Hunter, see Figure 2.28. (b) Glass rods, 90 mm long × 4 mm diameter, $N = 80$ (Adapted from Phani, K.K et al. 1987).

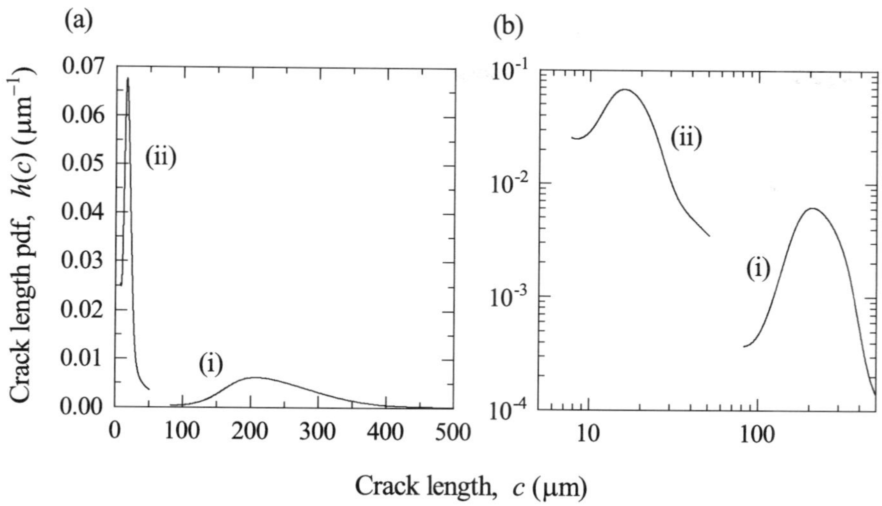

Figure 5.2 Plots of crack length pdf variations, $h(c)$, for glass components; determined from strength measurements of Figure 5.1. (a) Plot in linear coordinates. (b) Plot in logarithmic coordinates. Label (i) corresponds to Figure 5.1a, label (ii) corresponds to Figure 5.1b.

Figure 5.2a. In Figure 5.2b, the $h(c)$ variations are replotted, using logarithmic coordinates and labeled (i) and (ii) as before. The similarity of shape in the two responses is now apparent, as is the characteristic width of the crack length populations of a factor of approximately 6, consistent with the width of the strength domains, ≈ 2.5, and the Griffith equation, $c = (B/\sigma)^2$. Representations of crack length pdf $h(c)$ variations in logarithmic coordinates will be used throughout the book.

5.2.2 Alumina, Bending Loading

Figure 5.3 shows strength edf $Pr(\sigma)$ plots for four sets of polycrystalline alumina (Al_2O_3) strength measurements, using data derived from the published works cited. Figure 5.3a shows the behavior of tubular beams fabricated from a porous custom material, from the work of de Wit et al. (2017). Figure 5.3b and c show the behavior of rectangular section beams fabricated from commercial materials from the works of Quinn (1989), and Cook and DelRio (2019b), respectively; The observations of Quinn (1987) are similar to Figure 5.3c. Figure 5.3d shows the behavior of cylinders and plates fabricated from commercial material and loaded in biaxial flexure from the work of Simpatico et al. (1999). Symbols represent individual strength measurements. Lines represent visual best fits. All sets of data are well described by sigmoidal behavior, although there is a clear tendency to non-zero $\mathrm{d}Pr(\sigma)/\mathrm{d}\sigma$ derivative behavior at the data bounds. All sets of data exhibit strength domain widths σ_N/σ_1 of approximately 2.2. Detailed information regarding alumina microstructures can be found elsewhere (Kingery et al. 1975).

Figure 5.4 shows the crack length pdf $h(c)$ variations underlying the alumina strength behavior. As in Figure 5.2, the $h(c)$ variations are shown in two ways. In Figure 5.4a linear coordinates are used, in Figure 5.4b logarithmic coordinates are used. The curves are labeled (i)–(iv), corresponding to Figure 5.3a–d, respectively, and the area under each curve is 1. The B values used were (all in MPa m$^{1/2}$) (i) 1.0, (ii) and (iii) 3.0, and (iv) 3.5, taking into account the material microstructures. Consideration of Figures 5.3a and 5.3b shows that the strength distributions do not overlap. In this case, as the B values used were reversed in magnitude relative to the strengths, the crack length curves, (i) and (ii), overlap considerably. The physical interpretation is that although materials (i) and (ii) contained strength-limiting cracks of approximately the same size, the porosity in material (i) lead to decreased fracture resistance, and thence strengths, relative to material (ii). The strength-controlling flaws in (i) were likely bulk macroscopic pores (de Wit et al. 2017) and those in (ii) were likely surface and edge machining damage (Quinn 1989). Consideration of Figures 5.3b–d shows that the strength distributions barely overlap. In this case, as the B values used were similar, the crack length curves, (ii)–(iv), also barely overlap. The strength differences for these materials are simply and conventionally interpreted as differences in flaw size distributions. As in Figure 5.2b, the similarity in the shapes of of the crack length populations are made much more clearly evident through the use of logarithmic coordinates, as are the characteristic widths of the crack length populations of factors of approximately 5.

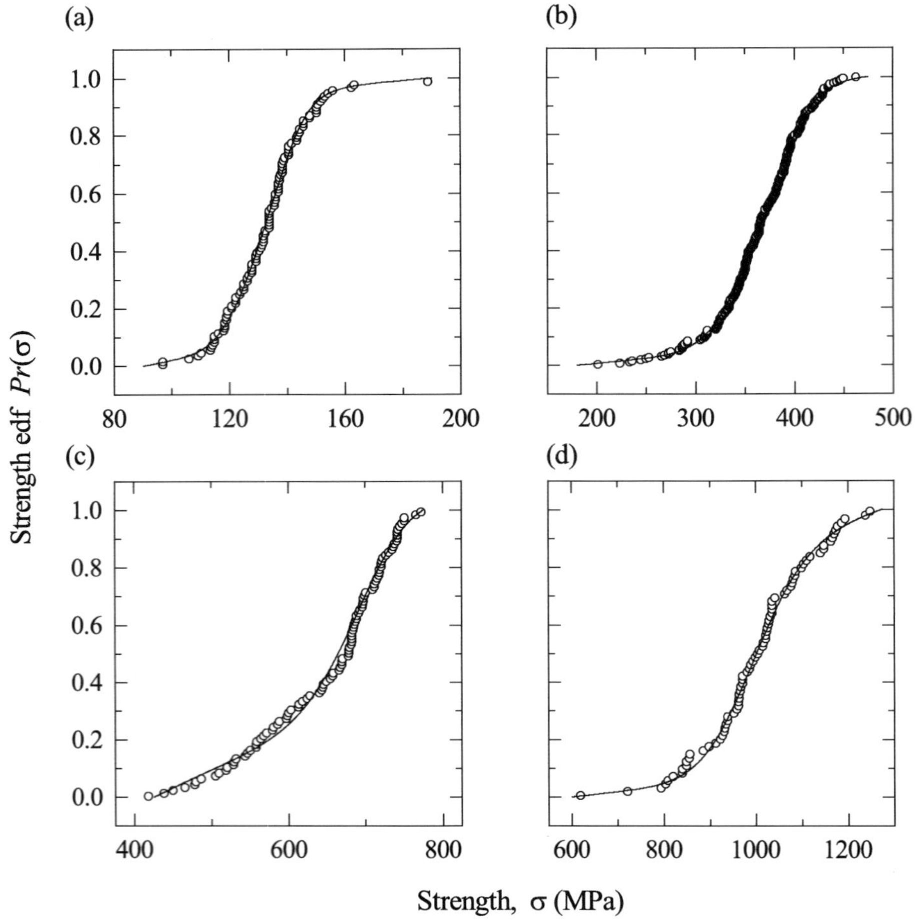

Figure 5.3 Plots of strength edf behavior, $Pr(\sigma)$, for alumina components. (a) Porous alumina tubular beams, 40 mm long × 1.5 mm diameter, N = 103 (Adapted from de Wit, P et al. 2017). (b) Alumina beams 50 mm long × 4 mm × 3 mm, N = 253 (Adapted from Quinn, G.D 1989). (c) Alumina beams, 32 mm long × 8 mm × 0.65 mm, N = 100 (Adapted from Cook, R.F et al. 2019b). (d) Alumina cylinders 20 mm diameter, or plates 20 mm × 20 mm edges, both 0.635 mm thick, N = 77 (Adapted from Simpatico, A et al. 1999).

5.2.3 Silicon Nitride, Bending Loading

Figure 5.5 shows strength edf $Pr(\sigma)$ plots for four sets of polycrystalline silicon nitride (Si_3N_4) strength measurements, using data derived from the published works cited. Figure 5.5a shows the behavior of rectangular section beams fabricated as a reaction bonded custom material, from the work of Quinn (1989). Figure 5.5b shows the behavior of rectangular section beams fabricated as a hot pressed commercial material, from the work of Easler et al. (1982); the observations of Evans and Jones (1978) are similar. Figure 5.5c and Figure 5.5d show the behavior of (probably) rectangular section beams fabricated as (probably) hot pressed materials, from the works of NIST (1996) (batch 2 in Figure 4.07) and Usami (Ito, as reported by Usami 1986), respectively (the strength values suggest the probable shapes and processing). Symbols represent individual strength measurements. Lines represent visual best fits. All sets of data, except Figure 5.5d, are well described by sigmoidal behavior, although there are clear tendencies to *convex* behavior (positive curvature) in Figures 5.5c and 5.5d. All sets of data exhibit strength domain widths σ_N/σ_1 of approximately 2. Detailed information regarding ceramic processing can be found elsewhere (Kingery et al. 1975).

Using logarithmic coordinates, Figure 5.6 shows the crack length pdf $h(c)$ variations underlying the silicon nitride strength behavior. The curves are labeled (i)–(iv), corresponding to Figures 5.5a–d, respectively, and the area under each curve is 1. The B values used were (all in MPa $m^{1/2}$) (i) 3.0, (ii) 3.5, and (iii) and (iv) 4.0, taking into account the material microstructures. Consideration of Figure 5.5 shows that the strength distributions overlap sequentially and as similar B values were used for all sets of components, the crack length populations also overlap sequentially. The strength-controlling flaws in (i) were likely the reported un-reacted zones of silicon (Quinn 1989) and in (ii)–(iv) likely edge flaws generated by machining, which sequentially decreased in size. The characteristic widths of the crack length populations (i)–(iii) are approximately factors of 4. Although the crack length $h(c)$ variation of sample (iv) exhibits a long crack tail similar in slope

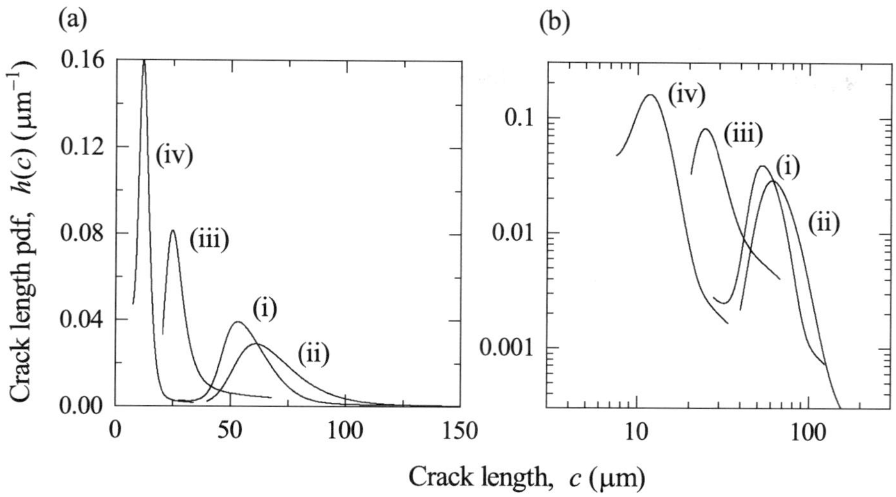

Figure 5.4 Plots of crack length pdf variations, $h(c)$, for alumina components; determined from strength measurements of Figure 5.3. (a) Plot in linear coordinates. (b) Plot in logarithmic coordinates. Label (i) corresponds to Figure 5.3a, label (ii) corresponds to Figure 5.3b, label (iii) corresponds to Figure 5.3c, and label (iv) corresponds to Figure 5.3d. Note reversal of (i) and (ii).

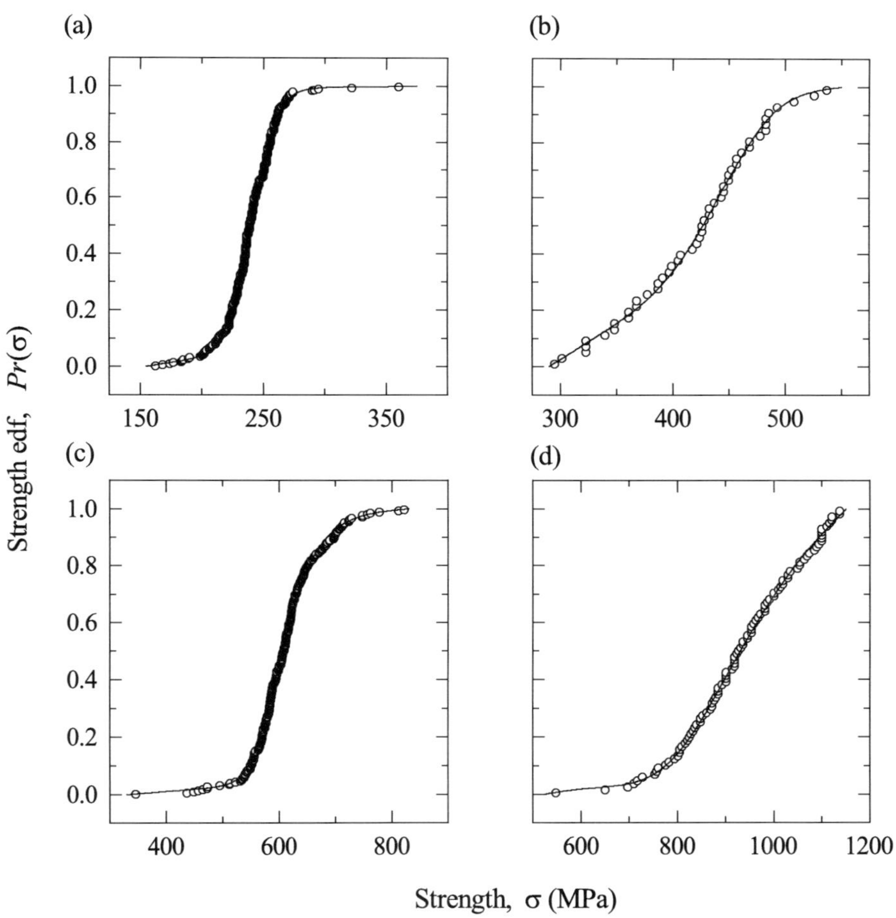

Figure 5.5 Plots of strength edf behavior, $Pr(\sigma)$, for silicon nitride components. (a) Reaction bonded beams, >40 mm long × 4 mm × 3 mm, $N = 269$ (Adapted from Quinn, G.D 1989). (b) Hot pressed beams, 30 mm long × 2.5 mm × 2.5 mm, $N = 49$ (Adapted from Easler, T.E et al. 1982). (c) $N = 240$ and (d) $N = 93$; unknown components, probably hot pressed (Adapted from NIST 1996; Usami, S et al. 1986).

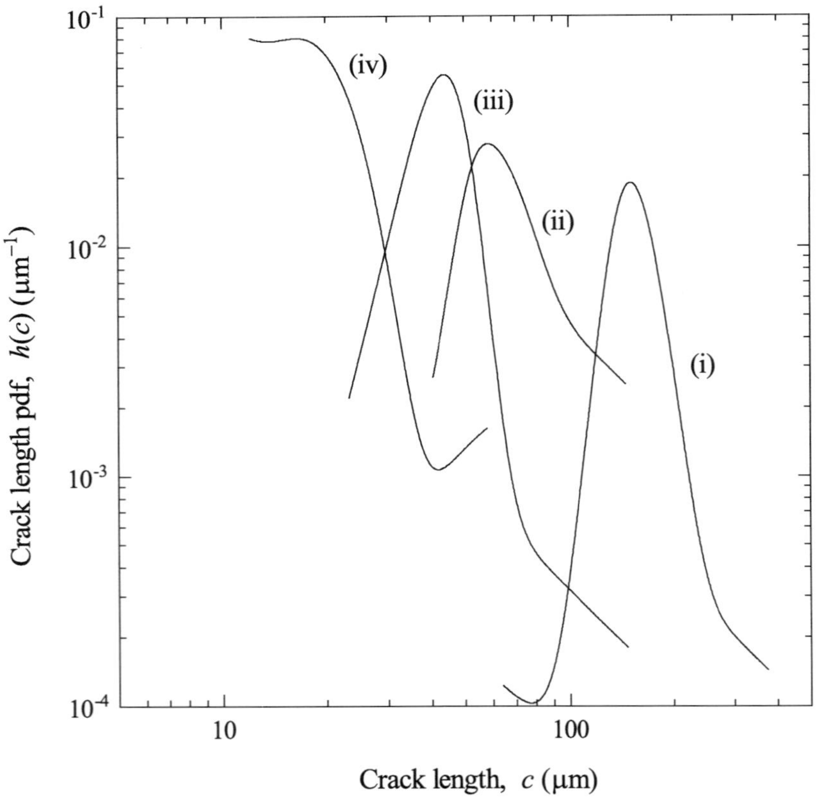

Figure 5.6 Plot in logarithmic coordinates of crack length pdf variations, $h(c)$, for silicon nitride components; determined from strength measurements of Figure 5.5. Label (i) corresponds to Figure 5.5a, label (ii) corresponds to Figure 5.5b, label (iii) corresponds to Figure 5.5c, and label (iv) corresponds to Figure 5.5d.

to (i)–(iii), the variation does not exhibit a well defined maximum, a consequence of the near convex strength distribution behavior (Figure 5.5d).

5.2.4 Porcelain, Bending Loading

Figure 5.7 shows strength edf $Pr(\sigma)$ plots for four sets of porcelain strength measurements, using data derived from the published works cited. Porcelain is a particulate composite ceramic material consisting of a glassy matrix encapsulating crystalline particles of mullite, quartz, and feldspar. Although there was debate up until the 1970s, it is now clear that the strength of porcelain is usually limited by cracks introduced in the microstructure by thermal expansion mismatch between the matrix and the quartz particles (Carty and Senapati 1998). Hence, an important technical advance is the replacement of at least some of the quartz by alumina particles, which do not lead to cracking and thus increase the material strength (and this is the basis of "electrical porcelain"). Figure 5.7a shows the behavior of cylindrical porcelain rods fabricated as a commercial material, from the work of Weibull (1939), in one of the earliest studies of strength distributions. Figure 5.7b shows the behavior of rectangular section beams custom fabricated as a control material, from the work of Akatsu et al. (2020). Figure 5.7c shows the behavior of (probably) rectangular section beams custom fabricated as a material with ultra-fine quartz particles, from the work of Kobayashi et al. (1994). Figure 5.7d shows the behavior of (probably) rectangular section beams custom fabricated as a material with a majority of the quartz particles replaced by alumina particles, from the work of Kobayashi et al. (2003). Symbols represent individual strength measurements. Lines represent visual best fits. The strength distribution data in Figure 5.7 are only weakly sigmoidal, exhibiting only small deviations from linear behavior and Figure 5.7a appears predominantly concave. The data exhibit strength domain widths σ_N/σ_1 of approximately 1.4, increasing slightly as the tendency to sigmoidal behavior increases.

Using logarithmic coordinates, Figure 5.8 shows the crack length pdf $h(c)$ variations underlying the porcelain strength behavior. The curves are labeled (i)–(iv), corresponding to Figure 5.7a–d, respectively, and the area under each curve is 1. The B values used were (all in MPa m$^{1/2}$) (i)–(iii) 1.0 and (iv) 1.5, taking into account the material microstructures. A feature of the $h(c)$ variations is the narrow characteristic widths of the crack length populations, approximately factors of 2. The microstructural variations in the example porcelains are clearly discernible in Figure 5.8. Curves (i) and (ii), associated

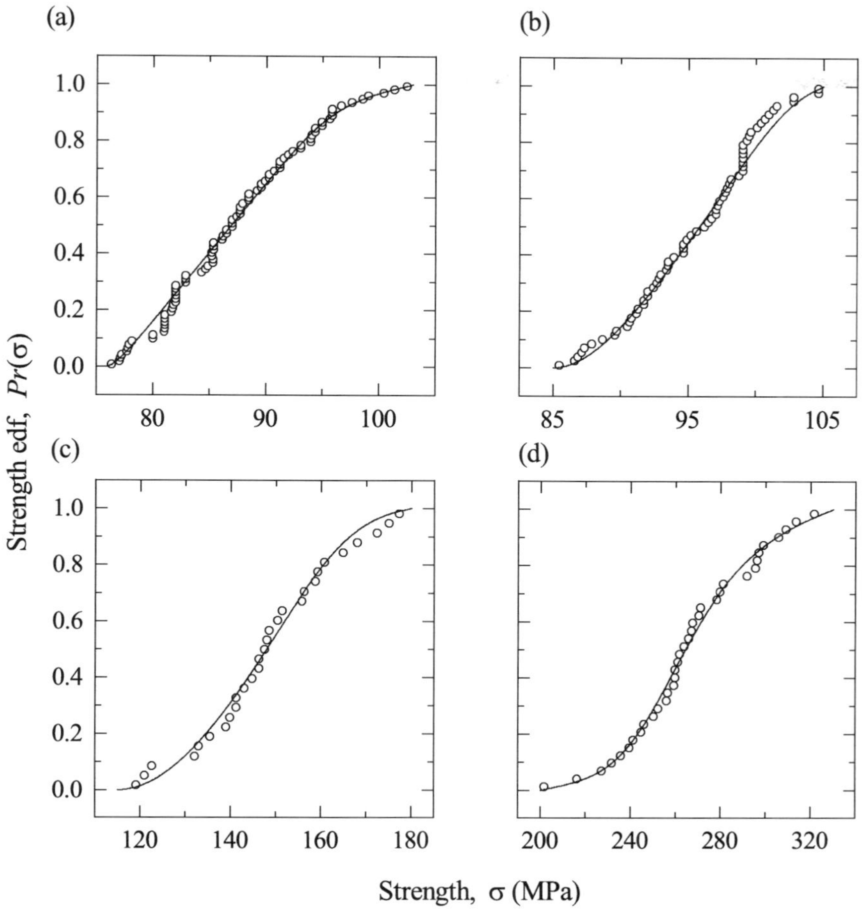

Figure 5.7 Plots of strength edf behavior, $Pr(\sigma)$, for porcelain components. (a) Cylindrical rods, >100 mm long × 19.2 mm diameter, $N = 80$ (Adapted from Weibull, W 1939). (b) Beams, 40 mm long × 8 mm × 4 mm, $N = 65$ (Adapted from Akatsu, T et al. 2020). (c) and (d) Beams, probably 80 mm long × 10 mm × 5 mm; (c) $N = 29$ and (d) $N = 36$ (Adapted from Kobayashi, Y et al. 1994, 2003).

with the conventional porcelains, show the largest cracks and the smallest crack population widths, consistent with pervasive strength-limiting cracks adjacent to similarly sized quartz particles generated in the fabrication process. Curve (iii), associated with a porcelain containing smaller quartz particles, shows smaller cracks and a slightly wider crack population, consistent with the smaller particle size and greater variability in the largest particle in a component. Curve (iv), associated with an aluminous porcelain containing very little quartz, shows the smallest cracks and a crack population width and shape similar to those considered earlier, suggesting that the strength-limiting flaw population was now associated with surface machining flaws rather than with bulk inclusions.

The crack length and microstructural interpretations of the strength distributions of porcelain probably also apply to another composite material, WC-Co, in which crystalline ceramic WC grains are imbedded in a metallic Co matrix. The strength distributions of such materials have been observed to exhibit sigmoidal shapes with strength domain widths of approximately 2, similar to Figure 5.7d, as in the early bending experiments of Gee (1984). However, more recent strength distributions of many other WC-Co materials have been observed to exhibit near convex, almost linear, behavior with strength domain widths of approximately 1.1–1.3, similar to Figures 5.7a–c, as in the tension and various bending experiments of Klünser et al. (2011), Jonke et al. (2017), and Klünser et al. (2020). The early experiments exhibited mean strengths of approximately 1.5 GPa and the later experiments exhibited mean strengths of approximately 3.5 GPa. An interpretation of these observations is that: The components in the early experiments contained the largest cracks and a crack population width and shape similar to those considered earlier, in alumina for example, suggesting that the strength-limiting crack population was associated with surface machining flaws. The components in the more recent experiments contained the smallest cracks and a crack population width and shape similar to those considered here in porcelain, suggesting that the strength-limiting flaw populations were associated with particles. Small strengths and very narrow strength distributions were observed for the composite materials with the largest WC particle sizes, consistent with this interpretation.

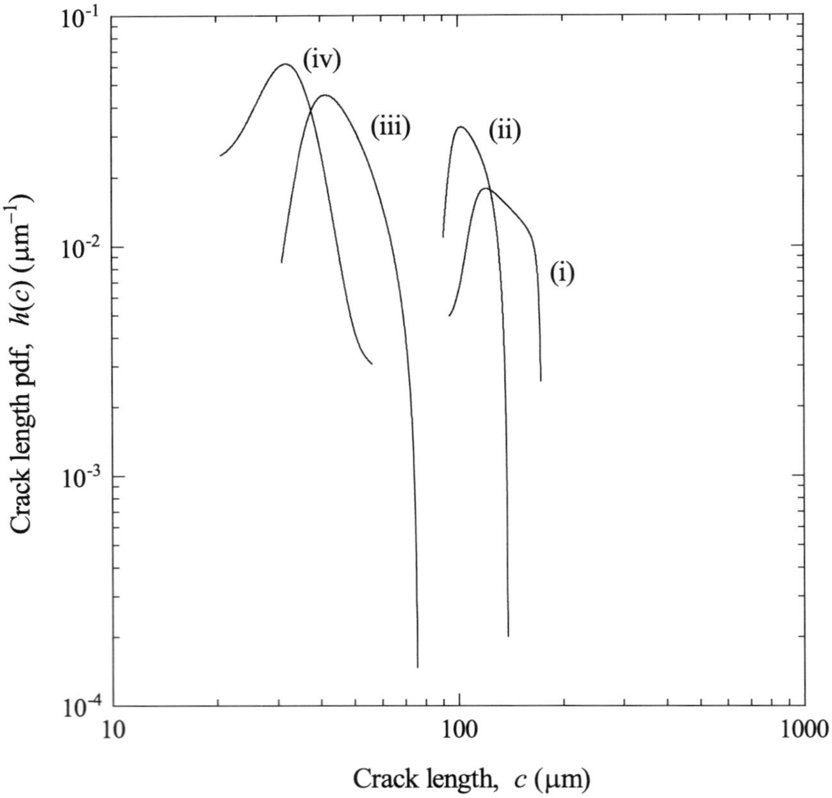

Figure 5.8 Plot in logarithmic coordinates of crack length pdf variations, $h(c)$, for porcelain components; determined from strength measurements of Figure 5.7. Label (i) corresponds to Figure 5.7a, label (ii) corresponds to Figure 5.7b, label (iii) corresponds to Figure 5.7c, and label (iv) corresponds to Figure 5.7d.

5.2.5 Silicon, Bending and Tension Loading

Figure 5.9 shows strength edf $Pr(\sigma)$ plots for four sets of silicon strength measurements, using data derived from the published works cited. Silicon is a semiconducting material used in two forms. Single crystal silicon (SCS) is the primary form, and used extensively in the electronics industry as the active layer and substrate material in microelectronic devices that are fabricated as small, flat, multi-millimeter scale "chips." Polycrystalline silicon (polysilicon) is the secondary form, and used in the microelectromechanical systems (MEMS) industry as the structural material in MEMS devices that are fabricated as small, nearly flat, sub-millimeter scale mechanisms. Strength-controlling flaws in microelectronic chips are usually associated with edge damage remnant from the dicing (sawing) process used to separate the chips. Strength-controlling flaws in MEMS components are usually associated with surface damage remnant from the etching process used to form the components. In polysilicon MEMS, grain boundaries significantly affect etch damage morphology, leading to component side-wall grooves. Figure 5.9a shows the behavior of commercially fabricated SCS chips measured in bending, in a data set slightly extended from earlier work (Cook and DelRio 2019b). The sub-gigapascal strengths of these chips is considerably less than the gigapascal strengths of the MEMS components. Figure 5.9b shows the behavior of polysilicon tensile bars fabricated in a multi-user MEMS facility, from the work of Yuan (as cited by LaVan and Bucheit 1999). Figure 5.9c shows the behavior of custom fabricated SCS tensile bars, from the work of Gaither et al. (2013). Figure 5.9d shows the behavior of polysilicon tensile bars fabricated in another multi-user MEMS facility, from the work of Boyce (2010). The $Pr(\sigma)$ and underlying crack length behavior of Figures 5.9c and 5.9d were considered earlier (Cook and DelRio 2019a). Symbols represent individual strength measurements. Lines represent visual best fits. All sets of data are well described by sigmoidal behavior. The sets of data exhibit strength domain widths σ_N/σ_1 decreasing from approximately 5 to approximately 1.5 as the strength increases (a)–(d). Detailed information regarding strengths controlled by dicing of SCS and etching of polysilicon can be found elsewhere (DelRio et al. 2015; Cook et al. 2019; Cook 2021).

Using logarithmic coordinates, Figure 5.10 shows the crack length pdf $h(c)$ variations underlying the silicon strength behavior. The curves are labeled (i)–(iv), corresponding to Figure 5.9a–d, respectively, and the area under each curve is 1. The B values used were all 0.7 MPa m$^{1/2}$ (DelRio et al. 2015). A feature of the $h(c)$ variations is the clear separation between the larger cracks of the diced components (i) and the smaller cracks of the etched components (ii)–(iv), consistent with

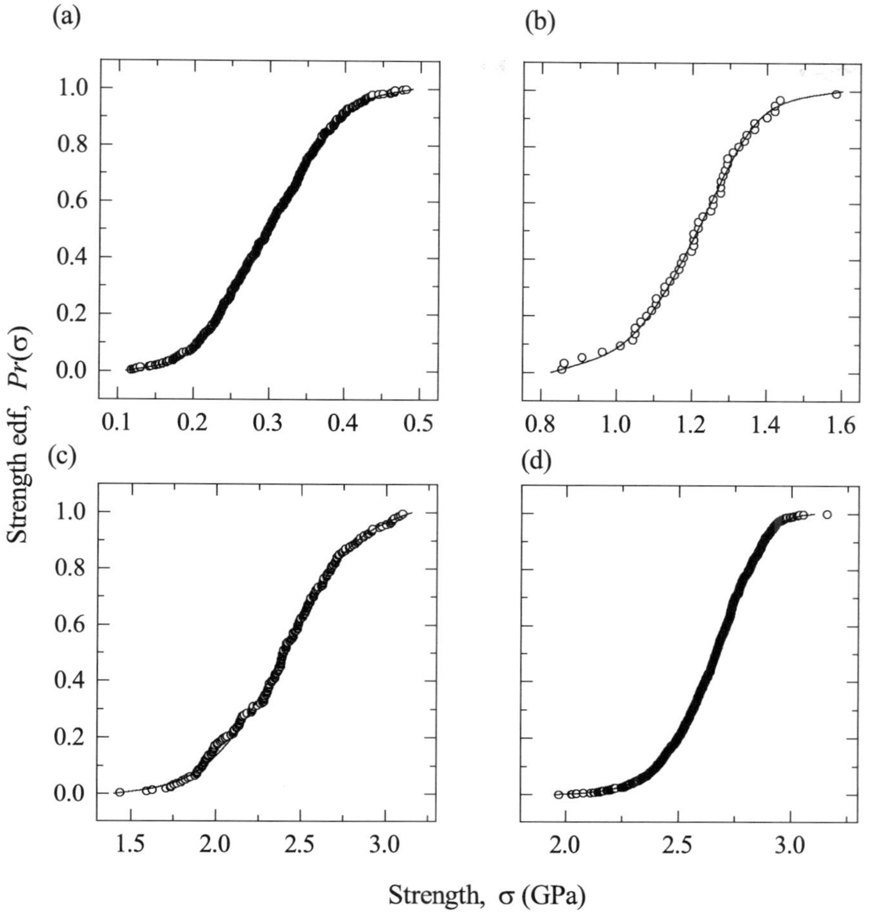

Figure 5.9 Plots of strength edf behavior, $Pr(\sigma)$, for silicon components. (a) Single crystal bend beams, 15 mm long × 11 mm × 0.73 mm, N = 482 (Adapted from Cook, R.F et al. 2019b). (b) Polycrystal tensile bars, 4 mm long × 600 μm × 3.5 μm, N = 48 (Adapted from LaVan, D.A et al. 1999). (c) Single crystal bars, 250 μm long × 25 μm × 8 μm, N = 209 (Adapted from Gaither, M.S et al. 2013). (d) Polycrystal bars, 250 μm long × 2.25 μm × 2.0 μm, N = 1008 (Adapted from Boyce, B.L 2010).

the strength observations, although the distinction between a crack and a "notch" and any associated plastic deformation for such small scale MEMS flaws is somewhat tenuous (DelRio et al. 2015). Figures 5.9b and 5.9d make clear the noted difference in strength between components fabricated in the two major MEMS multi-user facilities (LaVan and Bucheit 1999; DelRio et al. 2015). Figure 5.10 interprets this difference as a difference in flaw size, curves (ii) and (iv), but it is possible of course that the flaw populations generated in the two facilities are similar and that the fracture criterion B differs, smaller for (ii), as implemented for Si_3N_4 in Figure 4.9.

5.2.6 Fibers, Tensile Loading

Figure 5.11 shows strength edf, $Pr(\sigma)$, plots for four sets of fiber tensile strength measurements, using data derived from the published works cited. Three of the fibers, flax (largely cellulose), silicon carbide (SiC), and sapphire (single crystal Al_2O_3), Figure 5.11a, Figure 5.11b, Figure 5.11c, respectively, were intended for application as the reinforcing elements in fiber composites. One fiber, silicate glass, Figure 5.11d, was intended for application in optical communications. Flax fibers are obtained from flax plants in a hierarchical structure of tows, bundles, and individual or elementary fibers (Lamon et al. 2016). Figure 5.11a shows the shows the behavior of elementary flax fibers, from the work of Andersons et al. (2008). Figure 5.11b shows the shows the behavior of commercial SiC fibers intended for metal matrix composite reinforcements, from the work of Goda and Fukunaga (1986). Figure 5.11c shows the shows the behavior of commercial sapphire fibers from the work of Kotchick et al. (1975). The mechanical integrity of glass fibers in long lengths (kilometers) is critical to their use in optical communications applications. Figure 5.11d shows the shows the behavior of long samples of commercial glass fibers from the work of Kurkjian et al. (1976). Symbols represent individual strength measurements. Lines represent visual best fits. The strength distribution data in Figures 5.11a–5.11c are only weakly sigmoidal, exhibiting only small deviations

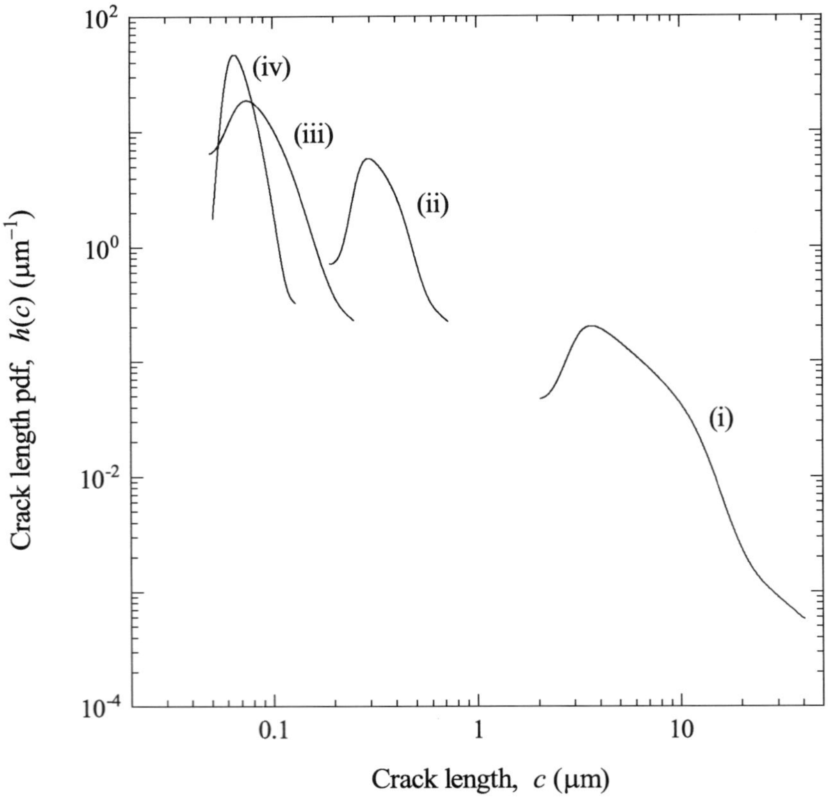

Figure 5.10 Plot in logarithmic coordinates of crack length pdf variations, $h(c)$, for silicon components; determined from strength measurements of Figure 5.9. Label (i) corresponds to Figure 5.9a, label (ii) corresponds to Figure 5.9b, label (iii) corresponds to Figure 5.9c, and label (iv) corresponds to Figure 5.9d.

from linear behavior and Figure 5.7d exhibits a clear tendency to non-zero $\mathrm{d}Pr(\sigma)/\mathrm{d}\sigma$ derivative linear behavior at the data bounds. The data exhibit strength domain widths σ_N/σ_1 of approximately 1.8–10 and should be compared with the strength distribution behavior of porcelain (Figure 5.7), which exhibited similar shapes but much smaller σ_N/σ_1 ratios.

A consequence of the wide strength strength distributions is observed in the crack length pdf $h(c)$ variations, shown using logarithmic coordinates in Figure 5.12. The curves are labeled (i)–(iv), corresponding to Figures 5.11a–d, respectively, and the area under each curve is 1. The B values used were (i) 1.0, (ii) 3.0, (iii) 1.5, and (iv) 0.5, (in MPa $m^{1/2}$) taking into account the materials. A feature of the $h(c)$ variations is the broad widths of the crack length populations, ranging from 5 to 100. Curve (iii) is a peaked function similar to those observed earlier. Curve (iv) exhibits a peak at the smallest crack lengths, but also exhibits an extremely long tail. Curves (i) and (ii) do not exhibit well-defined peaks, and curve (ii) does not even exhibit a clear maximum. Again, these variations should be contrasted with those of porcelain (Figure 5.8), which exhibited narrow crack length populations. An interpretation of the difference between the strength and flaw behavior of the porcelain and fiber systems is as follows: In the former case, the flaws were largely intrinsic and determined by invariant, microstructural influences. In the latter case, the flaws were largely extrinsic and determined by variable, surface influences. Hence, although the porcelain microstructures were inhomogeneous, the extreme flaws in each component were of similar size. The fiber microstructures were homogeneous, but therefore sensitive to the details of erratic contact and processing defects.

5.2.7 Shells, Flexure Loading

Figure 5.13 shows strength edf, $Pr(\sigma)$, plots for four sets of shell or shell-like component flexure strength measurements, using data derived from the published works cited. One set of measurements was conducted on full spherical shells, two sets of measurements were conducted on hemispherical shells, and a final set of measurements was conducted on a spherical C structure. The shells were sub-millimeter in scale and were formed of ceramic multilayers intended as nuclear fuel encapsulants, for which the multilayer structure is known as tri-isotropic (TRISO). The C structure was larger, multi-millimeter in scale and formed of a dense ceramic intended for use in bearing applications. The predominant structural material in

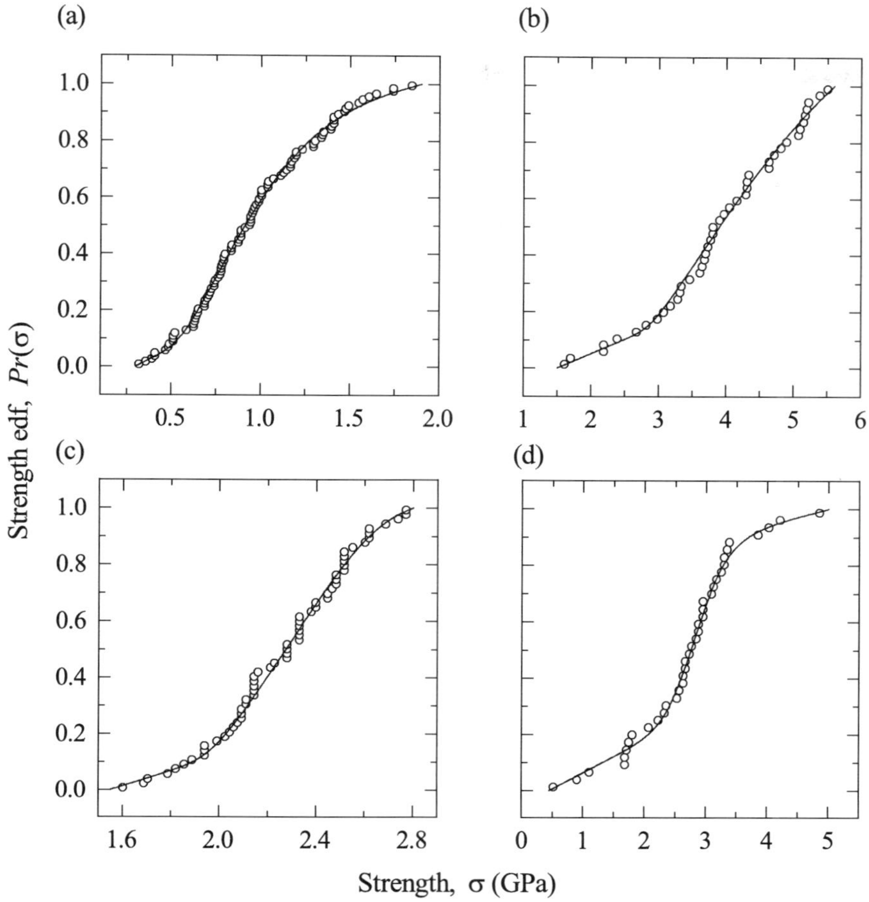

Figure 5.11 Plots of strength edf behavior, $Pr(\sigma)$, for fiber components. (a) Flax fibers, 5 mm long × ≈ 15 μm diameter, $N = 97$ (Adapted from Andersons, J et al. 2008). (b) SiC fibers, 10 mm long × ≈ 15 μm diameter, $N = 43$ (Adapted from Goda, K et al. 1986). (c) Sapphire fibers, 75 mm long × 250 μm diameter, $N = 61$ (Adapted from Kotchick, D.M et al. 1975). (d) Glass fibers, 20 m long × 110 μm diameter, $N = 38$ (Adapted from Kurkjian, C.R et al. 1976).

the shells was SiC. The C structure was formed of Si_3N_4. Schematic diagrams of the structures and loading for strength measurements is shown in Figure 5.14. Figure 5.13a shows the behavior of the full spherical shell structure, from the work of vanRooyen et al. (2010). The shell diameter was 800 μm and the shell thickness was 30 μm. Figures 5.13b and 5.13c show the behavior of the half-shell structures, from the works of Hong et al. (2007) and Xie et al. (2019). The shell diameters were 800 μm and 980 μm, and the shell thicknesses were 30 μm and 21 μm, respectively. Figure 5.13d shows the behavior of the C structure from the work of Wereszczak et al. (2007). The C structure diameter was 12.7 mm and the wall thickness was 2.54 mm. Symbols represent individual strength measurements. Lines represent visual best fits. The strength distribution data in Figure 5.13 are clearly sigmoidal, although exhibiting some degrees of asymmetry and tendencies to non-zero $dPr(\sigma)/d\sigma$ derivative linear behavior at the data bounds. The data exhibit strength domain widths σ_N/σ_1 of 2–3 and should be compared with the strength distribution behavior of alumina, Figure 5.3, which exhibited similar shapes and σ_N/σ_1 ratios.

Using logarithmic coordinates, Figure 5.15 shows the crack length pdf $h(c)$ variations underlying the spherical flexure strength behavior. The curves are labeled (i)–(iv), corresponding to Figure 5.15a–d, respectively, and the area under each curve is 1. The B values used were (i)–(iii) 3 MPa $m^{1/2}$ and (iv) 4.5 MPa $m^{1/2}$ appropriate to these structural ceramics. All materials exhibit clear peaks in the flaw size populations, at dimensions much smaller than those of the component thicknesses. Again, these variations should be compared with those of alumina, Figure 5.4.

In more detail, the strength and flaw distributions of Figures 5.13d and 5.15(iv) represent the behavior of hot-pressed, dense, bearing grade polycrystalline Si_3N_4 machined into "C-sphere" shaped components. In these components, a notch is milled in one side of a spherical particle such that the exterior surface of the resulting C is placed in tension when the particle is loaded in diametral compression. As such, these "particle" specimens are more closely related to extended bending components than true particles. The components are somewhat similar to the diametrally loaded C-ring, components investigated by Shelleman et al. (1991) as part of a study of silicon carbide including O-ring, bending, and pressurized

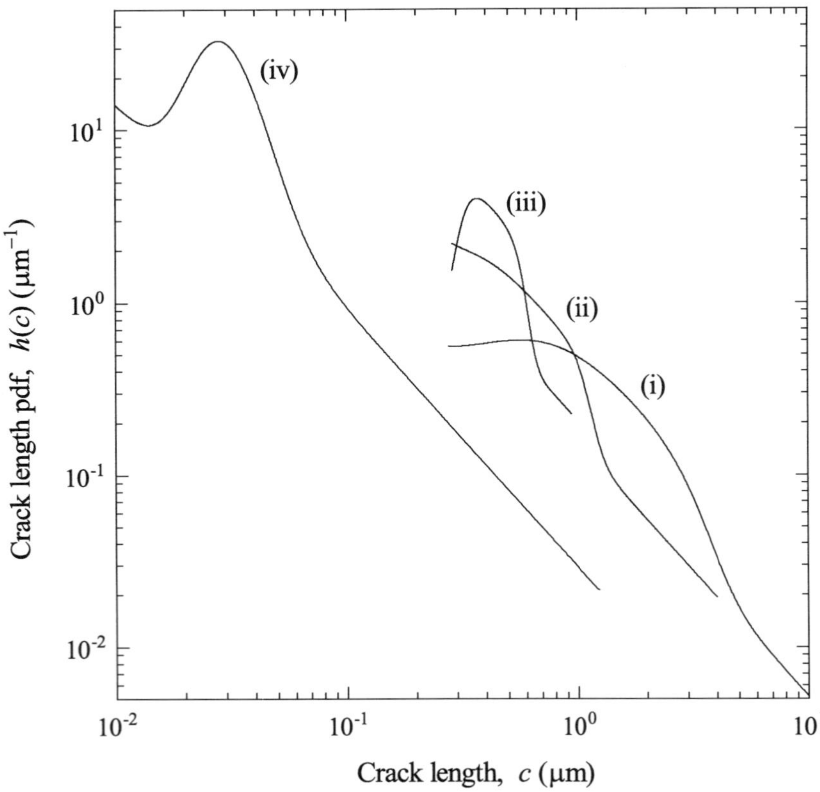

Figure 5.12 Plot in logarithmic coordinates of crack length pdf variations, *h*(*c*), for fiber components; determined from strength measurements of Figure 5.11. Label (i) corresponds to Figure 5.11a, label (ii) corresponds to Figure 5.11b, label (iii) corresponds to Figure 5.11c, and label (iv) corresponds to Figure 5.11d.

tube components, and by de With (1984) on diametrally loaded alumina O-ring components (that exhibited a clear size effect). The strength distribution here of the commercial grade of Si_3N_4 tested in the C configuration reflect the extended geometry and closely resembles bending component strength data shown earlier. The strength of the materials is about 1 GPa, reflecting the strong covalent bonding and hence large fracture resistance of Si_3N_4, and the small flaws on the highly polished bearing particle surfaces.

5.2.8 Columns, Compressive Loading

Figure 5.16 shows strength edf, $Pr(\sigma)$, plots for two sets of compressive strength measurements of columns of brittle materials, using data derived from the published works cited. Figure 5.16a shows the behavior of a porous phosphate material intended for use as a biomaterial scaffold, from the work of Meininger et al. (2016). Figure 5.16b shows the shows the behavior of concrete material used in bridge construction, from the work of Al-Manaseer et al. (2011). Symbols represent individual strength measurements. Lines represent visual best fits. Both sets of data are well described by sigmoidal behavior and Figure 5.16a exhibits a clear tendency to non-zero $\mathrm{d}Pr(\sigma)/\mathrm{d}\sigma$ derivative linear behavior at the data bounds. The data exhibit strength domain widths σ_N/σ_1 of approximately 2. The data are comparable to the glass data of Figure 5.1.

Using logarithmic coordinates, Figure 5.17 shows the crack length pdf $h(c)$ variations underlying the compressive strength behavior. The curves are labeled (i) and (ii), corresponding to Figures 5.16a and 5.16b, respectively, and the area under each curve is 1. The B values used were (i) 0.3 MPa m$^{1/2}$ and (ii) 1.5 MPa m$^{1/2}$. Both materials exhibit clear peaks in the flaw size populations, consistent with the scale of open porosity in the biomaterial scaffold, about 50 μm, and the scale of aggregate-related defects in concrete, sub-mm. In this case, the clear separation of the flaw populations gives rise to barely overlapping strength distributions as the fracture criteria scale with flaw size, whereas in Figure 5.1 the fracture criterion was invariant and strengths scaled directly with flaws.

5.2.9 Materials Survey Summary

This section has surveyed in detail the strength edf $Pr(\sigma)$ behavior of a wide variety of brittle materials—amorphous, crystalline, polycrystalline, and multi-phase—in a wide variety of loading geometries—primarily bending, but also tension, compression, and flexure. The predominant behavior is sigmoidal, strength edf responses characterized by mid-domain inflection points. Many responses exhibited ideal "sigmoidal" behavior of an inflection point centered in the strength domain and near zero derivatives at the domain bounds. Observed variations from the ideal included: an inflection point not centered in the domain, leading to an asymmetric edf that was sometimes near convex or near concave; non-zero derivative regions at one or both bounds, such that the response resembled a reversed "Z" more than an "S"; or, an ill-defined inflection point, leading to a weakly sigmoidal shape that was near linear. The strength domain widths σ_N/σ_1 were typically about 2, although the observed range was 1.4–5. There was a slight tendency to weaker sigmoidal behavior with smaller domain widths. The crack length pdf $h(c)$ behavior deconvoluted from the strength responses was typically a well

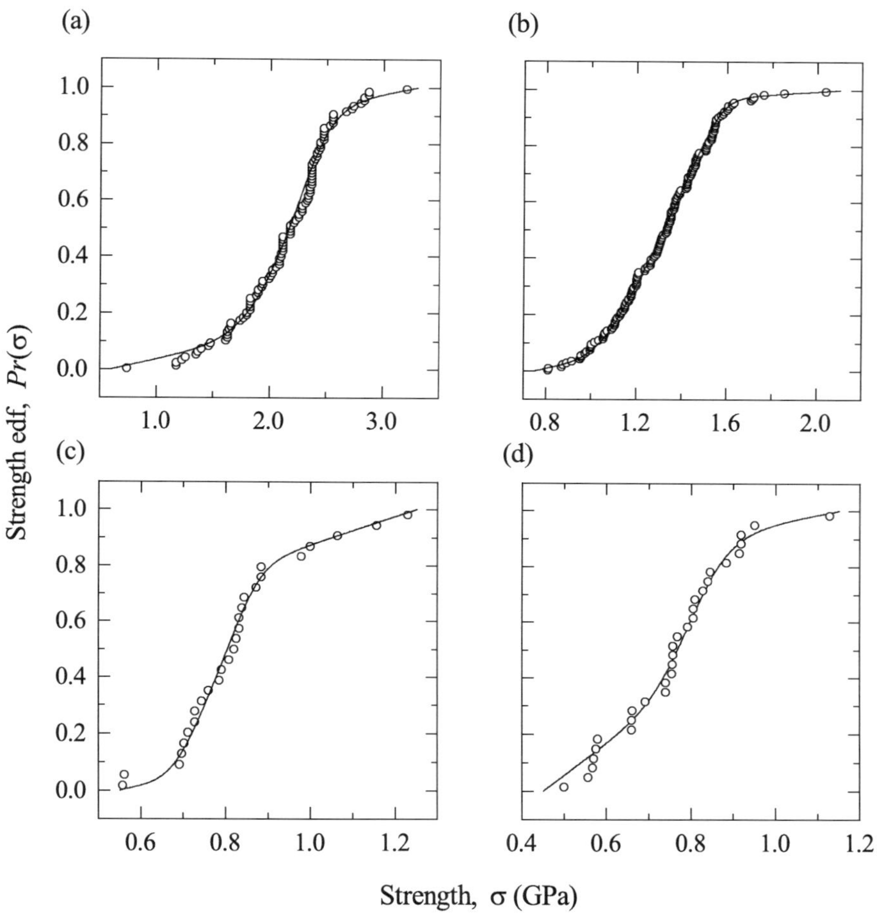

Figure 5.13 Plots of strength edf behavior, $Pr(\sigma)$, for shell components. (a) Spherical SiC shells, 800 μm diameter, $N = 102$ (Adapted from Van Rooyen, G.T (2010)). (b) Hemispherical SiC shells, 800 μm diameter, $N = 43$ (Adapted from Xie, X. (2019)). (c) Hemispherical SiC shells, 980 μm diameter, $N = 38$ (Adapted from Hong, S.G et al., (2007)). (d) Si_3N_4 C shaped components 12.7 mm diameter, $N = 30$ (Adapted from Wereszczak, A.A et al. 2007).

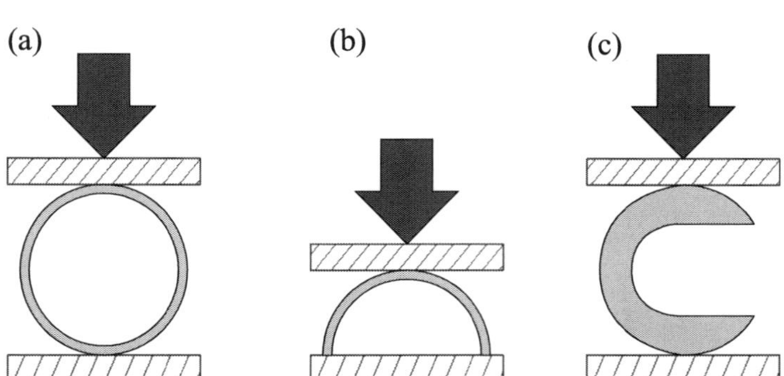

Figure 5.14 Schematic cross-section diagrams illustrating the loading configurations of shell structures. (a) Spherical. (b) Hemispherical. (c) C shaped.

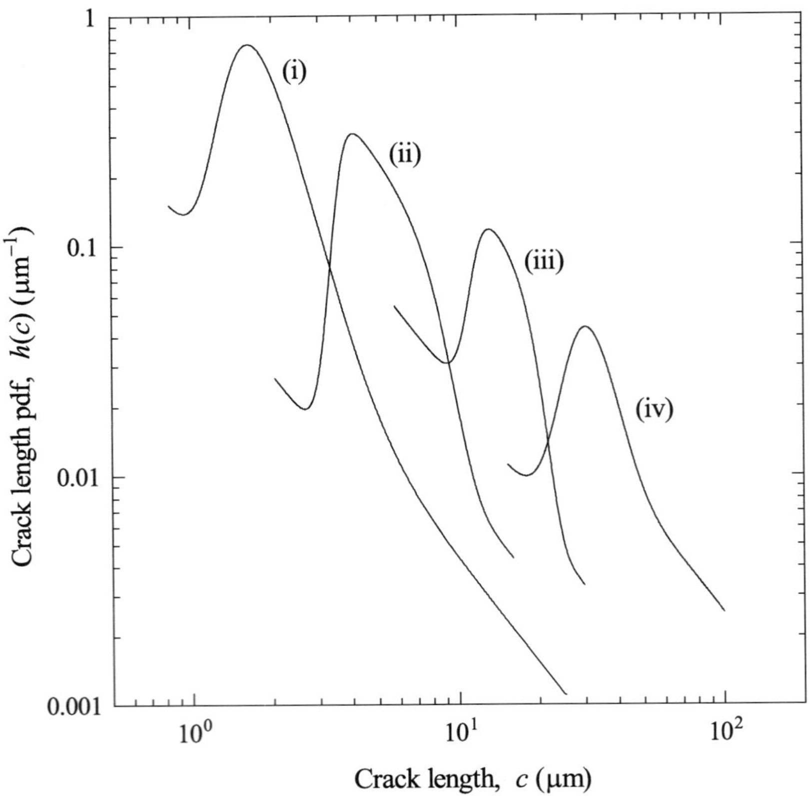

Figure 5.15 Plot in logarithmic coordinates of crack length pdf variations, $h(c)$, for shell components; determined from strength measurements of Figure 5.13. Label (i) corresponds to Figure 5.13a, label (ii) corresponds to Figure 5.13b, label (iii) corresponds to Figure 5.13c, and label (iv) corresponds to Figure 5.13d.

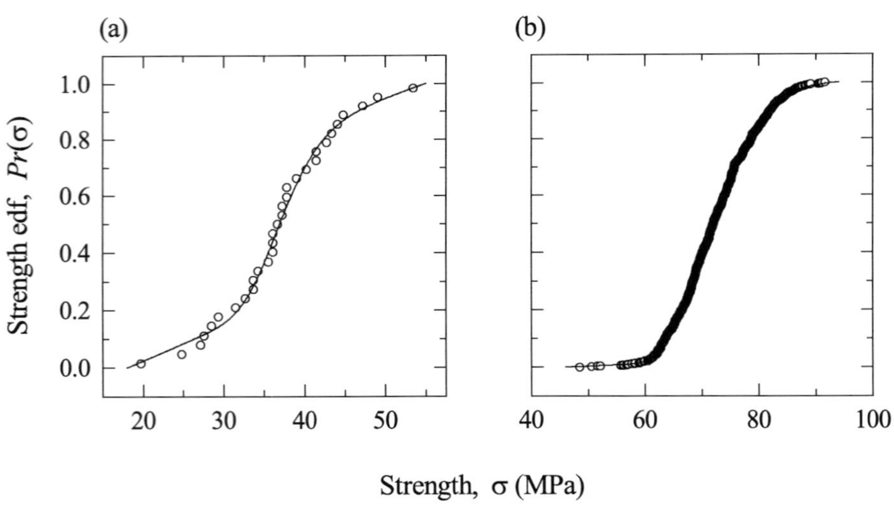

Figure 5.16 Plots of strength edf behavior, $Pr(\sigma)$, for two sets of brittle components tested in compression. (a) Phosphate biomaterial compression columns, 12 mm tall × 6 mm × 6 mm, $N = 31$ (Adapted from Meininger, S et al. 2016). (b) Concrete compression columns, 200 mm or 300 mm tall × 100 mm or 150 mm diameter, $N = 1025$ (Adapted from Al-Manaseer, A et al. 2011).

formed peak, often extending into a long, light tail at large crack lengths. Plotting of $h(c)$ responses in logarithmic coordinates enabled a clear assessments of relative widths and shapes of pdf behavior that nearly always differed significantly in absolute scale. A graphical summary of the various crack length behaviors observed in this section is given in Figure 5.18, showing one or two representative pdf $h(c)$ variations for each system. Note the range of crack length domains exhibited

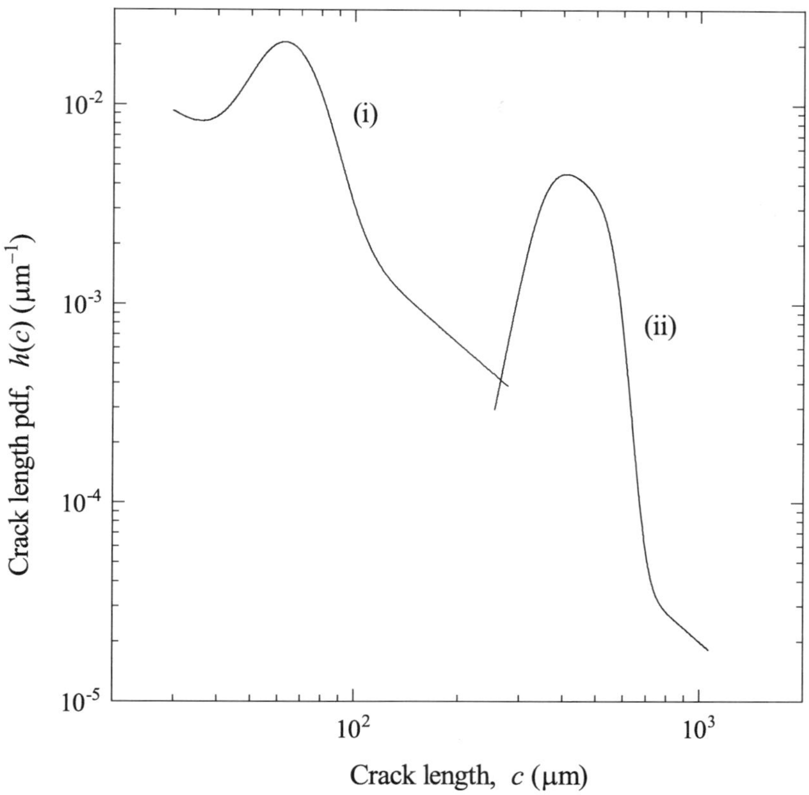

Figure 5.17 Plot in logarithmic coordinates of crack length pdf variations, $h(c)$, for components tested in compression; determined from strength measurements of Figure 5.16. Label (i) corresponds to Figure 5.16a, label (ii) corresponds to Figure 5.16b.

and the overall similarity in size and shape (the decreasing location is simply a consequence of imposed normalization, i.e. the c^{-1} trend is a geometrical imposition). The logarithmic coordinates highlight that crack length domains for these systems are less common for small cracks—it is more difficult to manufacture a small crack-large strength component. The relative domain positions of $h(c)$ responses were, of course, dependent on the choice of the fracture criterion parameter B linking strengths and crack lengths. Values in the range $B = 0.5$ MPa m$^{1/2}$–4.5 MPa m$^{1/2}$, selected on the basis of known brittle materials characteristics, generated expected and explainable $h(c)$ behavior. Examples include glass bottles contain larger flaws than glass optical fibers, sawn silicon components contain larger flaws than etched silicon MEMS devices, and porcelain containing misfit quartz particles contains larger flaws than porcelain containing better fitting alumina particles. A conclusion to be drawn from this survey is that extended components of brittle materials, those likely to contain many cracks and flaws, usually exhibit sigmoidal strength $Pr(\sigma)$ distributions. Analysis of such distributions within the probabilistic strength-crack length framework developed previously shows that the underlying crack lengths are described by probability densities $h(c)$ with clearly identifiable peaks and sometimes extended large crack tails.

It is important to remember that the $h(c)$ responses determined from strength measurements describe the probability density of the extreme value distribution of a sample of components. It is also important to remember that extended components need not be physically large (e.g. Si MEMS, TRISO shells, optical fibers), simply analytically large in containing many flaws. For extended components, the sample distribution $h(c)$ is not the population distribution $f(c)$, for two reasons. Extended components each contain k flaws, but only one flaw in each component, the extreme flaw, is sensed in a strength test. The domain of flaws sensed in a sample of strengths is thus contracted to the extreme flaws in each component. Samples contain N components, but Nk is usually a small fraction of the total population of flaws, $\lambda\Omega$. The domain of flaws sensed in a sample of strengths is thus truncated to the most probable flaws in the population. These domain restriction issues were explored in detail in Chapter 3, leading to conclusions that large N and small k provided the best sample configuration for determining $f(c)$ from $h(c)$. The experimental data in this section were all formed from large N samples, such that estimation of the characteristic feature in $h(c)$, the peak, was unambiguous. However, the samples all consisted of large k components, such that the relation between $h(c)$ and the population $f(c)$ was unknown. As $Pr(\sigma)$ and thus $\hat{H}(\sigma)$ are greater than $F(\sigma)$, the peak position in $h(c)$ must be an upper bound to the peak position in $f(c)$, independent of extreme value effects giving rise to sample strength distribution increases relative to the population strength distribution. If samples

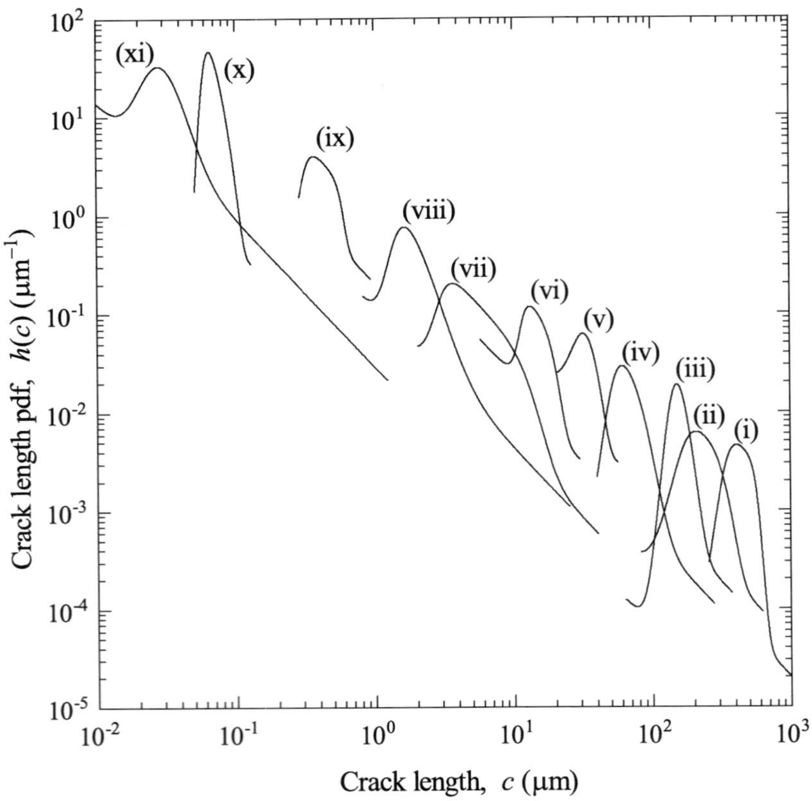

Figure 5.18 Plot in logarithmic coordinates of crack length pdf variations, $h(c)$, for a range of materials and loading configurations representative of the extended component data analyzed in this section. (i) concrete columns; (ii) glass bottles; (iii) polycrystalline Si_3N_4 bend beams; (iv) polycrystalline Al_2O_3 bend beams; (v) porcelain bend beams; (vi) polcrystalline SiC hemi-spherical shells; (vii) single crystal Si bend beams; (viii) polcrystalline SiC spherical shells; (ix) single crystal Al_2O_3 (sapphire) fibers; (x) polycrystalline Si MEMS tensile bars; (xi) glass fibers. Note the extremely large range of flaw sizes.

of different sized components (different k) are taken from the population, more information can be gained. In particular, the nature of the extreme value size effects, specifically stochastic vs deterministic, and the likely position and form of the crack length population $f(c)$, can be determined. The next section examines experimental observations of size effects in extended components, utilizing selected data from this survey.

5.3 Size Effects

5.3.1 Stochastic

Stochastic extreme value size effects influence sample strength distributions only through the probability that a component contains an element exhibiting a particular extreme strength, selected from the invariant population. The guiding equation in assessment of stochastic size effects is the relation between strength distributions of samples of different sized components, Eq. (3.36):

$$H_2(\sigma) = 1 - [1 - H_1(\sigma)]^{k_2/k_1}. \tag{5.1}$$

$H_1(\sigma)$ and $H_2(\sigma)$ are the strength distributions of the different samples and k_1 and k_2 are the different sizes of the components, expressed as numbers of fundamental volume elements, and flaws, in the components. It is noted that the the exponent includes only the ratio k_2/k_1 of the numbers of flaws in the different sized components. For $k_2 > k_1$, $H_2 > H_1$. There are several points to note from Eq. (5.1). First it is explicit that the strength domains of $H_1(\sigma)$ and $H_2(\sigma)$ are identical, and by extension identical with the strength domain of the population, $F(\sigma)$, $[\sigma_{th}, \sigma_u]$. In particular, the lower bound threshold, σ_{th}, is in common. Second, by the definition of H, $dH(\sigma)/d\sigma > 0$ throughout the domain and thus for $H_2 > H_1$, $dH_2(\sigma)/d\sigma > dH_1(\sigma)/d\sigma$ for $\sigma \to \sigma_{th}$. Strength distributions of larger components will thus appear "steeper" (truncation effects usually eliminate observations of $\sigma \to \sigma_u$). The fully sufficient condition for two sample strength distributions to be related by stochastic extreme value effects is demonstration that they obey Eq. (5.1); a necessary condition, useful for initial experimental assessment, is thus that the distributions appear as similarly shaped functions "fanned out" from a common threshold. Third, systems that obey Eq. (5.1) imply the existence of a single underlying population of *independent* strength-controlling flaws, regardless of the details (shape, location) of that population. Two stochastic size effect strength examples are considered below for two sets of extended components, each exhibiting different strength distribution shapes and locations, and thus reflecting different underlying crack length populations.

Figure 5.19 shows strength edf $Pr(\sigma)$ plots for four samples of polycrystalline alumina rectangular section components measured in bending, from Quinn (1989) using data derived from the published work. The bold solid line is identical to that in Figure 5.3b and represents the best fit to the strength distribution of 4 mm × 3 mm components tested using a 20 mm inner bending span. Symbols represent individual strength measurements of components of three other sizes, represented by different symbols. The least strengths were observed for 8 mm × 6 mm components tested using a 40 mm inner bending span, intermediate strengths were observed for 6 mm × 3 mm components tested using a 20 mm inner bending span, and greatest strengths were observed for 2 mm × 1.5 mm components tested using a 10 mm inner bending span. The strength distributions appear as similarly shaped (sigmoidal) functions fanned out from a common threshold, implying the possibility of relation by stochastic extreme value effects. The fine lines in Figure 5.19 are visual best fits to the symbol data, using Eq. (5.1) and taking $H_1(\sigma)$ as the base 4 mm × 3 mm data. The data are well described by the analysis, using size ratios, k_2/k_1, of (i) 3.0, (ii) 1.5, and (iii) 0.75, describing the lesser, intermediate, and greater strengths, respectively. The conclusion from Figure 5.19 is that fracture strength of the polycrystalline alumina is described by stochastic size effects.

Using logarithmic coordinates, Figure 5.20 shows the crack length pdf $h(c)$ variations underlying the strength behavior. The bold solid line and the fine lines labeled (i), (ii), and (iii) correspond to the same lines in Figure 5.19. As all responses were derived from a common base, identical B values were used, all responses extend over a common domain, and the area under all curves is 1. As anticipated from Figure 5.19, the $h(c)$ variations exhibit peaks at decreasing crack lengths as the size of the component decreases (i) to (iii), although the peaks become less pronounced. Alternatively, and perhaps more physically, the $h(c)$ variations exhibit peaks at increasing crack lengths as the size of the component increases and the peaks

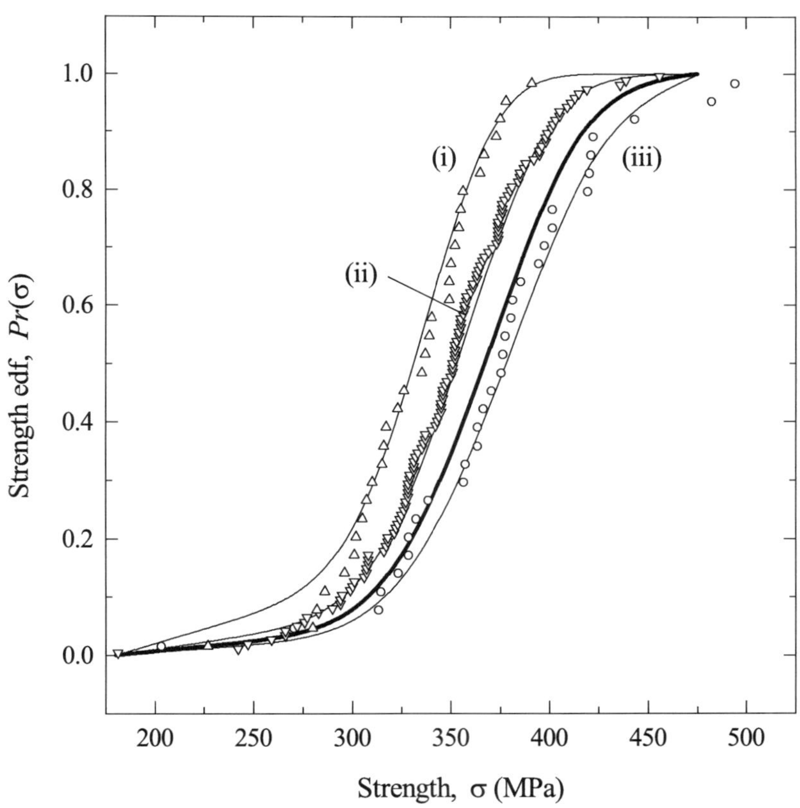

Figure 5.19 Plots of strength edf behavior, $Pr(\sigma)$, for four samples of a polycrystalline alumina material tested in bending, increasing component size left to right, N_{tot} = 448 (Adapted from Quinn, G.D 1989). The responses originating from a common threshold indicate a stochastic size effect.

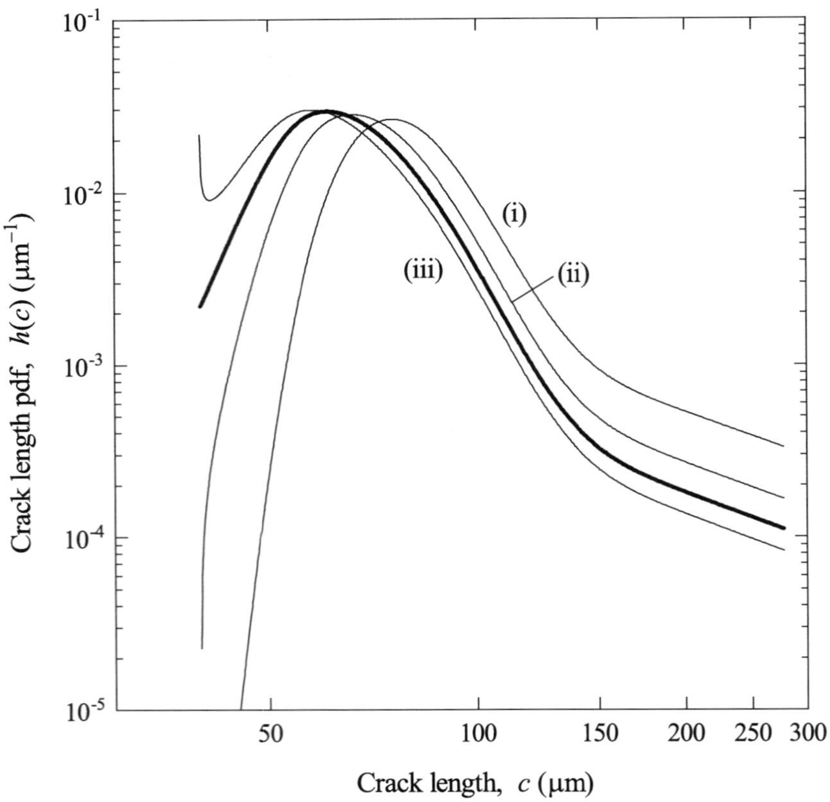

Figure 5.20 Plot in logarithmic coordinates of crack length pdf variations, $h(c)$, for a polycrystalline alumina material tested in bending; determined from strength measurements of Figure 5.19. Labels (i)–(iii) correspond to those in Figure 5.19.

become more pronounced. This behavior reflects a greater chance of occurrence of a large flaw in a large component and thus component and sample strengths tending to the lower bound threshold.

Stochastic size effects, as discussed and described in Chapter 3, are small and subtle. The peak in the crack length density of Figure 5.20 varies in position by about a factor of 1.3, consistent with the variation in the median strengths in Figure 5.19 by about a factor of 1.15. The means and standard deviations of the four samples of strengths in Figure 5.19, in order of increasing size, are: (372 ± 57) MPa, (364 ± 45) MPa, (347 ± 43) MPa, and (329 ± 35) MPa. The means overlap by all but the smallest standard deviation; conventional inspection would not suggest a substantial difference as the means vary by a factor of 1.13 and relative standard deviations are > 0.10. Use of standard error ($N_{tot} \approx 650$) does show the estimated means are significantly different; the above analysis also shows the distributions are different, but provides greater insight (similar to consideration of the batches of silicon nitride material in Chapter 4).

Figure 5.21 shows strength edf $Pr(\sigma)$ plots for three samples of flax fiber components measured in tension, from Andersons et al. (2008) using data derived from the published work. The bold solid line is identical to that in Figure 5.11a and represents the best fit to the strength distribution of 5 mm long fibers. Symbols represent individual strength measurements of fibers of two other sizes, represented by different symbols. The least strengths were observed for 20 mm long fibers and intermediate strengths were observed for 10 mm long fibers. The strength distributions appear as similarly shaped (weak sigmoidal, near concave) functions fanned out from a common threshold, implying the possibility of relation by stochastic extreme value effects. The fine lines in Figure 5.21 are visual best fits to the symbol data, using Eq. (5.1) and taking $H_1(\sigma)$ as the base 5 mm data. The data are well described by the analysis, using size ratios, k_2/k_1, of (i) 2.5 and (ii) 1.25, describing the lesser and intermediate strengths, respectively. The conclusion from Figure 5.21 is that fracture strength of the flax fibers is described by stochastic size effects.

Using logarithmic coordinates, Figure 5.22 shows the crack length pdf $h(c)$ variations underlying the strength behavior. The bold solid line and the fine lines labeled (i) and (ii) correspond to the same lines in Figure 5.21. As all responses were derived from a common base, identical B values were used, all responses extend over a common domain, and the area under all curves is 1. As anticipated from Figure 5.21, the $h(c)$ variations exhibit peaks at decreasing crack lengths as the length of

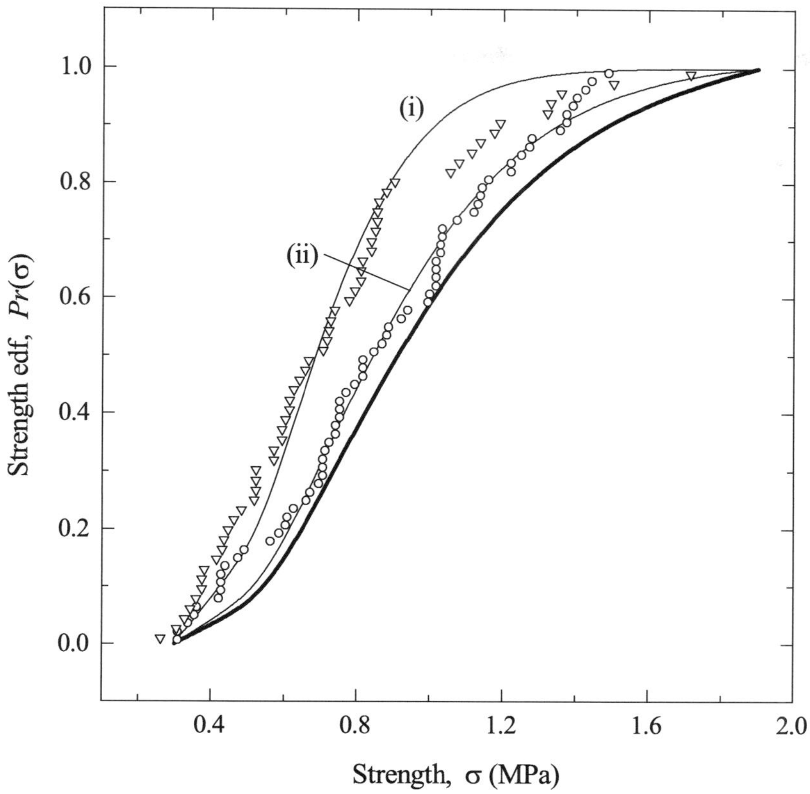

Figure 5.21 Plots of strength edf behavior, $Pr(\sigma)$, for three samples of flax fibers tested in tension, increasing fiber length left to right, N_{tot} = 225 (Adapted from Andersons, J et al. 2008). The responses originating from a common threshold indicate a stochastic size effect.

the fiber decreases: bold line, (i), and (ii), although, as above, the peaks become less pronounced. Alternatively, and perhaps more physically, the $h(c)$ variations exhibit peaks at increasing crack lengths as the length of the fiber increases and the peaks become more pronounced. This behavior reflects a greater chance of occurrence of a large flaw in a longer fiber and thus fiber and sample strengths tending to the lower bound threshold.

Once again, observed stochastic size effects are small and subtle. The peak in the crack length density of Figure 5.22 varies in position by about a factor of 3, consistent with the variation in the median strengths in Figure 5.21 by about a factor of 1.7. The means and standard deviations of the four samples of strengths in Figure 5.21, in order of decreasing size, are: (0.96 ± 0.35) GPa, (0.87 ± 0.32) GPa, and (0.74 ± 0.32) GPa. The means overlap in all cases; conventional inspection would not suggest a substantial difference as the means vary by a factor of 1.3 and relative standard deviations are > 0.3. Use of standard error ($N_{tot} \approx 230$) does show the estimated means are significantly different; the above analysis also shows the distributions are different, but again provides greater insight.

The two forms of strength distributions considered above, symmetric and asymmetric sigmoids, provided clear evidence of stochastic size effects—as component size increased, sample strength distributions tended towards smaller strengths consistent with stochastic analyses. The tendency reflected a greater fraction within a sample of components containing larger extreme flaws, which in turn was a consequence of larger components containing greater numbers of flaws. A central issue in both materials processing and component design is then identifying the element of "component size" that controls the number of flaws. As discussed in Section 3.3, depending on the nature of the flaws the number of flaws in a component may scale with a component length, a component area, or the component volume. In the experiments discussed here, the relative numbers of flaws in components of different sizes was determined as the ratio k_2/k_1. Comparison of this ratio with the ratio of selected measures of component size can provide the required scaling. The simplest case is one-dimensional: Strength-controlling flaws are assumed to be arrayed along a length dimension l of components, which are loaded in tension along l. Examples here included fibers and silicon chips and MEMS bars. If the length of the components is varied from sample to sample, l_1 and l_2 say, stochastic size effects should scale linearly, as $k_2/k_1 \sim l_2/l_1$.

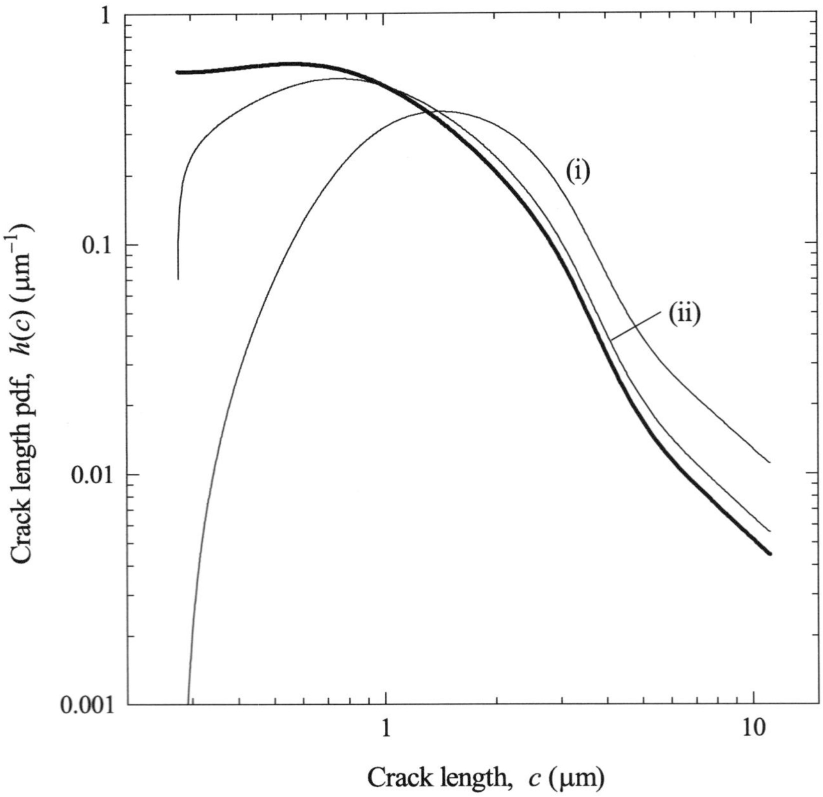

Figure 5.22 Plot in logarithmic coordinates of crack length pdf variations, $h(c)$, for flax fibers tested in tension; determined from strength measurements of Figure 5.21. Labels (i) and (ii) correspond to those in Figure 5.21.

Generalized versions of these parameters that enable comparison of multiple one-dimensional systems are k/k_{ref} and l/l_{ref}. Values of k and l are compared to a single reference configuration k_{ref} and l_{ref} within a given system and comparison between systems is thus relative to common reference point (1,1). Figure 5.23 shows a plot in logarithmic coordinates of the relative number of flaws k/k_{ref} vs the the relative size l/l_{ref} for several one-dimensional systems that exhibited stochastic size effects. The solid line has a slope of 1, consistent with ideal linear scaling. The dashed lines are guides to the eye of slopes 2 and 3. The solid symbols include measurements from the flax fibers of Figure 5.21 from the work of Andersons et al. (2008), glass fibers from the work of Kurkjian et al. (1976), including Figure 5.11d, and silicon MEMS bars from the work of Saleh et al. (2014) and Cook et al. (2019, 2021), including material similar to Figure 5.9d (Boyce 2010). Overall, there is agreement with the expected scaling over more than a factor of 30 in relative size.

Comparisons with systems that are not one dimensional are enabled by a further generalization that expresses the stressed volume of a component as $V = l^3$. In this way, the plotting scheme of Figure 5.23 is retained, allowing simple assessments of the nature of stochastic scaling and comparisons between systems of different physical dimensions. For example, if strength-controlling flaws are distributed throughout the volume of a material, as in the case of pores distributed throughout silicon nitride (Chao and Shetty 1992), strength distributions of components with different test volumes would exhibit k/k_{ref} vs l/l_{ref} behavior of slope 3. However, if strength-controlling flaws are distributed along a length of the fabricated components, as in the case of the sawn edges of the silicon chips tested in uniaxial bending, strength distributions of components with different volumes would exhibit k/k_{ref} vs l/l_{ref} behavior of slope less than 1. Similarly, if strength-controlling flaws are distributed over an area of the fabricated components, as in the case of the machined surfaces of the alumina cylinders and plates tested in biaxial flexure, strength distributions of components with different volumes would exhibit k/k_{ref} vs l/l_{ref} behavior of slope less than 2. The key here is that although a volume is tested, strength measurements are only sensitive to a single dimension, length or area, describing that volume. In the case of fibers, in which the area transverse to the length is invariant, the slope is exactly 1. The open symbols in Figure 5.23 represent analysis in this way of the responses of the alumina bend bars shown in Figure 5.19 from the work of Quinn (1989). The responses are clearly not of slope 3 and are grouped in with slope 1 fiber responses. The implication is that the strength-controlling flaws in the alumina

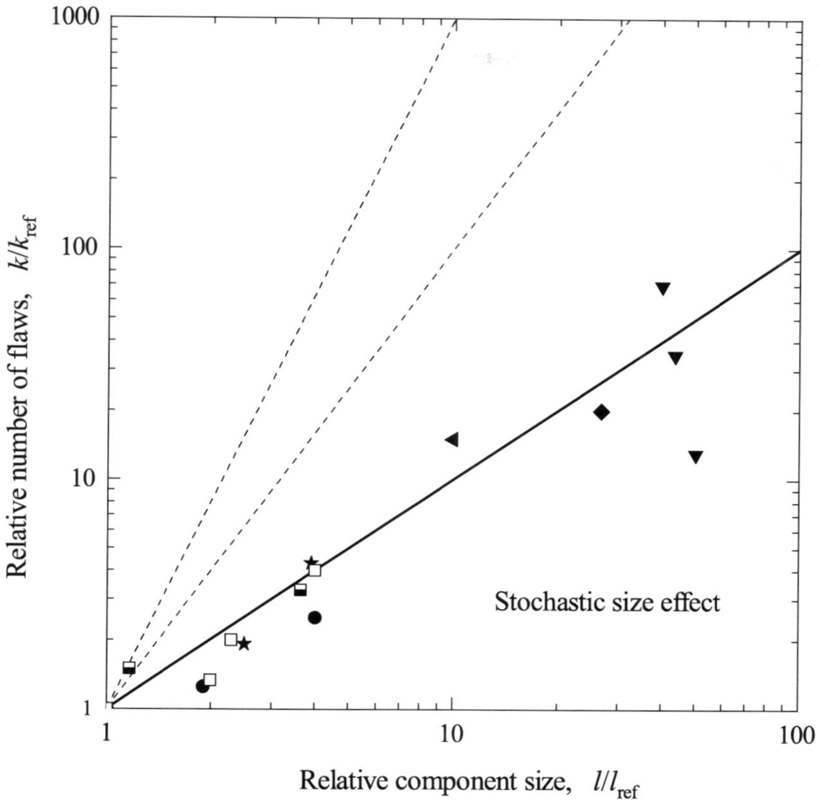

Figure 5.23 Plot of number of flaws of flaws k vs component size l for extended components exhibiting stochastic size effects on strength distributions. Lines are guides to the eye indicating scaling behavior; most systems behave linearly.

were not volume-distributed flaws, but more likely face- and edge-distributed flaws arising from the machining of the bend bars, giving rise to a linear stochastic size effect on strength as the bar and test fixture size increased. The upper half-open symbol in Figure 5.23 represents analysis of the responses of germanium biaxial flexure components from the work of Adler and Mihora (1992). As noted by them, the responses are consistent with linear stochastic size effect scaling, rather than the anticipated area or volume scaling. It is likely that the strength-controlling flaws were surface scratches with potency determined by length (and thus test fixture and component linear dimension) (Cook 2017). The lower half-open symbol in Figure 5.23 represents analysis of the responses of the alumina flexure components shown in Figure 5.19 from the work of Simpatico et al. (1999). Although the component and test fixture size increase was small, the response is consistent, overall scatter notwithstanding, with slope 2 and an area stochastic size effect on strength.

However, it must be noted that, despite extensive research for this book, there are very few points on Figure 5.23. Despite anticipation of stochastic size effects and study of one-dimensional systems since the earliest works (Weibull 1939; Epstein 1948; Epstein and Brooks 1948), definitive, quantifiable demonstrations of stochastic size effects are rare and largely unnoticed. The paucity of examples is due in part to a lack of analysis and the effects of distorting and obscuring linearized plotting schemes. In addition, in many cases the application of such plotting schemes leads to the mistaken impression that observed size effects on strength distributions are due to stochastic behavior, when in fact they are due to deterministic behavior. Examples of deterministic size effects on strength behavior of extended components are considered in the next section.

5.3.2 Deterministic

Deterministic extreme value size effects influence sample strength distributions through both the probability that a component contains an element exhibiting a particular extreme strength *and* determination by component size of the population of strengths from which the element is selected. In comparison of stochastic and deterministic failure behavior, basic concepts such as the fundamental volume element, the relation between crack length and strength, and component size expressed as k, the number of volume elements and cracks, remain. The lack of an invariant flaw population in deterministic systems, however, prevents detailed specification of a strength distribution H, as in Eq. (5.1). In particular, the constraint of an invariant strength domain, and thus an invariant strength threshold, is removed. It is anticipated that $H = H(\sigma, k)$ is greater for larger k.

Nevertheless, a useful equation in consideration of deterministic size effects on strength distributions is a generalization of Eq. (3.31) that underlies Eq. (5.1):

$$H(\sigma, k) = 1 - [1 - F(\sigma, k)]^k, \tag{5.2}$$

where $F(\sigma, k)$ is a size- (k-) dependent population given by the product

$$F(\sigma, k) = F(\sigma, 1)\Delta F(k). \tag{5.3}$$

$F(\sigma, 1)$ may be regarded similarly to the base elemental ($k = 1$) population discussed earlier, $F(\sigma, 1) \approx F(\sigma)$. $\Delta F(k)$ is a deterministic, component size dependent, perturbation that modifies the base population. In particular, $\Delta F(k)$ modifies the strength population domain. Thus, Eq. (5.2) and Eq. (5.3) enable deterministic strength distributions to be regarded to first approximation as related by changes in location, quantified by $\Delta F(k)$, and the familiar changes in shape, quantified by k. Two deterministic size effect strength examples are considered below for extended components, each arising from different size variations: the components themselves and the loading configuration. Detailed analyses of deterministic size effects on strength are considered in Chapters 8 and 9. Here attention is focused on establishing the major experimental observations, using Eq. (5.2) and Eq. (5.3) as simple physical descriptors of the major phenomena.

Figure 5.24 shows strength edf $Pr(\sigma)$ plots for four samples of polycrystalline diamond (5 % N_2) rectangular section components measured in tension, from Peng et al. (2007) using data derived from the published work. Symbols represent individual strength measurements of components of four sizes. Solid lines represent unconstrained visual best fits. In order of decreasing strength, the component sizes were 100 μm long × 5 μm × 1 μm, 200 μm long × 20 μm × 0.5 μm, 200 μm long × 20 μm × 1 μm, and 400 μm long × 40 μm × 1 μm. All sets of data are well described by sigmoidal behavior and a

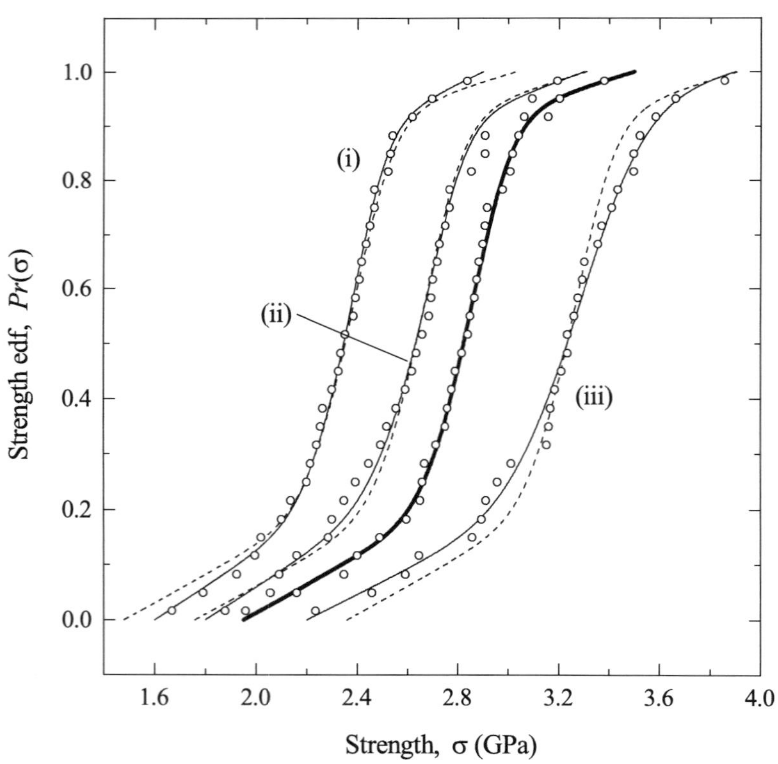

Figure 5.24 Plots of strength edf behavior, $Pr(\sigma)$, for four samples of polycrystalline diamond films tested in tension, increasing component size left to right, $N_{tot} = 120$ (Adapted from Peng, B et al. (2007)). The responses do not originate from a common threshold and therefore indicate a deterministic size effect.

tendency to non-zero $dPr(\sigma)/d\sigma$ derivative behavior at the data bounds. The data exhibit strength domain widths σ_N/σ_1 of approximately 1.7. The strength distributions appear as near identically shaped (sigmoidal) functions translated along the strength axis, implying deterministic extreme value size effects. The bold solid line represents a mid-size reference set of data. The dashed lines represent strength-translated versions of this line to match the medians of the other data sets. In all cases, the translated lines are close to the best fit lines, although there is a clear tendency for slight domain expansion with decreasing component size and increasing strength, curves (i), (ii), and (iii), respectively.

Using logarithmic coordinates, Figure 5.25 shows the crack length pdf $h(c)$ variations underlying the strength behavior. The bold solid line and the fine lines labeled (i), (ii), and (iii) correspond to the same lines in Figure 5.24. As all responses were derived from the same material, identical $B = 1$ MPa $m^{1/2}$ values were used. The area under all curves is 1. The responses extend over different domains, and, as anticipated from Figure 5.24, the $h(c)$ variations exhibit peaks and domains at decreasing crack lengths as the size of the component decreases (i) to (iii). As perhaps also anticipated, the responses are of very similar shape and appear as diagonally translated versions, in clear distinction to Figure 5.20. Other diamond compositions (undoped, 10, 20 % N_2) exhibited similar deterministic strength behavior.

The deterministic size effects demonstrated here are not small or subtle. The peak in the crack length density of Figure 5.25 varies in position by about a factor of 2, consistent with the variation in the median strengths in Figure 5.24 by about a factor of 1.4. The means and standard deviations of the four samples of strengths in Figure 5.24, in order of increasing size, are: (3.2 ± 0.4) GPa, (2.8 ± 0.3) GPa, (2.6 ± 0.3) GPa, and (2.3 ± 0.3) GPa. The means barely overlap within the standard deviations; conventional inspection would suggest a substantial strength difference as the means vary by a factor of 1.5 and relative standard deviations are ≈ 0.15. The other diamond compositions exhibited similar strength variations. Without analysis beyond the simple means and standard deviations, however, it is very difficult to judge whether any of the strength variations arise from stochastic or deterministic size effects.

Figure 5.26 shows strength edf $Pr(\sigma)$ plots for three samples of commercial polycrystalline alumina components measured in biaxial flexure, from Simpatico et al. (1999) using data derived from the published work. The material in each sample was the same and only the test configuration differed. The bold solid line (i) is identical to that in Figure 5.3d and

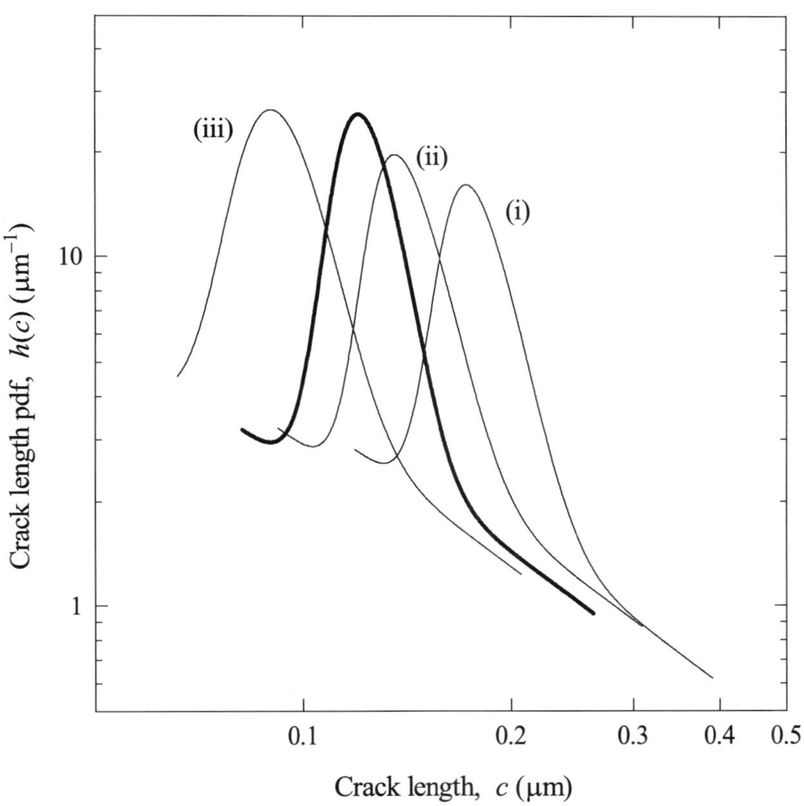

Figure 5.25 Plot in logarithmic coordinates of crack length pdf variations, $h(c)$, for polycrystalline diamond films tested in tension; determined from strength measurements of Figure 5.24. Labels (i)–(iii) correspond to those in Figure 5.24.

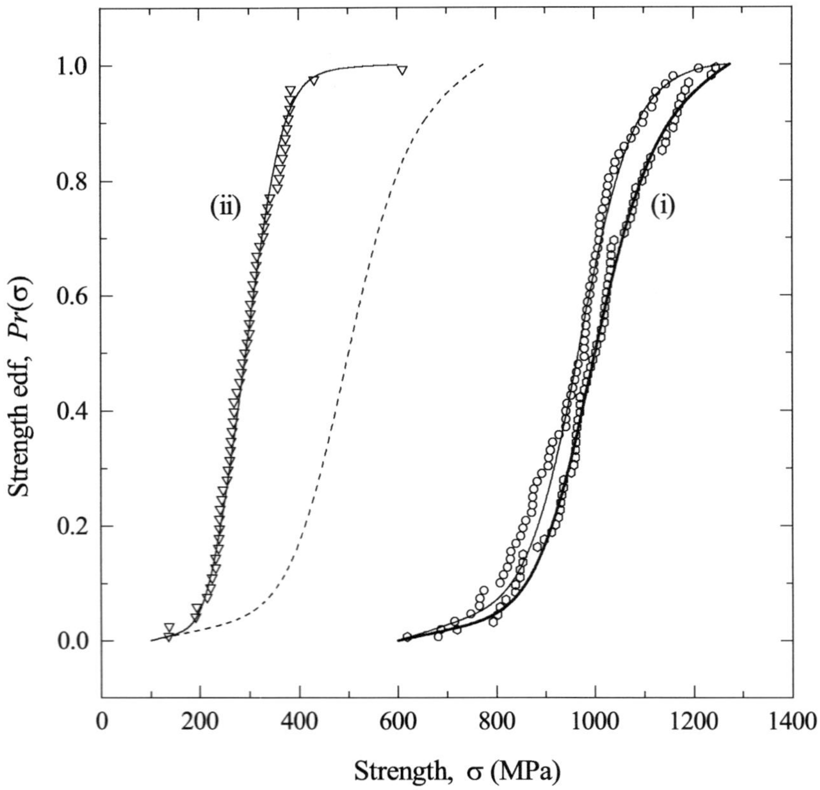

Figure 5.26 Plots of strength edf behavior, $Pr(\sigma)$, for three samples of a polycrystalline alumina material tested in flexure, increasing component size left to right, N_{tot} = 224 (Adapted from Simpatico, A et al. 1999). The larger strength responses originate from a common threshold indicating a stochastic size effect; the smaller strength response does not originate from the same threshold and therefore indicates a deterministic size effect.

represents the best fit to the strength distribution of 20 mm (external plan dimension) thin cylinder and square plate components tested using a common ball-on-ring biaxial flexure geometry. Symbols represent individual strength measurements of two other sets of components. The group of symbols at larger strengths represent 25 mm cylinder and square components similarly tested in biaxial flexure using a larger ball-on-ring geometry. The fine line at larger strengths is a visual best fit of $H_2(\sigma)$ to these data, assuming a stochastic size effect and Eq. (5.1), taking $H_1(\sigma)$ as the base 20 mm data. The larger component data are well described by the analysis, using a size ratio of $k_2/k_1 = 1.5$. The conclusion here is that these overlapping strength distributions of alumina are well described by stochastic size effects.

The group of symbols at smaller strengths in Figure 5.26 represent 20 mm components tested using the less common hydraulic burst biaxial flexure geometry. The hydraulic loading alters the spatial stress distribution in the component significantly, placing almost the entire flexure face of the component in biaxial tension, as opposed to tension localized in the face center of the ball-on-ring configuration. As a consequence, the effective size of the component is increased significantly, up to a factor of approximately 300 in area, and, perhaps more significantly, by a similar factor of 300 in volume (Simpatico et al. 1999). It is thus expected that the hydraulic burst configuration loads more, and different, flaws than the ball-on-ring geometry. The significantly smaller strengths and observations of failure from volume rather than surface flaws are consistent with this expectation and with a deterministic size effect. The dashed line in Figure 5.26 is a strength-translated version of the bold line to match the threshold of the hydraulic burst data. The dashed line does not match the hydraulic burst strength distribution in either magnitude or shape, suggesting that both ΔF and k influences in Eq. (5.3) are active in the size effect. The fine line at smaller strengths (ii) in Figure 5.26 represents an unconstrained visual best fit to the hydraulic burst data, corresponding approximately to a domain shift of $\Delta F \approx \sigma \rightarrow \sigma - 500$ MPa relative to the ball-on-ring tests and $k \approx 15$. The conclusion from these data is that the well separated fracture strength distributions of alumina in these two different test configurations are described by deterministic size effects.

Using logarithmic coordinates, Figure 5.27 shows the crack length pdf $h(c)$ variations underlying the strength behavior. The bold and fine lines labeled (i) and (ii) correspond to the same lines in Figure 5.26. As all responses were derived from

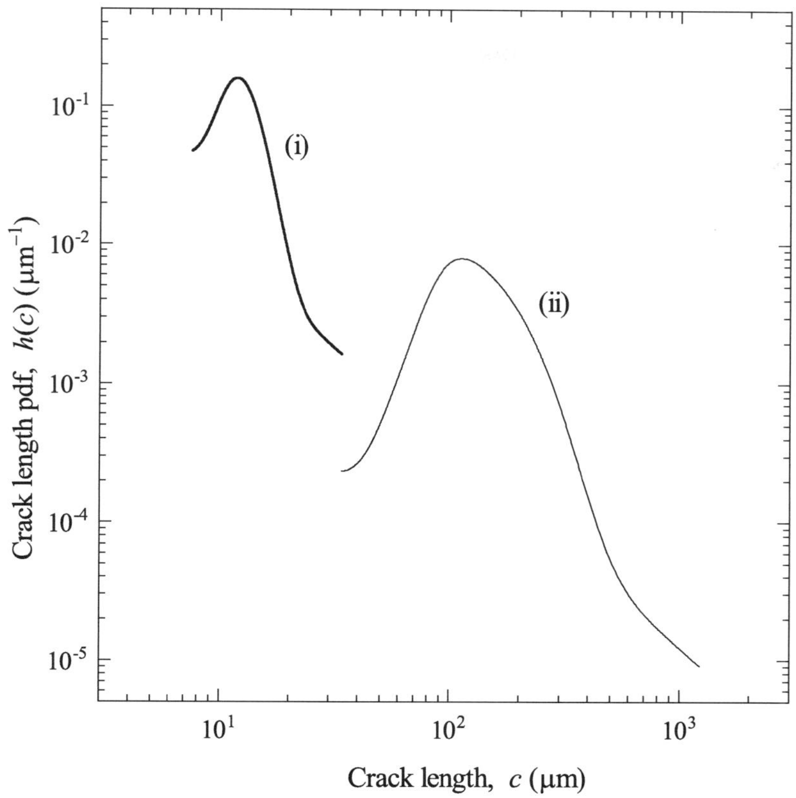

Figure 5.27 Plot in logarithmic coordinates of crack length pdf variations, $h(c)$, for polycrystalline alumina tested in flexure; determined from strength measurements of Figure 5.26. Labels (i) and (ii) correspond to those in Figure 5.26.

the same material, identical $B = 3.5$ MPa m$^{1/2}$ values were used. The area under each curve is 1. The responses extend over different domains, and, as anticipated from Figure 5.26, the $h(c)$ variations exhibit peaks and domains at increasing crack lengths as the effective size of the component increases (i) to (ii). As perhaps also anticipated, the responses are of very similar shape, but not scale, and appear as diagonally translated versions, in clear distinction to Figure 5.20.

Once again, the deterministic size effects demonstrated here are not small or subtle. The peak in the crack length density of Figure 5.27 varies in position by about a factor of 10, consistent with the variation in the median strengths in Figure 5.26 by about a factor of 3. The means and standard deviations of the three samples of strengths in Figure 5.24, in order of increasing size, are: (998 ± 120) MPa, (952 ± 113) MPa, and (296 ± 75) MPa. For the first two sets, tested in ball-on-ring geometry, the different sizes lead to a very small, almost indistinguishable, mean strength difference, consistent with stochastic size effects. The last set, tested in the hydraulic burst geometry that generates a much larger effective component size, exhibited a very small mean strength and a large strength difference from the first two sets, consistent with deterministic size effects. Other alumina and AlN components also exhibited small mean strengths in the hydraulic burst geometry (Simpatico et al. 1999). Although Simpatico et al. analyzed the above alumina strength data in terms of strength distributions, a primary assumption and motivation of the analysis was the extension of experimental assessment of stochastic size effects to extremely large component sizes. The use of linearized coordinates for analyzing strength distributions completely obscured the fact that the large strength difference observed was due to deterministic size effects. The observations here reinforce the point that without using unbiased coordinates, it is very difficult to judge whether strength variations arise from stochastic or deterministic size effects.

The two sets of strength distributions considered here, consisting of near-symmetric sigmoids, provided clear evidence of deterministic size effects—as component size increased, sample strength distributions contracted and translated to smaller strengths. Such behavior is inconsistent with stochastic size effects. The translation reflected the appearance of new, larger extreme flaws within a sample as component size increased. The issue in materials processing and component design in this case is then identifying the element of "component size" that controls the size of the flaws. As discussed above in the context of stochastic size effects, depending on the nature of the flaws the number of flaws in a component may scale with component length, area, or volume. The same is true of the size of the flaws and thus the related strengths of components controlled by deterministic size effects. In the experiments and analyses here, the relative lower bound strengths of samples of different size components is given by σ_{L1}/σ_{L2}. The strengths σ_{L1} and σ_{L2} are the empirically fit lower bounds of distributions, as specified in Eq. (4.1) and used throughout. The ratio of the conjugate crack length upper bounds is then given by $c_{U2}/c_{U1} = (\sigma_{L1}/\sigma_{L2})^2$. For stochastic size effects, the crack length ratio is 1 by definition. For deterministic size effects, comparison of the crack length ratio c_{U2}/c_{U1} with the ratio of component sizes l_2/l_1 discussed above can provide the required scaling. The simplest case is one-dimensional: Strength-controlling flaw sizes scale linearly with length dimension l of components that are loaded in tension along l. However, as component dimensions and component features transverse to l that control flaw size may not scale with l, e.g. fiber diameters, surface roughness on films, there is reason to expect sublinear scaling.

As above, generalized versions of the crack length and size parameter ratios c/c_{ref} and l/l_{ref} enable comparison of multiple systems. Values of c and l are compared to a single reference configuration c_{ref} and l_{ref} within a given system and comparison between systems is thus relative to common reference point (1,1). Multi-dimensional systems, e.g. bend bars, are included as above by expressing the stressed volume of a component as $V = l^3$. Figure 5.28 shows a plot in logarithmic coordinates of the relative size of flaws c/c_{ref} vs the relative component size l/l_{ref} for several extended component systems that exhibited deterministic size effects. The medians of strength distribution best fits were matched to reference distribution fits and the ratios of the resulting fit thresholds used to determined c/c_{ref}. The solid line is a guide to the eye and has slope 1/2. The solid symbols indicate measurements from the diamond films of Figure 5.24 from the work of Peng et al. (2007). The open symbols include measurements from the convex strength distributions describing tensile tests of Al_2O_3 and SiC fibers from the work of Goda and Fukunaga (1986), measurements from the sigmoidal strength distributions describing flexure tests of alumina cylinders and plates from the work of Simpatico et al. (1999), measurements from the sigmoidal strength distributions describing bending tests of Si beams from the work of Namazu et al. (2000), measurements from the weakly sigmoidal strength distributions describing flexure and burst tests of SiC shells from the work of Hong et al. (2007) (somewhat similar to Simpatico et al. (1999)), measurements from the linear strength distributions describing tensile tests

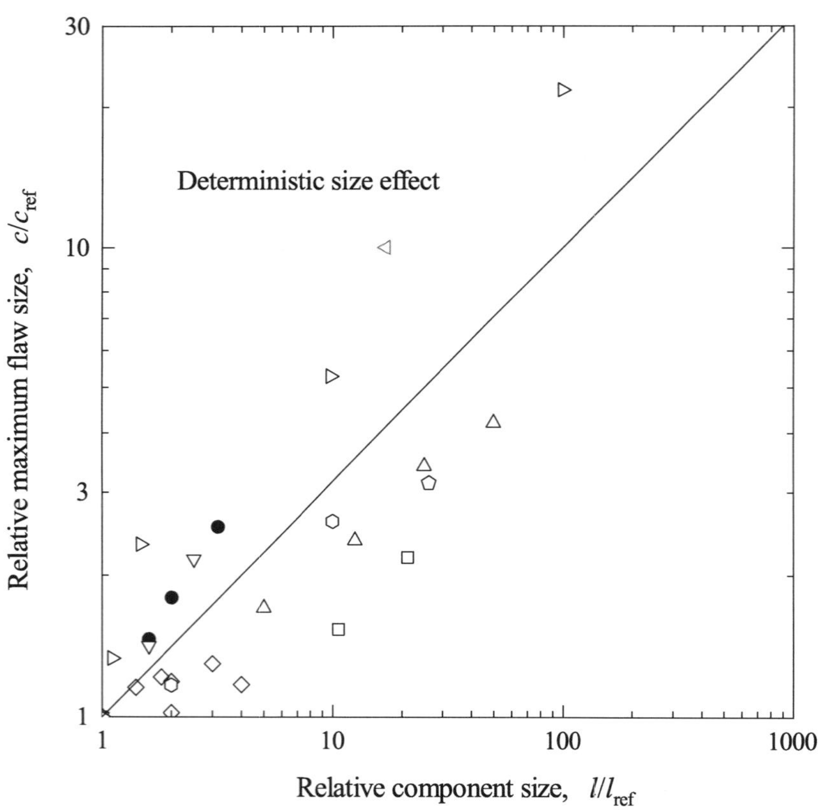

Figure 5.28 Plot of maximum size of flaws c vs component size l for extended components exhibiting deterministic size effects on strength distributions. Line is guide to the eye indicating scaling behavior; most systems behave sublinearly.

of C fibers from the work of Naito et al. (2012), and measurements from the weakly sigmoidal and linear strength distributions describing bending tests of concrete beams from the work of Koide et al. and Zi et al. as cited by Lei and Yu (2016). Overall, maximum crack sizes scale weakly, sublinearly, with component size in deterministic systems. A large crack size change in Figure 5.28, a factor of 5, responsible for more than halving the threshold strength, required a factor of 10 increase in component size. Deterministic size effects on strength are typically greater than stochastic size effects, but still require very large changes in component size. (The limited number of points on Figure 5.28 is in large part due to the requirement here that strength distributions contained sufficient information to enable estimation of threshold strengths from fits to the distribution. Estimating the threshold from the smallest observed strength leads to great uncertainty).

It must be noted, again, that in many prior works the application of distorting and obscuring linearized plotting schemes has impeded identification of deterministic size effects on strength distributions. For example, in the extensive studies of Namazu et al. (2000) and Peng et al. (2007), length, area, and volume scaling of stochastic effects of component size on strength were considered, but only in linearized coordinates. Namazu et al. made no conclusion regarding scaling and Peng et al. concluded that volume scaling was the best description. In both cases, unbiased assessment of the strength data, as in Figure 5.24, would have identified that the size effects were deterministic, not stochastic, and could have been interpreted in terms of the associated extensive microscopy observations.

5.3.3 Size Effect Summary

This section has surveyed in detail the two major size effects—stochastic and deterministic—on strength edf $Pr(\sigma)$ behavior of extended components. Stochastic size effects influence the strength distributions of components by altering the probability that the extreme, largest, strength-controlling flaw in a component is of a certain size. Such a component contains independent flaws extracted from an invariant population that describes the probability distribution of flaws in components of all sizes. As a consequence of the population invariance and flaw independence, strength distributions determined by stochastic extreme value size effects exhibit invariant strength domains and a fixed analytic relationship between samples of components of different sizes. A key experimental identifier for stochastic size effects is an invariant lower bound threshold strength exhibited by all samples. Such behavior was demonstrated here in analysis of polycrystalline alumina bend beams and flax tensile fibers. These and additional data derived from other extended component strength studies showed that stochastic size effects in these systems were consistent with the number of flaws in components scaling linearly with component size. The invariant threshold and the linear scaling lead to small variations in strength with component size in stochastic systems. Deterministic size effects influence the strength distributions of components by determining the population of flaws as a function of component size. The extreme, largest, strength-controlling flaw in a component is still controlled by probabilistic selection and the flaws are still independent, but the population is not component size invariant. As a consequence of the population variation, strength distributions controlled by deterministic extreme value size effects exhibit varying strength domains and no fixed analytic relationship between samples of components of different sizes. A key experimental identifier for deterministic size effects is a threshold strength that varies with component size. Such behavior was demonstrated here in analysis of polycrystalline diamond tensile bars and alumina flexure plates and thin cylinders. These and additional data derived from other extended component strength studies showed that deterministic size effects in these systems were consistent with the maximum size of flaws in components scaling sublinearly with component size. In deterministic systems, the varying threshold leads to large variations in strength with component size. A conclusion to be drawn from this section is that although extended components of brittle materials clearly exhibit both stochastic and deterministic size effects on strength distributions across a wide range of materials and forms of $Pr(\sigma)$, definitive demonstration of such effects is not frequent.

5.4 Discussion and Summary

This Chapter 5, has provided a firm foundation for consideration of particle strength distributions. The Chapter has surveyed in qualitative and quantitative detail the brittle fracture strength distributions of "conventional" extended components. Such components are usually physically large, such that they contain many strength-controlling cracks and the survey considered a wide variety of materials, loading configurations, and size effects in the strength distributions of samples of extended components. The quantitative aspect of the survey drew heavily and repeatedly on the probabilistic framework and procedures developed and demonstrated in Chapters 3 and 4 to relate observed strength distributions to underlying crack probability densities. Broad characterizations of extended component strength distribution behavior against which to view particle strength distribution behavior are that (i) extended component strength distributions are predominantly sigmoidal, although linear, convex, and concave perturbations are observed; (ii) the relative width of the sample strength domains of extended component distributions is approximately 2; (iii) the conjugate extreme crack length density functions of samples of extended components are usually well formed peaks with relative widths of approximately 4; (iv) samples of extended components of different sizes exhibited both stochastic and deterministic size effects with the former exhibiting small variations in strength and a fixed threshold and the latter exhibiting large variations in strength and a variable threshold; (v) no significant trend in these characteristics was observed with material type (amorphous, crystalline, polycrystalline, composite, porous) or loading configuration (tension, compression, bending). Particles are similar to extended components in that they are formed from the same range of materials. However, particles are expected to differ from extended components in that they will usually be physically small, and therefore contain few strength-controlling cracks, and the loading configuration will be restricted, to diametral compression. These two physical factors are likely to significantly influence size effects in particle strength distributions.

A qualitative description of size effects in extended components, representing in pictorial form the quantitative description used here in Section 5.3, provides additional physical insight into stochastic and deterministic effects and an additional foundation for consideration of particle strength distributions. Figure 5.29 is a schematic diagram of a sample of $N = 5$ planar tensile components. The loading leading to tensile stress is indicated by arrows. The components are of uniform size and each consists of $k = 10$ fundamental volume elements. Each fundamental volume element contains (by definition) one flaw, shown as a bold line. Not all flaws are the same size (for illustration, orientations shown also differ) as the flaw sizes reflect an overall distributed, but invariant flaw population. The components are regarded as extended, as every volume element and flaw experiences maximum, uniform, stress. Failure ensues from the largest, extreme flaw within the ensemble of flaws contained by each component. The volume element of the failure flaw is shown shaded and failure is independent of location within a component. The distribution of strengths for the sample reflects the extreme value distribution of flaws. The majority of flaws within the ensembles and the overall population of flaws are not sensed by strength measurements, which are biased by extreme value effects toward the large flaw, small strength threshold.

Figure 5.30 illustrates stochastic size effects and is a schematic diagram of a similar sample of planar tensile components. In this case the components are larger and consist of $k = 20$ fundamental volume elements and flaws drawn from the same invariant population as depicted in Figure 5.29. The volume element of the failure flaw is shown shaded and failure is again independent of location within a component. Consideration of the failure flaws shows them to be slightly larger than those in Figure 5.29. Similar consideration of the non-failure flaws shows them to be slightly smaller. Both effects are a consequence of a greater number of stochastic selections. The extreme value distribution of flaws thus includes more large flaws and the resulting distribution of strengths for the sample thus includes more small strengths. The majority of (non-failure) flaws within the ensembles and the overall population of flaws are again not sensed by strength measurements, which are now more biased toward the large flaw, small strength threshold. Hence, for purely stochastic selection reasons, the strength distribution tends to smaller values as the component size increases.

Deterministic size effects occur in several ways, but all involve more physically based processes. The simplest deterministic size effect is illustrated in Figure 5.31, which is a schematic diagram of a sample of planar tensile components similar

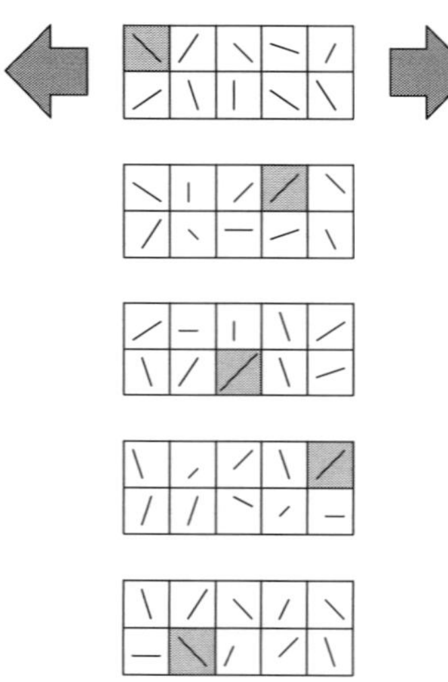

Figure 5.29 Schematic diagram illustrating extreme value effects in brittle failure of a sample of extended components. Applied tensile stress is indicated by arrows. Sample size $N = 5$. Component size $k = 10$. Flaws indicated by bold lines; strength-controlling extreme flaws responsible for failure indicated shaded. Note that failure occurs at largest flaws, independent of location.

to those in Figure 5.30. In this case the components still consist of $k = 20$ fundamental volume elements but here the flaws are drawn from a different population, larger than those depicted in Figure 5.29. The volume element of the failure flaw is again shown shaded and is independent of location within a component. For this population, both the failure flaws and the non-failure flaws are larger than those in Figure 5.29. The extreme value distribution of flaws thus includes larger flaws and the resulting distribution of strengths for the sample thus includes smaller strengths. In this case, as the populations of flaws in Figures 5.31 and 5.29 do not overlap, the resulting strength distributions would also not overlap. As extreme value effects still apply, overlapping sample strength distributions require overlapping domains of flaws in the large flaw regions. This effect appears to be common, e.g. Figure 5.24.

Another deterministic size effect is illustrated in Figure 5.32 and is a schematic diagram of a sample of planar tensile components. In this case, the components consist of $k = 5$ fundamental volume elements and both the volume elements and the flaws are larger than those depicted in Figure 5.29, although the components are physically the same size. The volume element of the failure flaw is again shown shaded and is independent of location within a component. As in Figure 5.31 both the failure flaws and the non-failure flaws are larger than those in Figure 5.29. The extreme value distribution of flaws thus includes larger flaws and the resulting distribution of strengths for the sample thus includes smaller strengths. In this case, as the populations of flaws in Figures 5.32 and 5.29 do not overlap, the resulting strength distributions would also not overlap. Comparison of Figures 5.31 and 5.32 suggests that interaction or coalescence of adjacent flaws in adjacent volume elements is a potential mechanism for increasing the sizes of both flaws and volume elements. This mechanism, unlike the deterministic size effect above, leads to decreases in strengths with increases in component size by removing flaw independence. Such ideas have been considered by Hunt (1978) and Todinov (2001). Retaining flaw size and flaw independence while increasing the volume element size is equivalent to reducing flaw spatial density or making flaws more sparse, and has the effect of increasing the fraction of large strengths within an invariant strength domain. The limit of this

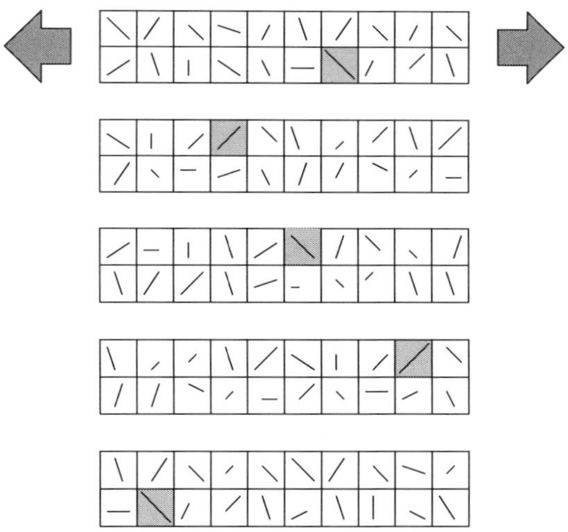

Figure 5.30 Schematic diagram illustrating stochastic extreme value effects in brittle failure of a sample of extended components. Applied tensile stress is indicated by arrows. Sample size $N = 5$. Component size $k = 20$ is double that of Figure 5.29. Flaws indicated by bold lines; strength-controlling extreme flaws responsible for failure indicated shaded. Note that failure flaws are slightly larger than those in Figure 5.29.

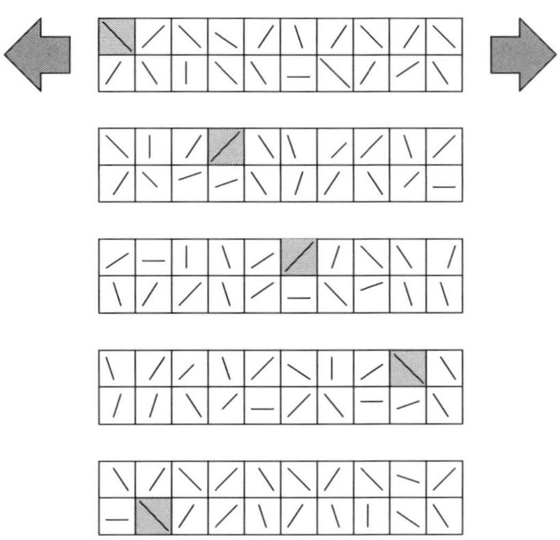

Figure 5.31 Schematic diagram illustrating deterministic extreme value effects in brittle failure of a sample of extended components. Applied tensile stress is indicated by arrows. Sample size $N = 5$. Component size $k = 20$ is double that of Figure 5.29. Flaws indicated by bold lines; strength-controlling extreme flaws responsible for failure indicated shaded. Note that all flaws are larger than those in Figure 5.29 and Figure 5.30.

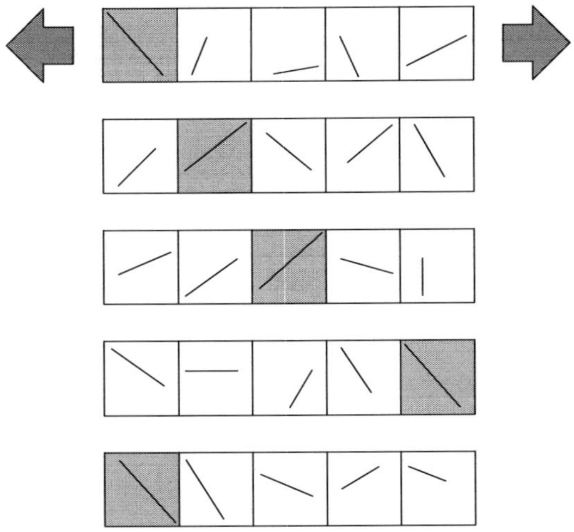

Figure 5.32 Schematic diagram illustrating deterministic extreme value effects in brittle failure of a sample of extended components. Applied tensile stress is indicated by arrows. Sample size $N = 5$. Component size $k = 5$ is half that of Figure 5.29. Flaws indicated by bold lines; strength-controlling extreme flaws responsible for failure indicated shaded. Note that all flaws are larger than those in Figure 5.29.

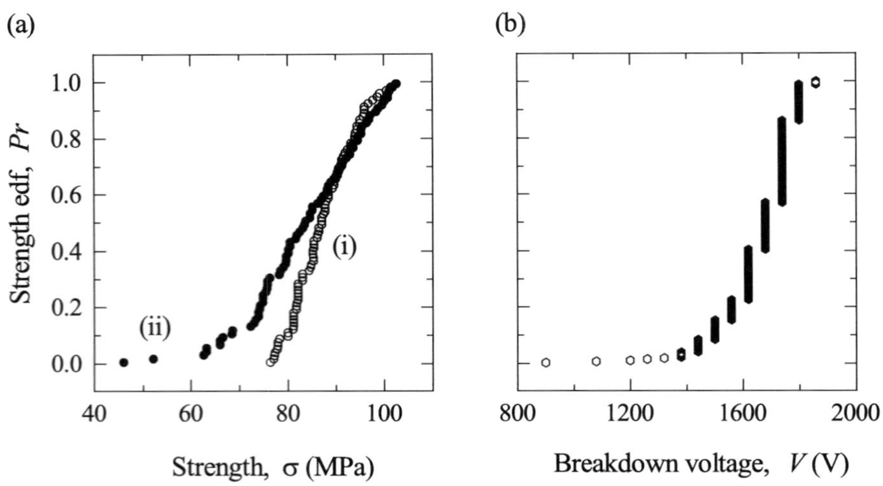

Figure 5.33 Plots of strength edf behavior, *Pr*, from early works considering strength distributions. (a) Samples of two porcelain materials tested in bending, $N_{tot} = 166$ (Adapted from Weibull, W 1939). (b) Sample of paper capacitor material tested electrically, $N_{tot} = 250$ (Adapted from Epstein, B et al., 1948).

effect is one or two flaws per component, such that the sample strength distribution approaches that of the population, and this is an expectation for some particle systems.

Before closing this survey it is useful to place the observations here in the context of the early works. Figure 5.33 shows mechanical and electrical strength edf plots for (a) two samples of porcelain components measured mechanically in bending, from Weibull et al. (1939) and (b) a sample of paper capacitor material measured electrically to breakdown, from Epstein and Brooks (1948), in both cases using data derived from the published work. Symbols represent individual strength measurements. Data (i) in (a) are identical to those in Figure 5.7a. The plot formats in Figure 5.33 are identical to those used throughout this chapter. The data in (a) were originally presented in linearized form. The data in (b) were originally presented in tabular form. None of the three data sets would be regarded as overtly sigmoidal; one is nearly linear, (a)(i), two are convex, (a)(ii) and (b), but none stand out from the range of edf responses presented here. The data in (a) were presented as part of a survey of many different types of distributions and were described by data-shifted versions of Eq. (4.10)c. The data in (b) were presented as an example of the need to consider asymmetric probability densities and were described by a data-shifted version of Eq. (4.10)a. It is perhaps ironic that both sets of data in (a) were presented in linearized coordinates: (a)(i) is near linear in such coordinates, but not sigmoidal as the coordinates were intended to portray; (a)(ii) is non-linear in such coordinates, but linear in absolute coordinates, as is the case with some of the data sets examined here, e.g. Figure 5.7. (It is not recorded whether the analyst, Miss Wiwica Weibull, noted these behaviors.) The long small-strength tails in Figure 5.33 were observed here in Figure 5.5 and may well represent two concurrent, differently sized, flaw populations, leading to "bimodal" effects (if widely separated) as suggested by Goda and Fukunaga (1986). (This possibility has been explicitly incorporated into the fitting routine here, as y_1 and y_2 are independent elements of the fitting function—see Chapter 4). Consideration of the early works reinforces the point that assumed distributions and linearized plotting coordinates may obscure information more clearly apparent in unbiased absolute coordinates.

The survey of strength distributions of extended components presented in this Chapter 5, provides background and foundation for an equivalent survey of strength distributions of particles presented in the following chapter, Chapter 6. Identical analytical procedures are implemented to relate strengths to underlying crack lengths and similar behavior regarding material and size effects will be investigated. In contrast to the components considered here, particles are often small and tensile stress generated by loading is usually localized—strength behavior is expected to reflect these differences.

References

Akatsu, T., Tomiyasu, M., Shingae, T., and Kamochi, N. (2020). Strengthening in porcelain reinforced with alumina particles. *Journal of the Ceramic Society of Japan* 128: 1045–1054.

Al-Manaseer, A., Nadeem, M.S., Magenti, R., and Lee, P. (2011). Strength unit weight and elasticity of concrete cylinders for the Benicia Martinez bridge. Report No. CA 10-1862 California Department of Transportation, Sacramento, California.

Andersons, J., Sparnins, E., Porike, E., and Joffe, R. (2008). Strength distribution of elementary flax fibres due to mechanical defects. In *International Inorganic-Bonded Fiber Composites Conference* 247–253.

Boyce, B.L. (2010). A sequential tensile method for rapid characterization of extreme-value behavior in microfabricated materials. *Experimental Mechanics* 50: 993–997.

Carty, W.M. and Senapati, U. (1998). Porcelain–raw materials, processing, phase evolution, and mechanical behavior. *Journal of the American Ceramic Society* 81: 3–20.

Cook, R.F. (2017). Fracture mechanics of sharp scratch strength of polycrystalline alumina. *Journal of the American Ceramic Society* 100: 1146–1160.

Cook, R.F. (2020). Single particle strength distributions: Heavy tails and extreme values. http://doi.org/10.5281/zenodo.4024618 (accessed July 4, 2021).

Cook, R.F. (2021). Edge chipping at small scales and strengths of diced components. http://doi.org/10.5281/zenodo.4924668 (accessed February 12, 2022).

Cook, R.F. and DelRio, F.W. (2019a). Material flaw populations and component strength distributions in the context of the Weibull function. *Experimental Mechanics* 59: 279–293.

Cook, R.F. and DelRio, F.W. (2019b). Determination of ceramic flaw populations from component strengths. *Journal of the American Ceramic Society* 102: 4794–4808. (typographical error in Eq. (9)).

Cook, R.F., DelRio, F.W., and Boyce, B. L. (2019). Predicting strength distributions of MEMS structures using flaw size and spatial density. *Microsystems & Nanoengineering* 5: 1–12.

DelRio, F.W., Cook, R.F., and Boyce, B.L. (2015). Fracture strength of micro-and nano-scale silicon components. *Applied Physics Reviews* 2: 021303.

de Wit, P., van Daalen, F.S., and Benes, N.E. (2017). The mechanical strength of a ceramic porous hollow fiber. *Journal of Membrane Science* 524: 721–728.

de With, G. (1984). Note on the use of the diametral compression test for the strength measurement of ceramics. *Journal of Materials Science Letters* 3: 1000–1002.

Easler, T.E., Bradt, R.C., and Tressler, R.E. (1982). Effects of oxidation and oxidation under load on strength distributions of Si_3N_4. *Journal of the American Ceramic Society* 65: 317–320.

Epstein, B. (1948). Statistical aspects of fracture problems. *Journal of Applied Physics* 19: 140–147.

Epstein, B. and Brooks, H. (1948). Application of the theory of extreme values in fracture problems. *Journal of the American Statistical Association* 43: 403–412.

Evans, A.G. and Jones, R.L. (1978). Evaluation of a fundamental approach for the statistical analysis of fracture. *Journal of the American Ceramic Society* 61: 156–160.

Gaither, M.S., Gates, R.S. Kirkpatrick, R., Cook, R.F., and DelRio, F.W. (2013). Etching process effects on surface structure, fracture strength, and reliability of single-crystal silicon theta-like specimens. *Journal of Microelectromechanical Systems* 22: 589–602.

Gee, M.G. (1984). Brittle fracture of hardmetals: Dependence of strength on defect size distribution. *International Journal of Mechanical Sciences* 26: 85–91.

Goda, K. and Fukunaga, H. (1986). The evaluation of the strength distribution of silicon carbide and alumina fibres by a multi-modal Weibull distribution. *Journal of Materials Science* 21: 4475–4480.

Hong, S.G., Byun, T.S., Lowden, R.A., Snead, L.L., and Katoh, Y. (2007). Evaluation of the fracture strength for silicon carbide layers in the tri-isotropic-coated fuel particle. *Journal of the American Ceramic Society* 90: 184–191.

Hunt, R.A. (1978). A theory of the statistical linking of microcracks consistent with classical reliability theory. *Acta Metallurgica* 26: 1443–1452.

Jonke, M., Klünsner, T., Supancic, P., Harrer, W., Glätzle, J., Barbist, R., and Ebner, R. (2017). Strength of WC-Co hard metals as a function of the effectively loaded volume. *International Journal of Refractory Metals and Hard Materials* 64: 219–224.

Kingery, W.D., Bowen, H.K., and Uhlmann, D.R. (1975). *Introduction to Ceramics*. Wiley.

Klünsner, T., Wurster, S., Supancic, P., Ebner, R., Jenko, M., Glätzle, J., Püschel, A., and Pippan, R. (2011). Effect of specimen size on the tensile strength of WC-Co hard metal. *Acta Materialia* 59: 4244–4252.

Klünsner, T., Lube, T., Gettinger, C., Walch, L., and Pippan, R. (2020). Influence of WC-Co hard metal microstructure on defect density, initiation and propagation kinetics of fatigue cracks starting at intrinsic and artificial defects under a negative stress ratio. *Acta Materialia* 188: 30–39.

Kobayashi, Y., Ohira, O., and Isoyama, H. (2003). Effect of cristobalite formation on bending strength of alumina-strengthened porcelain bodies. *Journal of the Ceramic Society of Japan* 111: 122–125.

Kobayashi, Y., Ohira, O., Satoh, T., and Kato, E. (1994). Effect of quartz on the sintering and bending strength of the porcelain bodies in quartz-feldspar-kaolin system. *Journal of the Ceramic Society of Japan* 102: 99–104.

Kotchick, D.M., Hink, R.C., and Tressler, R.E. (1975). Gauge length and surface damage effects on the strength distributions of silicon carbide and sapphire filaments. *Journal of Composite Materials* 9: 327–336.

Kurkjian, C.R., Albarino, R.V., Krause, J.T., Vazirani, H.N., DiMarcello, F.V., Torza, S., and Schonhorn, H. (1976). Strength of 0.04–50-m lengths of coated fused silica fibers. *Applied Physics Letters* 28: 588–590.

Lamon, J., R'Mili, M., and Reveron, H. (2016). Investigation of statistical distributions of fracture strengths for flax fibre using the tow-based approach. *Journal of Materials Science* 51: 8687–8698.

LaVan, D.A. and Buchheit, T.E. (1999) Strength of polysilicon for MEMS devices. *MEMS Reliability for Critical and Space Applications* 3880: 40–44. International Society for Optics and Photonics.

Lei, W.S. and Yu, Z. (2016). A statistical approach to scaling size effect on strength of concrete incorporating spatial distribution of flaws. *Construction and Building Materials* 122: 702–713.

Meininger, S., Mandal, S., Kumar, A., Groll, J., Basu, B., and Gbureck, U. (2016). Strength reliability and *in vitro* degradation of three-dimensional powder printed strontium-substituted magnesium phosphate scaffolds. *Acta Biomaterialia* 31: 401–411.

NIST (1996). *Ceramic Strength*. https://www.itl.nist.gov/div898/handbook/eda/section4/eda42a.htm (accessed January 29, 2022).

Peng, B., Li, C., Moldovan, N., Espinosa, H.D., Xiao, X., Auciello, O., and Carlisle, J.A (2007). Fracture size effect in ultrananocrystalline diamond: Applicability of Weibull theory. *Journal of Materials Research* 22: 913–925.

Phani, K.K. and De, A.K. (1987). A flaw distribution function for failure analysis of brittle materials. *Journal of Applied Physics* 62: 4433–4437.

Preston, F.W. (1937). Concerning the strength of the weakest bottles: Applicability of the "normal" curve of errors to statistical analyses of strength tests of glassware. *Journal of the American Ceramic Society* 20: 329–336.

Quinn, G.D. (1987). Delayed failure of a commercial vitreous bonded alumina. *Journal of Materials Science* 22: 2309–2318.

Quinn, G.D. (1989). Flexure strength of advanced ceramics-A round robin exercise. MTL TR 89–62. US Army Materials Technology Laboratory, Watertown MA.

Saleh, M.E., Beuth, J.L., and de Boer, M.P. (2014). Validated prediction of the strength size effect in polycrystalline silicon using the three-parameter Weibull function. *Journal of the American Ceramic Society* 97: 3982–3990.

Shelleman, D.L., Jadaan, O.M., Conway Jr., J.C., and Mecholsky Jr., J.J. (1991). Prediction of the strength of ceramic tubular components: Part II—Experimental verification. *Journal of Testing and Evaluation* 19: 192–200.

Simpatico, A., Cannon, W.R., and Matthewson, M.J. (1999). Comparison of hydraulic-burst and ball-on-ring tests for measuring biaxial strength. *Journal of the American Ceramic Society* 82: 2737–2744.

Todinov, M.T. (2001). Estimating the probabilities of triggering brittle fracture associated with the defects in the materials. *Materials Science and Engineering: A* 302: 235–245.

Usami, S., Kimoto, H., Takahashi, I., and Shida, S (1986). Strength of ceramic materials containing small flaws. *Engineering Fracture Mechanics* 23: 745–761.

Van Rooyen, G.T., Du Preez, R., De Villiers, J. and Cromarty, R. (2010). The fracture strength of TRISO-coated particles determined by compression testing between soft aluminium anvils. *Journal of Nuclear Materials* 403: 126–134.

Weibull, W. (1939). A statistical theory of the strength of materials. *Proceedings of the Royal Swedish Institute of Engineering Research* 151: 1–45.

Wereszczak, A.A., Kirkland, T.P., and Jadaan, O.M. (2007). Strength measurement of ceramic spheres using a diametrally compressed "C-sphere" specimen. *Journal of the American Ceramic Society* 90: 1843–1849.

Xie, X., Liu, B., Guo, Y., Liu, R., Zhao, X., Ni, N., Guo, F. and Xiao, P. (2019). Effect of hydrothermal corrosion on the fracture strength of SiC layer in tristructural-isotropic fuel particles. *Journal of the American Ceramic Society* 102: 5555–5564.

6

Survey of Particle Strength Distributions

This chapter uses the analysis of Chapter 4 to survey the strength distributions observed for representative particle systems and compares and contrasts particle behavior with that of extended components in terms of strengths and flaw sizes. The goal of this chapter is to demonstrate the clear difference in extreme value behavior of particle strengths relative to extended component strengths. That is, the focus is on the science and physics of particle failure. Detailed engineering aspects such as particle size effects and particle material effects are not considered here.

6.1 Introduction

Analyses and methodology for determination of crack length distributions underlying strength behavior of brittle components were considered in detail in Chapter 4. The precision of the methodology was verified by its application to several sets of computationally generated data. In Chapter 5, extensive application of the methodology was demonstrated in a comprehensive survey of a wide variety of strength distributions describing "conventional" extended structural components. The survey made clear the pervasive nature of sigmoidal strength distributions exhibited by such components, and the underlying peaked populations of the strength-controlling crack lengths. Although material-to-material variations in the detailed shapes of the sigmoidal distributions were observed, the survey of Chapter 5 provides a firm foundation to assess the strength and crack length behavior of other components. In particular, in this chapter, attention is focused on the brittle fracture strength and crack length behavior of *particles*. Particles are considered here to be compact objects that are loaded in diametral compression such that a localized zone of tension is generated in the center of a particle (described in detail in Chapter 2). As a consequence of the localization, only a few strength-controlling flaws and associated cracks are exposed to significant stress on loading. Typically, particles are physically small, although this is not a necessary condition: The localisation of stress characterizes a particle, just as the spatially distributed stress generated on loading characterizes an extended component, which need not be physically large (described in detail in Chapter 5). In consideration of strength distributions, the major implication of the particle loading configuration is that the distributions are likely to be influenced only weakly by stochastic extreme value size effects.

This chapter surveys experimental observations of strength distributions, and thus crack lengths, of particles—observations that extend back over 50 years. The observations are directly analogous to those of extended components and are presented in three sections. In this current section, analyses of particle strength data are illustrated through consideration of a prototypical example of a particle material, alumina. The strength and crack length characteristics of alumina particles are considered in the context of the crack length characteristics of extended alumina components of similar strength. The example will demonstrate the techniques used for analysis of particle strength in this chapter and throughout. The example will also illustrate the clear difference between the strength distribution of a sample of extended components, typically sigmoidal, and the strength distribution of a sample of particles, typically concave. In detail, in order to accommodate and compare the concave behavior all particle $Pr(\sigma)$ strength edf data will be fit by the *bilinear* interpolated continuum form (unless otherwise noted). The bilinear form is described by Eqs. (4.1)–(4.5) with y_2 suppressed. The strength fits will be deconvoluted into crack length pdf variations as described in Chapter 4 and notated as $h(c)$. The B values used in deconvolution will be based on materials considerations and constrained such the largest inferred crack length is smaller than the particle size. As in Chapter 5, all fitting and deconvolution steps are performed computationally and unless required for discussion, parameters associated with fitting steps and uncertainties are not given. In analogy with Section 5.2, the next

Particle Strengths: Extreme Value Distributions in Fracture, First Edition. Robert F. Cook.

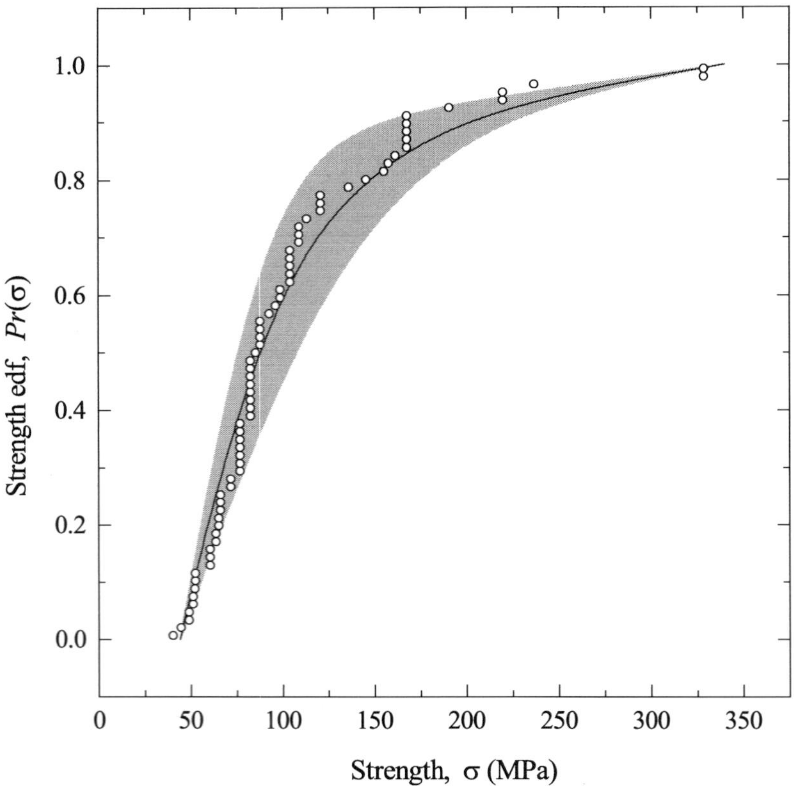

Figure 6.1 Plot of strength edf behavior, $Pr(\sigma)$, for alumina particles. Particle diameter, D = 1.29 (estimated $\pm$ 0.15) mm, number of measurements, N = 73 (Adapted from Huang, H et al. 1995). Solid line shows concave best fit, shaded band shows bounds.

section considers the detailed behavior of typical particle materials: alumina, quartz, limestone, and rock, illustrating a range of phenomena. This is followed by three sections surveying the behavior of particles classified by size. These sections are less detailed than the previous section and "size" here simply refers to the classification system, not to any size effects. Stochastic and deterministic size effects on particle strengths are considered in detail in Chapters 7–9 and the survey here provides context for such considerations by including many single-size examples of particle strengths of similar materials. Particles of both natural and engineered materials are considered; fabrication details can be found in the cited works.

Figure 6.1 shows a strength edf $Pr(\sigma)$ plot for a sample of polycrystalline alumina abrasive particles (Al_2O_3, often referred to as corundum), approximately 1 mm in diameter tested in a rotating roller configuration, from the work of Huang et al. (1993, 1995), using data derived from the published works cited. Here and following, as in earlier chapters: details of component and sample sizes, diameters of particles and numbers of measurements, respectively, are provided in the figure caption; $Pr(\sigma)$ plots are provided in an unbiassed format in linear coordinates, the abscissa and ordinate extending from approximately the minimum observed strength, σ_1 to approximately the maximum observed strength, σ_N, and from -0.1 to 1.1, respectively. Symbols represent individual strength measurements. The line in Figure 6.1 represents represents the visual best fit to data and the shaded band represent bounds on the fit encompassing most of the observations. The data are well described by concave behavior, and the data set exhibits a strength domain width σ_N/σ_1 of approximately 7. Comparison with earlier results, shows that both the shape and the extent of the strength response are very different from those of alumina extended components, Figure 5.3b, and similar to the carbonaceous sand particles of Figure 4.11.

Figure 6.2 shows as a bold line the crack length pdf $h(c)$ variation underlying the alumina particle strength behavior. The $h(c)$ variation was determined using $B = 1.5$ MPa m$^{1/2}$. The fine line shows the $h(c)$ variation from Figure 5.4 describing alumina extended component behavior. As might be anticipated from the strength responses, the two crack length pdf $h(c)$ variations are very different. The particle crack population extends over a wide crack length domain with little variation. The extended component crack population is concentrated in a narrow crack length domain and exhibits substantial variation. The dotted line in Figure 6.2 indicates zero, and the particle pdf $h(c)$ is non-zero at the upper domain bound, suggesting that the particle crack population is "heavy tailed" (see Chapter 3). The particle crack length population tail behavior is more clearly observed in Figure 6.3, which shows an expanded view and includes crack length uncertainty bounds (shown shaded) derived from the strength fit bounds. Within experimental uncertainty, the strength-controlling crack population

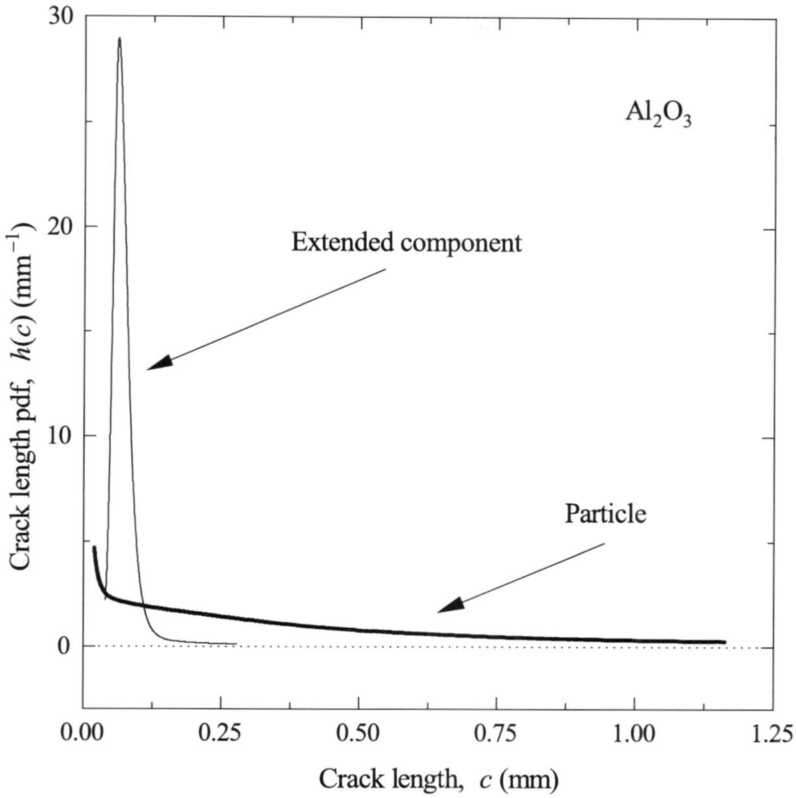

Figure 6.2 Plot of crack length pdf $h(c)$ variations for alumina. Particle response shown as bold line; determined from Figure 6.1. Extended component response shown as fine line; determined from data in Chapter 5 (Adapted from Quinn, G.D 1989).

density of the particles does not reach zero at the upper bound of the crack length domain, similar to that observed earlier for carbonaceous sand (Figure 4.12). Particle crack length populations are heavy tailed and lead to particle strength distributions that are concave. A comparison of alumina particle and extended component crack length population densities $h(c)$ is shown in logarithmic coordinates of Figure 6.4. Particle behavior is represented by the bold line and shaded band indicating uncertainty; extended component behavior is represented by the fine line. The extended component behavior is peaked with a clear mode. The particle behavior is an undulating decrease from a maximum at the lower bound and no intermediate mode. The increased domain extent of the particle crack length population relative to the extended component population is clear. These comparative traits should be considered in comparisons of the resulting strength distribution of extended components (Chapter 5) and particles (this chapter and following).

6.2 Materials Comparisons

6.2.1 Alumina

Due to their superior structural properties, some of the most studied engineered particles consist of alumina-based compounds (Brecker 1974; Wong et al. 1987; Bertrand et al. 1988; Shipway and Hutchings 1993; Huang et al. 1993, 1995; Verspui et al. 1997; Tavares and King, 1998; Salman et al. 2002; Müller 2013; Fedorov and Gulyaeva 2019). The primary constituent of alumina is Al_2O_3, properly termed corundum in its single-crystal form (sometimes also known as sapphire or ruby when naturally or artificially doped for color). The term alumina is predominantly reserved for engineered polycrystalline Al_2O_3 materials. These materials also usually contain deliberately added grain-boundary related microstructural control agents in the form of small amounts of other oxides or glasses. Bauxite is a natural sedimentary rock consisting primarily of aluminum-based minerals, along with some iron and silicate minerals. Bauxite is the principal ore used to form Al_2O_3 powder for aluminum metal production and (less so) for fabrication of polycrystalline alumina components, either extended or particles. Unless otherwise required, for brevity the term alumina will be used here; bauxite derived particles are considered in Chapter 8.

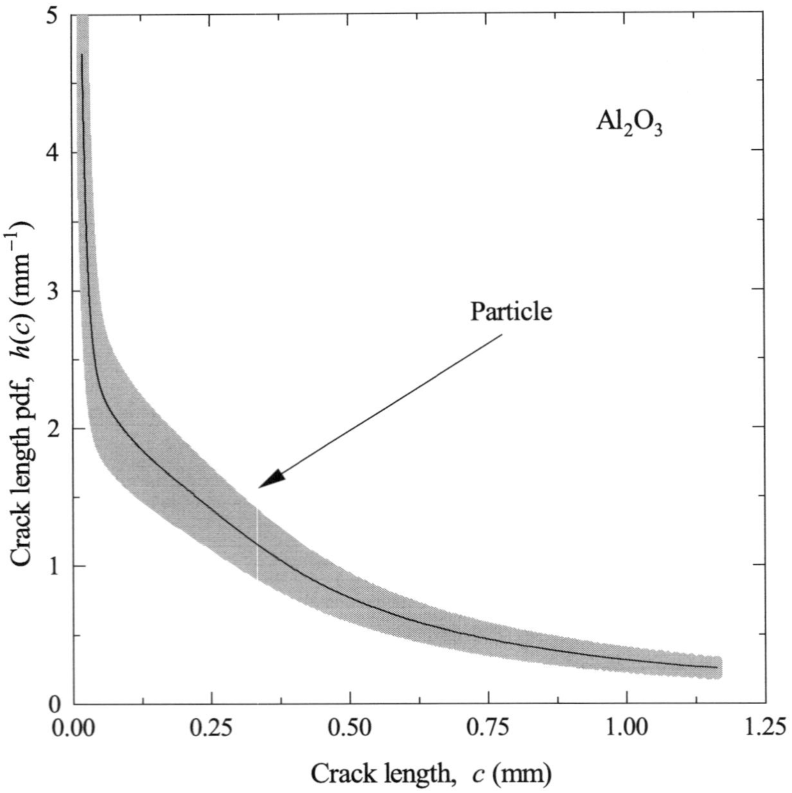

Figure 6.3 Plot on expanded scale of crack length pdf $h(c)$ variation for alumina particles. Solid line shows concave best fit, shaded band shows bounds; determined from Figures 6.1 and 6.2.

Figure 6.5 shows strength edf $Pr(\sigma)$ plots for four sets of mm-scale alumina particle strength measurements, using data derived from the published works cited. Figure 6.5a shows the behavior of particles tested in a drop weight configuration, from the work of Tavares and King (1998). Figure 6.5b shows the behavior of particles tested in a rotating roller configuration from Brecker (1974), in some of the earliest work. Figure 6.5c shows the behavior of particles tested in a conventional quasi-static compression platen configuration, from Bertrand et al. (1988), also in early work. Figure 6.5d repeats the results from Huang et al. (1995) shown above. Symbols represent individual strength measurements. Lines represent visual best fits. All sets of data are well described by concave behavior, although there are instances of very weak strength observations separate from the main distribution; this behavior is discussed and analyzed below in terms of perturbations to the predominant concave trends. The data exhibit strength domain widths σ_N/σ_1 from 3–10.

Using logarithmic coordinates, Figure 6.6 shows the crack length pdf $h(c)$ variations underlying the alumina particle strength behavior. The curves are labeled (i)–(iv), corresponding to Figure 6.5a–d, respectively, and the area under each curve is 1. The B values used were (all in MPa m$^{1/2}$) (i) 0.7, (ii) 1.5, (iii) 0.6, and (iv) 1.5, taking into account the particle sizes. Consideration of Figure 6.5 shows that the strength distributions overlap. Although a range of B values were used, the crack length curves also overlap as a consequence of the similar particle sizes. The unlabeled curve in Figure 6.6 is that for extended alumina components from Figure 5.4. A feature in common for the particle $h(c)$ variations of Figure 6.6 is the well-developed large crack tail, exhibited particularly by curve (iii). In some cases, the $h(c)$ variations exhibit a peak in population density, curves (i)–(iii), although in one case not very well formed, (ii). These peaks are extremely broad relative to that for the extended components. (As in earlier comparisons, the similarity in the shapes of the crack length populations is made clearly evident through the use of logarithmic coordinates.) Another feature in common for the $h(c)$ variations is that the B values used for the particles were all less than that (3.0) used for the polycrystalline alumina extended components. Taken together, the implications of the strength and crack length analysis are that the alumina particles contained a greater fraction of large cracks in material that was less resistant to fracture than its extended component analogue. Cracked grain boundary facets observed in images of the particles (Bertrand et al. 1988; Huang et al. 1995) support these implications.

For commercial reasons, the effects of particle shape were considered by both Brecker (1974) and Huang et al. (1993, 1995). The earliest use of a counter-rotating roller crushing apparatus to measure particle strength appears to be by Brecker

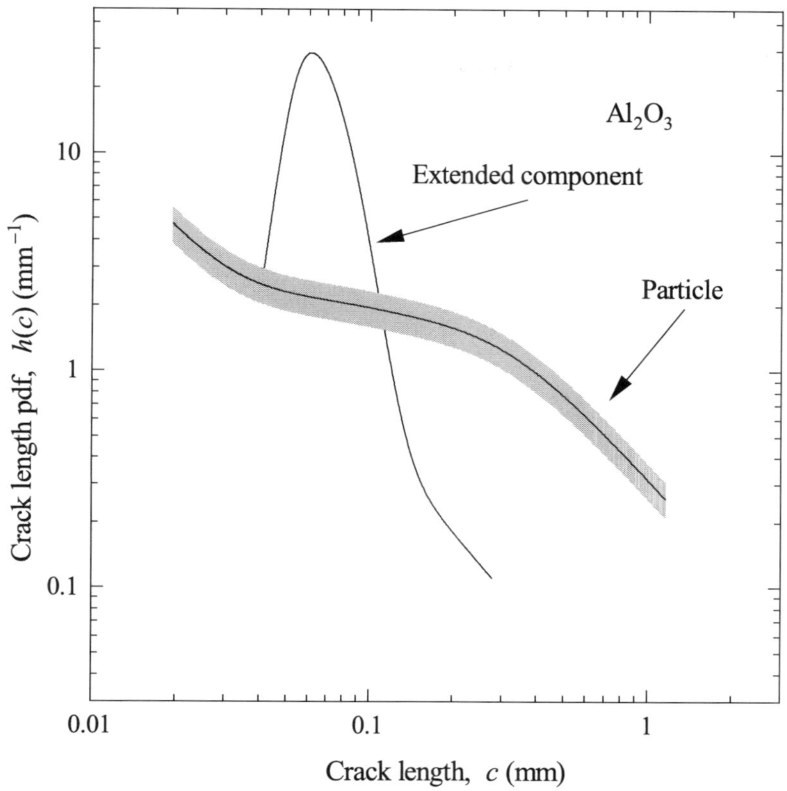

Figure 6.4 Plot in logarithmic coordinates of crack length pdf $h(c)$ variations for alumina. Particle response shown as bold line and shaded band, extended component response shown as fine line; determined from Figure 6.2.

(1974), who was motivated by optimizing the condition of abrasive particles used in industrial grinding applications, particularly for metals. Brecker mostly examined alumina-based materials but also considered SiC and examined the effects of deliberate alteration of particle shape. Brecker deliberately "ground" flats on alumina particles, leading to significant strengthening, about a factor of five increase in the failure force. These ideas were all pursued by Huang et al. 20 years later. Huang et al. (1993, 1995) used both a counter-rotating roller crushing apparatus and a conventional platen apparatus to measure the strength distributions of several different materials of abrasive particles, including SiC and a series of polycrystalline alumina particles several mm in size (although termed "corundum" by Huang et al., micrographs clearly show polycrystalline structures). The intent was to identify, long-lived sharp particles for cutting and machining manufacture operations (the intent of Bertrand et al. 1988 was similar). Huang et al. "rounded" brown alumina particles by ball-milling, leading to a slight strengthening, about a factor of 1.5 in failure force. The implication is that particle surface flats introduced by grinding and milling altered the geometry of the particles such that the central biaxial tension was decreased during testing; clearly a deterministic effect, but on stress not flaw population. The shapes of the strength distributions of the as-received and ground particles were similar, supporting the idea that the flaw populations were not altered by grinding or milling.

6.2.2 Quartz

Quartz is crystalline SiO_2 and is the most common mineral found at the earth's surface. Most beach sand is composed of quartz particles. The chemical name for SiO_2 is silica. (Fused silica is an amorphous glass that rarely occurs in nature but is commonly used in the semiconductor or MEMS industries, where it is often confusingly referred to as "fused quartz.") The term quartz will be used here.

Figure 6.7 shows strength edf $Pr(\sigma)$ plots for four sets of mm-scale quartz particle strength measurements, using data derived from the published works cited. Figures 6.7a and 6.7b show the behavior of particles tested in drop weight configurations, from the work of King and Bourgeois (1993) and Tavares and King (1998), respectively. Figures 6.7c and 6.7d show the behavior of particles tested in conventional quasi-static compression platen configurations from the work of Wang and Coop (2016) and McDowell (2002), respectively. Symbols represent individual strength measurements. Lines represent

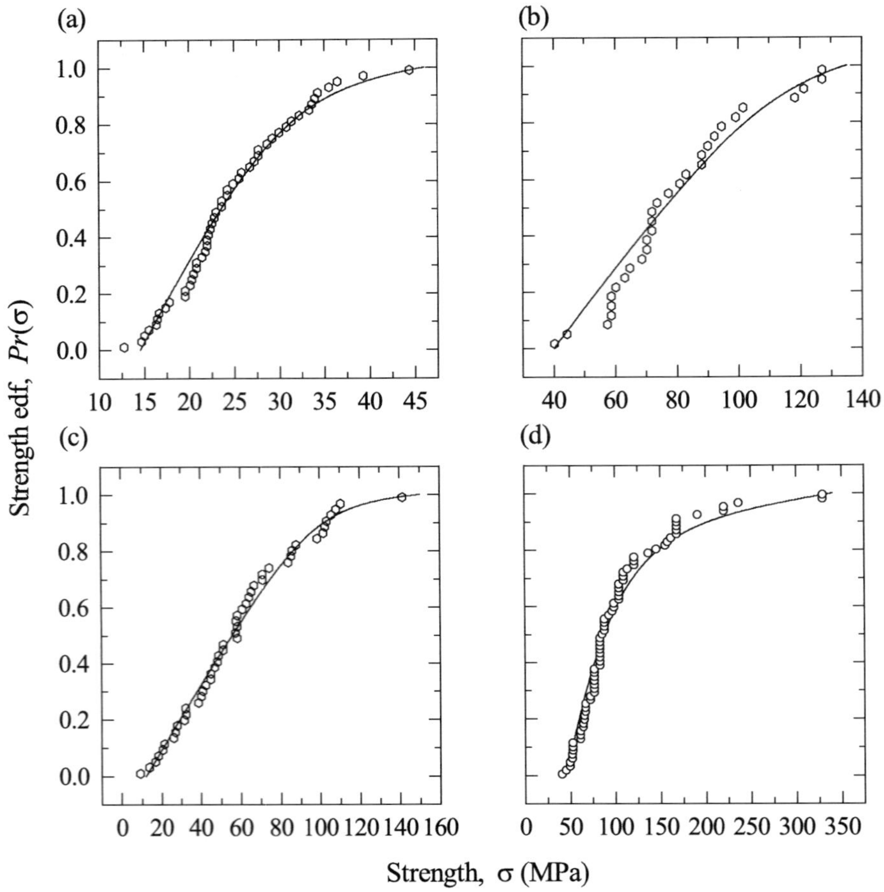

Figure 6.5 Plots of strength edf behavior, $Pr(\sigma)$, for alumina particles. (a) Particle diameter, $D = 2.0-2.8$ mm, number of measurements, $N = 50$ (Adapted from Tavares, L.M et al. 1998). (b) $D = 1.6$ mm, $N = 31$ (Adapted from Brecker, J.N 1974). (c) $D = 2.4$ mm $\pm$ 0.1 mm, $N = 48$ (Adapted from Bertrand, P.T et al. 1988). (d) $D = 1.29$ mm (estimated $\pm$ 0.15 mm), $N = 73$ (Adapted from Huang, H et al. 1995).

visual best fits. All sets of data are well described by concave behavior, although there are some deviations from the fitted lines at both large strengths and small strengths (the latter is discussed below). The data exhibit strength domain widths σ_N/σ_1 of approximately 4.

Using logarithmic coordinates, Figure 6.8 shows the crack length pdf $h(c)$ variations underlying the quartz particle strength behavior. The curves are labeled (i)–(iv), corresponding to Figures 6.7a–d, respectively, and the area under each curve is 1. The B values used were (all in MPa $m^{1/2}$) (i) 0.4 and (ii)–(iv) 0.5, taking into account the material and the particle sizes. Consideration of Figure 6.7 shows that the strength distributions overlap sequentially. As a limited range of B values was used, the crack length curves also overlap sequentially. The unlabeled curves shown in gray in Figure 6.8 are those for extended glass components from Figure 5.2. Many of the characteristics noted above are observed here in the particle $h(c)$ variations of Figure 6.8: a feature in common is the well-developed large crack tail, exhibited particularly by curve (i); in some cases the $h(c)$ variations exhibit a peak in population density, curves (ii)–(iv), although in one case not very well formed, (iv); and, the peaks are extremely broad relative to those for the extended components. Another feature in common for the $h(c)$ variations is that the B values used for the particles were all comparable, approximately 0.5 MPa $m^{1/2}$, and comparable to those observed for extended components of silica and other glasses. The implication in this case is that although the quartz particles contained a greater fraction of large cracks than those observed in extended components, the material exhibited approximately the same resistance to fracture as its extended component analogue.

It is interesting to compare the motivations of the works cited above. King and Bourgeois (1993) and Tavares and King (1998) were concerned with optimizing industrial comminution of particles in mining. The early work of King and Bourgeois (1993) described the development and application of an ultra-fast load cell in instrumented impact of particles by a falling metal sphere. King and Bourgeois used the resulting measurements to determine the distribution of specific failure energies, E_m, for a range of materials. In an extensive follow-up work using the same apparatus, Tavares and King

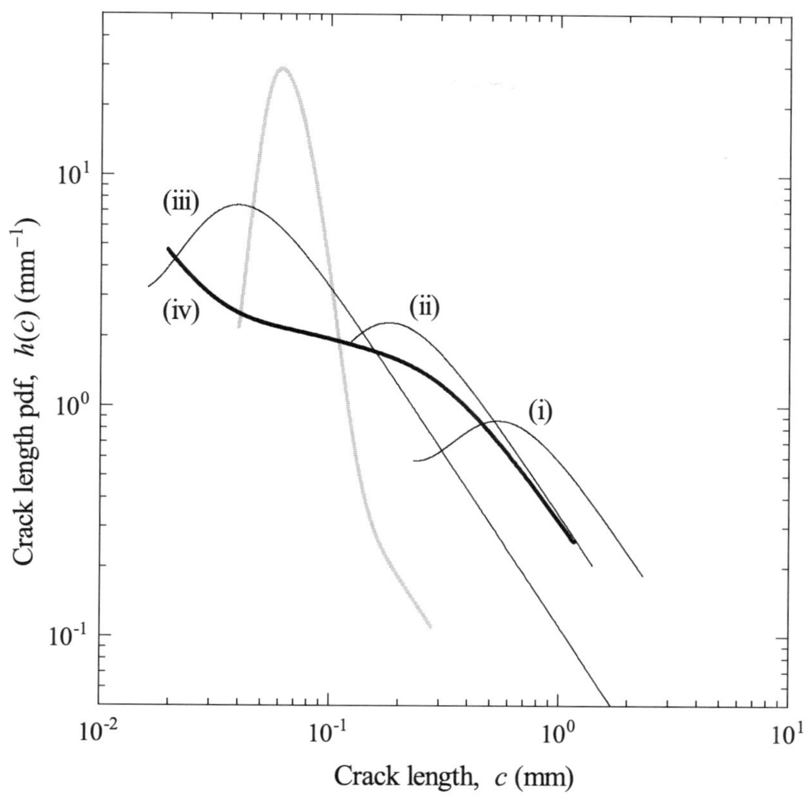

Figure 6.6 Plot in logarithmic coordinates of crack length pdf variations, $h(c)$, for alumina particles; determined from strength measurements of Figure 6.5. Label (i) corresponds to Figure 6.5a, label (ii) corresponds to Figure 6.5b, label (iii) corresponds to Figure 6.5c, and label (iv) corresponds to Figure 6.5d.

(1998) determined specific failure energies for a wide range of materials, examined particle shape effects, and discussed the many applications of particle strength measurement—including comminution efficiency in materials processing. Whereas Brecker (1974) and Huang et al. (1993, 1995) were concerned with prolonging the lifetimes of intact particles, King and Bourgeois (1993) and Tavares and King (1998) were concerned with efficient destruction of particles. These disparate engineering goals are considered in Chapter 13. McDowell (2002) and Wang and Coop (2016) were concerned with compaction of particle aggregates in geotechnical applications, in particular, the effects of individual particle fracture on the overall force-displacement behavior of aggregates of sand.

6.2.3 Limestone

Limestone and marble are both composed primarily of the mineral $CaCO_3$; limestone is a sedimentary rock and is typically porous, and marble is a metamorphic rock and dense (and was initially limestone). In particle form there is often little difference, although "limestone" may contain a microstructure based on animal shells. Calcareous or carbonaceous sands contain significant fractions of $CaCO_3$ shell fragment particles. The term limestone will be used here for brevity.

Figure 6.9 shows strength edf $Pr(\sigma)$ plots for four sets of limestone particle strength measurements, using data derived from the published works cited. The particles ranged in size from multi- to sub-millimeter. Figures 6.9a and 6.9b show the behavior of particles tested in conventional quasi-static compression platen configurations, from the work of Wang et al. (2015) and Rozenblat et al. (2011), respectively. Figures 6.9c and 6.9d show the behavior of particles tested in drop weight configurations from the work of Barrios et al. (2011) and King and Bourgeois (1993), respectively. Symbols represent individual strength measurements. Lines represent visual best fits. All sets of data are well described by concave behavior, although there are some very weak strength observations separate from the main distribution (discussed below). The data exhibit strength domain widths σ_N/σ_1 of 4–8.

Using logarithmic coordinates, Figure 6.10 shows the crack length pdf $h(c)$ variations underlying the limestone particle strength behavior. The curves are labeled (i)–(iv), corresponding to Figures 6.9a–d, respectively, and the area under each curve is 1. The B values used were (all in MPa $m^{1/2}$) (ii) 0.1 and (i), (iii), and (iv) 0.25, taking into account the material

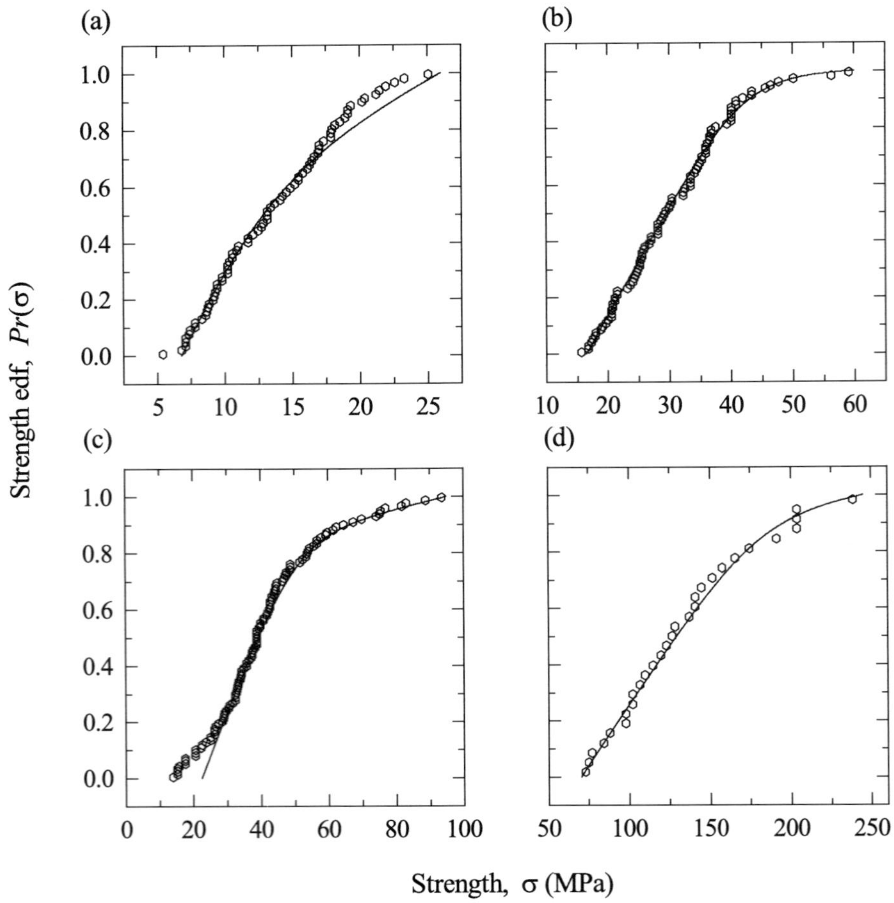

Figure 6.7 Plots of strength edf behavior, $Pr(\sigma)$, for quartz particles. (a) Particle diameter, D = 3.35 mm–4.00 mm, number of measurements, N = 73 (Adapted from King, R.P et al. 1993). (b) D = 1.0 mm–1.18 mm, N = 106 (Adapted from Tavares, L.M et al. 1998). (c) D = 1.18 mm–2.36 mm, N = 88 (Adapted from Wang, W et al. 2016). (d) D = 0.3 mm–0.6 mm, N = 29 (Adapted from McDowell, G.R 2002).

and the particle sizes. Consideration of Figure 6.9 shows that the strength distributions overlap sequentially. As a limited range of B values was used, the crack length curves also overlap sequentially. Many of the characteristics noted above are observed here in the particle $h(c)$ variations of Figure 6.10: a feature in common is the well-developed large crack tail, exhibited particularly by curve (ii); in some cases the $h(c)$ variations exhibit a peak in population density, curves (i), (ii), and (iv); and, the peaks are extremely broad relative to those for extended components.

In this case also motivations diverged: Wang et al. (2015) were concerned with particle stability in geotechnical engineering applications, whereas Rozenblat et al. (2011), Barrios et al. (2011), and King and Bourgeois (1993) were concerned with comminution efficiency in materials processing. The concave behavior in Figure 6.9 is similar to that observed by Vogel and Peukert (2002) in failure energy results obtained using an impact mill in comminution studies on limestone and other materials.

6.2.4 Rock

A rock here is taken to be a naturally formed material composed of mineral crystals. The minerals are usually silicates (e.g. quartz) or aluminosilicates (e.g., feldspar, mica), often with calcium, magnesium, or sodium constituents. Carbonates (e.g. calcite, dolomite), sulphates (e.g. gypsum), and halides (e.g. salt), are also common, as is the carbide coal. A common hard rock is granite, a composite of quartz, feldspar, and mica. A common soft rock is limestone, polycrystalline calcite. Ceramics are engineered materials composed of similar crystals (e.g. alumina, porcelain). Extensive details can be found elsewhere (Deer et al. 1966; Kingery et al. 1975). For brevity, naturally formed mineral assemblages will be referred to collectively as rock, and distinctions made as necessary.

Figure 6.11 shows strength edf $Pr(\sigma)$ plots for two contrasting sets of rock particle strength measurements, using data derived from the published works cited. The particles ranged in size from millimeter to multi-millimeter and represented

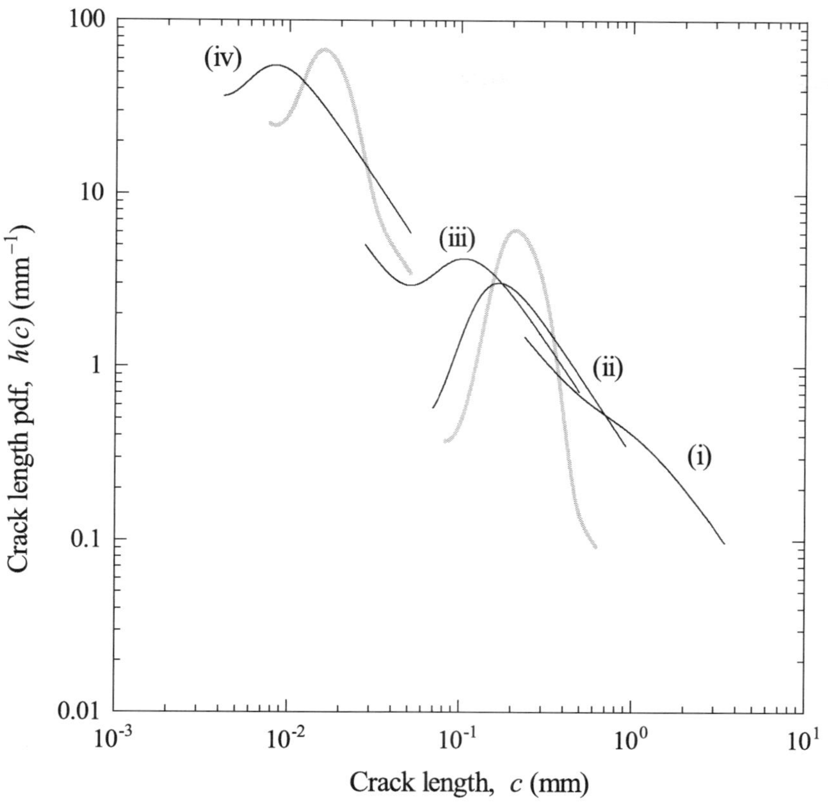

Figure 6.8 Plot in logarithmic coordinates of crack length pdf variations, $h(c)$, for quartz particles; determined from strength measurements of Figure 6.7. Label (i) corresponds to Figure 6.7a, label (ii) corresponds to Figure 6.7b, label (iii) corresponds to Figure 6.7c, and label (iv) corresponds to Figure 6.7d.

"soft" rock, limestone-based and coal, and "hard" rock, granite- and dolomite-based, intended for use as railway ballast. All were tested in conventional quasi-static compression platen configurations. Figure 6.11a shows the behavior of limestone-based rock particles from the work of Hu et al. (2011) (as cited by Ovalle et al. 2014). Figure 6.11b shows the behavior of coal-based particles from the work of Dong et al. (2018). Figure 6.11c and Figure 6.11c show the behavior of granite and dolomite rock particles from the work of Lim et al. (2004) and Koohmishi and Palassi (2016), respectively (the latter used a localized probe in diametral loading reminiscent of the HO simulations). Symbols represent individual strength measurements. Lines represent visual best fits. All sets of data are well described by concave behavior. The data exhibit strength domain widths σ_N/σ_1 of 4–10.

Using logarithmic coordinates, Figure 6.12 shows the crack length pdf $h(c)$ variations underlying the rock particle strength behavior. The curves are labeled (i)–(iv), corresponding to Figure 6.11a–d, respectively, and the area under each curve is 1. The B values used were (all in MPa $m^{1/2}$) (i) 0.05 and (ii) 0.06, and (iii) and (iv) 1.0, taking into account the materials and the particle sizes. Consideration of Figure 6.9 shows that the strength distributions overlap sequentially. As a very wide range of B values was used, although the crack length curves also overlap sequentially, the sequence is reversed. The soft rock particles exhibited small strength-controlling cracks, but the material resistance to fracture was very small, such that the strengths were small. Conversely, the hard rock particles exhibited large strength-controlling cracks, but the resistance to fracture was large, such that the strengths were large. The predominant characteristic of the $h(c)$ variations is that in only one case is a peak observed; a common characteristic is the well-developed large crack tail.

6.2.5 Threshold perturbations

A feature noted in some strength edf observations is the appearance adjacent to the threshold of a few small strength values that do not appear to be part of an overall concave trend. Fitting edf data including such strengths leads to a small dilemma: A strict analytical approach would remove the suspect data and recalculate the edf so as to optimise for a selected simple fitting form. A strict experimental approach would equally weight all data and recalculate the fit in a more

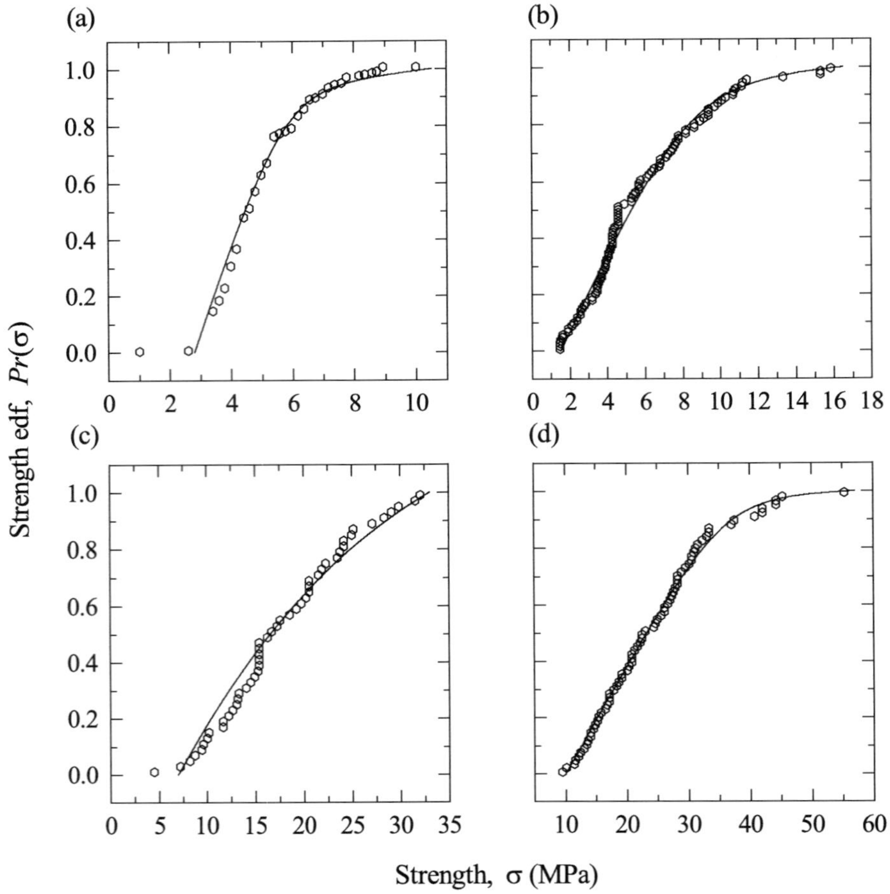

Figure 6.9 Plots of strength edf behavior, $Pr(\sigma)$, for limestone particles. (a) Particle diameter, $D = 28–33$ mm, number of measurements, $N = 31$ (Adapted from Wang, Y et al. 2015). (b) $D = 4–5$ mm, $N = 97$ (Adapted from Rozenblat, Y et al. 2011). (c) $D = 2.36–2.83$ mm, $N = 50$ (Adapted from Barrios, G.K.P et al., 2011). (d) $D = 0.5–0.7$ mm, $N = 72$ (Adapted from King, R.P et al. 1993).

complex and inclusive fitting form. Here, the fitting approach to these strengths has been an analytical and experimental middle course: neither censoring (removing) such strengths from a presented edf nor explicitly fitting to such strengths. The middle course thus avoids discarding information while retaining analytic simplicity. However, in order to enable assessment of presented strength fits in terms of inferred crack length distributions, the likely effects of data omitted from a fit must be known. Such effects are investigated here for a test example that takes advantage of the full trilinear fitting analysis developed in Chapter 4.

Figure 6.13 shows a strength edf $Pr(\sigma)$ plot for salt particles from the work of Rozenblat et al. (2011), using data derived from the published work. The particles were multi-millimeter in scale and tested in a conventional quasi-static compression platen configuration. Symbols represent individual strength measurements. Adjacent to the threshold there are a few strength observations that depart from the overall trend leading to a small region of reduced slope. The solid line in Figure 6.13 is a visual fit to the data using the bilinear concave form used above. The fit omits the small strength observations adjacent to the threshold, but otherwise intersects most data. Using logarithmic coordinates, Figure 6.14 shows as a solid line the crack length pdf $h(c)$ variation underlying this concave fit. The B value used was 0.1 MPa $m^{1/2}$. The deconvoluted $h(c)$ variation is an undulating decreasing trend with a weak intermediate maximum. The shaded band in Figure 6.13 represents the bounds of a visual fit to the strength data using the trilinear sigmoidal form used extensively in Chapter 5. The bounds encompass all the data, including the small strength observations, thus slightly increasing the fitted strength domain. Figure 6.14 shows as a shaded band the crack length pdf $h(c)$ variation underlying the sigmoidal fit using the same B value. The deconvoluted $h(c)$ variation is an undulating decreasing trend with slightly greater domain and range and a slightly more pronounced maximum than the solid line. The shaded band describing particle behavior deviates from the solid line only at large crack lengths—the position and nature of the peak remain much the same.

For comparison, also shown in Figure 6.14 is the similarly located $h(c)$ response underlying the sigmoidal strength behavior of the glass bottles from Chapter 2 and Chapter 5. The particle responses are clearly distinct from that of the extended

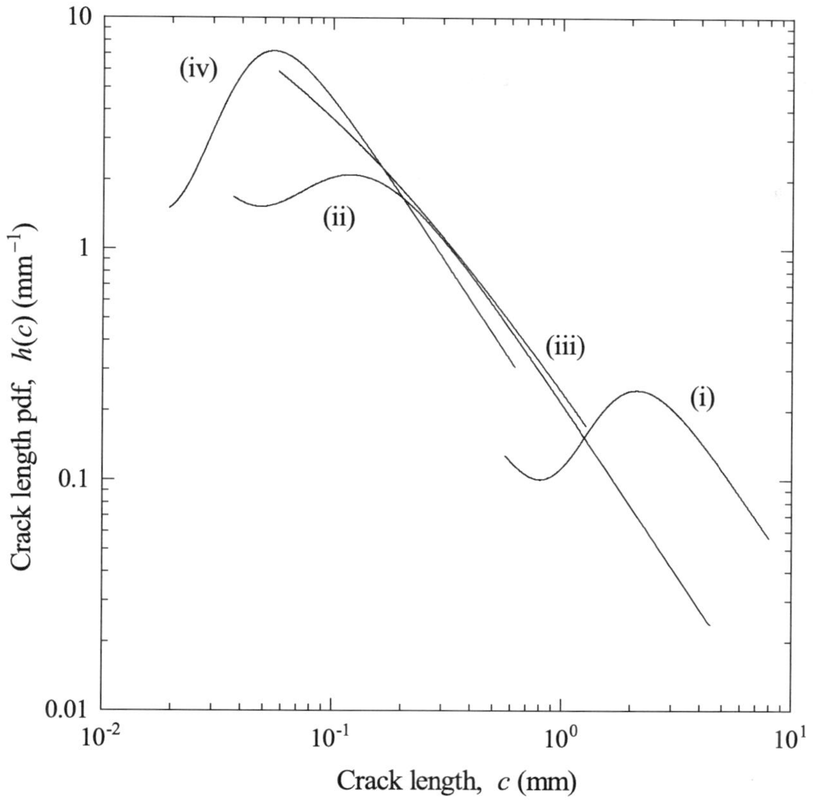

Figure 6.10 Plot in logarithmic coordinates of crack length pdf variations, $h(c)$, for limestone particles; determined from strength measurements of Figure 6.9. Label (i) corresponds to Figure 6.9a, label (ii) corresponds to Figure 6.9b, label (iii) corresponds to Figure 6.9c, and label (iv) corresponds to Figure 6.9d.

components, for both forms of the strength fit. Although there are expected similarities in shape at large crack lengths, the particle responses are broader and have less well developed peaks. Nevertheless, although the additional strengths in Figure 6.13 form a small fraction of the overall $Pr(\sigma)$ behavior, the visual effect on the inferred $h(c)$ variation is significant, a consequence of the magnifying effects of the $c \sim \sigma^{-2}$ relation and logarithmic coordinates. Although there are exceptions (e.g. Figure 6.7c), small strength perturbations of the threshold are usually minor. Description of particle $Pr(\sigma)$ responses as concave thus enables the best overall description of behavior; retention of the plotted data enables easy recognition that in some cases such descriptions lead to upper bounds on crack length distributions.

6.3 Size Comparisons

The previous section surveyed the strength behavior of four common particle materials—alumina, quartz, limestone, and rock—noting that the particle strength distributions (usually concave over broad domains) were not the same shape or width as those of extended components (usually sigmoidal over restricted domains, Chapter 5). This section continues the survey of particle strength behavior, classifying particles by size—small, medium, and large. Small particles are classified here as those with diameter $D < 1$ mm; the strengths of such particles are infrequently reported. Medium particles are those with $1 < D < 10$ mm and the strengths of such particles are the most commonly reported; medium particles were the majority of those considered in the previous section. Large particles are those with $D > 10$ mm; often terms such as fragments, aggregates, stones, or rocks, are used to refer to such particles.

6.3.1 Small Particles

Figure 6.15 shows strength edf $Pr(\sigma)$ plots for four sets of small particle strength measurements, using data derived from the published works cited. The particles ranged in size from sub-micrometer to sub-millimeter and represented engineered and natural soft and hard particles. Symbols represent individual strength measurements. Lines represent visual best fits. In

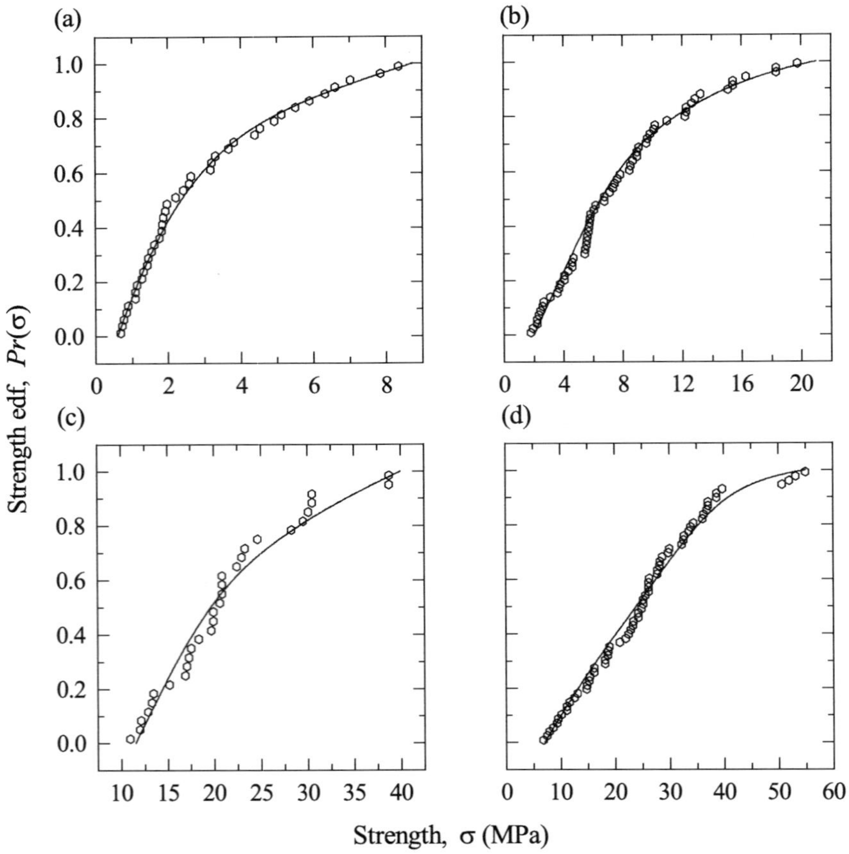

Figure 6.11 Plots of strength edf behavior, $Pr(\sigma)$, for rock particles. (a) Particle diameter, D = 7–15 mm, number of measurements, N = 40 Adapted from (Hu, W et al. 2011; Ovalle, C et al. 2014). (b) D = 0.2–1.0 mm, N = 62 (Adapted from Dong, J et al. 2018). (c) D = 10–14 mm, N = 30 (Adapted from Lim, W.L et al. 2004). (d) D = 25–37.5 mm, N = 64 (Adapted from Koohmishi, M et al. 2016).

order to demonstrate the ability of commercial instruments to perform mechanical testing at small scales, Ribas et al. (2014) determined the strength distributions of particles, about 45 μm in diameter, from a diverse set of six materials. The responses of the small soft particles are shown in Figure 6.15a for rice husk ash (smaller strengths) and limestone (larger); coal shale particles behaved similarly. Motivated by the geological importance of particle fracture during compaction of sandstones and sands, Brzesowsky et al. (2011) determined the strength distributions of a set of four different sized sub-millimeter sand (predominantly quartz) particles with a conventional testing apparatus. The strengths of the 378 μm (smaller strengths) and 115 μm (larger) particles are shown in Figure 6.15b; intermediate sized particles exhibited intermediate strengths. The symbols in Figure 6.15b represent strength values smaller than those reported by Brzesowsky et al. (2011) as the HO equation was used to determine the representative particle center stress at failure. The values do not affect the observations here. The responses of the small hard particles determined by Ribas et al. (2014) are shown in Figure 6.15c for quartz (smaller strengths) and SiC (larger); blast furnace slag (predominantly aluminosilicate oxide) particles behaved similarly.

The strength distributions of Figures 6.15a–6.15c are clearly concave, as observed earlier for the majority of brittle particles. Those in Figure 6.15d are sigmoidal and reflect a different deformation and failure mechanism. Failure measurements using a custom instrument were performed by Pejchal et al. (2017, 2018) to determine the strength distributions of (20 to 60) μm SiO_2 and amorphous or nanocrystalline near-eutectic Al_2O_3-ZrO_2-SiO_2 ceramic particles (a-Eu or nc-Eu). The response of the a-Eu particles is shown in Figure 6.15d (smaller strengths); the responses of SiO_2 and nc-Eu were similar. In order to explore potential ductility in ceramic particles, Herre et al. (2017) used commercial instrumentation in an electron microscope to determine strength distributions of TiO_2 particles, about 600 nm in diameter. The response of the rutile phase is shown in Figure 6.15d (larger strengths); the responses of other TiO_2 polymorphs were similar. Images of failed particles from Pejchal et al. (2017, 2018) and Herre et al. (2017) show extensive plastic deformation in which the contact flats are comparable in size to the particle diameter and extensive fine fragmentation originating beneath the contact, in addition to meridional cracks. Similar strength behavior and images were obtained by Dang et al. (2019) on (7 to 17)

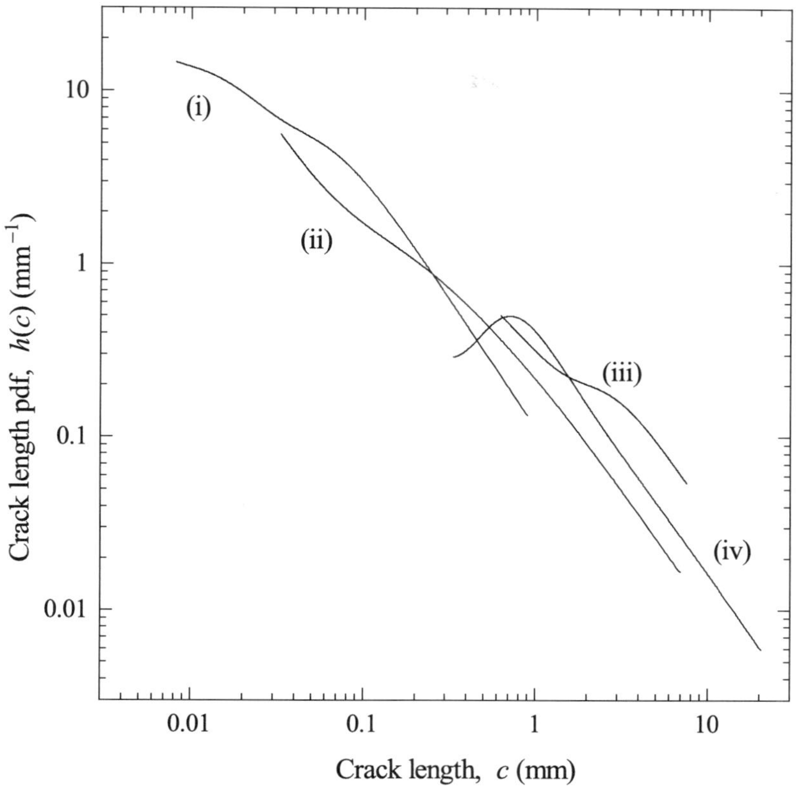

Figure 6.12 Plot in logarithmic coordinates of crack length pdf variations, *h*(*c*), for rock particles; determined from strength measurements of Figure 6.11. Label (i) corresponds to Figure 6.11a, label (ii) corresponds to Figure 6.11b, label (iii) corresponds to Figure 6.11c, and label (iv) corresponds to Figure 6.11d.

μm Li-Ni-Mn-Co-O particles using a flat punch indentation system. The implication in these cases is that the deformation state of the particles was not primarily elastic at failure and that particle failure was significantly influenced by distributed plastic deformation events in addition to fracture. Sections of micrometer-sized pores were observed on fracture surfaces by Pejchal et al. (2018) and it is possible that distributed porosity also influenced particle failure in the Eu ceramic. The sigmoidal strength responses of Figure 6.15d are similar to those of extended components (Chapter 5), consistent with a distributed failure mechanism.

Large strengths have been reported in other small particle systems. As part of a larger study of particle size degradation during erosive machining (Slikkerveer et al. 2000), Verspui et al. (1997) determined the strength distribution of 44 μm Al_2O_3 particles with a custom testing apparatus. Strengths of up to 3 GPa were reported as part of concave strength distribution that was extremely broad, $\sigma_N/\sigma_1 \approx 10^3$. The large fraction of very small strengths in the distribution, in addition to the large strengths, suggests that there were significant concurrent flaw population effects. As part of a study on coal breakage, Sikong et al. (1990) observed near GPa strengths for (5–50) μm particles of quartz and (10–200) MPa strengths of similarly sized particles of feldspar, limestone, marble, gypsum, and coal. Images of small particles of quartz and coal exhibited significant plastic deformation on loading, similar to that described above, and the smallest particles deformed without fracture, supporting a change in failure mechanism for very small particles. Watkins and Prado (2015) produced glass particles (70–150) μm in diameter by a flame spheroidization process and demonstrated strengths in the (150–1050) MPa domain that were particle size dependent. The strength distributions were near-linear, weakly concave, similar to the silica sand particles in Figure 6.15b and exhibited deterministic size effects also similar to Figure 6.15b. The deterministic behavior of both sets of particles will be studied in Chapter 9.

Using logarithmic coordinates, Figure 6.16 shows the crack length pdf $h(c)$ variations underlying the small particle strength behavior. The curves are labeled (i)–(iii), corresponding to Figures 6.16a–c, respectively, and the area under each curve is 1. The B values used were (all in MPa $m^{1/2}$) (i) 0.025, (ii) 0.5, and (iii) 0.25 and 0.5 taking into account the material and the particle sizes. Consideration of Figure 6.15 shows that the strength distributions overlap sequentially. As a very wide range of B values was used, the crack length curves overlap almost completely, a consequence of the similar particle

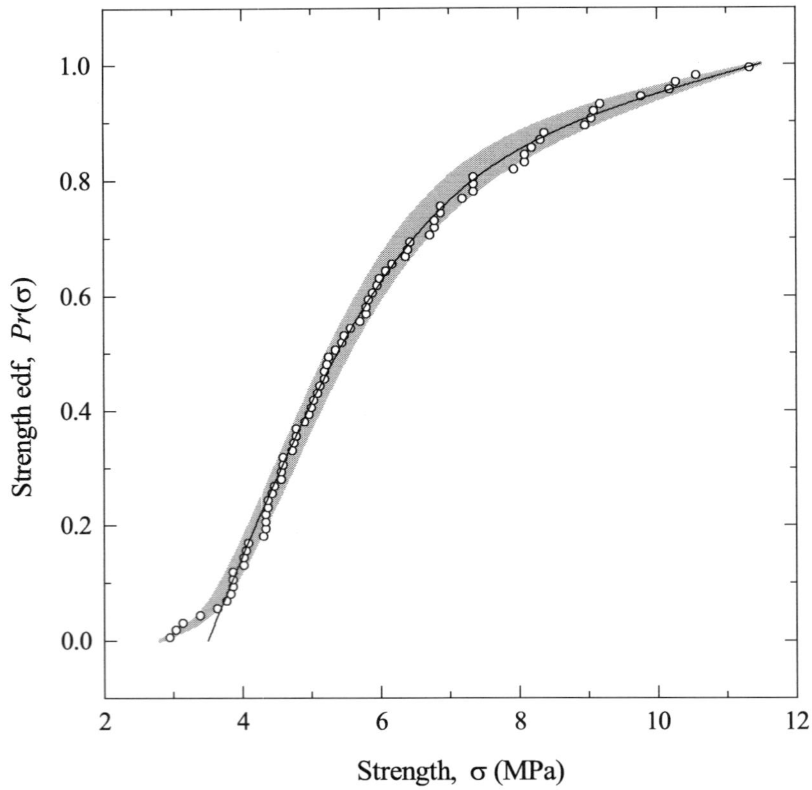

Figure 6.13 Plot of strength edf behavior, $Pr(\sigma)$, for salt particles, particle diameter, $D = 2.36 - 3.35$ mm, number of specimens, $N = 80$ (Adapted from Rozenblat, Y et al. 2011).

sizes. Note that the very small B values for small coal and limestone particles used here are similar to those used above for medium sized particles. The dashed gray curves Figure 6.16 are evaluated from the strength measurements of rutile and a-Eu from Figure 6.15d using the established fracture analysis and B values of 0.7 MPa $m^{1/2}$ (small cracks, rutile) and 1.0 (large, a-Eu), consistent with the materials and particle sizes. The responses are peaked and do not resemble the other small particle responses, reinforcing the idea that failure in these systems was not caused primarily by a localized fracture event. The mode flaw size for the a-Eu (about 1 μm) is comparable to the pore sizes observed in this material.

6.3.2 Medium Particles

In order to emphasize similarity in the forms of particle strength distributions, the strength behavior of a range of medium sized particles is presented here without analysis. The strength data vary in particle material and test configuration and provide context for the stochastic and deterministic size effects to be studied in Chapters 7 and 9. In addition, the data emphasize the discriminating abilities of the unbiased edf presentation format.

Figure 6.17 shows strength edf $Pr(\sigma)$ plots for four sets of medium particle strength measurements, using data derived from the published works cited. The particles ranged in size from (1.4–4) mm and represent engineered and natural particles. Symbols represent individual strength measurements; the open symbols represent the following first named particle material and the solid symbols represent the second named. Figure 6.17a shows strength measurements of Cu ore and Au ore particles from the drop weight tests of King and Bourgeois (1993). Figure 6.17b shows strength measurements of Masado sand (a degraded granite) and silica particles from the custom compression platen tests of Nakata et al. (2001a, 2001b). Figure 6.17c shows strength measurements of fused and tabular alumina particles from the compression platen tests of Bertrand et al. (1988). Figure 6.17d shows strength measurements of SiC and alumina particles from the rotating roller tests of Huang et al. (1993, 1995). In all cases, the predominant form of the strength edf is concave, although there are clear, small strength perturbations adjacent to the threshold in some systems. In many cases, the threshold strength is very small, leading to very large domain widths, and often $\sigma_N/\sigma_1 \gg 50$, similar to the strengths of sand and coral shown in

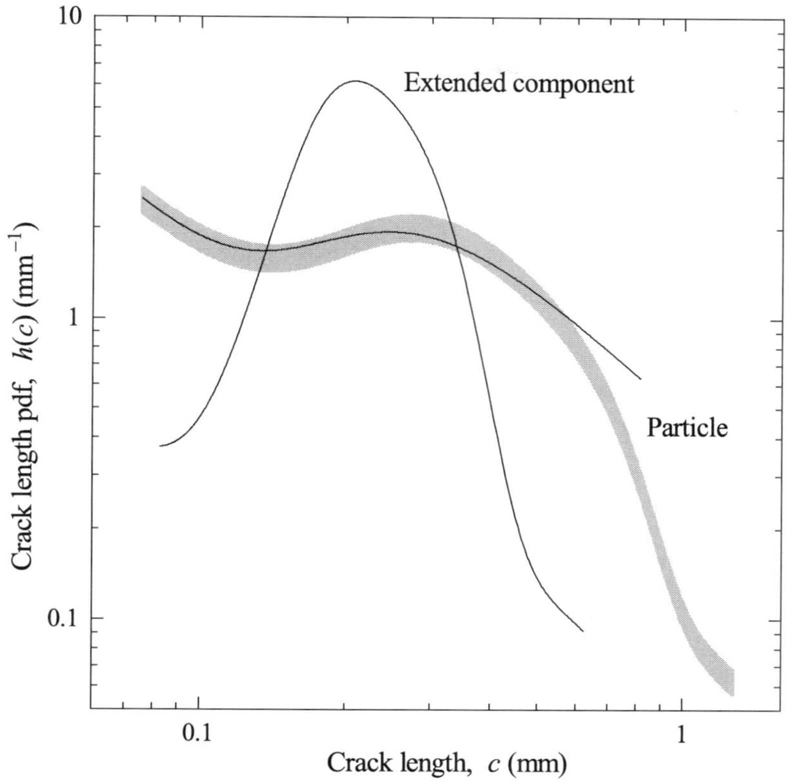

Figure 6.14 Plot in logarithmic coordinates of crack length pdf variations, $h(c)$, for salt particles; determined from strength measurements of Figure 6.13.

Chapter 2. The two materials in each edf plot are clearly distinguishable—a finding that would be difficult to reach using an assumed linearized plotting scheme or measures of central tendency such as mean and standard deviation.

It is interesting to note that in an extensive study of contact shape effects on strength, Todisco et al. (2017) reported similar data for 2.9 mm sand particles crushed between platens, but also reported that similar results were given in the original HO (1966) work on rock particles. The concave failure distribution often extends beyond considerations of strength. Rozenblat et al. (2011) showed similarly shaped failure force distributions for mm-scale potash and basalt particles. Salman et al. (2002, 2003) showed similarly shaped distributions in impact failure velocity observations of 5 mm alumina and fertilizer particles.

6.3.3 Large Particles

As in the previous section, the strength behavior of large sized particles is presented here without analysis. The strength data represent the behavior of two sets of rocks tested in conventional platen compression configurations and provide context for stochastic size effects to be studied in Chapter 7.

Figure 6.18 shows strength edf $Pr(\sigma)$ plots for two sets of large particle strength measurements, using data derived from the published works cited. The particles ranged in size from (7–50) mm and represent various rock particles. Symbols represent individual strength measurements; the open symbols represent the following first named particle material and the solid symbols represent the second named. Figure 6.18a shows strength measurements of calcareous rock and quartzite shale rock from quarries in France from the work of Hu et al. (2011) and Ovalle et al. (2014) considering particle breakage effects in aggregate deformation. Figure 6.18b shows strength measurements of basalt (an igneous rock) and dolomite (a sedimentary rock) from quarries in Iran from the work of Koohmishi and Palassi (2016) considering the effects of particle shape on railway ballast effectiveness.

Again, the predominant form of the strength edf is concave and the threshold strengths are small, leading to very large domain widths, $\sigma_N/\sigma_1 \approx 10$. The two materials in each edf plot are clearly distinguishable, although in this case measures of central tendency such as mean and standard deviation would reveal distinctions.

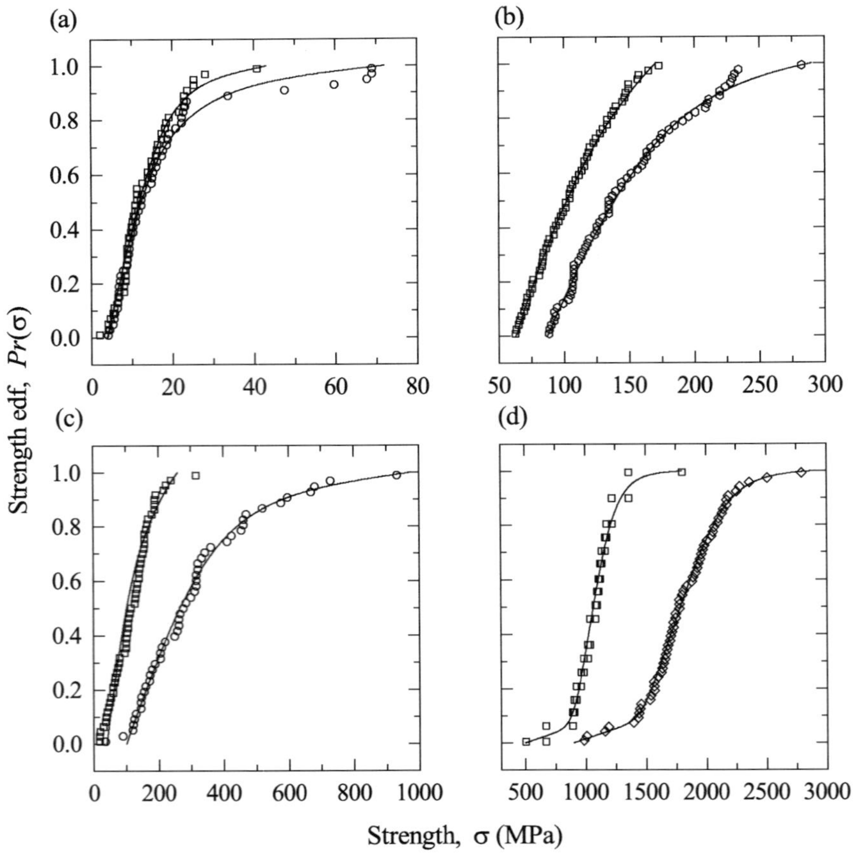

Figure 6.15 Plots of representative strength edf behavior, $Pr(\sigma)$, for small ($D < 1$ mm) particles. (a) Smaller strength, rice husk ash; larger strength, limestone. Particle diameter, $D = (45 \pm 3)$ μm, number of measurements, $N_{\text{tot}} = 100$ (Adapted from Ribas, L et al. 2014). (b) Silica sand. Smaller strength $D = (378 \pm 22)$ μm, larger strength, $D = (115 \pm 9)$ μm, $N_{\text{tot}} = 123$ (Adapted from Brzesowsky, R.H et al. 2011). (c) Smaller strength, quartz; larger strength, SiC. Particle diameter, $D = (45 \pm 3)$ μm, $N_{\text{tot}} = 104$ (Adapted from Ribas, L et al. 2014). (d) Smaller strength, amorphous Eucor ceramic, $D = (25$ to $35)$ μm (Adapted from Pejchal, V et al. 2018) $N = 36$; larger strength, rutile (TiO_2), $D = (533 \pm 22)$ nm, $N = 60$ (Adapted from Herre, P et al. 2017).

6.4 Summary and Discussion

The survey here of particle strength distributions encompassed all material types—amorphous, crystalline, polycrystalline, composite, organic, inorganic, engineered, and natural—across a wide range of particle sizes—from sub-micrometer to tens of millimeters. Two features distinguished particle behavior from that of the equally broad survey in Chapter 5 of strength distributions of extended components. Particle strength edf $Pr(\sigma)$ variations are predominantly concave, as opposed to predominantly sigmoidal for extended components, and the relative widths of particle strength distribution domains predominantly exceed 10, as opposed to predominantly less than 10 for extended components. Particle strength distributions differ markedly from extended component strength distributions.

The clearest distinguishing feature of particle strength edf variation $Pr(\sigma)$ is the linear increasing behavior extending from the threshold strength to mid-domain strengths. In quantitative edf terms, $Pr(\sigma)$ extends linearly from 0 at $\sigma = \sigma_1$ to > 0.5 at $\sigma \approx (\sigma_N - \sigma_1)/2$. In the relative terms used for forward analysis (Chapter 3), $H(\mu)$ extends linearly from $H(\mu = 0) = 0$ to $H(\mu \approx 0.5) > 0.5$. In the absolute terms used for reverse analysis (Chapter 4), the continuum expression for the discrete strength data $H(\sigma)$ for small strengths is

$$H(\sigma) = \beta(\sigma - \sigma_{\text{L}}), \tag{6.1}$$

where σ_{L} is a lower limit strength fitting parameter (usually very close to σ_1) and β is a slope parameter (related to a_1 of Chapter 4). Using the relationships between the strength and crack length probabilities and the Griffith equation (Chapter 3), Eq. (6.1) becomes

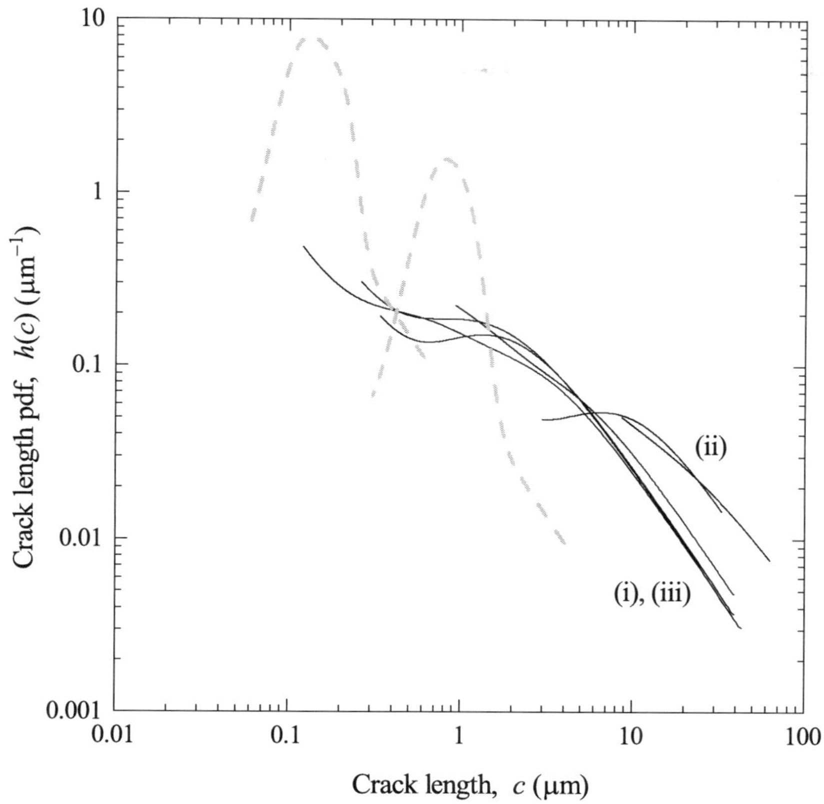

Figure 6.16 Plot in logarithmic coordinates of crack length pdf variations, $h(c)$, for small particles; determined from strength measurements of Figure 6.15. Label (i) corresponds to Figure 6.15a, label (ii) corresponds to Figure 6.15b, and label (iii) corresponds to Figure 6.15c. The dashed responses correspond to a crack length interpretation of Figure 6.15d.

$$H(c) = 1 - H(\sigma) = 1 - \beta B(c^{-1/2} - c_U{}^{-1/2}), \tag{6.2}$$

where c_U is the upper bound of the strength controlling crack length population conjugate to the strength lower bound (Chapter 4). Differentiation of this expression gives the behavior of the crack length probability density for large cracks:

$$h(c) = dH(c)/dc = (\beta B/2)c^{-3/2}. \tag{6.3}$$

Eq. (6.3) provides a convenient reference for comparisons of crack length probability behavior. (The emphasis on large crack behavior is consistent with the engineering importance of threshold strengths. Small crack, large strength behavior is far more variable for both particles and extended components and can be described by a number of equations—see Chapter 4.)

A graphical summary of the various crack length behaviors observed in this Chapter is given in Figure 6.19a, showing representative pdf $h(c)$ variations for particle systems. The logarithmic scale of the plot is identical to that of the similar plot for extended components, Figure 5.18, $10^7 \times 10^5$. The dashed line in Figure 6.19a is a guide to the eye in the $c^{-3/2}$ form of Eq. (6.3). Note the range of crack length domains exhibited and the overall similarity in size and shape of the responses, particularly the large crack tails that exhibit the slope of the dashed line. Figure 6.19b reproduces Figure 5.18, converting the crack length units to mm. The location of the domain of crack lengths exhibited by the set of extended components is slightly smaller than that exhibited by the set of particles. The dashed line in Figure 6.19b is also a guide to the eye in the $c^{-3/2}$ form, describing the large crack behavior of a few extended components. (The relative c^{-1} locations of the $h(c)$ responses in both Figures 6.19a and 6.19b is a geometrical consequence of normalization.)

The crack length domain positions of $h(c)$ responses for particles were, as for extended components, dependent on the choice of the fracture criterion parameter B linking strengths and crack lengths. Values in the range $B = 0.025$–1.5 MPa $m^{1/2}$ were used for particles. The values were selected on the basis of known brittle materials characteristics and, more particularly here, constrained by particle size, such that the largest inferred crack length in a strength-controlling population was smaller than the particle size in the sample. These values were on average less than those used to describe the behavior of extended components (0.5–4 MPa $m^{1/2}$) and reflect a greater susceptibility of particle materials to brittle failure, relative to the extended components examined. It is important to remember that although material toughness is a

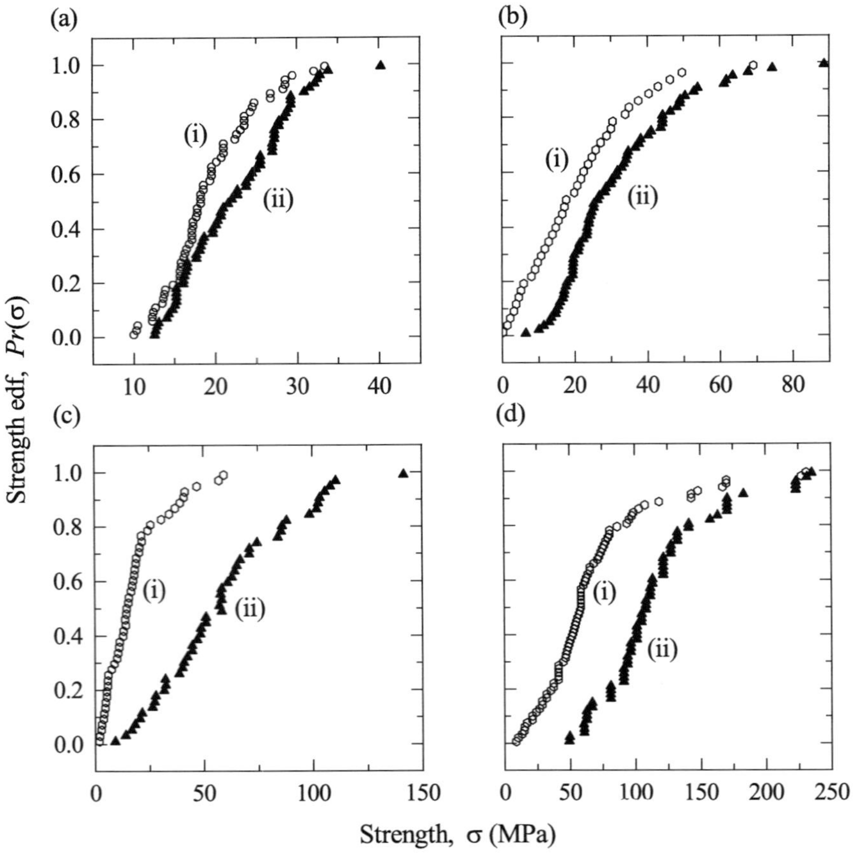

Figure 6.17 Plots of representative strength edf behavior, $Pr(\sigma)$, for medium ($1 < D < 10$) mm particles, the most common size. (a) (i) Cu ore; (ii) Au ore. Particle diameter, D = (3.35–4.00) mm, number of measurements, N_{tot} = 124 (Adapted from King, R.P et al. 1993). (b) (i) Masado sand (degraded granite); (ii) silica. D = (1.4–1.7) mm, N = 108. (Adapted from Nakata, Y et al., 2001b). (c) (i) Fused alumina, $D = (2.6 \pm 0.1)$ mm; (ii) Tabular alumina, $D = (2.4 \pm 0.1)$ mm, N_{tot} = 97 (Adapted from Bertrand P.T et al. 1988). (d) (i) SiC; (ii) sintered alumina. D = 1.85 (estimated $\pm$ 0.25) mm, N = 139 (Adapted from Huang, H et al. 1995).

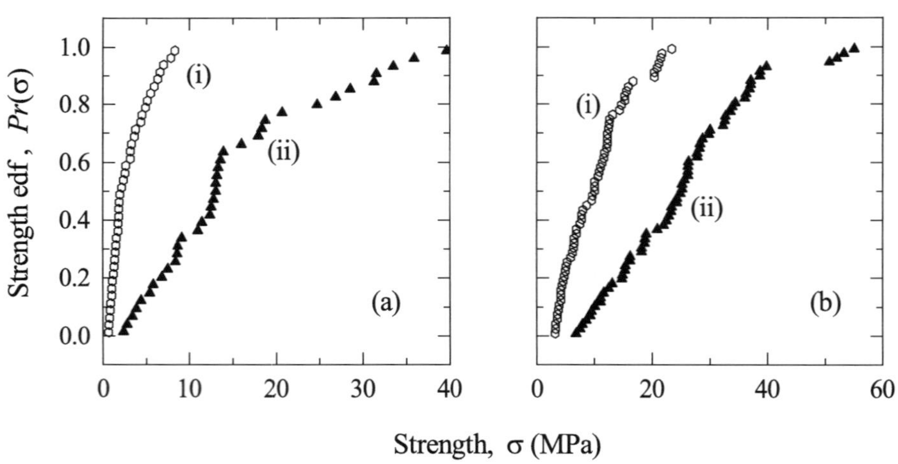

Figure 6.18 Plots of representative strength edf behavior, $Pr(\sigma)$, for large ($D > 10$ mm) particles. (a) (i) Calcareous rock, particle diameter, D = 7–15 mm, number of measurements, N = 40; (ii) quartzitic rock, D = 15–30 mm, N = 37; (Adapted from Ovalle, C et al. 2014). (b) (i) Basalt, D = 50–62.5 mm, N = 61; (ii) dolomite, D = 25–37.5 mm, N = 64 (Adapted from Koohmishi, M et al. 2016).

significant factor in determining B, strength-controlling flaw physical attributes, e.g. shape, residual stress state, are also factors. Such factors are considered in Chapter 12. Nevertheless, the relative B values for the particles lead to expected and explainable $h(c)$ behavior. Examples include: rocks contain larger flaws than alumina abrasives, which in turn contain

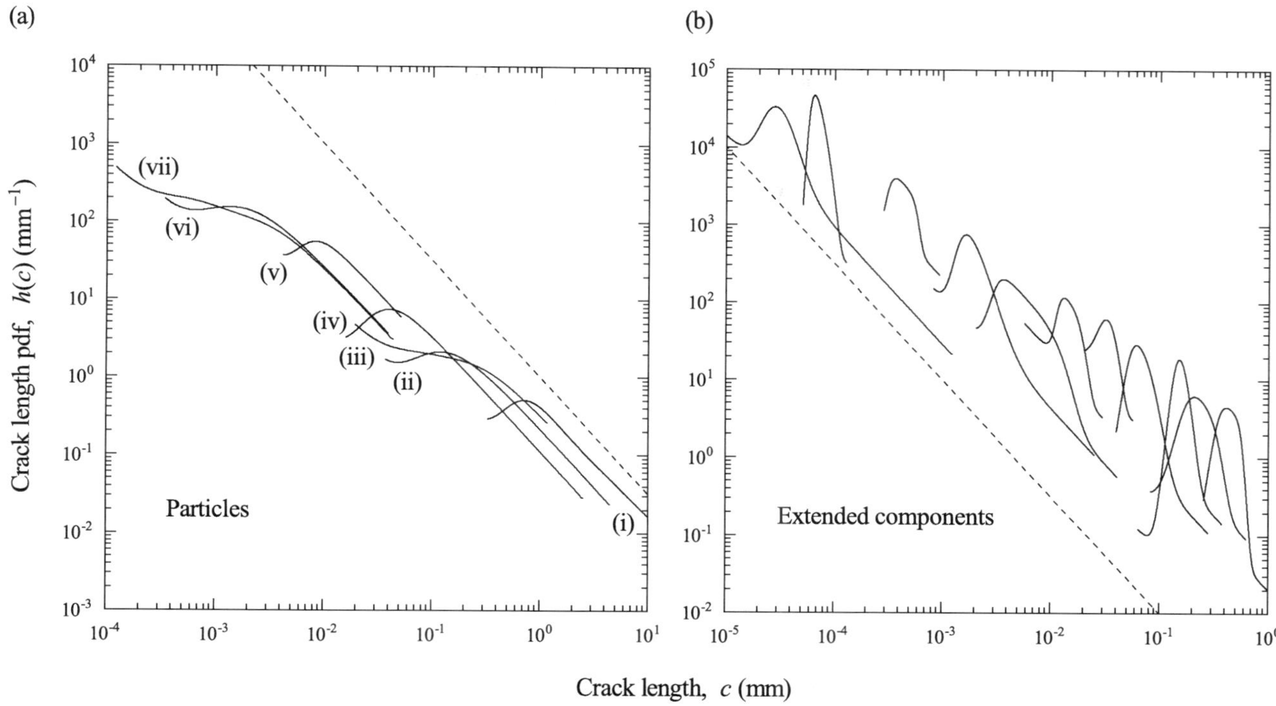

Figure 6.19 (a) Plot in logarithmic coordinates of crack length pdf variations, $h(c)$, for a range of materials and loading configurations representative of the particle data analyzed in this chapter. (i) rock; (ii) limestone; (iii) alumina; (iv) alumina; (v) quartz; (vi) coal; (vii) limestone. Note the extremely large range of flaw sizes. (b) Reproduction of the similar plot, Figure 5.18, for extended components.

larger flaws than coal dust. It is also important to remember that the B value does not change the shape of the crack length population deconvoluted from strength measurements.

Comparison of Figures 6.19a and 6.19b shows that the domains of strength controlling crack lengths in particles are wider than those of extended components. As a consequence, the amplitude variations of particle crack length probability densities are smaller, such that peaks, where present, are broad and not very well defined, and are often absent. The predominant feature of the crack length probability densities for particles is the heavy tail. The opposite is true for extended components: the domains are more compact, well-defined peaks are the predominant feature, and, where present, tails are light. Strength controlling crack length populations in particles differ markedly from those of extended components.

A qualitative description of strength controlling cracks in particles and extended components, representing in pictorial form the above quantitative description provides additional physical insight. Figure 6.20a is a schematic diagram of a sample of $N = 5$ particles loaded in diametral compression. The diagram is in direct analogy to that of extended components loaded in tension, Figure 5.29, reproduced here as Figure 6.20b. The particles are of uniform size, consisting of 10 fundamental volume elements, as are the tensile components. In Figure 6.20a the compressive loading is indicated by arrows and leads to localized zones of tensile stress at the center of the particles, indicated by shading. Each fundamental volume element contains one flaw, shown as a bold line. Not all flaws are the same size (orientations shown also differ) and the flaw sizes reflect an overall distributed, but invariant flaw population that is identical in the particles and the extended components. In the particles, only the central volume element and flaw experience the maximum tensile stress and thus $k = 1$. Failure thus ensues from the central flaw within the ensemble of flaws contained by each particle, independent of the size of the flaw at that location. As the size of the strength-controlling flaw at the center of a particle is determined by the probability distribution of flaw sizes in the population, the distribution of strengths for a sample of particles reflects the entire distribution of flaws. All flaws within the ensembles and the overall population of flaws are sensed by strength measurements, which are unbiased by extreme value effects in Figure 6.20a. The opposite effect occurs in the tensile components. Failure occurs at the largest flaw, shown shaded, in the ensemble of $k = 10$, independent of location. Most flaws are not sensed by strength measurements, which are biased by extreme value effects, Figure 6.20b.

The particles in diametral loading in Figure 6.20a have regions of central tension comparable in size to the fundamental volume elements containing single flaws. The strength measurements of small, $D < 1$ mm, particles reviewed here, Section 6.3.1, provide the best chance of exhibiting this configuration and enabling direct assessment of the flaw population as each

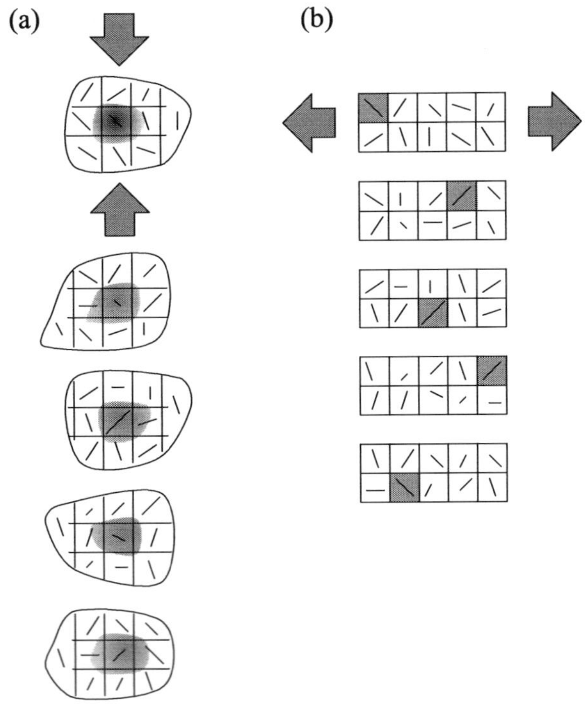

Figure 6.20 Schematic diagrams illustrating extreme value effects in brittle failure of samples of (a) particles in diametral compression and (b) extended components in tension. Applied loading is indicated by arrows. In each case, sample size $N = 5$, component size = 10. Flaws indicated by bold lines; strength-controlling flaws responsible for failure indicated shaded. Note that in (a) failure occurs at center, independent of flaw size, $k = 1$ and in (b) failure occurs at largest flaws, independent of location, $k = 10$.

particle contains a single strength-containing flaw. Hence, the heavy tailed flaw probability densities shown by the solid lines in Figure 6.16 can be interpreted as representing the crack length populations, $f(c) \approx h(c)$. However, the relative size of the zone of central tension and the particle size is fixed. If the size of a fundamental volume element is invariant, as particle increases the tensile zone will encompass a greater numbers of flaws. The implication is that medium particles, $1 < D < 10$ mm, the most common size, contain several flaws within the tensile zone and the strength distributions for medium particles presented here were a reflection of stochastic extreme value size effects. A follow on implication is that large particles, $D > 10$ mm, contain many flaws within the central tensile zone and the strength distributions for large particles presented here were most likely controlled by stochastic extreme value size effects. In addition, fabrication methods generate flaws in increasingly different ways as component size changes (Ashby 1999), a phenomenon explored in detail for MEMS components (DelRio et al. 2015). Hence, all the particle strength distributions presented here may also reflect superposed deterministic extreme value size effects. Distinguishing and characterizing stochastic and deterministic flaw size effects thus requires strength measurements on different size particles and this is the subject of the following three chapters.

Chapter 7 investigates stochastic size effects in particle strength distributions. Such effects are evident as different size particles randomly sample a single flaw population. Chapters 8 and 9 investigate deterministic size effects in particle strength distributions. Such effects are evident as different size particles systematically sample multiple flaw populations. The distinctions in behavior are not absolute, but, as will be observed, purely stochastic behavior is infrequently observed relative to deterministic behavior. Chapter 7 is analogous to Section 5.3.1 and Chapter 9 is analogous to Section 5.3.2. Chapter 8 is a "case study" of deterministic size effects on strength as a consequence of the evolution of flaw populations during fabrication of ceramic particles.

References

Ashby, M.F. (1999). *Materials Selection in Mechanical Design*. Butterworth-Heinemann.

Barrios, G.K.P., de Carvalho, R.M., and Tavares, L.M. (2011). Modeling breakage of monodispersed particles in unconfined beds. *Minerals Engineering* 24: 308–318 .

Bertrand, P.T., Laurich-McIntyre, S.E., and Bradt, R.C. (1988). Strengths of fused and tabular alumina refractory grains. *American Ceramic Society Bulletin* 67: 1217–1222.

Brecker, J.N. (1974). The fracture strength of abrasive grains. *Journal of Engineering for Industry* 96: 1253–1257.

Brzesowsky, R.H., Spiers, C.J., Peach, C.J., and Hangx, S.J.T. (2011). Failure behavior of single sand grains: Theory versus experiment. *Journal of Geophysical Research* 116: B06205.

Deer, W.A., Howie, R.A., and Zussman, J. (1966). *An Introduction to the Rock Forming Minerals*. Longman.

DelRio, F.W., Cook, R.F., and Boyce, B.L. (2015). Fracture strength of micro-and nano-scale silicon components. *Applied Physics Reviews* 2: 021303.

Dong, J., Cheng, Y., Hu, B., Hao, C., Tu, Q., and Liu, Z. (2018). Experimental study of the mechanical properties of intact and tectonic coal via compression of a single particle. *Powder Technology* 325: 412–419.

Fedorov, A.V. and Gulyaeva, Y.K. (2019). Strength statistics for porous alumina. *Powder Technology* 343: 783–791.

Herre, P., Romeis, S., Mačković, M., Przybilla, T., Paul, J., Schwenger, J., Torun, B., Grundmeier, G., Spiecker, E., and Peukert, W. (2017). Deformation behavior of nanocrystalline titania particles accessed by complementary in situ electron microscopy techniques. *Journal of the American Ceramic Society* 100: 5709–5722.

Hu, W., Dano, C., Hicher, P.Y., Le Touzo, J.Y., Derkx, F., and Merliot, E. (2011). Effect of sample size on the behavior of granular materials. *Geotechnical Testing Journal* 34: 186–197.

Huang, H., Huang, Q.K., Zhu, X.H., and Hu, X.Z. (1993). An experimental investigation of the strengths of individual brown corundum abrasive grains. *Scripta Metallurgica* 29: 299–304.

Huang, H., Zhu, X.H., Huang, Q.K., and Hu, X.Z. (1995). Weibull strength distributions and fracture characteristics of abrasive materials. *Engineering Fracture Mechanics* 52: 15–24.

King, R.P. and Bourgeois, F. (1993). Measurement of fracture energy during single-particle fracture. *Minerals Engineering* 6: 353–367.

Kingery, W.D., Bowen, H.K., and Uhlmann, D.R. (1975). *Introduction to Ceramics*. Wiley.

Koohmishi, M. and Palassi, M. (2016). Evaluation of the strength of railway ballast using point load test for various size fractions and particle shapes. *Rock Mechanics and Rock Engineering* 49: 2655–2664.

Lim, W.L., McDowell, G.R., and Collop, A.C. (2004). The application of Weibull statistics to the strength of railway ballast. *Granular Matter* 6: 229–237.

McDowell, G.R. (2002). On the yielding and plastic compression of sand. *Soils and Foundations* 42: 139–145.

Müller, P., Seeger, M., and Tomas, J. (2013). Compression and breakage behavior of γ-Al_2O_3 granules. *Powder Technology* 237: 125–133.

Nakata, Y., Hyodo, M., Hyde, A. F. L., Kato, Y., and Murata, H. (2001a). Microscopic particle crushing of sand subjected to high pressure one-dimensional compression. *Soils and Foundations* 41: 69–82.

Nakata, Y., Kato, Y., Hyodo, M., Hyde, A.F.L., and Murata, H. (2001b). One-dimensional compression behavior of uniformly graded sand related to single particle crushing strength. *Soils and Foundations* 41: 39–51.

Ovalle, C., Frossard, E., Dano, C., Hu, W., Maiolino, S., and Hicher, P.Y. (2014). The effect of size on the strength of coarse rock aggregates and large rockfill samples through experimental data. *Acta Mechanica* 225: 2199–2216.

Pejchal, V., Žagar, G., Charvet, R., Dénéréaz, C., and Mortensen, A. (2017). Compression testing spherical particles for strength: Theory of the meridian crack test and implementation for microscopic fused quartz. *Journal of Mechanics and Physics of Solids* 99: 70–92.

Pejchal, V., Fornabaio, M., Žagar, G., Riesen, G., Martin, R.G., Medřický, J., Chráska, T., and Mortensen, A. (2018). Meridian crack test strength of plasma-sprayed amorphous and nanocrystalline ceramic microparticles. *Acta Materialia* 145: 278–289.

Quinn, G.D. (1989). Flexure strength of advanced ceramics-A round robin exercise. MTL TR 89-62. US Army Materials Technology Laboratory, Watertown MA.

Ribas, L., Cordeiro, G.C., Toledo Filho, R.D., and Tavares, L.M. (2014). Measuring the strength of irregularly-shaped fine particles in a microcompression tester. *Minerals Engineering* 65: 149–155.

Rozenblat, Y., Portnikov, D., Levy, A., Kalman, H., Aman, S., and Tomas, J. (2011). Strength distribution of particles under compression. *Powder Technology* 208: 215–224.

Salman, A.D., Biggs, C.A., Fu, J., Angyal, I., Szabó, M., and Hounslow, M.J. (2002). An experimental investigation of particle fragmentation using single particle impact studies. *Powder Technology* 128: 36–46.

Salman, A.D., Fu, J., Gorham, D.A., and Hounslow, M.J. (2003). Impact breakage of fertiliser granules. *Powder Technology* 130: 359–366.

Shipway, P.H. and Hutchings, I.M. (1993). Fracture of brittle spheres under compression and impact loading II. Results for lead-glass and sapphire spheres. *Philosophical Magazine A* 67: 1405–1421.

Sikong, L., Hashimoto, H., and Yashima, S. (1990). Breakage behavior of fine particles of brittle minerals and coals. *Powder Technology* 61: 51–57.

Slikkerveer, P.J., in't Veld, H. Verspui, M., de With, B., and Reefman, D. (2000). Alumina particle degradation during solid particle impact on glass. *Journal of the American Ceramic Society* 83: 2263–2266.

Tavares, L.M. and King, R.P. (1998). Single-particle fracture under impact loading. *International Journal of Mineral Processing* 54: 1–28.

Todisco, M.C., Wang, W., Coop, M.R., and Senetakis, K. (2017). Multiple contact compression tests on sand particles. *Soils and Foundations* 57: 126–140.

Verspui, M.A., de With, G., and Dekkers, E.C.A. (1997). A crusher for single particle testing. *Reviews of Scientific Instruments* 68: 1553–1556 .

Vogel, L. and Peukert, W. (2002). Characterisation of grinding-relevant particle properties by inverting a population balance model. *Particles & Particle System Characterisation* 19: 149–157.

Wang, W. and Coop, M.R. (2016). An investigation of breakage behaviour of single sand particles using a high-speed microscope camera. *Géotechnique* 66: 984–998.

Wang, Y., Dan, W., Xu, Y., and Xi, Y. (2015). Fractal and morphological characteristics of single marble particle crushing in uniaxial compression tests. *Advances in Materials Science and Engineering* 2015: 537692.

Watkins, I.G. and Prado, M. (2015). Mechanical properties of glass microspheres. *Procedia Materials Science* 8: 1057–1065.

Wong, J.Y., Laurich-McIntyre, S.E., Khaund, A.K., and Bradt, R.C. (1987). Strengths of green and fired spherical aluminosilicate aggregates. *Journal of the American Ceramic Society* 70: 785–791.

7

Stochastic Scaling of Particle Strength Distributions

This chapter uses the analysis of Chapter 4 to survey in detail the strength distributions of particle systems that demonstrate stochastic size effects. Such systems are based on an invariant population of flaws that is modified by particle size effects to form extreme value flaw distributions that determine particle strength distributions. The strength distributions are closely linked and in particular are characterized by a common lower bound strength threshold. Scaling of strength distributions by stochastic size effects is demonstrated for concave and sigmoidal distributions.

7.1 Introduction

Particle strength edf $Pr(\sigma)$ variations are usually concave and wide, as shown in the extensive survey of Chapter 6 and the few earlier examples in Chapter 2. Analysis within the probabilistic fracture framework developed in Chapters 3 and 4 showed that the observed $Pr(\sigma)$ behavior is a consequence of underlying strength-controlling flaw distributions $h(c)$ that are predominantly heavy tailed with weakly defined peaks. In samples of particles from a material, the strength-controlling flaws form extreme value distributions based on the ensembles of flaws within each particle. In turn, the ensembles are formed by selection from the overall material flaw population $f(c)$. *Stochastic* size effects occur if particle size determines only the *probability* that a particular flaw size selected from an *invariant* population appears within the ensembles, and thus determines the extreme value distribution. The extreme value distribution of interest here is that of the largest cracks in each ensemble, as this determines the strength distribution of the sample. This Chapter 7 surveys and investigates in a quantitative manner stochastic size effects in strength behavior of particle systems. The chapter is analogous to Section 5.3.1 in Chapter 5 that investigated stochastic size effects in extended components.

A pictorial, qualitative description of stochastic size effects in particles is shown in Figure 7.1. The description is similar to those representing size and geometry effects in extended components and particles in Figures 5.29, 5.30, and 6.20 and schematizes the quantitative analyses of flaw distributions developed in Chapter 3. Figure 7.1a is a schematic diagram of a planar particle loaded in diametral compression, similar to Figure 6.20a. The particle consists of 9 fundamental volume elements, and thus $D \approx 3$, and the compressive loading, indicated by arrows, leads to a localized zone of tensile stress at the center of the particle, indicated by shading. Each fundamental volume element contains one flaw, shown as a bold line, and the flaw sizes reflect an invariant flaw population. In the particle of Figure 7.1a, only the central volume element and flaw experience the maximum tensile stress and thus $k = 1$. Failure thus ensues from the central flaw, indicated by a hatched element, within the ensemble of flaws contained by the particle. As the flaw sizes at all locations in the particles, including the centers, are determined by the probability distribution of flaw sizes in the population, the distribution of strengths for a sample of like particles is thus unbiased by extreme value effects and reflects the entire distribution of flaws.

Figures 7.1b and 7.1c are schematic diagrams of larger particles loaded in diametral compression with $D \approx 6$ and $D \approx 9$ and consisting of 36 and 81 fundamental volume elements, respectively. Again, each fundamental volume element contains one flaw, and the flaw sizes reflect the same invariant flaw population represented in Figure 7.1a. Diametral compression loading leads to localized zones of tensile stress at the center of each particle that scale with particle size (about one third) and are indicated by shading. In the particle of Figure 7.1b, the central four volume elements and flaws experience the maximum tensile stress and thus $k = 4$ and in the particle of Figure 7.1c the central nine volume elements and flaws experience the maximum tensile stress and thus $k = 9$. As the flaw sizes within the tensile zones represent stochastic

Particle Strengths: Extreme Value Distributions in Fracture, First Edition. Robert F. Cook.

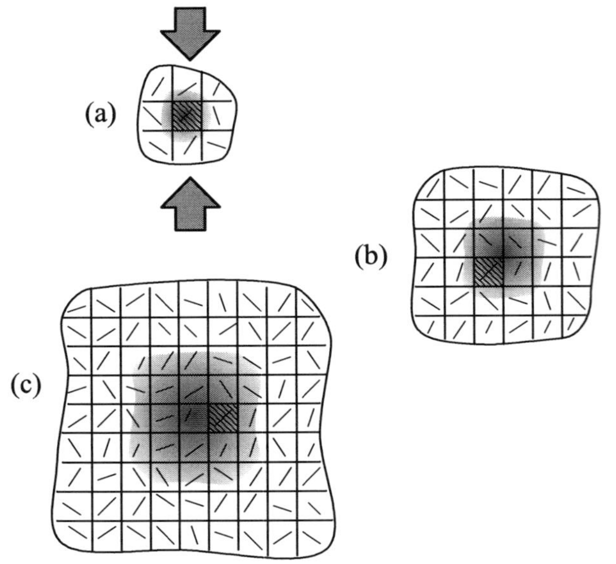

Figure 7.1 Schematic diagrams illustrating extreme value effects in brittle failure of particles in diametral compression. Particle diameter (a) $D = 3$, (b) $D = 6$, (c) $D = 9$. Applied loading is indicated by arrows in (a). Tensile zones indicated by shading. Flaws indicated by bold lines; strength-controlling flaws responsible for failure indicated hatched. Note that in (a) failure occurs at center, independent of flaw size, $k = 1$ and in (b) $k = 4$ and (c) $k = 9$, failure occurs at largest flaws, independent of location in tensile zone.

selections from the population of flaws, the size of the largest (or smallest) flaw in each zone approaches that of the largest (or smallest) flaw in the population as the zone size increases. As failure ensues from the largest or maximal extreme flaw within the ensemble of flaws contained by the tensile zone, failure tends to occur from larger flaws as particle size increases. Hence, in these cases, failure location *is* dependent on the size of the flaw within the central tensile zone, as it is the location of the largest flaw within the zone, again indicated by a hatched element. The distributions of strengths for samples of these larger particles are biased by extreme value effects—more so for the $k = 9$ particles than for the $k = 4$ particles—toward the large flaw, small strength threshold. The comparison of Figure 7.1b and Figure 7.1c is analogous to the comparison between Figure 5.29 and Figure 5.30. Note that the above discussion makes a clear distinction between (spatial) particle size D and (numerical) ensemble size k.

Hence, as for the extended components in Chapter 5, for purely stochastic selection reasons strength distributions of particles tend to smaller values of σ as the particle size k increases. The equation describing stochastic size effects is the relation between strength distributions of samples of different sized components, Eq. (3.26), here particles:

$$H_2(\sigma) = 1 - [1 - H_1(\sigma)]^{k_2/k_1}. \tag{7.1}$$

$H_1(\sigma)$ and $H_2(\sigma)$ are strength distributions of samples of different sized particles and k_1 and k_2 are the sizes expressed as the numbers of *stressed* fundamental volume elements, and flaws, in the particles. As noted earlier regarding Eq. (7.1): (i) The strength domains of $H_1(\sigma)$ and $H_2(\sigma)$ are identical, and identical with the strength domain $[\sigma_{th}, \sigma_u]$ of the population, $F(\sigma)$. In particular, the lower bound threshold, σ_{th}, is common to all distributions. (ii) Strength distributions of larger components are steeper, although truncation effects usually eliminate observations for $\sigma \to \sigma_u$. (iii) Systems that obey Eq. (7.1) imply the existence of a single underlying population of *independent* strength-controlling flaws, regardless of the details (shape, location) of that population. The majority of (non-failure) flaws within particle ensembles and the overall population of flaws are not sensed by strength measurements. The fully sufficient condition for the strength distributions of two samples of particles to be related by stochastic extreme value effects is demonstration that they obey Eq. (7.1). A consequence of the domain and derivative conditions is that such strength distributions appear as similarly shaped functions fanned out from the threshold.

It is important to note that the exponent in Eq. (7.1) provides only the ratio k_2/k_1 of the numbers of flaws effectively stressed in strength tests of particles. In particular, for particles of different sizes, D_2 and D_1, demonstration of stochastic behavior does not specify the absolute values of k_2 or k_1 or enable the specification of an absolute size relation $k(D)$. However, as in consideration of stochastic behavior of extended components of different sizes (Section 5.3), the scaling of the ratio k_2/k_1 with size is informative. Here, the ratio k_2/k_1 obtained from strength measurements can be compared with the ratio D_2/D_1, obtained independently from particle size measurements. For example, the schematic diagrams of the particles in Figure 7.1 exhibit $D_1 : D_2 : D_3$ of $1 : 2 : 3$. As drawn and described above as two-dimensional planar particles, the proportion $k_1 : k_2 : k_3$ is $1 : 4 : 9$. If Figure 7.1 is taken to represent a cross-section of three-dimensional particles, the proportion $k_1 : k_2 : k_3$ is 1 : 8 : 27. Such behavior is easily assessed for various particle systems and will be discussed in Section 7.4.

In addition, particle systems provide a potential natural internal constraint on interpretation of strength behavior. The largest flaw inferred from strength tests on a sample of particles cannot be larger than the particle size. For the example above, in which D_1 is the smallest particle, expressing this constraint in strength terms gives

$$(B/\sigma_{\rm th})^2 \leq D_1, \tag{7.2}$$

where B is the coefficient relating strength and crack length in the Griffith equation, $\sigma = Bc^{-1/2}$. As $\sigma_{\rm th}$ is measured (as estimated by $\sigma_{\rm L}$) and if D_1 is known, this constraint effectively sets the material-flaw composite parameter B. In Chapters 2 and 6, this procedure was implemented and the resulting B values used to interpret material properties (e.g. porous limestone exhibited B values less than dense alumina). In other previous works considering silicon MEMS components, an invariant value of B based on material (polysilicon) considerations was used to infer crack lengths, independent of component size (Cook et al. 2019, 2021; DelRio et al. 2020). In these works, stress concentrating effects in components with notches, fillets, or extremely small dimensions enabled $k_1 = 1$ to be set for such components, such that k_2, k_3, and so on could be compared with independent measurements of component dimensions and flaw spacing. Here, an invariant value of B for a particle system was determined using the constraint of Eq. (7.2), setting D_1 as the smallest particle in a set of particle samples. A further interpretation is that in the limit of equality of Eq. (7.2), $k_1 = 1$ for the smallest particle and thus the response of the smallest particles under this constraint can be taken to approximate the population, $H_1(\sigma) \approx F(\sigma)$ and $h_1(c) \approx f(c)$.

Figure 7.2 shows a strength edf $Pr(\sigma)$ plot for samples of quartz (crystalline SiO_2) particles approximately 1.4 mm (i) and 3.35 mm (ii) in diameter, tested in a drop weight configuration from the work of King and Bourgeois (1993), using data derived from the published work. The results were originally reported in terms of specific failure energy and were motivated by considerations of minimizing the work required to crush particles in ore refinement. (As earlier, details of component and sample sizes are provided in the figure caption.) Symbols represent individual strength measurements; different symbols represent different particle sizes, D. The lower solid line in Figure 7.2 represents a visual best fit, taken here

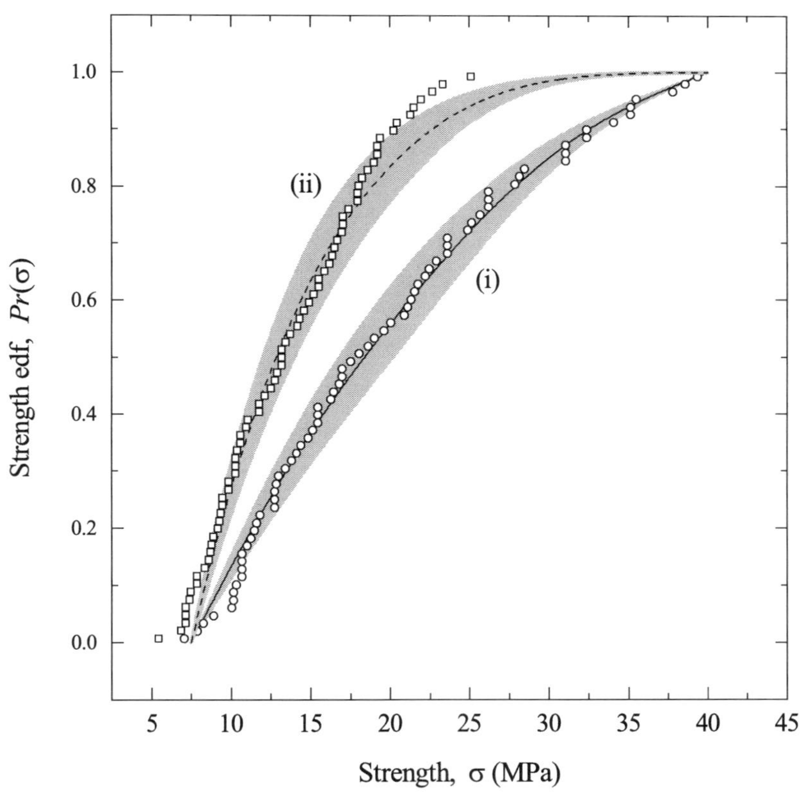

Figure 7.2 Plot of strength edf behavior, $Pr(\sigma)$, for quartz particles. Particle diameter, (i) D = 1.4 mm–1.7 mm and (ii) D = 3.35 mm–4.00 mm, number of measurements, $N_{\rm tot}$ = 147 (Adapted from King, R.P et al., 1993). Solid line shows concave best fit, dashed line shows best fit constrained by stochastic size effect, shaded bands show bounds.

as $H_1(\sigma)$, to the $D_1 = 1.4$ mm data and the shaded band represents bounds on the fit encompassing most of the observations. The data are well described by concave behavior and the data set exhibits a strength domain width σ_N/σ_1 of approximately 8. Comparison with the survey results of Chapter 6 shows that both the shape and the extent of the strength response are typical for particles. The upper dashed line in Figure 7.2 represents a constrained fit, taken here as $H_2(\sigma)$, to the $D_2 = 3.35$ mm data, using Eq. (7.1) and k_2/k_1 as a fitting parameter. The shaded band represents the bounds on $H_2(\sigma)$ derived from those of $H_1(\sigma)$. In this case, a value of $k_2/k_1 = 2.2$ provided a fit to the behavior of the larger particles encompassing most of the strength observations. The conclusion from Figure 7.2 is that the strength distributions of samples of the two sizes of particles are related by a stochastic size effect and a common flaw population. Similar strength results and conclusions can be drawn from the experiments of Wang and Coop (2016) in which 1 mm and 2 mm quartz particles were examined in a custom compression apparatus.

Figure 7.3 shows as a solid line (i) the crack length pdf $h(c)$ variation underlying the smaller D_1 quartz particle strength behavior described by $H_1(\sigma)$. The $h(c)$ variation was determined using $B = 0.25$ MPa m$^{1/2}$ and Eq. (7.2) as a constraint. Comparison with the results in Chapter 6 indicates that the tail-dominated shape, weakly defined peak, and wide extent of the crack length response are typical for particles. The discussion above suggests that this response approximates the overall population of flaws, $f(c)$, applicable to all sizes of particle of this material. The dashed line (ii) in Figure 7.3 represents the crack length pdf $h(c)$ variation underlying the larger D_2 quartz particle strength behavior described by $H_2(\sigma)$. The $h(c)$ variation was determined using $B = 0.25$ MPa m$^{1/2}$ and extends over the same crack length domain as that of the smaller particle response (i) approximating the population $f(c)$. The dashed line $h(c)$ response (ii) represents the continuum prediction for the strength-controlling crack lengths of the larger particles and reflects the extreme value bias towards fewer small cracks and more large cracks. As large strengths in large particles are not frequently observed, the crack length domain conjugate to the experimentally observed strengths is truncated relative to the population crack length domain. The truncated crack length domain here is indicated by the grayed shaded line and does not include small cracks. The shape of the shaded line is similar to that of response (i) and other directly fit crack length responses shown in Chapter 6. Although the dashed line prediction follows analytically from Eq. (7.1), the complete response is not accessible experimentally.

The presentation formats of Figure 7.2 and Figure 7.3 will be used throughout this Chapter in surveying stochastic extreme value size effects in strength behavior of particles. In the example above, it is noted that the strength distributions

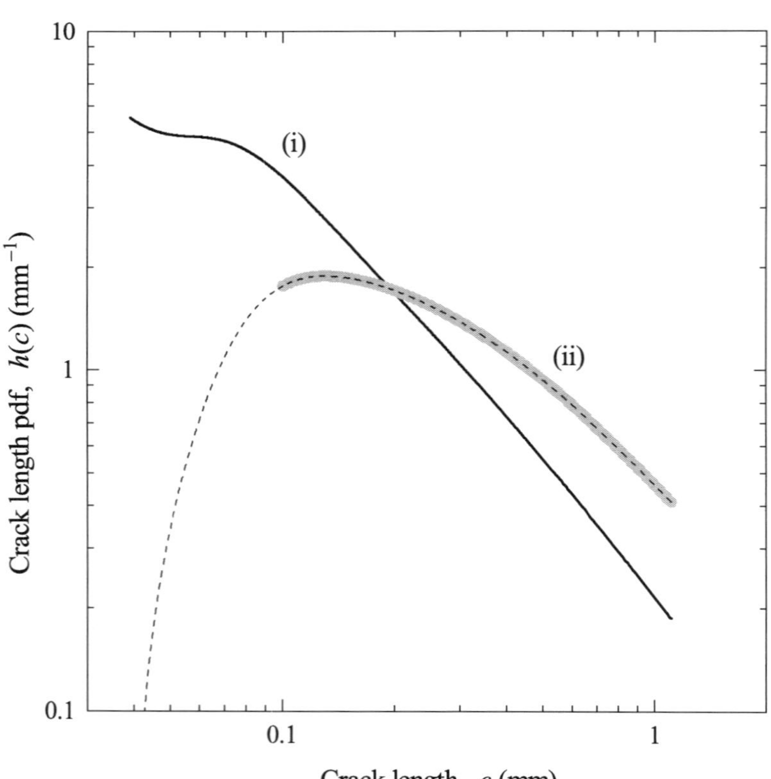

Figure 7.3 Plot in logarithmic coordinates of crack length pdf $h(c)$ variations for quartz particles. Solid line (i) $D = 1.4$ mm–1.7 mm and dashed and grayed lines (ii) $D = 3.35$ mm–4.00 mm; determined from Figure 7.2.

and particle sizes were related by $k_2/k_1 = 2.2 \approx D_2/D_1 = 2.4$. A goal of the survey is to use observations such as this to provide an experimental assessment of the scaling relation $k(D)$ for particles.

7.2 Concave Stochastic Distributions

7.2.1 Alumina

Figure 7.4 shows strength edf $Pr(\sigma)$ plots for samples of alumina (polycrystalline Al_2O_3) particles 1.2–3.8 mm in diameter, tested in a conventional compression platen apparatus from the work of Bertrand et al. (1988), using data derived from the published work. Distinct from the study on quartz above, the goal in this case was to identify optimum particles for use in refractory bricks for high temperature manufacturing. The particles were commercial materials, differing by microstructure, which consisted of 50 μm-scale grains and included intra- and inter-granular porosity. Four grades of alumina in four size classifications were examined, all exhibiting similar behavior. Tabular (a) and fused (b) microstructures are considered here. Symbols represent individual strength measurements; different symbols represent different particle sizes, D. The lower solid lines in Figure 7.4 represent visual best fits, taken here as $H_1(\sigma)$, to the (a) $D_1 = 1.2$ mm–2.4 mm data and (b) 1.2 mm data. The data, although not completely separated by particle size, are well described by concave behavior and the data sets exhibit very wide strength domains σ_N/σ_1 of 20–50. In these cases, the maximum/minimum strength ratios are extremely large as the minimum strengths are near zero. The upper dashed line in Figure 7.4a represents a constrained fit, taken here as $H_2(\sigma)$, to the $D_2 = 3.8$ mm data, using Eq. (7.1) and k_2/k_1 as a fitting parameter. The upper dashed lines in Figure 7.4b represent similarly constrained fits to the $D = 2.6$ and 2.0 mm (lower) and 3.2 mm (upper) data. The constrained fits provide descriptions of the larger particle strength distributions over the observed strength domains. The conclusion from Figure 7.4 is that the strength distributions of samples of the multiple sizes of particles are related by stochastic size effects and common flaw populations. The fitted values of k_2/k_1 were in the range 3–8 and are discussed in Section 7.4, along with other similar measurements from the data discussed below.

Figure 7.5 shows as solid lines (i) the crack length pdf $h(c)$ variations underlying the smaller D_1 alumina particle strength behaviors described by $H_1(\sigma)$. The $h(c)$ variations were determined using (a) $B = 0.4$ MPa m$^{1/2}$ and (b) $B = 0.12$ MPa m$^{1/2}$ and Eq. (7.2) as a constraint. The discussion above suggests that these responses approximate the overall populations of flaws, $f(c)$, applicable to all sizes of particle of these materials. The different B values point to the differences

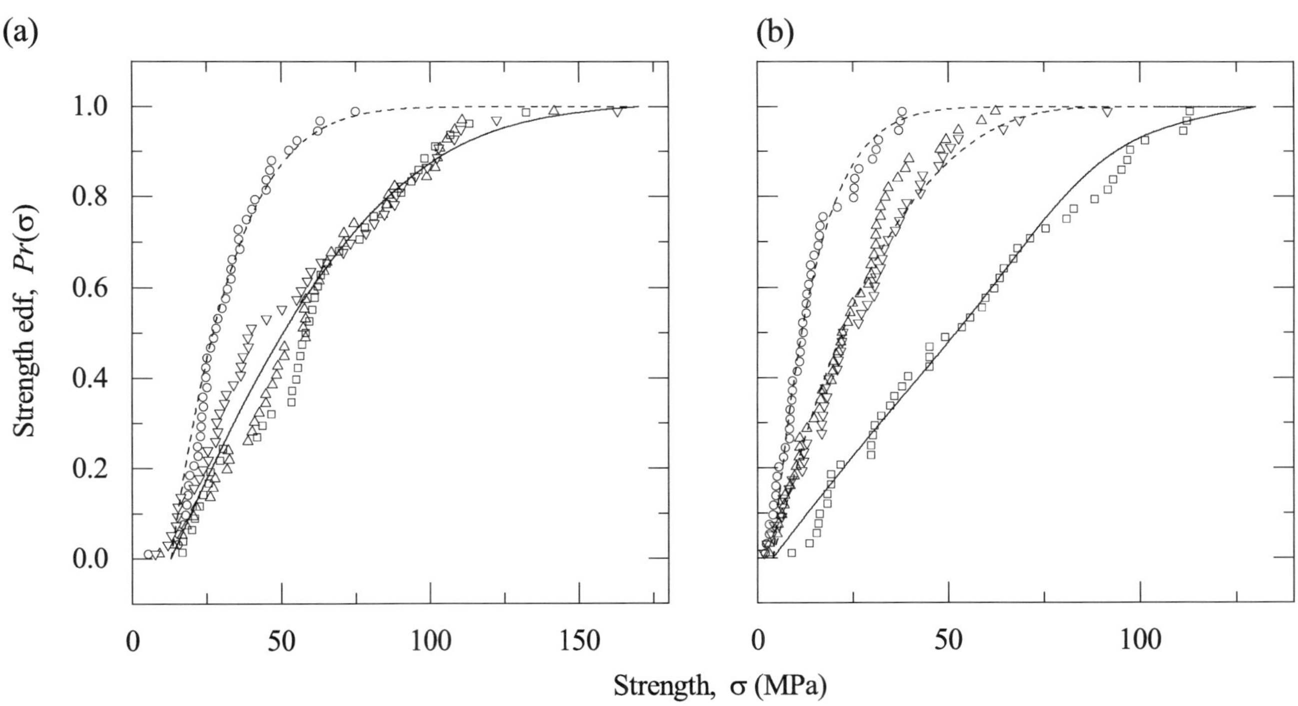

Figure 7.4 Plots of strength edf behavior, $Pr(\sigma)$, for alumina particles. (a) D = (3.8, 2.4, 2.0, and 1.2) mm and (b) D = (3.2, 2.6, 2.0, and 1.2) mm, all size uncertainties ± 0.1 mm, N_{tot} = 370 (Bertrand, P.T et al. 1988). Solid lines show concave best fits, dashed lines show best fits constrained by stochastic size effects.

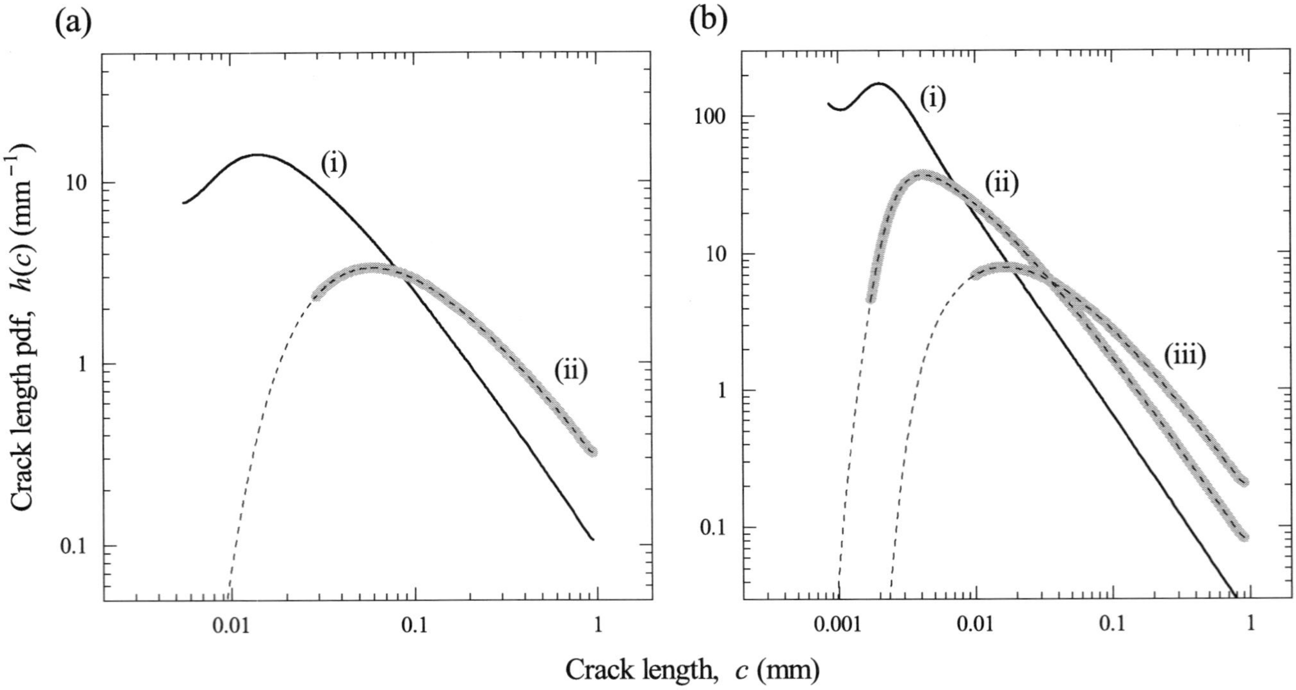

Figure 7.5 Plots in logarithmic coordinates of crack length pdf $h(c)$ variations for alumina particles. Solid lines (i) $D = 1.2$ mm and dashed and grayed lines (ii) and (iii) $D = 2.0$ mm–3.8 mm; determined from Figure 7.4.

in the materials; the tabular material was more fracture resistant than the fused material. The dashed lines (ii) and (iii) represent the crack length pdf $h(c)$ variations underlying the larger alumina particle strength behaviors described by $H_2(\sigma)$. These $h(c)$ variations were determined using the same B values and extend over the same crack length domains as those of the smaller particle responses (i) approximating the populations $f(c)$. As above, the dashed line $h(c)$ responses (ii) and (iii) represent the continuum predictions for the strength-controlling crack lengths of larger particles and reflect the extreme value bias toward fewer small cracks and more large cracks. Also as above, large strengths in large particles were not frequently observed. The crack length domains conjugate to the experimentally observed strengths are truncated relative to the population crack length domains. The truncated crack length domains here are indicated by the grayed shaded lines and do not include small cracks.

7.2.2 Limestone

This section examines stochastic size effects in the strength distributions of three sets of limestone (predominantly polycrystalline $CaCO_3$) particles. Similar to the study on quartz above, the goal in the first study was to aid in minimizing the work required to crush particles in ore refinement. The goal in the second and third studies was to aid in interpretation of the deformation of particulate aggregates in geotechnical applications (e.g. foundations). The studies are examined in order of increasing width of the crack length population domain. Figure 7.6 shows a strength edf $Pr(\sigma)$ plot for samples of limestone particles approximately 0.5 mm (i) and 4 mm (ii) in diameter, tested in a drop weight configuration from the work of King and Bourgeois (1993), using data derived from the published work. Symbols represent individual strength measurements; different symbols represent different particle sizes, D. The lower solid line represents a visual best fit, taken here as $H_1(\sigma)$, to the $D_1 = 0.5$ mm data. The data are well described by concave behavior and the data set exhibits a strength domain width σ_N/σ_1 of approximately 5. The upper dashed line represents a constrained fit, taken here as $H_2(\sigma)$, to the $D_2 = 4$ mm data, using Eq. (7.1) and k_2/k_1 as a fitting parameter. The conclusion from Figure 7.6 is that the strength distributions of samples of the two sizes of particles are related by a stochastic size effect and a common flaw population. Similar strength results and conclusions can be drawn from the experiments of Wang et al. (2015) in which 33 mm and 6 mm limestone particles were examined.

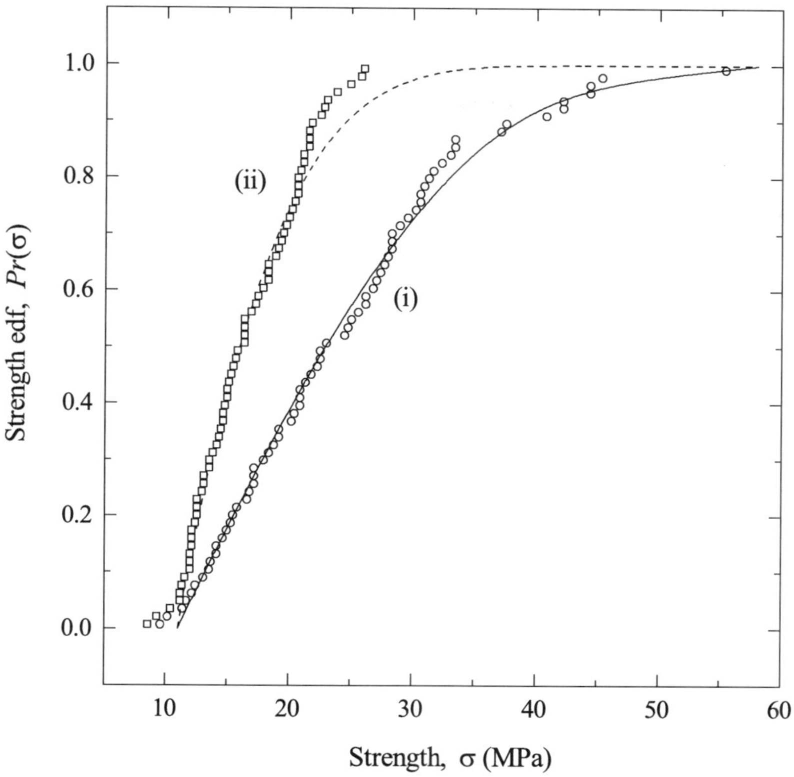

Figure 7.6 Plot of strength edf behavior, $Pr(\sigma)$, for limestone particles. (i) D = 0.5 mm–0.7 mm and (ii) D = 4.00 mm–4.75 mm, N_{tot} = 144 (Adapted from King, R.P et al., 1993). Solid line shows concave best fit, dashed line shows best fit constrained by stochastic size effect.

Figure 7.7 shows as a solid line (i) the crack length pdf $h(c)$ variation underlying the smaller D_1 limestone particle strength behavior described by $H_1(\sigma)$. The $h(c)$ variation was determined using $B = 0.25$ MPa m$^{1/2}$ and Eq. (7.2) as a constraint. The discussion above suggests that this response approximates the overall population of flaws, $f(c)$, applicable to all sizes of particle of this material. The dashed line (ii) represents the crack length pdf $h(c)$ variation underlying the larger D_2 limestone particle strength behavior described by $H_2(\sigma)$. The $h(c)$ variation was determined using $B = 0.25$ MPa m$^{1/2}$ and extends over the same crack length domain as that of the smaller particle response (i) approximating the population $f(c)$. The dashed line $h(c)$ response (ii) represents the continuum prediction for the strength-controlling crack lengths of the larger particles and reflects the extreme value bias toward fewer small cracks and more large cracks. As large strengths in large particles were not frequently observed, the crack length domain conjugate to the experimentally observed strengths is truncated relative to the population crack length domain. The truncated crack length domain here is indicated by the grayed shaded line and does not include small cracks.

Figure 7.8 shows a strength edf $Pr(\sigma)$ plot for samples of limestone particles 7–50 mm in diameter, tested in a conventional compression platen apparatus from the work of Hu et al. (2011, as cited by Ovalle et al. 2014), using data derived from the published work. Symbols represent individual strength measurements; different symbols represent different particle sizes, D. The lower solid line represents a visual best fit, taken here as $H_1(\sigma)$, to the D_1 = 7 mm and 15 mm data. The data, although not completely separated by particle size, are well described by concave behavior and the data sets exhibit strength domains σ_N/σ_1 of approximately 8. The upper dashed lines represent constrained fits, taken here as $H_2(\sigma)$, to the D_2 = 25 mm and 50 mm data, using Eq. (7.1) and k_2/k_1 as a fitting parameter. The constrained fits provide descriptions of the larger particle strength distributions over the observed strength domains. The conclusion from Figure 7.8 is that the strength distributions of samples of the multiple sizes of particles are related by stochastic size effects and common flaw populations.

Figure 7.9 shows as a solid line (i) the crack length pdf $h(c)$ variation underlying the smaller D_1 limestone particle strength behaviors described by $H_1(\sigma)$. The $h(c)$ variation was determined using $B = 0.07$ MPa m$^{1/2}$ and Eq. (7.2) as a constraint.

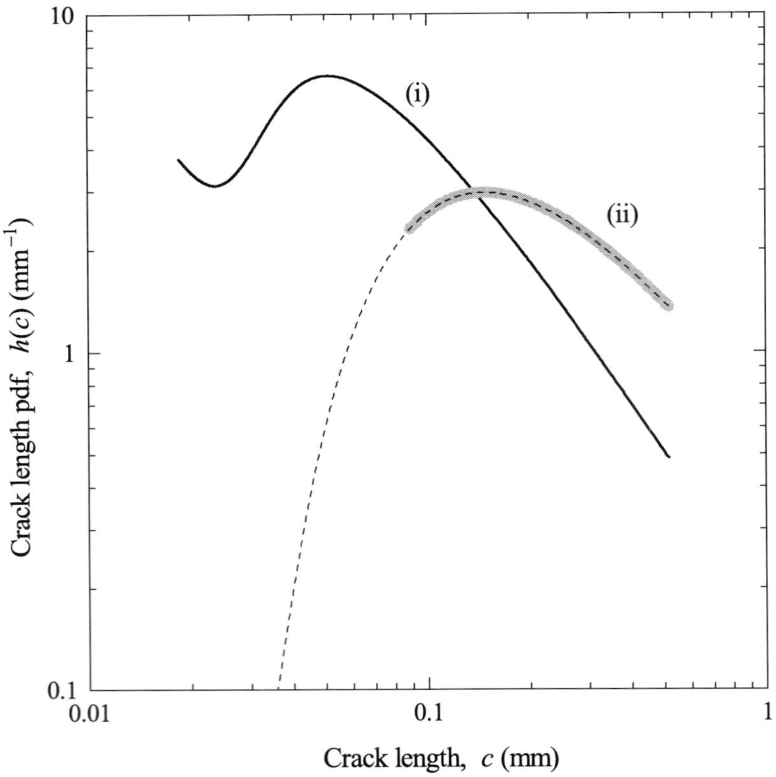

Figure 7.7 Plot in logarithmic coordinates of crack length pdf $h(c)$ variations for limestone particles. Solid line (i) $D = 0.5–0.7$ mm and dashed and grayed lines (ii) $D = 4.00–4.75$ mm; determined from Figure 7.6.

The discussion above suggests that this response approximates the overall population of flaws, $f(c)$, applicable to all sizes of particle of this material. The dashed lines (ii) and (iii) represent the crack length pdf $h(c)$ variations underlying the larger limestone particle strength behaviors described by $H_2(\sigma)$. These $h(c)$ variations were determined using the same B value and extend over the same crack length domain as that of the smaller particle response (i) approximating the populations $f(c)$. As above, the dashed line $h(c)$ responses (ii) and (iii) represent the continuum predictions for the strength-controlling crack lengths of larger particles and reflect the extreme value bias toward fewer small cracks and more large cracks. Also as above, large strengths in large particles were not frequently observed. The crack length domains conjugate to the experimentally observed strengths are truncated relative to the population crack length domains. The truncated crack length domains here are indicated by the grayed shaded lines and do not include small cracks.

A striking example of strength distributions formed by stochastic size effects can be observed in the experiments by McDowell and Amon (2000) on Quiou sand particles. Quiou sand is a calcareous, limestone-based sand. Motivated by considerations of compaction of soil grain aggregates, McDowell and Amon used a custom compression apparatus to test the strengths of carefully graded Quiou sand particles from 1 mm to 16 mm in size. Figure 7.10 shows the resulting strength edf $Pr(\sigma)$ plots for samples of Quiou sand particles, using data derived from the published work. The data are particularly well suited to assess stochastic size effects: The strength distributions are wide, the cumulative data sets exhibit a strength domain σ_N/σ_1 of approximately 300, and the data are well separated by particle size. Symbols represent individual strength measurements; different symbols represent different particle sizes, D, identified in the figure. The solid line represents a visual best fit, taken here as $H_1(\sigma)$, to the $D_1 = 1$ mm data. The data are well described by concave behavior. The dashed lines for the $D = 2$ mm, 4 mm, and combined 8 mm and 16 mm data represent constrained fits, taken here as $H_2(\sigma)$, using Eq. (7.1) and k_2/k_1 as a fitting parameter. The constrained fits provide descriptions of the larger particle strength distributions over the increasingly reduced strength domains. The conclusion from Figure 7.10 is that the strength distributions of samples of the multiple sizes of particles are related by stochastic size effects and a common flaw population.

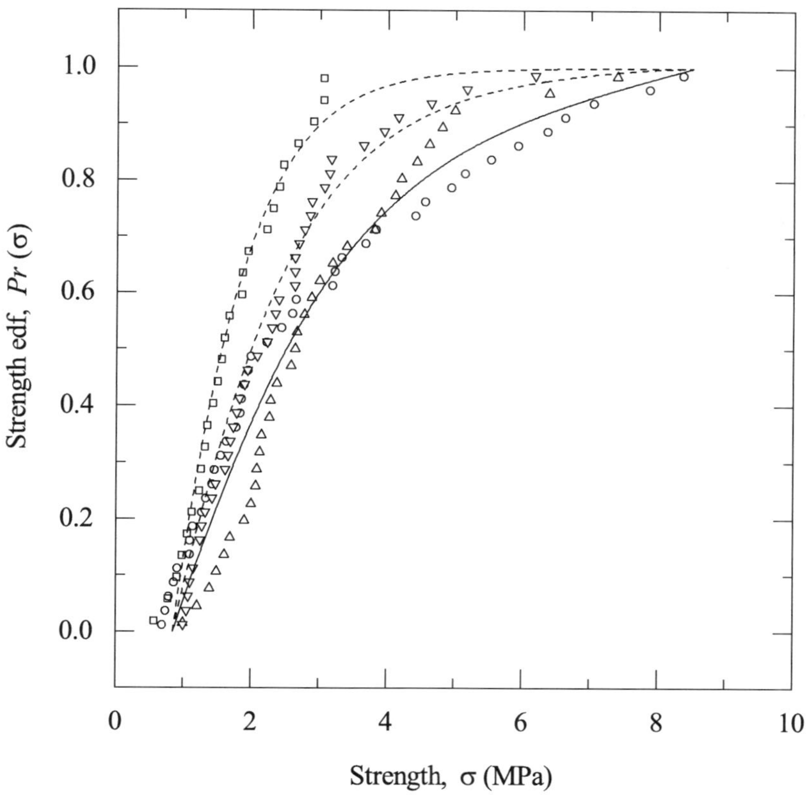

Figure 7.8 Plot of strength edf behavior, $Pr(\sigma)$, for limestone particles. (a) D = (7–15) mm, (15–25) mm, (25–50) mm, (50–80) mm, N_{tot} = 139 (Adapted from Hu, W et al., 2011; Ovalle C et al., 2014). Solid line shows concave best fit, dashed lines show best fits constrained by stochastic size effects.

Figure 7.11 shows as a solid line (i) the crack length pdf $h(c)$ variation underlying the 1 mm Quiou sand particle strength behavior described by $H_1(\sigma)$. The $h(c)$ variation was determined using $B = 0.009$ MPa $m^{1/2}$ and Eq. (7.2) as a constraint. The discussion above suggests that this response approximates the overall population of flaws, $f(c)$, applicable to all sizes of particle of this material and the small value of B is consistent with a very small fracture resistance. The dashed lines (ii)–(iv) represent the crack length pdf $h(c)$ variations underlying the larger Quiou sand particle strength behaviors described by $H_2(\sigma)$. These $h(c)$ variations were determined using the same B value and extend over the same crack length domain as that of the smaller particle response (i) approximating the populations $f(c)$. As above, the dashed line $h(c)$ responses (ii)–(iv) represent the continuum predictions for the strength-controlling crack lengths of larger particles and reflect the extreme value bias toward fewer small cracks and more large cracks. In this case in particular, large strengths in large particles were not observed. The crack length domains conjugate to the experimentally observed strengths are substantially truncated relative to the very wide population crack length domain. The truncated crack length domains here are indicated by the grayed shaded lines and do not include small cracks (and some large cracks for the largest particles).

7.2.3 Coral

A similar example of strength distributions formed by stochastic size effects to that of Quiou sand can be observed in the experiments by Shen et al. (2020) on coral sand particles. Coral sand is a porous material formed by marine organisms and is a mixture of $CaCO_3$ and $MgCO_3$. The study of Shen et al. was motivated by evaluation of coral sand as a constituent of cement mortar. Shen et al. (2020) used a conventional compression platen apparatus to test the strengths of graded coral particles from 1.5 mm to 11 mm in size (nominally 2, 4, and 8 mm). Figure 7.12 shows the resulting strength edf $Pr(\sigma)$

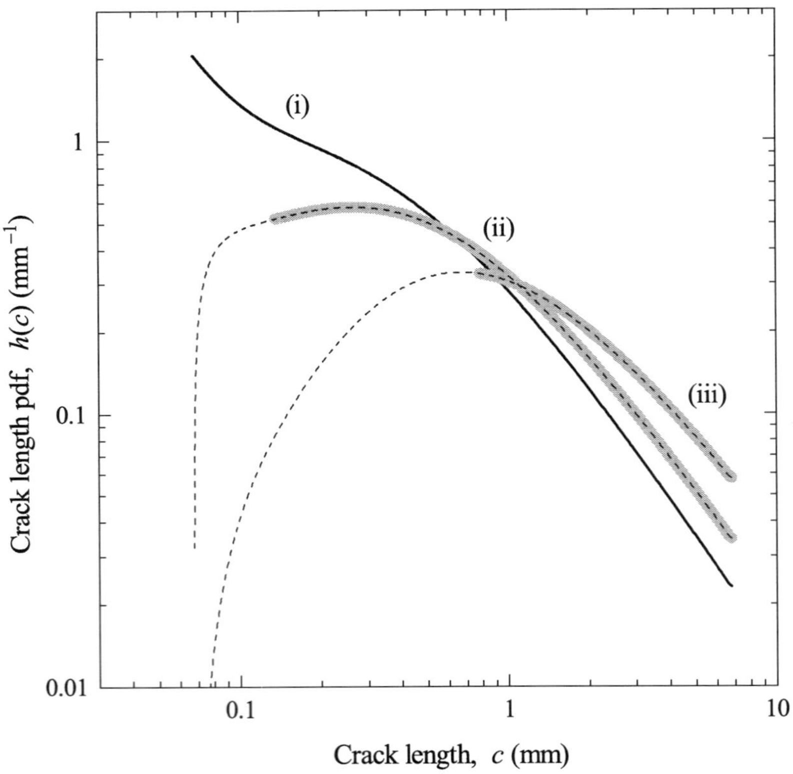

Figure 7.9 Plot in logarithmic coordinates of crack length pdf $h(c)$ variations for limestone particles. Solid line (i) $D = 7$ and 15 mm and dashed and grayed lines (ii) and (iii) $D = 25$ mm and 50 mm; determined from Figure 7.8.

plots for samples of coral particles, using data derived from the published work, repeating the unanalyzed data presented in Chapter 2. As with the previous study on limestone-based Quiou sand, the coral sand data are particularly well suited to assess stochastic size effects. The strength distributions are wide and the data are well separated by particle size. Symbols represent individual strength measurements; different symbols represent different particle sizes, increasing D shown as Figure 7.12abc, respectively. The solid line represents a visual best fit, taken here as $H_1(\sigma)$, to the $D_1 = 2$ mm data. The data are well described by concave behavior. The dashed lines for the $D = 4$ mm and 8 mm data represent constrained fits, taken here as $H_2(\sigma)$, using Eq. (7.1) and k_2/k_1 as a fitting parameter. The constrained fits provide descriptions of the larger particle strength distributions over the increasingly reduced strength domains. The conclusion from Figure 7.14 is that the strength distributions of samples of the multiple sizes of particles are related by stochastic size effects and a common flaw population.

Figure 7.13 shows as a solid line (i) the crack length pdf $h(c)$ variation underlying the 2 mm coral sand particle strength behavior described by $H_1(\sigma)$. The $h(c)$ variation was determined using $B = 0.06$ MPa m$^{1/2}$ and Eq. (7.2) as a constraint. The discussion here suggests that this response approximates the overall population of flaws, $f(c)$, applicable to all sizes of particle of this material. The dashed lines (ii) and (iii) represent the crack length pdf $h(c)$ variations underlying the larger silica sand particle strength behaviors described by $H_2(\sigma)$. These $h(c)$ variations were determined using the same B value and extend over the same crack length domain as that of the smaller particle response (i) approximating the populations $f(c)$. As above, the dashed line $h(c)$ responses (ii) and (iii) represent the continuum predictions for the strength-controlling crack lengths of larger particles and reflect the extreme value bias toward fewer small cracks and more large cracks. As in the case of the Quiou sand, large strengths in large coral sand particles were not observed. The crack length domains conjugate to the experimentally observed strengths are substantially truncated relative to the very wide population crack length domain. The truncated crack length domains here are indicated by the grayed shaded lines and do not include small cracks.

7.2.4 Quartz and Quartzite

Another striking example of strength distributions formed by stochastic size effects can be observed in the experiments by Nakata et al. (2001a, 2001b) on silica sand particles. As in the previous study on Quiou sand by McDowell and Amon

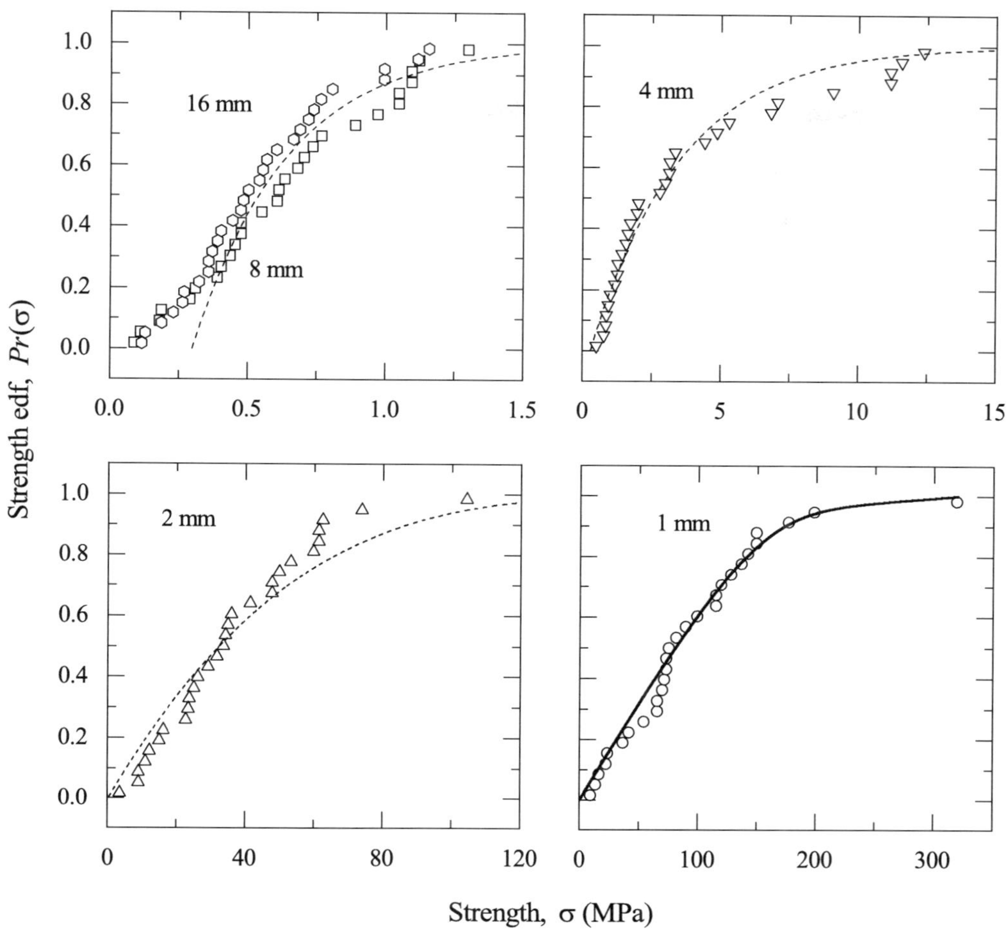

Figure 7.10 Plot of strength edf behavior, $Pr(\sigma)$, for Quiou sand (limestone) particles. D = (0.66–0.996, 1.43–2.01, 3.33–4.41, 7.22–8.5, and 14.7–16.3) mm, nominally (1, 2, 4, 8, and 16) mm, N_{tot} = 146 (Adapted from McDowell, G.R et al., 2000). Solid line shows concave best fit, dashed lines show best fits constrained by stochastic size effects.

(2000), Nakata et al. were motivated by considerations of compaction of soil grain aggregates. Nakata et al. (2001b) used a conventional compression platen apparatus to test the strengths of carefully graded silica (quartz, crystalline SiO_2) sand particles from 0.1 mm to 1.5 mm in size. Figure 7.14 shows the resulting strength edf $Pr(\sigma)$ plots for samples of silica sand particles, using data derived from the published work, repeating the unanalyzed data presented in Chapter 2. As with the previous study on limestone-based Quiou sand, the silica sand data are particularly well suited to assess stochastic size effects. The strength distributions are wide and the data are well separated by particle size. Symbols represent individual strength measurements; different symbols represent different particle sizes, increasing D shown as Figure 7.14a, 7.14b, 7.14c, respectively. The solid line represents a visual best fit, taken here as $H_1(\sigma)$, to the $D_1 = 0.1$ mm data. The data are well described by concave behavior. The dashed lines for the $D = 0.4$ mm and 1.5 mm data represent constrained fits, taken here as $H_2(\sigma)$, using Eq. (7.1) and k_2/k_1 as a fitting parameter. The constrained fits provide descriptions of the larger particle strength distributions over the increasingly reduced strength domains. The conclusion from Figure 7.14 is that the strength distributions of samples of the multiple sizes of particles are related by stochastic size effects and a common flaw population.

Figure 7.15 shows as a solid line (i) the crack length pdf $h(c)$ variation underlying the 0.1 mm silica sand particle strength behavior described by $H_1(\sigma)$. The $h(c)$ variation was determined using $B = 0.15$ MPa m$^{1/2}$ and Eq. (7.2) as a constraint. The discussion above suggests that this response approximates the overall population of flaws, $f(c)$, applicable to all sizes of particle of this material. The dashed lines (ii) and (iii) represent the crack length pdf $h(c)$ variations underlying the larger silica sand particle strength behaviors described by $H_2(\sigma)$. These $h(c)$ variations were determined using the same B value and extend over the same crack length domain as that of the smaller particle response (i) approximating the populations

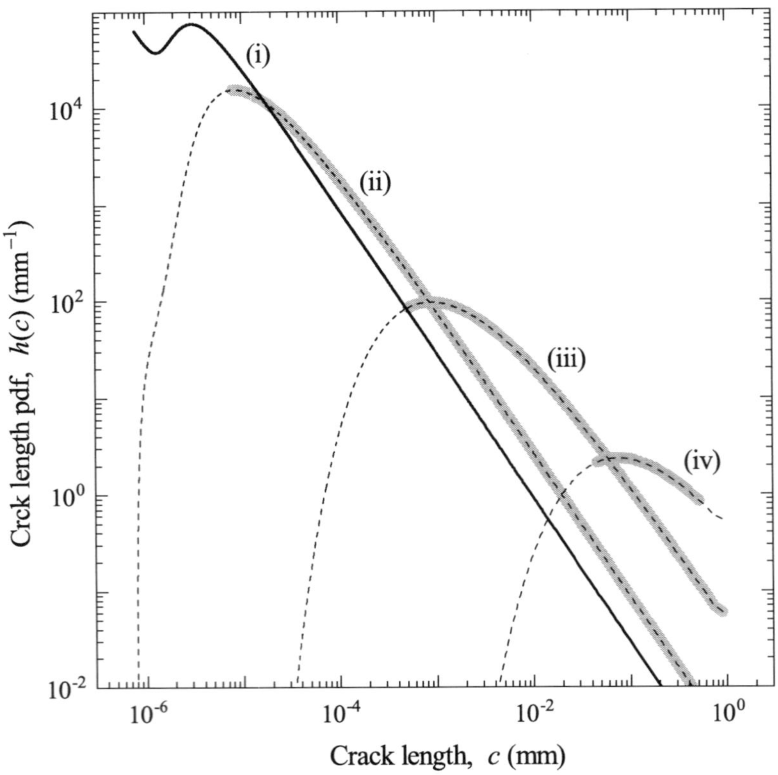

Figure 7.11 Plot in logarithmic coordinates of crack length pdf $h(c)$ variations for Quiou sand (limestone) particles. Solid line (i) $D = 1$ mm and dashed and grayed lines (ii) and (iv) $D = 2$ mm, 4 mm, and 8 mm and 16 mm; determined from Figure 7.10.

$f(c)$. As above, the dashed line $h(c)$ responses (ii) and (iii) represent the continuum predictions for the strength-controlling crack lengths of larger particles and reflect the extreme value bias toward fewer small cracks and more large cracks. As in the case of the Quiou sand, large strengths in large silica sand particles were not observed. The crack length domains conjugate to the experimentally observed strengths are substantially truncated relative to the very wide population crack length domain. The truncated crack length domains here are indicated by the grayed shaded lines and do not include small cracks.

Figure 7.16 shows strength edf $Pr(\sigma)$ plots for samples of quartzite (polycrystalline quartz) particles, 15–70 mm in diameter, tested in a conventional compression platen apparatus from the work of Ovalle et al. (2014), using data derived from the published work. Symbols represent individual strength measurements; different symbols represent different particle sizes, D. The solid line represents a visual best fit, taken here as $H_1(\sigma)$, to the $D_1 = 15$ mm data. The data are well described by concave behavior. The dashed lines for the $D = 30$ mm and 40 mm data represent constrained fits, taken here as $H_2(\sigma)$, using Eq. (7.1) and k_2/k_1 as a fitting parameter. The constrained fits provide descriptions of the larger particle strength distributions. The conclusion from Figure 7.16 is that the strength distributions of samples of the multiple sizes of particles are related by stochastic size effects and a common flaw population.

Figure 7.17 shows as a solid line (i) the crack length pdf $h(c)$ variation underlying the 15 mm quartzite particle strength behavior described by $H_1(\sigma)$. The $h(c)$ variation was determined using $B = 0.24$ MPa m$^{1/2}$ and Eq. (7.2) as a constraint. The discussion above suggests that this response approximates the overall population of flaws, $f(c)$, applicable to all sizes of particle of this material. The dashed lines (ii) and (iii) represent the crack length pdf $h(c)$ variations underlying the larger quartzite particle strength behaviors described by $H_2(\sigma)$. These $h(c)$ variations were determined using the same B value and extend over the same crack length domain as that of the smaller particle response (i) approximating the populations $f(c)$. As above, the dashed line $h(c)$ responses (ii) and (iii) represent the continuum predictions for the strength-controlling crack lengths of larger particles and reflect the extreme value bias toward fewer small cracks and more large cracks. Large strengths in large particles were not observed and thus the crack length domains conjugate to the experimentally observed

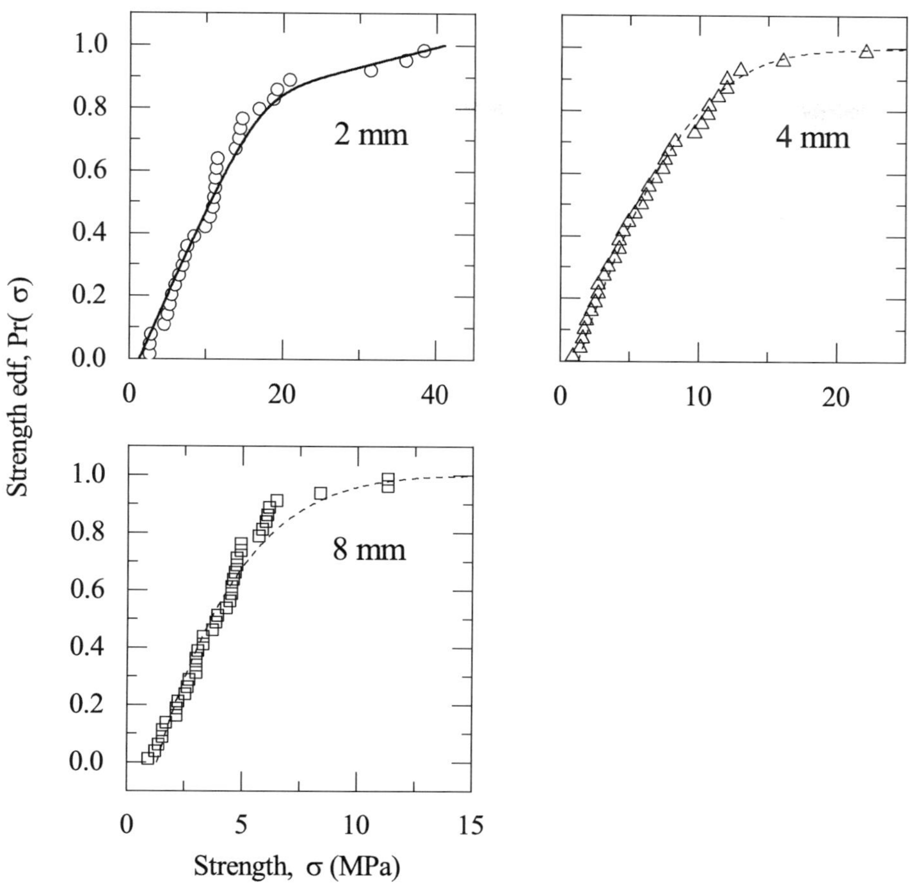

Figure 7.12 Plots of strength edf behavior, $Pr(\sigma)$, for coral sand particles. D = 1.53–1.95 mm, 2.94–4.42 mm, and 6.28–11.11 mm (nominally 2, 4, 8 mm), N_{tot} = 107 (Adapted from Shen, J et al. 2020). Solid line shows concave best fit, dashed lines show best fits constrained by stochastic size effects.

strengths are truncated relative to the population crack length domain. The truncated crack length domains are indicated by the grayed shaded lines and do not include small cracks.

7.2.5 Basalt

Figure 7.18 shows strength edf $Pr(\sigma)$ plots for samples of basalt particles. Basalt is a silica poor, aluminosilicate rich volcanic rock. The particles considered here were 1–5 mm in diameter, tested in a conventional compression platen apparatus from the work of Rozenblat et al. (2011), using data derived from the published work. Symbols represent strength measurements censored into deciles from the full edf measurements; different symbols represent different particle sizes, D. The solid line represents a visual best fit, taken here as $H_1(\sigma)$, to the $D_1 = 1$ mm data. The data are well described by concave behavior. The dashed lines for the $D = 1.6$ mm (2 and 3.15) mm combined, and 5 mm data represent constrained fits, taken here as $H_2(\sigma)$, using Eq. (7.1) and k_2/k_1 as a fitting parameter. The constrained fits provide descriptions of the larger particle strength distributions. The conclusion from Figure 7.18 is that the strength distributions of samples of the multiple sizes of particles are related by stochastic size effects and a common flaw population.

Figure 7.19 shows as a solid line (i) the crack length pdf $h(c)$ variation underlying the 1 mm basalt particle strength behavior described by $H_1(\sigma)$. The $h(c)$ variation was determined using $B = 0.1$ MPa m$^{1/2}$ and Eq. (7.2) as a constraint. The discussion above suggests that this response approximates the overall population of flaws, $f(c)$, applicable to all sizes of particle of this material. The dashed lines (ii)–(iv) represent the crack length pdf $h(c)$ variations underlying the larger basalt particle strength behaviors described by $H_2(\sigma)$. These $h(c)$ variations were determined using the same B value and extend over the same crack length domain as that of the smaller particle response (i) approximating the populations $f(c)$. As above,

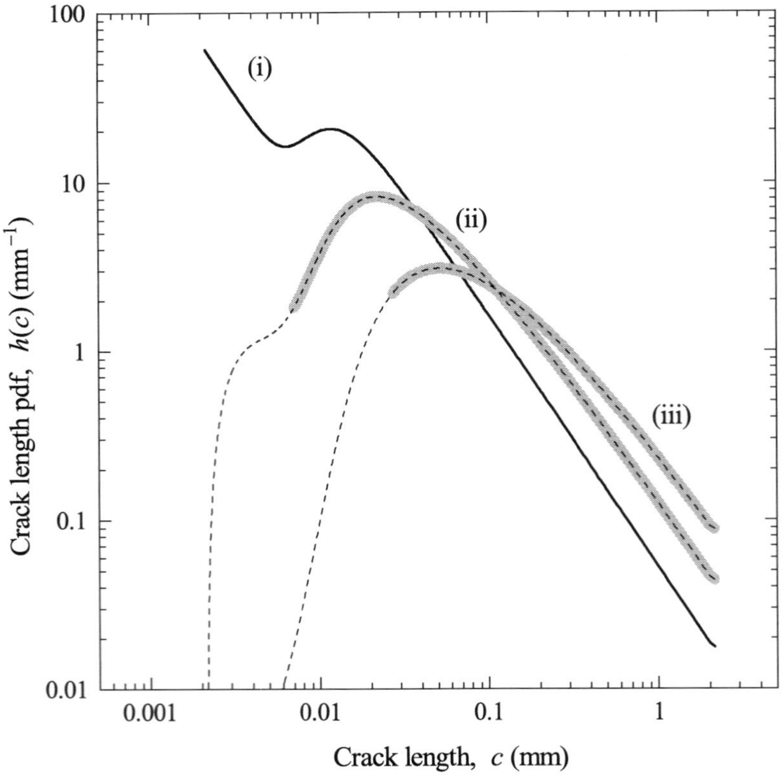

Figure 7.13 Plot in logarithmic coordinates of crack length pdf $h(c)$ variations for coral sand particles. Solid line (i) $D = 2$ mm and dashed and grayed lines (ii) and (iii) $D = 4$ mm and 8 mm; determined from Figure 7.12.

the dashed line $h(c)$ responses (ii)–(iv) represent the continuum predictions for the strength-controlling crack lengths of larger particles and reflect the extreme value bias towards fewer small cracks and more large cracks. Large strengths in large particles were not observed and thus the crack length domains conjugate to the experimentally observed strengths are substantially truncated relative to the population crack length domain. The truncated crack length domains are indicated by the grayed shaded lines and do not include small cracks.

7.3 Sigmoidal Stochastic Distributions

The vast majority of particle strength distributions are concave, Chapter 6, and many examples of concave distributions exhibiting stochastic behavior were examined in the previous section. However, as noted in Chapter 6, there are a few particle systems that exhibit sigmoidal distributions. This section examines two particle systems that exhibit sigmoidal distributions and stochastic behavior. Both systems are from the work of Rozenblat et al. (2011).

7.3.1 Fertilizer

Commercial and domestic plant fertilizers are often formed of clay particles coated with active compounds. A common fertilizer variety is GNP (Granular Nitrogen and Phosphate), often formed as millimeter-sized particles. Figure 7.20 shows strength edf $Pr(\sigma)$ plots for samples of GNP particles, 2–4.75 mm in diameter, tested in a conventional compression platen apparatus from the work of Rozenblat et al. (2011), using data derived from the published work. Symbols represent strength measurements censored into deciles from the full edf measurements, except for the 2.36 mm particles for which the symbols represent individual strength measurements (data also reported by Liburkin et al. 2015); different symbols represent different particle sizes, D. The solid line represents a visual best fit, taken here as $H_1(\sigma)$, to the $D_1 = 2$ mm data. The data are well described by sigmoidal behavior and the data set exhibits a strength domain width σ_N/σ_1 of approximately 6. The

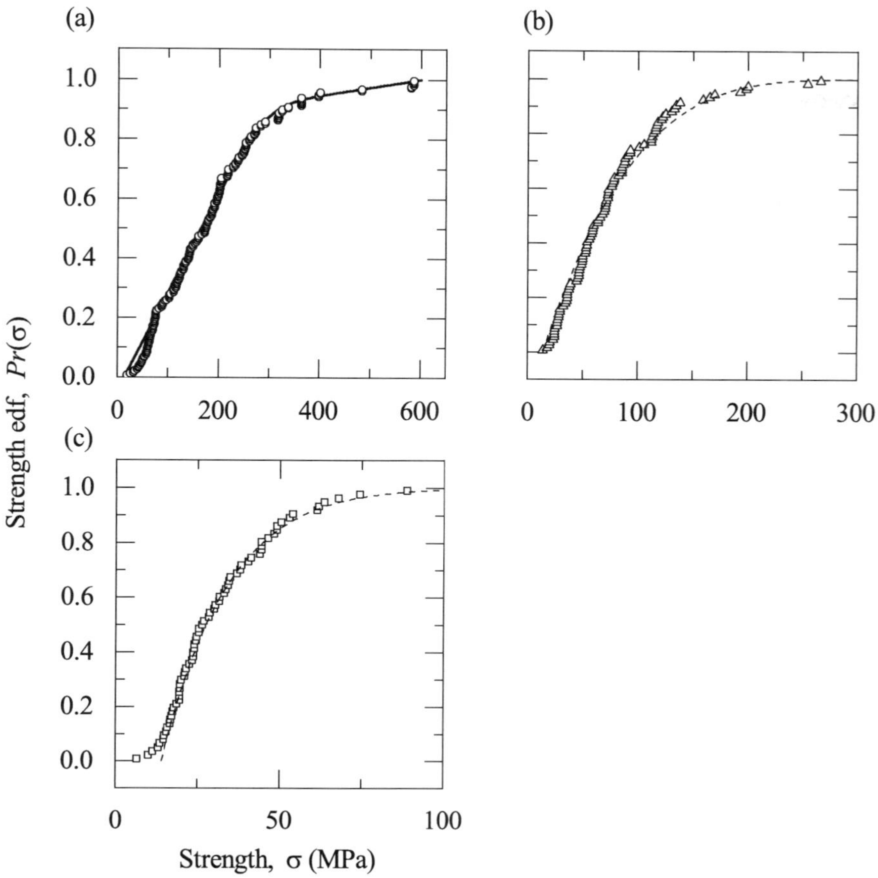

Figure 7.14 Plots of strength edf behavior, $Pr(\sigma)$, for silica sand particles. D = (a) 0.12–0.4 mm, (b) 0.4–0.9 mm, and (c) 1.1–1.8 mm, N_{tot} = 368 (Adapted from Nakata, Y et al. 2001b). Solid line shows concave best fit, dashed lines show best fits constrained by stochastic size effects.

dashed lines for the larger particle data represent constrained fits, taken here as $H_2(\sigma)$, using Eq. (7.1) and k_2/k_1 as a fitting parameter. The constrained fits provide descriptions of the larger particle strength distributions. The conclusion from Figure 7.20 is that the strength distributions of samples of the multiple sizes of particles are related by stochastic size effects and a common flaw population that results in sigmoidal strength distribution behavior.

Figure 7.21 shows as a solid line (i) the crack length pdf $h(c)$ variation underlying the 2 mm GNP particle strength behavior described by $H_1(\sigma)$. The $h(c)$ variation was determined using $B = 0.11$ MPa m$^{1/2}$ and Eq. (7.2) as a constraint. The discussion above suggests that this response approximates the overall population of flaws, $f(c)$, applicable to all sizes of particle of this material. The dashed lines (ii)–(iv) represent the crack length pdf $h(c)$ variations underlying the larger GNP particle strength behaviors described by $H_2(\sigma)$. These $h(c)$ variations were determined using the same B value and extend over the same crack length domain as that of the smaller particle response (i) approximating the populations $f(c)$. As above, the dashed line $h(c)$ responses (ii)–(iv) represent the continuum predictions for the strength-controlling crack lengths of larger particles and reflect the extreme value bias toward fewer small cracks and more large cracks.

The best fit and predicted $h(c)$ responses of these fertilizer particles are very different from those of the particle systems discussed in the previous section, for example Figure 7.17, although all are based on stochastic size effects on strength. Figure 7.21 is plotted in the same 1 : 1 aspect ratio as the other pdf plots in this chapter, in this case $10^3 : 10^3$, and emphasizes the narrow strength-controlling crack length domains of the fertilizer particles, leading to the narrow sigmoidal strength distributions. Although large strengths in large particles were not observed here, the effect was not great, and thus the crack length domains conjugate to the experimentally observed strengths were not very truncated relative to the population crack length domain. The truncated crack length domains are indicated by the grayed shaded lines in Figure 7.21. Both the narrow strength distribution and the sigmoidal shape lead to less truncation of the strength and crack length distributions as the

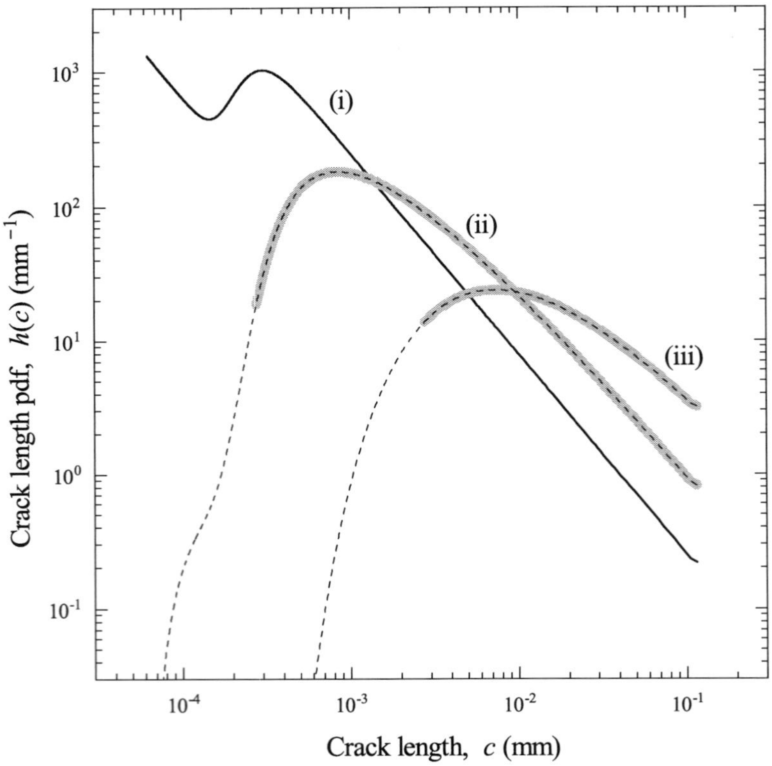

Figure 7.15 Plot in logarithmic coordinates of crack length pdf $h(c)$ variations for silica sand particles. Solid line (i) $D = 0.12$ mm and dashed and grayed lines (ii) and (iii) $D = 0.4$ mm and 1.1 mm; determined from Figure 7.14.

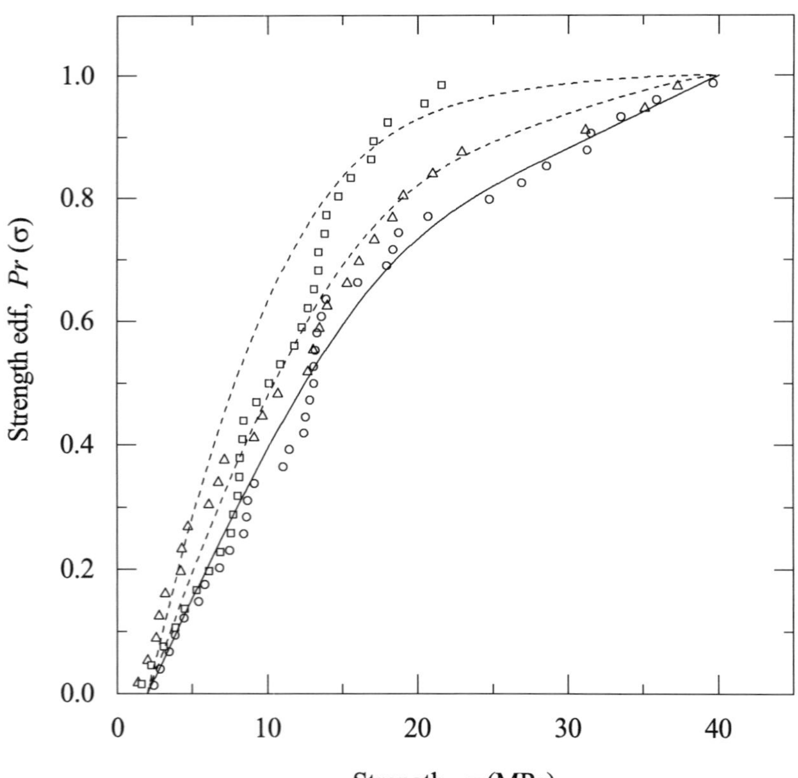

Figure 7.16 Plot of strength edf behavior, $Pr(\sigma)$, for quartzite particles. $D = (15–30, 30–40,$ and $40–70)$ mm, $N_{tot} = 98$ (Adapted from Ovalle, C et al. 2014). Solid line shows concave best fit, dashed lines show best fits constrained by stochastic size effects.

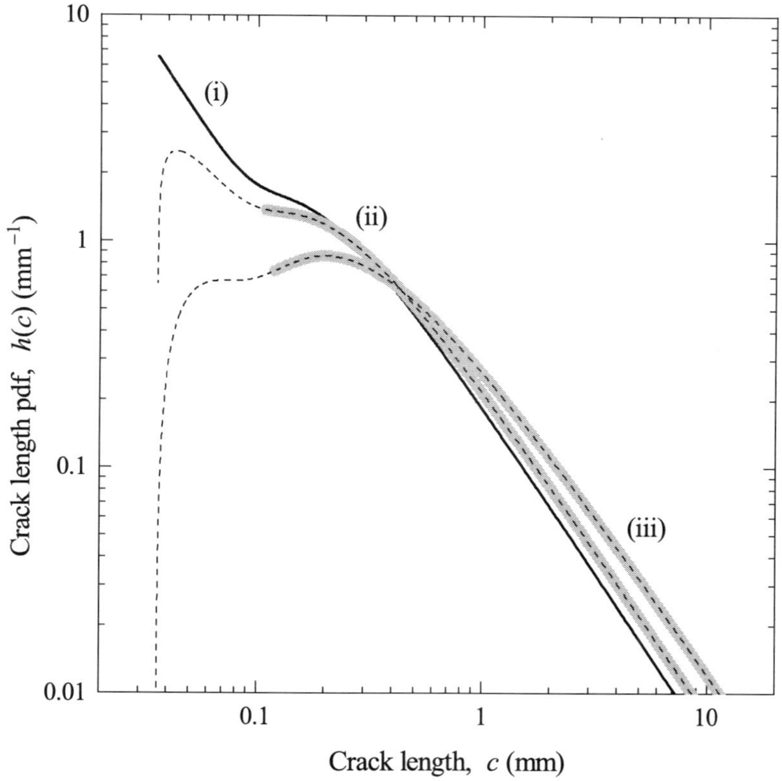

Figure 7.17 Plot in logarithmic coordinates of crack length pdf $h(c)$ variations for quartzite particles. Solid line (i) D = 15 mm and dashed and grayed lines (ii) and (iii) D = 30 mm and 40 mm; determined from Figure 7.16.

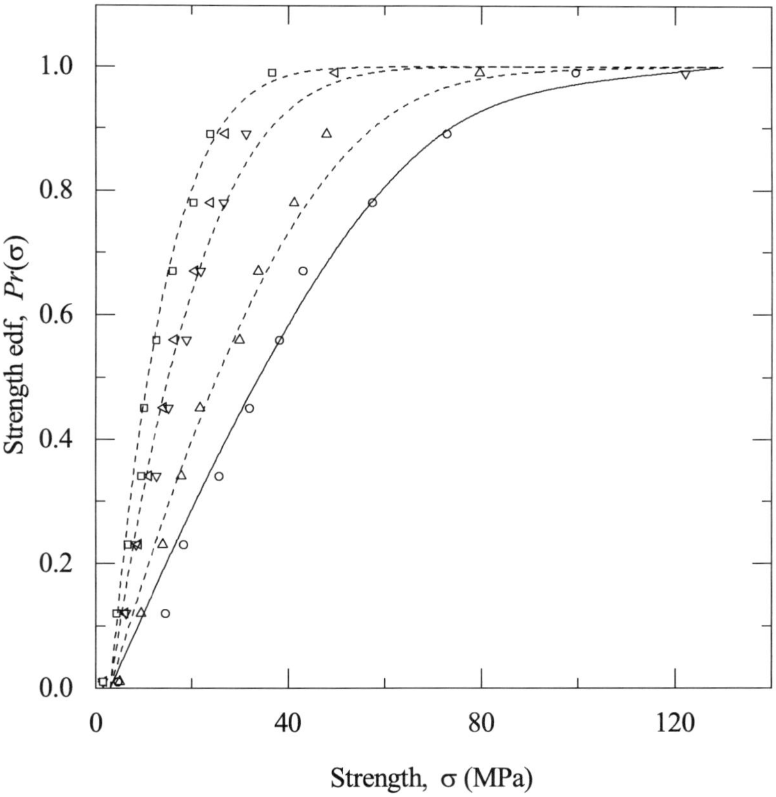

Figure 7.18 Plot of strength edf behavior, $Pr(\sigma)$, for basalt particles. D = (1.0–1.6, 1.6–2.0, 2.0–2.5, 3.15–4.0, and 4.0–5.0) mm, N_{tot} = 50 (Adapted from Rozenblat, Y et al. 2011). Solid line shows concave best fit, dashed lines show best fits constrained by stochastic size effects.

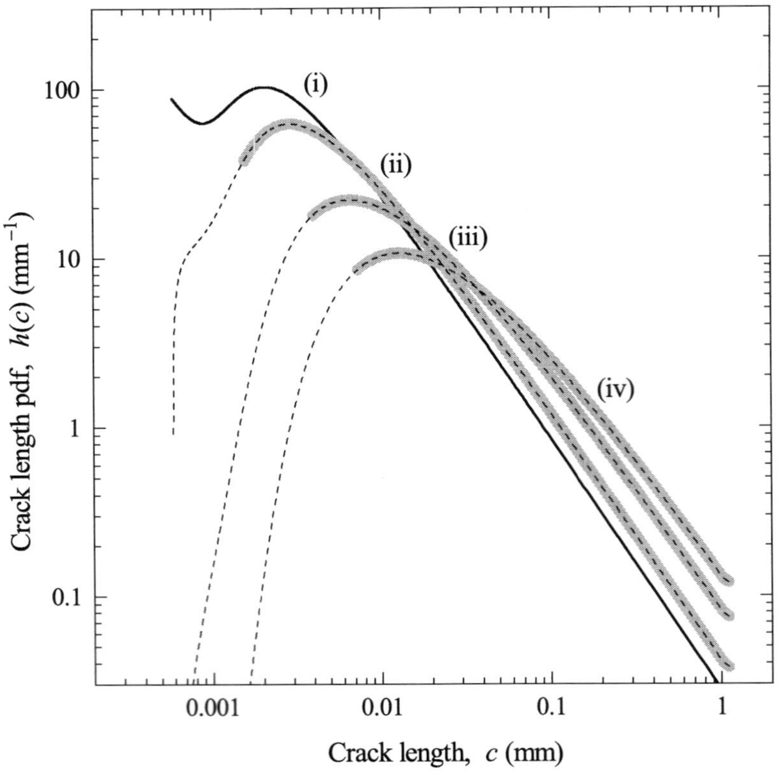

Figure 7.19 Plot in logarithmic coordinates of crack length pdf $h(c)$ variations for basalt particles. Solid line (i) $D = 1$ mm and dashed and grayed lines (ii)–(iv) $D = 1.6$ mm, (2 and 3.15) mm, and 4 mm; determined from Figure 7.18.

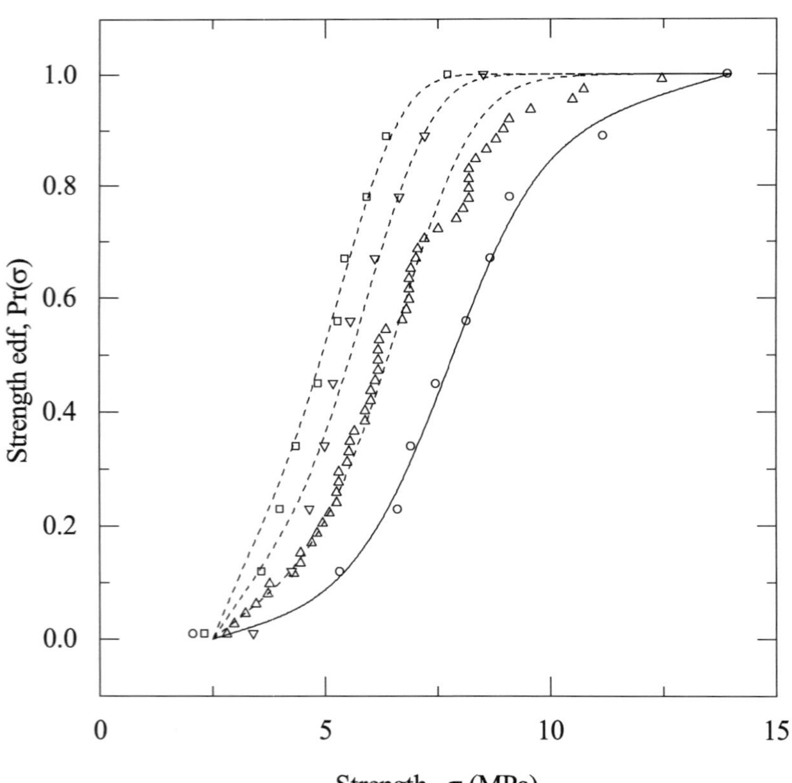

Figure 7.20 Plot of strength edf behavior, $Pr(\sigma)$, for GNP fertilizer particles. $D = (2.00–2.36, 2.36–3.35, 3.35–4.00$, and $4.00–4.75)$ mm, $N_{tot} = 86$ (Adapted from Rozenblat, Y et al. 2011). Solid line shows sigmoidal best fit, dashed lines show best fits constrained by stochastic size effects.

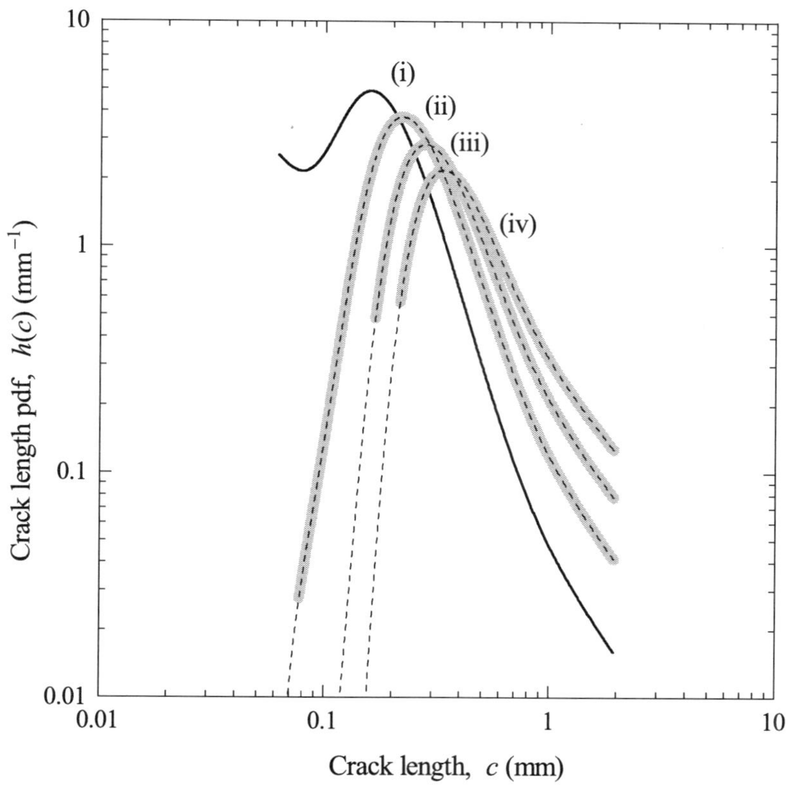

Figure 7.21 Plot in logarithmic coordinates of crack length pdf $h(c)$ variations for GNP fertilizer particles. Solid line (i) D = 2 mm and dashed and grayed lines (ii)–(iv) D = (2.36, 3.35, and 4) mm; determined from Figure 7.20.

size of the components (in this case, particles) increases. This effect is clear in the relatively longer overlaid grayed lines in Figure 7.21. Overall, the particle responses of Figures 7.20 and 7.21 closely resemble those of the extended components in Chapter 5.

7.3.2 Glass

Figure 7.22 shows strength edf $Pr(\sigma)$ plots for samples of silicate glass spherical particles, 0.7–3.3 mm in diameter, tested in a conventional compression platen apparatus from the work of Rozenblat et al. (2011), using data derived from the published work. Symbols represent strength measurements censored into deciles from the full edf measurements for the 2.85 mm data and individual strength measurements for the 0.7 mm data; different symbols represent different particle sizes, D. The solid line represents a visual best fit, taken here as $H_1(\sigma)$, to the $D_1 = 0.7$ mm data. The data are well described by sigmoidal behavior. The dashed line for the $D = 2.85$ mm data represents a constrained fit, taken here as $H_2(\sigma)$, using Eq. (7.1) and k_2/k_1 as a fitting parameter. The constrained fit provides a description of the larger particle strength distribution. The conclusion from Figure 7.22 is that the strength distributions of samples of both sizes of particles are related by stochastic size effects and a common flaw population.

Figure 7.23 shows as a solid line (i) the crack length pdf $h(c)$ variation underlying the 2.85 mm glass particle strength behavior described by $H_1(\sigma)$. The $h(c)$ variation was determined using $B = 1.0$ MPa m$^{1/2}$ and Eq. (7.2) as a constraint. The discussion above suggests that this response approximates the overall population of flaws, $f(c)$, applicable to all sizes of particle of this material. The dashed line (ii) represents the crack length pdf $h(c)$ variation underlying the larger glass particle strength behavior described by $H_2(\sigma)$. This $h(c)$ variation was determined using the same B value and extends over the same crack length domain as that of the smaller particle response (i) approximating the populations $f(c)$. As earlier, the dashed line $h(c)$ response (ii) represents the continuum predictions for the strength-controlling crack lengths of the larger particles and reflects the extreme value bias toward fewer small cracks and more large cracks. Large strengths in the large particles were not observed and thus the crack length domain conjugate to the experimentally observed strengths is

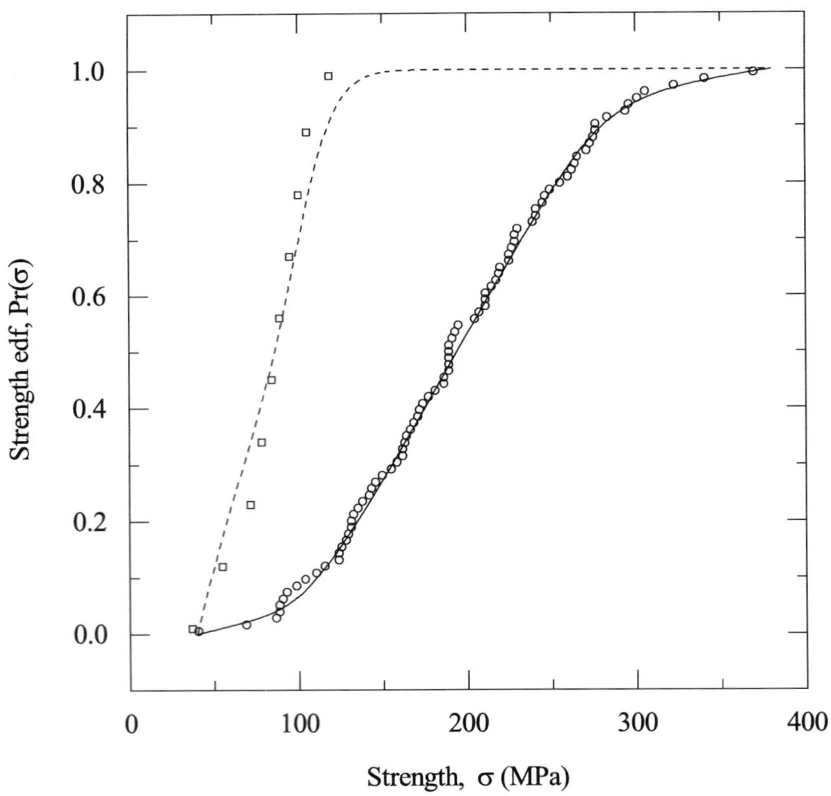

Figure 7.22 Plot of strength edf behavior, $Pr(\sigma)$, for glass particles. D = (0.71–1.00 and 2.85–3.3) mm, N_{tot} = 98 (Adapted from Rozenblat, Y et al. 2011). Solid line shows sigmoidal best fit, dashed line shows best fit constrained by stochastic size effects.

truncated relative to the population crack length domain. The truncated crack length domain is indicated by the grayed shaded line and does not include small cracks.

Extensive investigations of strength distributions of glass particles in Chapter 9 will show that the stochastic size effects presented here are in fact infrequently observed. Glass particles tend to exhibit deterministic size effects.

7.4 Summary and Discussion

This chapter has surveyed experimental observations of particle strength distributions that exhibit stochastic size effects. In such effects, a number of particles are sampled from a material containing a population of flaws characterized by a spatial density λ and crack length probability density function $f(c)$. Each particle, size D, contains a number of cracks, k, forming an ensemble selected from $f(c)$. The ensembles of cracks in the sample of particles are probabilistically identical and described by $f(c)$. However, the largest cracks in each ensemble form an extreme value distribution, characterized by a crack length probability density function $h(c)$. In stochastic size effects, the cracks are independent such that $f(c)$ is invariant and the relationship between the cumulative distribution functions of $h(c)$ and $f(c)$, $H(c)$ and $F(c)$ respectively, are given by a simple power law in k. As the strength of a particle is determined by the largest crack in the ensemble of cracks contained by the particle, the strength distribution of a sample of particles, $H(\sigma)$ is determined by $H(c)$. Hence, $H(\sigma)$ is related by the same simple power law relation to $F(\sigma)$, the invariant strength distribution characterizing the material. Experimentally, stochastic size effects are demonstrated by showing that samples of particles of different D exhibit strength distributions that are consistent with the same relation, Eq. (7.1). Such behavior was demonstrated here in an extensive survey of a broad range of particle systems, including natural (e.g. rock, coral) and engineered (e.g. alumina, fertilizer) materials, and a wide domain of particle sizes (0.1–80 mm). Stochastic behavior was observed for both the predominant, concave, form of particle strength distributions and the less frequently observed sigmoidal form. In addition to exploring the phenomena of stochastic size effects in particles, goals of the survey were the elucidation of the

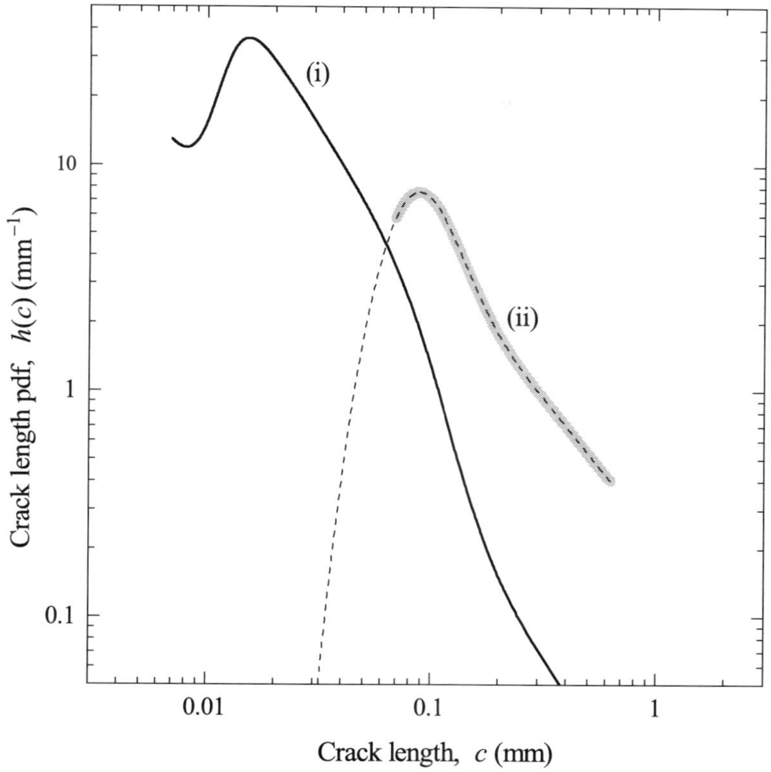

Figure 7.23 Plot in logarithmic coordinates of crack length pdf $h(c)$ variations for glass particles. Solid line (i) $D = 0.7$ mm and dashed and grayed line (ii) $D = 2.85$ mm; determined from Figure 7.22.

form of $k(D)$, thus providing insight into λ, and determination of the parameter B linking particle strength σ and crack length c.

Figure 7.24 summarizes the behavior of strength-controlling crack length populations for particles with strength distributions linked by stochastic extreme value size effects. Crack length populations from quartz (Figure 7.15) and limestone (Figure 7.9) and shown in Figure 7.24 using the equi-axed logarithmic plotting scheme used earlier. The fine solid lines in Figure 7.24 represent the strength-controlling crack length populations of the smallest particles in each case, 0.1 mm and 7 mm respectively, and approximate the invariant flaw populations of the two materials. In changing materials, from quartz to limestone for example, the material flaw population changes, indicated by the lower arrow. For a given material, increasing the particle size contracts and truncates the strength-controlling crack population, such that there are greater proportions of larger cracks, indicated by the upper arrows. The extent of contraction depends on the nature of the crack density λ, and thus the form of $k(D)$, and the ratios of the particle sizes D measured. As the ratios of the largest/smallest particles for both materials in Figure 7.24 were both ≈ 5, the greater relative contraction for quartz implies a greater $k(D)$ variation relative to limestone. Note that contraction occurs for small cracks only: the large crack domain bound is invariant, and is reflected in an invariant threshold strength that is determined by the material parameter B. The particle material dependence of the $k(D)$ variation and the value of B can be evaluated from the results of the strength distribution survey encapsulated in plots such as Figure 7.24.

Following the discussions in Chapters 3 and 5 regarding the nature of strength controlling flaws, the number of flaws in a particle may scale with a particle linear dimension, a particle area, or the particle volume. In the experiments discussed here, the relative numbers of flaws in particles of different sizes was determined as the ratio k_2/k_1. Comparison of this ratio with the ratio of selected measures of particle size can provide the relationship $k(D)$. Here, the simplest case might be three-dimensional: Strength-controlling flaws are assumed to be arrayed uniformly throughout the volume of the tensile stress zone at the center of a particle in diametral compression, as depicted in the schematic diagram of particles, Figure 7.1. As the volume of the zone scales as D^3, if particle size is varied from sample to sample, D_1 and D_2 say, stochastic size effects should scale in a cubic manner, as $k_2/k_1 \sim (D_2/D_1)^3$. Generalized versions of these parameters that enable comparison

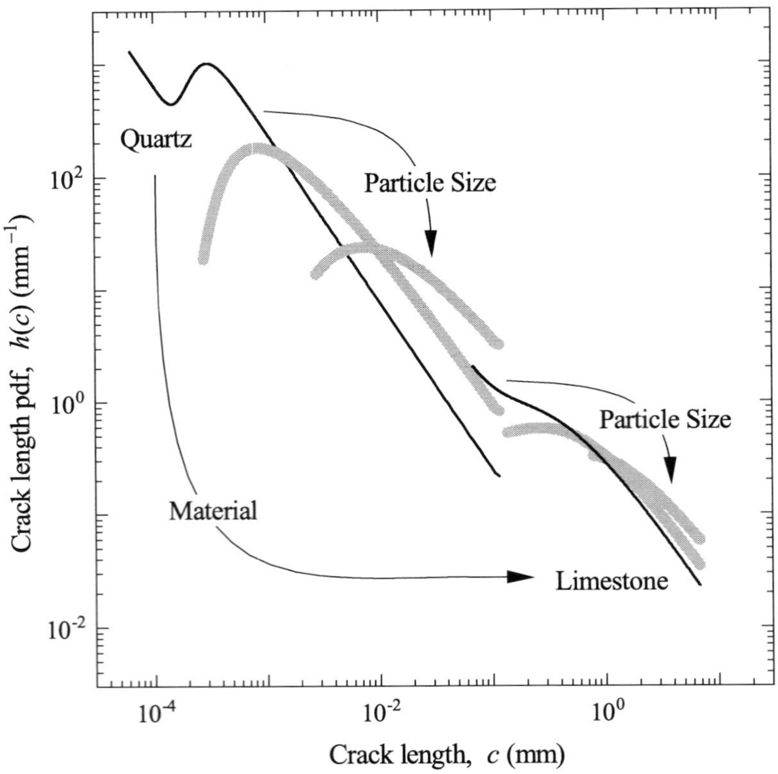

Figure 7.24 Plot in logarithmic coordinates of crack length pdf $h(c)$ variations for limestone and quartz particles repeated from Figure 7.9 and Figure 7.15. Particle sizes are limestone 7–80 mm and quartz 0.1–1.1 mm. Solid lines indicate material flaw populations sensed by small particles, grayed lines indicate extreme value flaw populations controlling strength populations of large particles.

of multiple particle (and other component) systems are k/k_{ref} and D/D_{ref}. Values of k and D are compared to a single reference configuration k_{ref} and D_{ref} within a given system and comparison between systems is thus relative to common reference point (1,1). In this way, the plotting scheme of Figure 5.23 is retained, allowing simple assessments of the nature of stochastic scaling and comparisons between particles and other components. For example, if strength-controlling flaws are distributed throughout a volume of a material, as discussed above, strength distributions of particles with different test volumes would exhibit (logarithmic) k/k_{ref} vs D/D_{ref} behavior of slope 3. However, if strength-controlling flaws are distributed along a length dimension of a particle, for example along the diametral loading axis, strength distributions of particles with different volumes would exhibit k/k_{ref} vs D/D_{ref} behavior of slope less than 1. Similarly, if strength-controlling flaws are distributed over an area within a particle, for example over the equatorial plane, strength distributions of particles with different volumes would exhibit k/k_{ref} vs D/D_{ref} behavior of slope less than 2. The key here, as before, is that although a volume is tested, strength measurements are only sensitive to a single dimension, length or area, describing that volume.

Figure 7.25 shows a plot of the relative number of flaws k/k_{ref} vs relative particle size D/D_{ref} obtained from analysis of the stochastic strength behavior of particles in this chapter. The solid symbols represent individual particle strength distributions; the common point at (1,1) is omitted. The error bars represent the uncertainties in relative particle sizes determined from the cited ranges. The solid line is of slope 1 and the dashed lines are of slope 2 and 3. The open symbols repeat the extended component data of Figure 5.23. A major feature of Figure 7.25 is that most observations are in a small relative particle size domain, $D/D_{\text{ref}} < 5$. This feature points to the fact that is extremely difficult experimentally to measure the strengths of identically fabricated particles over a wide domain of sizes, a feature in common with many extended components. A second noticeable feature is that most of the responses are weakly grouped around a line of slope 1, a feature that is also in common with extended components. Distinct from extended components, some particle systems displayed no size effect at all (e.g. alumina) and some clearly displayed a size effect approaching volume scaling (slope 3, e.g. Quiou sand).

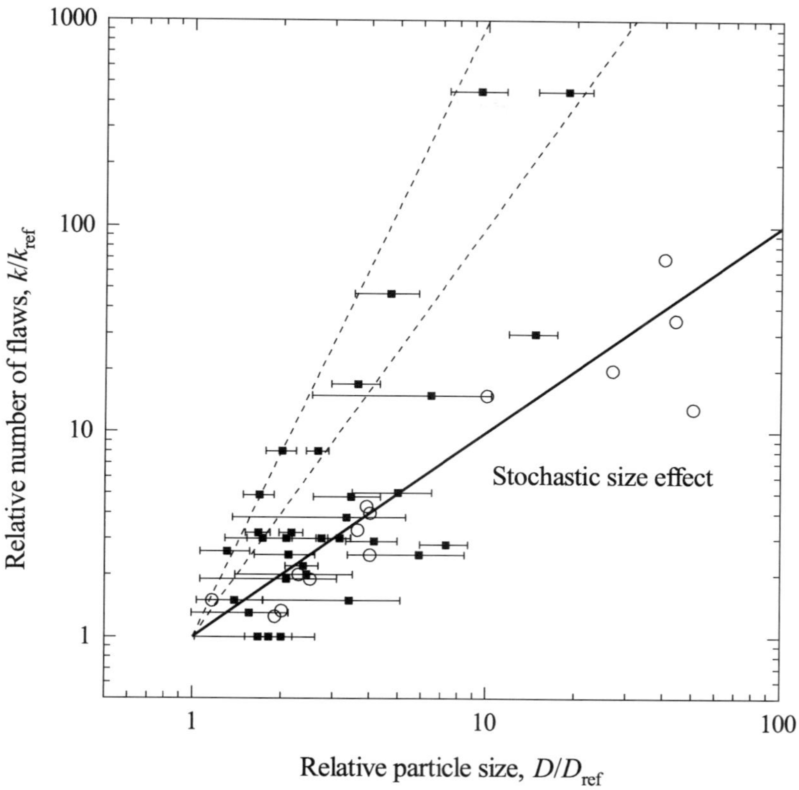

Figure 7.25 Plot of number of flaws of flaws k vs particle size D for particles exhibiting stochastic size effects on strength distributions. Lines are guides to the eye indicating scaling behavior; most systems behave linearly, although there are a few that approach cubic scaling.

Overall, however, although the experimental evidence for stochastic size effects in many strength distributions of particles is substantial and significant, as shown throughout this chapter, the evidence in support of a particular scaling law is weak (Figure 7.25). The predominant trend is that of linear scaling, suggesting that a linear array of flaws that increases in length as particle size is increased is responsible for stochastic size effects on strength. Such linear scaling is consistent with the elongated, approximately linear, zone of transverse tensile stress generated on a particle axis during loading, Figures 2.19–2.22. Combining this more restricted stress distribution view with the linear stochastic failure ideas of Figure 5.29 enables modification of Figure 7.1 so as to reflect linear scaling of particle strengths. The key is limited transverse extension of the stress field with increasing particle size, as indicated in the schematic diagram of Figure 7.26. In the diagram, both particle size D and the number of stressed volume elements k increase in the ratio 1 : 2 : 3, such that $k \sim D^1$. More detailed consideration, including the few $k \sim D^3$ observations require greater consideration of particle shape.

Again, it must be noted that, despite extensive research for this book, there are very few points on Figure 7.25. Despite anticipation of stochastic size effects and study of scaling of particle systems since the early works of the Bradt group (Kschinka et al. 1986; Wong et al. 1987), definitive, quantifiable demonstrations of stochastic size effects in particle strengths are rare and largely unnoticed. As with extended components, the paucity of examples is due in part to a lack of analysis and the effects of distorting and obscuring linearized plotting schemes. In addition, in many cases the application of such plotting schemes leads to the mistaken impression that observed size effects on strength distributions are due to stochastic behavior, when in fact they are due to deterministic behavior. Examples of deterministic size effects on strength behavior of particles is considered in Chapters 8 and 9.

In addition to the invariance of the flaw population attributes λ and $f(c)$ in stochastic size effects, it is implicit that the property B controlling the relationship between strength and flaw size is also invariant. B is a composite property that derives from both material fracture resistance and flaw geometry. In selecting values of B to deconvolute strength distributions of extended components in terms of underlying crack length distributions (Chapter 5), guidance was provided primarily by considerations of known material fracture resistance or toughness. The absolute value of B does not influence the forms of

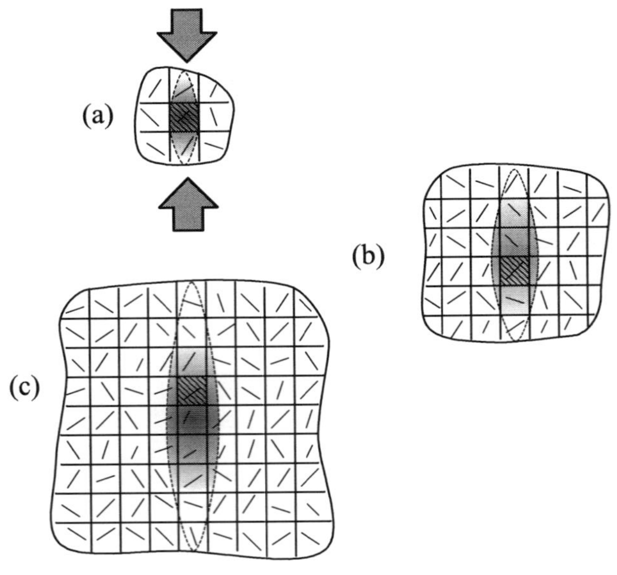

Figure 7.26 Schematic diagrams illustrating extreme value effects in brittle failure of particles in diametral compression. Particle diameter (a) $D = 3$, (b) $D = 6$, (c) $D = 9$. Applied loading is indicated by arrows in (a). Tensile zones restricted to axes and scaling linearly with particle dimensions indicated by shading and dashed outlines. Flaws indicated by bold lines; strength-controlling flaws responsible for failure indicated hatched. Note that in (a) failure occurs at center, independent of flaw size, $k = 1$ and in (b) $k = 2$ and (c) $k = 3$, failure occurs at largest flaws, independent of location in tensile zone.

deconvoluted crack length distributions or the relationships between crack length distributions describing different sized components. Hence, such consideration for extended components enabled correspondence to be maintained between the sizes of the deconvoluted cracks and independent observations. A brittle extended component typically exhibits strengths of a few hundred MPa arising from crack lengths of a few tens of μm. In selecting values of B to deconvolute strength distributions of particles, material guidance is secondary. Particle size places a constraint on the upper bound value of B such the largest deconvoluted crack length is smaller than the smallest particle size. As a consequence, the allowed values of B for particles can be interpreted in terms of characterizing particle materials and compared with those used to characterize extended component materials.

Using values determined from the strength-crack length analyses of the experiments of Chapters 5 and 6, and this chapter 7, Figure 7.27 shows a semi-logarithmic plot of the failure resistance parameter B for extended components and particles. The values are separately ranked for each component type; extended components are indicated by open symbols, particles are indicated by filled symbols. A conventional MPa $m^{1/2}$ scale is shown on the right of the plot. As a guide to the eye, a horizontal line is drawn at $B = 0.7$ MPa $m^{1/2}$, about the toughness of glass in inert conditions. Both extended components and particles exhibit considerable ranges about this value, about a factor of 10 for extended components, predominantly greater than 0.7 MPa $m^{1/2}$, and about a factor of 100 for particles, predominantly less than 0.7 MPa $m^{1/2}$. Although the ranges overlap, particles are usually less failure resistant than extended components. The overlap region for both component types includes silicates and silicate glasses in the range 0.4 MPa $m^{1/2}$–1.0 MPa $m^{1/2}$. The extremes for each component type are marked, porous alumina to dense Si_3N_4 for extended components and porous limestone to dense Al_2O_3 for particles. As a brittle particle of size 1 mm typically exhibits strengths of a few tens of MPa, a more particle-appropriate unit for B is MPa $mm^{1/2}$, noting that 1 MPa $m^{1/2} \approx 31.6$ MPa $mm^{1/2}$ such that 0.7 MPa $m^{1/2} \approx 22$ MPa $mm^{1/2}$. A MPa $mm^{1/2}$ scale is shown on the left side of the plot. The relative rankings of particles and extended components and the overlap about a common set of materials in Figure 7.27 further supports and validates (Chapter 4) the physical basis and accuracy of the analysis techniques developed and applied here.

This chapter has examined the closest relationship between strength distributions of different sized particles: that in which the ensembles of flaws in particles are selected stochastically from an invariant population, $f(c)$, and particle size determines the extreme value distribution of the largest flaws in a sample of particles, $h(c)$. Both the domain and forms of $h(c)$ are prescribed by this relationship, such that the domain and forms of the conjugate strength distributions of different sized particles are also prescribed, specifically by Eq. (7.1). The ensuing behavior gives rise to stochastic extreme value size effects in strength distributions of particles, the subject of this chapter. The following two Chapters, 8 and 9, relax the constraints leading to stochastic size effects and examine *deterministic* size effects in which a flaw population is not invariant. In these cases, particle size *determines* the flaw population from which the ensembles of flaws in different sized particles are selected. The following chapter, Chapter 8, considers a simple case in which the domains of the flaw populations for different sized particles have a common bound. The system examined is that of bauxite-based ceramic particles intended

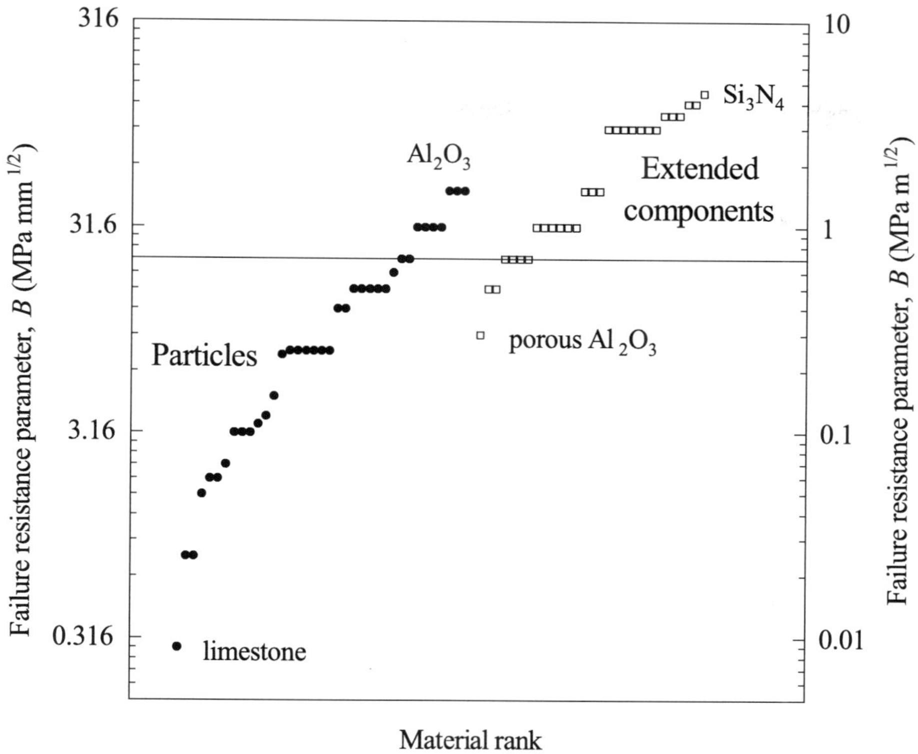

Figure 7.27 Plot of ranked failure resistance parameter *B* for extended components and particles using values determined from strength-crack length analyses of experimental observations of Chapters 5, 6, and 7. Extended component values are in agreement with materials considerations; particle values are constrained by particle size.

for use in proppant beds during oil extraction (Wong et al. 1987). The chapter describes the evolution of particle strength during particle fabrication and may be considered a "case study" of the analyses developed here in a materials science and engineering application. Chapter 9 is a broad survey of deterministic size effects in particles.

References

Bertrand, P.T., Laurich-McIntyre, S.E., and Bradt, R.C. (1988). Strengths of fused and tabular alumina refractory grains. *American Ceramic Society Bulletin* 67: 1217–1221.

Cook, R.F., Boyce, B.L., Friedman, L.H., and DelRio, F.W. (2021). High-throughput bend-strengths of ultra-small polysilicon MEMS components. *Applied Physics Letters* 118: 201601.

Cook, R.F., DelRio, F.W., and Boyce, B. L. (2019). Predicting strength distributions of MEMS structures using flaw size and spatial density. *Microsystems & Nanoengineering* 5: 1–12.

DelRio, F.W., Boyce, B.L., Benzing, J.T., Friedman, L.H., and Cook, R.F. (2020). Shoulder fillet effects in strength distributions of microelectromechanical system components. *Journal of Micromechanics and Microengineering* 30: 125013.

Hu, W., Dano, C., Hicher, P.Y., Le Touzo, J.Y., Derkx, F., and Merliot, E. (2011). Effect of sample size on the behavior of granular materials. *Geotechnical Testing Journal* 34: 186–197.

King, R.P. and Bourgeois, F. (1993). Measurement of fracture energy during single-particle fracture. *Minerals Engineering* 6: 353–367.

Kschinka, B.A., Perrella, S., Nguyen, H., and Bradt, R.C. (1986). Strengths of glass spheres in compression. *Journal of the American Ceramic Society* 69: 467–472.

Liburkin, R., Portnikov, D., and Kalman, H. (2015). Comparing particle breakage in an uniaxial confined compression test to single particle crush tests–model and experimental results. *Powder Technology* 284: 344–354.

McDowell, G.R. and Amon, A. (2000). The application of Weibull statistics to the fracture of soil particles. *Soils and Foundations* 40: 133–141.

Nakata, Y., Hyodo, M., Hyde, A. F. L., Kato, Y., and Murata, H. (2001a). Microscopic particle crushing of sand subjected to high pressure one-dimensional compression. *Soils and Foundations* 41: 69–82.

Nakata, Y., Kato, Y., Hyodo, M., Hyde, A.F.L., and Murata, H. (2001b). One-dimensional compression behavior of uniformly graded sand related to single particle crushing strength. *Soils and Foundations* 41: 39–51.

Ovalle, C., Frossard, E., Dano, C., Hu, W., Maiolino, S., and Hicher, P.Y. (2014). The effect of size on the strength of coarse rock aggregates and large rockfill samples through experimental data. *Acta Mechanica* 225: 2199–2216.

Rozenblat, Y., Portnikov, D., Levy, A., Kalman, H., Aman, S., and Tomas, J. (2011). Strength distribution of particles under compression. *Powder Technology* 208: 215–224.

Shen, J., Dongsheng, X.U., Liu, Z., and Wei, H. (2020). Effect of particle characteristics stress on the mechanical properties of cement mortar with coral sand. *Construction and Building Materials* 260: 119836.

Wang, W. and Coop, M.R. (2016). An investigation of breakage behaviour of single sand particles using a high-speed microscope camera. *Géotechnique* 66: 984–998.

Wang, Y., Dan, W., Xu, Y., and Xi, Y. (2015). Fractal and morphological characteristics of single marble particle crushing in uniaxial compression tests. *Advances in Materials Science and Engineering* 2015: 537692.

Wong, J.Y., Laurich-McIntyre, S.E., Khaund, A.K., and Bradt, R.C. (1987). Strengths of green and fired spherical aluminosilicate aggregates. *Journal of the American Ceramic Society* 70: 785–791.

8

Case Study: Strength Evolution in Ceramic Particles

This chapter uses the analysis of Chapter 4 to examine in detail the strength distributions of ceramic bauxite and mullite particles. Such particles demonstrate flaw evolution on transformation from the unfired ("green") state (in this case bauxite) to the fired state (in this case mullite) and deterministic size effects. The particles are those used in proppant beds in the "fracking" operation of oil and gas extraction. Analysis of the strength evolution data illustrates the deterministic size effects of flaw domain truncation. This chapter has the approach of a detailed case study.

8.1 Introduction

A ceramic material fabrication process typically consists of two stages: the first stage assembles and combines the raw materials and forms an intermediate ceramic "green body" and the second stage heat treats the green body and generates the final ceramic "fired body" (Kingery et al. 1975). In the first stage, mineral powders, usually oxides, are mixed into a liquid binder to form a malleable dough that is formed into a shape suitable for handling and eventual finishing. This is the ceramic green body, containing μm-scale mineral powder particles held together by liquid capillary forces and surficial secondary bonding forces and typically including about 50 % relative porosity by volume. Due to the weak bonding and included porosity, ceramic green bodies are usually not very strong. In the second stage, the green body is heated in a furnace, initially at low temperatures to burn off residual liquid and often organic constituents of the binder. This heating stage is followed by heating at high temperatures to generate phase transformations between the mineral powder particles and, most importantly, to generate consolidation of the green body. The consolidation stage is known as "sintering." Sintering is a thermally activated process that reduces the internal surface area and thus surface energy of the body and generates solid primary bonds between the atoms of the original powders. The reduction in relative porosity on sintering is usually substantial, typically to a value less than 10 %, and often to less than 5 %. In this state, the ceramic material has a relative density of greater than 95 % and the ceramic microstructure contains only isolated pores. This is the fired ceramic body, usually about 50 % smaller in volume than the green body. Due to strong internal bonding and much reduced porosity, fired ceramic components can be very strong.

Optimization of fired ceramic component strength leads to consideration of the evolution of strength during fabrication and hence to consideration of the evolution of strength-controlling flaws during the processing sequence described earlier. In particular, in the terms used in this book, relevant questions are "Does the population of flaws in the green state, characterized by $f_{\mathrm{green}}(c)$, determine the population of flaws in the fired state, characterized by $f_{\mathrm{fired}}(c)$?" and "Do the distributions of strengths $Pr_{\mathrm{green}}(\sigma)$ and $Pr_{\mathrm{fired}}(\sigma)$ reflecting these flaw populations exhibit stochastic or deterministic size effects?" Answering the first of these questions was the goal of a study by Wong et al. (1987), in which green and fired ceramic particles were used as the test vehicles. This chapter re-examines the results of Wong et al. using current terms and analysis and arrives at the same answer as Wong et al. for the particles studied of "yes" to the first question. Here, as a consequence of interpretation in unbiased coordinates, the analysis is more detailed and the answer more definitive. Based on the new analysis, the second question is addressed, arriving at the answer "yes, deterministic size effects." Analyses and physical interpretations are then developed to describe the observed particle strength distribution phenomena in terms of underlying flaw populations and deterministic size effects.

The fired and sintered ceramic particles studied by Wong et al. had materials engineering application as synthetic proppants for use in well-based hydrocarbon extraction—hydraulic "fracking." In such wells, water is pumped underground

Particle Strengths: Extreme Value Distributions in Fracture, First Edition. Robert F. Cook.

into bore holes at high pressure, leading to fracture of sub-surface hydrocarbon bearing rock. Proppants are subsequently pumped into the bore holes to "prop" the fractured rock faces and fissures open and enable hydrocarbon extraction (Wang and Elsworth, 2018). Proppants must fulfill many requirements, including maintaining mechanical integrity in the confining pressure, several tens of MPa, and corrosive conditions, acidic brine, of the sub-surface rock environment—and be cost-effective. Many materials, including natural sands and synthetic ceramic- and composite-based systems, have been used and developed as proppants (Cooke 1977; Volk et al. 1981; Breval et al. 1987; Komarneni et al. 1987; Gourav et al. 2012; Liang et al. 2016; Mocciaro et al. 2018; and Feng et al. 2021). As many tonnes of proppants are required per well, proppant fabrication cost is an issue and manufacturing by simple heat treatment of plentiful minerals has long been a focus (Cooke 1977; Breval et al. 1987)—in particular, the mixing, granulation, and firing of bauxite ore to form mullite particles—and this was the approach of Wong et al. (1987). Bauxite is a naturally occurring clay mineral that is rich in aluminum hydroxides and also contains silica and iron oxides (Gow and Lozej, 1993). It is the raw ore material used in the production of aluminum metal by chemical means. In addition, bauxite can be processed into a ceramic by thermal means, as by Wong et al.

The starting material used by Wong et al. was a bauxite powder, milled to generate 5–8 μm sized sub-particles that were fabricated into two sets of mm-scale particles. In the first set, water and starch were used to bind the sub-particles into spherical agglomerates formed by repeated collisions and tumbling in a rotary mixer. The agglomerates were then dried to form particles in the green state screened into four size samples. In the second set, half the green state particles of each size were kiln fired, leading to sintering, shrinkage, and near-complete transformation to aluminosilicate mullite ($3Al_2O_3{\cdot}2SiO_2$). Particles were selected from the second set for sphericity and 15 % relative diameter shrinkage on firing. The emphasis of Wong et al. was to identify the evolution of defects from the green state to the sintered state rather than proppant development. Wong et al. noted the earlier work of Kapur and Fuerstenau (1967) and Capes (1971) on particle agglomerates of sub-particles, and of Kschinka et al. (1986) and Kendall et al. (1986) on strength distributions. In a broader context, very early work by Jewett et al. (1961) had considered the effects of sub-particle size on the strengths of green-state pelletized iron ore agglomerates. The particle strength studies of Kschinka et al. (1986), Shipway and Hutchings (1993c), and Watkins and Prado (2015) on glass, and Luscher et al. (2007) on kaolinite and bauxite agglomerates (2007) were at least partially motivated by proppant applications. The results from some of these other studies are examined in this book, Chapters 2 and 9. Wong et al. used a conventional compression platen apparatus to measure the strengths of the selected green and fired particles.

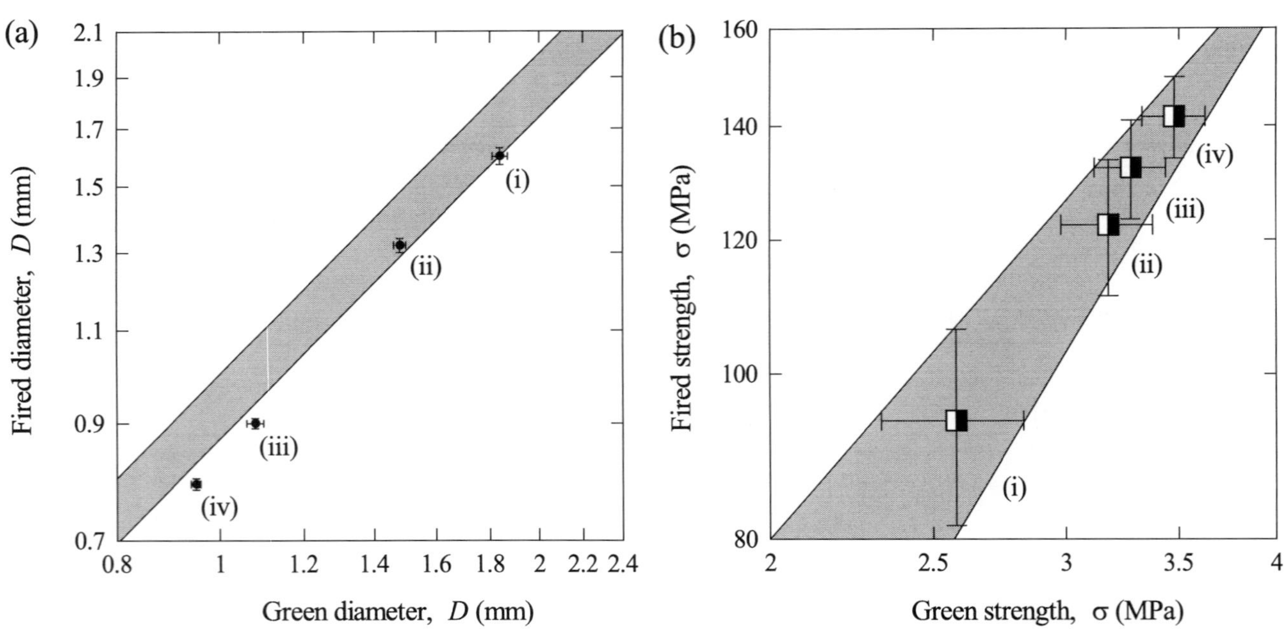

Figure 8.1 (a) Means and standard deviations of selected spherical particles sizes for green bauxite and fired mullite particles. Numbering scheme (i)–(iv) is from largest to smallest mean particle size in each state. (b) Means and standard deviations of strengths of size-selected spherical particles of green bauxite and fired mullite from (a). Numbering scheme is as in (a), indicating that the smallest particles have the greatest mean strength and least strength dispersion. (Adapted from Wong, J.Y et al., 1987).

Figure 8.1a shows a 3 × 3 logarithmic plot of the means and standard deviations of the ceramic particle diameters in the green and fired states as classified by Wong et al. The numbering scheme (i)–(iv), from largest to smallest particle sample, for both green and fired particles, is that of Wong et al. (and largely contrary to the scheme used in this book). The upper solid line indicates no shrinkage on firing and the lower solid line of the gray band indicates 15 % relative diameter shrinkage, corresponding to ≈ 50% relative volume shrinkage. The agreement of the measurements with the 15 % diameter shrinkage trend points to the well-matched nature of the particles selected by Wong et al. Figure 8.1b shows a 2 × 2 logarithmic plot of the means and standard deviations of the ceramic particle strengths in the green and fired states as reported by Wong et al. The slightly reduced plot dimensions indicate that strength variation (about a factor of 1.4) was less than that of the selected particle diameter variation (about a factor of 2). Distinct from the particle sizes, in which fired and green diameters were comparable, the fired strengths are about a factor of 40 greater than the green strengths. Note that the same numbering scheme is used and that mean strengths *increase* in order (i)–(iv) from largest to smallest particles and that the standard deviations *decrease* in order (i)–(iv) from largest to smallest particles. As equi-axed logarithmic coordinates are used, the comparable dimensions of the standard deviations for the green (horizontal) and fired (vertical) strengths for each particle size indicate comparable relative strength dispersions at each size. The upper and lower solid lines and the shaded band are empirical power-law fits to the exterior bounds of the standard deviations of the strengths. The shaded band encompasses the means and standard deviation strength limits nearly identically for all particle sizes, indicating a systematic monotonic trend in strength behavior with particle size.

The correlations of means and standard deviations of strengths in the green and fired states for well-matched particle sizes is a clear indication that the populations of strength-controlling flaws in the green state are related to the populations of strength-controlling flaws in the fired state. In the terms used in this book, $h_{\text{green}}(c)$ is related to $h_{\text{fired}}(c)$. There is a clear dependence on particle size, implying that the strengths reflect extreme value size effects on flaws selected from underlying flaw populations, $f_{\text{green}}(c)$ and $f_{\text{fired}}(c)$, respectively. Further, the dispersions of strengths and thus strength-controlling flaws increase as particle size D increases—the opposite of that observed for stochastic size effects (Chapters 5 and 7), in which strengths contracted to a common threshold as D increased. The overall conclusions to be drawn from Figure 8.1 is that an underlying green state particle flaw population $f_{\text{green}}(c)$ probably generates an underlying fired state particle flaw population $f_{\text{fired}}(c)$, and that both are modified by *deterministic* extreme value size effects in generating strength-controlling flaw populations as a function of particle size. The following section uses the analyses of Chapters 4 and 5 to address strength distributions and flaw populations of the green and fired particles.

8.2 Strength and Flaw Size Observations

Figure 8.2 shows strength edf $Pr(\sigma)$ plots for samples of (a) green bauxite and (b) fired mullite particles 0.79–1.84 mm in diameter from the work of Wong et al. (1987), using data derived from the published work. Symbols represent individual strength measurements. Different symbols represent different particle sizes, D; particle sizes decrease left to right, consistent with Figure 8.1. The magnitudes of the strengths are consistent with those of similar materials: Green strengths are greater than those of the 10 mm diameter iron ore agglomerate particles (0.01–0.07 MPa) of Jewett et al. (1961) and the limestone, sand, and cement agglomerates (Kapur and Fuerstenau 1967; Capes 1971; Vallet and Charmet 1995) shown in Figure 2.26, and comparable to the titania agglomerates (Kendall 1988) shown in Figure 3.2, and the alumina agglomerates of 70 mm bend specimens (1 MPa–12 MPa) and 2 mm particles (3 MPa–10 MPa) of Tanaka et al. (1994) and Satone et al. (2017), respectively. The fired strengths are comparable to those of dense alumina particles (Tavares and King 1998; Brecker 1974; Bertrand et al. 1988; Huang et al. 1995) shown in Figure 6.5. Two clear features of Figure 8.2 are that the strength distribution data converge at large strengths in each state (green and fired) and that the relative shapes of the distributions are not significantly altered by firing. The solid lines in Figure 8.2 represent visual best fits to the individual data sets. The data for the largest particles in each state were fit by the concave bi-linear strength description. The fits to the data for the remaining smaller particles were fit by the sigmoidal tri-linear strength description. The fits were constrained such that the upper bound strengths for each state were fixed: 5.0 MPa for the green particles and 225 MPa for the fired particles. The fits were unconstrained otherwise with regard to shapes and lower bound strength thresholds. The data are well separated by particle size except at the upper bounds and are well described by concave and sigmoidal behavior. The data sets exhibit strength domains σ_N/σ_1 of approximately 5, somewhat narrow for particles. The increase in mean and decrease in standard deviation of strength with decrease in particle size, Figure 8.1b, is shown in Figure 8.2 to arise from increase in the threshold strength with decrease in particle size, with little other change to the strength distributions.

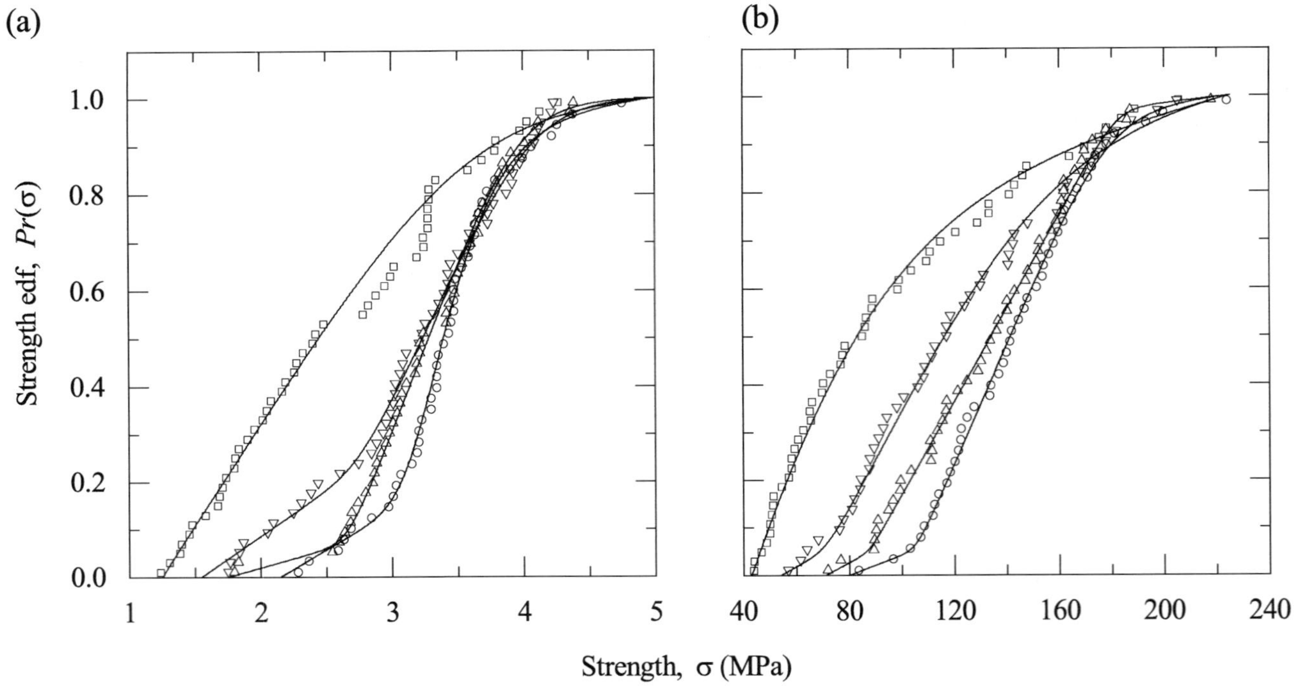

Figure 8.2 Strength distributions of size selected ceramic particles of (a) green bauxite and (b) fired mullite (Adapted from Wong, J.Y et al., 1987). Different symbols indicate different sizes. Decreasing particle size left to right. (a) $D = (1.84 \pm 0.03, 1.48 \pm 0.02, 1.08 \pm 0.02$, and $0.95 \pm 0.01)$ mm. (b) $D = (1.60 \pm 0.03, 1.32 \pm 0.02, 0.90 \pm 0.01$, and $0.79 \pm 0.01)$ mm. Total number of observations, $N_{tot} = 380$.

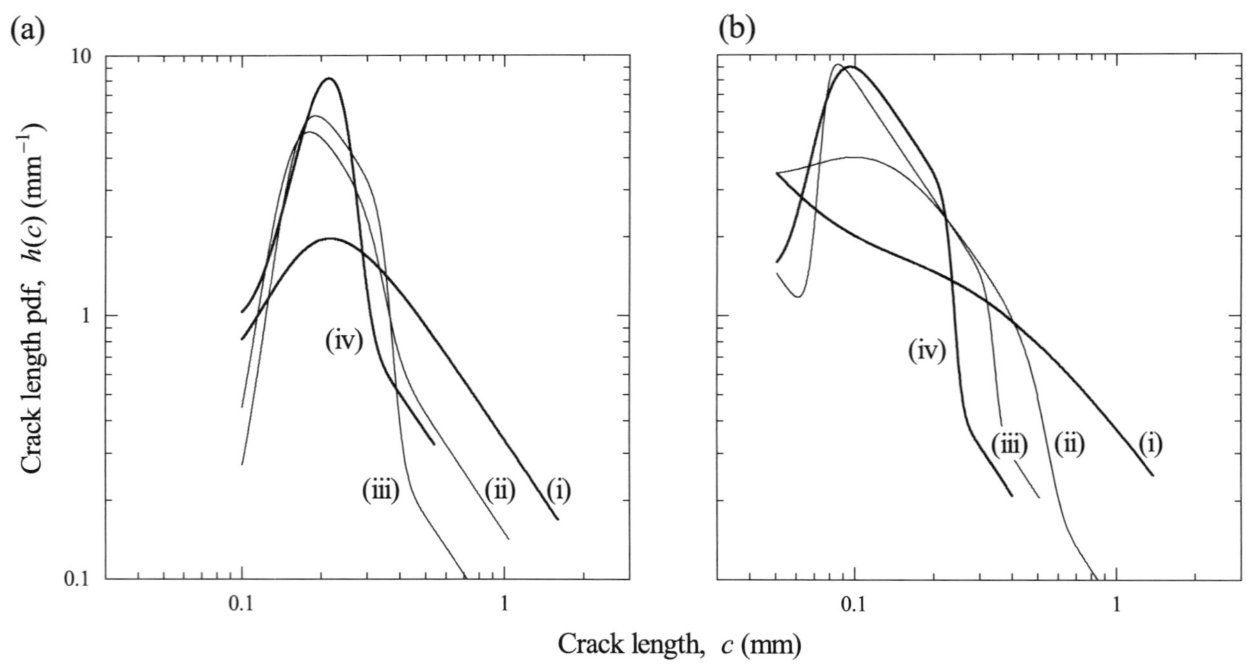

Figure 8.3 Plots in logarithmic coordinates of crack length pdf $h(c)$ variations for (a) green bauxite and (b) fired mullite particles. Particle sizes decrease in order (i)–(iv) and bold lines (i) and (iv) indicate ranges of behavior for both states. Responses determined from Figure 8.2.

Using logarithmic coordinates, Figure 8.3 shows the crack length pdf $h(c)$ variations underlying the bauxite and mullite particle strength behavior. The curves are labeled (i)–(iv), corresponding to the decreasing particle size numbering scheme and the area under each curve is 1. The B values used were (a) green bauxite, 0.05 MPa m$^{1/2}$ and (b) fired mullite 1.6

MPa $m^{1/2}$, taking into account the particle sizes, Eq. (7.2), such that the inferred largest crack was smaller than the particle size in each case (implementing this constraint for the largest particle in both states was sufficient). These values (1.6 MPa $mm^{1/2}$ and 51 MPa $mm^{1/2}$, respectively) are consistent with those inferred for particles displaying stochastic size effects, Figure 7.27. As the strength distributions overlap, so do the strength-controlling crack length populations. The extremes of the particle sizes, (i) and (iv), for each state are shown bold and highlight the major points. In both cases, the crack length populations for the largest particle sizes have the widest domains and least variations. As particle size decreases, the crack population domains decrease and modes become well defined as peaks narrow within the domains. As the upper bounds to the strength fits for individual particle sizes in each state were constrained to common invariant values, the lower bounds of the crack length populations are also common invariant values: a factor of two larger for the green particles relative to the fired particles, 0.10 mm vs 0.05 mm. Variations in the forms of the crack length populations are thus relative to these values: the mode crack lengths (at curvature minima) decreased similarly, approximately 0.2 mm vs 0.1 mm, but the characteristic peak widths (between inflection points), increased from approximately 0.2 mm to 0.5 mm.

The crack length distribution parameters are shown in greater detail as functions of particle size in Figure 8.4 for both (a) green and (b) fired particles. The dashed lines indicate the particle size in each case—upper bounds to all parameters. The upper and lower solid lines indicate the crack length domain bounds, calibrated to be less than the particle sizes for the upper bounds and invariant for the lower bounds. The central bold lines indicate the crack length modes. The shaded bands indicate the crack length populations peak bounds. In the green state, the weakly decreasing central population peak width and decrease in the large crack tail with decreasing particle size are clear. In the fired state, the shifted and decreasing central population peak width and weakly decreasing large crack tail with decreasing particle size are also clear. The similarity in the shapes of the shaded bands in Figure 8.4 point to a relationship between the green and fired flaw populations. Quantification of the relation between the crack length peak widths in the green and fired states, the widths of the shaded bands, makes this clear. Figure 8.5 shows a logarithmic plot of the fired state particle and crack length distribution dimensions as a function of the conjugate green state parameters. The solid line indicates ideal agreement. The filled symbols indicate particle size, from Figure 8.1b, the half-filled symbols indicate mode crack length, related to the means in Figure 8.1b, and the open symbols indicate the crack length population bounds. Over a factor of 10 in dimensions, these quantities are in agreement. Of perhaps most interest here are the shaded symbols, indicating the variation of the crack population distribution widths, taken from Figure 8.4. These data fall within the overall agreement, but more importantly exhibit a monotonic (albeit, non-linear) trend as an isolated group (as indicated by the line). Taken together, the data in Figure 8.5 imply that

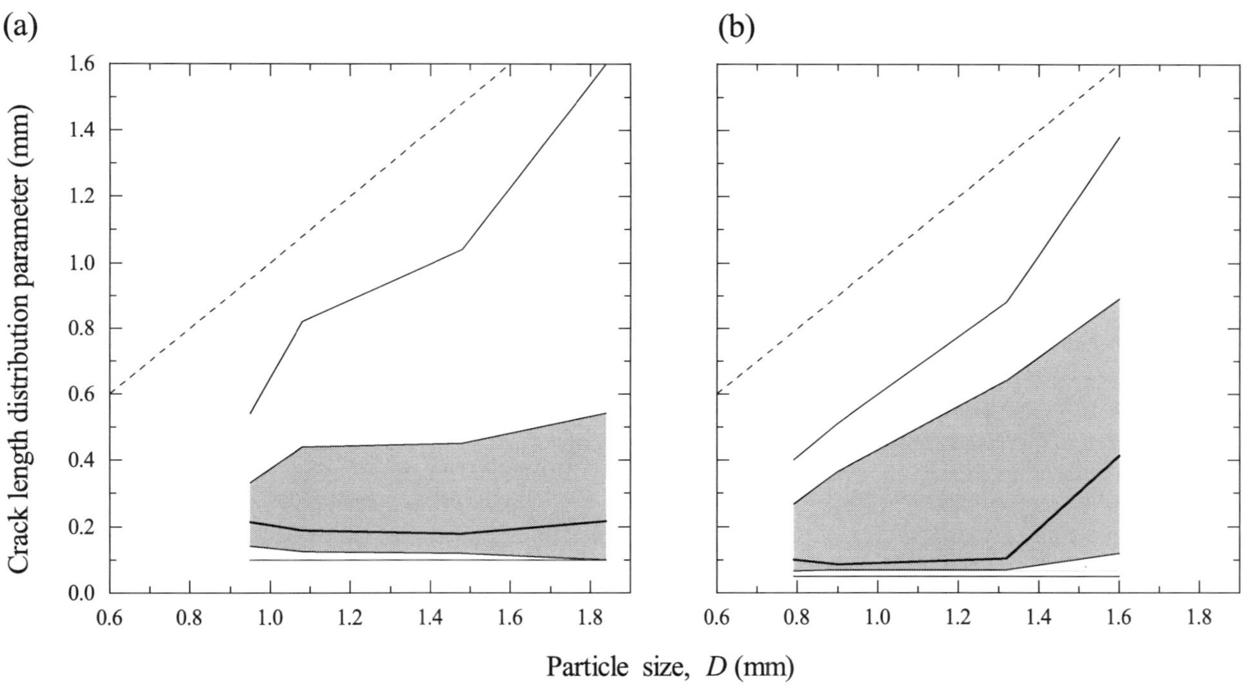

Figure 8.4 Crack length distribution parameters for (a) green bauxite and (b) fired mullite particles. Dashed lines indicate upper bounds, fine solid lines indicate crack distribution bounds, bold solid line indicates mode behavior, shaded bands represent peak bounds. Responses determined from Figure 8.3.

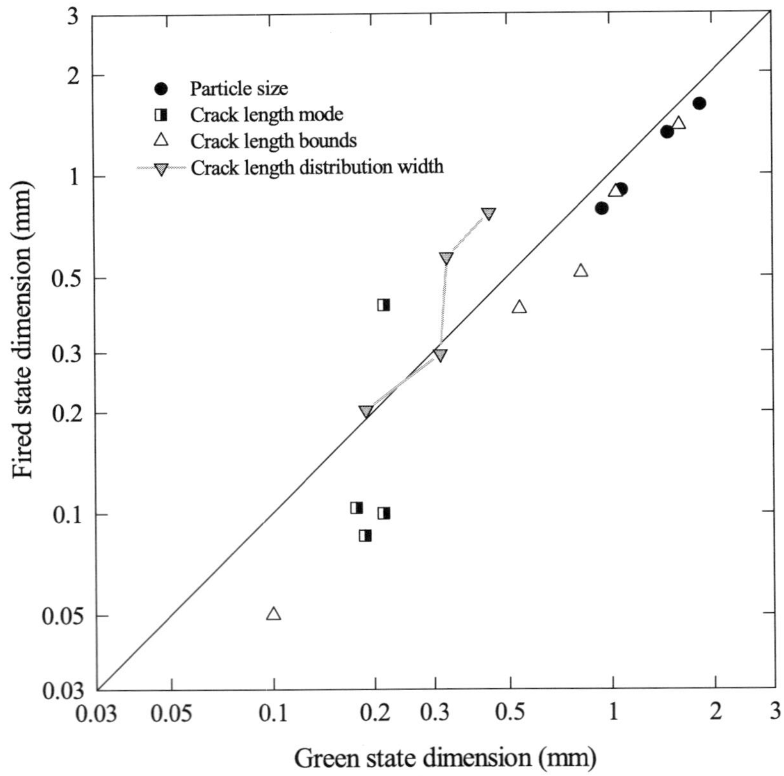

Figure 8.5 Plot in logarithmic coordinates of particle size and crack distribution dimensions for fired mullite vs conjugate green bauxite particles. The correlation between the two states is clear.

the flaw populations established in the green state by tumbling aggregation were passed with modification into the fired state. The modifications included a slight broadening of the peak in the flaw population towards smaller cracks, such that the modes and lower bound crack lengths were both reduced by factors of approximately 2. The firing also had the effect of altering the materials such the failure resistance of the particles, B, increased by a factor of approximately 30, thereby increasing strengths. In both green and fired states, the upper bound crack lengths significantly increased with particle size, indicating a deterministic size effect on the crack population. Analysis and quantification of this size effect is the subject of the next section.

8.3 Strength and Flaw Size Analysis

The physical phenomena underlying the strength distributions of the green and fired particles in Figure 8.2 have much in common with those of particles exhibiting stochastic size effects in Chapter 7. In both cases, an individual particle contains an ensemble of flaws. On loading in diametral compression, a zone of tensile stress is generated at the center of a particle and failure ensues from the largest flaw in the zone. The distribution of strengths arising from a sample of similarly sized particles is an extreme value distribution reflecting the distribution of largest flaws from each ensemble. However, clear differences between the strength distribution behavior exhibited in Figure 8.2 and Chapter 7 point to differences in the relation between the ensembles as a function of particle size. In stochastic failure, a lower bound threshold is common to strength distributions for particles of all sizes. Here, the upper bound strength is common to strength distributions of particles of different sizes. In stochastic failure, the smallest particles exhibit the widest domain of strengths, and strength domains contract toward the threshold as particle size is increased. Here, the largest particles exhibit the widest domains of strengths, and strength domains contract toward the upper bound as particle size is decreased. Interpretation of stochastic failure assumes that all ensembles of flaws represent a single invariant population and that particle size controls the probability of extreme flaws selected from this population. Here, it appears that the ensembles of flaws derive from multiple populations and that particle size controls the form of a population, leading to deterministic failure. Just as the common lower bound strength in stochastic systems suggests a linkage between the extreme value flaw distributions of different sized particles, the common upper bound strength in the deterministic system here suggests a linkage between flaw distributions. The smallest particles in stochastic systems exhibit the widest strength domains and thus closest estimation of the underlying flaw population. Similarly, as the largest particles in the deterministic systems here exhibit the widest strength domains, closest estimation of base flaw populations can be obtained from the behavior of large particles.

Combining these observations in consideration of deterministic size effects leads to a particle size- (D-) dependent extreme value crack length pdf described by the product

$$h(c, D) = f(c, \Omega)\Delta f(D). \tag{8.1}$$

Equation (8.1) is similar to the product of Eq. (5.3) describing the cdf of deterministic strengths. Here, $f(c, \Omega)$ may be regarded similarly to the base elemental flaw population discussed earlier, $f(c, \Omega) \approx f(c)$. In the case of strengths, Eq. (5.3), the notation $F(\sigma, 1)$ was used for the base cdf to indicate that the cdf applied to minimal $k = 1$ components. In the case of crack lengths, Eq. (8.1), the notation $f(c, \Omega)$ for the base pdf indicates that the pdf applies to maximal $D \approx \Omega$ components. $\Delta f(D)$ is a deterministic, particle size dependent, perturbation that modifies the base population. In particular, $\Delta f(D)$ modifies the crack length population domain. Here, $f(c, \Omega)$ will be taken as the $h(c)$ function describing the largest particles in the green and fired states, responses (i) in Figure 8.3a and Figure 8.3b. $\Delta f(D)$ will be given by a generalized step function,

$$\Delta f(D) = \Delta f_i(c, a_i, b_i, c_i) = \begin{cases} a_i & : c < c_i \\ b_i & : c \geq c_i \end{cases}, \tag{8.2}$$

where the index i indicates the step function acting at the crack length c_i. At c_i the function steps between a_i and b_i. A familiar example is the Heaviside unit step function centered at 0, for which $a = 0$, $b = 1$, and $c = 0$. Here, the parameters a_i, b_i, and c_i depend on D.

The above analysis is able to describe the major effect of particle size observed here for the bauxite and mullite particles: increases in strength thresholds and contraction of strength distributions with decreasing particle size, Figure 8.2. Figure 8.6a shows as a dashed line the crack length pdf variation $h(c)$ over the inferred domain, 0.1 mm–1.6 mm, for the largest green bauxite particles, $D = 1.84$ mm. The solid line shows a modification of this pdf using Eq. (8.1) and a single perturbing function of the form Eq. (8.2) with $a_1 = 1$, $b_1 = 0$, and $c_1 = 0.8$ mm. The original function is truncated (and normalized) such that there are no strength controlling cracks longer than 0.8 mm. Figure 8.6b shows as dashed and solid lines the resultant strength cdf variations. The lower bound threshold strength associated with the modified crack length population is greater, reflecting the absence of larger cracks, but otherwise the strength distribution is unaltered relative to the original strength behavior in shape or the upper bound. Perturbations of the behavior in Figure 8.6a are illustrated in Figure 8.7, which again shows as a dashed line the crack length pdf variation for the largest green bauxite particles. The solid line shows a normalized modification of this pdf using Eq. (8.1) and three perturbing functions of the form Eq. (8.2) with: $a_1 = 1$, $b_1 = 0$, $c_1 = 1.2$ mm; $a_2 = 1$, $b_2 = 0.2$, $c_2 = 0.6$ mm; $a_3 = 0.25$, $b_3 = 1$, $c_3 = 0.15$ mm. The perturbing functions have effect from large to small crack lengths (right to left in Figure 8.7) and truncate the original distribution, shown as the dark shaded region, or suppress the original distribution, the light shaded regions. This multi-region form will be used to perturb the base crack length distributions of the green and fired particles so as describe the deterministic size effects on the strength distributions.

Figure 8.8 and Figure 8.9 are analogous to Figure 8.2 and Figure 8.3 with a major exception: Instead of representing independent fits to experimental observations, the later figures represent predictions of strength and crack length behavior. The predictions are derived from the base crack length populations modified by deterministic size effects—this is *forward analysis* (Chapter 3). The prediction process begins with the crack length populations inferred from experimental strength measurements of the largest green and fired particles. These "base" populations are indicated as the bold lines (i) in Figure 8.9 and are reproduced from Figure 8.3. The resulting strength distribution behaviors are shown as the left-most responses in Figure 8.8, and reproduce those from Figure 8.2. The multi-region truncation and suppression perturbation method outlined above was then implemented on the base crack length populations. The perturbations lead to narrowing consistent with the measured particle sizes and agreement with the measured strengths. The truncated and suppressed crack length populations are indicated by the fine and bold lines (ii) and (iv) in Figure 8.9 (for ease of visualization, (iii) is not shown). The piece-wise continuous nature of the perturbed responses is clear, as is the overall topological similarity of the responses to those in Figure 8.3. The resultant simulated strength distribution responses are indicated by the central and right variations in Figure 8.8. The similarity with the "penguin" like shape of the experimental responses of Figure 8.2 is

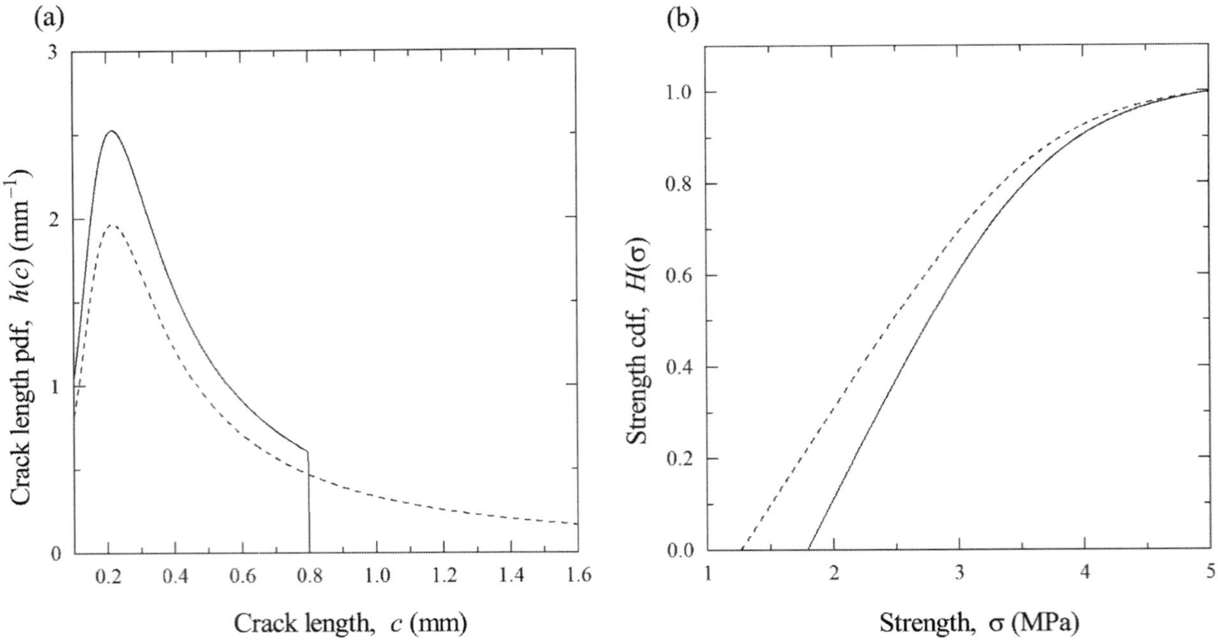

Figure 8.6 (a) Plot of crack length pdf variations for green bauxite particles from Figure 8.3, dashed line, and simply truncated and re-normalized pdf representing a deterministic size effect, solid line. (b) Plot of resultant concave strength cdf variations from (a) showing increased lower bound threshold and invariant upper bound strengths on truncation.

clear ((iii) is again omitted). (As $H(\sigma, D)$ predictions reflect integration of $h(c, D)$, steps in the piecewise continuous curves in Figure 8.9 are reflected as changes in derivative in Figure 8.8. These features could be removed easily by moving average smoothing.) The parameters a_i, b_i, and c_i underlying the strength distributions varied systematically with D, as expected for deterministic size effects. In particular, the narrowing flaw distributions were characterized by decreased c_1 and c_2 and increased c_3. The crack lengths of the simulations were in quantitative agreement with those observed experimentally, c_1 agreed with the bounds (open symbols in Figure 8.5), and $(c_2 - c_3)$ agreed with the peak widths (shaded symbols in Figure 8.5). The strength distribution variations with particle size, in both green and fired states, are well described by a deterministic size effect.

8.4 Summary and Discussion

This chapter has examined in detail the strength distributions and underlying crack lengths of two related series of ceramic particles. The series were related by fabrication: one series consisted of green state bauxite particles and the other series consisted of fired state mullite particles formed from the first series. The particle sizes of each of the four members of the two series were matched by careful selection so that a member of the fired state series represented an approximate 50 % volume reduction relative to the conjugate member of the green state series. Particle diameters spanned factors of approximately 2 across each series. The strengths of the particles were typical of green agglomerates and fired ceramics. The fired particles were about a factor of 40 stronger than the green particles. Both series of particles exhibited strength distribution behavior consistent with deterministic size effects on the strength controlling crack lengths. Specifically, the lower bound threshold strengths in each series increased with decreasing particle size, such that the strength distributions contracted toward invariant upper bounds. As a consequence, the mean strengths increased and the strength dispersions decreased as particle size decreased. The underlying strength-controlling crack lengths were deconvoluted from the strength measurements using the probabilistic analyses developed earlier. For both series, crack length populations narrowed with decreasing particle size, primarily due to truncation and suppression of the proportions of large crack lengths. The fired state crack lengths exhibited wider peaked distributions than the green state crack lengths, primarily due to increases in the proportion of small crack lengths. The detailed findings regarding crack length distributions are in agreement with the

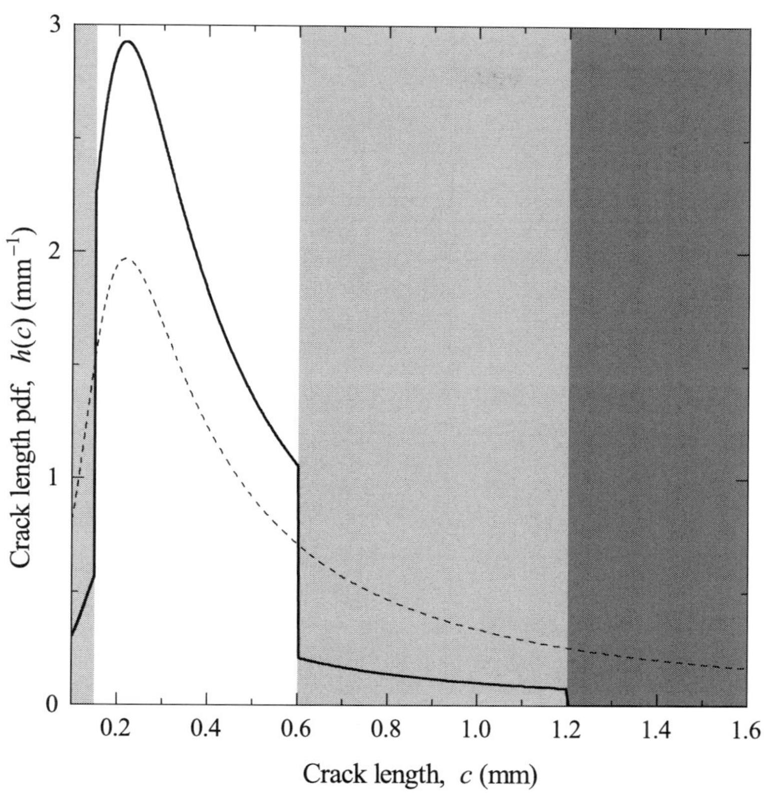

Figure 8.7 Plot of crack length pdf variations for green bauxite particles from Figure 8.3, dashed line, and multiply truncated, suppressed and re-normalized pdf, solid line. The alteration represents a deterministic narrowing of the crack length population with changing particle size: dark shaded region indicates cracks removed from the population and light shaded regions indicates crack population suppression.

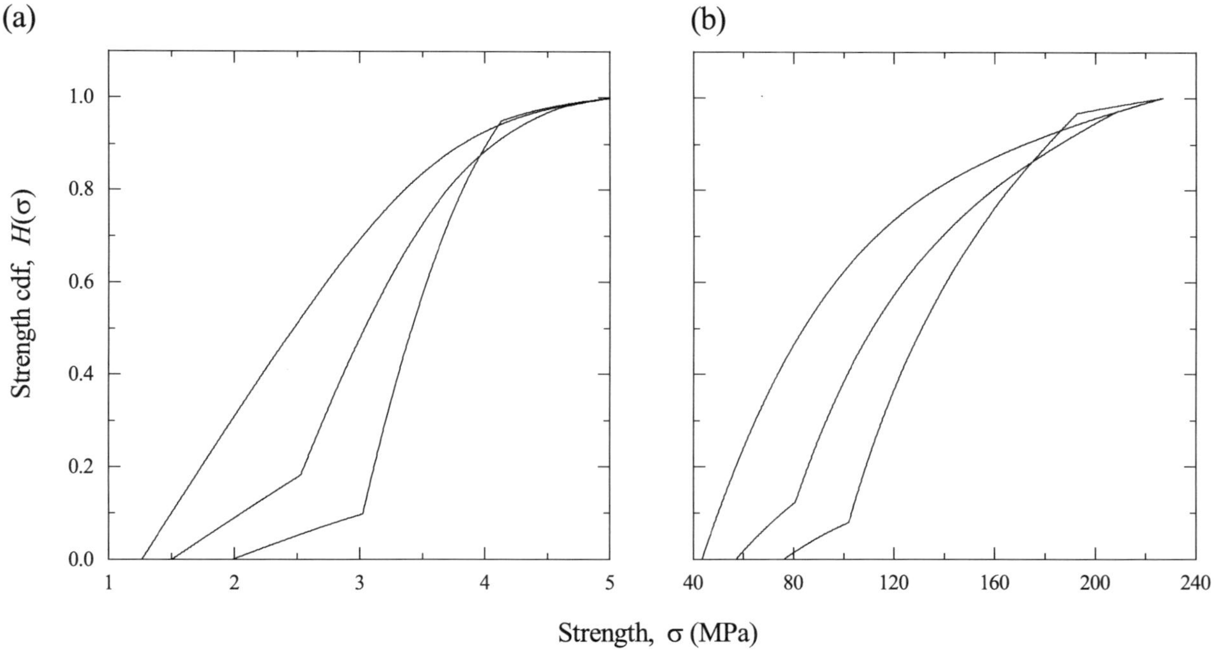

Figure 8.8 Simulated strength distributions of ceramic particles of (a) green bauxite and (b) fired mullite. Decreasing particle size left to right. The simulations are conjugate to the crack length populations in Figure 8.9 and are based on deterministic size effects acting on base, large particle, populations.

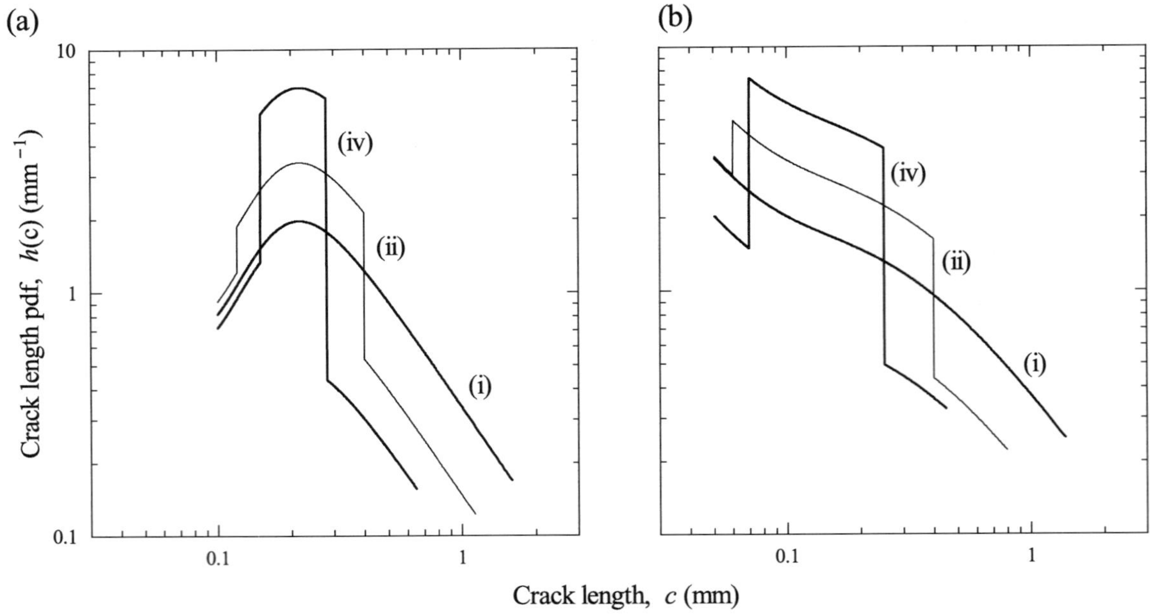

Figure 8.9 Simulated crack length populations of ceramic particles of (a) green bauxite and (b) fired mullite. Decreasing particle size in order (i), (ii), and (iv). The conjugate strengths are shown in Figure 8.8. The simulations are based on deterministic size effects acting to narrow the base, large particle, populations (i).

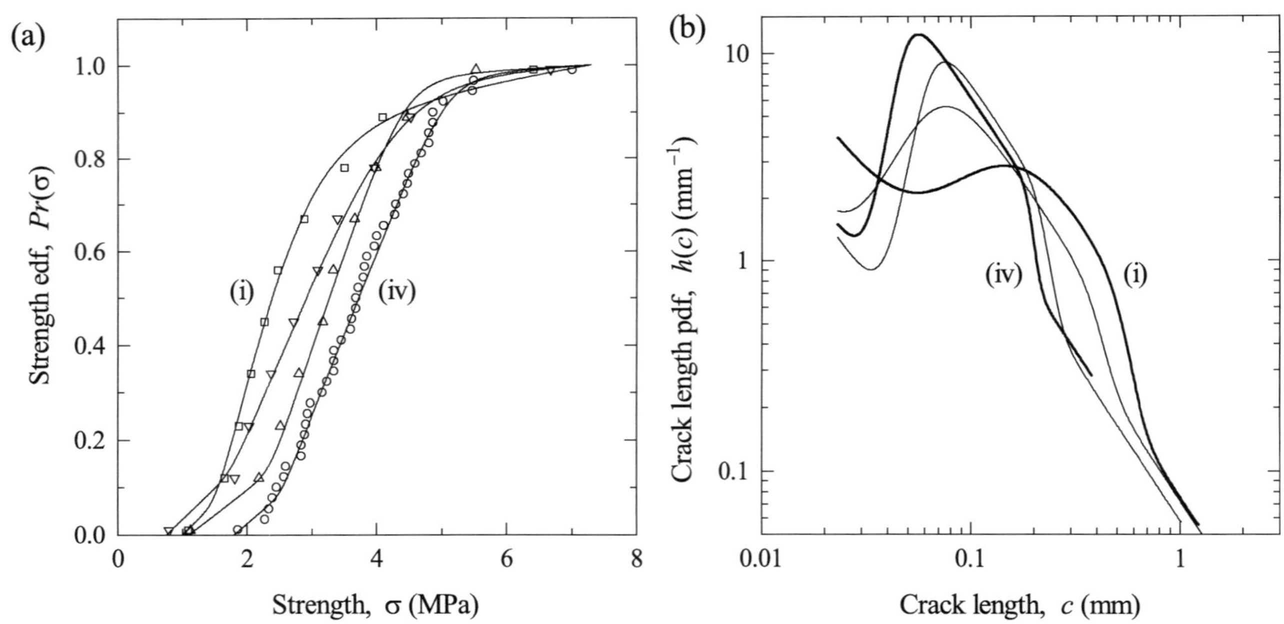

Figure 8.10 (a) Strength distributions of potash particles (Adapted from Rozenblat, Y et al., 2011). Different symbols indicate different sizes. Decreasing particle size left to right. (a) D = 1.4 mm–2 mm (i), 2 mm–2.36 mm, 2.36 mm–3.35 mm, and 3.35 mm–4 mm (iv). N_{tot} = 76. (b) Plot in logarithmic coordinates of crack length pdf $h(c)$ variations for potash particles. Bold lines (i) and (iv) indicate range of behavior. Responses determined from (a).

assessment from the original work of of Wong et al. (1987): the green state flaw distributions are related to, and probably determined, the fired state flaw distributions. An assessment arrived at by Wong et al. through consideration of the strength dispersions in assumed linearized form.

Similar strength distribution behavior, consistent with deterministic size effects on strength controlling crack lengths, was exhibited by potash and iron ore particles. Potash particles are considered first. Potash is a common commercial and domestic fertilizer, supplying potassium in ion-accessible form to plants. Potash particles consist of sub-particles containing primarily KCl, K_2SO_4, and K_2CO_3, usually mined from deep salt deposits and tumbled with an organic binder to form spherical particles similar to the green ceramics above (and the fertilizer of Figure 7.20). Figure 8.10a shows strength edf $Pr(\sigma)$ plots for samples of potash particles 1.4–4 mm in diameter from the work of Rozenblat et al. (2011), using data derived from the published work. Strengths were measured using a conventional compression platen apparatus. Symbols represent individual strength measurements. Different symbols represent different particle sizes, D; particle sizes decrease left to right,(i)–(iv). The magnitudes of the strengths are similar to those of the green bauxite particles above and those of similar agglomerate materials in the works cited. As above, the strength distributions in Figure 8.10a converge at large strengths and the relative shapes of the distributions do not alter significantly with particle size. The solid lines in Figure 8.10a represent visual best fits to the individual data sets by the sigmoidal tri-linear strength description. The fits were constrained such that the upper bound strengths for each size was fixed at 7.3 MPa. The fits were unconstrained otherwise. The data are well separated by particle size except at the upper bounds and are well described by sigmoidal behavior. The data sets exhibit strength domains σ_N/σ_1 of approximately 5. As a group, there is an increase in the threshold strength with decrease in particle size, with little other change to the strength distributions.

Using logarithmic coordinates, Figure 8.10b shows the crack length pdf $h(c)$ variations underlying the strength behavior. The bold curves labeled (i) and (iv) correspond to the range of particle sizes and the area under each curve is 1. The B value used was 0.035 MPa $m^{1/2}$ taking into account the particle sizes, Eq. (7.2), consistent with earlier values. As the strength distributions overlap, so do the strength-controlling crack length populations. The crack length population for the largest particle size has the widest domain and least variation and as particle size decreased the crack population domains decreased and modes became well defined as peaks narrowed within the domains. As the upper bound to the strength fits for individual particle sizes was constrained to a common value, the lower bounds of the crack length populations are also in common. Overall, the behavior is very similar to that of the green ceramic particles, implying a similar deterministic size effect based on flaw distribution truncation.

Iron ore particles are now considered. Iron ore is frequently formed into pellets or particles to facilitate air flow through the packed ore in blast furnaces for steel production. The particles also facilitate ore handling and transport. The particles are formed by high-temperature thermal processing and contain about 70 % iron oxide by relative volume, with the remainder other oxides and limestone and clay as flux and binding agents. The nearly spherical particles are typically > 10 mm in size but otherwise are similar to the fired mullite particles above. In order to develop and refine models for iron ore particle failure during handling and transport, Gustaffsson et al. (2013a) used a conventional platen apparatus to measure the strength distribution of a sample of commercial (Swedish) iron ore particles. Tavares et al. (2018) used a platen apparatus and impact testing to measure the specific failure energy of a set of four sizes of commercial (Brazilian) iron ore particles. Similar measurements were performed by Cavalcanti and Tavares (2018) using a platen apparatus and by Gustafsson et al. (2017) using a split Hopkinson bar impact apparatus, the latter as part of a larger program measuring the properties of ensembles of iron ore particles under shear and confined compression (Gustafsson et al. 2009, 2013bc).

Figure 8.11a shows strength edf $Pr(\sigma)$ plots for samples of iron ore particles 9–19 mm in diameter from the work of Tavares et al. (2018), using data derived from the published work. Symbols represent individual strength measurements. Different symbols represent different particle sizes, D; particle sizes decrease left to right, (i)–(iv). The magnitudes of the strengths are intermediate between those of the green bauxite, potash, and similar agglomerate materials cited earlier and the fired mullite particles. (The strengths of 11 mm iron ore particles reported by Gustaffson et al. (2013a) were similar.) As above, the strength distributions in Figure 8.11a converge at large strengths and the relative shapes of the distributions do not alter significantly with particle size. In particular, the dominant central strength edf responses for all four sizes of particles in Figure 8.11a are clearly linear and of the same slope, such that the central linear responses appear simply translated. The central linear responses are similar in shape to those exhibited by the small particles of Figure 6.15 but more importantly the appearance of translation is similar to that exhibited in the deterministic size effects of the polycrystalline diamond thin films of Figure 5.24. Distinct from the two systems considered earlier, the strengths converge at the small strength threshold, but the predominant central translated responses indicate that this is a deterministic rather than stochastic system. The lines in Figure 8.11a represent visual best fits to the individual data sets by the sigmoidal tri-linear strength description. The fits were constrained such for each particle size the lower and upper bound strengths were 4.5 MPa and 38 MPa, respectively, and the central slope was 67 MPa^{-1}. The fits were unconstrained otherwise. The data are well separated by particle size

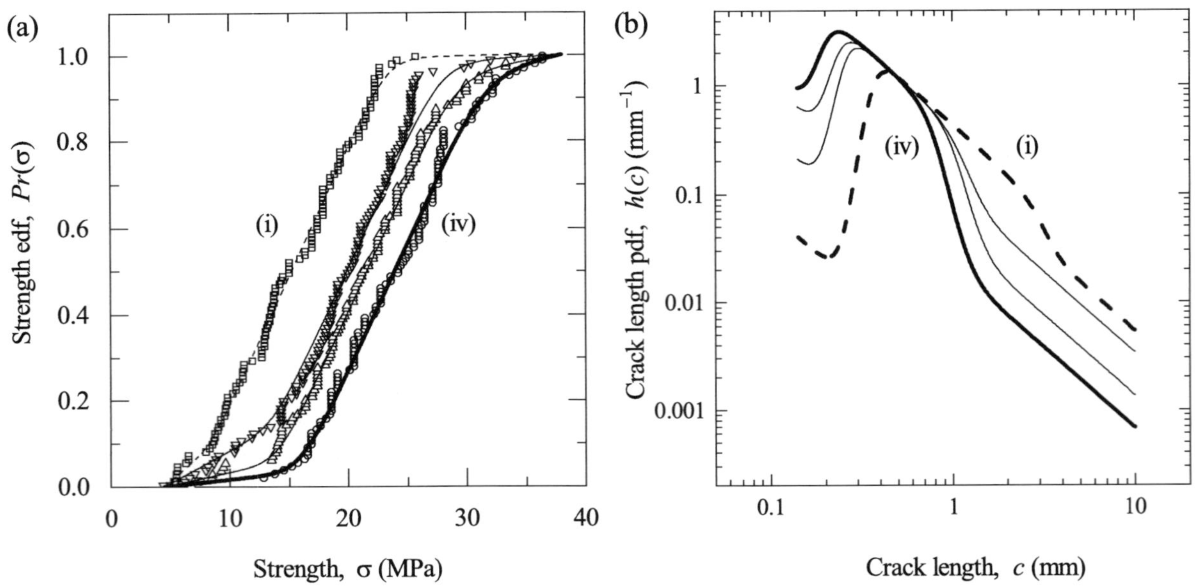

Figure 8.11 (a) Strength distributions of iron ore particles (Adapted from Tavares, L.M et al., 2018). Different symbols indicate different sizes. Decreasing particle size left to right. (a) D = 9 mm−12.5 mm (i), 12.5 mm−14 mm, 14 mm−16 mm, and 16 mm−19 mm (iv), N_{tot} = 474. (b) Plot in logarithmic coordinates of crack length pdf $h(c)$ variations for iron ore particles. Bold lines (i) and (iv) indicate range of behavior. Responses determined from (a).

and are well described by sigmoidal behavior. The data sets exhibit strength domains σ_N/σ_1 of approximately 8. Bounds are indicated by dashed and solid bold lines.

Using logarithmic coordinates, Figure 8.11b shows the crack length pdf $h(c)$ variations underlying the strength behavior. The dashed and bold curves labeled (i) and (iv) correspond to the range of particle sizes as in Figure 8.11a and the area under each curve is 1. The B value used was 0.45 MPa m$^{1/2}$ taking into account the particle sizes, Eq. (7.2), consistent with earlier values. As the strength distributions overlap on a fixed domain, so do the strength-controlling crack length populations, in the form of overlapping peaks. A feature of the pdf responses of Figure 8.11b is the shared responses adjacent to the central peaks, reflecting the common central slope in the strength responses. The crack length population for the largest particle size (dashed line) has the widest peak and least variation. As particle size decreased the crack population peak widths decreased and modes became well defined as peaks narrowed within the domains, typified by the smallest particle size (bold line). As the upper bound to the strength fits for individual particle sizes was constrained to a common value, the lower bounds of the crack length populations are also in common. Overall, the behavior is very similar to that of the green ceramic particles, implying a similar deterministic size effect based on flaw distribution suppression at small crack lengths.

The four sets of strength distributions considered above, consisting of concave and asymmetric sigmoid shapes, provided clear evidence of deterministic size effects—as particle size increased, sample strength distributions expanded and shifted to smaller strengths. Such behavior is inconsistent with stochastic size effects. The expansion and shifts reflected the appearance of new or more frequent, larger extreme flaws within a sample with increasing particle size. As discussed in Chapters 5 and 6, generalized ratios of the maximum extreme crack lengths and component sizes (relative to the behavior of a small, reference component) enable comparison of multiple systems. In this case, the ratios enable consideration of the deterministic scaling of the maximum crack length with particle size. (For stochastic systems maximum crack length is invariant.) Here, values of maximum crack lengths c in samples of particles size D, inferred from analysis of strength distributions, are compared to a single reference configuration c_{ref} and D_{ref} within a given system. Comparison of c/c_{ref} and D/D_{ref} between systems is thus relative to common reference point (1,1).

Figure 8.12 shows a plot in logarithmic coordinates of the relative size of flaws c/c_{ref} vs the relative particle size D/D_{ref} for the systems studied in this chapter. The crack lengths were determined from fits to the strength distributions (e.g. the upper bounds in Figure 8.10b or maximum curvature points in Figure 8.11b). The bars represent uncertainties determined from the particle size ranges given in the cited works. The solid line is a guide to the eye and has slope 2. The shaded band indicates the range of responses observed for extended components over the same generalized component size domain, Figure 5.28.

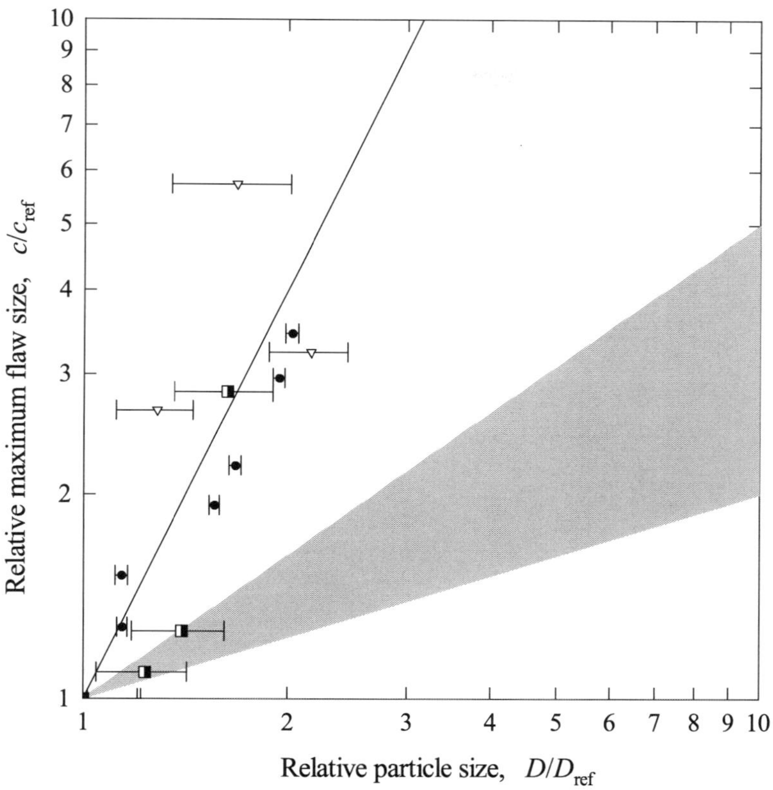

Figure 8.12 Logarithmic plot of maximum size of flaws c vs particle size D for particles exhibiting deterministic size effects on strength distributions. The particle systems were those considered in this chapter and exhibited invariant strength maxima. The line is of slope 2 and a guide to the eye; the shaded band indicates the behavior of extended components.

The solid symbols with small uncertainty bars indicate measurements from the manually size-selected bauxite and mullite particle systems of Figure 8.2 and Figure 8.3. The open symbols indicate measurements from the sieved potash particle system of Figure 8.10. The semi-filled symbols indicate measurements from the iron ore particle system of Figure 8.11. A major point to note in Figure 8.12 is that the overall domain of observed particle sizes is small, a factor of 2, compared with nearly 100 for extended components, Figure 5.28. It is also important to note that the deterministic systems examined here were limited to those that were related by a common minimum flaw size, and therefore common maximum strength. Nevertheless, it is clear that the maximum crack lengths in the deterministic particle systems here exhibited a much greater dependence on size than the extended components. A more thorough examination of scaling effects in deterministic particle systems is given in Chapter 9.

A physical interpretation of the strength and flaw size phenomena exhibited in Figures 8.2, 8.3, and 8.10 is based on the mechanism of particle formation during the rotary milling and tumbling fabrication process (also known as "balling"). In this process, very small sub-particles of powder agglomerate with the aid of an added binder to form larger particles. In this case the particles were size selected for mechanical testing. The process is illustrated in the schematic diagram of Figure 8.13, in which the initial sub-particles are shown at the top left and increasing particle size is shown in the direction top to bottom. Note that this direction follows that of particle enlargement and is the opposite of that used above in size classification. Two mechanisms of particle formation and enlargement are considered that give rise to the two limits of strength behavior: (i) upper bound strengths that are particle size invariant and (ii) lower bound strengths that decrease with increasing particle size. The two mechanisms give rise to deterministic size effects on strength as they lead to lower bound flaw sizes that are a particle size invariant and upper bound flaw sizes that increase with particle size.

The first mechanism of particle formation is one of uniform aggregation, shown in the left side column of Figure 8.13. In this mechanism the sub-particles, shown as shaded discs, bind together incrementally to form near close packed arrays (shown in 2-D in Figure 8.13). The arrays constitute the particles, indicated by curved outlines, and the sub-particles bind to the arrays in small numbers such that the array structure is maintained as the particle size increases. Within such uniform particles, the largest unbonded regions occur at the interstices between the sub-particles. The interstices are also uniform in

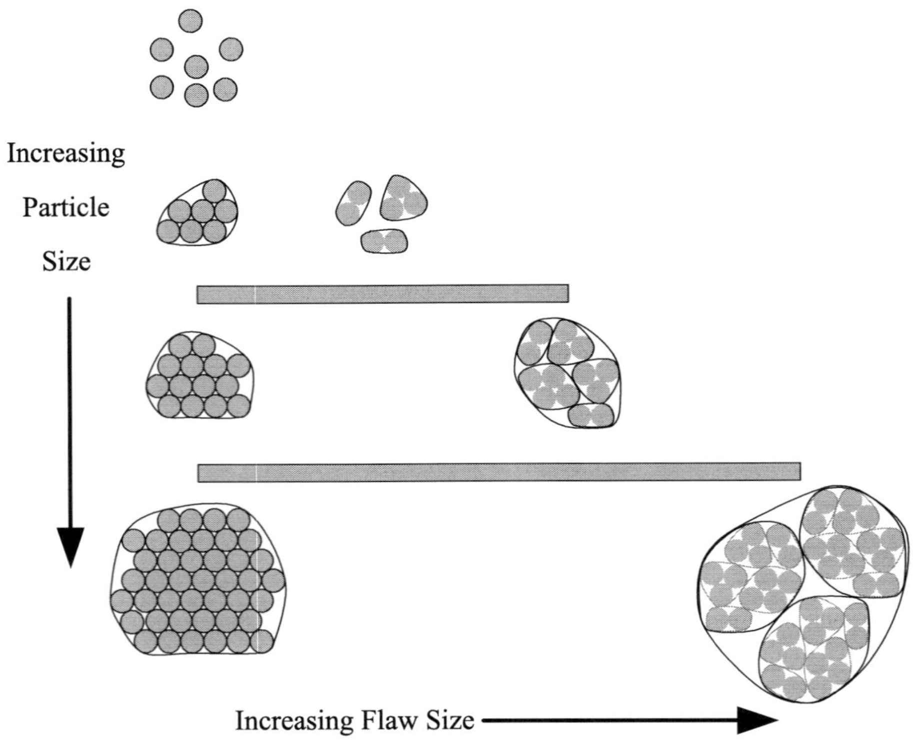

Figure 8.13 Schematic diagram illustrating particle formation and growth by aggregation of sub particles. Discrete sub particle shown at top left. Left column shows uniform aggregation, leading to a particle-size invariant flaw size. Right diagonal shows hierarchical aggregation, leading to increasing flaw size with increasing particle size. The growth mechanisms account for the deterministic size effects on particle strength: the shaded bands represent flaw size domains at fixed particle size.

size and are the largest "flaws" within the particles. Loading of a particle in diametral compression places the central region of the particle in tension and thus leads to fracture from one of the central interstices. As illustrated in the left column of Figure 8.13, the size of the interstices does not alter with particle size, as it is determined by sub-particle packing. Hence, flaw size, and thus strength, is invariant with regard to particle size. As the sub-particle interstices are the lower bound flaws, this invariant strength is an upper bound for all particle sizes.

The second mechanism of particle formation is one of hierarchical aggregation, shown on the right side diagonal of Figure 8.13. In this mechanism, the sub-particles initially bind to form small clusters. Several clusters then bind to form a larger cluster, which in turn binds with several other similar cluster to form a yet larger cluster and so on in a multi-generational manner. The clusters constitute the particles, again indicated by the curved outlines. The number of sub-particles in each cluster generation in Figure 8.13 is the same as that in the adjacent array. However, it is clear that the local close packing of the sub-particles in the first cluster generation is not maintained as the particle size increases. Within such hierarchical particles, the interstices are not uniform in size and the largest unbonded regions occur at the interstices between the largest clusters. Hence, the largest "flaws" within the particles scale with particle size and thus particle strength decreases with increasing particle size. As the hierarchical cluster interstices are the upper bound flaws, this decreasing strength is the lower bound for a given particle size.

In Figure 8.13, increasing flaw size for a given particle size is shown in the direction left to right, corresponding to transition in the particle formation mechanism from uniform aggregation of sub-particles, left, to hierarchical aggregation, right. Variations between these two mechanisms lead to domains of largest, extreme flaws within samples of similarly sized particles. The flaw size domains for the two largest particles, bottom, are indicated by shaded horizontal bands. The lower bounds of these bands are identical and the upper bounds increase with particle size, in agreement with experimental observations. An implication is that the mean strengths of such structures should decrease with sub-particle size D_{sub} and exhibit a dependence somewhat greater than $D_{sub}^{-1/2}$. This dependence takes into account the increase in flaw size with sub-particle size and the decrease in fracture resistance with sub-particle size. Studies of green pellet strengths

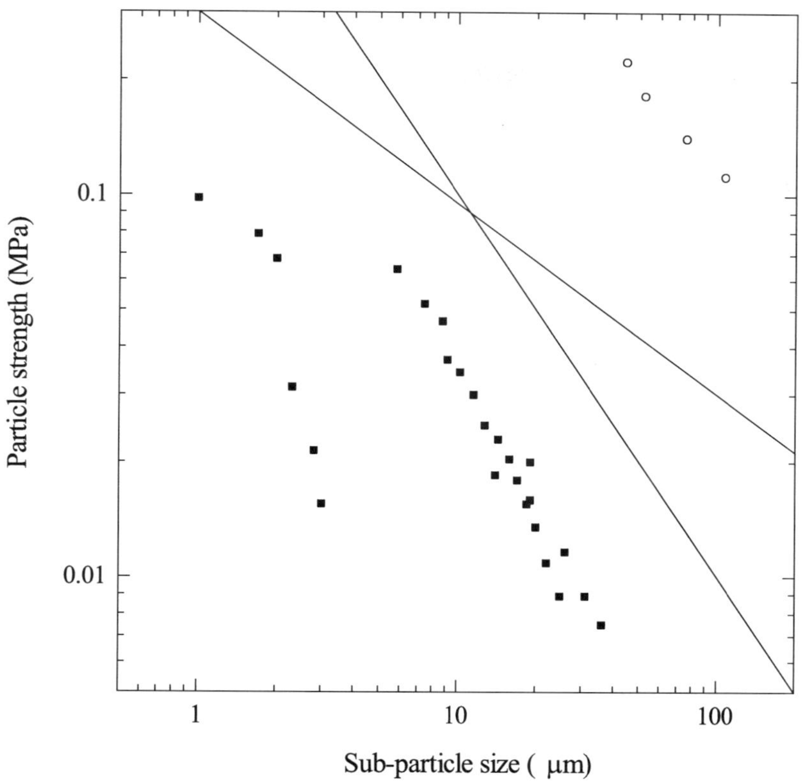

Figure 8.14 Logarithmic plot of green strengths of iron ore particles as a function of constituent sub-particle size. Different symbols represent different compositions (adapted from Jewett, R.P et al., 1961; Patra, S et al., 2017). Solid lines are guides to the eye of slope −0.5 and −1.

of iron ore particles are consistent with this expectation. Figure 8.14 shows the green strengths of iron ore particles tested in conventional platen systems, from the works of Jewett (1961) (filled symbols) and Patra et al. (2017) (open symbols), using data derived from the published work. The particles or "pellets" were formed by balling sub-particles of various sizes and chemical compositions. Different groups of symbols represent the mean strengths of samples of pellets, grouped by material at the specified sub-particle sizes. Particle sizes were approximately 10 mm. The solid lines are guides to the eye of slopes −0.5 and −1. The variations of strengths with sub-particle size are consistent with the expected trend. The above considerations of green and fired particles of matched sizes suggest that the behavior of Figure 8.14 is transmitted into the sintered state from the agglomerate state, leading to similar fracture and strength behavior.

The detailed fracture mechanics of hierarchical aggregated structures developed earlier (Cook 1989, 1993), includes examination of the behavior of flaws intermediate in size between the bounds. Such fracture mechanics applies directly to consideration of the bauxite and mullite particles considered here in proppant applications. Such applications require large strengths and small densities of particles, suggesting that a microstructure intermediate between uniform and hierarchical might be optimum in application. The materials engineering goal would then be to refine the balling and sintering procedures so as to generate particles at a mid-point in the domains of Figure 8.13. Fracture mechanics of aggregated structures, including time-dependent failure (Cook 1993) is also applicable to consideration of particles imbedded in rock fissure walls in both hydrocarbon extraction (Volk 1981) and groundwater motion (Sun 2019).

Chapter 8 has examined in detail the strength behavior of a particle system composed of two related materials, bauxite and mullite. The two materials and the particles considered were related by heat treatment that transformed green bauxite agglomerate particles into sintered mullite ceramic particles. The examination applied the analysis of Chapter 4 to deconvolute underlying flaw populations from strength distribution measurements and placed the flaw populations in the context of those describing extended components and other particles, Chapters 5 and 6. As opposed to the invariant populations characterizing stochastic size effects, Chapter 7, the strength and flaw size behavior here exhibited deterministic size effects, in which particle size determines multiple flaw populations in a system. In the bauxite and mullite systems here, the multiple flaw populations were related by common lower bounds for each particle material, leading to common upper bound strengths. In Chapter 9, fully deterministic size effects on particle strengths are considered, in which the bounds on strengths are independent.

References

Bertrand, P.T., Laurich-McIntyre, S.E., and Bradt, R.C. (1988). Strengths of fused and tabular alumina refractory grains. *American Ceramic Society Bulletin* 67: 1217–1221.

Brecker, J.N. (1974). The fracture strength of abrasive grains. *Journal of Engineering for Industry* 96: 1253–1257.

Breval, E., Jennings, J.S., Komarneni, S., Macmillan, N.H., and Lunghofer, E.P. (1987). Microstructure, strength and environmental degradation of proppants. *Journal of Materials Science* 22: 2124–2134.

Capes, C.E. (1971). The correlation of agglomerate strength with size. *Powder Technology* 5: 119–125.

Cook, R.F. (1989). Effective-medium theory for the fracture of fractal porous media. *Physical Review B* 39: 2811–2814.

Cook, R.F. (1993). Theory of time-dependent failure for fractal porous aggregates. In: *Defects and Processes in the Solid State: Geoscience Applications The McLaren Volume* (ed. J.N. Boland and J.D. Fitz Gerald), 229–242. Elsevier Science Publishers, B.V., Amsterdam.

Cooke, C.E. (1977). Fracturing with a high-strength proppant. *Journal of Petroleum Technology* 29: 1222–1226.

Feng, Y.C., Ma, C.Y., Deng, J.G., Li, X.R., Chu, M.M., Hui, C., and Luo, Y.Y. (2021). A comprehensive review of ultralow-weight proppant technology. *Petroleum Science* 18: 807–826.

Gaurav, A., Dao, E.K., and Mohanty, K.K. (2012). Evaluation of ultra-light-weight proppants for shale fracturing. *Journal of Petroleum Science and Engineering* 92: 82–88.

Gow, N. and Lozej, G. (1993). Bauxite. *Geoscience Canada* 20: 9–16.

Huang, H., Zhu, X.H., Huang, Q.K., and Hu, X.Z. (1995). Weibull strength distributions and fracture characteristics of abrasive materials. *Engineering Fracture Mechanics* 52: 15–24.

Jewett, R.P., Wood, C.E., and Hansen, J.P. (1961). Effect of particle size upon the green strength of iron oxide pellets. Bureau of Mines Report of Investigations 5762. US Department of Interior. Washington.

Kapur, P.C. and Fuerstenau, D.W. (1967). Dry strength of pelletized spheres. *Journal of the American Ceramic Society* 50: 14–18.

Kendall, K. (1988). Agglomerate strength. *Powder Metallurgy* 31: 28–31.

Kendall, K., Alford, N.M., Tan, S.R., and Birchall, J.D. (1986). Influence of toughness on Weibull modulus of ceramic bending strength. *Journal of Materials Research* 1: 120–123.

Kingery, W.D., Bowen, H.K., and Uhlmann, D.R. (1975). *Introduction to Ceramics*. Wiley.

Komarneni, S., Breval, E., Jennings, J.S., Macmillan, N.J., and Lunghofer, E.P. (1987). Mechanisms of environmental degradation of proppants. *Journal of Materials Science Letters* 6: 263–266.

Kschinka, B.A., Perrella, S., Nguyen, H., and Bradt, R.C. (1986). Strengths of glass spheres in compression. *Journal of the American Ceramic Society* 69: 467–472.

Liang, F., Sayed, M., Al-Muntasheri, G.A., Chang, F.F., and Li, L. (2016). A comprehensive review on proppant technologies. *Petroleum* 2: 26–39.

Luscher, W.G., Hellmann, J.R., Segall, A.E., Shelleman, D.L., and Scheetz, B.E. (2007). A critical review of the diametral compression method for determining the tensile strength of spherical aggregates. *Journal of Testing and Evaluation* 35: 624–629.

Mocciaro, A., Lombardi, M.B., and Scian, A.N. (2018). Effect of raw material milling on ceramic proppants properties. *Applied Clay Science* 153: 90–94.

Patra, S., Kumar, A., and Rayasam, V. (2017). The effect of particle size on green pellet properties of iron ore fines. *Journal of Mining and Metallurgy A: Mining* 53: 31-41.

Rozenblat, Y., Portnikov, D., Levy, A., Kalman, H., Aman, S., and Tomas, J. (2011). Strength distribution of particles under compression. *Powder Technology* 208: 215–224.

Shipway, P.H. and Hutchings, I.M. (1993). Attrition of brittle spheres by fracture under compression and impact loading. *Powder Technology* 76: 23–30.

Sun, J. (2019). Hard particle force in a soft fracture. *Scientific Reports* 9: 3065.

Tanaka, H., Fukai, S., Uchida, N., Uematsu, K., Sakamoto, A., and Nagao, Y. (1994). Effect of moisture on the structure and fracture strength of ceramic green bodies. *Journal of the American Ceramic Society* 77: 3077–3080.

Tavares, L.M. and King, R.P. (1998). Single-particle fracture under impact loading. *International Journal of Mineral Processing* 54: 1–28.

Vallet, D. and Charmet, J.C. (1995). Mechanical behaviour of brittle cement grains. *Journal of Materials Science* 30: 2962–2967.

Volk, L.J., Raible, C.J., Carroll, H.B., and Spears, J.S. (1981). Embedment of high strength proppant into low-permeability reservoir rock. In SPE/DOE *Low Permeability Gas Reservoirs Symposium*. OnePetro.

Wang, J. and Elsworth, D. (2018). Role of proppant distribution on the evolution of hydraulic fracture conductivity. *Journal of Petroleum Science and Engineering* 166: 249–262.

Watkins, I.G. and Prado, M. (2015). Mechanical properties of glass microspheres. *Procedia Materials Science* 8: 1057–1065.

Wong, J.Y., Laurich-McIntyre, S.E., Khaund, A.K., and Bradt, R.C. (1987). Strengths of green and fired spherical aluminosilicate aggregates. *Journal of the American Ceramic Society* 70: 785–791.

9

Deterministic Scaling of Particle Strength Distributions

This chapter uses the analysis of Chapter 4 to examine in detail the strength distributions of particle systems that demonstrate deterministic size effects. Such systems comprise the majority of particle systems at all sizes and exhibit predominantly concave strength distributions, although sigmoidal and linear distributions are also observed. Many of the particle materials that exhibited stochastic size effects considered in Chapter 7 are represented here. Analyses are developed to describe and predict linear and concave deterministic strength distributions as functions of particle size. The analyses are based on forward and reverse application of the probabilistic framework developed in Chapter 3 and include extreme value size effects and particle size dependent flaw populations.

9.1 Introduction

Particle strength edf $Pr(\sigma)$ variations are usually concave and wide, as shown in the extensive surveys of Chapters 6, 7, and 8, and the few earlier examples in Chapter 2. Analyses within the probabilistic fracture framework developed in Chapters 3 and 4 showed that the observed $Pr(\sigma)$ behavior is a consequence of underlying strength-controlling flaw distributions $h(c)$ that are predominantly heavy tailed with weakly defined peaks. In samples of particles from a material, the strength-controlling flaws form extreme value distributions based on the ensembles of flaws within each particle. In turn, the ensembles are formed by selection from a flaw population f. In Chapter 7, the flaw population was considered an invariant function of crack length c alone, $f(c)$, characterizing the overall material. The invariance lead to stochastic size effects, in which particle size D determined only the probability that a particular flaw size selected from the population appeared within an ensemble, and thus determined the extreme value distribution. The extreme value distribution of interest was that of the largest cracks in each ensemble, as this determined the strength distribution of the sample. In contrast, the flaw population can be considered a function of crack length and particle size, $f(c, D)$, characterizing the particular *particle size* sampled from the material. Thus, the flaw population is not an invariant material property, but an extensive sample property that depends on particle size. The dependence leads to *deterministic* size effects, in which particle size D determines both the probability that a particular flaw size selected from the population appears within an ensemble of flaws in a particle and the population. Hence, the extreme value distribution of largest flaws in a sample and thence the strength distribution of a sample exhibits a much greater particle size effect. In particular, the material constraints on the domains, including the lower bound thresholds, and shapes of strength distributions are greatly weakened by deterministic effects. This Chapter surveys and investigates in a quantitative manner deterministic size effects in strength behavior of particle systems. The chapter is analogous to Section 5.3.2 in Chapter 5 that investigated deterministic size effects in extended components.

A pictorial, qualitative description of deterministic size effects in particles is shown in Figure 9.1. The description is similar to those representing size and geometry effects in extended components and particles in Figures 5.29–5.31 and Figure 7.1, and schematizes the quantitative analyses of flaw distributions developed in Chapter 3. In particular, Figure 9.1 represents a deterministic size effect in which the crack density λ is invariant and the crack population domain upper bound increases with particle size. In current notation, this effect corresponds to both component size k and maximum crack length c_{max} increasing with D, and $f(c)$ altering accordingly. For ease of visualization, detailed variations in c are not shown in Figure 9.1 and increases in c_{max} are represented simply as populations of small, medium, and large cracks. Figure 9.1a is a schematic diagram of a planar particle loaded in diametral compression, similar to Figure 7.1a. The particle consists of 9 fundamental volume elements and the compressive loading, indicated by arrows, leads to a localized zone of

Particle Strengths: Extreme Value Distributions in Fracture, First Edition. Robert F. Cook.

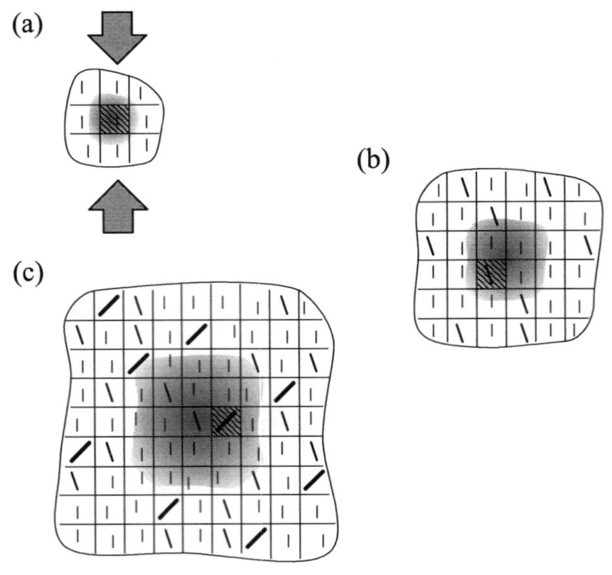

Figure 9.1 Schematic diagrams illustrating deterministic extreme value effects in brittle failure of particles in diametral compression. Flaw population alters with particle size. Particle diameter (a) $D = 3$, (b) $D = 6$, (c) $D = 9$. Applied loading is indicated by arrows in (a). Tensile zones indicated by shading. Flaws indicated by bold lines; strength-controlling flaws responsible for failure indicated hatched. Note that in (a) failure occurs at center, independent of flaw size, $k = 1$ and in (b) $k = 4$ and (c) $k = 9$, failure occurs at largest flaws, independent of location in tensile zone.

tensile stress at the center of the particle, indicated by shading. Each fundamental volume element contains one flaw, shown as a fine line representing a "base" small flaw population. In the particle of Figure 9.1a, only the central volume element and flaw experience the maximum tensile stress and thus $k = 1$. Failure thus ensues from the central flaw, indicated by a hatched element, within the ensemble of flaws contained by the particle. The flaw sizes at all locations, including the centers, in all similarly sized particles are determined by the probability distribution of flaw sizes for that particle size. Hence, the distribution of strengths for a sample of like particles is thus unbiased by extreme value effects and reflects the entire distribution of flaws.

Figures 9.1b and 9.1c are schematic diagrams of larger particles loaded in diametral compression, consisting of 36 and 81 fundamental volume elements, respectively. The volume elements are invariant in size, and, again, each fundamental volume element contains one flaw. In these cases, however, the flaw population alters with particle size so as to include larger cracks as the particle size increases. The larger cracks are shown as inclined bolder lines. In Figure 9.1b approximately one third of the additional cracks are medium sized, and in Figure 9.1c approximately one fifth of the additional cracks are large. Relative to Figure 9.1a, in Figures 9.1b and 9.1c $c_{\max}$ is increased and $f(c)$ has been altered by an extended large crack tail. Diametral compression loading leads to localized zones of tensile stress at the centers of each particle that scale with particle size (about one third) and are indicated by shading. In the particle of Figure 9.1b, the central four volume elements and flaws experience the maximum tensile stress and thus $k = 4$ and in the particle of Figure 9.1c the central nine volume elements and flaws experience the maximum tensile stress and thus $k = 9$. For the example alterations to the flaw populations, stochastic selection is extremely likely to include a medium flaw in the tensile zone of the particle in Figure 9.1b and a large flaw in the tensile zone of the particle in Figure 9.1c, and both cases of flaws are shown. Failure thus ensues from these flaws, indicated by hatched elements, within the ensembles of flaws contained by the particles. The distributions of strengths for samples of these larger particles are biased by extreme value effects—more so for the $k = 9$ particles than for the $k = 4$ particles—toward the medium or large flaw lower bound strength thresholds. The comparison of Figures 9.1b and 9.1c is analogous to the comparison between Figure 5.29 and a combination of Figure 5.30 (increased k) and Figure 5.31 (increased $c_{\max}$). In experimental terms, particles described by Figure 9.1 are expected to exhibit smaller strength thresholds and smaller strength dispersions as particle size increases.

A pictorial description of an alternative deterministic size effect in particles is shown in Figure 9.2. The description is similar to that representing size and geometry effects in extended components in Figure 5.32 and the major effect is one of uniform enlargement. In particular, Figure 9.2 represents a deterministic size effect in which the crack density λ decreases and the crack population domain upper bound increases with particle size. In current notation, this effect corresponds to maximum crack length $c_{\max}$ increasing with D and $f(c)$ altering accordingly, but component size k increasingly moderately or not at all. For ease of visualization, variations in $f(c)$ are not shown in Figure 9.2 and increases in $c_{\max}$ are represented by small, medium, and large cracks. Figure 9.2a is a schematic diagram of a planar particle loaded in diametral compression, identical to Figure 9.1a. Figures 9.2b and 9.2c are schematic diagrams of larger particles loaded in diametral compression, each consisting of nine fundamental volume elements that vary in size with particle size and contain one flaw. The flaw

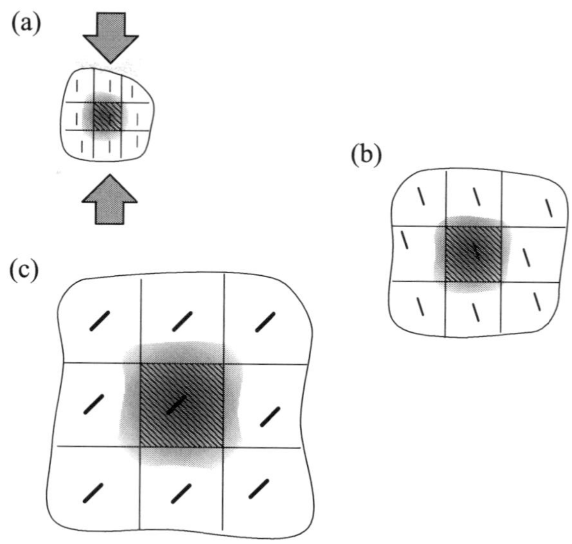

Figure 9.2 Schematic diagrams illustrating deterministic extreme value effects in brittle failure of particles in diametral compression. Flaw population and flaw spatial density alter with particle size, *cf* Figure 9.1. Particle diameter (a) $D = 3$, (b) $D = 6$, (c) $D = 9$. Applied loading is indicated by arrows in (a). Tensile zones indicated by shading. Flaws indicated by bold lines; strength-controlling flaws responsible for failure indicated hatched. Note that failure occurs at center, independent of flaw size, $k = 1$.

population alters with particle size so as to be based on larger cracks as the particle size increases. In Figure 9.1b all the cracks are medium sized and in Figure 9.1c all the cracks are large. Relative to Figure 9.1a, in Figures 9.1b and 9.1c c_{max} is increased and $f(c)$ may have been altered in shape. Diametral compression loading leads to localized zones of tensile stress at the centers of each particle that scale with particle size (about one third) and are indicated by shading. In this scheme, the central volume element and flaw in each particle experiences the maximum tensile stress and thus $k = 1$ in each case. Failure thus ensues from the central flaw, indicated by a hatched element as described above. The flaw sizes at all locations in all similarly sized particles, including the centers, are determined by the probability distribution of flaw sizes for that particle size. Hence, the distribution of strengths for a sample of like particles is thus unbiased by extreme value effects and reflects the entire distribution of flaws. The comparison of Figures 9.2a, 9.2b, and 9.2c is analogous to the comparison between Figures 5.29 and 5.32 (increased c_{max}, invariant $k = 1$). In experimental terms, particles described by Figure 9.2 are expected to exhibit smaller strength thresholds. If the relative shapes and domains of flaw populations are invariant with particle size, strength dispersions will decrease as particle size increases. In experimental terms, particle systems described by Figures 9.1 and 9.2 would be very difficult to distinguish by measurements of strength central tendencies (e.g. mean, variance). Such distinction requires examination of strength distributions exhibited by particles of different sizes.

The necessary condition for strengths σ of particle systems to be described as exhibiting stochastic size effects is that the strength distributions $H(\sigma)$ can be related by

$$H_2(\sigma) = 1 - [1 - H_1(\sigma)]^{k_2/k_1}, \tag{9.1}$$

as used earlier (Eq. (3.36), Eq. (5.1), Eq. (7.1)). $H_2(\sigma)$ and $H_1(\sigma)$ are the strength edf variations for samples of particles of two different sizes, D_2 and D_1, and the exponent k_2/k_1 is the ratio of the numbers of stressed flaws in the different sized particles. Implicit in Eq. (9.1) is that the domain of strengths $\sigma_L \leq \sigma \leq \sigma_U$ is identical for $H_2(\sigma)$ and $H_1(\sigma)$, although it is sufficient that Eq. (9.1) pertains and not necessary that samples of particles D_2 and D_1 exhibit strengths that fill the domain. This point is particularly so for $H(\sigma)$ variations that exhibit $\mathrm{d}H(\sigma)/\mathrm{d}\sigma \to 0$ at domain bounds. For concave $H(\sigma)$ variations, typical for particles, this is not an issue at the lower bound threshold strengths, as observed sample lower bounds σ_1 are usually comparable to fit lower bounds σ_L. As a consequence, it is usually a matter of simple inspection to reject a stochastic size effect description if the lower bounds of a series of strength variations, particularly concave variations, do not align. If the domains of the strength variations do align, Eq. (9.1) is a simple test that is usually also a matter of simple inspection to reject a stochastic size effect description (e.g. Figure 4.10). These criteria were used here (and earlier, e.g. Figure 8.12) to identify particle systems to be described as exhibiting deterministic size effects. Note that the internal system constraint of Eq. (7.2) applies to interpretation of both stochastic and deterministic particle strength distributions in terms of underlying crack lengths—crack length must be less than particle size. Analysis of strength-crack length relations for deterministic systems is considered in detail in Section 9.5, and develops an alternative to Eq. (9.1) based on the ideas of Eq. (8.1).

An example of deterministic size effects on strength behavior is exhibited in the strength edf $Pr(\sigma)$ plot for samples of limestone particles, Figure 9.3. The particles were approximately 2.4 mm to 63 mm in diameter, tested in a drop weight configuration in the work of Barrios et al. (2011) and the plot was constructed using data derived from the published work.

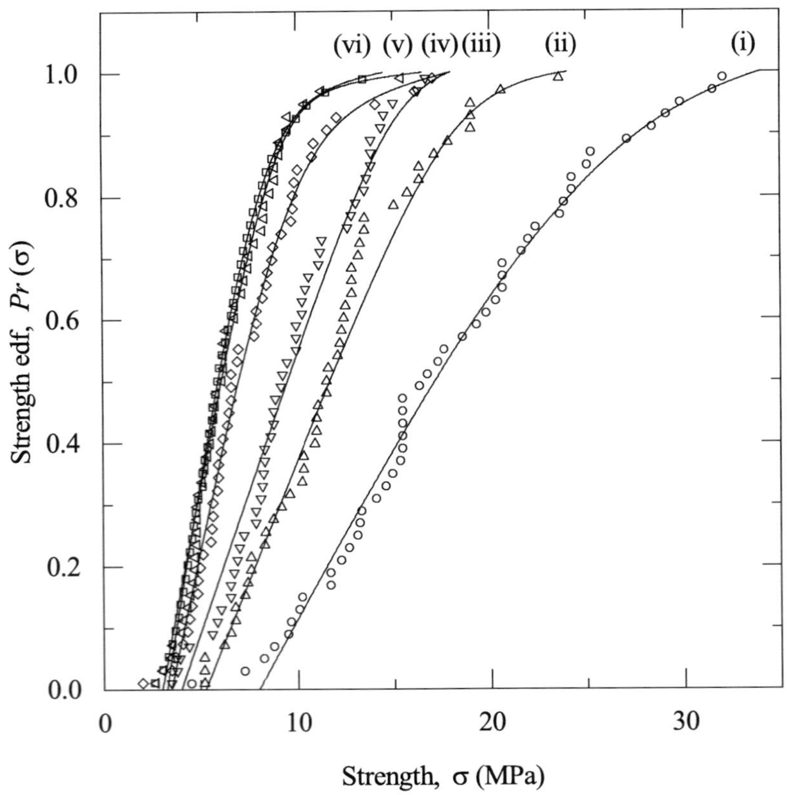

Figure 9.3 Plot of strength edf behavior, $Pr(\sigma)$, for limestone particles. Particle diameter, D = (2.36–2.83, 4.75–5.6, 9.5–11.2, 19.2–22.4, 37.5–45.5, and 53–63) mm, labeled (i)–(vi), number of measurements, N_{tot} = 293. (Adapted from Barrios G.K.P et al., 2011). Solid lines show independent concave best fits. The separated lower bound threshold strengths are indicative of deterministic size effects.

The results were originally reported in terms of specific failure energy and were motivated by considerations of particle confinement on comminution efficiency in materials processing. Samples of increasing particle size are labeled (i)–(vi). Symbols represent individual strength measurements; different symbols represent different particle sizes, D. Comparison with survey observations (Chapter 6) shows that both the shape and the extent of the strength responses are typical for particles. (As earlier, details of component and sample sizes are provided in the figure caption.) Lines represent independent visual best fits to the data using the bi-linear concave description developed earlier (Chapter 4). A clear feature of Figure 9.3 is that the data do not exhibit a common lower bound threshold strength. The threshold strength decreases with increasing particle size. The relative strength dispersions are approximately invariant with particle size, exhibiting strength domain widths σ_N/σ_1 of approximately 6. The strength responses thus appear similar, superficially, to those of stochastic systems (Chapter 7).

Figure 9.4 shows as solid lines the crack length pdf $h(c)$ variations underlying the limestone particle strength behavior. The $h(c)$ variations were determined using $B = 0.4$ MPa m$^{1/2}$ and particle size, Eq. (7.2), as a constraint. Comparison with the earlier survey (Chapter 6) and consideration of particles influenced by stochastic size effects (Chapter 7) shows that the tail-dominated shapes and wide extents are typical for particles but that the peaks are more well-defined than is usually observed. There is, however, a critical difference in interpretation of the $h(c)$ responses in Figure 9.4 and those presented earlier for stochastic particle systems. The solid lines in Figure 9.4 represent the full strength-controlling crack length distributions underlying the observed strength behavior of Figure 9.3. No truncation effects are overlaid as were required for stochastic systems (represented by grayed lines in Chapter 7). In stochastic systems, the strength responses for different particle sizes are not independent (there is an identical strength domain). Truncation of responses within the single inferred crack length domain is thus required to account for experimental strength observations. In deterministic systems, the strength responses for different particle sizes are independent (there are separate domains) and truncation of inferred crack length responses is not required. Hence, the entirety of the small-crack responses in Figure 9.4 reflects the active strength-controlling crack lengths for the large strength particles in each sample. Similarly, the large-crack behavior of the responses in Figure 9.4 reflects the variable strength-controlling crack lengths in the small strength particles in each sample, as indicated by the increasing crack length upper bounds with increasing particle size.

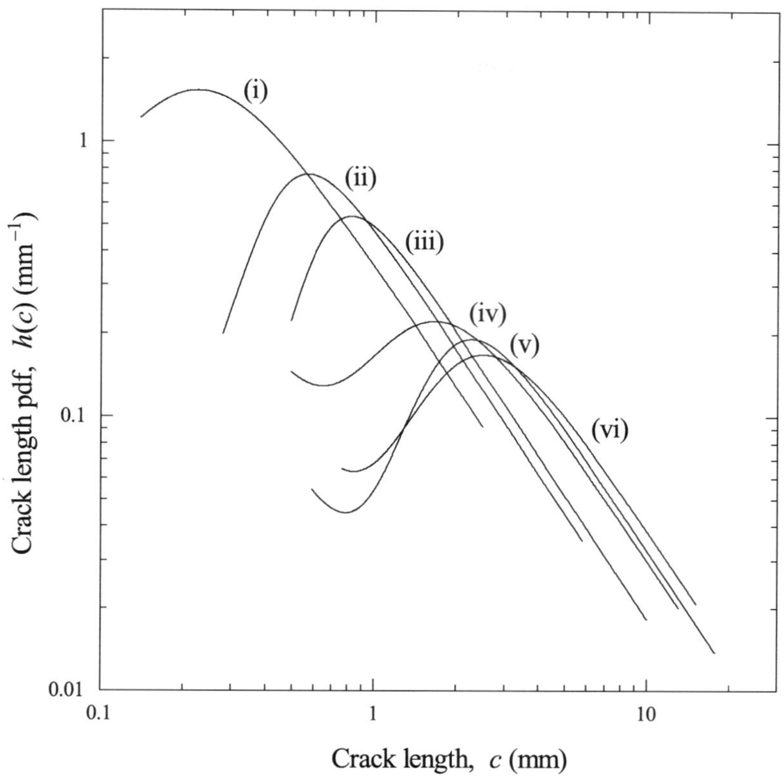

Figure 9.4 Plot in logarithmic coordinates of crack length pdf $h(c)$ variations for limestone particles. D = 2.4 mm–63 mm, increasing particle size responses labeled (i)–(vi); determined from Figure 9.3.

The presentation formats of Figures 9.3 and 9.4 will be used throughout this chapter in surveying deterministic extreme value size effects in strength behavior of particles. The format is similar to that used in surveying stochastic effects in Chapter 7. As in Chapter 7, the survey here is divided by the shape of the observed particle strength distributions, and then by material. In this case, the shapes are concave (the most common), sigmoidal, and linear (rare). In the limestone example above, it is noted that the maximum crack length range was approximately 6 and the particle sizes range was approximately 22. A goal of the survey is to use observations such as this to provide an experimental assessment of the scaling relation $c_{max}(D)$ for particles.

9.2 Concave Deterministic Distributions

9.2.1 Alumina

Figure 9.5 shows strength edf $Pr(\sigma)$ plots for samples of alumina (polycrystalline Al_2O_3) particles 0.7–2.6 mm in diameter, from the work of Huang et al. (1995) (see also Huang et al. 1993), using data derived from the published work. The motivation for the work was to prolong the useful life of intact abrasive particles used in industrial grinding applications. The particles were not natural ores or minerals, but particles engineered by heat treatment of mineral powders, based primarily on corundum (Al_2O_3, the main chemical constituent of bauxite). Huang et al. examined several abrasive particles, mostly Al_2O_3 based, but including SiC, and measured the strength distributions using a counter-rotating roller crushing apparatus. Figure 9.5 shows the strength behavior of a single grade of Al_2O_3 ("brown") in four different particle sizes. Symbols represent individual strength measurements. Different symbols represent different particle sizes, D; increasing particle size is labelled (i)–(iv). The data are well separated by particle size and exhibit wide strength domains σ_N/σ_1 of approximately 10. As above, a clear feature of Figure 9.5 is that the data exhibit threshold strengths that decrease with increasing particle size.

The works on alumina-based particles of Bertrand et al. (1988) (Chapter 7), Wong et al. (1987) (Chapter 8), and Huang et al. (1995) (here) illustrate a sequence of decreasing constraint on strength distributions. The sequence reflects underlying flaw populations that are increasingly unrelated. The measurements of Bertrand et al. demonstrated stochastic size effects

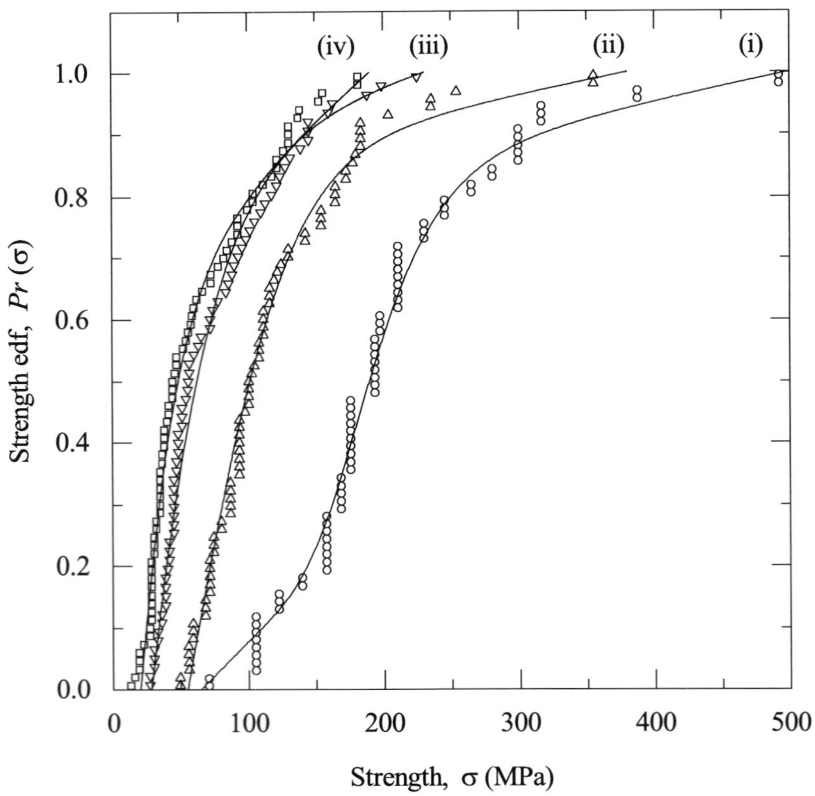

Figure 9.5 Plot of strength edf behavior, $Pr(\sigma)$, for alumina particles. $D = (0.7 \pm 0.15, 1.29 \pm 0.15, 1.85 \pm 0.25, 2.58 \pm 0.30)$ mm (uncertainties estimated), $N_{tot} = 583$ (Adapted from Huang, H et al., 1995). Solid lines show independent best fits. The separated lower bound threshold strengths are indicative of deterministic size effects.

on strength, in which the strength distributions for particles of different sizes were related by Eq. (9.1). The measurements of Wong et al. demonstrated constrained deterministic size effects on strength, in which the strength distributions for particles of different sizes were related by a common upper bound and truncation of a common flaw population. The measurements of Huang et al. demonstrate unconstrained deterministic size effects on strength, in which the strength distributions for particles of different sizes are freely determined. The underlying flaw populations thus evolve in sequence from a single population, to a set of perturbations of a single population, to multiple populations. The solid lines in Figure 9.5 thus represent unconstrained visual best fits to the data consistent with multiple flaw populations. The data are fit by a tri-linear sigmoidal function for the smallest particles and bi-linear concave functions for the larger particles (slight sigmoidal effects can also be observed in the smallest particle strength data of Figure 9.3). Deterministic size effects on strength are thus observed here as variations in both strength distribution domain and shape.

Figure 9.6 shows as solid lines the crack length pdf $h(c)$ variations underlying the alumina particle strength behaviors. The $h(c)$ variations were determined using $B = 1.0$ MPa m$^{1/2}$ and particle size and Eq. (7.2) as a constraint. Increasing particle size variations are labelled (i)–(iv). Consistent with deterministic effects, the crack length distributions vary in domain location and distribution shape as particle size changes. In particular, as particle size increases the crack length domains shift to larger crack lengths and the crack length distributions alter from a clearly peaked function (i) to a decreasing function with a weak inflection (iv). The decreasing amplitude peak, at small crack lengths, is associated with an increasing length tail at large crack lengths. In Figure 9.4 the predominant deterministic size effect appeared to be a shift of crack length distributions to larger crack lengths as particle size increased. Here in Figure 9.6 the crack length distribution is also shifted but the predominant deterministic size effects appears to be a suppression in crack length distribution peak as particle size increased.

9.2.2 Quartz

Figure 9.7 shows strength edf $Pr(\sigma)$ plots for samples of quartz (crystalline SiO_2) particles 0.25 mm–4.75 mm in diameter, from the work of Tavares and King (1998), using data derived from the published work. As above, the motivation for the

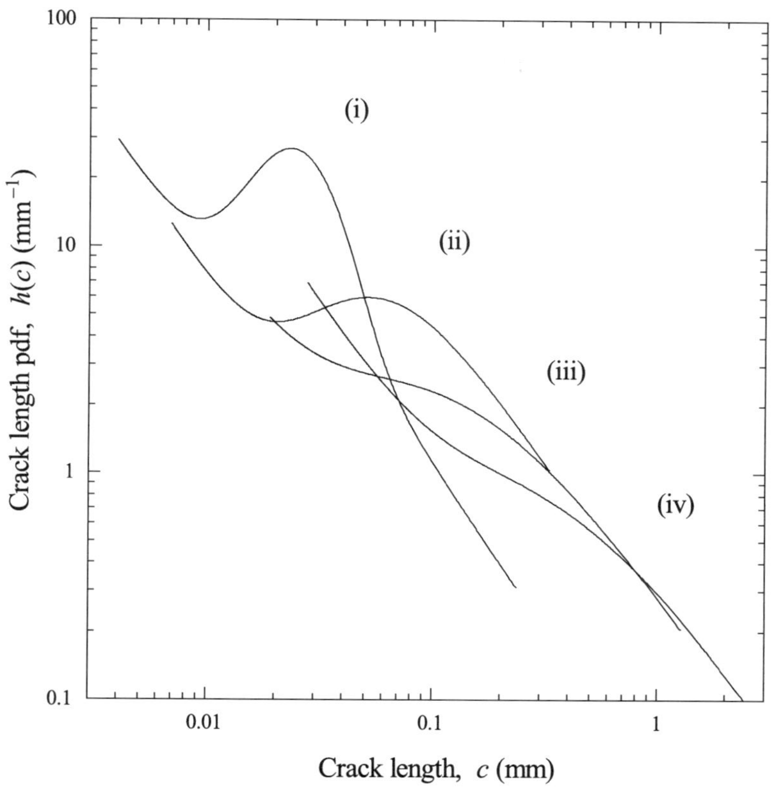

Figure 9.6 Plot in logarithmic coordinates of crack length pdf $h(c)$ variations for alumina particles. D = 0.7 mm–2.6 mm, increasing particle size responses labeled (i)–(iv); determined from Figure 9.5.

work was to improve comminution efficiency in materials processing. The strength distributions were determined in a drop weight configuration and originally reported in terms of specific failure energy. Symbols represent individual strength measurements. Different symbols represent different particle sizes, D; increasing particle size is labelled (i)–(v). The data are well separated by particle size and exhibit moderately wide strength domains σ_N/σ_1 of approximately 4. A particularly clear feature of Figure 9.7 is that the data exhibit threshold strengths that decrease with increasing particle size. Qualitatively, the strength responses are similar to those of alumina seen earlier in Figure 9.5. The solid lines in Figure 9.7 represent unconstrained visual best fits to the data consistent with multiple deterministic flaw populations. The data are fit by a tri-linear sigmoidal function for the smallest particles and bi-linear concave functions for the larger particles (as above, sigmoidal effects are observed in the smallest particle strength data). Deterministic size effects on strength are thus observed here as variations in both strength distribution domain and shape.

Figure 9.8 shows as solid lines the crack length pdf $h(c)$ variations underlying the quartz particle strength behaviors. The $h(c)$ variations were determined using $B = 0.45$ MPa m$^{1/2}$ and particle size and Eq. (7.2) as a constraint. Increasing particle size variations are labeled (i)–(v). Consistent with deterministic effects, the crack length distributions vary in domain location and distribution shape as particle size changes, although the distributions remain peaked as observed for limestone particles (Figure 9.3). In particular, as particle size increases the crack length domains shift to larger crack lengths and contract slightly so as to exhibit a decreasing length tail at large crack lengths.

Two further examples of deterministic size effects on strength behavior of quartz particles are exhibited in the strength edf $Pr(\sigma)$ and crack length pdf $h(c)$ plots of Figures 9.9 and 9.10. In both cases strengths were measured in conventional compression apparatus. The works were motivated by the geological importance of particle fracture during compaction of sandstones and sands and the compaction of particle aggregates in geotechnical applications. Figure 9.9a shows strength edf $Pr(\sigma)$ plots for samples of quartz particles 0.3–2 mm in diameter, from the work of McDowell (2002), using data derived from the published work. Symbols represent individual strength measurements. Different symbols represent different particle sizes, D; increasing particle size is labeled (i)–(iii). The data are well separated by particle size and exhibit moderately wide strength domains σ_N/σ_1 of approximately 5. A clear feature of Figure 9.9a is that the data exhibit threshold strengths that decrease with increasing particle size. Qualitatively, the strength responses are similar to those of limestone seen in Figure 9.3. The solid lines in Figure 9.9a represent unconstrained visual best fits to the data consistent with multiple deterministic flaw populations. The data are fit by bi-linear concave functions. Deterministic size effects on strength are thus

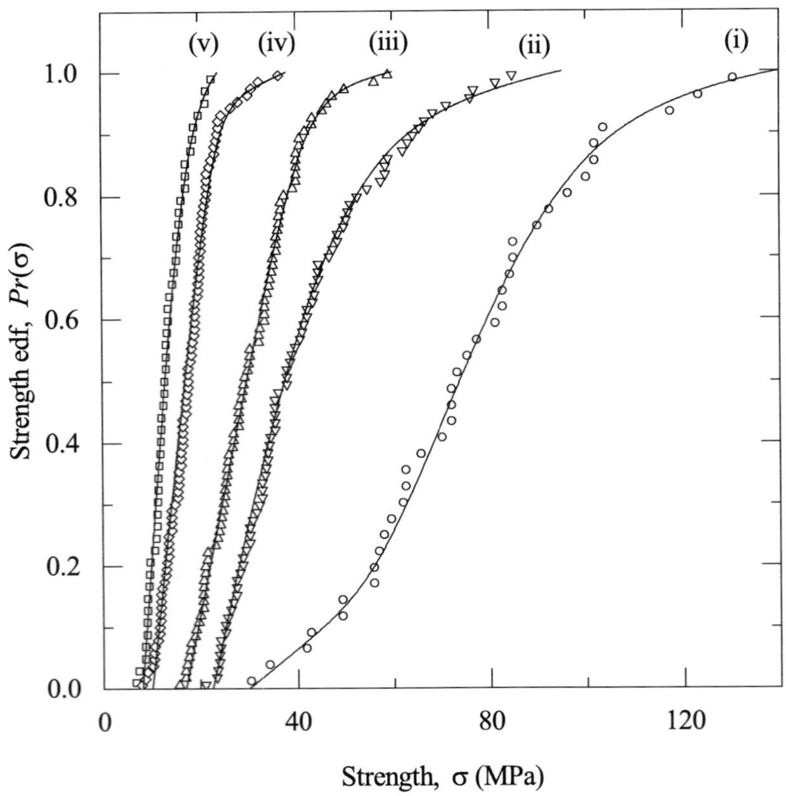

Figure 9.7 Plot of strength edf behavior, $Pr(\sigma)$, for quartz particles. D = (0.25–0.35, 0.5–0.7, 1–1.18, 2–2.8, and 4–4.75) mm, N_{tot} = 354 (adapted from Tavares, L.M et al. 1998). Solid lines show independent best fits. The separated lower bound threshold strengths are indicative of deterministic size effects.

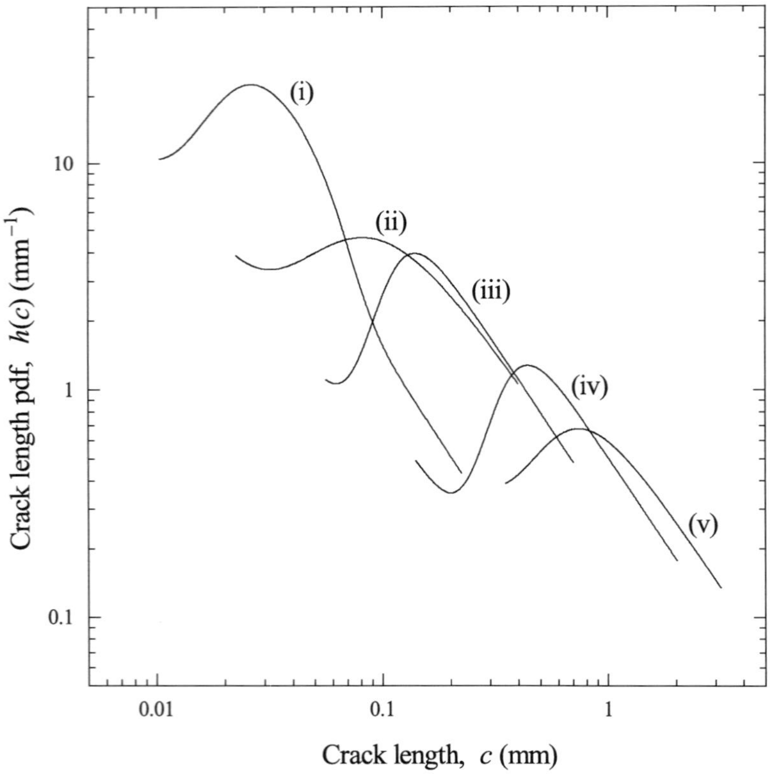

Figure 9.8 Plot in logarithmic coordinates of crack length pdf $h(c)$ variations for quartz particles. D = 0.25 mm–4.75 mm, increasing particle size responses labelled (i)–(v); determined from Figure 9.7.

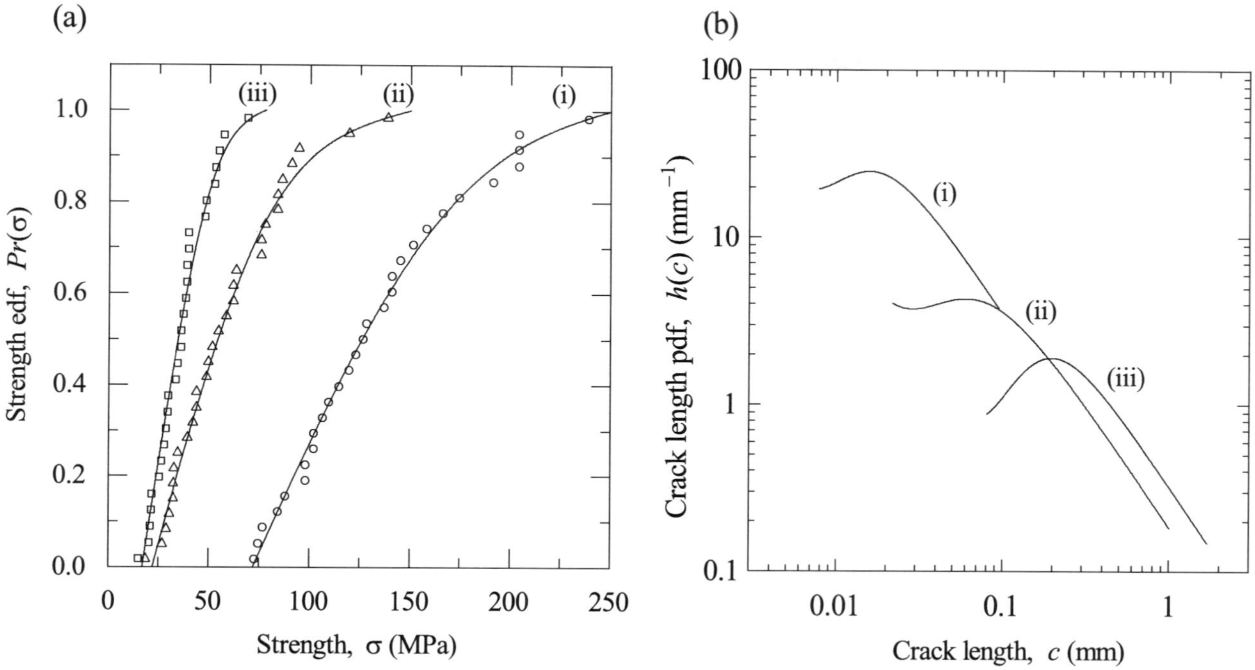

Figure 9.9 (a) Plot of strength edf behavior, $Pr(\sigma)$, for quartz particles. D = (0.3–0.6, 0.6–1.18, and 1.18–2.00) mm, N_{tot} = 87 (adapted from McDowell, G.R 2002). Solid lines show independent best fits. The separated lower bound threshold strengths are indicative of deterministic size effects. (b) Plot in logarithmic coordinates of crack length pdf $h(c)$ variations for quartz particles. Increasing particle size responses labeled (i)–(iii); determined from (a).

observed here as variations in the strength distribution domain. Figure 9.9b shows as solid lines the crack length pdf $h(c)$ variations underlying the quartz particle strength behavior. The $h(c)$ variations were determined using $B = 0.70$ MPa m$^{1/2}$ and particle size and Eq. (7.2) as a constraint. Increasing particle size variations are labeled (i)–(iii). Consistent with deterministic effects, the crack length distributions vary in domain location and in this case the distributions are all weakly peaked. In particular, as particle size increases the crack length domains shift to larger crack lengths. Similar domain shift behavior has been observed for related sandstone particles (Xiao et al. 2020), albeit for larger particles, (2.5–10) mm, and smaller strengths, (1–12) MPa.

Figure 9.10 shows in material context and in more detail, the strength and crack length responses of the small quartz sand particles of Figures 6.15b and 6.16, from the work of Brzesowsky et al. (2011), using data derived from the published work. Figure 9.10a shows strength edf $Pr(\sigma)$ plots for particles 115–378 μm in diameter. Symbols represent individual strength measurements. Different symbols represent different particle sizes, D; increasing particle size is labelled (i)–(iv). The data are very tightly clustered, despite a factor of over 3 in particle size, and exhibit narrow strength domains σ_N/σ_1 of approximately 4; the threshold strengths decrease similarly with increasing particle size. Qualitatively, the strength responses are somewhat similar to those of limestone as in Figure 9.3. The solid lines in Figure 9.10a represent unconstrained visual best fits to the data consistent with multiple deterministic flaw populations. The data are fit by bi-linear concave functions. Deterministic size effects on strength are thus observed here as variations in the strength distribution domain. Figure 9.10b shows as solid lines the crack length pdf $h(c)$ variations underlying the quartz sand particle strength behavior (note change in crack length scale). The $h(c)$ variations were determined using $B = 0.90$ MPa m$^{1/2}$ and particle size and Eq. (7.2) as a constraint. Increasing particle size variations are labeled (i)–(iv). Consistent with deterministic effects and the strength clustering, the crack length distributions vary weakly in domain location; the distributions are all peaked. In particular, as particle size increases the crack length domains shift to larger crack lengths.

9.2.3 Salt

Figure 9.11 shows strength edf $Pr(\sigma)$ plots for samples of salt (crystalline NaCl) particles 0.71 mm–4.00 mm in diameter, from the work of Rozenblat et al. 2011), using data derived from the published work. The motivation for the work was

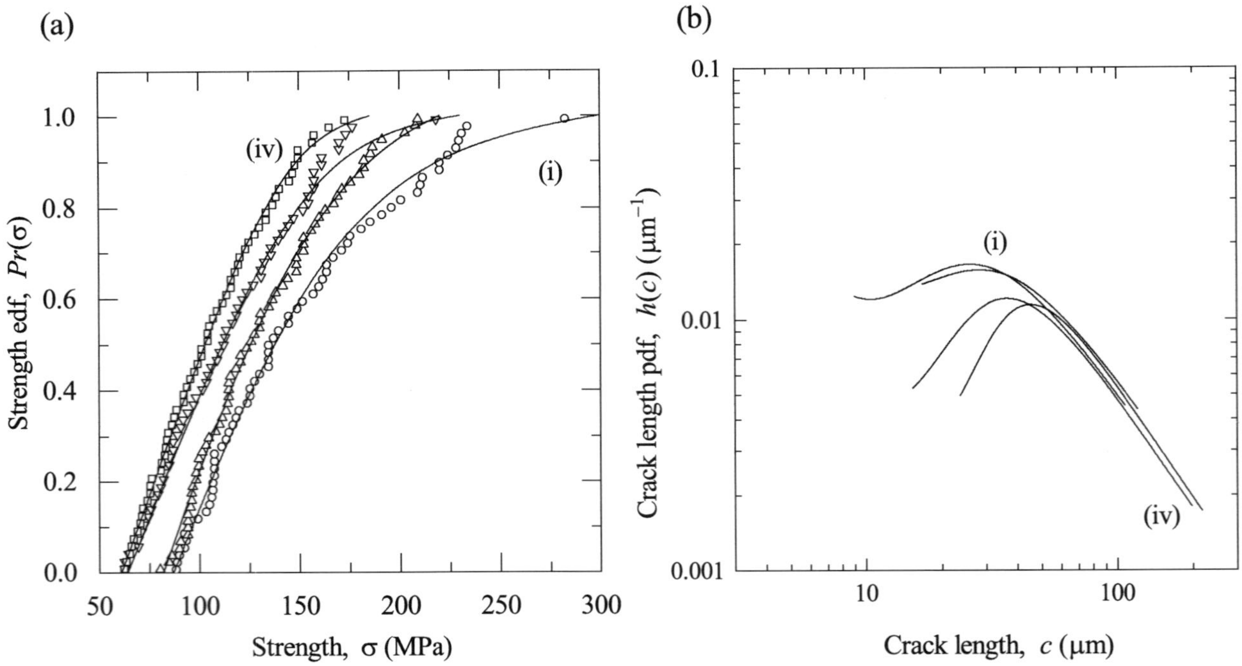

Figure 9.10 (a) Plot of strength edf behavior, $Pr(\sigma)$, for silica sand particles. $D = (115 \pm 9, 196 \pm 16, 275 \pm 25,$ and $378 \pm 22)$ μm, $N_{tot} = 262$ (Adapted from Brzesowsky, R.H et al., 2011). Solid lines show independent best fits. The separated lower bound threshold strengths are indicative of deterministic size effects. (b) Plot in logarithmic coordinates of crack length pdf $h(c)$ variations for sand particles. Increasing particle size responses labeled (i)–(iv); determined from (a).

to improve comminution efficiency in materials processing and strength distributions were determined in a conventional compression apparatus. Symbols represent individual strength measurements. Different symbols represent different particle sizes, D; increasing particle size is labeled (i)–(vi). The data, although compact, are well separated by particle size and exhibit moderately wide strength domains σ_N/σ_1 of approximately 6. A clear feature of Figure 9.11 is the data exhibit threshold strengths that decrease with increasing particle size. Qualitatively, the strength responses are similar to those of limestone seen in Figure 9.3. The solid lines in Figure 9.11 represent unconstrained visual best fits to the data consistent with multiple deterministic flaw populations. The data are fit by bi-linear concave functions. Deterministic size effects on strength are thus observed here as variations in strength distribution domain.

Figure 9.12 shows as solid lines the crack length pdf $h(c)$ variations underlying the salt particle strength behaviors. The $h(c)$ variations were determined using $B = 0.08$ MPa m$^{1/2}$ and particle size and Eq. (7.2) as a constraint. Increasing particle size variations are labeled (i)–(vi). Consistent with deterministic effects, the crack length distributions vary in domain location and distribution shape as particle size changes, although the distributions remain weakly peaked as observed for some quartz particles (Figure 9.9b). In particular, as particle size increases, the crack length domains shift to larger crack lengths and contract slightly so as to exhibit a decreasing length tail at large crack lengths.

9.2.4 Rock

Strength distribution studies of "large" rock particles, with diameters predominantly in the domain $D > 10$ mm, although infrequent, provide clear examples of material and size effects. Material effects were demonstrated in the survey of Chapter 6, Figure 6.18, in which the behavior of various large rock particles used in geotechnical applications (Ovalle et al. 2014) and in railway ballast applications (Kohhmishi and Palassi 2016) was considered. Stochastic size effects were demonstrated in Chapter 7, Figures 7.7–7.9, 7.12, 7.13, 7.16, and 7.17 in which the behavior of large limestone, coral, and quartzite particles in geotechnical applications (Hu 2011; Ovalle et al. 2014; Shen et al. 2020) was considered. Here, deterministic size effects are demonstrated for three large rock particle systems.

In an extensive study considering composition, microstructure, wear, and fracture properties of Brazilian quarry rocks for geotechnical applications, Tavares and das Neves (2008) used a ball-drop impact tester to measure specific failure energy distributions of rock particles of seven different sizes from four different quarries. The focus here is on rocks from two quarries, the Vigné (V series rocks) and the Santa Luzia (SL series rocks). In both cases, some particles were quite large (90

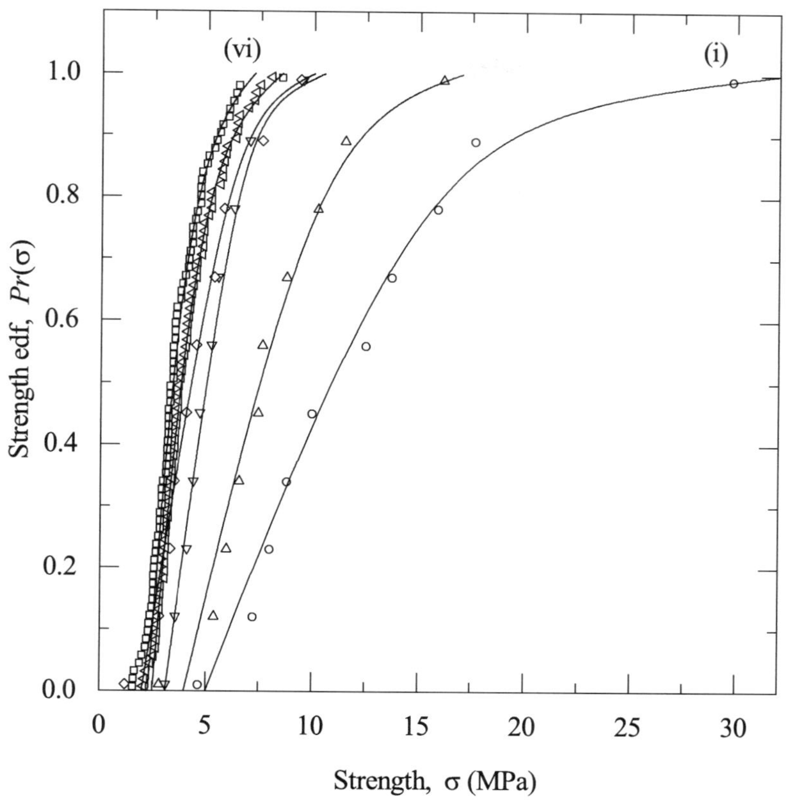

Figure 9.11 Plot of strength edf behavior, $Pr(\sigma)$, for salt particles. D = (0.71–1.00, 1.00–1.40, 1.40–2.00, 2.00–2.36, 2.36–3.35, and 3.35–4.00) mm, N_{tot} = 198 (adapted from Rozenblat, Y et al., 2011). Solid lines show independent best fits. The separated lower bound threshold strengths are indicative of deterministic size effects.

mm) and the size range was greater than a factor of 100. The V series rocks are considered first. Figure 9.13 shows strength edf $Pr(\sigma)$ plots for samples of V series rock particles 0.59–90 mm in diameter, from the work of Tavares and das Neves, using data derived from the published work. Symbols represent individual strength measurements. Different symbols represent different particle sizes, D; increasing particle size is labeled (i)–(vi). The data are well separated by particle size and exhibit narrow strength domains σ_N/σ_1 of approximately 3. A clear feature of Figure 9.13 is that the data exhibit threshold strengths that decrease with increasing particle size, and that the shape of the strength distribution of the smallest particles is distinguished from that of larger sized particles, in this case by exhibiting almost linear behavior. The solid lines in Figure 9.13 represent unconstrained visual best fits to the data consistent with multiple deterministic flaw populations. The data are fit by bi-linear concave functions. Deterministic size effects on strength are thus observed here as variations in both strength distribution domain and shape. Figure 9.14 shows as solid lines the crack length pdf $h(c)$ variations underlying the V series rock particle strength behaviors. The $h(c)$ variations were determined using $B = 0.75$ MPa m$^{1/2}$ and particle size and Eq. (7.2) as a constraint. Increasing particle size variations are labelled (i)–(vi). Consistent with deterministic effects, the crack length distributions vary in domain location and distribution shape as particle size changes. In particular, as particle size increases the crack length domains shift to larger crack lengths, expand slightly, and exhibit more defined peaks in addition to tails at large crack lengths.

Two further examples of deterministic size effects on strength behavior of rock particles are exhibited in the strength edf $Pr(\sigma)$ and crack length pdf $h(c)$ plots of Figures 9.15 and 9.16. In the first example, strengths measurements were motivated by increasing the longevity of rock used as railway ballast. Railway ballast is a layer of aggregated angular rocks beneath railroad tracks, providing elastic foundation, noise absorption, and fluid removal by percolation between the rock particles. Repeated train loading leads to ballast degradation by crushing of ballast particles, with clear performance and economic implications. In an extensive study, Lim et al. (2004) used a conventional loading platen geometry to observe and measure the strength behavior of six different ballast rocks separated into three different size classes. As an example of the results, Figure 9.15a shows strength edf $Pr(\sigma)$ plots for samples of rock particles 10–50 mm in diameter, using data derived from

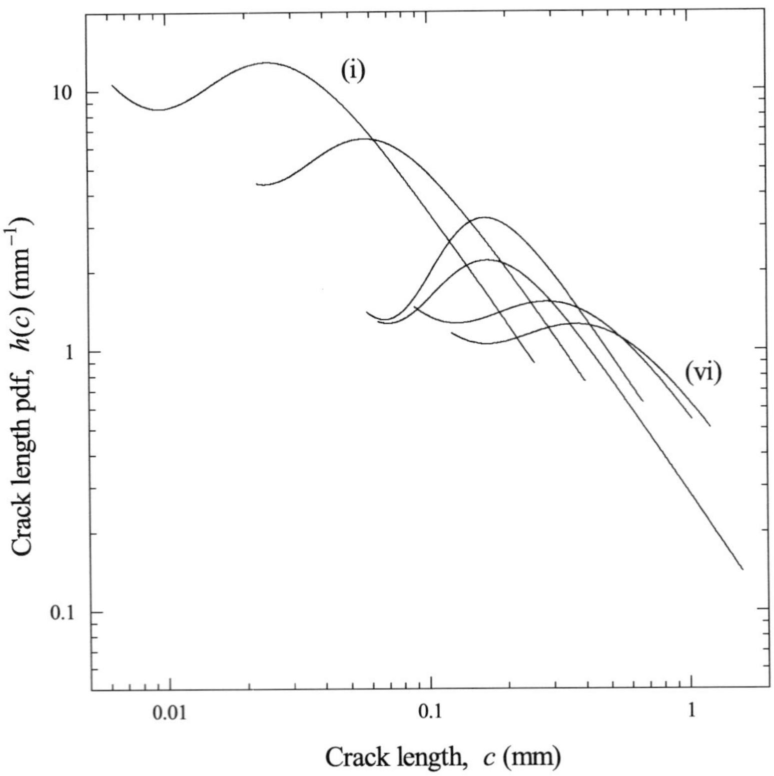

Figure 9.12 Plot in logarithmic coordinates of crack length pdf $h(c)$ variations for salt particles. D = 0.71 mm–4.00 mm, increasing particle size responses labeled (i)–(vi); determined from Figure 9.11.

the published work. Symbols represent individual strength measurements. Different symbols represent different particle sizes, D; increasing particle size is labeled (i)–(iii). The data are well separated by particle size and exhibit narrow strength domains σ_N/σ_1 of approximately 3. A feature of Figure 9.15a is that the data exhibit threshold strengths that decrease with increasing particle size. Qualitatively, the strength responses are similar to those of quartz in Figure 9.9a. The solid lines in Figure 9.15a represent unconstrained visual best fits to the data consistent with multiple deterministic flaw populations. The data are fit by bi-linear concave functions. Deterministic size effects on strength are thus observed here as variations in the strength distribution domain. Figure 9.15b shows as solid lines the crack length pdf $h(c)$ variations underlying the rock particle strength behavior. The $h(c)$ variations were determined using $B = 1.0$ MPa m$^{1/2}$ and particle size and Eq. (7.2) as a constraint. Increasing particle size variations are labeled (i)–(iii). Consistent with deterministic effects, the crack length distributions vary in domain location and in this case the distributions are all peaked. In particular, as particle size increases the crack length domains shift to larger crack lengths. In another extensive study, Koohmishi and Palassi (2016) (see Figure 6.18b) used localized probe diametral loading to observe and measure the strength behavior of four different ballast rocks separated into three different shape classes (reminiscent of the early Brecker 1974 study) and sieved into four different size classes. No significant effect of shape on strength was observed, although there was a clear effect of size and rock type.

The second example is the SL series quarry rocks considered by Tavares and das Neves (2008). Figure 9.16a shows strength edf $Pr(\sigma)$ plots for samples of rock particles 0.59–90 mm in diameter, using data derived from the published work. Symbols represent individual strength measurements. Different symbols represent different particle sizes, D; increasing particle size is labeled (i)–(v). The data are well separated by particle size and exhibit moderaetely wide strength domains σ_N/σ_1 of approximately 4. A feature of Figure 9.16a is that the data exhibit threshold strengths that decrease with increasing particle size. Qualitatively, the strength responses are similar to those of the V series rocks above, Figure 9.13a. The solid lines in Figure 9.16a represent unconstrained visual best fits to the data consistent with multiple deterministic flaw populations. The data are fit by bi-linear concave functions. Deterministic size effects on strength are thus observed here as variations in the strength distribution domain. Figure 9.16b shows as solid lines the crack length pdf $h(c)$ variations underlying the rock particle strength behavior. The $h(c)$ variations were determined using $B = 0.35$ MPa m$^{1/2}$ and particle size and Eq. (7.2) as a constraint. Increasing particle size variations are labeled (i)-(v). Consistent with deterministic effects, the crack length

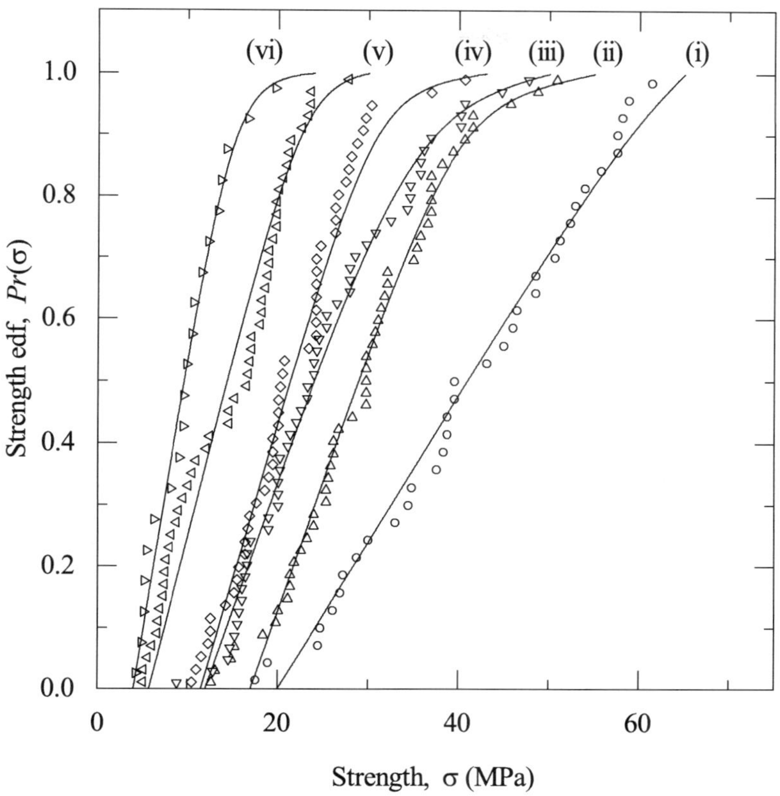

Figure 9.13 Plot of strength edf behavior, $Pr(\sigma)$, for V series rock particles. D = (0.59–0.70, 2.36–2.83, 4.75–5.6, 13.3–16, 37.5–45, and 75–90) mm, N_{tot} = 256 (adapted from Tavares, L.M et al (2008)). Solid lines show independent best fits. The separated lower bound threshold strengths are indicative of deterministic size effects.

distributions vary in domain location and in this case the distributions are nearly all peaked. Similar strength distributions and influences of particle size were reported by Lobo-Guerrero and Vallejo (2006) for two rocks types tested using opposing conical platens and by Nad and Saramak (2018) for three types of copper ore from Poland tested using conventional platens.

The implication of the results from the above studies is that very large, angular, fragments that might be regarded as "rocks" exhibit strength distributions identical to those of small, rounded, objects identified as "particles" (see images in Chapter 1). At all sizes, particles exhibit strength distributions clearly distinct from extended components (Chapter 5).

9.2.5 Coal

Figure 9.17 shows strength edf $Pr(\sigma)$ and crack length pdf $h(c)$ plots for samples of coal (polycrystalline C) particles 0.25–3.75 mm in diameter, from the work of Dong et al. (2018), using data derived from the published work. The motivation for the work was to improve efficiency in coal and hydrocarbon extraction in mining. Strength measurements were performed in a conventional compression platen apparatus. Symbols represent individual strength measurements. Different symbols represent different particle sizes, D; increasing particle size is labeled (i)–(iii). The data are well separated by particle size and exhibit wide strength domains σ_N/σ_1 of approximately 10. In terms of particle size and strength magnitude and dispersion, the data are similar to the coal particle strength observations of Wang et al. (2019) in Figure 2.26e. A feature of Figure 9.17a is that the data exhibit threshold strengths that decrease with increasing particle size. Qualitatively, the strength responses are similar to those of quartz in Figure 9.9. The solid lines in Figure 9.17a represent unconstrained visual best fits to the data consistent with multiple deterministic flaw populations. The data are fit by bi-linear concave functions. Deterministic size effects on strength are thus observed here as variations in strength distribution domain.

Figure 9.17b shows as solid lines the crack length pdf $h(c)$ variations underlying the coal particle strength behaviors. The $h(c)$ variations were determined using $B = 0.037$ MPa m$^{1/2}$ and particle size and Eq. (7.2) as a constraint. Increasing

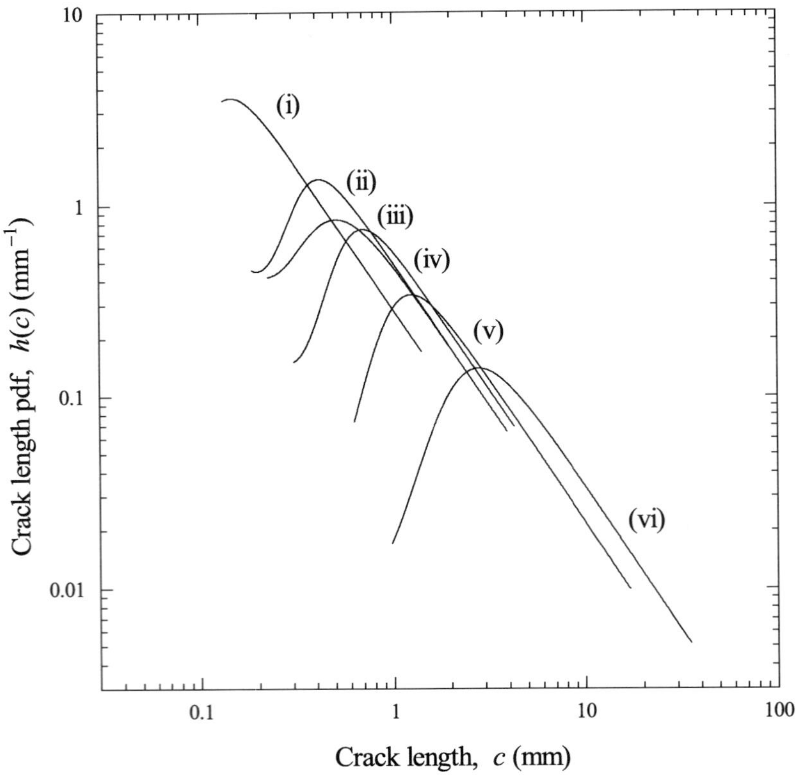

Figure 9.14 Plot in logarithmic coordinates of crack length pdf $h(c)$ variations for V series rock particles. D = 0.59 mm–90 mm, increasing particle size responses labeled (i)–(vi); determined from Figure 9.13.

particle size variations are labeled (i)–(iii). Consistent with deterministic effects, the crack length distributions vary in domain location as particle size changes, although the distributions are weakly peaked, as observed for some quartz particles (Figure 9.9b). In particular, as particle size increases the crack length domains shift to larger crack lengths and and in all cases are dominated by a long tail at large crack lengths.

9.2.6 Coral

Figure 9.18 shows strength edf $Pr(\sigma)$ and crack length pdf $h(c)$ plots for samples of coral particles 3–12 mm in diameter, from the work of Ma et al. (2019), using data derived from the published work. Coral is a calcareous, porous agglomerate generated by oceanic creatures and coral sands are widely used in oceanic engineering. The motivation for the work was to assess the efficiency of lightweight coral materials in aggregate structures. Strength measurements were performed in a conventional compression platen apparatus. Symbols represent individual strength measurements. Different symbols represent different particle sizes, D; increasing particle size is labeled (i)–(iii). The data, although clustered, are separated by particle size and exhibit wide strength domains σ_N/σ_1 of approximately 10. In terms of particle size and strength magnitude and dispersion, the data are similar to the coral particle strength observations of Shen et al. (2020) in Figures 2.26d, 2.29a, and 7.12. A feature of Figure 9.18a is that the data exhibit threshold strengths that decrease with increasing particle size. Qualitatively, the strength responses are somewhat similar to those of quartz in Figure 9.10a. The solid lines in Figure 9.18a represent unconstrained visual best fits to the data consistent with multiple deterministic flaw populations. The data are fit by bi-linear concave functions. Deterministic size effects on strength are thus observed here as variations in strength distribution domain, distinguishing the behavior from that of the stochastic coral system considered earlier (Figure 7.12).

Figure 9.18b shows as solid lines the crack length pdf $h(c)$ variations underlying the coral particle strength behaviors. The $h(c)$ variations were determined using $B = 0.08$ MPa m$^{1/2}$ and particle size and Eq. (7.2) as a constraint. Increasing

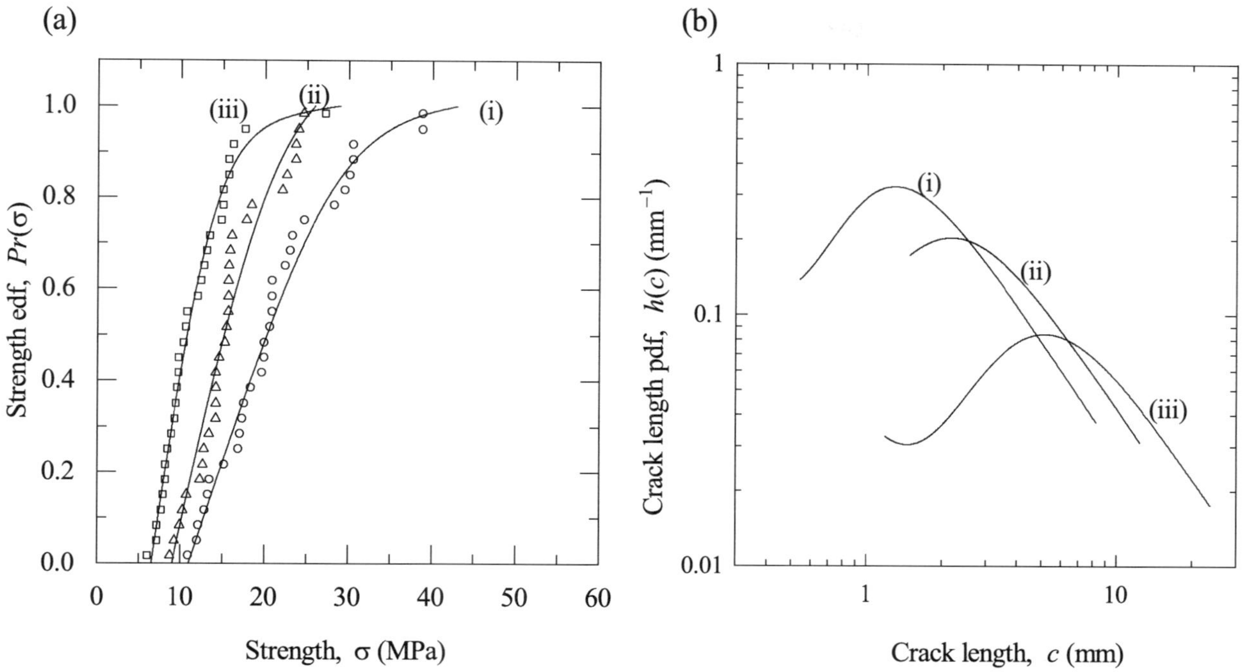

Figure 9.15 (a) Plot of strength edf behavior, $Pr(\sigma)$, for railway ballast rock particles. D = (10–14, 20–28, and 37.5–50) mm, N_{tot} = 90 (adapted from Lim, W.L et al., 2004). Solid lines show independent best fits. The separated lower bound threshold strengths are indicative of deterministic size effects. (b) Plot in logarithmic coordinates of crack length pdf $h(c)$ variations for rock particles. Increasing particle size responses labeled (i)–(iii); determined from (a).

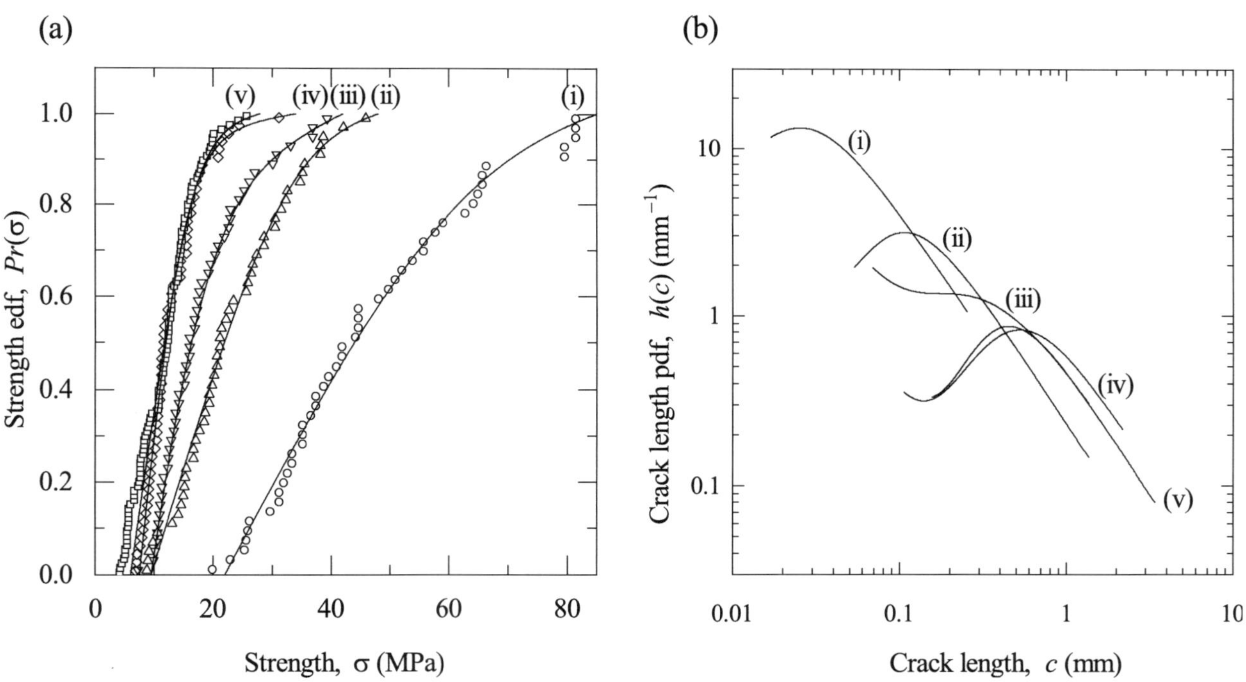

Figure 9.16 (a) Plot of strength edf behavior, $Pr(\sigma)$, for S series rock particles. D = (0.59–0.7, 1.18–1.4, 2.36–2.83, 4.75–5.6, 13.3–90) mm, N_{tot} = 317 (adapted from Tavares, L.M (2008)). Solid lines show independent best fits. The separated lower bound threshold strengths are indicative of deterministic size effects. (b) Plot in logarithmic coordinates of crack length pdf $h(c)$ variations for rock particles. Increasing particle size responses labeled (i)–(v); determined from (a).

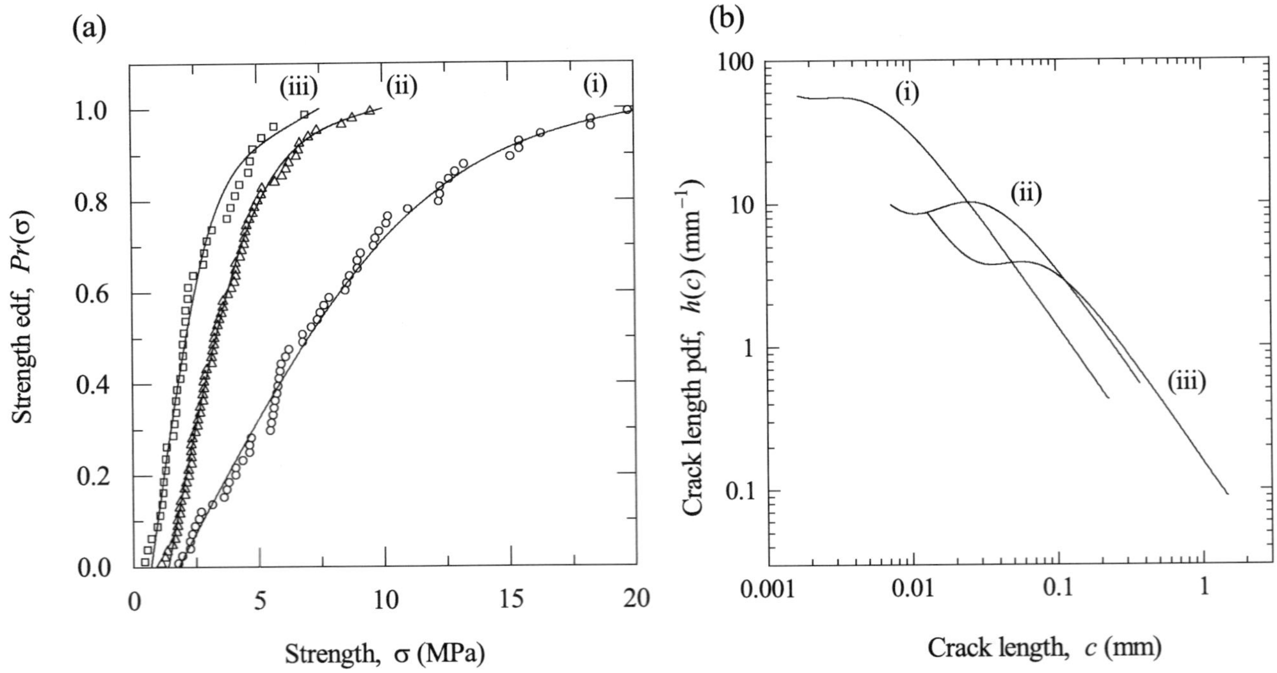

Figure 9.17 (a) Plot of strength edf behavior, $Pr(\sigma)$, for coal particles. D = (0.25–1, 1–2, and > 2) mm, N_{tot} = 175 (adapted from Dong, J et al., 2018). Solid lines show independent best fits. The separated lower bound threshold strengths are indicative of deterministic size effects. (b) Plot in logarithmic coordinates of crack length pdf $h(c)$ variations for coal particles. Increasing particle size responses labelled (i)–(iii); determined from (a).

particle size variations are labeled (i)–(iii). Consistent with deterministic effects, the crack length distributions vary in domain location as particle size changes, although the distributions are weakly peaked and barely separated. In all cases, the distributions are dominated by a long tail at large crack lengths.

9.3 Sigmoidal Deterministic Distributions

9.3.1 Glass

Glass (soda lime silicate) is the stereotypical brittle material and has extensive commercial use in sheet and moulded forms for architectural and domestic uses. Glass particles are also used commercially in many of the applications noted in Chapter 1 and elsewhere throughout, including grinding, proppant beds, structural reinforcement in composites, and as catalytic and biological agents. As a consequence, the strength of glass particles was measured in early work by Kschinka et al. (1986) using a conventional platen apparatus and more recently by Aman et al. (2010), Rozenblat et al. (2011), and Portnikov et al. (2013); also using a conventional platen apparatus, by Huang et al. (2014) and Shan et al. (2018) using a conventional platen apparatus and a split Hopkinson bar impact apparatus; and by Watkins and Prado (2015) on small particles using a custom platen apparatus. Many of these works and others have examined the failure mechanisms of glass particles, aided by glass transparency (Shipway and Hutchings 1993; Tavares and King 1998; Gorham and Salman 2005; Aman et al. 2010; Rozenblat et al. 2011; Huang et al. 2014; Paul et al. 2015; Parab et al. 2017; Pejchal et al. 2017; Shan et al. 2018). In most of these strength and observation studies, the motivation was to explore the fundamentals of fracture processes using a well studied material.

The overall strength behavior of glass particles is shown in the logarithmic plot of glass particle strength, σ, as a function of particle diameter, D, for a range of glass particle systems, Figure 9.19. The strengths were determined from Kschinka et al. (1986), Aman et al. (2010), Rozenblat et al. (2011), Portnikov et al. (2013), Huang et al. (2014), Shan et al. (2018), and Watkins and Prado (2015), using data derived from the published works cited. The plot is similar to those in Figure 2.26, with an expanded diameter axis. The bars in Figure 9.19 indicate the ranges of strengths observed at a given particle

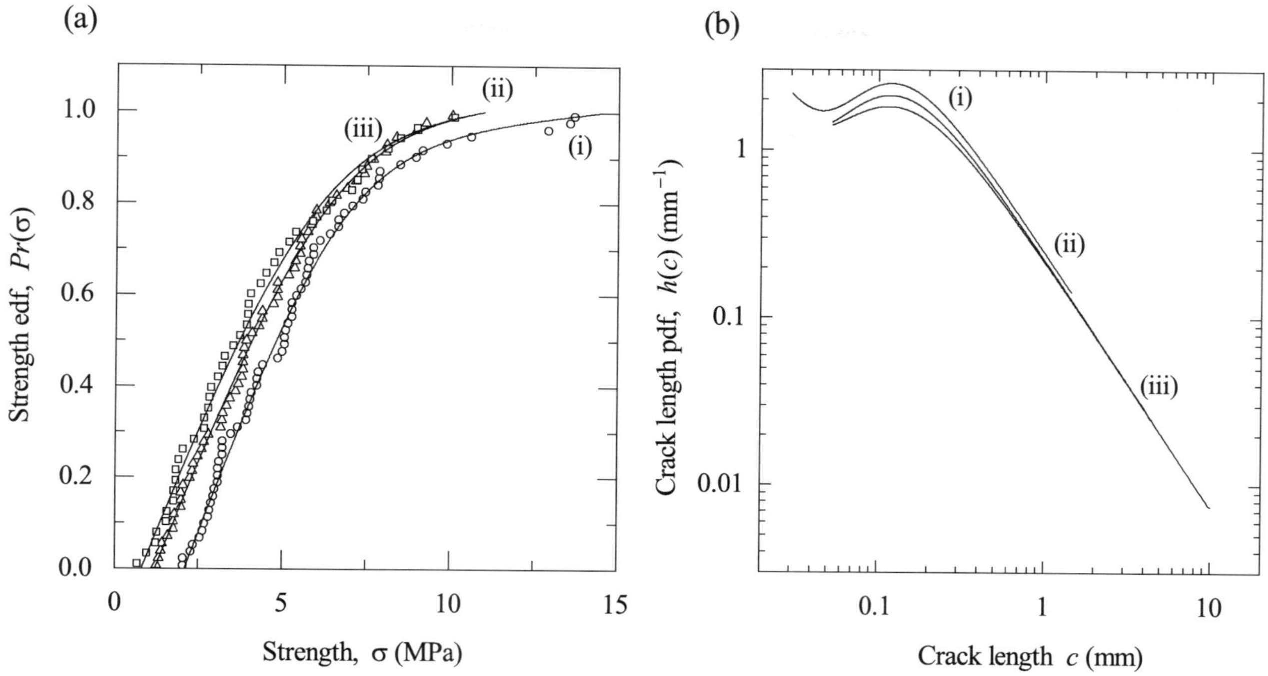

Figure 9.18 (a) Plot of strength edf behavior, $Pr(\sigma)$, for coral particles. D = (3–5, 6.5–8.5, and 10–12) mm, N_{tot} = 173 (adapted from Ma, L et al., 2019). Solid lines show independent best fits. The separated lower bound threshold strengths are indicative of deterministic size effects. (b) Plot in logarithmic coordinates of crack length pdf $h(c)$ variations for rock particles. Increasing particle size responses labeled (i)–(iii); determined from (a).

diameter and the ranges of diameters where cited. The symbols are a guide to the eye and represent the mid-points of the ranges. Different symbols represent different strength studies. As in Figure 2.26, the solid lines are power-law guides to the eye of slopes −0.5 and −1.0 and bound the trend in the observations over a factor of 10^3 in particle diameter. The range in strengths for a given particle diameter was approximately a factor of 5 and a typical individual particle study exhibited approximately a factor of 5 in mid-point strength range. Overall, Figure 9.19 indicates that the strength behavior of particles of glass (often referred to as glass "spheres") is similar to that of particles of other materials and that there is agreement between the many existing glass particle studies.

Figure 9.20 shows strength edf $Pr(\sigma)$ plots for samples of glass particles 4.4–24.7 mm in diameter, from the work of Huang et al. (2014), using data derived from the published work. Symbols represent individual strength measurements. Different symbols represent different particle sizes, D; increasing particle size is labeled (i)–(v). The data are well separated by particle size and exhibit narrow strength domains σ_N/σ_1 of approximately 2.5. A clear feature of Figure 9.20 is that the data exhibit threshold strengths that decrease with increasing particle size, indicative of deterministic size effects. This feature distinguishes these glass particle data from those considered in Chapter 7 (Rozenblat et al. 2011, Figure 7.22), in which an invariant threshold was indicative of stochastic size effects. Another clear feature of Figure 9.20 is that the absolute strength dispersions also decrease with increasing particle size. This feature distinguishes this deterministic strength behavior from that of the diamond films considered in Chapter 5 (Peng et al. 2007, Figure 5.24), in which deterministic behavior was apparent as simple strength distribution translation. The solid lines in Figure 9.20 represent unconstrained visual best fits to the data consistent with multiple deterministic flaw populations. The data are fit by trilinear sigmoidal functions. Deterministic size effects on strength are thus observed here as variations in strength distribution domain.

Figure 9.21 shows as solid lines the crack length pdf $h(c)$ variations underlying the glass particle strength behaviors. The $h(c)$ variations were determined using $B = 2.0$ MPa $m^{1/2}$ and particle size and Eq. (7.2) as a constraint. Increasing particle size variations are labeled (i)-(v). The crack length variations exhibit well defined peaks and consistent with deterministic effects the crack length distributions vary in domain location as particle size changes. In particular, as particle size increases the crack length domains shift to larger crack lengths. (The large B value and thus large crack lengths inferred here arise

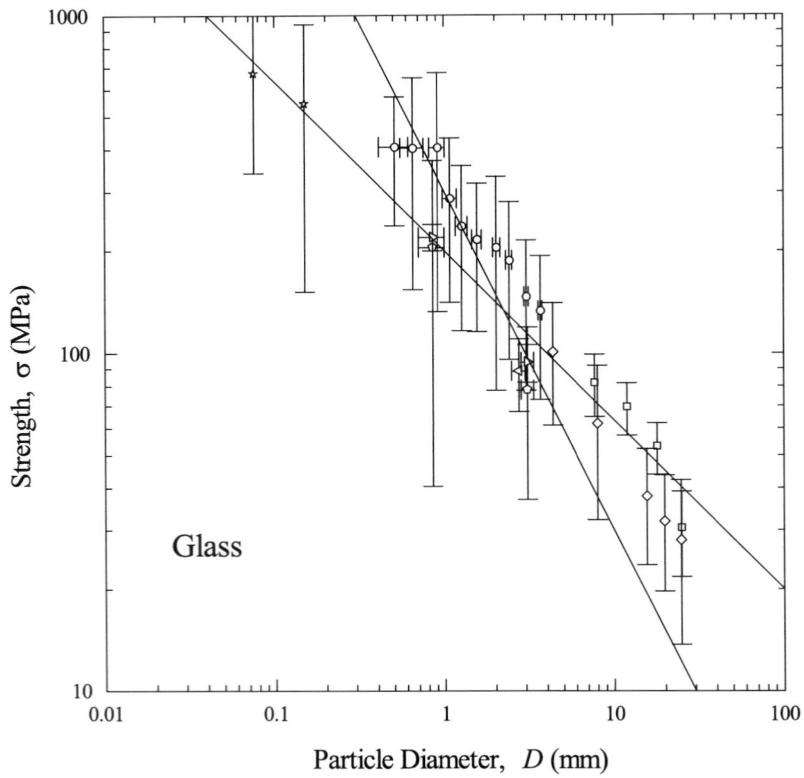

Figure 9.19 Logarithmic plot of strength σ vs diameter D for glass particle systems, showing measurements for samples of particles illustrating decreasing strength trend and dispersions. Data from Kschinka, B.A et al. (1986), Aman, S et al. (2010), Rozenblat, Y et al. (2011), Portnikov, D et al. (2013), Huang, J et al. (2014), Shan, J et al. (2018), and Watkins, I.G et al. (2015).

from consistent application of the particle size constraint used in earlier analyses, Chapters 6, 7, 8, and here. The value of B does not affect the form or relative positions of inferred $h(c)$ responses as in Figure 9.21, but the large values appear unphysical and are addressed in Section 9.6.)

Figure 9.22 shows strength edf $Pr(\sigma)$ plots for samples of glass particles 0.51–3.68 mm in diameter, from the work of Kschinka et al. (1986), using data derived from the published work. Different symbols represent different particle sizes, D; increasing particle size is indicated by the gray arrow (right to left). Solid symbols represent individual strength measurements for D = 2.41 mm particles. Open symbols represent strength values reconstructed from the powered exponential distribution parameters provided and the number of components N for samples of other D particles. The data are well separated by particle size and exhibit narrow strength domains σ_N/σ_1 of approximately 3. A clear feature of Figure 9.22 is that the absolute strength dispersions decrease with increasing particle size. Although not uniformly obvious at this scale, an additional feature of Figure 9.22 is that the data exhibit threshold strengths that decrease with increasing particle size, indicative of deterministic size effects. Hence, although the data here appear superficially similar to the fertilizer strength measurements of Figure 7.20, for example, the separated thresholds distinguish the glass particle data from the stochastic data considered in Chapter 7. The solid lines in Figure 9.22 represent unconstrained visual best fits to the data consistent with multiple deterministic flaw populations. The fits used tri-linear sigmoidal functions. Deterministic size effects on strength are thus observed here as variations in strength distribution domain.

Figure 9.23 shows as solid lines the crack length pdf $h(c)$ variations underlying the glass particle strength behaviors. The $h(c)$ variations were determined using B = 2.0 MPa m$^{1/2}$ and particle size and Eq. (7.2) as a constraint. Increasing particle size variations are indicated by the gray arrow (diagonally left to right). The crack length variations exhibit very well defined peaks and consistent with deterministic effects the crack length distributions vary in domain location as particle size changes. In particular, as particle size increases the crack length domains shift to larger crack lengths and the large crack tail contracts.

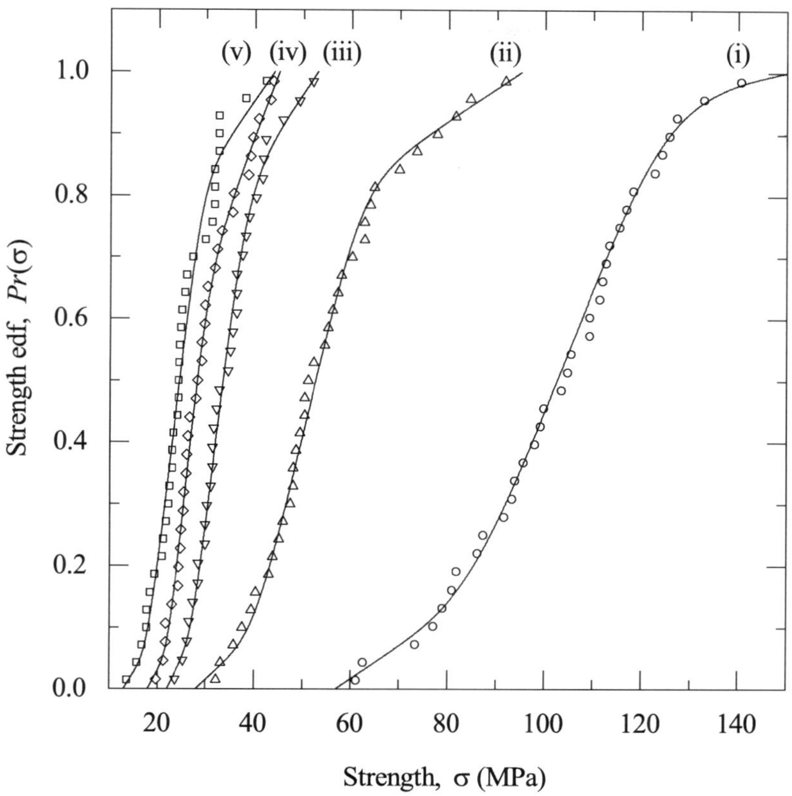

Figure 9.20 Plot of strength edf behavior, $Pr(\sigma)$, for glass particles. D = (4.36 ± 0.08, 8.02 ± 0.19, 15.77 ± 0.19, 19.78 ± 0.15, and 24.66 ± 0.22) mm, N_{tot} = 169 (adapted from Huang, J et al., 2014). Solid lines show independent sigmoidal best fits. The separated lower bound threshold strengths are indicative of deterministic size effects.

Two further examples of deterministic size effects on strength behavior of glass particles are exhibited in the strength edf $Pr(\sigma)$ and crack length pdf $h(c)$ plots of Figures 9.24 and 9.25. In the first example, Shan et al. (2018) repeated a sub-set of the experiments of Huang et al. (2014). The results are shown in Figure 9.24a as the strength edf $Pr(\sigma)$ plots for samples of glass particles 7–25 mm in diameter, using data derived from the published work. Symbols represent individual strength measurements. Different symbols represent different particle sizes, D; increasing particle size is labeled (i)–(iv). The data are well separated by particle size and exhibit narrow strength domains σ_N/σ_1 of approximately 2. A feature of Figure 9.24a is that the data exhibit threshold strengths that decrease with increasing particle size. The strength responses are similar to those of the earlier glass particles measurements of Figure 9.20. The solid lines in Figure 9.24a represent unconstrained visual best fits to the data consistent with multiple deterministic flaw populations. The data are fit by tri-linear sigmoidal functions. Deterministic size effects on strength are thus observed here as variations in the strength distribution domain. Figure 9.24b shows as solid lines the crack length pdf $h(c)$ variations underlying the glass particle strength behavior. The $h(c)$ variations were determined using $B = 3.0$ MPa m$^{1/2}$ and particle size and Eq. (7.2) as a constraint. Increasing particle size variations are labeled (i)-(iv). Consistent with deterministic effects, the crack length distributions vary in domain location and in this case the distributions are all peaked. In particular, as particle size increases the crack length domains shift to larger crack lengths.

The second example is a series of small glass particles considered by Watkins and Prado (2015). Figure 9.25a shows strength edf $Pr(\sigma)$ plots for samples of glass particles 75 μm and 150 μm in diameter, using data derived from the published work. Symbols represent individual strength measurements. Different symbols represent different particle sizes, D; increasing particle size is labeled (i) and (ii). The data are well separated by particle size and exhibit moderate strength domains σ_N/σ_1 of approximately 3 (particles of intermediate size exhibited strengths similar to the larger particles). A feature of Figure 9.25a is that the data exhibit threshold strengths that decrease with increasing particle size. The solid lines in Figure 9.25a represent unconstrained visual best fits to the data consistent with multiple deterministic flaw populations. The data are fit by tri-linear sigmoidal functions. Deterministic size effects on strength are thus observed here as variations

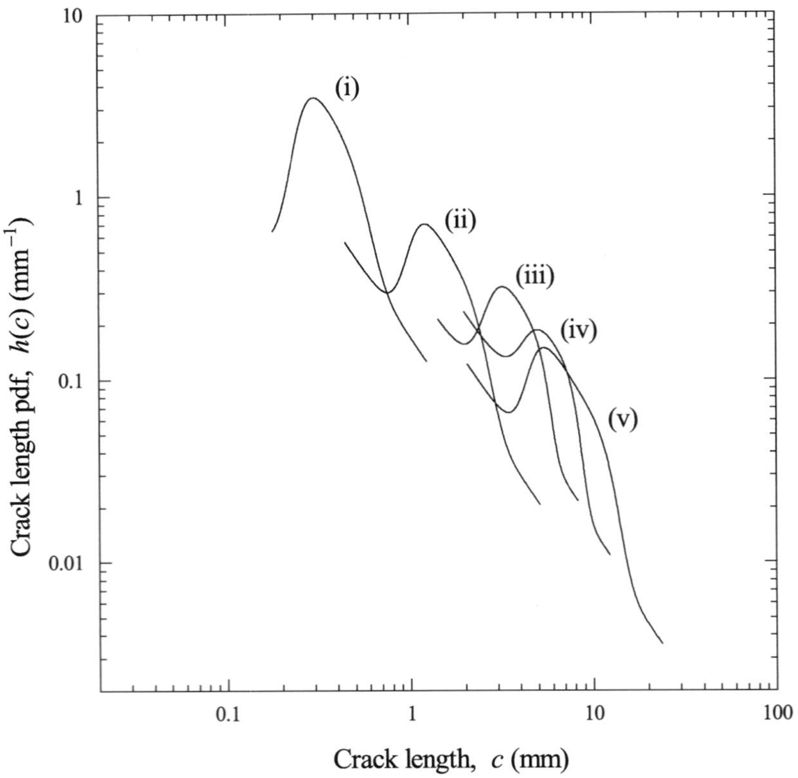

Figure 9.21 Plot in logarithmic coordinates of crack length pdf $h(c)$ variations for limestone particles. D = 4.36 mm–24.66 mm, increasing particle size responses labelled (i)–(v); determined from Figure 9.20.

in the strength distribution domain. Figure 9.25b shows as solid lines the crack length pdf $h(c)$ variations underlying the glass particle strength behavior. The $h(c)$ variations were determined using $B = 1.3$ MPa m$^{1/2}$ and particle size and Eq. (7.2) as a constraint. Increasing particle size variations are labeled (i) and (ii). Consistent with deterministic effects, the crack length distributions vary in domain location and shape.

9.3.2 Rock

A major economic factor in the mining and materials processing industries is energy consumption in comminution of rock, mineral, and ore particles, as discussed in Chapter 1. Hence, in many of the strength studies of rock and mineral particles considered in Chapters 6 and 7 and here, Chapter 9, improving the energy efficiency of comminution was the motivating factor. In those studies, the focus was on the effects of variations in particle size and particle material on post comminution strength. Here, in a study by Tavares (2005), particle size and material are fixed and the focus is on the effect of variation in comminution method on particle strength. Tavares compared the crushing and subsequent strength of Cu ore particles by (i) a conventional hammer mill, (ii) a conventional smooth roller mill, and (iii) and a high pressure roll grinder (HPRG). The crushed ore particles were 4.75–5.6 mm in size and were strength tested using a drop weight configuration.

Figure 9.26a shows strength edf $Pr(\sigma)$ plots for samples of Cu ore particles formed by the three techniques, using data derived from the published work. Symbols represent individual strength measurements. Different symbols represent different particle fabrication methods; the hammer mill and HPRG data are labeled (i) and (iii). The data are well separated by particle fabrication method and exhibit moderate strength domains σ_N/σ_1 of approximately 3. The strength variations are similar to those exhibited by Cu and Au ore in Figure 6.17a. A feature of Figure 9.26a is that the data exhibit threshold strengths that differ with fabrication method. In addition, the data appear to exhibit similarly shaped strength behavior. These features are similar to those exhibited by the diamond films of Figure 5.24. The solid line in Figure 9.26a represents an unconstrained visual best fit to the hammer mill (i) data by the tri-linear sigmoidal function.

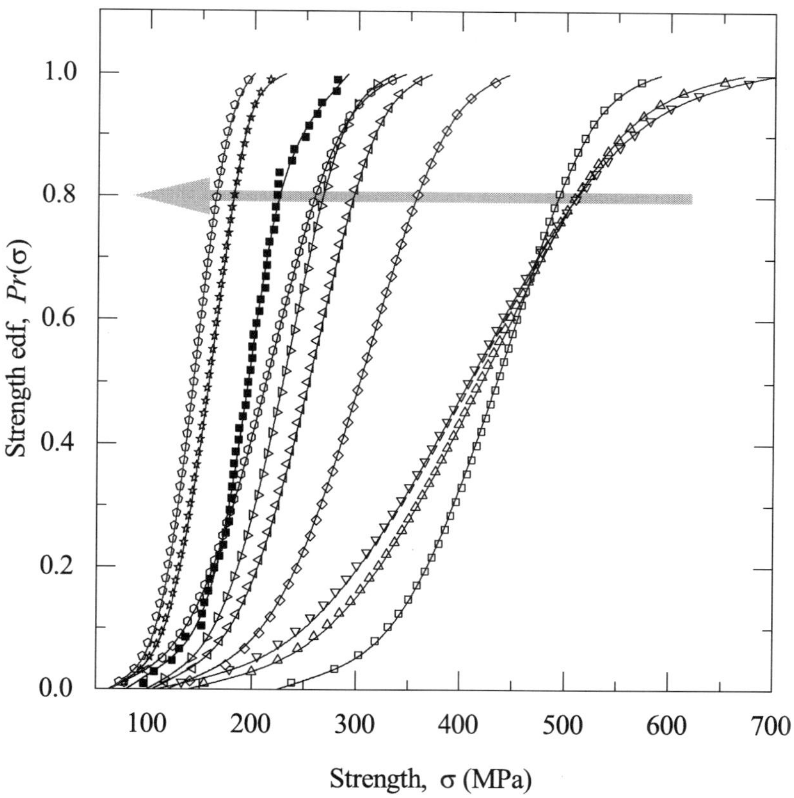

Figure 9.22 Plot of strength edf behavior, $Pr(\sigma)$, for glass particles. D = (0.51, 0.65, 0.91, 1.08,1.27, 1.56, 2.03, 2.41, 3.05, and 3.68) mm, all uncertainties ± 0.1 mm, increasing particle size responses labeled indicated by gray arrow, N_{tot} = 507 (adapted from Kschinka, B.A et al., 1986). Solid lines show independent sigmoidal best fits. The separated lower bound threshold strengths are indicative of deterministic size effects.

Following Figure 5.24, the dashed lines in Figure 9.26a represent simple translations in this fit to match the behavior of the smooth roller mill (ii) and HPRG (iii) strength data. The translated behavior describes the data well. Deterministic fabrication effects on strength are thus observed here as variations in the location of the strength distribution domain. Figure 9.26b shows as solid lines the crack length pdf $h(c)$ variations underlying the Cu ore particle strength behavior. The $h(c)$ variations were determined using $B = 0.3$ MPa m$^{1/2}$ and particle size and Eq. (7.2) as a constraint. Particle fabrication variations are labeled (i) and (iii). The very close sigmoidal strength variations result in very close peaked crack length populations, similar to Figure 5.25. Consistent with deterministic effects, the crack length distributions vary in domain location.

9.4 Linear Deterministic Distributions

Many samples of particles, and other components, exhibit linear strength edf $Pr(\sigma)$ variations. Some examples, plotted in unbiased coordinates, are shown in Chapter 5 (Figure 5.33) of predominately linear strength distribution behavior from some early studies of extended components. Two particle systems, cement and ice, are examined here that also exhibit linear strength distribution behavior. These systems, in which particle size was varied in strength measurements, exhibit deterministic size effects.

In quantitative edf terms, the distinguishing feature of linear strength edf variation is increase in $Pr(\sigma)$ from 0 at $\sigma = \sigma_1$ to 1 at σ_N with constant derivative (see Chapter 6). In the relative terms used for forward analysis (Chapter 3), $H(\mu)$ extends linearly from $H(\mu = 0) = 0$ to $H(1) = 1$. In the absolute terms used for reverse analysis (Chapter 4), the continuum expression for the discrete strength data $H(\sigma)$ is given by an extension of Eq. (6.1):

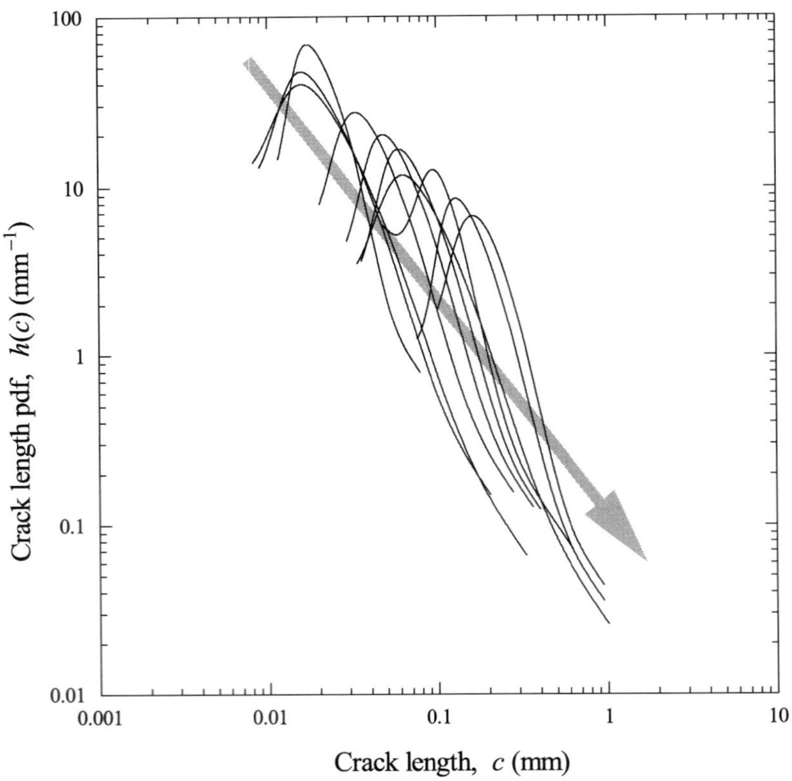

Figure 9.23 Plot in logarithmic coordinates of crack length pdf $h(c)$ variations for glass particles. D = 0.51 mm–3.68 mm, increasing particle size responses labeled indicated by gray arrow; determined from Figure 9.22.

$$
\begin{aligned}
H(\sigma) &= (\sigma - \sigma_L)/(\sigma_U - \sigma_L) \\
&= \beta(\sigma - \sigma_L), \qquad \sigma_L \le \sigma \le \sigma_U.
\end{aligned} \tag{9.2}
$$

σ_L is a lower limit strength fitting parameter (usually very close to σ_1) and σ_U is an upper limit strength fitting parameter (usually very close to σ_N). The derivative parameter β of Eq. (6.1) and Eq. (9.2) is now specified by the strength fitting parameters as

$$\beta = 1/(\sigma_U - \sigma_L). \tag{9.3}$$

The crack length pdf $h(c)$ follows as before,

$$h(c) = \mathrm{d}H(c)/\mathrm{d}c = (\beta B/2)c^{-3/2}, \qquad c_L \le \sigma \le c_U, \tag{9.4}$$

identical in form to Eq. (6.3). Eq. (9.4) however, pertains over the entire crack length domain, $c_L = (B/\sigma_U)^2$, $c_U = (B/\sigma_L)^2$, and the value of β is specified. Linear strength distributions of different sized particles of the same material are described by Eq. (9.2) with different σ_L, σ_U, and thus β parameters and underlying crack length populations given by Eq. (9.4) with identical B values. However, such multiple samples cannot be described simultaneously by variations of Eq. (9.2) and related by Eq. (9.1), indicating that linear strength distributions are a mark of deterministic size effects.

9.4.1 Cement

Concrete components are usually composites, consisting of a cement matrix containing rock and gravel particles ("aggregate") and metal reinforcing bars ("rebar"). Cement can be used on its own for non-structural applications, but is usually the binder for much stronger, load-bearing composites. Cement is formed by mixing limestone and calcium aluminosilicate minerals and clays and heating the mixture to react and sinter the materials to form cement clinkers, cement particles in the 1 mm to 10 mm size range. Clinkers facilitate handling, transport, and storage of cement, but must be ground into powder for mixing and reaction with water (and more lime) for use in concrete. Hence the strength of cement clinkers is critical

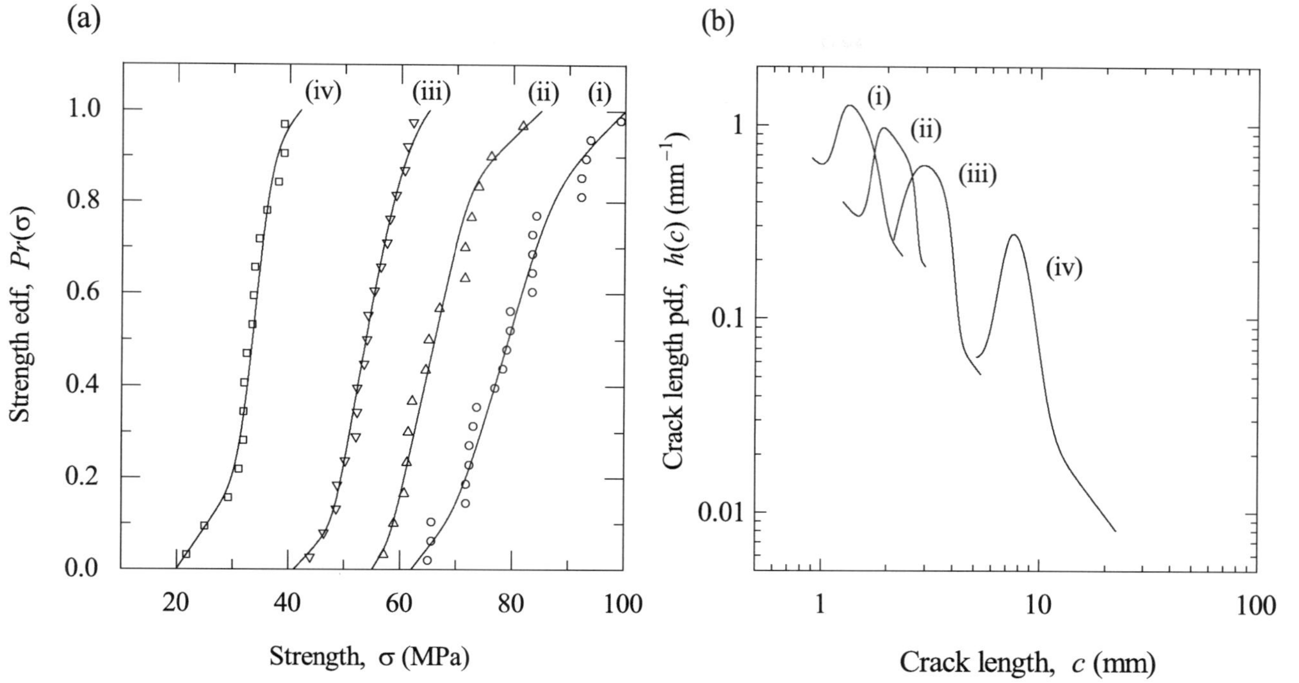

Figure 9.24 (a) Plot of strength edf behavior, $Pr(\sigma)$, for glass particles. $D = (7.71 \pm 0.14, 11.90 \pm 0.1, 17.88 \pm 0.15$, and $24.87 \pm 0.11)$ mm, $N_{tot} = 74$ (adapted from Shan, J et al., 2018). Solid lines show independent best fits. The separated lower bound threshold strengths are indicative of deterministic size effects. (b) Plot in logarithmic coordinates of crack length pdf $h(c)$ variations for glass particles. Increasing particle size responses labeled (i)–(iv); determined from (a).

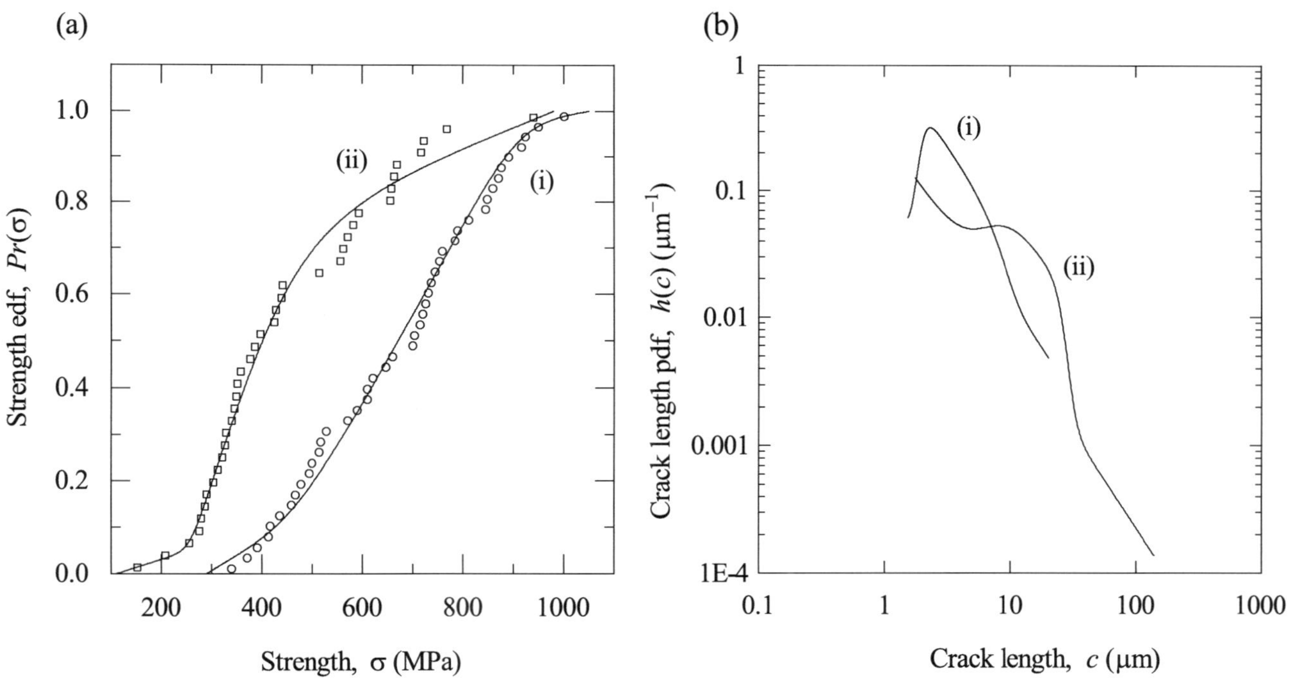

Figure 9.25 (a) Plot of strength edf behavior, $Pr(\sigma)$, for glass particles. $D = 75$ μm and 150 μm, $N_{tot} = 82$ (adapted from Watkins, I.G et al. 2015). Solid lines show independent best fits. The separated lower bound threshold strengths are indicative of deterministic size effects. (b) Plot in logarithmic coordinates of crack length pdf $h(c)$ variations for glass particles. Increasing particle size responses labeled (i)–(ii); determined from (a).

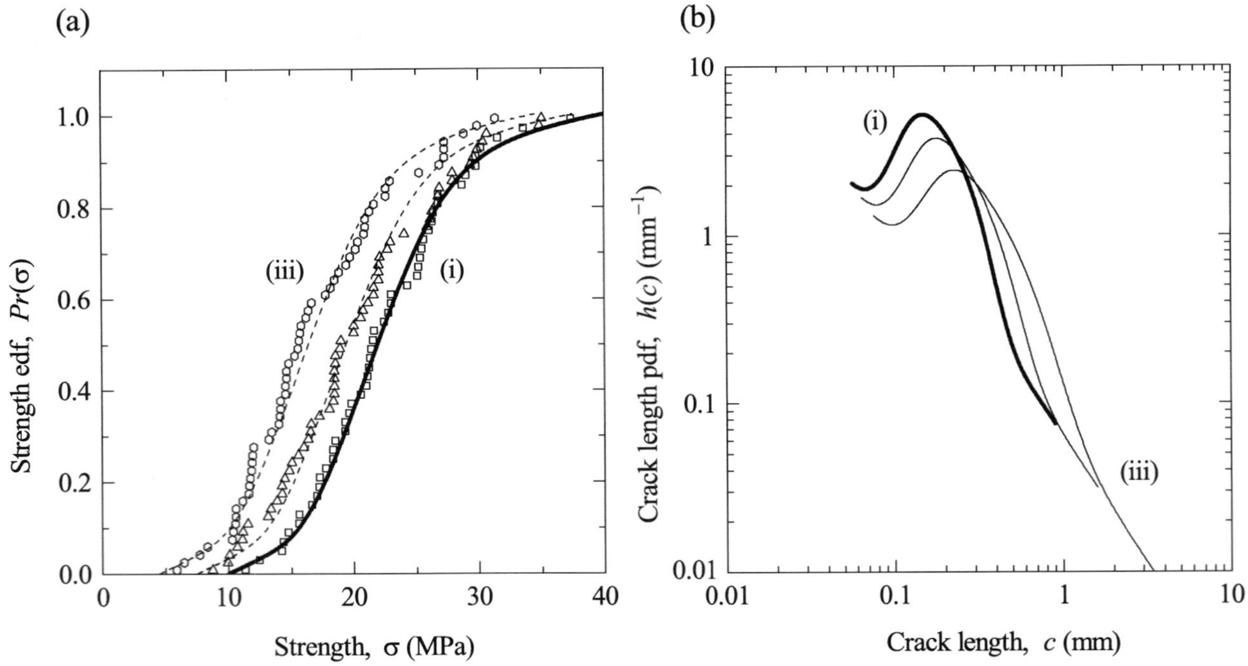

Figure 9.26 (a) Plot of strength edf behavior, $Pr(\sigma)$, for Cu ore particles. $D = 4.75-5.6$ mm, $N_{tot} = 170$ (Adapted from Tavares, L.M 2005). Solid lines show independent best fits. The separated lower bound threshold strengths are indicative of deterministic effects. (b) Plot in logarithmic coordinates of crack length pdf $h(c)$ variations for Cu ore particles. Different particle formation methods labeled (i)–(iii); determined from (a).

to the raw material pre-concrete stages, to avoid material loss by the generation of "fines" and early hydrolysis, and in the concrete production stage, to minimize the energy required for material grinding. The production of cement is a major carbon dioxide emitter and thus optimizing clinker mechanical performance can contribute to reductions in emissions and energy use.

Linear strength distributions have been exhibited by cement particles as shown in the strength edf $Pr(\sigma)$ and resulting crack length pdf $h(c)$ plots of Figure 9.27. Figure 9.27a shows strength behavior of samples of cement particles, 4–20 mm in diameter, measured using a compression platen apparatus in the work of May (1974), as cited in Jansen and Stoyan (2000), using data derived from the published work. Symbols represent individual strength measurements. Different symbols represent different particle sizes, D; increasing particle size is shown right to left. The results for the largest particles are shown as filled symbols; all others are open. The strengths are significantly greater (tens of megapascals) than those reported for cement clinkers by Vallet and Charmet (1995) (a few megapascals or less) and discussed earlier (Figure 2.26c). The data are well separated by particle size and exhibit moderate strength domains σ_N/σ_1 of approximately 3. A feature of Figure 9.27a is that the data exhibit threshold strengths that decrease with increasing particle size, as discussed earlier in this Chapter. However, perhaps the most distinctive feature of the data in Figure 9.27a, not in common with the studies above, is that the behavior of the strength distributions is near linear.

Figure 9.27b shows strength behavior of samples of cement particles, 0.5 mm and 8 mm in diameter, measured using a drop weight apparatus in the work of Tavares and Cerqueria (2006), using data derived from the published work. Symbols represent individual strength measurements. Different symbols represent different particle sizes, D; increasing particle size is shown right to left. The results for the smallest particles are shown as filled symbols; all others are open. The strengths of the larger particles are comparable to those reported in Figure 9.27a and the strengths of the smaller particles are significantly greater. The data are well separated by particle size and exhibit moderate strength domains σ_N/σ_1 of approximately 4. The data exhibit threshold strengths that decrease with increasing particle size, as discussed earlier, but again the most distinctive feature of the data in Figure 9.27b is that the behavior of the strength distributions is near linear.

The solid lines in Figures 9.27a and 9.27b represent unconstrained linear visual best fits to the data consistent with multiple deterministic flaw populations, Eq. (9.2). Deterministic size effects on strength are thus observed here as variations in the strength distribution domains, characterized by σ_L and σ_U separately, and the slope within domains, characterized

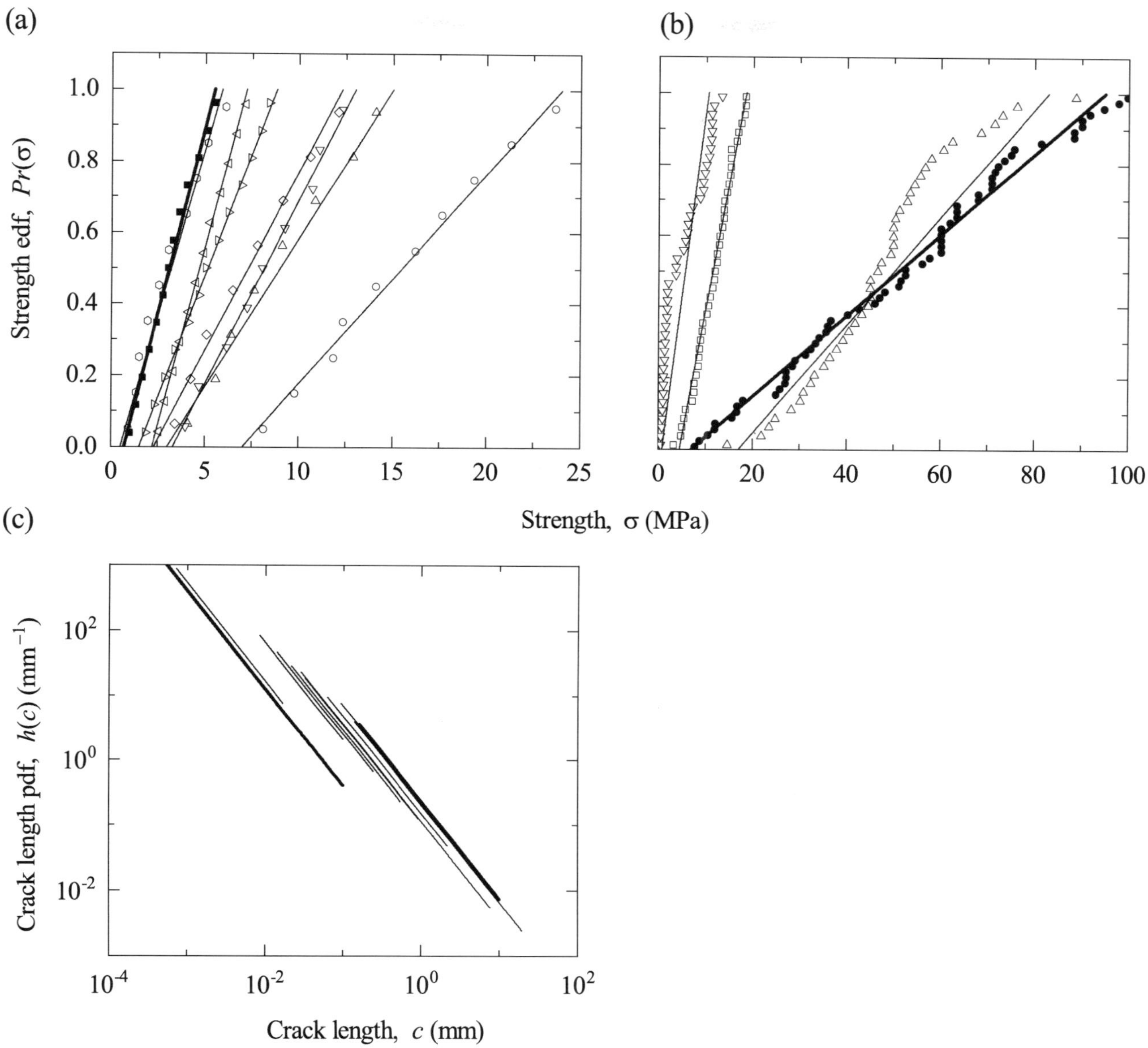

Figure 9.27 (a) Plot of strength edf behavior, $Pr(\sigma)$, for cement particles. D = (4, 5, 6.3, 10, 12, 16, 18, and 20) mm, N_{tot} = 103 (adapted from Jansen, U et al. 2000). (b) Plot of strength edf behavior, $Pr(\sigma)$, for cement particles. D = (0.5 to 0.7 and 8 to 9.5) mm, N_{tot} = 186 (adapted from Tavares, L.M et al. 2006). In both (a) and (b), solid lines show independent linear best fits and the separated lower bound threshold strengths are indicative of deterministic size effects. (c) Plot in logarithmic coordinates of crack length pdf $h(c)$ variations for cement particles; determined from (a) and (b).

by $\beta = 1/(\sigma_U - \sigma_L)$. The bounding behaviors of the strength distribution fits are shown as the bold solid lines. Figure 9.27c shows as solid lines the crack length pdf $h(c)$ variations underlying the cement particle strength behavior, Eq. (9.4). The $h(c)$ variations were determined using $B = 0.07$ MPa m$^{1/2}$ for both sets of samples and particle size and Eq. (7.2) as a constraint. In the logarithmic coordinates of Figure 9.27c, the variations of Eq. (9.4) appear as straight lines of slope $-3/2$. The responses of the bounding behavior are shown as bold solid lines and particle size increases left to right. Consistent with deterministic effects, the crack length distributions vary in domain location and in this case the distributions are all linear. In particular, as particle size increases the crack length domains shift to larger crack lengths.

9.4.2 Ice

Linear strength distributions have also been exhibited by ice particles, strictly ice "blocks" as shown in the strength edf $Pr(\sigma)$ and resulting crack length pdf $h(c)$ plots of Figure 9.28. Figure 9.28a shows strength behavior of samples of ice blocks,

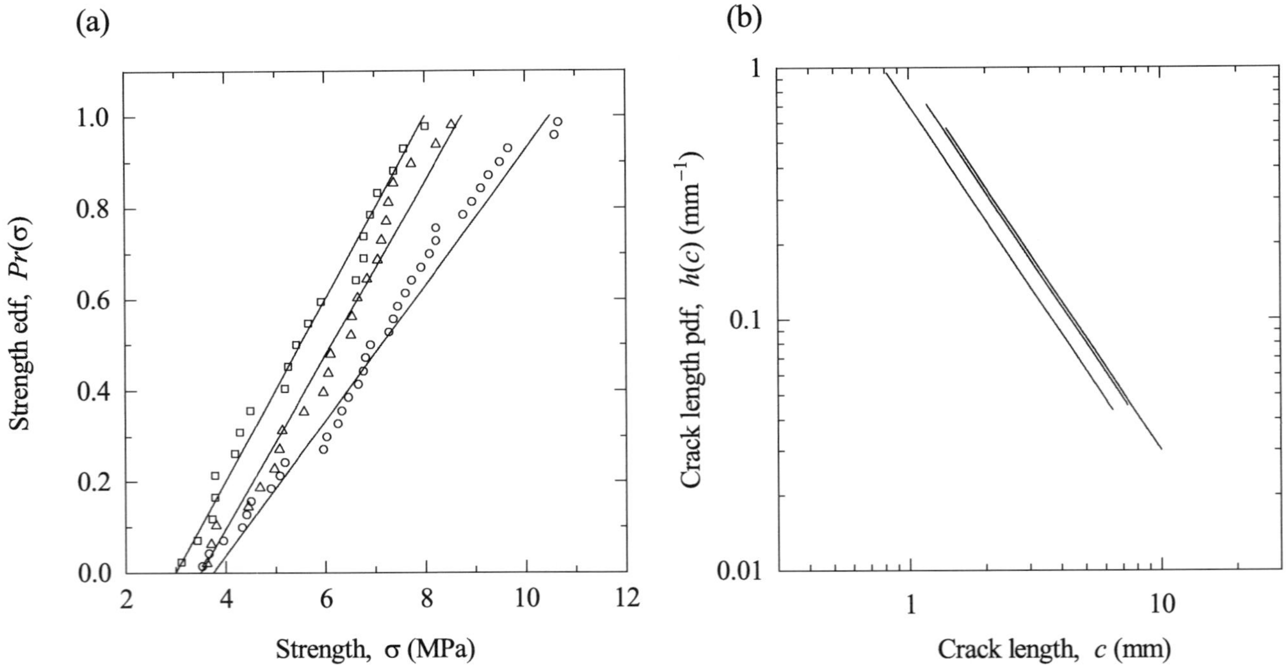

Figure 9.28 (a) Plot of strength edf behavior, $Pr(\sigma)$, for ice blocks. $D = (40, 70, \text{and } 150)$ mm, $N_{tot} = 111$ (adapted from Kuehn, G.A et al., 1993). Solid lines show independent linear best fits and the separated lower bound threshold strengths are indicative of deterministic size effects. (b) Plot in logarithmic coordinates of crack length pdf $h(c)$ variations for ice blocks; determined from (a).

40–150 mm in diameter, measured using a compression platen apparatus in the work of Kuehn et al. (1993), using data derived from the published work. Symbols represent individual strength measurements. Different symbols represent different particle sizes, D; increasing particle size is shown right to left. Results for 10 mm and 20 mm blocks were not significantly different from those of the 40 mm blocks. The data are separated by particle size and exhibit moderate strength domains σ_N/σ_1 of approximately 3. In common with the observations on cement particles above, features of Figure 9.28a are that the data exhibit threshold strengths that weakly decrease with increasing particle size and that the behavior of the strength distributions is near linear. The solid lines in Figure 9.28a represent unconstrained linear visual best fits to the data consistent with multiple deterministic flaw populations, Eq. (9.2). Deterministic size effects on strength are thus observed here as variations in the strength distribution domains and the slope within domains. Figure 9.27b shows as solid lines the crack length pdf $h(c)$ variations underlying the cement particle strength behavior, Eq. (9.4). The $h(c)$ variations were determined using $B = 0.3$ MPa m$^{1/2}$ and particle size and Eq. (7.2) as a constraint. In the logarithmic coordinates of Figure 9.28b, the variations of Eq. (9.4) appear as straight lines of slope $-3/2$ and block size increases left to right. Consistent with deterministic effects, the crack length distributions vary in domain location and in this case the distributions are all linear. In particular, as particle size increases the crack length domains shift to larger crack lengths.

9.5 Deterministic Strength and Flaw Size Analyses

An extensive survey of particle strength distributions controlled by deterministic size effects has been provided in the preceding sections. Such strength distributions are not described by Eq. (9.1), which is based on stochastic selection of flaws from a single, particle size invariant, flaw population. Instead, deterministic strength distributions are based on selection of flaws from multiple, particle size dependent, flaw populations. A feature of deterministic strength distributions, independent of distribution shape—concave, sigmoidal, or linear—is that the lower bound threshold strengths decrease with increasing particle size, reflecting the changing flaw populations. The implication is that the ensembles of flaws in larger particles in deterministic systems include larger flaws (the ensembles of flaws in stochastic systems are invariant and hence so is the strength threshold; see Chapter 7). Extreme value effects act equally in strength tests in both stochastic and deterministic systems, such that in both systems strength distributions are extreme value distributions: The distributions reflect the distribution of largest flaws from each ensemble. For all systems it is reasonable to assume that the size of the ensemble, here characterized by k, increases with increasing particle size, characterized by D, although the nature of the $k(D)$ scaling is *a priori* unknown. Hence, in both cases extreme value effects lead to strength distribution contraction as particle size increases. In the deterministic case, the size of the maximum flaw, here characterized by c_{max}, increases with increasing particle size, although the nature of the $c_{max}(D)$ scaling is also *a priori* unknown. Hence, in the deterministic case, strength distributions both contract and shift as particle size increases.

This section develops analyses for strength distributions of deterministic systems, similar to those describing stochastic systems culminating in Eq. (9.1). Two forms of strength distribution are considered, linear and concave. Linear distributions are considered initially, for three reasons. First, linear distributions are the baseline behavior for all cdf and edf variations. Convex, concave, and sigmoidal behaviors can all be viewed as variations on a linear background that extends in distribution space from (0,0) to (1,1). Hence, the basic physics of deterministic behavior can be elucidated with linear variations. Second, linearity enables tractable analyses in closed form, such that the effects of various parameter changes (e.g. k) are readily apparent. Third, as noted in the previous section, linear distributions describe a few sets of strength observations. Deterministic concave strength distributions are considered next. Such distributions constitute the majority of particle strength behavior. The linear deterministic strength distribution analysis takes a quasi-empirical reverse analysis approach to estimate underlying crack length populations from strength distributions (see Chapter 4). The concave deterministic strength distribution analysis takes a forward analysis approach to predict strength distributions from assumed crack length populations (see Chapter 3). Both analyses explicitly incorporate determination of flaw populations by particle size.

9.5.1 Linear Strength Distributions

The starting point for deterministic linear strength distribution analysis is Eq. (9.2) and inclusion of particle size D dependence for the equation parameters. Experimental observations show that the parameters σ_L and σ_U decrease with particle size D and the parameter β increases with D. A large particle size, D_Ω, is introduced that is an upper bound for the material. D_Ω characterizes the scaling behavior of the strength distribution parameters, here chosen to be simple inverse linear dependencies,

$$\sigma_L = \sigma_{L\Omega}(D_\Omega/D), \tag{9.5}$$

$$\sigma_U = \sigma_{U\Omega}(D_\Omega/D). \tag{9.6}$$

$\sigma_{L\Omega}$ is a lower limit strength bound for the largest particles ($D = D_\Omega$) in the system and $\sigma_{U\Omega}$ is the conjugate upper limit strength bound. An upper limit β_Ω to the derivative parameter of Eq. (9.3) is specified by the limiting strength bounds as

$$\beta_\Omega = 1/(\sigma_{U\Omega} - \sigma_{L\Omega}) \tag{9.7}$$

and Eq. (9.3) thus expressed with explicit linear particle size dependence as

$$\beta = \beta_\Omega(D/D_\Omega). \tag{9.8}$$

Eq. (9.2) can thus be re-expressed with explicit particle size dependence as

$$\begin{aligned} H(\sigma) &= \beta_\Omega(D/D_\Omega)\,[\sigma - \sigma_{L\Omega}(D_\Omega/D)] \\ &= \beta_\Omega\,[\sigma(D/D_\Omega) - \sigma_{L\Omega}]. \end{aligned} \tag{9.9}$$

The domain of this strength distribution is particle size dependent, $\sigma_{L\Omega}(D_\Omega/D) \le \sigma \le \sigma_{U\Omega}(D_\Omega/D)$. A limiting strength distribution $H_\Omega(\sigma)$ for Eq. (9.2) is specified by combining Eqs. (9.5) and (9.7) or by setting $D = D_\Omega$ in Eq. (9.9) to gain

$$H_\Omega(\sigma) = \beta_\Omega(\sigma - \sigma_{L\Omega}), \qquad \sigma_{L\Omega} \le \sigma \le \sigma_{U\Omega}, \tag{9.10}$$

noting that the domain of the limiting strength distribution is particle size independent.

Combining Eqs. (9.9) and (9.10) expresses the strength distribution for an arbitrary particle size in terms of the limiting distribution:

$$H(\sigma) = \left[\frac{\sigma(D/D_\Omega) - \sigma_{L\Omega}}{\sigma - \sigma_{L\Omega}}\right] H_\Omega(\sigma) \tag{9.11}$$

This equation, expressing the size dependence of deterministic strength distributions in terms of a limiting distribution, is analogous to Eq. (3.31) expressing the size dependence of stochastic strength distributions in terms of an invariant distribution. A key difference, noted earlier (Chapter 8), is that deterministic behavior is expressed in terms of the empirical strength distribution of a maximal component, whereas stochastic behavior is expressed in terms of the fundamental strength distribution of a minimal component. For two different size particles, D_1 and D_2, the strength distributions are

$$H_1(\sigma) = \left[\frac{\sigma(D_1/D_\Omega) - \sigma_{L\Omega}}{\sigma - \sigma_{L\Omega}}\right] H_\Omega(\sigma) \tag{9.12}$$

and

$$H_2(\sigma) = \left[\frac{\sigma(D_2/D_\Omega) - \sigma_{L\Omega}}{\sigma - \sigma_{L\Omega}}\right] H_\Omega(\sigma). \tag{9.13}$$

Combining Eqs. (9.12) and (9.13) to eliminate $H_\Omega(\sigma)$ expresses the linear strength distribution for one particle size in a deterministic system in terms of another:

$$H_2(\sigma) = \left[\frac{\sigma(D_2/D_\Omega) - \sigma_{L\Omega}}{\sigma(D_1/D_\Omega) - \sigma_{L\Omega}}\right] H_1(\sigma). \tag{9.14}$$

This expression is analogous to that for stochastic systems, Eq. (9.1). As might be anticipated for deterministic systems, there is a dependence on the particle sizes relative to an absolute value (D_Ω) and a dependence on the strength value relative to an absolute value ($\sigma_{L\Omega}$). The strength distributions are not related simply by a particle size ratio D_2/D_1 similar to that appearing in the stochastic formulation, k_2/k_1. In addition, the distributions do not have the same domain.

For large particles D_Ω, a limiting strength-controlling crack length population $h_\Omega(c)$ underlies the limiting strength distribution. From Eq. (9.4),

$$h_\Omega(c) = \mathrm{d}H_\Omega(c)/\mathrm{d}c = (\beta_\Omega B/2)c^{-3/2}, \qquad c_{L\Omega} \le \sigma \le c_{U\Omega}. \tag{9.15}$$

The crack length bounds on the limiting population are particle size independent and given by

$$c_{L\Omega} = (B/\sigma_{U\Omega})^2, \tag{9.16}$$

$$c_{U\Omega} = (B/\sigma_{L\Omega})^2. \tag{9.17}$$

For an arbitrary particle size $D \le D_\Omega$, the strength-controlling crack length population $h(c)$ underlying the strength distribution is thus

$$h(c) = (D/D_\Omega)h_\Omega(c), \qquad c_L \le \sigma \le c_U. \tag{9.18}$$

The crack length bounds on the population are particle size dependent and, in accord with intuition, decrease with particle size and are given by

$$c_L = (D/D_\Omega)^2 c_{L\Omega}, \tag{9.19}$$

$$c_U = (D/D_\Omega)^2 c_{U\Omega}. \tag{9.20}$$

An example of the deterministic linear strength distribution analysis is shown in Figure 9.29. The limiting strength distribution bounds were set as $\sigma_{L\Omega} = 1$ MPa and $\sigma_{U\Omega} = 5$ MPa, resulting in $\beta_\Omega = 0.25$ MPa^{-1}, and the limiting particle size was set as $D_\Omega = 10$ mm. The resulting linear strength distributions for particle sizes of D(i) = 1.25 mm, D(ii) = 2.5 mm, D(iii) = 5 mm, and D(iv) = 10 mm, using Eq. (9.9), are shown in Figure 9.29a. The distributions are similar to those of cement particles, Figure 9.27, as the bounding parameters are very similar. The bounding distributions are indicated in bold; particle size increases right to left. Note that in this example the convenient (but not required) choice of D(iv) = D_Ω was made so that $H(\sigma)$(iv) = $H_\Omega(\sigma)$. The $h(c)$ variations underlying the strength distributions were determined using $B = 0.1$ MPa

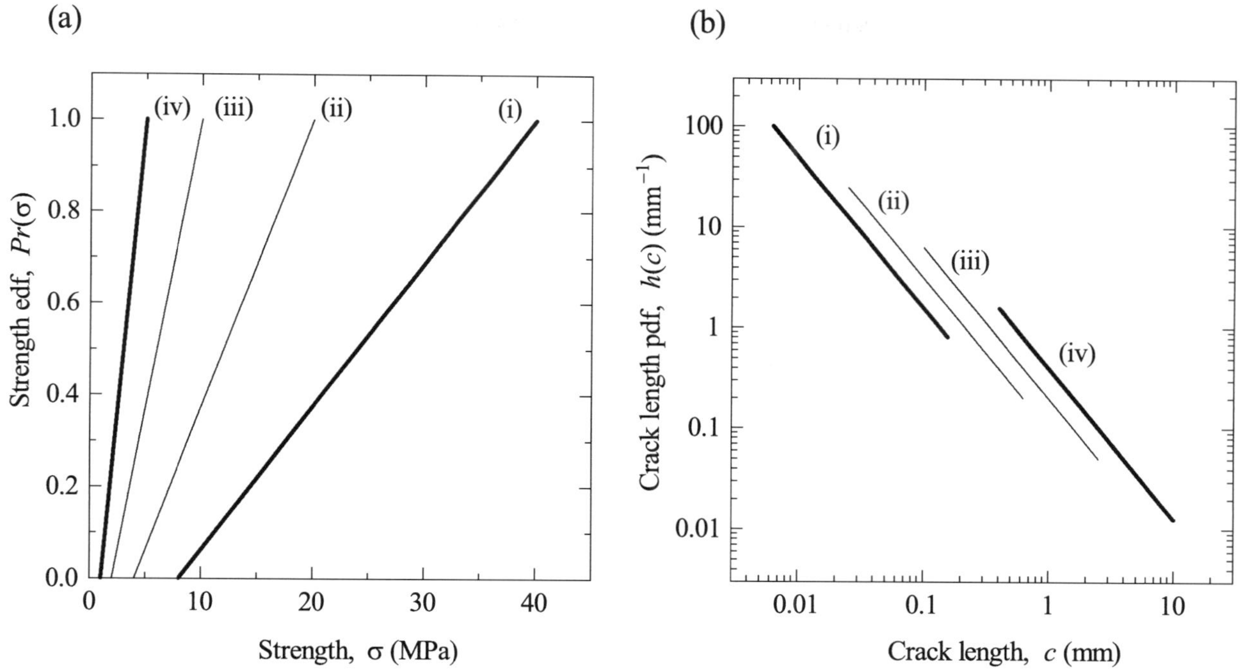

Figure 9.29 (a) Simulated deterministic linear strength distributions and (b) conjugate strength limiting crack length populations. (a) Increasing particle size right to left, (i)–(iv). (b) Increasing particle size left to right. The simulations derive from deterministic size effects acting on a base, large particle, population. Compare with Figure 9.27.

$m^{1/2}$ and Eqs. (9.15)–(9.20), noting that the particle size constraint is fulfilled by Eq. (9.17). The variations are shown in the logarithmic coordinates of Figure 9.29b. The variations appear as straight lines of slope $-3/2$. The responses of the bounding behavior are shown as bold solid lines and particle size increases left to right. Consistent with deterministic effects, the crack length distributions vary in domain location. In particular, as particle size increases the crack length domains shift to larger crack lengths.

The crack length pdf variations $h(c)$ of Figure 9.29b represent the distributions of strength controlling flaws that give rise to the particle strength edf variations of Figure 9.29a. The $h(c)$ variations are thus extreme value distributions of largest crack lengths, "selected" by strength testing from the ensembles of cracks within each of the particles. The distribution of crack lengths for each ensemble is identical (within samples of particles of identical size). In deterministic systems the ensemble distribution is described by a particle size dependent population pdf $f(c, D)$. In stochastic systems, $f(c)$ is independent of particle size. In both cases, f and h are connected by an extreme value relationship that depends on the size of the ensembles, k. The importance of this relationship is that flaw populations f, determined by particle manufacture, can be estimated from strength controlling flaw pdf variations h, assessed by particle strength measurements. The relationship is considered here for deterministic linear strength distribution behavior—this is *reverse analysis* (Chapter 4).

Analysis of ensemble and strength controlling flaw distributions begins with a re-statement of the extreme value relationship between the strength distribution of a population of fundamental elements, $F(\sigma, D)$, and the strength distribution of a sample of components, here particles, $H(\sigma, D)$, Eq. (3.31):

$$[1 - F(\sigma, D)]^k = [1 - H(\sigma, D)]\,. \tag{9.21}$$

Similar to Eq. (3.31), the fundamental elements contain $k = 1$ flaw each and the sampled particles contain k flaws (there are N particles in the sample, but this does not bear on extreme value calculations). Distinct from Eq. (3.31), a particle size dependence of $F(\sigma, D)$ is included, appropriate for deterministic systems. In stochastic systems, $F(\sigma)$ is independent of particle size. It is implicit that $k = k(D)$. Using the relationships, Eq. (3.29), between crack length and strength distributions, Eq. (9.21) can be re-expressed as

$$F(c, D)^k = H(c, D)$$
$$F(c, D) = H(c, D)^{1/k}, \tag{9.22}$$

and as $f(c) = \mathrm{d}F(c)/\mathrm{d}c$ and $h(c) = \mathrm{d}H(c)/\mathrm{d}c$,

$$f(c, D) = \frac{1}{k} H(c, D)^{(1-k)/k} h(c, D). \tag{9.23}$$

For the linear system here, combining Eq. (3.29) and Eq. (9.2) gives

$$H(c, D) = \beta(\sigma_U - \sigma) = \beta(Bc_L{}^{-1/2} - Bc^{-1/2}) \tag{9.24}$$

and thus Eq. (9.23) becomes

$$f(c, D) = \frac{1}{k}\left[\beta B(c_L{}^{-1/2} - c^{-1/2})\right]^{(1-k)/k}(\beta B/2)c^{-3/2}, \tag{9.25}$$

in which the crack length dependence is expressed explicitly and the particle size dependence is implicit in $\beta(D)$, $c_L(D)$, and $k(D)$. The important large crack asymptotic behavior of Eq. (9.25) is $f(c, D) \sim c^{-2}$ as $c \to c_U$. For small cracks, $f(c, D)$ approaches very large values as $c \to c_L$.

Figure 9.30 shows a logarithmic plot of deterministic crack length population behavior $f(c)$ from Eq. (9.25). The behavior uses the $h(c)$ information from Figure 9.29, approximating the cement response of Figure 9.27, and assuming a k variation of $k(D) = k_\Omega(D/D_\Omega)^2$, consistent with the schematic diagrams of Figure 9.1. In this case $k_\Omega = 64$ was selected. The behavior of $f(c, D)$ from Eq. (9.25), reflecting the entire ensemble of flaws in a particle, is shown as the fine dark lines. The behavior of $h(c, D)$ from Figure 9.29b, reflecting the extreme strength controlling flaws, is shown as bold grayed lines. The largest particle size (iv) has $D = 10$ mm $= D_\Omega$, and thus contains an ensemble of 64 flaws. $f(c)$ for this particle size exhibits a clear maximum at the lower bound crack length and then decreases significantly with increasing crack length before approaching the c^{-2} asymptote at large crack lengths. Most flaws in these large particles are small, close to the lower bound in size, and only a few are larger. $h(c)$ for this particle size is a straight line exhibiting $c^{-3/2}$ behavior that lies midway between the $f(c)$ maximum and the asymptote. The implication is clear and expected: In physical terms, the extreme value nature of strength testing forms a crack length distribution from the rare large flaws and essentially ignores the small flaws. In mathematical

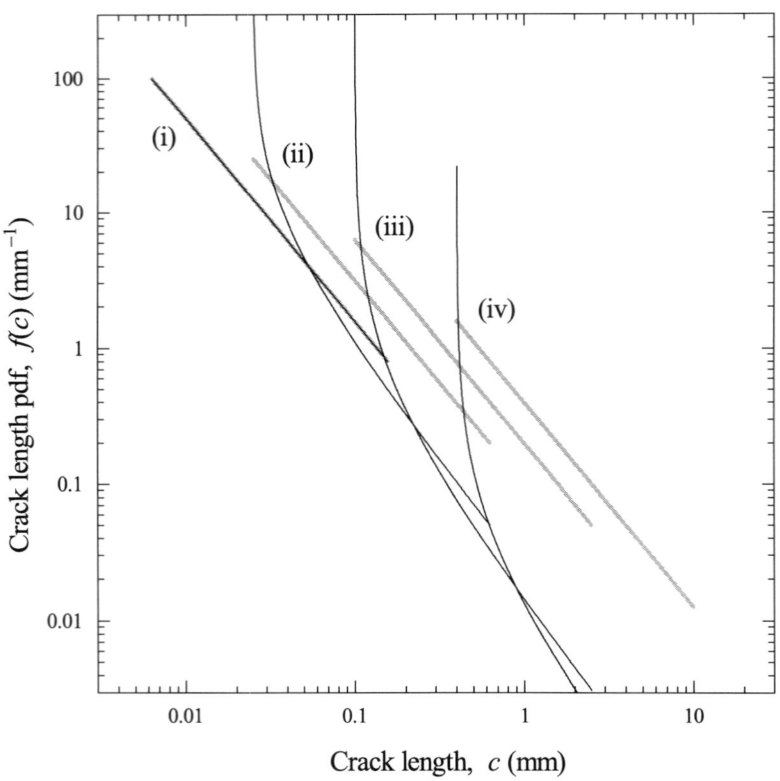

Figure 9.30 Simulated crack length populations, black lines, of particles giving rise to deterministic linear strength variations, Figure 9.29a. Increasing particle size left to right (i)–(iv). The extreme value flaw populations are shown as gray lines from Figure 9.29b.

terms, the area under the maximum peak is transferred to the area above the lower asymptote, so as to form a straight line. Similar, but increasingly less extreme, phenomena occur for the smaller particle sizes (iii) and (ii) that contain ensembles of 16 and 4 flaws, respectively. The smallest particle size (i) has $D = 1.25$ mm $= D_\Omega/8$, and thus contains $k = 1$ flaw. $f(c)$ for this particle size is identical to $h(c)$, Eq. (9.23), and the absence (or limit) of extreme value effects is reached.

9.5.2 Concave Strength Distributions

The deterministic concave strength distribution analysis developed here builds on three aspects of particle strength and crack length analyses developed earlier. The first is the probabilistic framework linking population pdf behavior f to population cdf behavior F by differentiation (or F to f by integration) and the Griffith equation, as developed in Chapter 3 and applied throughout. The second is population domain perturbation by truncation, first introduced in Chapter 5 in consideration of deterministic effects on strengths, and then applied in Chapter 8 in description of deterministic effects on crack lengths. The third is the extreme value framework linking population crack length behavior described by f or F to sample strength behavior described by h or H as mediated by component size k, as developed in Chapter 3 and applied throughout. These three aspects are applied here in sequence, using the concepts of particle size D dependent crack length bounds and component size and a limiting large particle size D_Ω, as developed in Section 9.5.1.

Analysis begins with the establishment of a limiting strength population cdf, $F_\Omega(\sigma)$. The domain of this cdf extends from a lower bound of $\sigma_{L\Omega}$ to an upper bound of $\sigma_{U\Omega}$ and encompasses the entirety of all strengths to be described, $\sigma_{L\Omega} \leq \sigma \leq \sigma_{U\Omega}$. $F_\Omega(\sigma)$ describes the distribution of strengths in limiting large sized particles $D = D_\Omega$ containing limiting large ensembles of fundamental volume elements, $k = k_\Omega$. It is convenient to establish $F_\Omega(\sigma)$ using the form of an observed strength cdf and modify the bounds accordingly. The strength population cdf has a conjugate crack length cdf, $F_\Omega(c)$, given by $F_\Omega(c) = 1 - F_\Omega(\sigma)$, using the inverted Griffith equation $c = (B/\sigma)^2$, constraining B by particle size as seen earlier. The related limiting crack length pdf, $f_\Omega(c)$, is given by $f_\Omega(c) = \mathrm{d}F_\Omega(c)/\mathrm{d}c$. The domain of the pdf is conjugate to that of strength and extends from a lower bound of $c_{L\Omega}$ to an upper bound of $c_{U\Omega}$ and encompasses the entirety of all crack lengths to be described, $c_{L\Omega} \leq \sigma \leq c_{U\Omega}$.

It is not necessary that limiting particles of size D_Ω or limiting distributions $F_\Omega(\sigma)$ or $f_\Omega(c)$ physically exist. However, $f_\Omega(c)$ provides the analytical basis for description of crack length and strength distributions observed or inferred for particles of arbitrary size $D \leq D_\Omega$. In particular, using the idea and notation of Eq. (8.1), the population of cracks, $f(c, D)$ for particles of size D is described by the product

$$f(c, D) = f_\Omega(c)\Delta f(D). \tag{9.26}$$

$\Delta f(D)$ is a deterministic, particle size dependent, perturbation that modifies the limiting, maximal population to generate a particle size dependent population pdf $f(c, D)$. As earlier, $\Delta f(D)$ is a step function,

$$\Delta f(D) = \Delta f(c, a, 0, c_U) = \begin{cases} a & : c < c_U) \\ 0 & : c \geq c_U) \end{cases}, \tag{9.27}$$

where the step function acts at the crack length c_U and steps between a and 0. Similar to the behavior shown in Figure 8.6a, the domain of the limiting population is effectively truncated at an upper bound crack length c_U that is particle size dependent, $c_U = c_U(D)$. The particle size dependent population cdf $F(c, D)$ is then given by integration of the pdf

$$F(c, D) = \int_{c_{L\Omega}}^{c} f(u, D)\mathrm{d}u, \tag{9.28}$$

where u is a dummy crack length variable. Note that the lower bound $c_{L\Omega}$ is unchanged from the limiting population value and Eqs. (9.26) and (9.27) lead to

$$\int_{c_U}^{c_{U\Omega}} f(u, D)\mathrm{d}u = 0. \tag{9.29}$$

Normalization requires

$$F(c_U, D) = \int_{c_{L\Omega}}^{c_U} f(u, D)\mathrm{d}u = 1, \tag{9.30}$$

setting the value of a in Eq. (9.27). All particles of size D contain distributions of cracks that are described by the cdf $F(c, D)$ and pdf $f(c, D)$. For two different size particles, D_1 and D_2, $F_1(c, D_1) \neq F_2(c, D_2)$, although F_1 and F_2 are related by the deterministic Eq. (9.26) (for stochastic systems $F_1 = F_2$).

Particles of size D each contain ensembles of $k = k(D)$ cracks; each ensemble described by $F(c, D)$, as is a sample of N such particles. Each ensemble (and therefore particle) contains a largest crack. The distribution of largest cracks in a sample is an extreme value distribution $H(c, D)$, given by $H(c, D) = F(c, D)^k$ (Eq. 3.29). Strength measurements are only sensitive to the largest crack in a component and hence the distribution of particle strengths in a sample is a also an extreme value distribution $H(\sigma, D)$, given by $H(\sigma, D) = 1 - H(c, D)$, using the Griffith equation $\sigma = Bc^{-1/2}$. The strength controlling crack length population pdf $h(c, D)$ is given by $h(c, D) = \mathrm{d}H(c, D)/\mathrm{d}c$. This entire process, leading to prediction of $H(\sigma, D)$ and $h(c, D)$ from an initial assumed crack length population is an example of *forward analysis*.

An example of the deterministic concave strength distribution analysis is shown in Figure 9.31 and Figure 9.32. (For easier comparison with experimental observations, the crack length behavior that is the basis for analysis is shown after the resulting strength behavior, as in the combination of Figures 8.8 and 8.9.) The limiting strength distribution bounds were set as $\sigma_{L\Omega} = 6.5$ MPa and $\sigma_{U\Omega} = 165$ MPa, resulting in limiting crack length bounds of $c_{L\Omega} = 5.9$ μm and $c_{U\Omega} = 3.8$ mm, using $B = 0.4$ MPa m$^{1/2}$. These parameters are similar to those describing quartz particles, Figures 9.7 and 9.8. The form of the limiting strength distribution and resulting limiting crack length population, $F_\Omega(\sigma)$ and $f_\Omega(c)$, were set by a concave fit to the $D = 0.25$ mm quartz particle response, Figure 9.7. The limiting large particle size was set as $D_\Omega = 4$ mm and the particle sizes considered were $D = (0.25, 0.5, 1, 2,$ and $4)$ mm. The upper bound truncation crack lengths for the particle populations was set as $c_U = 3(D/D_\Omega)$ mm. The number of fundamental elements and cracks for the particle populations was set as $k = 20(D/D_\Omega)$. The largest particles thus contained 20 cracks with an upper bound crack length of 3 mm. The smallest particles contained 1.25 cracks (on average) with an upper bound crack length of 0.19 mm. The resulting strength distributions are shown in Figure 9.31. Particle size increases right to left, (i)–(v). The distributions are truncated at $Pr(\sigma) = 0.99$, typical of experimental data and finite $N = 100$. The distributions are all concave and the resemblance to Figure 9.7, both qualitatively and quantitatively, is clear, including the decreasing strength threshold and strength dispersion with increasing particle size.

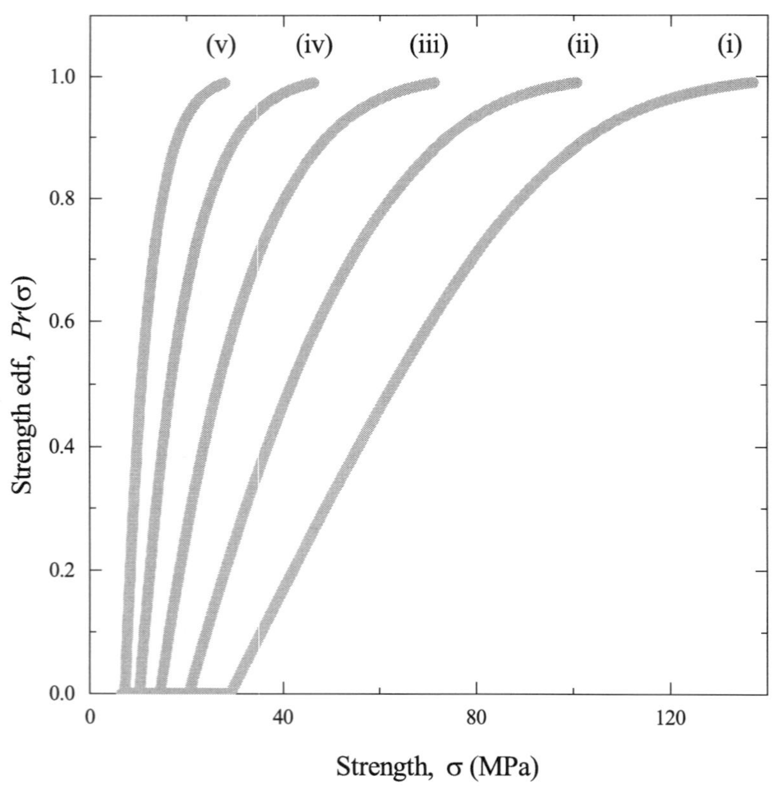

Figure 9.31 Simulated deterministic concave strength distributions. Increasing particle size right to left, (i)–(v). The simulations derive from deterministic size effects acting on a base, large particle, population. Compare with Figure 9.3.

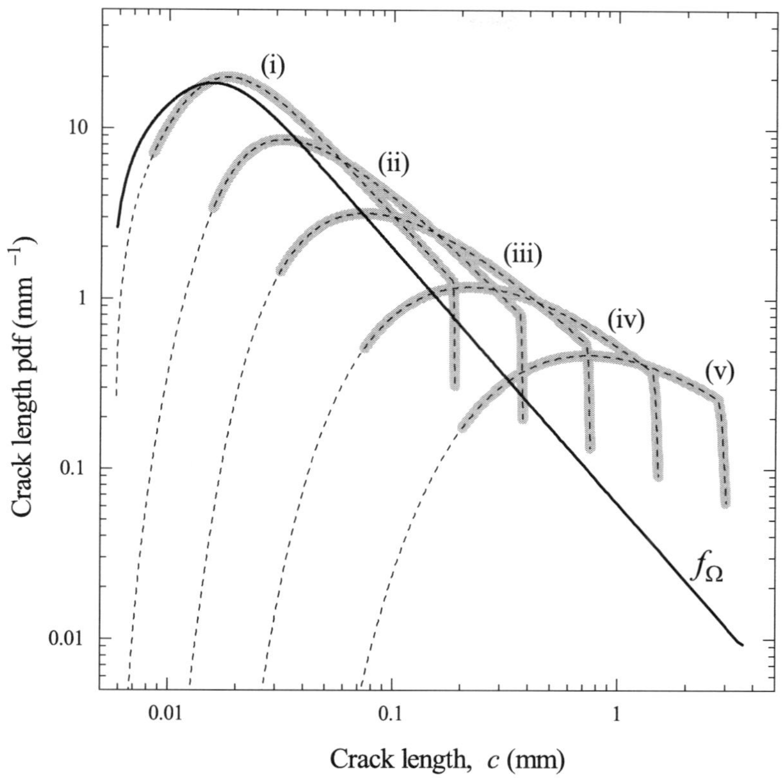

Figure 9.32 Plot in logarithmic coordinates of crack length pdf $h(c)$ variations conjugate to the strength simulation of Figure 9.31. Base population shown as fine solid line. Extreme value distributions shown as dashed lines. Increasing particle size left to right, (i)–(v). Experimentally attainable domains shown as gray lines.

Figure 9.32 shows a logarithmic plot of the strength controlling crack length populations underlying the strength behavior. The limiting strength population pdf $f_\Omega(c)$ is shown as the fine solid line and consists of a well-formed peak at small crack lengths and an extended large crack tail of slope $-3/2$. Following the convention of Chapter 7, the full strength controlling crack length populations, $h(c, D)$, determined by truncation and extreme value calculations from $f_\Omega(c)$, are shown as the fine dashed lines. The partial crack length populations, constrained by the expected observed strength domains, and sub sets of $h(c, D)$ are shown by the bold grayed lines. The crack length responses increase with particle size left to right. The response of the smallest particle size (i) most closely approximates the limiting population, with a peak at small crack lengths and a large crack tail. The response is slightly displaced from the limiting population due to $k = 1.25$ rather than $k = 1$, and the peak is constrained at small crack lengths due to the finite N effect. The $h(c, D)$ response differs most clearly from the $f_\Omega(c)$ behavior in the existence of a short downturn at large crack lengths associated with the particle size dependent threshold $c_U(D)$. The weakly defined peak and reduced tail features of the largest particle size (v) deviate most from the limiting population. The response is greatly displaced from the limiting population and again differs from the limiting behavior with the existence of a short downturn at large crack lengths. As in the linear responses of Figure 9.30, the $h(c, D)$ responses intersect the $f_\Omega(c)$ behavior. The intersections indicate many more large strength controlling crack lengths and many fewer small strength controlling crack lengths than in the limiting population. The detailed behavior of the particle size responses (ii)–(iv) are intermediate between responses (i) and (v), although there is a common feature of a short downturn at large crack lengths associated with particle size dependent thresholds. Overall, the clear and detailed resemblance of Figure 9.31 and Figure 9.32 to the experimental data surveyed in this chapter indicate that the analysis developed captures the physics and mathematical description of concave deterministic particle strength behavior.

9.6 Summary and Discussion

Chapter 9 has surveyed experimental observations of particle strength distributions that exhibit deterministic size effects. In such effects, particles sampled from a material exhibit flaw populations that depend on the particle size D. In particular, the crack length probability density function $f(c)$ depends on D, such that $f(c) = f(c, D)$. The spatial density of the population of flaws λ may also depend on D, such that $\lambda = \lambda(D)$. Many of the phenomena associated with particle systems that exhibit stochastic size effects (see Chapters 3 and 7) also pertain to deterministic particle systems: Each particle contains a number of cracks, k, forming an ensemble selected from the deterministic population $f(c, D)$. The ensembles of cracks in a sample of like-sized particles are probabilistically identical and described by $f(c, D)$. The largest cracks in each ensemble form an extreme value distribution, characterized by a crack length probability density function $h(c, D)$. In stochastic size effects, $f(c)$ is invariant and the relationship between the cumulative distribution functions of $h(c)$ and $f(c)$, $H(c)$ and $F(c)$ respectively, are given by a simple power law in k that extends to all particle sizes. In deterministic size effects, although $f(c, D)$, and thus $F(c, D)$, are not invariant, the relationship between the resultant cumulative distribution functions of $f(c, D)$ and $h(c, D)$, $H(c, D)$ and $F(c, D)$, remains a power law, as the relationship is based on extreme value principles. However, the relationship is restricted to a single particle size. A particularly clear factor is that the domain of $H(c, D)$ may depend on D. As the strength of a particle is determined by the largest crack in the ensemble of cracks contained by the particle, the strength distribution of a sample of particles, $H(\sigma, D)$ is determined by $H(c, D)$ and related to $F(c, D)$ as for stochastic systems. The critical distinction for deterministic systems is that the $H(\sigma, D)$ responses for different D, and thus k, values are not related by a power law. Experimentally, deterministic size effects are demonstrated by showing that samples of particles of different D exhibit strength distributions that are inconsistent with the relation Eq. (9.1). A clear inconsistency, following the point regarding the extreme value crack length domain, is that the strength domain is also dependent on D, most commonly observed in threshold strength variation.

Deterministic size effects in strength behavior were demonstrated here in an extensive survey of a broad range of particle systems. As in the particle systems exhibiting stochastic size effects, the deterministic systems included natural (e.g. salt, coal) and engineered (e.g. glass, cement) materials, and a wide domain of particle sizes (0.115–90 mm). Many of the particle materials that exhibited deterministic behavior here had also exhibited stochastic behavior in earlier chapters (e.g. alumina, limestone, coral, quartz, rock, and glass—see Chapter 7). Deterministic behavior was observed for both the predominant, concave, form of particle strength distributions and the less frequently observed sigmoidal and linear forms. In addition to exploring the phenomena of deterministic size effects in particles, goals of the survey were the elucidation of the form of $k(D)$, thus providing insight into λ, and determination of the parameter B linking particle strength σ and crack length c.

Figure 9.33 summarizes the behavior of strength-controlling crack length populations for particles with strength distributions that exhibited deterministic extreme value size effects. Crack length populations for salt, solid lines, (Figure 9.12) and limestone, dashed lines, (Figure 9.4) are shown in Figure 9.33 using the equi-axed logarithmic plotting scheme used earlier. The fine black lines in Figure 9.33 represent the strength-controlling crack length populations of the smallest particles in each case, 0.7 mm and 2.4 mm, respectively. The bold gray lines in Figure 9.33 represent the strength-controlling crack length populations of the largest particles in each case, 4 mm and 63 mm. In changing materials, from salt to limestone, the material flaw population changes, indicated by the lower arrow. For a given material, increasing the particle size shifts, contracts, and alters the shape of the strength-controlling crack population, indicated by the upper arrows. The shifts in crack populations between small and large particles leads to incomplete overlap of the strength controlling crack populations for a given material. In particular, for particles of different sizes, the large crack bounds of the strength controlling flaw populations are not coincident. The contractions and shape alterations for changes from small to large particles lead to centralization of population peaks within the contracted domains and much greater proportions of larger cracks in larger particles. The extent of contraction depends on the nature of the crack density λ, and thus the form of $k(D)$, and the ratios of the particle sizes D measured. The ratios of the largest/smallest particles for the materials in Figure 9.33 increased significantly from ≈ 6 for salt to ≈ 25 for limestone. The greater flaw population domain contraction for salt relative to limestone is consistent with a much more strongly increasing $k(D)$ function for salt. The large crack asymptotic behavior of $c^{-3/2}$ for individual systems is indicative of linear strength distribution behavior adjacent to strength thresholds. The overall c^{-1} behavior for the group of systems is indicative of $f(c, D)$ normalization. The relative positions of the crack populations for materials depends on the fracture resistance as quantified by the crack length-strength parameter B. The larger crack length domain positions for limestone here relative to salt imply a greater B value for limestone. The particle material dependence of the $k(D)$ variation and the value of B can be evaluated from the results of the strength distribution survey encapsulated in plots such as Figure 9.33.

The strength distributions considered in this chapter, consisting of concave, sigmoidal, and linear behavior, provided clear evidence of deterministic size effects—as particle size increased, sample strength distribution domains shifted to smaller strengths. Such behavior is inconsistent with stochastic size effects. The shifts reflected the appearance of new or more frequent, larger extreme flaws within a sample with increasing particle size. As discussed in Chapters 5, 6, and 8, generalized ratios of the maximum extreme crack lengths and particle sizes (relative to the behavior of a small, reference particle)

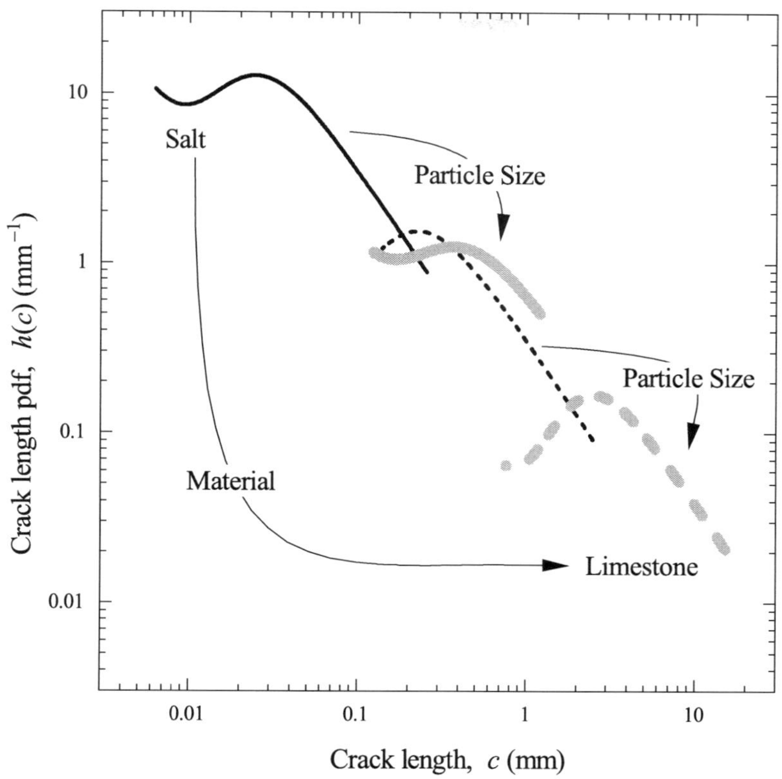

Figure 9.33 Plot in logarithmic coordinates of crack length pdf $h(c)$ variations for salt and limestone particles repeated from Figure 9.12 and Figure 9.4. Particle sizes are salt 0.4 mm and 4 mm and limestone 2.4 mm and 63 mm. Fine lines indicate material flaw populations sensed by small particles, bold lines indicate flaw populations sensed by large particles.

enable comparison of multiple particle systems and evaluation of the deterministic scaling of the maximum crack length with particle size. (For stochastic systems maximum crack length is invariant.) Here, values of maximum crack lengths c in samples of particles size D, inferred from analysis of strength distributions, are compared to a single reference configuration c_{ref} and D_{ref} within a given system. Comparison of c/c_{ref} and D/D_{ref} between systems is thus relative to common reference point (1, 1).

Figure 9.34 shows a plot in logarithmic coordinates of the relative size of flaws c/c_{ref} vs the relative particle size D/D_{ref} for the systems studied in this chapter. The format is the same as Figure 8.12. The crack lengths were determined from the lower bounds of the fits to the strength distributions. The bars represent uncertainties determined from the particle size ranges given in the cited works. The solid lines are guides to the eye of slopes 1 and 2. The shaded band indicates the range of responses observed for extended components over the same generalized component size domain, Figure 5.28. Note that the relative particle size domain is much greater here. The majority open symbols indicate measurements from particle systems that exhibited concave strength distributions. The filled symbols indicate measurements from glass particle systems that exhibited sigmoidal strength distributions. The partially filled symbols indicate measurements from particle systems that exhibited linear strength distributions. There are several points to note in Figure 9.34: The first point is that the vast majority of observed behavior is restricted to a small domain of relative particle sizes, approximately a factor of 5, compared with nearly 100 for extended components, Figure 5.28. There are exceptions, but the small relative particle size domain implies that it would be difficult to discern a clear trend in the dependence of maximum crack length with particle size. The second point is that there is a considerable diversity of responses, extending from an apparent dependence of maximum crack length on particle size of greater than $c \sim D^2$ to less than $c \sim D^{0.5}$. This diversity is largely restricted to the particle systems that exhibited concave strength distributions (open symbols). The third point is that the particle systems that exhibited sigmoidal and linear strength distributions (filled and partially filled symbols) exhibited dependencies of maximum crack length on particle size of approximately $c \sim D^2$.

The extensive survey here suggests that maximum crack lengths in deterministic particle systems exhibit much greater dependence on particle size than extended components, in agreement with the limited data of Chapter 8. The successful

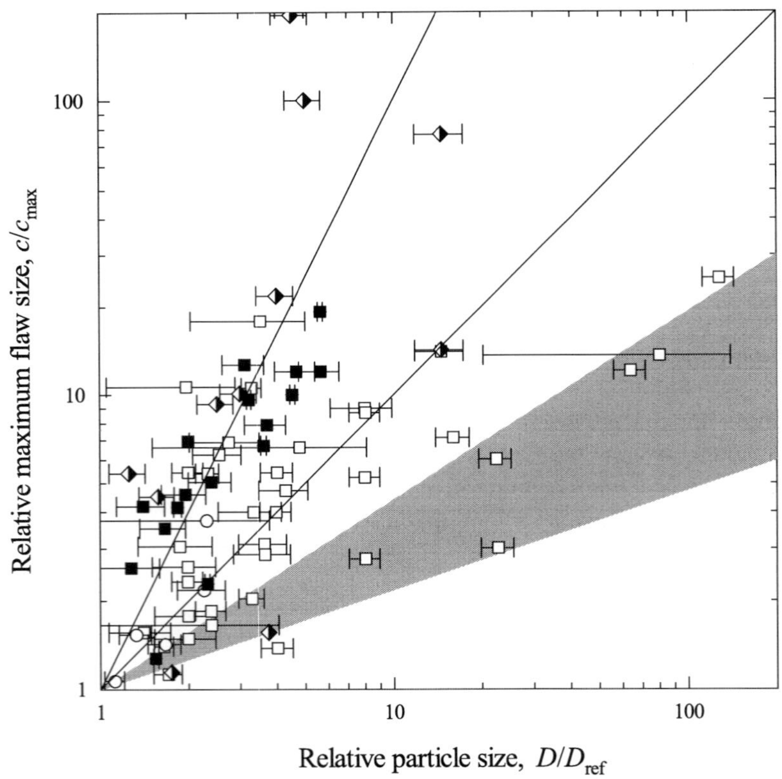

Figure 9.34 Logarithmic plot of maximum size of flaws c vs particle size D for particles exhibiting deterministic size effects on strength distributions. The particle systems were those considered in this chapter: open symbols are systems that exhibited concave strength behavior, filled symbols, sigmoidal behavior, and partially filled symbols, linear behavior. The lines are of slope 1 and 2 and are guides to the eye; the shaded band indicates the behavior of extended components.

descriptions of linear and sigmoidal strength distributions by the deterministic strength analyses developed in Section 9.5 are in part based on the $c \sim D^2$ (Eq. (9.20)) and $c \sim D^1$ scaling of Figure 9.34. Overall, however, although the experimental evidence for deterministic size effects in many strength distributions of particles is substantial and significant, as shown throughout this chapter, the evidence in support of a particular scaling law is weak. The overall trend is at best semi-quantitative, suggesting that maximum flaw size and particle size may both be determined by similar mechanisms in the particle fabrication process. In contrast to the limited number of stochastic particle systems from which to draw scaling information, there is a significant number of deterministic particle systems. However, data interpretation is still hampered by the effects of distorting and obscuring linearized plotting schemes and the application of such plotting schemes leading to the mistaken impression that observed size effects are due to stochastic behavior rather than deterministic behavior, as detailed here in consideration of the studies of Wong et al. (1987, ceramic particles, Chapter 8) and Kschinka et al. (1986, glass particles, Section 9.3.1).

In interpreting strength distribution data in terms of underlying crack length distributions, an invariant B value controlling the relationship between strength and flaw size was selected for each material. The selection was based on consistent implementation of the particle size constraint, Eq. (7.2). The invariant values, although not required, lead to simplifying consistency between different particle sizes of the same material. The particle size constraint is required, but may lead to inconsistency between materials if, as here, the upper bound B value is selected. A particular case is that of glass particles (Section 9.3.1), for which unphysically large values (1.3 MPa $m^{1/2}$–3 MPa $m^{1/2}$) were implemented, leading to very large upper bound crack lengths (that were still smaller than the particle size). A realistic B value for glass, consistent with the behavior of extended glass components and the behavior of particles of other materials is $B = 0.5$ MPa $m^{1/2}$. Note that implementation of such a value still meets the particle size constraint, does not alter the relative positions of crack length distributions, and thus does not alter the scaling behavior of Figure 9.34. Using the B values determined from the strength-crack length analyses of the experiments in this chapter and the amended values for glass, Figure 9.35 shows a semi-logarithmic plot in a format similar to Figure 7.27 of the failure resistance parameter B. The behaviors of extended

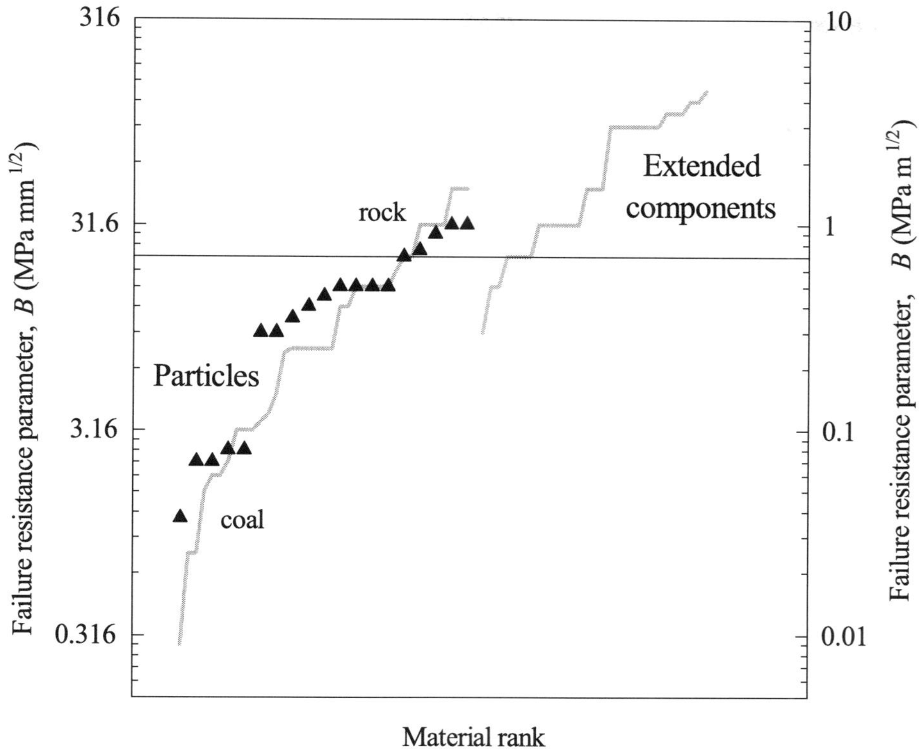

Figure 9.35 Plot of ranked failure resistance parameter B for deterministic particle systems using values determined from strength-crack length analyses of experimental observations in this chapter. Behaviors of extended components and stochastic particle systems are shown as gray lines. Extended component values are in agreement with materials considerations; particle values are constrained by particle size.

components and stochastic particles are shown as gray lines. The values for deterministic particles are indicated by filled symbols and extend over the range 0.04 MPa $m^{1/2}$–1.0 MPa $m^{1/2}$. The extremes are labeled, coal to rock. The range of B values for particle systems that exhibit deterministic size effects on strength is slightly smaller than that for particle systems that exhibit stochastic size effects. The values for both particle systems are less than those for extended components, implying that particles are weaker than their extended analogues independent of the flaw and strength scaling.

In addition to the extensive survey of experimental behavior, this chapter has developed two analyses describing the strength behavior of particle systems that are controlled by deterministic size effects. The first analysis describes deterministic variation of linear strength distributions (Section 9.5.1). The second analysis describes deterministic variation of concave strength distributions (Section 9.5.2). In both cases, the analyses include as central elements decreases in the lower bound strength threshold and the strength domain width with increasing particle size. In the first case, the analysis was applied to simulate the linear strength distribution of cement particles, Figure 9.29 and based on the reverse analysis of Chapter 4. In the second case, the analysis was applied to simulate the concave strength distribution of quartz particles, Figure 9.31 and based on the forward analysis of Chapter 3. In both cases, the similarity of the simulations to the observations suggests that the physical principles underlying the simulations were correct. The simulations are based on continuum representations of crack length and strength distributions. Extension to discrete representations (as in Chapter 3) is straightforward as the mathematical construction of the simulations is implemented simply in numerical form.

This chapter has examined a weak relationship between strength distributions of different sized particles: that in which the ensembles of flaws in particles are selected from multiple populations, $f(c, D)$, that depends on particle size. Particle size determines the extreme value distribution of the largest flaws in a sample of particles, $h(c, D)$, but the domain and form of $h(c, D)$ for one particle size are not simply related to those of another particle size. Hence, the domains and forms of the conjugate strength distributions of different sized particles are also not simply related, and specifically not related by Eq. (9.1). The ensuing behavior gives rise to deterministic extreme value size effects in strength distributions of particles, the subject of this chapter. This chapter and the previous three chapters have predominantly considered the strength behavior of particles of hard, stiff materials in which interpretations of brittle fracture—stochastic and deterministic—are straightforward. In studies of such particles, there is frequently sufficient data to warrant development and application of the appropriate analysis. The following two chapters, Chapters 10 and 11, move away from this survey then analyze approach to consider more compliant (often termed "soft") materials. For these materials, brittle fracture interpretations of strength are less straightforward and may not even be applicable and there is frequently insufficient data to permit detailed analysis. Chapter 10 surveys the behavior of agglomerate particles and Chapter 11 surveys the behavior of elastomer and gel particles.

References

Aman, S., Tomas, J., and Kalman, H. (2010). Breakage probability of irregularly shaped particles. *Chemical Engineering Science* 65: 1503–1512.

Barrios, G.K.P., de Carvalho, R.M., and Tavares, L.M. (2011). Modeling breakage of monodispersed particles in unconfined beds. *Minerals Engineering* 24: 308–318 .

Bertrand, P.T., Laurich-McIntyre, S.E., and Bradt, R.C. (1988). Strengths of fused and tabular alumina refractory grains. *American Ceramic Society Bulletin* 67: 1217–1221.

Brecker, J.N. (1974). The fracture strength of abrasive grains. *Journal of Engineering for Industry* 96: 1253–1257.

Brzesowsky, R.H., Spiers, C.J., Peach, C.J., and Hangx, S.J.T. (2011). Failure behavior of single sand grains: Theory versus experiment. *Journal of Geophysical Research* 116: B06205.

Dong, J., Cheng, Y., Hu, B., Hao, C., Tu, Q., and Liu, Z. (2018). Experimental study of the mechanical properties of intact and tectonic coal via compression of a single particle. *Powder Technology* 325: 412–419.

Gorham, D.A. and Salman, A.D. (2005). The failure of spherical particles under impact. *Wear* 258: 580–587.

Hu, W., Dano, C., Hicher, P.Y., Le Touzo, J.Y., Derkx, F., and Merliot, E. (2011). Effect of sample size on the behavior of granular materials. *Geotechnical Testing Journal* 34: 186–197.

Huang, J., Xu, S., Yi, H., and Hu, S. (2014). Size effect on the compression breakage strengths of glass particles. *Powder Technology* 268: 86–94.

Huang, H., Huang, Q.K., Zhu, X.H., and Hu, X.Z. (1993). An experimental investigation of the strengths of individual brown corundum abrasive grains. *Scripta Metallurgica* 29: 299–304.

Huang, H., Zhu, X.H., Huang, Q.K., and Hu, X.Z. (1995). Weibull strength distributions and fracture characteristics of abrasive materials. *Engineering Fracture Mechanics* 52: 15–24.

Jansen, U. and Stoyan, D. (2000). On the validity of the Weibull failure model for brittle particles. *Granular Matter* 2: 165–170.

Koohmishi, M. and Palassi, M. (2016). Evaluation of the strength of railway ballast using point load test for various size fractions and particle shapes. *Rock Mechanics and Rock Engineering* 49: 2655–2664.

Kschinka, B.A., Perrella, S., Nguyen, H., and Bradt, R.C. (1986). Strengths of glass spheres in compression. *Journal of the American Ceramic Society* 69: 467–472.

Kuehn, G.A., Schulson E.M., Jones D.E., and Zhang J. (1993). The compressive strength of ice cubes of different sizes. *Journal of Offshore Mechanics and Arctic Engineering* 115: 142–148.

Lim, W.L., McDowell, G.R., and Collop, A.C. (2004). The application of Weibull statistics to the strength of railway ballast. *Granular Matter* 6: 229–237.

Lobo-Guerrero, S. and Vallejo, L.E. (2006). Application of Weibull statistics to the tensile strength of rock aggregates. *Journal of Geotechnical and Geoenvironmental Engineering* 132: 786–790.

Ma, L., Li, Z., Wang, M., Wei, H., and Fan, P. (2019). Effects of size and loading rate on the mechanical properties of single coral particles. *Powder Technology* 342: 961–971.

McDowell, G.R. (2002). On the yielding and plastic compression of sand. *Soils and Foundations* 42: 139–145.

Nad, A. and Saramak, D. (2018). Comparative analysis of the strength distribution for irregular particles of carbonates, shale, and sandstone ore. *Minerals* 8: 37.

Ovalle, C., Frossard, E., Dano, C., Hu, W., Maiolino, S., and Hicher, P.Y. (2014). The effect of size on the strength of coarse rock aggregates and large rockfill samples through experimental data. *Acta Mechanica* 225: 2199–2216.

Parab, N., Guo, Z., Hudspeth, M.C., Claus, B.J., Fezzaa, K., Sun, T., and Chen, W.W. (2017). Fracture mechanisms of glass particles under dynamic compression. *International Journal of Impact Engineering* 106: 146–154.

Paul, J., Romeis, S., Mačković, M., Marthala, V.R.R., Herre, P., Przybilla, T., Hartmann, M., Spiecker, E., Schmidt, J., and Peukert, W. (2015). In situ cracking of silica beads in the SEM and TEM – Effect of particle size on structure–property correlations. *Powder Technology* 270: 337–347.

Pejchal, V., Žagar, G., Charvet, R., Dénéréaz, C., and Mortensen, A. (2017). Compression testing spherical particles for strength: Theory of the meridian crack test and implementation for microscopic fused quartz. *Journal of Mechanics and Physics of Solids* 99: 70–92.

Peng, B., Li, C., Moldovan, N., Espinosa, H.D., Xiao, X., Auciello, O., and Carlisle, J.A (2007). Fracture size effect in ultrananocrystalline diamond: Applicability of Weibull theory. *Journal of Materials Research* 22: 913–925.

Portnikov, D., Kalman, H., Aman, S., and Tomas, J. (2013). Investigating the testing procedure limits for measuring particle strength distribution. *Powder Technology* 237: 489–496.

Rozenblat, Y., Portnikov, D., Levy, A., Kalman, H., Aman, S., and Tomas, J. (2011). Strength distribution of particles under compression. *Powder Technology* 208: 215–224.

Shan, J., Xu, S., Liu, Y., Zhou, L., and Wang, P. (2018). Dynamic breakage of glass sphere subjected to impact loading. *Powder Technology* 330: 317–329.

Shen, J., Xu, D., Liu, Z., and Wei, H. (2020). Effect of particle characteristics stress on the mechanical properties of cement mortar with coral sand. *Construction and Building Materials* 260: 119836.

Shipway, P.H. and Hutchings, I.M. (1993). Attrition of brittle spheres by fracture under compression and impact loading. *Powder Technology*, 76: 23–30.

Tavares, L.M. (2005). Particle weakening in high-pressure roll grinding. *Minerals Engineering* 18: 651–657.

Tavares, L.M. and Cerqueria, M.C. (2006). Statistical analysis of impact-fracture characteristics and microstructure of industrial Portland cement clinkers. *Cement and Concrete Research* 36: 409–415.

Tavares, L.M. and das Neves, P.B. (2008). Microstructure of quarry rocks and relationships to particle breakage and crushing. *International Journal of Mineral Processing* 87: 28–41.

Tavares, L.M. and King, R.P. (1998). Single-particle fracture under impact loading. *International Journal of Mineral Processing* 54: 1–28.

Vallet, D. and Charmet, J.C. (1995). Mechanical behaviour of brittle cement grains. *Journal of Materials Science* 30: 2962–2967.

Watkins, I.G. and Prado, M. (2015). Mechanical properties of glass microspheres. *Procedia Materials Science* 8: 1057–1065.

Wong, J.Y., Laurich-McIntyre, S.E., Khaund, A.K., and Bradt, R.C. (1987). Strengths of green and fired spherical aluminosilicate aggregates. *Journal of the American Ceramic Society* 70: 785–791.

Xiao, Y., Meng, M., Daouadji, A., Chen, Q., Wu, Z., and Jiang, X. (2020). Effects of particle size on crushing and deformation behaviors of rockfill materials. *Geoscience Frontiers* 11: 375–388.

10

Agglomerate Particle Strengths

This chapter presents in survey form the strengths and strength distributions of particles that are formed as agglomerates of subparticles and places them in the context of the monolithic particles examined earlier. Agglomerate particles exhibit fracture strengths much less than their monolithic analogs as a consequence of two distinctive structural features: large strength-limiting flaws and fracture planes with fewer and weaker bonds. Such agglomerate particles include commonly encountered pharmaceutical tablets and food particles, and the less commonly encountered catalyst particles. The near-exponential effects of porosity on agglomerate particle strength are demonstrated. The detailed effects of pharmaceutical tablet shapes, intermediate between spherical and cylindrical particles, are analyzed. The similarity in strength behavior of edible agglomerates, tablets and food, to inedible agglomerates, catalysts and cement, is noted.

10.1 Introduction

An important class of particles is that of the *agglomerates*—particles formed by the assembly and bonding of large numbers of smaller *sub-particles* to form a single mass (following the nomenclature of Nichols et al. 2002). Although there are very many types of agglomerate particles, they share several common structural features, as illustrated in Figure 10.1. Figure 10.1a is a schematic diagram of an agglomerate particle with mottled surface shading to indicate that the particle is not monolithic and dense, as in the glass, salt, and rock particles considered here previously, but comprised of discrete sub-particles and pores. Such agglomerate particles have been considered previously in this book and include sand (Chapter 1), gypsum (plaster, Chapter 1), Al_2O_3 (Chapter 2), TiO_2 (Chapter 3), cement clinkers (Chapters 1, 7, and 8), iron ore pellets (Chapters 2 and 8), fertilizer and potash (Chapters 7 and 8), bauxite and mullite based ceramics (Chapter 8), and coral and limestone (Chapters 1, 7, and 9). Figure 10.1b shows a magnified diagram of the agglomerate. Several sub-particles are shown shaded and are packed together such that each sub-particle contacts only a few other sub-particles over localized contact areas. Pores, shown as open, occur between the sub-particles such that the overall particle contains significant porosity. The sub-particles are usually monolithic and dense materials (and usually crystalline) with material density ρ_0. The pores incorporated into an agglomerate particle lead to a particle density ρ less than this, $\rho < \rho_0$, such that the relative porosity of the particle $P = 1 - (\rho/\rho_0)$ is often in the range $0.1 < P < 0.5$, and sometimes greater. Figure 10.1c shows a further magnified diagram of a sub-particle contact. The area of contact interaction is shown as a dark shaded region bounded by menisci. In some systems, the contact interaction region is a physical material bridge between the sub-particles: the necks that form by high temperature diffusion of sub-particle material during sintering of ceramics; the necks that form by low temperature reaction of sub-particle material (typically with ambient moisture) during formation of cement and plaster; or, the necks that form by evaporation of water from saturated sub-particle assemblies to leave ionic crystal bridges, as in the NaCl bridges in sand. In contrast, in some systems the contact interaction region demarcates an area of surface forces between the sub-particles in close proximity: for example, the van der Waals and hydrogen bonding forces between sub-particles of fertilizer, potash, bauxite, pre-clinker cement, or pre-pellet iron ore, all of which are tumbled in order to induce such contacts and form particles from the respective sub-particles. In both forms of sub-particle interaction, material bridges and surface forces, it is clear that that the mechanical properties of the particle are significantly influenced by the nature of the sub-particle contacts.

Consideration of agglomerate particle strengths σ begins with a re-statement of the fundamental Griffith equation, $\sigma = (ER)^{1/2}/\psi c^{1/2}$, following the discussion of fracture in Section 3.1. All terms in this equation refer to the porous particle: E is

Particle Strengths: Extreme Value Distributions in Fracture, First Edition. Robert F. Cook.

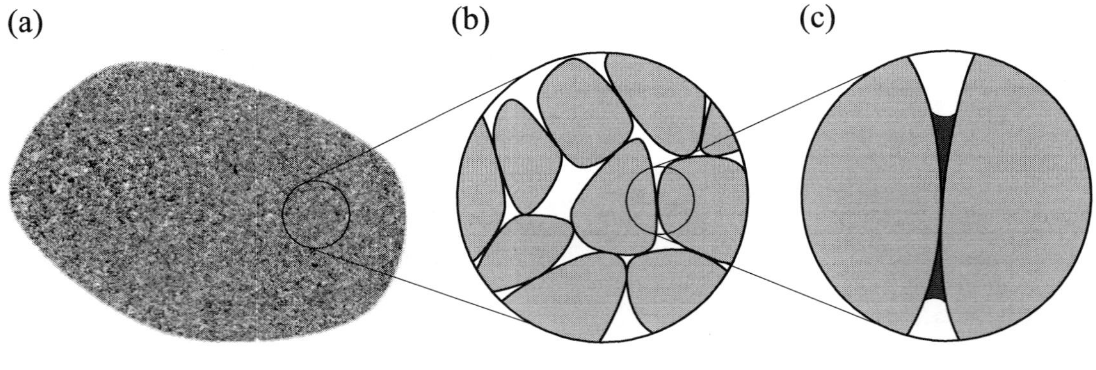

Figure 10.1 Schematic diagram of an agglomerate particle structure. (a) Particle, mottled to indicate an inhomogeneous microstructure consisting of sub-particles and pores. (b) Magnified view of the sub-particle array (shaded), showing pores (open) and localized sub-particle contacts. (c) Magnified view of a sub-particle contact, showing a sub-particle bridge (dark shading) that may be solid, fluid, or a region of surface force interaction.

the elastic Young's modulus, R is the fracture resistance, c is the strength-controlling crack length, and ψ is a dimensionless crack geometry term. The material properties of the dense sub-particles are elastic modulus E_0 and surface energy $R_0 = 2\gamma$. If the contacts between the sub-particles are material bridges formed of the same material as the sub-particles, simple bond density estimation with $P > 0$ leads to $E < E_0$ and $R < 2\gamma$ for the agglomerate particle. For small values of P, the pores are isolated and dilute and particle behavior is a linear decreasing function of porosity, $E \sim E_0(1 - P)$ and $R \sim 2\gamma(1 - P)$. Similarly, in this closed porosity domain the most potent strength-controlling flaw in a particle, an inter sub-particle pore as in Figure 10.1b, scales as a linear increasing function of porosity, $c \sim c_0(1 + P)$ and $\psi \sim \psi_0(1 + P)$. It is emphasized that c_0 and ψ_0 are scaling parameters associated with the geometry of the sub-particle assembly and are not sub-particle material properties. Combining these factors into the particle strength equation gives $\sigma \sim (2\gamma E_0)^{1/2}(1-P)/(\psi_0 c_0^{1/2})(1+P)^{3/2}$ and, recognizing that P is small, $\sigma \sim \sigma_0(1 - P)^{5/2}$, where σ_0 is a characteristic strength. Hence, agglomerate materials and particles with small values of porosity are expected to exhibit significant strength dependence on porosity, and in particular a much greater than linear decrease in σ with increasing P. (As noted by Hasselman 1963, for $P \to 0$ some systems exhibit linear behavior for both modulus and strength, $E \sim (1 - P)$ and $\sigma \sim (1 - P)$, implying that $\psi c^{1/2}$ is invariant in this limit for these systems.)

Conversely, for a structure in which the pores are not dilute, but interact to form an open, cellular, material, the contacts between the sub-particles are most likely to be bridges formed of material different from that of the sub-particles or to be adhesive contacts generated by surface forces. In this case, an upper bound to the elastic modulus behavior is that for an open cell foam, $E \sim E_0(1 - P)^2$ (Gibson and Ashby, 1997). The fracture resistance scales linearly with porosity as before, $R \sim 2\gamma_{\text{contact}}(1 - P)$, where the pre-factor is now the contact surface energy, $2\gamma_{\text{contact}} < 2\gamma$. In this porosity domain, the open cellular structure of the material contains many flaws, such that the size and geometry of the strength controlling flaw in a particle are determined by large ensemble extreme value effects. The crack length and geometry term are thus almost invariant with porosity and are fixed at population upper bound values, c_{max} and ψ_{max}. Again, it is emphasized that c_{max} and ψ_{max} are parameters associated with the sub-particle assembly and not sub-particle material properties. Combining this second set of factors into the particle strength equation gives $\sigma \sim (2\gamma_{\text{contact}}E_0)^{1/2}(1 - P)^{3/2}/(\psi_{\text{max}} c_{\text{max}}^{1/2})$, such that $\sigma \sim \sigma_{\text{cell}}(1 - P)^{3/2}$, where σ_{cell} is another characteristic strength and $\sigma_{\text{cell}} < \sigma_0$ (Gibson and Ashby arrive at identical scaling by related assumptions). Hence, agglomerate materials and particles with large values of porosity are also expected to exhibit strength dependence on porosity, and in particular a greater than linear decrease in σ with increasing P, although not as great as that as at small porosity. Although the two strength bounds discussed are well defined, most agglomerates contain a mix of open and closed porosity and the overall strength responses of agglomerate components, including particles, are thus expected to reflect transitions between the two bounds. The transitions should also exhibit steep dependencies of strength on porosity as the characteristic strength values, $\sigma_0 = (2\gamma E_0)^{1/2}/(\psi_0 c_0^{1/2})$ and $\sigma_{\text{cell}} = (2\gamma_{\text{contact}}E_0)^{1/2}/(\psi_{\text{max}} c_{\text{max}}^{1/2})$, are well separated and depend on different aspects of the agglomerate structure. Hence, the major conclusion from these considerations is that the strength of agglomerate components should decrease with increasing porosity with a greater than linear dependence, but that a universal behavior is precluded by variable microstructural effects.

The conclusion regarding porosity effects is borne out by consideration of the brittle fracture strengths of a range of porous materials tested in a variety of geometries. Figure 10.2 shows in semi-logarithmic coordinates the strength of several brittle

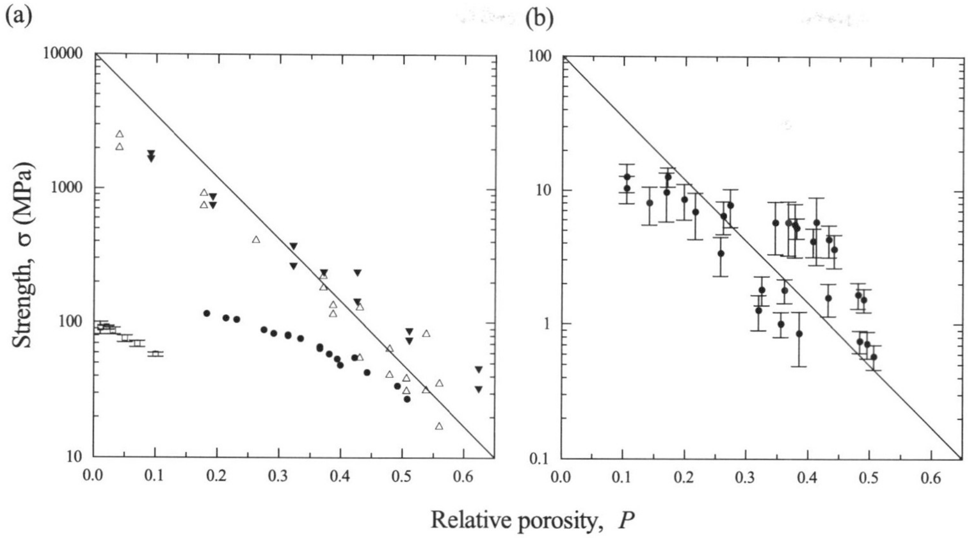

Figure 10.2 Semi-logarithmic plots of strength vs porosity for ceramic and glass materials. (a) Alumina and zirconia blocks tested in uniform compression, upper open and closed symbols. Cement blocks tested in uniform compression, lower solid symbols. Glass cylinders tested in diametral compression, lower open symbols. (Adapted from Ryshkewitch, E 1953, Powers, T.C 1958, Niesz, D.E 1965) (b) Aluminosilicate spherical particles tested in diametral compression (adapted from Moreno-Maroto, J.M et al. 2017).

materials as a function of porosity, using data derived from the published works cited. Measurements from some of the earliest works in this area are shown in Figure 10.2a. The upper open and closed symbols represent the behavior of sintered Al_2O_3 and ZrO_2, respectively, from the work of Ryshkewitch (1953), in which deliberate additions of fugitive hydrogen peroxide were used to create controlled pore structures with characteristic scales of (0.2–0.7) mm. 12 mm cubes of material were tested in uniform compression (as extended components, Chapter 5) using a conventional platen geometry. The domain of porosity extended over approximately 0.03–0.63 and the consequent range of observed strengths extended from approximately 200 MPa–20 MPa. The strength decrease was large and non-linear. As noted by Ryshkewitch, the behavior is well described empirically by an exponential decrease, $\sigma \sim \exp(-bP)$, indicated by the solid straight line in semi-logarithmic coordinates using $b \approx 10.6$. The line was fit to the Al_2O_3 data and subsequently used to predict, successfully, the ZrO_2 response based on independent measurements of the strength of dense ZrO_2. A similar, successful prediction was later made by Chamberlain (1978) using dense limestone as the basis for predicting the exponential decrease in the strength of coral with increasing porosity.

The lower solid symbols in Figure 10.2a represent the behavior of hardened cement paste from the work of Powers (1958). In cement hardening, gradual hydration of the calcium aluminosilicate cement sub-particles forms a gel that eliminates pores between the sub-particles within the paste. (The cement porosity values of Figure 10.2a neglect residual water and unreacted sub-particles.) 50 mm cubes of material were tested in uniform compression using a conventional platen geometry. The strengths decrease with increasing porosity, although not as markedly as the earlier ceramic materials. Powers noted that the strengths varied as approximately $(1 - P)^3$. The lower open symbols in Figure 10.2a represent the behavior of sintered glass, from the work of Niesz (1965), in which solid necks formed between (5–20) μm sub-particles of glass at high temperature, eliminating porosity with increasing heat treatment time. The domain of porosity studied was limited to less than 0.1, as this is the domain applicable to most commercial ceramics. Of perhaps more direct relevance to this book, the specimens were 12.7 mm diameter × 9.5 mm thick cylinders tested in diametral compression using a conventional compression platen geometry and interpreted via the "Brazil" strength equation, Eq. (2.22). The strengths decrease with increasing porosity, although, again, not as markedly as the earlier ceramic materials. Niesz noted that the strengths varied as approximately $\sigma \sim \exp(-bP)$, although as can be observed here, the value of b would be less than that used above.

Two historical points are worth noting in the context of the early works cited. First, in a Discussion appended to the Ryshkewitch (1953) work, Duckworth states that bend and tension tests on porcelain "gave essentially the same same strength-porosity characteristic as a compression test ... " and were in agreement with Ryshkewitch, although with a value of $b = 7$. An earlier work by Duckworth (1951) does indeed include tables of porcelain strengths that decrease with porosity, although the variation is restricted to four measurements over the 0–0.28 P domain and the strength decrease is only about a factor of 4 (more like the glass observations). Duckworth included the equation $\sigma \sim \exp(-bP)$, which Ryshkewitch did not, and extended the discussion to consider what is now recognized as a performance metric in design with porous materials, the strength/density ratio or specific strength σ/ρ (Ashby 1999). It is, however, unclear that co-citation as Ryshkewitch-Duckworth is warranted in reference to strength-porosity variations. Second, Niesz (1965) cites Rudnick et al. (1963) and Spriggs et al. (1964) in connection with the then near contemporary application of the cylinder-based diametral compression test for evaluating ceramic strength. Although the test was demonstrated as effective on a variety of brittle materials, largely by removing the need for gripping specimens, application to ceramics did not subsequently become widespread (see Snajdr 1964 for an example of implementation at the time; Darvell 1990 for an historical compendium). The large volume of material required and the recognition that beam bend tests were effective measures of brittle tensile strength (Duckworth 1951) essentially restricted the diametral test to rocks (already large, e.g. Jaeger 1966, 1967) and tablets (already approximately cylindrical, see next Section). The disc-like "theta" specimen, also developed in the 1960s and that also enabled tensile strength measurements by compressive loading, also did not become widely used. The theta specimens required relatively large volumes of material and were difficult to fabricate (although see Gaither et al. 2013 for a small-scale MEMS implementation in which these factors were overcome).

Figure 10.2b shows the strength of sintered ceramic particles, from the work of Moreno-Maroto et al. (2017). The particles were formed from industrial waste materials, granite and marble sawing sludge and rejected clay, and a polymer. The goal of the work was to assess the usefulness of the waste materials in the fabrication of light weight aggregate. The inorganic waste materials were milled to generate (4–20) μm sub-particles, combined with various amounts of the polymer, and formed into (9–9.5) mm spheres. The spheres were then fired for various times and temperatures to generate a series of particles with varying porosity. The strengths of the fired spherical particles were measured using a conventional compression platen geometry and the HO equation, Eq. (2.41). The symbols in Figure 10.2b represent the mean and standard deviation of 25 particle strength measurements at the porosity indicated. The solid line is a guide to the eye of the same form and slope ($b \approx 10.6$) as in Figure 10.2a—a similar guide will be repeated with this slope in subsequent plots. Although there is considerable scatter, preventing identification of a clear functional trend, the ceramic particle strengths clearly decrease with increasing porosity, and decreased sintering time and temperature, and are not inconsistent with the previous exponential behavior. The domains, ranges, and scatter of the data in Figure 10.2 are similar to those included in the extensive strength-porosity compilation of Li and Aubertin (2003). The compilation extended over ceramics, metals, and geological materials, and required a complex multi-parameter empirical expression to fit the diversity of responses observed. In the domain of large porosity, $P \approx (0.5\text{–}0.7)$, engineered ceramic structures for use in bone repair also demonstrated significant decreases in strength with increasing porosity (Roohani-Esfahani et al. 2016).

The porosity in the agglomerate blocks, cylinders, and spheres of Figure 10.2 was eliminated from the ceramic, glass, or cement compacts by diffusion or reaction of sub-particle material to fill the pore space. In these cases, the sub-particle contacts were physical necks of material. Decreases in porosity lead to increases in strength that in many systems could be described by an exponential function. These observations provide background for the following sections that examine the strengths of three common agglomerate particle systems—pharmaceuticals and related materials, food, and chemical catalysts. In the first, the particles are cylinder or tablet shaped and the sub-particle contacts are regions of surface forces. In the second, the particles are irregularly shaped and the sub-particles are predominantly plant cells held together by extended organic bonding contacts. Most foods are foams, although some food particles are monolithic crystals. In the third, the particles are spherical and the sub-particles are held together by solid necks of material in a manner very similar to the ceramics examined here and in Chapter 8. The first two systems are predominantly organic and the third is inorganic.

10.2 Pharmaceuticals

Pharmaceutical tablets are usually formed by die compaction of powders, Figure 10.3. The dies usually have a circular internal shape, typically (10–15) mm in diameter, and are filled with (0.5–1) g of powder composed of (10–300) μm sub-particles that is compacted and bound with (100–300) MPa of pressure (corresponding to tens of kN compaction forces). The die faces may be flat so as to generate a cylinder, Figures 10.3a and 10.3b, or doubly convex so as to generate a classic tablet shape, Figures 10.3d and 10.3e. In both cases, the circular forms are ideal for diametral compression strength testing, Figures 10.3c and 10.3f. In addition to an active pharmaceutical ingredient (API), the powders usually contain additives to aid in compaction, binding, color, taste, and API release. Such pharmaceutically inactive additives are termed excipients. A key element in tablet manufacturing is to optimize binding in compaction so that tablets are strong enough for storage and transport but not so strong that API release is inhibited. The variation of tablet strength with compaction pressure and

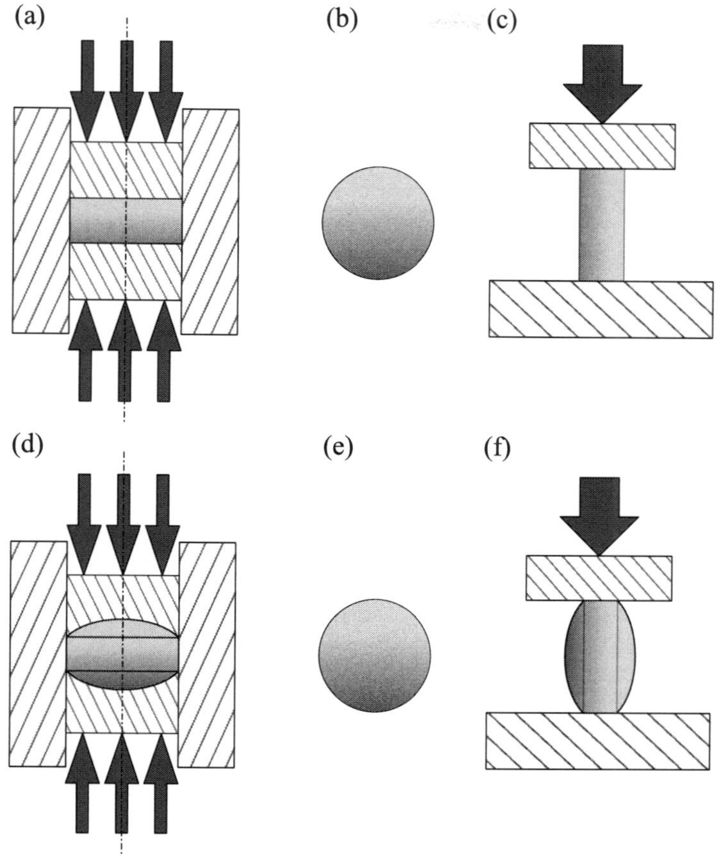

Figure 10.3 Schematic diagrams of pharmaceutical powder compaction to form circular outline tablets that are tested mechanically in diametral compression. (a) Compaction by a planar faced circular die to form a cylindrical tablet. (b) Circular elevation of a cylindrical tablet. (c) Rectangular elevation, perpendicular to (b) of a cylindrical tablet in diametral compression. (d) Compaction by a curved faced circular die to form a doubly convex tablet. (e) Circular elevation of a doubly convex tablet. (f) Rectangular elevation, perpendicular to (e) of a doubly convex tablet in diametral compression.

subsequent environmental exposure is thus a focus of pharmaceutical manufacturing. Tablet strength has most frequently been measured by diametral compression of cylindrical tablets in the classic "Brazilian" thin cylinder geometry. Tablets for strength testing are usually composed entirely of excipients, most frequently lactose, sucrose, microcrystalline cellulose, or gypsum (calcium sulphate—plaster), although sometimes real pharmaceutical tablets (e.g. aspirin) are studied. Prominent contributions to tablet strength measurement have been made by Newton, Stanley, and colleagues from the 1970s onward.

10.2.1 Porosity

The early tablet strength tests of Fell and Newton (1970) and Newton et al. (1971) established the major features of pharmaceutical tablet strength measurements. The tests focused on an important engineering aspect of tablet fabrication, the relation between compaction pressure used in tablet formation and the resultant tablet strength. A slightly earlier result that is key to interpreting such measurements is the relation between compaction pressure and resultant tablet porosity. Figure 10.4a shows a plot of resultant porosity as a function of compaction pressure for gypsum plaster cylinders 31.8 mm × 1.27 mm from the work of Soroka and Sereda (1968), using data derived from the published work. The compaction pressure domain spans approximately (20–1200) MPa and the conjugate porosity range is approximately 0.5–0.05. The scale used in Figure 10.4a is identical to that for Figure 10.2, but the independent and dependent variables are interchanged. Symbols represent measurements by two methods of agglomerate porosity at the compaction pressure indicated and the solid line is an empirical fit of the form $P \sim -\ln(\text{pressure})$. Combining this finding with the observation from Figure 10.2 that agglomerate strength varies as $\sigma \sim \exp(-P)$ implies that the expected strength-pressure variation should be linear, $\sigma \sim$ pressure.

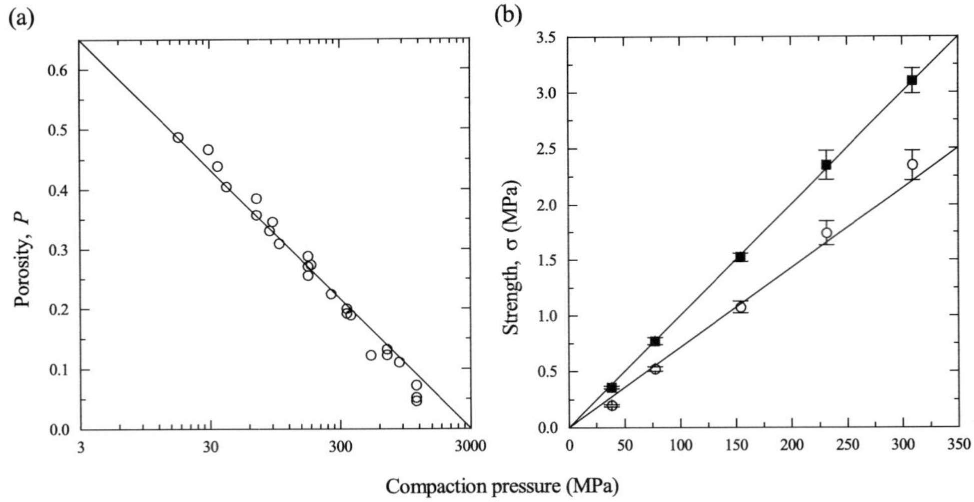

Figure 10.4 (a) Semi-logarithmic plot of relative porosity vs compaction pressure for gypsum (plaster) cylinders (adapted from Soroka, I et al. 1968). (b) Linear plot of strength vs compaction pressure for lactose cylinders, spray-dried (filled symbols) and crystalline (open symbols) (adapted from Fell, J.T et al. 1970). Note different scales in (a) and (b).

Figure 10.4b shows a plot of lactose tablet strength as a function of compaction pressure for plane faced cylinders, (Figure 10.3a, 10.3b, and 10.3c) 12.7 mm × 3 mm, from the work of Fell and Newton (1970), using data derived from the published work. The tablets were tested using a conventional compression platen geometry and failure forces varied from 10 N–130 N. Failure strengths were determined using the Brazil equation, Eq. (2.22). The solid symbols are for spray dried lactose and the open symbols are for crystalline lactose; both were composed of ≈30 μm sub-particles. Symbols represent means and standard deviations of about 5 measurements. A similar study by Newton et al. (1971) using many more measurements generated nearly identical results. It is to be noted that tablet strengths are about a factor of 100 less than the compacting pressures. The relationship between strength and pressure is indeed linear, consistent with the independent porosity-based measurements. Padded platens led to both greater forces (up to nearly 400 N) and force variability at failure, perhaps for the same reasons as the observations of Brecker (1974) on alumina particles with ground flats (Chapter 6). Decreasing the mass of powder loaded into a die prior to compaction to a fixed pressure lead to decreased strengths and increased strength variability, suggesting inhomogeneity in pore closure within a tablet as sub-particle constraint was reduced. The predominant failure mode was meridional fracture as in Figures 1.9b, 1.10b, and 1.11b.

In the intervening years since Fell and Newton's works, similar observations of failure loads in the range (50–220) N have been made on flat cylinder tablets formed from (20 to 700) μm sub-particle powders using (5 to 12.5) mm dies and (8 to 700) MPa compaction pressures to generate cylinders (3–9.5) mm thick. The inferred strengths are of course also similar and include: lactose, starch, and microcrystalline cellulose, (0.2–5) MPa (McKenna and McCafferty 1982); gypsum, lactose, microcrystalline cellulose, aspirin, and benzoic acid, (0.5–10) MPa (Jarosz and Parrot 1982); aspirin (0.6–1.4) MPa (Kennerley et al. 1982); lactose variants, gypsum, and microcrystalline cellulose, failure forces up to 220 N (Vromans et al. 1985); sucrose, (0.2 to 4) MPa (Es-Saheb 1996); lactose, sucrose, cellulose, Na bicarbonate, NaCl, and gypsum, (0.5 to 5) MPa (Adolfsson et al. 1997); microcrystalline cellulose variants, (0.5–5) MPa (Obae et al. 1999); salt, sucrose, cellulose, lactose, and paracetamol, (1 to 4) MPa (Sonnergaard 2002); lactose, microcrystalline cellulose, aspirin, gypsum, magnesium carbonate, paracetamol, and starch (0.5–10) MPa (Mohammed et al. 2005); microcrystalline cellulose, gypsum, and starch (0.2–5) MPa (Almaya and Aburub 2008); and lactose-cellulose bilayers (0.1–1.6) MPa (Chang et al. 2017). Most of these works demonstrated the key engineering result that greater compaction pressures result in greater strengths and many demonstrated a linear pressure-strength relationship; a few demonstrated the underlying logarithmic porosity-compaction pressure relationship. Some additional measurements included direct tension (Jarosz and Parrot 1982), thick samples (Es-Saheb 1996), and emphases on sub-particle contact dissolution (Adolfsson et al. 1997) and aspect ratio effects (Obae et al. 1999). Some studies were focused on compaction plasticity measurements (e.g. Mohammed et al. 2005; Almaya and Aburub 2008) as in Roberts and Rowe (1986, 1987a, 1987b).

The tablet compaction process increases the number of sub-particles per volume in the tablet and therefore increases the number of inter sub-particle contacts, probably as van der Waals or hydrogen bonding between the uncharged, unreacted sub-particles. Hence compacted tablet strength is often observed to decrease significantly with increasing porosity, as in the largely extended component geometries of Figure 10.2. Figure 10.5 shows examples of tablet strength variation with porosity, arranged in order of decreasing strength, using data derived from the published works cited. The data encompass (a) cellulose and starch (lower symbols) (Wu et al. 2005), (b) cellulose (Keleş et al. 2015) (filled symbols), (Shang et al. 2013a) and (van Veen et al. 2005) (open symbols), and (He et al. 2007) (half-filled symbols), (c) cellulose and lactose (lower symbols) (Yohannes et al. 2015), and (d) NaCl and sucrose (open symbols) (Fichtner et al. 2005). The tablets were all cylinders, (8–13) mm × (2–6) mm, tested in diametral compression, and the strengths evaluated using the Brazil equation. The plots are all semi-logarithmic at the same relative scale and the solid lines are all guides to the eye, characterizing the exponential dependence of strength on porosity noted above, $\sigma \sim \exp(-bP)$, using the same value of $b = 10.6$. The data are all linear in these coordinates, indicating agreement with an exponential trend, and the value of b used describes the overall behavior. The commonality of responses, relative to the diversity of Figure 10.2, probably reflects the commonality of fabrication method (compaction) and the resulting agglomerate microstructures. The fracture implication is that the extreme value strength-controlling largest crack in tablets varies exponentially with porosity as $c_{max} \sim \exp(2P)$. (If the sub-particles

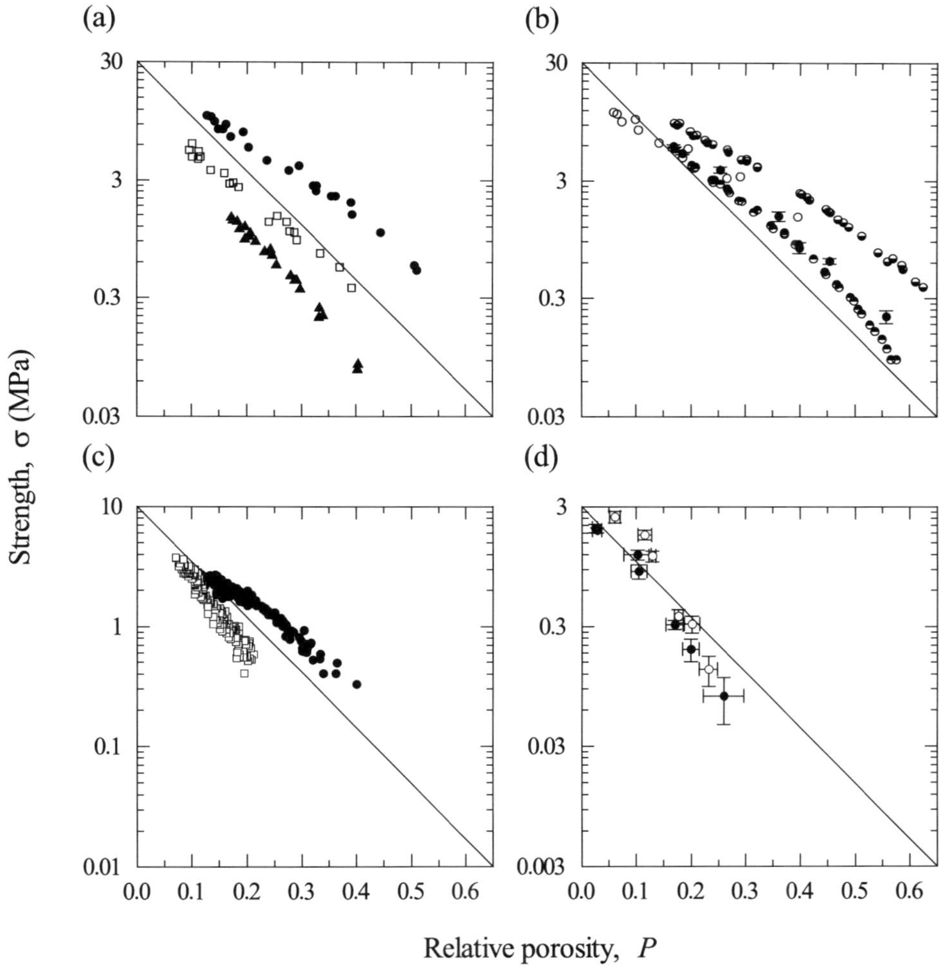

Figure 10.5 Semi-logarithmic plot of strength vs relative porosity for several pharmaceutical excipient materials. Tested as cylindrical particles in diametral compression. (a) Cellulose and starch (adapted from Wu et al. 2005). (b) Cellulose (adapted from Keleş, O et al. 2015; Shang, C et al. 2013a; Van Veen, B et al. 2005 and He, X et al. 2007), (c) Cellulose and lactose (lower symbols) (adapted from Yohannes, B et al. 2015). (d) NaCl and sucrose (open symbols) (adapted from Fichtner, F et al. 2005).

are deliberately combined at very large compaction pressures so as to remove the exponential strength-porosity trends of Figure 10.5, the linearity of the tablet strength-compaction pressure trend is also removed, e.g. Pazesh et al. 2019).

10.2.2 Shape

Tablet strength testing was greatly extended by Pitt, Newton, and Stanley in the 1980s, through the introduction of a strength formulation for tablets with the common shape of doubly convex exteriors, Figures 10.3d, 10.3e, and 10.3f. Figure 10.6 is a diagram of a doubly convex tablet, detailing the dimensional parameters. Figure 10.6a shows the circular outline elevation, using the coordinate system of Figure 2.18 and showing the diameter D, applied similarly to previous usage to characterize particle size. Figure 10.6b is a diagram of the perpendicular elevation, showing the central cylindrical barrel and the two spherical segment caps. The overall tablet thickness is T, composed of the barrel width W and two spherical segment heights h, $T = W + 2h$. The radii of the spherical caps, R are equal and $R \geq D/2$. The order of the dimensional parameters is usually $W < T < D < R$. The dimensions in the elevation Figure 10.6b are related by

$$T = W + 2R - (4R^2 - D^2)^{1/2}, \tag{10.1}$$

such that only two out of the three parameters T, W, R are independent once D is specified. The volume V of the doubly convex tablet is given by

$$V = 2\pi h^2(3R - h)/3 + \pi D^2 W/4, \tag{10.2}$$

such that the density of a tablet ρ can be specified from the tablet mass m and shape by $\rho = m/V$. (Note that the coordinate system in Figure 10.6 is rotated from that of Pitt et al. 1989a, and that the use of R is different from that of Chapter 2.)

It is convenient to describe the shapes of doubly convex tablets by independent ratios of the dimensions in Figure 10.6. In particular, the ratio W/D describes the aspect ratio of the central cylindrical barrel. The ratio D/R describes the relative curvature of the spherical caps (noting that the cap curvature is approximately $1/R$.) $D/R = 0$ characterizes planar faced cylindrical particles, in which case $T = W$. The combination $D/R = 2$ and $W/D = 0$ characterizes spherical particles. Figure 10.7 shows variations of these ratios for doubly convex tablets; rows (a) and (b) circular, row (c) rectangular. The compaction direction is perpendicular to the first diagram in each row. In row (a), barrel thickness $W/D = 0.2$ and cap curvature D/R increases left to right, $D/R = 0, 1.2, 1.9$. In row (b), $W/D = 0.4$ and $D/R = 0, 1.2, 1.9$. The progression from a thin cylinder, (a) left, to a near sphere, (b) right), is clear. In row (c), one outline perpendicular to the compaction direction is described by $W/D = 0.4$ and $D/R = 1.9$, but the tablet is elongated in the other direction ($W/D \gg 1$) to form a "caplet".

Similar to the mechanical analysis sequence of Chapter 2, the first step in developing a strength formulation for tablets is assessment of the stress distribution within diametrally loaded tablets, Figure 10.3c and f. Pitt et al. (1989a) performed such

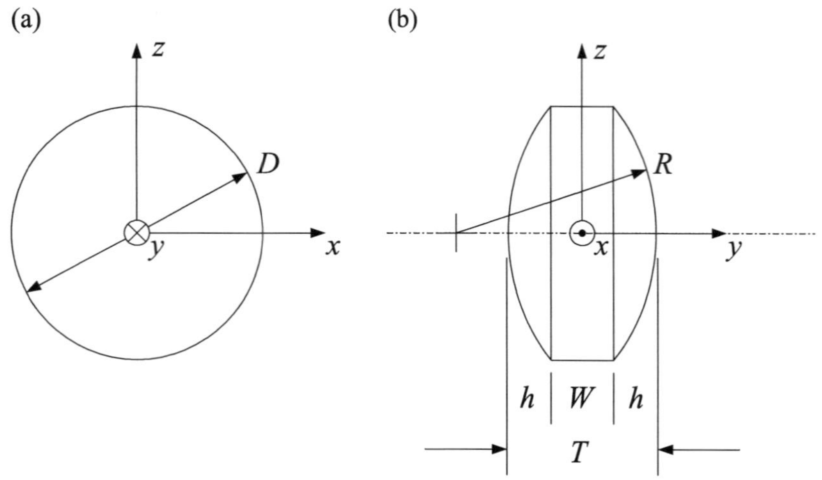

Figure 10.6 Diagrams illustrating the coordinate system and dimensions used in description of doubly convex tablets. (a) Front elevation, indicating circular outline diameter D. (b) Side elevation, indicating total thickness T, internal barrel width W, and external cap radii, R. Cap widths h are also indicated.

an assessment using a photoelastic technique and large-scale ($D = 50$ mm) epoxy models of circular tablets, with shape parameters in the ranges $0.1 \leq W/D \leq 0.3$ and $0 \leq D/R \leq 1.43$. The technique was validated by measurements on a thin planar faced cylinder, $W/D = 0.1$ and $D/R = 0$, loaded in diametral compression by padded platens. Figure 10.8 shows the results of stress measurements within the cylinder on the central xz plane (Figure 10.6a), using the same format as Figure 2.20 (some changed axis labels). The dashed lines represent predictions from the analysis of Hondros, repeated from Figure 2.20, for localized diametral pressure loading of a cylinder, as in the experiments. The symbols represent the measurements for two stress components (changing the normalization to σ_{mean}), σ_{zz} (open symbols) and σ_{xx} (filled symbols). Figure 10.8a shows stress variation in the x direction and Figure 10.8b shows variation in the z direction. The agreement between the predictions and experimental measurements, both the axial compressive stress and the transverse tensile stress is clear and validates the procedure. (Similar agreement was observed for pressure variation in a loaded sphere, measured by fluorescence; see Figure 2.11.) Of most interest in tablet strength measurements is the variation of the tensile stress σ_{xx} with tablet shape, as this determines fracture behavior. Figure 10.9 shows the variation of σ_{xx} in the z direction in the xz plane for several variants of the epoxy model tablets. The diagram is similar to Figure 10.8 with magnified stress scale and the dashed line is repeated from Figure 10.8, indicative of planar cylinder behavior. The symbols represent experimental measurements from an epoxy tablet model with $W/D = 0.1$ and $D/R = 1.43$ (similar to Figure 10.7a right). Following Pitt et al., for ease of visualization the stress values are exaggerated by a factor of T/W, about a factor of 5 in this case. The solid line is a fit to these data of the same form but reduced amplitude as the dashed line. The shaded band encompasses similar results from models with different shapes in the W/D and D/R (and T/W) ranges examined. The conclusion to be drawn from Figure 10.9 is that the maximum strength controlling tensile stress in tablets $\sigma_{xx}^{\max}$ is reduced from that in cylinders of the same W/D ratio. The reduction was expressed by Pitt et al. as a modification of the Brazil equation, Eq. (2.22),

$$\sigma_{xx}^{\max} = \frac{F}{\pi D^2} I(W/D, D/R), \tag{10.3}$$

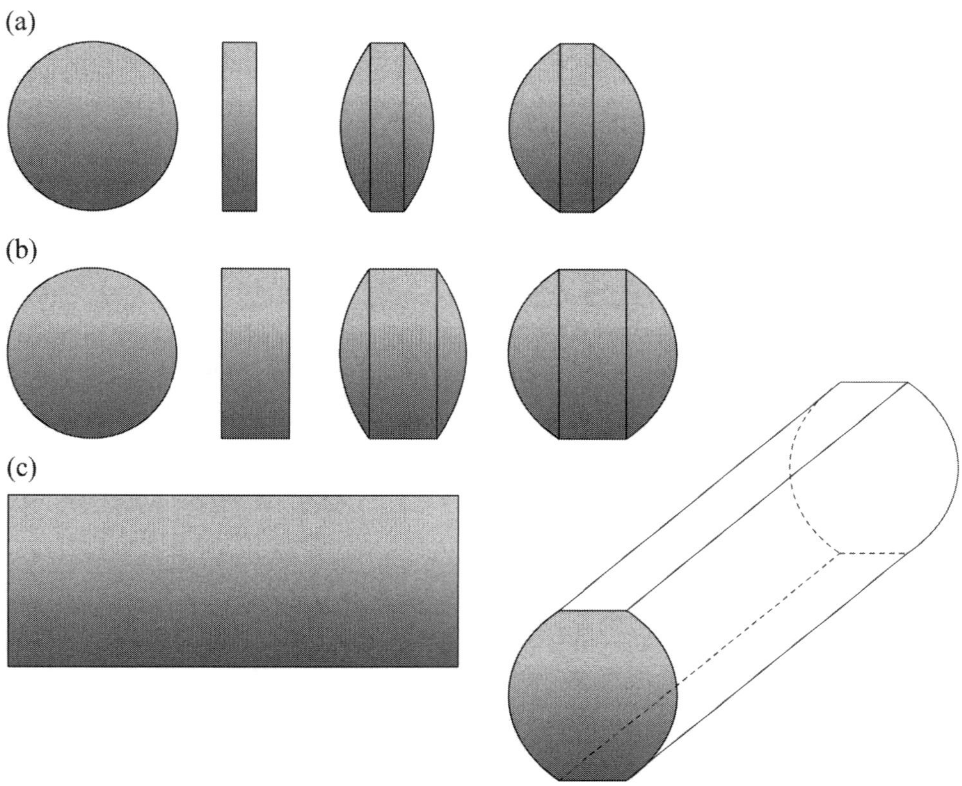

Figure 10.7 Diagrams illustrating variations in shape of doubly convex tablets. Left column, front elevation outlines, (a) and (b) circular, (c) rectangular. (a) Side elevations of cylindrical and doubly convex tablets, decreasing cap radii left to right. (b) Side elevations of wider barrel cylindrical and doubly convex tablets, decreasing cap radii left to right. (c) Right, perspective view of elongated doubly convex tablet.

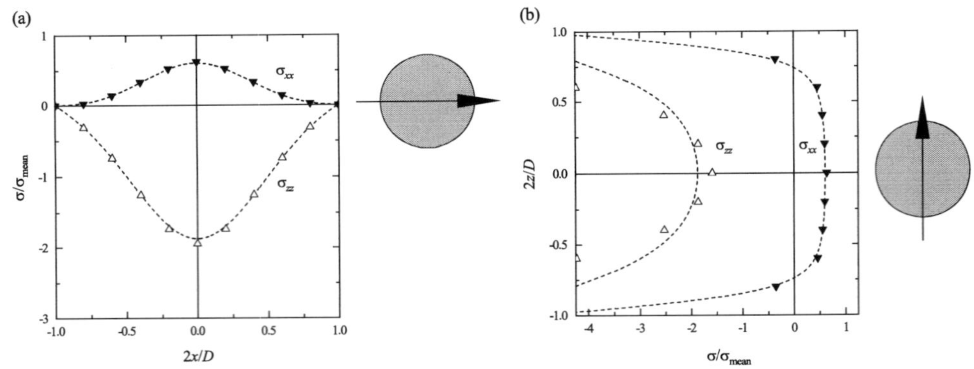

Figure 10.8 Plots of the variations in axial σ_{zz} and transverse σ_{xx} stresses in a cylinder loaded in diametral compression by a distributed pressure , Figure 2.18b. (a) Equatorial ($z = 0$) plane. (b) Axial ($x = 0$) plane. Dashed lines, analytical solutions. Symbols, photoelastic measurements. (Adapted from Pitt, K.G et al. 1989a.) Schematic diagrams of plot directions shown on right.

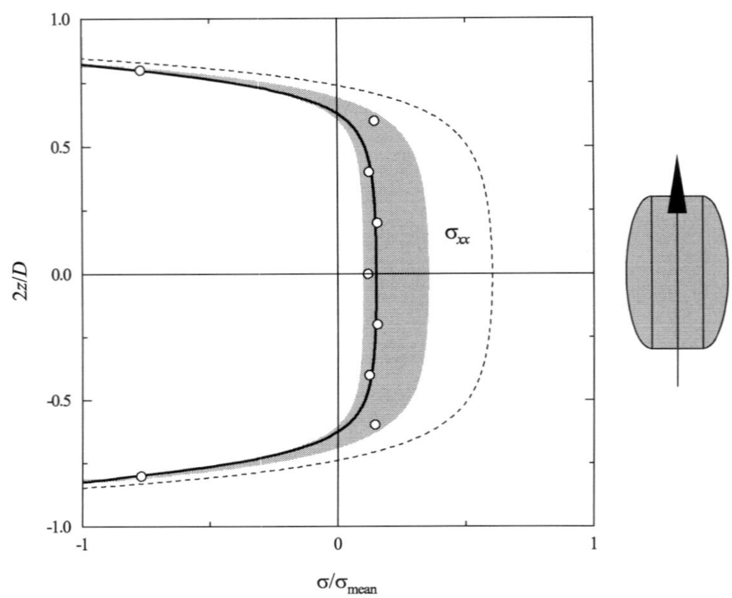

Figure 10.9 Plot of the variations in transverse σ_{xx} stress in doubly convex tablets loaded in diametral compression, Figure 10.3f. Dashed line, analytical solution for a cylinder. Symbols and solid line, photoelastic measurements for a single tablet. (Adapted from Pitt, K.G et al. 1989a.) Shaded band, range of photoelastic measurements for tablets of different aspect ratios, Figure 10.7.

where F is the applied force and $I(W/D, D/R) \leq 2D/W$ (the Hertzian limit) is a stress concentration factor that depends on tablet geometry (notation altered). Based on the experimental stress measurements, Pitt et al. (1989a) provided I values for a few combinations of W/D and D/R.

The second step taken by Pitt and colleagues in the development of a strength formulation for tablets was to build on the finding of Eq. (10.3) and calibrate by experimental measurements a similar expression for strength. In philosophy, this approach was similar to that of Hiramatsu and Oka (1966) in development of the similar HO equation, Eq. 2.41, with some exceptions. Hiramatsu and Oka used the results of finite element analyses to estimate a single value of I for near-spherical irregular shaped rocks characterized by a single size dimension D. The value of I (≈ 2.8) was then validated by comparison of strength measurements on planar cylinder rock specimens and irregular shaped rock specimens. The validation results of Hiramatsu and Oka and others are shown in Figure 2.23. The approach taken by Pitt et al. (1988) was to measure the strengths of model tablets of various shapes and use regression to arrive at an expression for I that was calibrated by strength measurements on planar cylinder specimens. Pitt et al. (1988) used wet cast gypsum (plaster) as $D = 12.5$ mm circular tablet and cylinder specimens with shape parameters in the ranges $0.06 \leq W/D \leq 0.3$ and $0 \leq D/R \leq 1.43$. The strength σ of the control cylinder specimens, evaluated using the Brazil equation, Eq. (2.22), was (mean ± standard deviation) (5.25 ± 0.53) MPa. The strengths σ of the test tablet specimens were evaluated using Eq. (10.3). The regressed expression for I developed by Pitt et al. was of the form

$$I = [a(T/D) + b(T/W) + c(W/D) + d]^{-1}, \qquad (10.4)$$

where a, b, c, d were specified coefficients (noting that I now refers to strength and that different notation is used). The resulting I values were in the range $0.3 \leq I \leq 3$. Figure 10.10 shows a plot of gypsum tablet strength as a function of failure force for the range of tablet geometries exam examined by Pitt et al. (1988), using data derived from the published work. Symbols represent means and standard deviations of strength measurements of 20 replicates of a tablet geometry that failed at the mean force indicated. The shaded band represents the mean and standard deviation of the control strength of the $W/D = T/D = 0.2$ planar cylinder specimens. The solid lines represent the mean strengths of $W/D = T/D = 0.1$, 0.2, and 0.3 planar cylinder specimens. The average response of the doubly convex tablet strengths over the shape domain examined, which covers most common tablets, was (5.44 ± 0.76) MPa. The expression for I, Eq. (10.4), describes tablet strengths that agree with the strengths of the control well within experimental uncertainty.

Although Eq. (10.4) was successful in the original application, the form of I omits two important constraints. The first, as noted in consideration of Eq. (10.1), is that the formulation is over-specified and omits the constraint that only two of the four coefficients in Eq. (10.4) are independent. The over-specification leads to an underestimate of the uncertainty in tablet strengths determined using Eq. (10.4) and therefore overestimates precision. Second, as noted in consideration of Eq. (10.3), the coefficients in the formulation are unconstrained and do not necessarily lead to the required convergence of $I \to 1/[0.5(W/D)]$ as $(D/R) \to 0$, and thus $T \to W$. The lack of constraint leads to unspecified variations in the values of tablet strengths determined using Eq. (10.4) and therefore degrades accuracy. The lack of constraints in Eq. (10.4) was noted by Shang et al. (2013a) in an extensive study of shape effects in microcrystalline cellulose tablet strengths. In addition to variations in W/D and D/R as studied by Pitt et al. (1988), Shang et al. also varied the compaction pressure of both doubly convex tablets and the planar faced cylinder controls. Shang et al. used an expression for I given in terms of the aspect ratios of the entire tablet and the central barrel:

$$I = [a(T/D) + b(W/D)]^{-1}, \qquad (10.5)$$

which is not over-specified. Equation (10.5) was fit to the doubly convex tablet strength data using the compaction pressure dependent cylinder strengths as controls and a and b as unconstrained empirical parameters. Eq. (10.5) was also fit to the tablet strength data using the cylinder controls with a and b constrained by the convergence criterion $a + b = 0.5$, such that

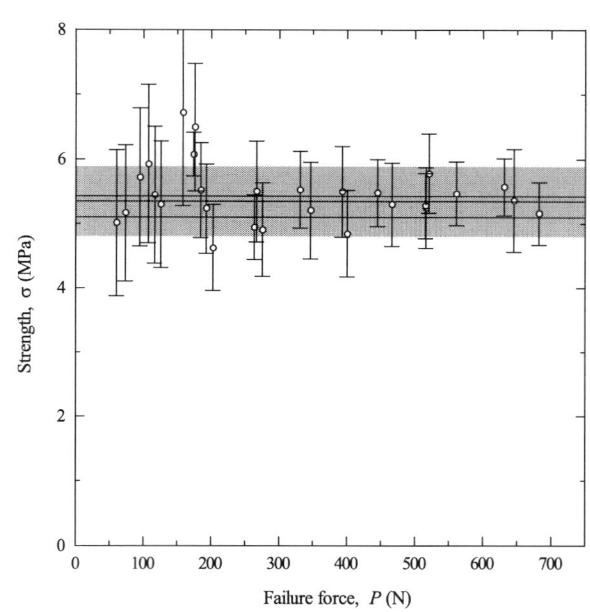

Figure 10.10 Plot of strength of doubly convex gypsum tablets as a function of failure force using calibrated strength formulation, Eqs. (10.3) and (10.4) (adapted from Pitt, K.G et al. 1988). Symbols represent means and standard deviations of 20 measurements of tablets of a fixed shape. Shaded band and solid lines represent strengths of control cylindrical tablets.

$I = 1/[0.5(W/D)]$ for $T = W$. For the microcrystalline cellulose tablet strengths examined by Shang et al., the difference between the constrained and unconstrained $(a + b)$ sums was about 3 %, much less than the strength dispersions that were typically 10 % or greater. When applied to the Pitt et al. (1988) gypsum data, the strength value averaged over all shapes was (5.53 ± 1.07) MPa, using the constrained Shang et al. formulation, Eq. (10.5) with $a + b = 0.5$. This value is in agreement with both the cylinder control measurement and the original Pitt et al. tablet estimate and the revised estimated uncertainty is (as expected) somewhat greater.

Test and control strength observations were compared in Figure 2.23 using the strengths of spherical particles determined using the HO equation and the strengths of cylinder specimens of the same materials determined using the Brazil equation. The Pitt et al. (1988), Shang et al. (2013a), and other tablet and cylinder observations can be used in the same manner using data derived from the published works cited. Figure 10.11 compares the test strengths of tablet particles determined using the constrained Shang et al. equation, Eq. (10.5), with control strengths of cylinder specimens determined using the Brazil equation. The plot uses logarithmic coordinates as in Figure 2.23 and the solid line indicates ideal agreement. The solid symbol in Figure 10.11 represents the mean and standard deviation of the gypsum tablet and cylinder strength data of Figure 10.10. The open symbols in Figure 10.11 represent individual strength measurements of microcrystalline cellulose tablet and cylinder strengths determined from the data of Shang et al., using data derived from the published work. The shaded and half-filled symbols in Figure 10.11 represent individual strength measurements of aspirin tablet and cylinder strengths determined from the data of Pitt et al. (1989b, 1990), using data derived from the published works cited. For the Shang et al. (2013a) and Pitt et al. (1989b, 1990) data, the multiple control strengths represent specimens generated at multiple compaction pressures; multiple test strengths at each control strength represent strength measurements on tablets of different shapes. (The single control strength of the Pitt et al. (1988) measurements represents a single compaction pressure and the mean and standard deviation of the resulting test strengths of different shaped tablets.) The agreement in Figure 10.11 validates the idea behind the original Pitt et al. (1988) strength expression and the precision and accuracy of the re-formulation and calibration by Shang et al. (2013a).

The validated strength expression of Eq. (10.5) can be used in an applied manner in the engineering of tablets and in a more fundamental manner to investigate the materials science of tablet strengths. An engineering application is illustrated in Figure 10.12 by aspirin tablet failure force and strength from the work of Pitt et al. (1990), using data derived from the published work. Figure 10.12a shows failure force as a function of compaction force for $D = 12.5$ mm aspirin tablets with relative curvature $D/R = 0.5$. The aspect ratios $W/D = 0.1, 0.2, 0.3$ are shown as different symbols, bottom to top, and, as might be expected, wider tablets have greater failure forces. Figure 10.12b shows the converted failure strengths, Eq. (10.5), as a function of compaction stress. The data are significantly converged, such that the $W/D = 0.3$ and 0.2 responses are coincident and the $W/D = 0.1$ data slightly separated to smaller values. The implication is that

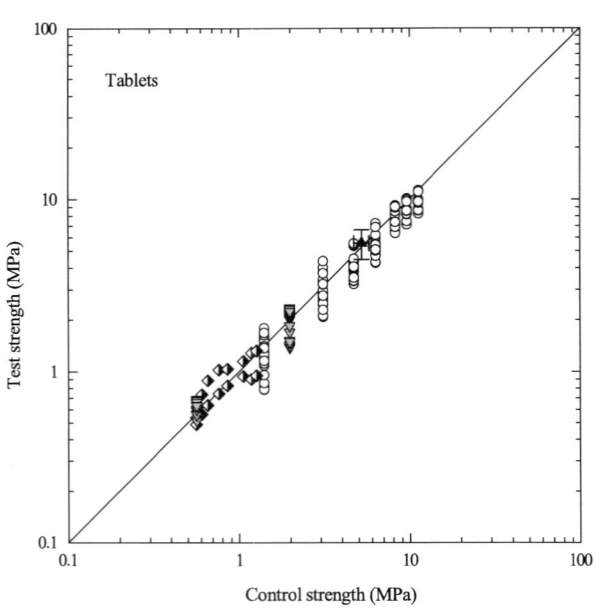

Figure 10.11 Plot demonstrating correlation of doubly convex tablet strength measurements determined by the constrained Shang et al. expression, Eq. (10.5), with cylindrical tablet strength measurements determined by the Brazil expression, Eq. (2.22). The open symbols represent calibration experiments adapted from Shang, C et al. (2013a).

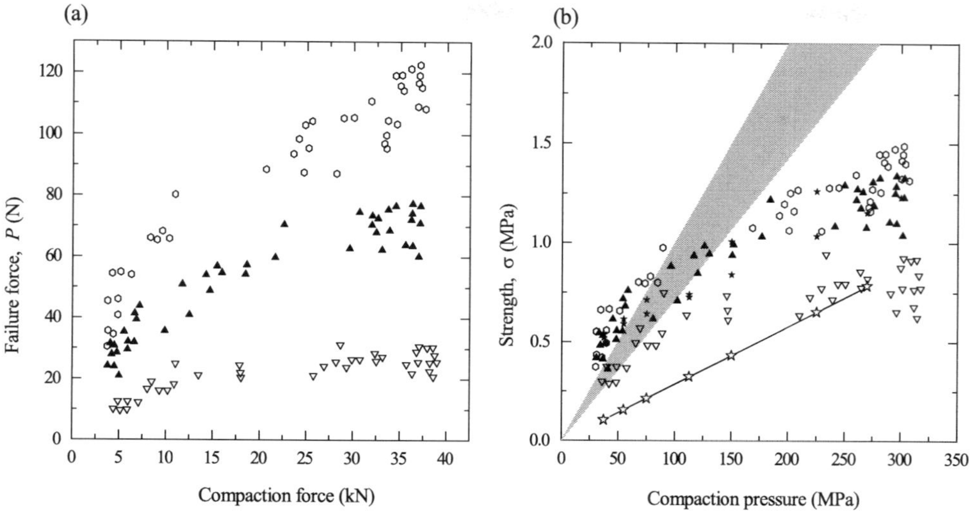

Figure 10.12 Plots demonstrating application of doubly convex tablet strength formulation. (a) Raw failure force vs compaction force data for three different thickness aspirin tablets (adapted from Pitt, K.G et al. 1990). (b) Strength vs compaction stress results after analysis: the data have converged to almost a single response. Starred symbols and line indicate cylindrical tablet aspirin responses. Shaded band indicates lactose cylindrical tablet response. Adapted from Pitt, K.G et al. 1990.

the thinner tablets are weaker. This conclusion is supported by within-material measurements on $W/D = 0.3, 0.2, 0.1$ aspirin cylinders, in which thinner cylinders were observed to be weaker (open starred connected symbols) than thicker cylinders (filled starred symbols). Between-material comparison is demonstrated by the shaded band, which represents the response of the lactose cylinders from Figure 10.4. The smaller values and non-linear dependence on compaction pressure of the aspirin tablet strengths compared with those of lactose cylinders are clear. Graphs such as Figure 10.12, connecting extensive and intensive failure properties, enable design and selection of materials and processes in tablet engineering.

A materials science application of tablet strength analysis is illustrated in Figure 10.13. Microcrystalline cellulose doubly convex tablet measurements from Shang et al. (2013a) and Newton et al. (2000a) are analyzed, using data derived from the published works cited. Specifically, the porosity of the tablets was determined using Eqs. (10.1) and (10.2), the specified geometry parameters D, W, R, and the masses of powder used in tablet fabrication. The porosity values for all tablets were determined from a common value of sub-particle density estimated from porosity bounds given in Newton et al. The strengths of the tablets were determined using the geometry parameters in Eq. (10.5) and the specified failure forces or strengths. The resulting strength vs porosity plots are shown in Figure 10.13a from Shang et al., in which symbols represent individual strength measurements, and Figure 10.13b from Newton et al., in which symbols represent the means of 10 strength measurements. The same semi-logarithmic scales as in earlier plots of extended component and cylinder tablet behavior are used, Figures 10.2 and 10.5, and the solid lines representing negative exponential behavior have the same slopes. In both cases, the strengths increase with decreasing porosity, although with decreased slopes relative to the reference exponential trend. The data exhibit overlap at approximate porosity of 0.45 and strength of 1 MPa, although diverge at that point: the Shang et al. data exhibit greater strengths with less dispersion and greater porosity dependence than the Newton et al. data. In both cases, the predominant fabrication parameter that determined porosity and thus strength was compaction pressure. Compaction pressures were varied in the ranges (25–250) MPa, Shang et al., and (20–112) MPa, Newton et al.; isolated groups of data at discrete compaction pressures are visible in Figure 10.13a and an arrow indicates the overall compaction pressure trajectory in Figure 10.13b.

A secondary fabrication factor that determined strength was tablet curvature. Relative curvatures, D/R, for the doubly convex tablets were varied in the ranges 0.35–1.84, Shang et al., and 0.25–1.43, Newton et al., and both works included planar tablet $D/R = 0$ measurements. An arrow indicates the overall curvature direction trajectory in Figure 10.13b and planar measurements are indicated as filled symbols. The relationships between compaction die pressure and curvature (and thus tablet curvature) and resulting tablet porosity and strength are not straightforward, as powder mass is not held invariant and the pressure and curvature axes in Figure 10.13b are thus not orthogonal. The somewhat counter-intuitive result was

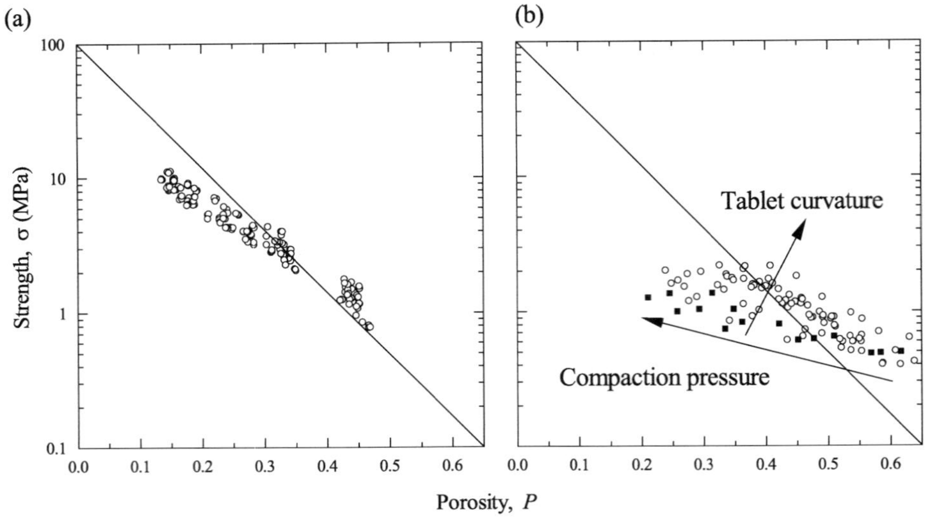

Figure 10.13 Plots of microcellulose doubly convex tablet strengths vs porosity (a) adapted from Shang, C et al. (2013a) and (b) adapted from Newton, J.M et al. (2000a). In (a) symbols represent individual strength measurements. In (b) symbols represent means of 10 strength measurements; filled symbols represent cylindrical tablets. Solid line is an empirical guide to the eye. Arrows indicate directions of increasing tablet fabrication parameters and apply to both plots.

observed for both sets of data: as tablet curvature was increased at constant die pressure, the porosity *and* strength both increased, as indicated by the direction of the curvature trajectory. In greater quantitative detail, Figure 10.14 re-plots the strength measurements of Figure 10.13 (using the same logarithmic scale) as a function of tablet curvature. Solid symbols, Newton et al., open symbols, Shang et al., and symbols with bars, ranges of measurements from Es-Saheb (1996), using data derived from the published work. The effects of die compaction pressure and curvature are now separated and the wide ranges of strengths reflect the compaction pressure domains; for a given curvature, tablet strength increased monotonically with pressure. There is no significant trend in either mean strength or strength distribution with curvature. The increasing porosity trajectory in Figure 10.14 is diagonally downward to the right and the increasing strength trajectory is diagonally upward to the right.

The implication of Figures 10.13 and 10.14 is that strength controlling flaw generation in compacted powder tablets is an inhomogeneous process that depends on the die pressure *and* shape. These parameters control the motion of sub-particles and flaw formation during compaction of the powder array. Such ideas have been discussed since the earliest tablet strength measurements (Newton et al. 1971, 2000a; Pitt et al. 1989a, 1990; Sonnergaard 1999), as it is clear that porosity or relative density alone is insufficient to completely characterize tablet strength. Flaw generation is influenced by both dilatational and shear strains in the sub-particle array. The importance of the relative motion of sub-particles in controlling flaws and strength is also made clear in the experiments of He et al. (2007); see Figure 10.5b. The lower half-filled symbols represent the strengths of tablets compacted from sub-particles that were lubricated in the pre-compaction rolling process—the strengths are clearly less than those of the control particles shown as the upper half-filled symbols. There are a large number of studies (e.g. McKenna and McCafferty 1982; Roberts and Rowe 1986; Roberts et al. 1997; Sonnergaard 1999; Almaya and Aburub 2008; Hooper et al. 2016; and Hernández et al. 2019) characterizing the deformation of many powders (e.g. lactose, microcrystalline cellulose, starch, NaCl, gypsum, ascorbic acid, paracetamol, ibuprofen, and mannitol) during compaction. However, they all relate the scalars of deformation, densification (negative dilatation) and pressure (negative hydrostatic stress). Often the behavior is presented using a linearized form of the exponential Heckel equation, $P \sim \exp(-\text{pressure})$ in current notation (although it is rarely plotted this way, but see e.g. Van Veen et al. 2005; Arida and Al-Tabakha 2008). Whether this is the best description or not (Sonnengaard, 1999), such experiments probably cannot provide complete insight into flaw generation on compaction in tablets as the measurements are restricted to characterization of porosity alone. The experiments here make clear, however, that much insight can be gained by measuring failure forces of tablets fabricated with identical included masses of powder and different die curvatures, in addition to different compaction pressures, as in this way shear and dilatation are separated. Such experiments could be combined with local density and properties measurements, as in the mechanical probe experiments of Cabiscol et al. (2018), the refractive index measurements of May et al. (2013), or the X-ray tomographic measurements of Akseli et al.

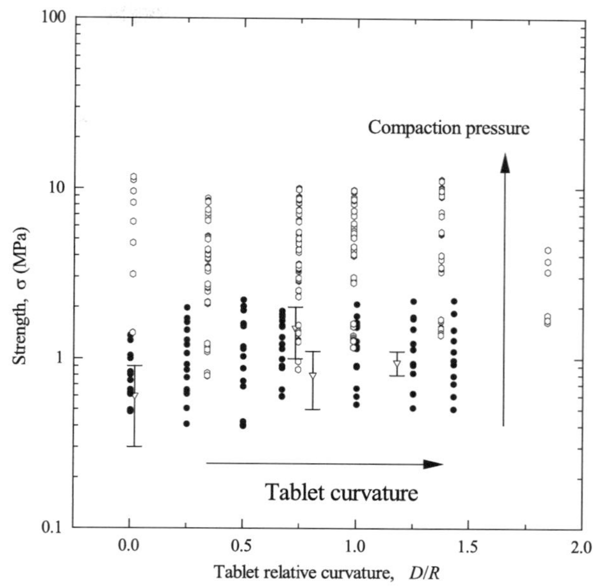

Figure 10.14 Replot of microcellulose doubly convex tablet strength data from Figure 10.13 as strength vs tablet relative curvature. (adapted from Shang, C et al. (2013a); Newton, J.M et al. (2000a); Es-Saheb, M.H.H (1996)). Arrows indicate directions of increasing tablet fabrication parameters. As tablet mass is increased with increasing tablet curvature, the strength is invariant.

(2011), all of which reveal inhomogeneity within tablets. In addition, models of compaction, such as by elastic-plastic finite element analyses (Han et al. 2008), or by discrete element models (Siiriä et al. 2011) can also provide insight into flaw generation.

It is important to remember that measurements of compaction deformation are only the first step in developing a greater understanding of tablet strength. Relating failure force to the stresses generated in a loaded tablet, either experimentally, as in the photoelastic measurements of Pitt et al. (1989a) or analytically, as in the elastic-plastic finite element analysis of Shang et al. (2013b), is an equally important second step. The failure of agglomerate tablets loaded in geometries other than diametral compression is likely to be controlled by different populations of strength-controlling flaws. Elongated tablets (Figure 10.7c), tested in axial compression exhibited strengths that were a simple 2/3 multiple of the typical strengths of circular tablets (0.1–3) MPa, consistent with elastic finite element analysis (Pitt and Heasly 2013). In both finite element analyses of Pitt and Heasley and Shang et al., the maximum principal stress (in thin cylinders usually σ_{xx}, in tablets often σ_{yy}) was used as the failure criterion, re-affirming that brittle fracture from internal bulk flaws was the failure mechanism. However, elongated cellulose tablets, loaded as uniaxial beams in bending, exhibited larger strengths, Newton et al. (2000b) and Podczeck et al. (2006), (0.2 to 12) MPa, as did biaxially flexed sorbitol cylinders (4 to 20) MPa (Podczeck 2007). The implication of the uniaxial and biaxial tests, more typical of extended components (Chapter 5), is that surface flaws are less potent than bulk flaws in these powder compacts. The similarity of the failure patterns of cylinder tablets with breaking lines (Podczeck et al. 2014, 2015) to those of alumina beams with scratches (Cook 2017) highlights the dominance of surface flaws in some tablet loading geometries.

10.2.3 Distributions

A few works considering tablet-like particle strengths have presented edf $Pr(\sigma)$ data. Figure 10.15 shows strength edf $Pr(\sigma)$ plots for four sets of agglomerate pharmaceutical tablet and particle strength measurements, using data derived from the published works cited. The plots are arranged in order of approximate increasing strength and the symbols represent individual strength measurements obtained in conventional compression platen geometries. Tablets were in the $D = (10–18)$ mm range. Figure 10.15a shows strength measurements of effervescent zinc tablets from Sonnergaard (2002); strengths were evaluated using the Brazil equation. Figure 10.15b shows strength measurements of aspirin tablets from Kennerley et al. (1982) and Stanley (2001); filled symbols represent measurements of the entire sample, open symbols represent measurements of a single batch. Figure 10.15c shows strength measurements of microcrystalline cellulose tablets from Keleṣ et al. (2015). Figure 10.15d shows strength measurements of paracetamol clusters from Ålander et al. (2004). The clusters were composed of 1–12, predominantly 6–8, crystals (the sub-particles) precipitated in acetone to form form particles sieved into the range (280–450) μm; strengths were evaluated using the HO equation. The tablet responses, Figure 10.15a–c, were all sigmoidal and exhibited narrow strength distributions, σ_N/σ_1, of 2 or less, both characteristics very similar to those

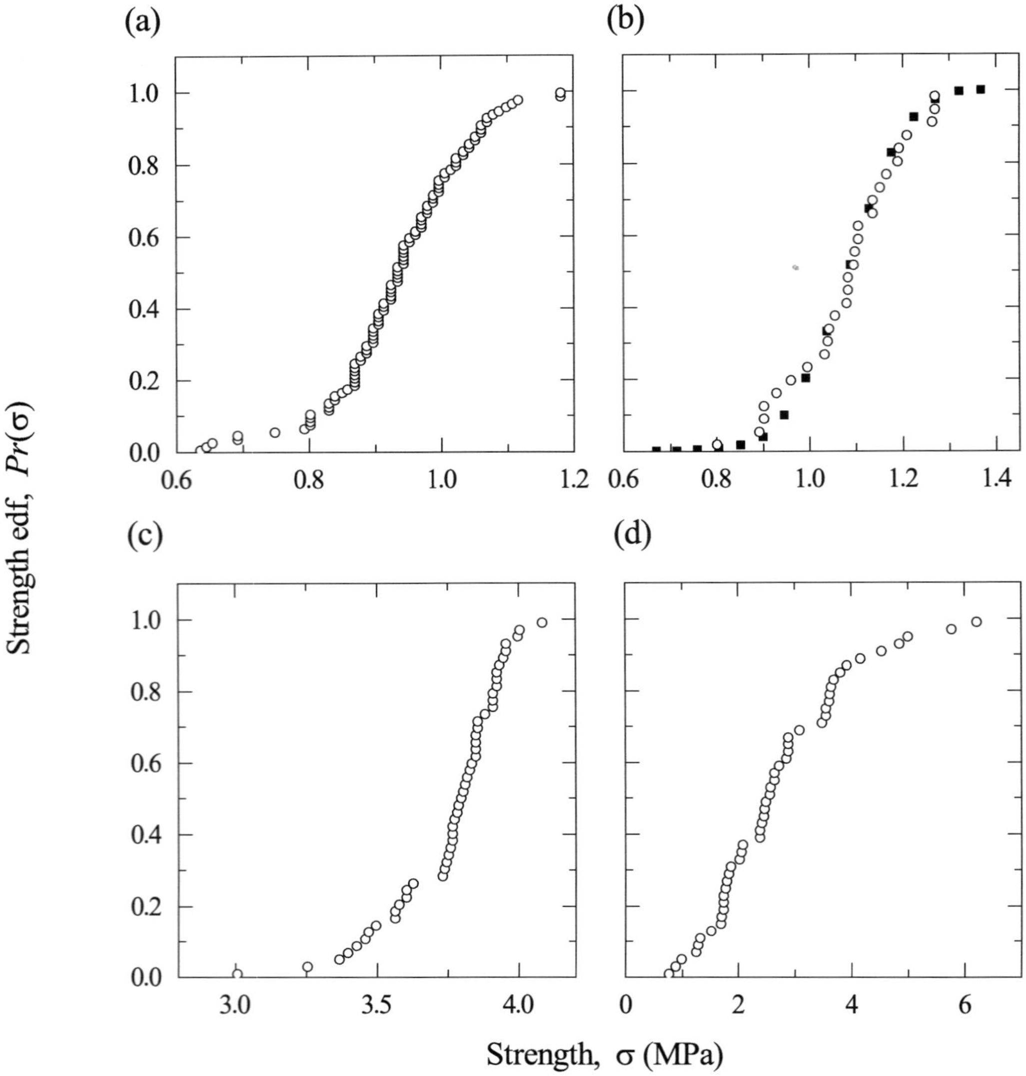

Figure 10.15 Strength edf behavior for four compacted agglomerate tablet systems. (a) effervescent zinc tablets, diameter $D = 18$ mm, number in sample $N = 100$ (adapted from Sonnergaard, J.M 2002). (b) aspirin tablets $D = 10$ mm, $N = 16$ and, $N = 28$ (adapted from Kennerley, J.W et al. 1982; Stanley, P 2001). (c) microcrystalline cellulose tablets $D = 10$ mm, $N = 51$ (adapted from Keles, O et al. 2015). (d) paracetamol clusters, $D \approx 360$ μm, $N = 50$ (adapted from Alander, E.M et al. 2004).

of the extended components of Chapter 5. The small cluster response, Figure 10.15d, was almost completely concave and exhibited a wide strength distribution, σ_N/σ_1, of approximately 6, both characteristics very similar to those of the particles of Chapter 6. Additional strength edf $Pr(\sigma)$ data are included in some of the works cited: Sonnergaard characterized a variety of commercial and custom tablets, Keleş et al. characterized the influence of porosity, and Ålander et al. characterized the influence of the precipitating environment. The additional data support the above observations. Otherwise, very little research appears in this area. The work of Keleş et al., now examined in detail, may provide an explanation for this absence.

Figure 10.16 shows the complete set of microcrystalline cellulose strength edf $Pr(\sigma)$ behavior from Keleş et al. (2015) using data derived from the published work. The plot coordinates are linear. The data appear right to left in order of increasing porosity and decreasing strength; the mean values and overall trend were shown left to right in Figure 10.5b. The symbols in Figure 10.16 represent individual strength measurements. The lines represent unconstrained best fits using the three parameter sigmoidal strength distribution formulation. The significant negative exponential effects of porosity are evident in that strength decreases from approximately 6 MPa to 0.2 MPa as porosity increases from approximately 0.17 to 0.56.

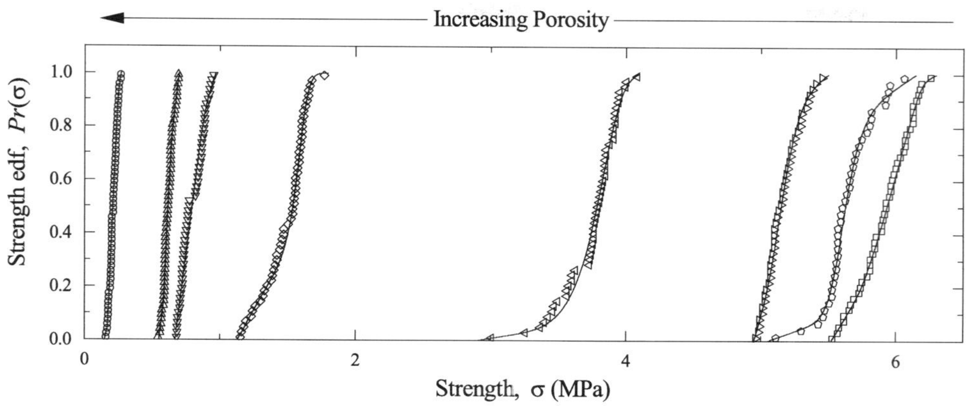

Figure 10.16 Strength edf behavior for several compacted microcellulose cylindrical tablet systems (adapted from Keles, O et al. 2015). The arrow indicates increasing tablet porosity, controlled by changing tablet mass and compaction pressure. Symbols represent individual strength measurements. Lines indicate unconstrained best fits. The strength distributions appear almost linear at this scale and are significantly dependent on porosity.

Although at this shared strength scale some distributions exhibit sigmoidal shapes (e.g. the center distribution, also used in Figure 10.15c), many do not, appearing as near straight. An assessment of distribution shapes is made more easily in Figure 10.17, in which the width of each distribution is scaled similarly and decreasing porosity indicated in the sequence Figure 10.17a–h. The symbols and lines in Figure 10.17 are identical to those in Figure 10.16. In this format the sigmoidal shapes of the edf responses are evident, although in some cases, the departure from a concave or linear response is weak. The asymmetry of the sigmoidal shapes is clear. Consideration of the strength scales in Figure 10.17 highlights the decrease in strength dispersion noted in Figure 10.5 and noted by Keleş et al., relative strength distribution widths, σ_N/σ_1, decrease from approximately 2, Figure 10.17a, to approximately 0.1, Figure 10.17h, as porosity decreases. A possible explanation for the relatively infrequent publication of tablet strength distributions is thus that the widths of the distributions are small relative to the changes in mean strengths caused by changes in porosity.

The strength variations in Figures 10.16 and 10.17 represent deterministic effects—there is certainly no common threshold strength. The underlying strength controlling flaw populations derive from different, but related, materials: the different porosity levels were obtained with different masses of compacted powder and different compaction pressures (Keleş et al. 2015). The flaw populations can be deconvoluted as described earlier (Chapter 4) and Figure 10.18 shows as solid lines the crack length pdf $h(c)$ variations underlying the cellulose tablet strength behavior. The direction of increasing porosity is indicated. For simplicity (see section 10.1), the $h(c)$ variations were determined using a linear dependence of B on porosity, $B \sim P$ between the bounds of (0.0033 and 0.001) MPa m$^{1/2}$. Sub-particle size (approximately 180 μm) and Eq. (7.2) were used as constraints. The widely separated near sigmoidal strength variations of Figure 10.16 result in widely separated peaked crack length populations, somewhat similar to Figure 5.25 and Figure 9.24. Consistent with deterministic effects, the crack length distributions vary in domain location and shape. Specifically, the crack populations in the least porous materials are narrow and nearly coincident, at small sub-μm lengths. For greater levels of porosity, the populations separate and increase in width and average length, to reach approximately (20 to 40) μm for the most porous material. On the basis of microstructural considerations such as depicted in Figure 10.1, it seems probable that these flaw populations, deconvoluted from strength tests of samples of tablets, are extreme value distributions. As the sub-particles and the related interstices are much smaller than the compacted agglomerate tablets, the stressed regions in the tablets are likely to contain many interstitial flaws, of which only the largest, most extreme, is sensed in a strength test. As such, although much more widely separated, the strength distributions in Figure 10.16 and Figure 10.18 are probably related by the same deterministic description as developed in Figure 9.31 and Figure 9.32. However, impressions can be deceiving with regard to visual flaw identification, as clearly demonstrated in consideration of the apparent "flaws" in side-walls of MEMS devices (Cook et al. 2019). This point is considered in Section 10.5. It is noted that the strength distribution spacings in Figure 10.16 are reversed from those of the underlying crack lengths in Figure 10.18 and that the prominent central strength distribution in Figure 10.16 is (now) not apparent in the set of crack length distributions.

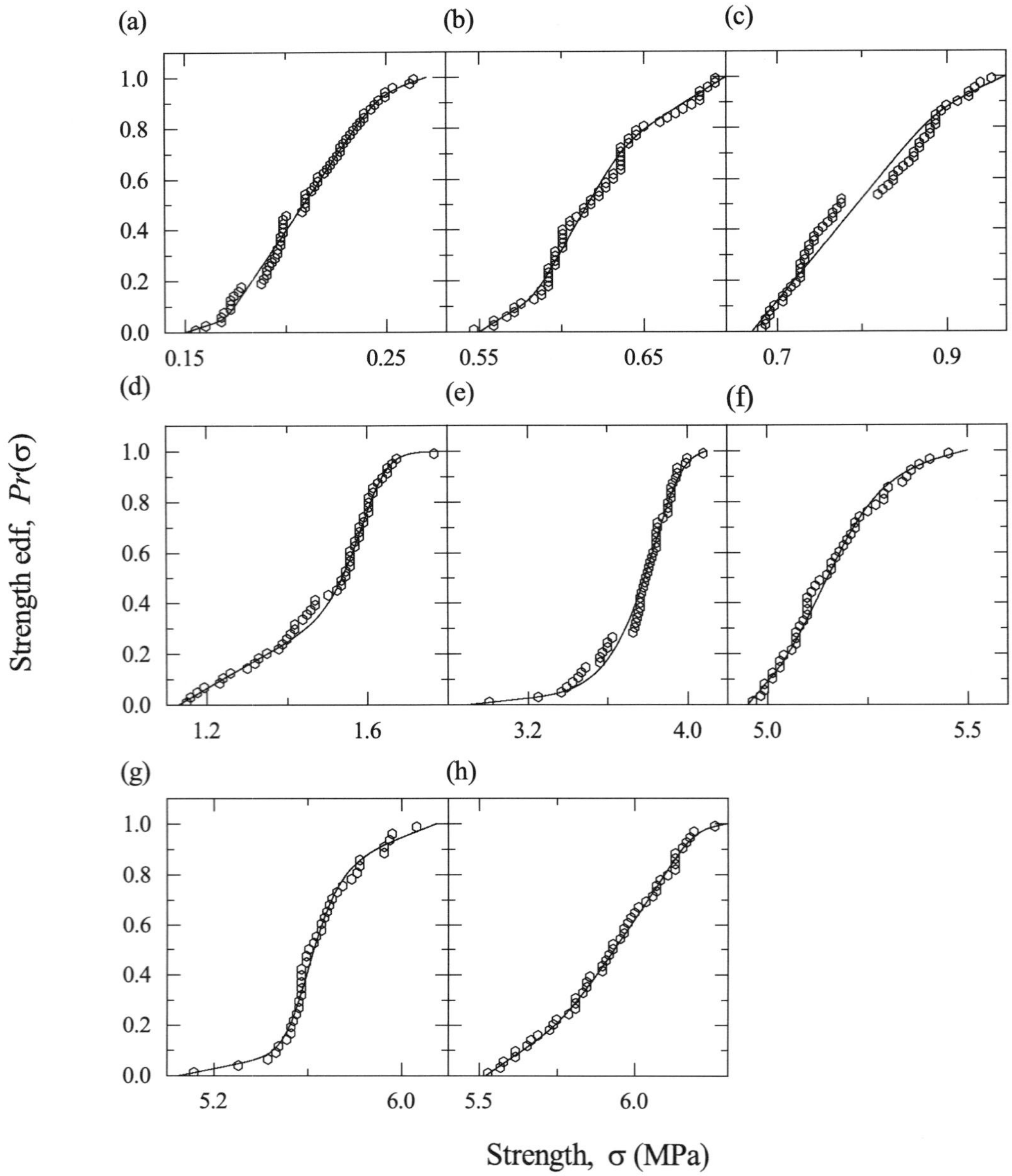

Figure 10.17 Replot of microcellulose cylindrical tablet strength data from Figure 10.16 using individually scaled strength axes. The common, weakly sigmoidal shape of the strength distribution measurements and best fits is now clear.

10.3 Foods

The application of excessive bite force through the jaw to strong (or "hard" or "tough") particles of foods can lead to chipping of teeth. Such chipping and fracture is of interest to dentists and zoologists to explain current-day human and animal behavior and to anthropologists and paleontologists to explain and reconstruct previous behavior. In order to infer the diets of living and fossil mammals and homonims, Constantino et al. (2010) and Lee et al. (2011) performed experiments and developed fracture mechanics models to analyze cracks in teeth. Among the many conclusions was that some species (e.g. orangutans) frequently eat "fallback" foods, not part of their regular diet, in the form of large, hard particles (e.g. seeds) leading to tooth fracture and some species (e.g. gorillas) rarely eat such foods and rarely exhibit tooth fracture or chipping. In addition, models of tooth fracture suggested that humans should exhibit smaller resistance to tooth cracking relative to other mammals and hominims in terms of critical bite forces. The focus of Constantino et al. and Lee et al. was on teeth.

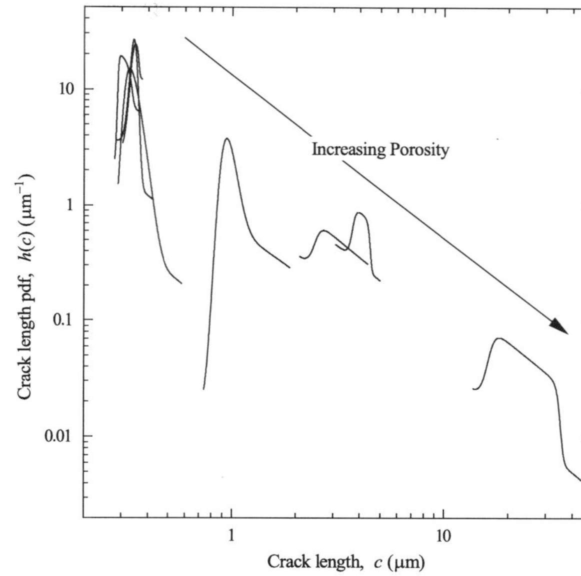

Figure 10.18 Plot in logarithmic coordinates of crack length pdf $h(c)$ variations for microcellulose tablets. Note the extremely narrow populations. Arrow indicates increasing porosity. Determined from Figure 10.17.

Here the focus is on the food (or other) particles that humans might eat and examines the strength distributions of such particles in the context of tooth fracture.

Figure 10.19 shows strength edf $Pr(\sigma)$ plots of food particles in approximate order of increasing strength, using data derived from the published works cited. All were tested using conventional platen apparatus. Figure 10.19a shows the behavior of corn flakes, rice krispies, and pasta from the work of McDowell and Humphreys (2002). Figure 10.19b shows the behavior of mustard seeds and peppercorns from the work of Singh et al. (2016). Figure 10.19c shows the behavior of cellulose from the work of Pitchumani et al. 2004). Figure 10.19d shows the behavior of sugar and cake decorations from the works of Rozenblat et al. (2011), Stefanou and Sulem (2016), and Singh et al. (2016). In Figure 10.19a and Figure 10.19d the data display the frequently observed particle strength characteristics of very broad, concave, strength distributions with maximum/minimum strength ratios $\sigma_N/\sigma_1 > 10$. In Figure 10.19b and Figure 10.19c the data display the extended component characteristics of narrower sigmoidal strength distributions with maximum/minimum strength ratios σ_N/σ_1 of 2–6. The data in Figure 10.19 can be divided into three groups: (a) porous particles, formed from aqueous mixtures containing smaller sub-particles of corn, rice, or wheat and then dried (not dissimilar to the bauxite, cement, or iron ore particles above); (b) nearly fully dense, composite structures formed as particles by plants; and (c) and (d) dense particles formed by crystallization of saccharide molecules, also by plants. There is a clear difference in strength between the food particles formed from sub-particles (a), about 1 MPa, and the food particles formed as composite seeds or as crystals (b), (c), and (d), 5–20 MPa. Consistent with the insights into porous particle failure, it is noted that the dense microcrystalline cellulose particles are stronger, 10 MPa to 20 MPa, Figure 10.19c, than the porous cellulose tablets considered above, 1 MPa to 5 MPa.

Using the observations of Constantino et al. and Lee et al., that human bite forces necessary for tooth cracking are about 500 N, the upper bound strength data of Figure 10.19 and the HO equation suggest that a 17 mm cornflake, an 8 mm peppercorn, or a 4 mm chunk of sugar could lead to a cracked tooth. The first particle is unlikely to be encountered, but the second and third are very common, although the broad domain of strengths suggests that fracture conditions are unlikely to be frequent for these food particles. Conversely, the significantly stronger particles of sand, about 100 MPa (Chapters 6 and 7) suggest that accidental biting of "grit" particles could easily lead to tooth fracture in both humans and other animals. Figure 9.11 shows that salt particles are comparable to the sugar particles of Figure 10.19d and could thus also lead to dietary tooth fracture. Similarly, Figure 9.28 shows that ice particles are only slightly less strong than sugar and salt, implying that it is possible, although unlikely, that tooth fracture could result from biting an ice block. At the other extreme, the strengths of tablets, (1-5) MPa, shown above to be consistent with other engineering particle measurements, are similar to the strengths of processed cereals, Figure 10.19a and consistent with chewability by humans.

As a side note, whole coconuts behave as almost ideal spherical particles in diametral compression, exhibiting planar meridional fracture in both the longitudinal and latitudinal coconut orientation (Schmier et al. 2020). As the failure forces are approximately 3500 N, opening a coconut by human hand is near impossible, although using the HO equation and

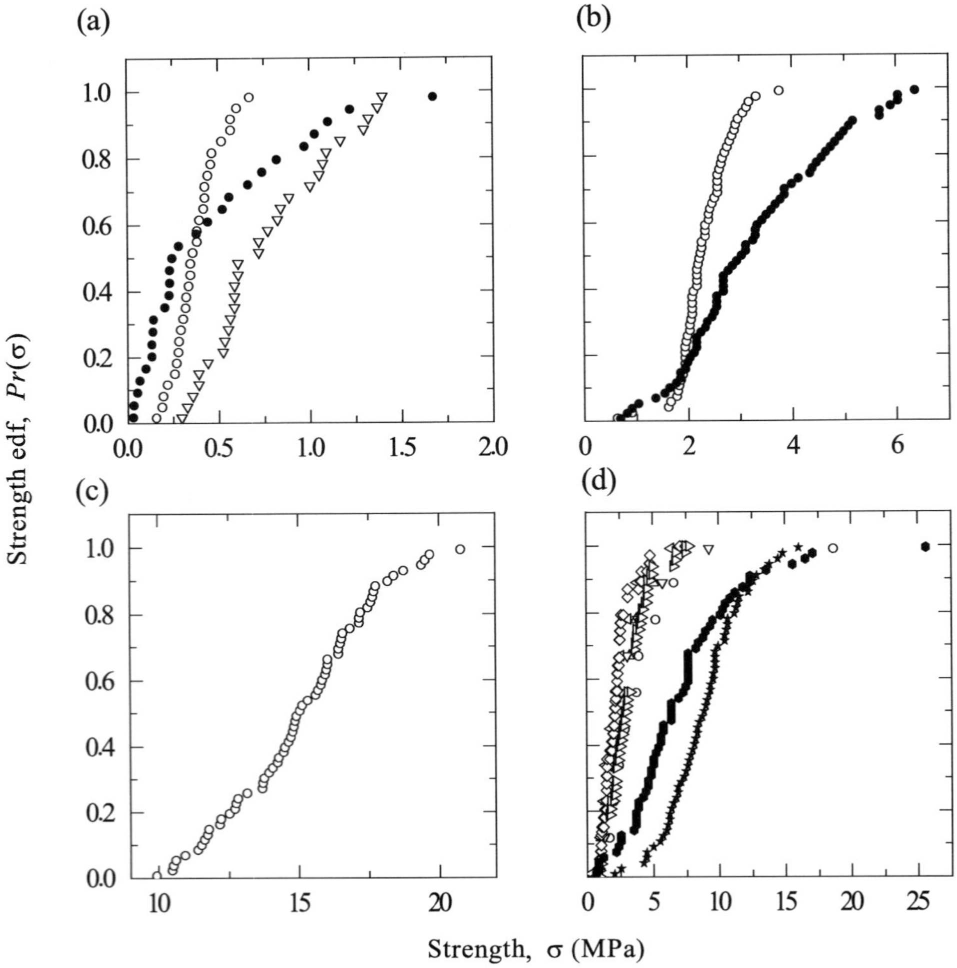

Figure 10.19 Strength behavior of food particles. (a) Corn flakes (filled symbols), rice krispies (open), and pasta (half-filled), irregular $D = (5-20)$ mm, $N_{\text{tot}} = 87$ (adapted from McDowell, G.R et al. 2002). (b) Mustard seeds (open symbols) and peppercorns (filled), $D = (1.6-2.6$ and $3.4-5.4)$ mm, $N_{\text{tot}} = 125$ (adapted from Singh, D et al. 2016). (c) Cellulose, $D = (0.75$ to $1.2)$ mm, $N = 64$ (adapted from Pitchumani, R et al. (2004)) (d) Sugar (open symbols) $D = (1.6-2, 2.5-3.15, 3.15-4, 4-5, 5-6.3, 6.3-8)$ mm $N_{\text{tot}} = 182$ and unrefined sugar and cake decorations (solid sybols) (0.8–2.2 and 1.2–2.0) mm, $N_{\text{tot}} = 21$ (adapted from Rozenblat, Y 2011; Stefanou, I et al. 2016; Singh, D et al. 2016).

$D \approx 100$ mm shows that the characteristic stress for whole coconut failure is only about 0.35 MPa. This value is comparable to the characteristic stress for whole pod failure of $D \approx 5$ mm canola seed pods that exhibit failure forces of about 8 N (Ghajarjazi et al. 2015) and thus stresses of about 0.28 MPa—easily and safely opened by human hands and jaws. These values are far less than the strengths estimated for isolated sections of shell or pod walls, of approximately 50 MPa in bending and 200 MPa in compression, for coconuts and other hard shelled fruits and seeds with particle-like shapes (Williamson and Lucas 1995; Schmier et al. 2020).

10.4 Catalysts

In many industrial chemical engineering processes, reactant gases or liquids are passed over a surface that catalyzes a chemical reaction, such that the reacted fluid product contains new chemical species or has had unwanted species removed. The catalytic surface is often formed by the interior and exterior surfaces of medium to large particles containing small-scale porosity and packed into an array or "bed," either fixed or mobile ("fluidized"), through which the reactants are passed, and from which the products are collected. Mechanical integrity of such catalyst particles is critical, particularly in fixed bed reactors, as broken or crushed particles can generate "fines," impeding flow through the particle bed and reducing reactor performance (similar to the concerns regarding air flow through beds of iron ore pellets in blast furnaces, explained earlier). Fracture of catalyst particles may also reduce their catalytic ability. Thus, a major requirement for catalyst and other particles with similar functions is that particle strength is sufficient for particles lower in the bed to support the support the weight of overlying particles higher in the bed. As catalytic function increases with porosity and, as demonstrated in Section 10.2, strength decreases with porosity, optimizing catalyst bed performance requires compromise in design. As a consequence, several measurements of the crushing strengths of catalyst particles have been performed, with a focus on zeolites and alumina.

Figure 10.20 shows edf $Pr(\sigma)$ plots of catalyst particles and some related porous alumina particles in approximate order of increasing strength, using data derived from the published works cited. All data, except those in Figure 10.20c, were obtained in conventional uniaxial compression apparatus, Figure 1.07a. The data in Figure 10.20c were obtained in a biaxial compression apparatus, Figure 1.07i. The symbols represent individual failure measurements. Figure 10.20a shows the behavior of three zeolite materials with $D \approx 6$ mm from the work of Li et al. (1999, 2000). Figure 10.20b shows the behavior of two alumina materials with $D \approx 5$ mm also from the work of Li et al. (1999, 2000); the open and filled symbols in each case indicate different materials. Relative porosity for all materials was $P \approx 0.5$. Similar results were reported by Wu et al. (2006, 2007). Figure 10.20a and Figure 10.20b illustrate a major feature of the mechanical behavior of porous catalyst particles: the strengths are small relative to dense materials of similar composition (see Figure 8.2 and Figure 9.5), a result observed by Li et al. for several other catalyst materials manufactured by various processes and loaded in various geometries.

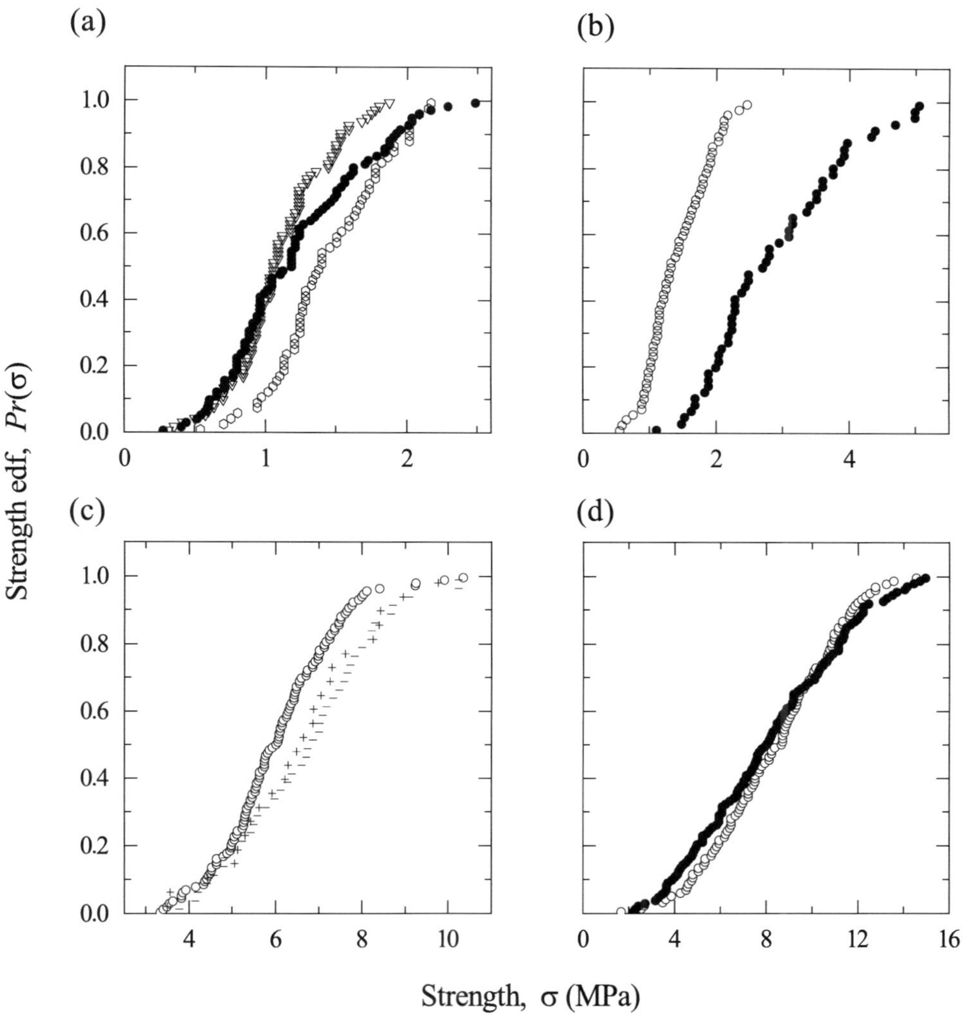

Figure 10.20 Strength behavior of catalyst and related particles. (a) Zeolite, particle size $D = (5.75 \pm 0.17, 6.02 \pm 0.14$ and $5.5–6.5)$ mm, sample size $N_{tot} = 236$ (adapted from Li, Y et al. 1999, 2000). (b) Alumina, $D = (4.43 \pm 0.14$ and $4.5–6.5)$ mm, $N_{tot} = 116$ (adapted from Li, Y et al. 1999, 2000). (c) Alumina, particles tested in mixed mode loading. $D = 1.97$ mm, $N_{tot} = 185$ (adapted from Satone, M et al. 2017). (d) Alumina, particles tested fresh and after use. $D = (1.7–2.0)$ mm, $N_{tot} = 225$. (adapted from Subero-Couroyer, C et al. 2003).

Figure 10.20c and Figure 10.20d show strength behaviors of porous particles that are relevant in catalyst applications. Figure 10.20c shows the behavior of organic gel derived alumina particles with $D \approx 2$ mm from the work of Satone et al. (2017). The particles were fired at a low temperature, enough to remove the organic gel binder, and had a relative porosity of $P \approx 0.4$. The particles were tested by application of two perpendicular loading axes, such that the direction of the net applied force could be varied from 0° to a selected particle axis, conventional uniaxial loading, to 45° to that axis, equibiaxial loading. The magnitude of the net force was used in strength calculations. The − and + symbols in Figure 10.20c represent the two loading mode limits of 0°–5° and 40°–45°. The open symbols in Figure 10.20c represent all the intermediate strength measurements, 5°–40°. There is no significant difference between any of these strength distributions, implying that the mode of loading (uniaxial, biaxial, or mixed) is not a large factor in determining particle strength. From a catalyst design perspective, the performance of low-lying catalyst particles in supporting the over burden of many higher lying particles can be estimated from simple uniaxial strength tests. Figure 10.20d shows the behavior of alumina particles used as catalyst carriers with $D = (1.7–2)$ mm from the work of Subero-Couroyer et al. (2003). The filled symbols in Figure 10.20d represent the strengths of particles that had been treated with metallic catalyst and used in a reforming refinery for 6–7 years. The open symbols represent the strengths of fresh carrier beads. There is no significant difference between these strength distributions, implying that carrier bead strength does not degrade with use. From a catalyst design and reliability estimation perspective, initial strength distributions and identification of initial strength controlling flaws (from the sol-gel fabrication process) are sufficient.

Other catalyst related particle strength measurements are similar to those in Figure 10.20: Antonyuk et al. (2005) (porosity $P = 0.45$) and Russell et al. (2014) (0.51 porosity) reported strength values of approximately 5 MPa and (2–7) MPa, respectively, for zeolite catalyst particles; Antonyuk et al. (2005), (0.68 porosity) Müller et al. (2013), (0.65-0.75 porosity) and Fedorov and Gulyaeva (2019) (0.73 porosity) reported values of approximately 12 MPa, (9–25) MPa, and (5–40) MPa for alumina catalyst particles, respectively. Somewhat greater values were reported by Samimi et al. (2015), (37–91) MPa, for alumina catalyst support pellets that were compacted, approximately (400–1300) MPa, prior to firing, leading to relative porosity of $P \approx 0.3$.

The predominant shape characteristic of the strength distributions in Figure 10.20 is sigmoidal, although there are clear tendencies to linear behavior, characteristics that are also observed in the works of Antonyuk et al., Russell et al., Müller et al., Fedorov and Gulyaeva, and Samimi et al. The relative widths of the strength distributions in Figure 10.20 are in the range $\sigma_N/\sigma_1 = 3–8$. This range is greater than the factor of 2 exhibited by the well-formed sigmoidal responses of extended components (Chapter 5) and some pharmaceutical tablets (Section 10.2.3), but comparable to the sigmoidal responses of fertilizer particles (Section 7.3), bauxite and mullite particles (Section 8.2), potash and iron ore particles (Section 8.4), and some food particles (Section 10.3). The similarity of agglomerate particle strength distributions to those of extended components, and the implications for the underlying flaw distributions, is considered in the discussion

10.5 Discussion and Summary

This chapter has surveyed experimental observations of the strengths of agglomerate particles—particles composed of collections of smaller sub-particles. Two distinct features characterize the structures of such agglomerate particles: The first is that the sub-particle array is usually not fully dense, incorporating porosity into the particle through interstices between the sub-particles. The second is that bonding between the sub-particles is usually restricted to localized areas at sub-particle contacts and inter sub-particle bonding is usually weaker and not the same as intra sub-particle bonding. As a consequence of these features, agglomerate particles exhibit fracture strengths much less than materials that are their dense analogs with compositions similar to those of the sub-particles. The strengths of agglomerate particle are decreased by both of their distinctive structural features: Agglomerate particles contain large, strength-limiting flaws, the sub-particle interstices and related internal pore structures, and fracture planes in agglomerate materials contain fewer and weaker bonds, the sub-particle contacts. In some cases, the sub-particle contacts in agglomerate particles consist of sub-particle material that has diffused to the contacts. This material is thus identical to that of the sub-particles and forms bridges and necks of material well bonded to the sub-particles. Such structures are typified by agglomerates of glasses and ceramics sintered at high temperatures. In some cases, the sub-particle contacts consist of reaction products formed by the sub-particle material and environmental species, often moisture. These products are thus related to but not identical to the material of the sub-particles and form bridges and necks of material moderately well bonded to the sub-particles. Such structures are typified by agglomerates of cements cured in the presence of water and ceramics fired in oxidizing environments at low temperatures. In some cases, the sub-particle contacts involve no transfer or introduction of solid material but consist of regions of surface force interactions, van der Waals bonding, hydrogen bonding, and water capillary meniscus bonding. These forces are very short range and weak, but pervasive within the sub-particle array. Such structures are typified by pressure compacted powders such as pharmaceutical tablets, some foods, and green ceramics. All three of these agglomerate structures exhibit significant decreases in strength with increasing porosity, consistent with scaling behavior based on fracture mechanics considerations. Empirically, agglomerate components, whether extended or particles, often exhibit negative exponential dependence of strength σ on relative porosity P, $\sigma \sim \exp(-bP)$, where b is an empirical exponent of order 10. Hence at a relative porosity of $P \approx 0.1$, the typical transition from a closed to an open porosity structure, the strength has decreased to about a third of the value exhibited by dense material. At a relative porosity of $P \approx 0.3$, the onset of the transition from a "solid" to a "foam", the strength has typically decreased by over an order of magnitude relative to the dense material.

A important class of agglomerate particles are pharmaceutical tablets, formed by the compaction of mixtures of pharmaceutically active and inactive (excipient) sub-particle powders (Muzzio et al. 2002). A commonly observed characteristic of such powders, important in commercial manufacture of tablets, is that tablet porosity decreases approximately logarithmically with compaction pressure, $P \sim -\ln(\text{pressure})$. As a consequence of the above negative exponential dependence of strength on porosity, tablet strength usually varies linearly with compaction pressure, $\sigma \sim$ pressure. A minor complication in quantifying this behavior is that tablets, although convenient for strength testing in diametral compression, are neither cylinders nor spheres, but intermediate doubly convex shapes. Empirical calibration of doubly convex tablet strength as a function of shape has enabled compacted tablet materials comparisons and identification of tablet shape factors (e.g. tablet curvature) that influence tablet strength and thus the underlying strength-controlling flaws. The sub-particles in pharmaceutical tablets are bound by surface forces, such that although the compaction pressures for tablets are typically several hundred MPa, tablet strengths are small, typically in the range (0.3–10) MPa, and 1 MPa is common.

In the absence of compaction, agglomerate particles formed of smaller sub-particles held together by localized forces can be very weak indeed, as noted in the early works considering agglomerate particles. Binding between sub-particles by van der Waals attraction or cementitious bonds was considered for limestone agglomerates (Kapur and Fuerstenau 1967), by NaCl salt bridges in sand pellets (Capes 1971), and by liquid bridges and capillaries in granular materials broadly considered (Pietsch 1968). The strengths of the agglomerate limestone and sand particles, approximately 0.1 MPa, are illustrated in Figure 2.26. At the "strong" end of the weak agglomerate range, Ålander et al. (2004) used a platen apparatus to measure strengths of (0.2 to 6) MPa for approximately 0.5 mm polycrystalline paracetamol agglomerates formed by crystallization from solution; see Figure 10.15d. Presumably the substituent crystals, about 0.1 mm in size, were bound within the particles by primary, grain-boundary, bonds and the small strength predominantly reflects particle porosity. Similarly, Kirsch et al. (2011) used a delicate custom apparatus to measure the failure forces of individual bridges generated in humid conditions between millimeter-scale sub-particles of urea. Characteristic strengths in tension of about 0.6 MPa were observed. Knoop et al. (2016) used acoustic methods to measure strengths of (0.4 to 4) MPa for (1 to 3) mm agglomerates of glass sub-particles, about 0.1 mm in size, held together by van der Waals forces and water capillary bridges. Direct platen measurements by Wollborn et al. (2017) of similar agglomerates with slightly different glass surface preparation exhibited strengths of (0.03 to 0.17) MPa. Adi et al. (2011) also used a direct platen test to measure the strengths of spray-dried mannitol agglomerates formed into 1.5 mm particles from 3 μm substituents, with the intention of assessing aerosol use. Strengths of (0.003 to 0.183) MPa were observed and correlated with agglomerate vapor dispersion. Das et al. (2013) were similarly concerned and inferred strengths of (0.001 to 0.1) MPa from agglomerate size measurements of (18 to 24) μm lactose substituents for aerosol use. At the weakest end of the agglomerate strength range, Mueller et al. (2017) formed 3 mm, approximately 1.4 mg, particles from (40 to 90) μm volcanic ash or glass substituents, deliberately bound with NaCl bridges, in order to mimic and model volcanic plume activity. Particle impact measurements showed increased fragmentation energy required for particles containing greater amounts of NaCl and inferred particle strengths of approximately (0.01 to 0.03) MPa. Extensive strength edf $Pr(\sigma)$ data in unbiased form were presented in the studies of the stronger agglomerate particles by Ålander et al., Knoop et al., and Wollborn et al. In the first case, the data are identical to the characteristic concave particle strength distributions of Chapter 6, reflecting heavy tail flaw populations. In the next two cases the data are very similar to the near linear strength distributions of the small alumina particles in Chapter 9. As above with tablets, the strengths of weak porous agglomerate particles are consistent with other particle measurements.

Reporting of agglomerate particle strength distributions is not pervasive as it is for dense monolithic particles (Chapters 6–9). In addition, estimates of the parameter B linking strength and crack length are extremely uncertain. Hence, application of reverse analyses as described in Chapter 4 for determination of crack length populations from strength distributions of agglomerates is limited, e.g. Figure 10.18. However, simulations of strength edf $Pr(\sigma)$ behavior for samples of particles, starting from material crack populations, can provide insight. Such simulations are based on forward analyses as described

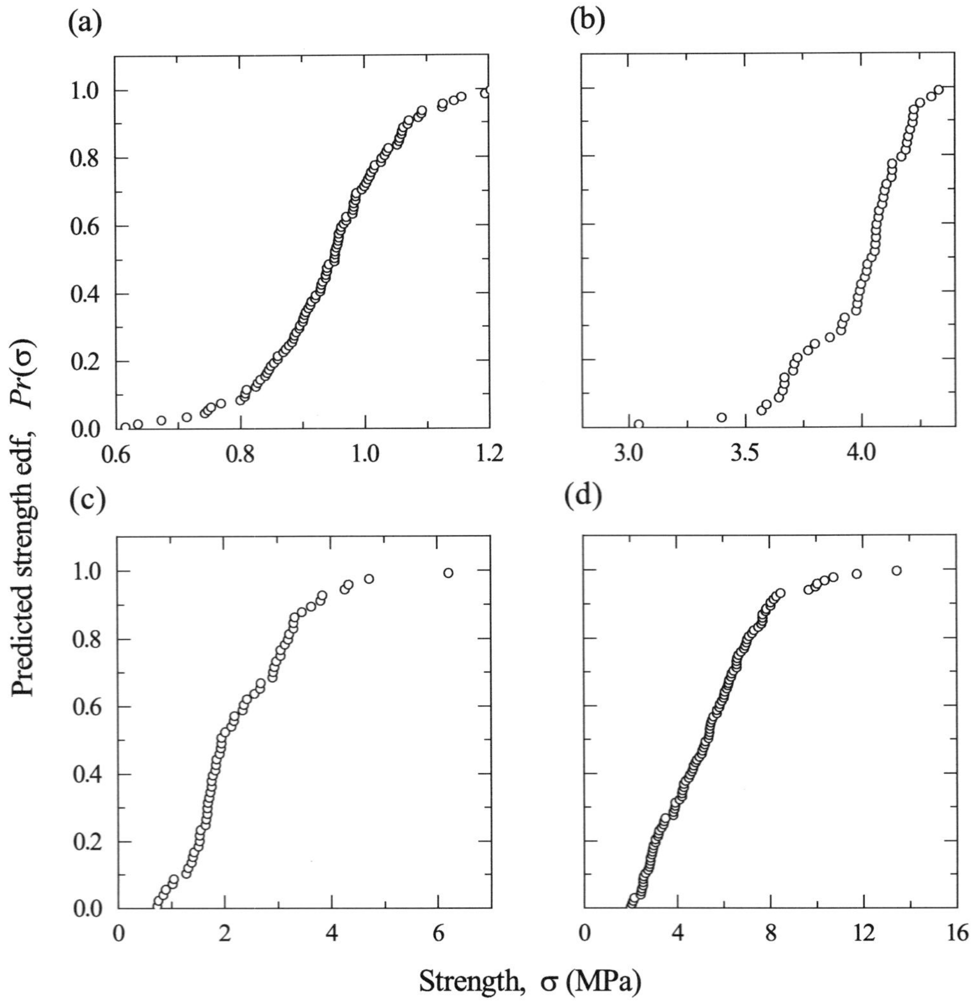

Figure 10.21 Simulated strength edf $Pr(\sigma)$ responses developed from initial crack length populations. Simulation parameters ($\omega, \kappa, f_0, f_1, \eta, k, N, \sigma_1$/MPa) and comparison experimental data are (a) (0.2, 20, 0.06, 0.01, 2.3, 1.5, 100, 0.6), Figure 10.15a, (b) (0.15, 20, 0.06, 0.01, 1.5, 1.5, 51, 3.0), Figure 10.15c, (c) (0.03, 15, 0.04, 0.01, 25, 1.5, 62, 0.7), Figure 10.19b, and (d) (0.04, 25, 0.06, 0.01, 45, 2.5, 110, 2.0), Figure 10.20d. The simulations well describe the experimental observations.

in Chapter 3 and are sensitive to the shape of the crack population, the width of the strength distribution, and the number of cracks in each particle. Simulations of strength distributions of glass bottles and coral particles were demonstrated in Chapter 3, Figure 3.17, and closely matched the experimental observations presented in Chapter 2, Figures 2.28 and 2.29. Figure 10.21 shows the simulated strength edf $Pr(\sigma)$ responses of the agglomerate particles from Figures 10.15a, 10.15c, 10.19b, and 10.20d. The plots cover the diversity of behavior observed for the three groups of materials considered in this chapter and are shown in approximate order of increasing strength. The plots use the same format and scales as those for the experimental data. The edf variations were simulated by forward analysis from an intensive crack population $f(\nu)$ and relative strength domain η, modified by the extensive component and sample parameters, k, N, and σ_1. In all cases, the crack population was described by a perturbed beta distribution, as in Figure 3.7c. The full set of parameters is given by $(\omega, \kappa, f_0, f_1, \eta, k, N, \sigma_1)$, described in detail in Chapter 3; numerical values are given in the caption to Figure 10.21. As before, the simulations match the experimental observations extremely well.

The simulations of Figure 10.21 are based on physical principles and thus an implication is that the underlying parameters are physically interpretable. This implication is aided by the fact that the parameter values for the simulations are quite restricted: The number of components tested in a sample N and the minimum strength observed σ_1 are set by experiment. A lower bound to η, (σ_N/σ_1), is also set by experiment. Although the perturbed beta distribution is extremely flexible, the crack population shape in practice is restricted to variations between Figures 3.6b and 3.7b in order to match experimental strength distribution shapes. The restriction constrains ω, κ, f_0, and f_1. Hence, as might be expected, the parameter that has the greatest freedom—the average number of cracks k in the ensemble of cracks in a component—has a great effect on simulated strength distributions, here for agglomerate particles. Of perhaps some surprise is that the numbers of cracks/particle used in the simulations of Figure 10.21 are small, $k = 1.5$ or 2.5. The strength distributions of the experimental observations of Figures 10.15a, 10.15c, 10.19b, and 10.20d very closely reflect the base flaw population, not an extreme value distribution of a sample from the population. This observation, that k is small, has been a feature of the recent development and application of the probabilistic framework used in the simulations (Cook and DelRio 2019a, 2019b; Cook et al. 2019, 2021; DelRio et al. 2020). The similar finding here supports the earlier observations that not all pores in a Si_3N_4 ceramic were possible strength controlling flaws (Chao and Shetty 1992) and that not all grain boundary grooves in Si MEMS sidewalls were possible strength controlling flaws (Cook et al. 2019). Small values of k were also required in the simulations of glass bottles and coral particles of Figure 3.17.

This chapter has surveyed the strength properties of agglomerate particles. Although such particles are characterized by appreciable porosity, they exhibit failure behavior that is classically brittle. Deformation on application of a force, usually in a diametral compression geometry, is small and linear elastic until fracture and failure ensue at peak force (although there are sometimes deviations, Sonnergaard 2013). Continued deformation leads to extended supported force and erratic behavior as the particles are crushed into several fragments. Although occurring at smaller stresses and strengths, the behavior is identical to that of the dense particles of the previous chapters. The porosity in these particles was typically of order 0.2 and unfilled, and the strains at failure in the solid fraction were typically 0.01 or less. The following chapter, Chapter 11, extends consideration of porous particles to those in which the porosity is 0.95 or greater and filled with fluid, usually water, and the strains at failure are typically 0.5. Although deformation prior to failure for these particles is elastic, it is large and non-linear, and such particles are not usually regarded as classically brittle.

References

Adi, S., Adi, H., Chan, H.-K., Finlay, W.H., Tong, Z., Yang, R., and Yu, A. (2011). Agglomerate strength and dispersion of pharmaceutical powders. *Journal of Aerosol Science* 42: 285–294.

Adolfsson, Å, Olsson, H., and Nyström, C. (1997). Effect of particle size and compaction load on interparticulate bonding structure for some pharmaceutical materials studied by compaction and strength characterisation in butanol. *European Journal of Pharmaceutics and Biopharmaceutics* 44: 243–251.

Akseli, I., Iyer, S., Lee, H.P., and Cuitiño, A.M. (2011). A quantitative correlation of the effect of density distributions in roller-compacted ribbons on the mechanical properties of tablets using ultrasonics and X-ray tomography. *AAPS Pharmscitech*, 12: 834–853.

Ålander, E.M., Uusi-Penttilä, M.S., and Rasmuson, Å.C. (2004). Agglomeration of Paracetamol during crystallization in pure and mixed solvents. *Industrial and Engineering Chemistry Research* 43: 629–637.

Almaya, A. and Aburub, A. (2008). Effect of particle size on compaction of materials with different deformation mechanisms with and without lubricants. *AAPS Pharmscitech* 9: 414–418.

Ashby, M.F. (1999). *Materials Selection in Mechanical Design*. Butterworth-Heinemann.

Arida, A.I. and Al-Tabakha, M.M. (2008). Compaction mechanism and tablet strength of Cellactose. *Jordan Journal of Pharmaceutical Sciences* 1: 71–82.

Brecker, J.N. (1974). The fracture strength of abrasive grains. *Journal of Engineering for Industry* 96: 1253–1257.

Cabiscol, R., Finke, J.H., Zetzener, H., and Kwade, A. (2018). Characterization of mechanical property distributions on tablet surfaces. *Pharmaceutics* 10: 184.

Capes, C.E. (1971). The correlation of agglomerate strength with size. *Powder Technology* 5: 119–125.

Chamberlain, J.A. (1978). Mechanical properties of coral skeleton: Compressive strength and its adaptive significance. *Paleobiology* 4: 419–435.

Chang, S.-Y., Li, J.-X., and Sun C.C. (2017). Tensile and shear methods for measuring strength of bilayer tablets. *International Journal of Pharmaceutics* 523: 121–126.

Chao, L.-K. and Shetty, D.K. (1992). Extreme-value statistics analysis of fracture strengths of a sintered silicon nitride failing from pores. *Journal of the American Ceramic Society* 75: 2116–2124.

Constantino, P.J., Lee, J.J.-W., Chai, H., Zipfel, B., Ziscovici, C., Lawn, B.R., and Lucas, P.W. (2010). Tooth chipping can reveal the diet and bite forces of fossil hominins. *Biology Letters* 6: 826–829.

Cook, R.F. (2017). Fracture mechanics of sharp scratch strength of polycrystalline alumina. *Journal of the American Ceramic Society* 100: 1146–1160.

Cook, R.F., Boyce, B.L., Friedman, L.H., and DelRio, F.W. (2021). High-throughput bend-strengths of ultra-small polysilicon MEMS components. *Applied Physics Letters* 118: 201601.

Cook, R.F. and DelRio, F.W.(2019a). Material flaw populations and component strength distributions in the context of the Weibull function. *Experimental Mechanics* 59: 279–293.

Cook, R.F. and DelRio, F.W. (2019b). Determination of ceramic flaw populations from component strengths. *Journal of the American Ceramic Society* 102: 4794–4808. (typographical error in Eq. (9)).

Cook, R.F., DelRio, F.W., and Boyce, B. L. (2019). Predicting strength distributions of MEMS structures using flaw size and spatial density. *Microsystems & Nanoengineering* 5: 1–12.

Darvell, B.W. (1990). Review Uniaxial compression tests and the validity of indirect tensile strength. *Journal of Materials Science* 25: 757–780.

Das, S.C., Behara, S.R.B., Morton, D.A.V., Larson, I., and Stewart, P.J. (2013). Importance of particle size and shape on the tensile strength distribution and de-agglomeration of cohesive powders. *Powder Technology* 249: 297–303.

Duckworth, W.H. (1951). Precise tensile properties of ceramic bodies. *Journal of the American Ceramic Society* 34: 1–9.

Es-Saheb, M.H.H. (1996). Tensile fracture characteristics of double convex-faced cylindrical powder compacts. *Journal of Materials Science* 31: 214–223.

Fell, J.T. and Newton, J.M. (1970). Determination of tablet strength by the diametral-compression test. *Journal of Pharmaceutical Sciences* 59: 688–691.

Fichtner, F., Rasmuson, Å., and Alderborn, G. (2005). Particle size distribution and evolution in tablet structure during and after compaction. *International Journal Pharmaceutics* 292: 211–225.

Gaither, M.S., Gates, R.S. Kirkpatrick, R., Cook, R.F., and DelRio, F.W. (2013). Etching process effects on surface structure, fracture strength, and reliability of single-crystal silicon theta-like specimens. *Journal of Microelectromechanical Systems* 22: 589–602.

Ghajarjazi, E., Azadbakht, M., and Ghaderi-Far, F. (2015). Determination of the mechanical properties of unbroken canola pods. *Agricultural Engineering International: CIGR Journal* 17: 392–403.

Gibson, L.J. and Ashby, M.F. (1997) *Cellular Solids: Structure and Properties, 2nd ed.* Cambridge University Press.

Han, L.H., Elliott, J.A., Bentham, A.C., Mills, A., Amidon, G.E., and Hancock, B.C. (2008). A modified Drucker-Prager Cap model for die compaction simulation of pharmaceutical powders. *International Journal of Solids and Structures* 45: 3088–3106.

Hasselman, D.P.H. (1963). Relation between effects of porosity on strength and on Young's modulus of elasticity of polycrystalline materials. *Journal of the American Ceramic Society* 46: 564–565.

He, X., Secreast, P.J., and Amidon, G.E. (2007). Mechanistic study of the effect of roller compaction and lubricant on tablet mechanical strength. *Journal of Pharmaceutical Sciences* 96: 1342–1355.

Hernández, O.C., Baltazar, E.H., González, E.A., and Contreras, L.M.M. (2019). Production of directly compressible excipients with mannitol by wet granulation: Rheological, compressibility and compactibility characterization. *Farmacia* 67: 973–985.

Hiramatsu, Y. and Oka, Y. (1966). Determination of the tensile strength of rock by a compression test of an irregular test piece. *International Journal of Rock Mechanics and Mining Sciences & Geomechanics Abstracts* 3: 89–99.

Hondros, G. (1959). The evaluation of Poisson's ratio and the modulus of materials of low tensile resistance by the Brazilian (indirect tensile) test with particular reference to concrete. *Australian Journal of Applied Science* 10: 243–268.

Hooper, D., Clarke, F.C., Mitchell, J.C., and Snowden, M.J. (2016). A modern approach to the Heckel Equation: The effect of compaction pressure on the yield pressure of ibuprofen and its sodium salt. *Journal of Nanomedicine and Nanotechnology* 7: 381. doi:10.4172/2157-7439.1000381.

Jaeger, J.C. and Hoskins, E.R. (1966). Rock failure under the confined Brazilian test. *Journal of Geophysical Research* 71: 2651–2659.

Jaeger, J.C. (1967). Failure of rocks under tensile conditions. *International Journal of Rock Mechanics and Mining Sciences* 4: 219–227.

Jarosz, P.J. and Parrott, E.L. (1982). Factors influencing axial and radial tensile strengths of tablets. *Journal of Pharmaceutical Sciences* 71: 607–614.

Kapur, P.C. and Fuerstenau, D.W. (1967). Dry strength of pelletized spheres. *Journal of the American Ceramic Society* 50: 14–18.

Keleș, Ö., Barcenas, N.P., Sprys, D.H., and Bowman, K.J. (2015). Effect of porosity on strength distribution of microcrystalline cellulose. *Pharmaceutical Science and Technology* 16: 1455–1464.

Kennerley, J.W., Newton, J.M., and Stanley, P. (1982). A modified Weibull treatment for the analysis of strength-test data from non-identical brittle specimens. *Journal of Materials Science* 17: 2947–2954.

Kirsch, R., Bröckel, U., Brendel, L., and Török, J. (2011). Measuring tensile, shear and torsional strength of solid bridges between particles in the millimeter regime. *Granular Matter* 13: 517–523.

Knoop, C., Todorova, Z., Tomas, J., and Fritsching, U. (2016). Agglomerate fragmentation in high-intensity acoustic standing wave fields. *Powder Technology* 291: 214–222.

Lee, J.J.-W., Constantino, P.J., Lucas, P.W., and Lawn, B.R. (2011). Fracture in teeth—a diagnostic for inferring bite force and tooth function. *Biological Reviews* 86: 959–974.

Li, L. and Aubertin, M. (2003). A general relationship between porosity and uniaxial strength of engineering materials. *Canadian Journal of Civil Engineering* 30: 644–658.

Li, Y., Wu, Do., Chang, L., Shi, Y., Wu, Di., and Fang, Z. (1999). A model for the bulk crushing strength of spherical catalysts. *Industrial and Engineering Chemistry Research* 38: 1911–1916.

Li, Y., Wu, Do., Zhang, J., Chang, L., Wu, Di., Fang, Z., and Shi, Y. (2000). Measurement and statistics of single pellet mechanical strength of differently shaped catalysts. *Powder Technology* 113: 176–184.

May, R.K., Su, K.E., Han, L., Zhong, S., Elliott, J.A., Gladden, L.F., Evans, M., Shen, Y., and Zeitler, J.A. (2013). Hardness and density distributions of pharmaceutical tablets measured by terahertz pulsed imaging. *Journal of Pharmaceutical Sciences* 102: 2179–2186.

McDowell, G.R. and Humphreys, A. (2002). Yielding of granular materials. *Granular Matter* 4: 1–8.

McKenna, A. and McCafferty, D.F. (1982). Effect of particle size on the compaction mechanism and tensile strength of tablets. *Journal of Pharmacy and Pharmacology* 34: 347–351.

Mohammed, H., Briscoe, B.J., and Pitt, K.G. (2005). The interrelationship between the compaction behaviour and the mechanical strength of pure pharmaceutical tablets. *Chemical Engineering Science* 60: 3941–3947.

Moreno-Maroto, J.M., González-Corrochano, B., Alonso-Azcárate, J., Rodríguez, L. and Acosta, A. (2017). Development of lightweight aggregates from stone cutting sludge, plastic wastes and sepiolite rejections for agricultural and environmental purposes. *Journal of Environmental Management* 200: 229–242.

Mueller, S.B., Kueppers, U., Ametsbichler, J., Cimarelli, C., Merrison, J. P., Poret, M., Wadsworth, F.B., and Dingwell, D. B. (2017). Stability of volcanic ash aggregates and break-up processes. *Scientific Reports* 7: 7440.

Muzzio, F.J., Shinbrot, T., and Glasser, B.J. (2002). Powder technology in the pharmaceutical industry: The need to catch up fast. *Powder Technology* 124: 1–7.

Newton, J.M., Rowley, G., Fell, J.T., Peacock, D.G., and Ridgway, K. (1971). Computer analysis of the relation between tablet strength and compaction pressure. *Journal of Pharmacy and Pharmacology* 23(Supplement 1): 195S–201S.

Newton, J.M., Haririan, I., and Podczeck, F. (2000a). The influence of punch curvature on the mechanical properties of compacted powders. *Powder Technology* 107: 79–83.

Newton, J.M., Haririan, I., and Podczeck, F. (2000b). The determination of the mechanical properties of elongated tablets of varying cross section. *European Journal of Pharmaceutics and Biopharmaceutics* 49: 59–64.

Nichols, G., Byard, S., Bloxham, M.J., Botterill, J., Dawson, N.J., Dennis, A., Diart, V., North, N.C., and Sherwood, J.D. (2002). A review of the terms agglomerate and aggregate with a recommendation for nomenclature used in powder and particle characterization. *Journal of Pharmaceutical Sciences* 91: 2103–2109.

Niesz, D.E. (1965). *Effect of Porosity on the Strength of a Sintered Glass Powder*. Ph.D. thesis. The Ohio State University.

Obae, K., Iijima, H., and Imada, K. (1999). Morphological effect of microcrystalline cellulose particles on tablet tensile strength. *International Journal of Pharmaceutics* 182: 155–164.

Pazesh, S., Persson, A.S., and Alderborn, G. (2019). Atypical compaction behaviour of disordered lactose explained by a shift in type of compact fracture pattern. *International Journal of Pharmaceutics: X* 1: 100037.

Pietsch, W.B. (1968). Tensile strength of granular materials. *Nature* 217: 736–737.

Pitchumani, R., Zhupanska, O., Meesters, G.M.H., and Scarlett, B. (2004). Measurement and characterization of particle strength using a new robotic compression tester. *Powder Technology* 143–144: 56–64.

Pitt, K.J. and Heasley, M.G. (2013). Determination of the tensile strength of elongated tablets. *Powder Technology* 238: 169–175.

Pitt, K.G., Newton, J.M., and Stanley, P. (1988). Tensile fracture of doubly convex cylindrical discs under diametral loading. *Journal of Materials Science* 23: 2723–2728.

Pitt, K.G., Newton, J.M., and Stanley, P. (1989a). Stress distributions in doubly convex cylindrical discs under diametral loading. *Journal of Physics D: Applied Physics* 22: 1114–1127.

Pitt, K.G., Newton, J.M., and Stanley, P. (1990). Effects of Compaction Variables on Porosity and Material Tensile Strength of Convex-faced Aspirin Tablets. *Journal of Pharmacy and Pharmacology* 43: 219–225.

Pitt, K.G., Newton, J.M., Richardson, R., and Stanley, P. (1989b). The material tensile strength of convex-faced aspirin tablets. *Journal of Pharmacy and Pharmacology* 41: 289–292.

Podczeck, F. (2007). Investigations into the mechanical strength anisotropy of Sorbitol Instant compacts made by uniaxial compression. *Advanced Powder Technology* 18: 361–379.

Podczeck, F., Drake, K.R., Newton, J.M., and Haririan, I. (2006). The strength of bilayered tablets. *European Journal of Pharmaceutical Sciences* 29: 361–366.

Podczeck, F., Newton, J.M., and Fromme, P. (2014). Theoretical investigations into the influence of the position of a breaking line on the tensile failure of flat, round, bevel-edged tablets using finite element methodology (FEM) and its practical relevance for industrial tablet strength testing. *International Journal of Pharmaceutics* 477: 306–316.

Podczeck, F., Newton, J.M., and Fromme, P. (2015). The bending strength of tablets with a breaking line–Comparison of the results of an elastic and a "brittle cracking"İ finite element model with experimental findings. *International Journal of Pharmaceutics* 495: 485–499.

Powers, T.C. (1958). Structure and physical properties of hardened Portland cement paste. *Journal of the American Ceramic Society* 41: 1–6.

Roberts, R.J. and Rowe, R.C. (1986). The effect of the relationship between punch velocity and particle size on the compaction behaviour of materials with varying deformation mechanisms. *Journal of Pharmacy and Pharmacology* 38: 567–571.

Roberts, R.J. and Rowe, R.C. (1987a). Brittle/ductile behaviour in pharmaceutical materials used in tabletting. *International Journal of Pharmaceutics* 36: 205–209.

Roberts, R.J. and Rowe, R.C. (1987b). The compaction of pharmaceutical and other model materials-a pragmatic approach. *Chemical Engineering Science* 42: 903–911.

Roberts, R.J., Rowe, R.C. and Kendall, K. (1997). The influence of lateral stresses on brittle–ductile transitions in the die-compaction of sodium chloride. *Journal of Materials Science* 32: 4183–4187.

Roohani-Esfahani, S.I., Newman, P., and Zreiqat, H. (2016). Design and fabrication of 3D printed scaffolds with a mechanical strength comparable to cortical bone to repair large bone defects. *Scientific Reports* 6: 19468.

Rozenblat, Y., Portnikov, D., Levy, A., Kalman, H., Aman, S., and Tomas, J. (2011). Strength distribution of particles under compression. *Powder Technology* 208: 215–224.

Ryshkewitch, E. (1953). Compression strength of porous sintered alumina and zirconia: 9th communication to ceramography. *Journal of the American Ceramic Society* 36: 65–68.

Samimi, A., Zakeri, M., Maleki, B., and Mohebbi-Kalhori, D. (2015). Experimental and statistical assessments of the mechanical strength reliability of gamma alumina catalyst supports. *Particuology* 21: 74–81.

Satone, M., Iimura, K., Teraoka, T., Hanafusa, T., Hisatani, S., Nishiwaki, M., and Suzuki, M. (2017). Analysis of granule fracture under biaxial compression. *Ceramics International* 43: 16835–16842.

Schmier, S., Jentzsch, M., Speck, T., and Thielen, M. (2020). Fracture mechanics of the endocarp of Cocos nucifera. *Materials & Design* 195: 108944.

Shang, C., Sinka, I. C., Jayaraman, B., and Pan, J. (2013a). Break force and tensile strength relationships for curved faced tablets subject to diametrical compression. *International Journal of Pharmaceutics* 442: 57–64.

Shang, C., Sinka, I C., and Pan, J. (2013b). Modeling of the break force of tablets under diametrical compression. *International Journal of Pharmaceutics* 445: 99–107.

Siiriä, S.M., Antikainen, O., Heinämäki, J., and Yliruusi, J. (2011). 3D simulation of internal tablet strength during tableting. *AAPS PharmSciTech* 12: 593–603.

Singh, D., McGlinchey, D., and Crapper, M. (2016). Breakage functions of particles of four different materials subjected to uniaxial compression. *Particulate Science and Technology* 34: 494–501.

Snajdr, E.A. (1964). Fracture stress of UO_2-tungsten cermets evaluated by the diametral tensile test. Interim Report No. 1. General Electric Co. Hanford Atomic Products Operation.

Sonnergaard, J.M. (1999). A critical evaluation of the Heckel equation. *International Journal of Pharmaceutics* 193: 63–71.

Sonnergaard, J.M.(2002). Distribution of crushing strength of tablets. *European Journal of Pharmaceutics and Biopharmaceutics* 53: 353–359.

Sonnergaard, J.M. (2013). A new brittleness index for compacted tablets. *Journal of Pharmaceutical Sciences* 102: 4347–4352.

Soroka, I. and Sereda, P.J. (1968). Interrelation of hardness, modulus of elasticity, and porosity in various gypsum systems. *Journal of the American Ceramic Society* 51: 337–340.

Stanley, P. (2001). Mechanical strength testing of compacted powders. *International Journal of Pharmaceutics* 227: 27–38.

Stefanou, I. and Sulem, J. (2016). Existence of a threshold for brittle grains crushing strength: two-versus three-parameter Weibull distribution fitting. *Granular Matter* 18: 14.

Subero-Couroyer, C., Ghadir, M., Brunard, N., and Kolenda, F. (2003). Weibull analysis of quasi-static crushing strength of catalyst particles. *Chemical Engineering Research and Design* 81: 953–962.

Van Veen, B., Bolhuis, G.K., Wu, Y.S., Zuurman, K., and Frijlink, H.W. (2005). Compaction mechanism and tablet strength of unlubricated and lubricated (silicified) microcrystalline cellulose. *European Journal of Pharmaceutics and Biopharmaceutics* 59: 133–138.

Vromans, H., De Boer, A.H., Bolhuis, G.K., Lerk, C.F., Kussendrager, K.D., and Bosch, H. (1985). Studies on tableting properties of lactose. Part 2. Consolidation and compaction of different types of crystalline lactose. *Pharmaceutisch Weekblad Scientific Edition* 7: 186–193.

Williamson, L. and Lucas, P. (1995). The effect of moisture content on the mechanical properties of a seed shell. *Journal of Materials Science* 30: 162–166.

Wollborn, T., Schwed, M.F., and Fritsching, U. (2017). Direct tensile tests on particulate agglomerates for the determination of tensile strength and interparticle bond forces. *Advanced Powder Technology* 28: 2177–2185.

Wu, C.Y., Best, S.M., Bentham, A.C., Hancock, B.C., and Bonfield, W. (2005). A simple predictive model for the tensile strength of binary tablets. *European Journal of Pharmaceutical Sciences* 25: 331–336.

Wu, D.F., Zhou, J.C., and Li, Y.D. (2006). Distribution of the mechanical strength of solid catalysts. *Chemical Engineering Research and Design* 84: 1152–1157.

Wu, D., Zhou, J. and Li, Y. (2007). Mechanical strength of solid catalysts: Recent developments and future prospects. *AIChE Journal* 53: 2618–2629.

Yohannes, B., Gonzalez, M., Abebe, A., Sprockel, O., Nikfar, F., Kang, S. and Cuitino, A.M. (2015). The role of fine particles on compaction and tensile strength of pharmaceutical powders. *Powder Technology* 274: 372–378.

11

Compliant Particles

This chapter considers the deformation and strength of spherical particles of hydrogel materials—materials composed of sparse polymer networks containing large volume fractions of water. The mechanical behavior of particles of such materials is marked by significant compliance and large, non-linear deformations. The axial and transverse deformation characteristics of spherical hydrogel particles in diametral compression are detailed and compared to those of structurally similar elastomeric rubber spheres. Experimental measurements show that both hydrogel and rubber spheres are characterized by constant volume deformation in diametral compression. The strengths of hydrogel particles are shown to be deformation rate dependent and are analyzed by deterministic extreme value effects, in this case controlled by testing rate not particle size.

11.1 Introduction–Hydrogel Particles

Compliant components exhibit large, reversible, deformations under the actions of small loads. For compliant particles loaded in diametral compression large axial and transverse displacements are generated by small axial forces. The prototypical compliant particle is the elastomeric "rubber ball." It is nearly impossible to fracture a rubber ball by generation of transverse strain and stress at the center of the ball through diametral compression. The ball simply flattens to a "pancake" geometry. However, as a consequence of the generated central tension, particles of other compliant materials do fracture at large deformations on loading in diametral compression. One such group of materials is the *hydrogels*. In this chapter the fracture strengths of compliant particles of hydrogel materials are considered, placing the pre-failure deformations of hydrogel particles in context with those of elastomer and rubber particles. (Hydrogels are often referred to as "soft" materials, although compliance is meant—hydrogels, like rubber, very rarely exhibit plastic deformation, and, in fact, are very hard.)

Hydrogel materials exhibit great potential for application in many technologies requiring localized and controlled liquid storage and transfer. In operation, devices containing hydrogels can absorb or release liquids, often selectively, many times the volume of the hydrogel component. Potential is particularly clear in bioengineering (Drury and Mooney 2003; Okay 2009; Oyen 2014; Gibbs et al. 2016), but also in technologies as diverse as agriculture (Abd El-Rehim et al. 2006) and soft robotics (Banerjee et al. 2018). Hydrogels consist of sparse, but three-dimensionally connected, networks of cross-linked polymer chains encapsulating a large number of water molecules in the liquid state. The volume fraction of water (the "hydro" component) in a hydrogel is typically greater than 99 % such that the volume fraction of the polymer network (the "gel" component) is very small, less than 1 %, and volume expansion ratios from a dry polymer state to a completely saturated hydrogel state are usually much greater than 100 (Bertrand et al. 2016; Aangenendt et al. 2020). An example of hydrogel swelling is shown in Figure 11.1, in which the effect of hydrolization on commercial poly(acrylamide) hydrogel particles (often domestically encountered as "water beads") is shown. The small particles on the left are in the dry state and have diameters D of approximately 2 mm, 3 mm, and 5 mm (upper to lower). The large particles on the right are in the swollen saturated state, after 3 days immersion in distilled water. These particles have diameters D of approximately 10 mm, 15 mm, and nearly 40 mm (upper to lower). The sparse nature of the polymer network generated by the absorption of water is clear in comparison of the left and right images.

The large water fractions in hydrogels lead to great biocompatibility and hence potential as tissue engineering scaffolds, cell encapsulates, or drug delivery vehicles in bioengineering. However, the sparse polymer fractions lead to often limiting mechanical properties as hydrogels are typically extremely compliant, exhibiting elastic moduli $\approx$ 100 kPa, and extremely

Particle Strengths: Extreme Value Distributions in Fracture, First Edition. Robert F. Cook.

Figure 11.1 Images of arrays of dry (left) and water saturated (right) hydrogel particles. Initial particle sizes are approximately 2 mm, 3 mm, and 5 mm (upper to lower) and saturated particle sizes are 10 mm, 15 mm, and 40 mm (upper to lower). The saturated particles consist of a sparse polymer network encapsulating water. *Source:* Robert F. Cook.

weak, exhibiting failure stresses ≈ 0.3 MPa. Deformation in most hydrogels is dominated by poroelastic (Wang 2000) and viscoelastic (Findley et al. 1976) effects and some hydrogels may exhibit (engineering) failure strains extending to ≈ 10. In order to take advantage of the liquid capture abilities, many microstructural modification schemes have been implemented to improve mechanical properties of base or "conventional" hydrogel materials composed of single polymer networks. Examples include conventional hydrogels formed by poly(acrylamide), poly(ethylene glycol) (PEG), or biological macromolecules such as agar or gelatin. A range of compositions and structures is shown in the compilation of Hua et al. (2018). The microstructural modifications generate "advanced" hydrogels, including double network gels, with sacrificial networks (Gong 2010) and composite gels, with particle or fiber reinforcements (Galli et al. 2011; Tonsomboon and Oyen 2013; Ji and Kim 2021). The mechanical behavior of advanced gels is commonly distinguished from conventional gels by increased failure strains and consequent increased works of failure (Cook and Oyen 2021), although it is unclear that the mechanical failure properties of the advanced gels are of direct physiological applicability. In addition to a diversity of hydrogel modification schemes, a range of test methods has been implemented to assess hydrogel mechanical properties. These methods include: (i) those focused on small-scale deformation, such as rheometry (Kong et al. 2002), constrained swelling (Bhattacharyya et al. 2020), and indentation (Galli et al. 2009); (ii) those focused on extended component deformation and failure, such as conventional tension and compression geometries (Normand et al. 2000; Gu et al. 2003), and compression and adhesion of spheres and shells (e.g. Liu 2006; Shi et al. 2012 James et al. 2020); and (iii) those focused on fracture, such as the trouser tear (Tonsomboon et al. 2017) and planar tension geometries (Long and Hui 2016) (the latter is commonly referred to as "pure shear" when applied to elastomers, Rivlin and Thomas 1953). Overviews of the macroscopic mechanics of swelling, deformation, and fracture are given in Zhao (2014), Liu et al. (2015), Long and Hui (2016), and Fennell and Huyghe (2019). The range of test methods and analyses has enabled a broad picture of hydrogel mechanical properties to emerge, including assessments of microstructural effects, such as those giving rise to the large failure strains noted above.

However, in order to guide development of hydrogel microstructures for specific applications, more precise assessments of hydrogel mechanical failure and strength are required. As highlighted in a recent Perspective considering hydrogels for cartilage replacement (Cook and Oyen 2021), the impressive works of failure attained at large strains for some advanced hydrogels are likely not of physiological relevance and different tests are required. This chapter extends the range of quantitative test methods applicable to hydrogels by interpreting strength in terms of underlying flaw population characteristics, specifically flaw size distribution. The work here is based on the analysis of Chapter 4 and a recent study considering hydrogel particle failure by James et al. (2020): James et al. described spherical compression testing of conventional poly(acrylamide) hydrogel particles in the $D = 17$ mm saturated state and demonstrated that the strength edf at a given applied deformation rate was sigmoidal and contracted as the rate decreased, although, as will be shown, not to a common threshold. The similarity of the results to the deterministic extreme value strength behavior considered in Chapter 9 is clear and interpreting strengths in this manner is a goal of this Chapter. As in Chapter 8, consideration of a single work is used here in the manner of a case study to illustrate application of the strength and flaw size analyses developed. As strength-controlling flaw populations are expected to be influenced directly by microstructure, the methods described here will provide guidance in advanced hydrogel development and in determining hydrogel performance measures, including reliability (similar to that provided in Chapter 8 for ceramic development).

A feature of the observations of James et al. was that the force-displacement behavior during loading in diametral compression of the hydrogel particles was non-linear and extended. An example is shown in Figure 11.2. Distinct from the particle force-displacement traces shown in Chapter 2, the convex behavior for the hydrogel sphere (similar to those shown in Figure 11.1) extends throughout the loading cycle, with no tendency to linear behavior after an initial convex region, as in Figures 2.5 and 2.6. Linear force-displacement behavior was shown in Chapter 2 to result from localized plastic deformation or compaction at the particle poles during loading, leading to force-displacement hysteresis on unloading. No such hysteresis was observed by James et al. and the force-displacement behavior of Figure 11.2 was completely reversible, although non-linearly elastic, for load-unload cycles up to the peak force. Another distinction between the particle force-displacement traces shown in Figure 11.2 and Chapter 2 is the extent of deformation. The relative deformation or characteristic strain, w/D, for the particle represented in Figure 11.2 is approximately 0.5, a factor of 10–100 greater than those shown in Chapter 2. However, a similarity between the particle force-displacement traces of Figure 11.2 and the majority of those in Chapter 2 is the behavior at peak force. In both cases, the force subsequently decreased precipitously to zero, indicative of complete failure of the particle. The force decrease and subsequent fragmentation are faster than a single video frame—the hydrogel particles are classically brittle. James et al. showed that the brittle mode of failure was independent of applied deformation rate. In addition, James et al. showed that although the peak forces, and thus particle strengths, exhibited dispersion, the median strength increased with deformation rate. The implication is that mechanisms in these hydrogel particles underlying the large deformations preceding peak force influence particle fracture criteria and this behavior is addressed in this chapter.

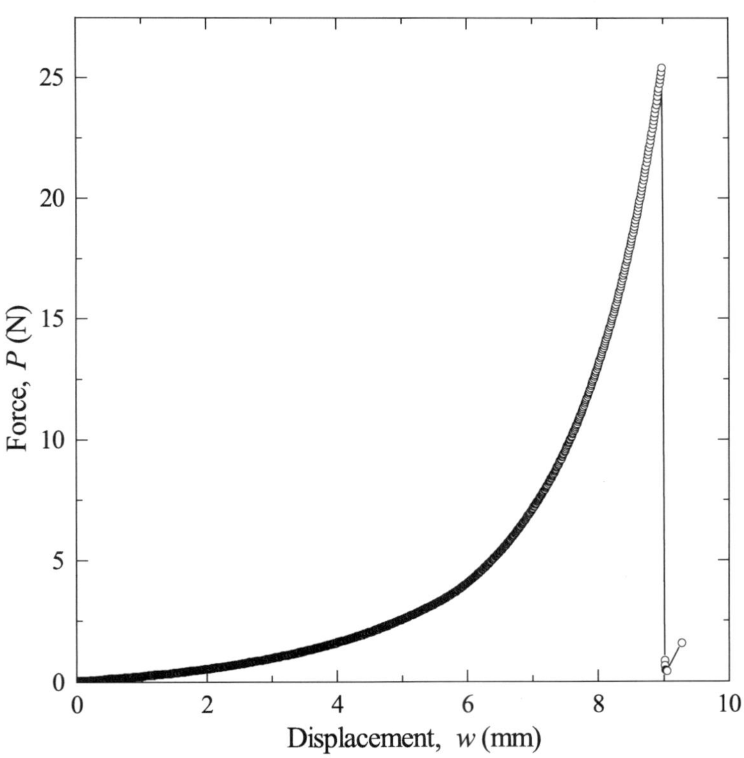

Figure 11.2 Plot of force-displacement behavior for a $D = 17$ mm diameter hydrogel particle (adapted from James, J.D et al. 2020). The behavior is significantly non-linear to peak load, at which point the particle fails abruptly in a brittle manner.

Large deformations generated by loading spherical particles in diametral compression have been studied previously, most notably by Durelli and Chen (1973) and Tatara and colleagues (Tatara 1991, 1993; Tatara et al. 1991; Shima et al. 1993). Durelli and Chen used a photoelastic technique, also used by Kilcast et al. (1984) and subsequently Rosenthal (2016) to measure stress distributions in indented gelatin (see also Henderson et al. 1985 for gelatin indentation) and by Pitt et al. (1989) in studying small deformations in diametrally compressed agglomerate tablets (Chapter 10). Tatara and colleagues generated both experimental data and descriptive analyses. A common example of large deformation in an elastic sphere, similar to that observed in the above studies by Durelli and Chen and Tatara and colleagues, is shown in Figure 11.3. In Figure 11.3a a polyurethane foam sphere $D = 70$ mm is shown in the unloaded state between two platens. In Figure 11.3b, the platens are displaced, loading and deforming the sphere in compression in the axial dimension such that $w/D \approx 0.5$. As a consequence, both the contact radius between the deformed sphere and the platens and the transverse radius at the sphere equator have increased significantly. Release of the platens leads to full recovery of the sphere. Schematic diagrams of the deformation in Figure 11.3 are shown in Figure 11.4, similar to Figure 2.10. In Figure 11.4a, the undeformed sphere, diameter $D = 2R$, is shown between two platens. In Figure 11.4b, the sphere is deformed by axial translation of the platens, leading to compressive force P and displacement $w = 2\delta$. The contact radius is dimension a and the equatorial radius is dimension b. In the small deformation limit of Herztian contact considered in Chapter 2, the axial and transverse deformations were infinitesimal, $\delta/R \ll 1$, $a/R \ll 1$, and $b/R \approx 1$. For the large deformations shown in Figures 11.3 and 11.4, the axial and

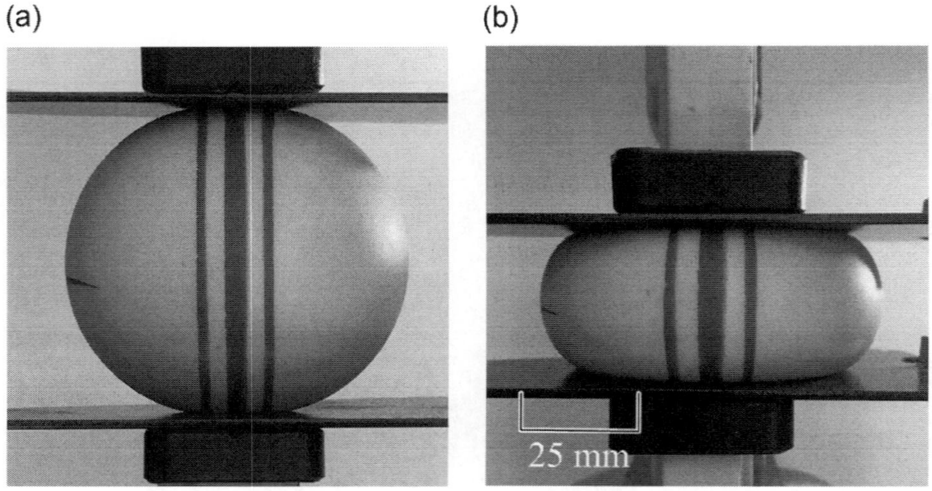

Figure 11.3 Images of an elastomer ball between two platens. (a) Free. (b) Compressed to approximately 0.5 of the original height by displaced and force-supporting platens. Note the increased contact and equatorial radii. *Source:* Robert F. Cook.

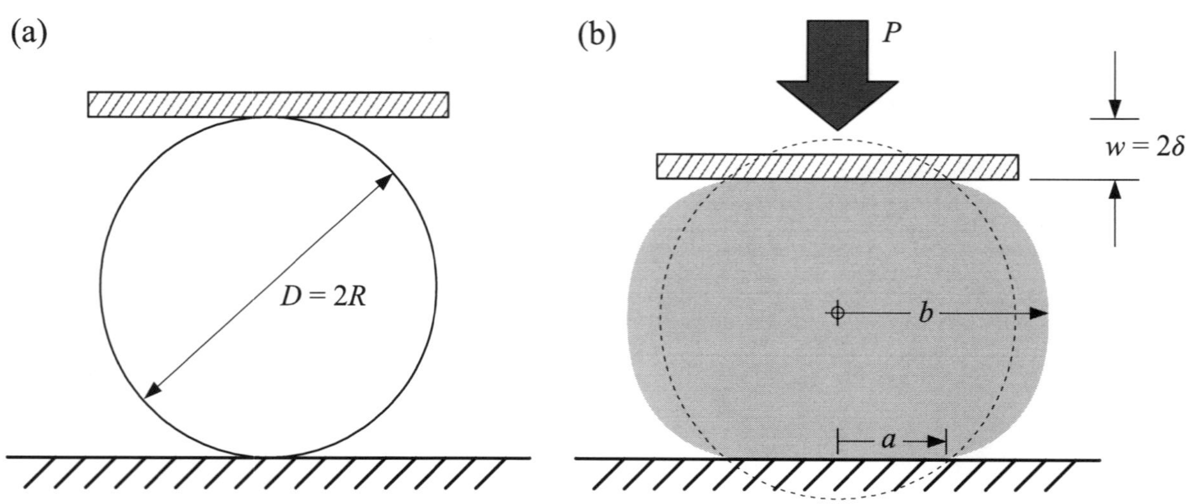

Figure 11.4 Schematic cross-section diagrams illustrating large deformation of a sphere in diametral compression. Loading platens shown hatched. (a) Unloaded, initial sphere diameter D. (b) Loaded by force P so as to generate axial displacement w, contact radius a, and equatorial radius b. Deformed sphere shown shaded. Original sphere shown as dashed outline. Compare with Figure 11.3.

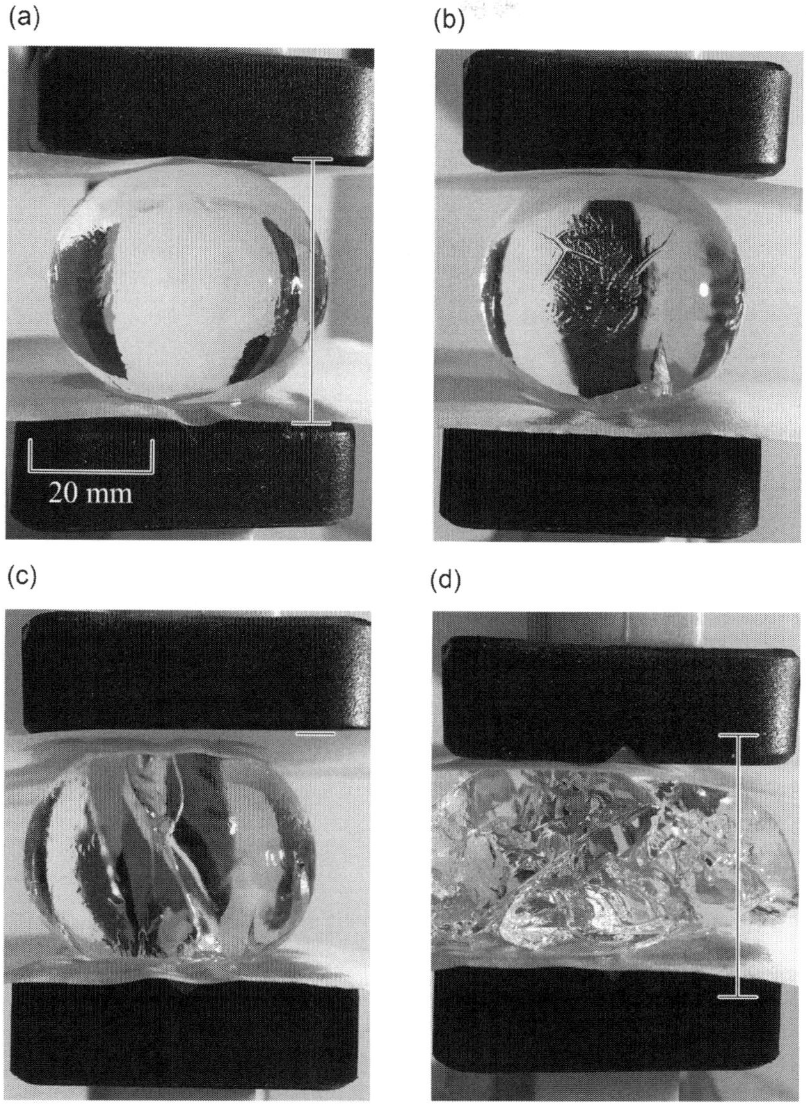

Figure 11.5 Sequence of images illustrating hydrogel sphere failure. (a) Nearly free; note slightly oblate shape induced by gravity. Bar at right indicates initial height. (b) Incipient failure. Note crack at lower right. (c) Mid-failure. Note meridional crack through axis. (d) Final failure. Particle has fragmented. Bar at right indicates initial height. Note platen displacement. *Source:* Robert F. Cook.

transverse deformations are finite, $\delta/R \approx 0.5$, $a/R \approx 0.5$, and $b/R \approx 1.5$. In order to provide background and context for the similarly finite deformation of hydrogels prior to failure in diametral compression, the deformation characteristics of similarly loaded compliant particles are considered in the section 11.2.

Figure 11.5 shows a sequence of images illustrating hydrogel particle failure. Figure 11.5a shows an unloaded $D = 40$ mm particle, the vertical bar at right indicates the approximate platen spacing. Inspection of Figure 11.5a indicates that the particle is oblate, contracted at the poles and bulging at the equator, $b/R \approx 1.1$, a consequence of the gravitational body force deriving from the mass of contained water acting on the compliant polymer network. A further consequence is that it is very difficult to set an initial position of the particle so that the contact radius is near zero, usually the initial contact conditions are $\delta/R \approx 0$, $a/R \approx 0.1$. (The particles recover their near-spherical shape on return to water.) Figure 11.5b shows the particle at incipient failure, a small crack is visible near the base of the particle. Figure 11.5c shows the particle at a later stage in the failure process, the crack appears to extend throughout the particle. Figure 11.5d shows the particle at complete failure, the material has fragmented and no significant force is supported, in distinction to the extended post-failure supported force behavior exhibited by the particles of Figure 2.7. The vertical bar at right indicates the original platen spacing and provides a measure of the finite deformation required to break the particle. As might be anticipated, the slightly oblate initial conditions are not observed to affect the failure conditions significantly. The platen displacement rate imposed in Figure 11.5 was quasi-static, the entire loading process extended over 20 min, such that in this case $w/D \approx 0.3$ at failure. The transition from Figures 11.5b to 11.5c took about 10 s; the transition from Figure 11.5c to Figure 11.5c took less than 0.1 s. The effects of displacement rate on failure strain and strength are considered in section 11.3.

11.2 Deformation

11.2.1 Axial

Figure 11.6 shows a logarithmic plot of experimental axial force-displacement behavior for several finite deformation sphere systems. To facilitate comparison between different sphere diameters, the relative displacement w/D is used. The data included are (in order of decreasing supported force): polyurethane rubber, diameter $D = 174$ mm (Durelli and Chen 1973), natural rubber, $D = 10$ mm (Shima et al. 1993), polyurethane rubber, $D = 48$ mm (Lin et al. 2008), water saturated polyacrylamide hydrogel, $D = 17$ mm (James et al. 2020), polyurethane rubber, $D = 0.27$ mm (Liu et al. 1998), and water saturated polysaccharide hydrogel, $D = 0.46$ mm (Andrei et al. 1996), using data derived from the published works cited. Symbols

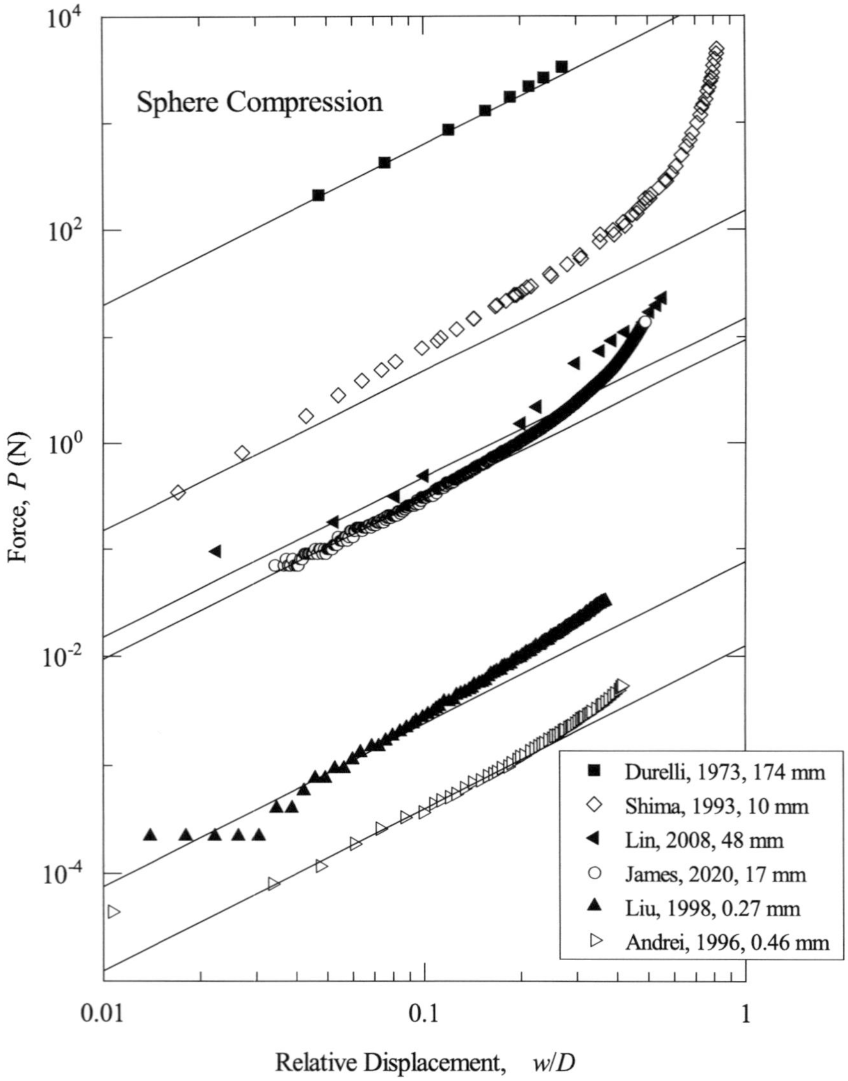

Figure 11.6 Logarithmic plot of force vs relative displacement for a number of spherical particle systems. The three upper sets of symbols and the lower filled set represent elastomers, the two lower sets of open symbols represent hydrogels. Note the similarity in behavior across a wide domain of particle sizes. Lines are guides to the eye representing small scale elastic behavior. First authors of works noted.

represent individual observations. As discussed in Chapter 2, the axial force-displacement behavior for small displacements in the Hertzian limit is $P \sim MD^2(w/D)^{3/2}$. The lines are guides to the eye consistent with Hertzian behavior of slope 3/2.

For subsequent, larger, displacements, two modifications are required to the scaling analysis of Chapter 2. First, the characteristic strain, for the entire deformed sphere, is now $\varepsilon \sim w/D$, reflecting the finite dimensions. Second, the contact radius and displacement are now related by a circular relation, $a^2 \sim \delta R(1 - \delta/2R)$ and thus $a^2 \sim wD(1 - w/2D)$, also reflecting the finite sphere and finite contact radius. Combining these modified relations into the unaltered expressions in Chapter 2 for stress and elastic behavior gives the force-displacement response for sphere compression as $P \sim Mw^2(1 - w/2D)$. Relative to infinitesimal Hertzian contact, this response has a slightly greater dependence of force with displacement for compression of a finite sphere, but somewhat less than fully quadratic; a result and expression consistent with the finite element analysis of Zhang et al. (2007). Note that the modified relations and the resulting force-displacement expression reflect refinements to the geometrical non-linearity of the system. At very large displacements, both geometric and material non-linearity need to be taken into account. The material of interest here is that of an elastically incompressible solid. Such materials conserve volume under non-hydrostatic loading and exhibit significant stiffening on deformation—rubber is the canonical example (Treloar 1975). The hyperelastic stiffening can be characterized by increases in modulus. Although details of the increases depend on the specific deformation geometry, the overall increasing trend can be represented by a power-law series in strain, $M \sim M_0 + k_1\varepsilon + k_2\varepsilon^2 + k_3\varepsilon^3 + \ldots$, where k_i are geometry-dependent series expansion coefficients. For the case here, combining the series description of material non-linearity and the above characteristic strain into the geometrical non-linear force-displacement response gives the force-displacement expression $P \sim M_0w^2[1 + M_1(w/D) + M_2(w/D)^2 + M_3(w/D)^3 + \ldots]$, where M_i are empirical factors. Hence, for large compressive deformations of a volume-conserving sphere, a rapidly increasing force-displacement response is predicted that also displays increasing stiffness with displacement. The greater powers of w/D dominate at large displacements such that for the expansion written here the response could be expected to approach $P \sim (w/D)^5$ for a highly deformed sphere. The overall force-displacement response during sphere compression is thus predicted to consist of three regions:

(i) At very small displacements, $w/D \ll 0.1$, the force variation is described by the Hertzian response, $P \sim (w/D)^{3/2}$. The material behaves linearly elastically and the deformations are infinitesimal relative to the sphere radius.
(ii) At moderate displacements, $0.1 < (w/D) < 0.3$, the force variation is approximately quadratic, $P \sim (w/D)^2$. The material continues to behave linearly elastically but the deformations are now finite relative to the sphere radius.
(iii) At large displacements, $(w/D) > 0.3$, the force variation is very strongly increasing and not well characterized by a single power law, although the response is expected to asymptotically approach $P \sim (w/D)^5$ or a greater variation at large displacements. The material behaves hyperelastically and the deformations are finite relative to the sphere radius.

All systems in Figure 11.6 are well described by Hertzian behavior at small relative displacements. However, over most of the experimental relative displacement domain, all systems exhibit slightly greater power law behavior, indicative of the above moderate displacement analysis. The observations of Shima et al. (1993) and James et al. (2020) exhibit rapidly increasing forces at large relative displacements, consistent with the varying power-law response predicted from the above large displacement analysis. These latter two sets of observations clearly display the three regions of behavior identified earlier. The rapid increase in force at large displacements is also evident in data, not shown, reported for similar rubber spheres (Tatara et al. 1991; Shima et al. 1993) but mostly absent from data reported for a porous, compressible sponge (Lin et al. 2008). The behavior of this and other systems, not described by the above considerations, are discussed below. The data of Figure 11.6, describing spheres varying in size by approximately a factor of 10^3 and supporting forces varying in magnitude by approximately a factor of 10^7, are well described by the above scaling laws. A mechanics-based analysis (Tatara 1991, 1993), implementing physical principles similar to those used here, provides a full description of the system including continuous quantitative force behavior over the entire displacement range. The analysis is complete, but of some complexity and requires iterative numerical solution.

An assumption implicit in the above analyses is that force and displacement are in near equilibrium such that observations are not greatly influenced by time. This assumption requires some caution. Figure 11.7 shows force-displacement load-unload cycle behavior for two hydrogel systems, (a) 0.46 mm water saturated polysaccharide hydrogel particles (Andrei et al. 1996) as above, and (b) 1 mm water saturated alginate hydrogel particles (David et al. 2006), using data derived from the published works cited. In both cases, there is clear hysteresis, although, as with small-scale indentation of rubber (Cook and Oyen 2007), the hysteresis is small, about 0.1 of the peak force. Similarly, Figure 11.8 shows constant displacement force-time relaxation behavior for two hydrogel systems, (a) Andrei et al. (1996) and (b) James et al. (2020), using data derived from the published works cited. In both cases, there is clear relaxation toward equilibrium values of about 0.7 of the peak

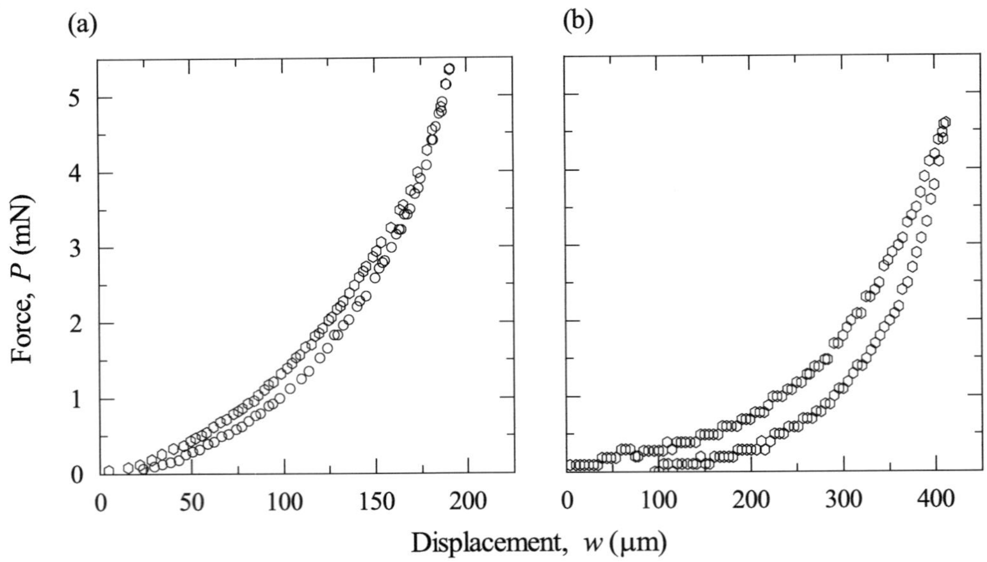

Figure 11.7 Plots of force-displacement behavior for two hydrogel particle systems illustrating hysteresis. (a) 0.46 mm particle (modified from Andrei et al. 1996). (b) 1 mm particle (adapted from David et al. 2006).

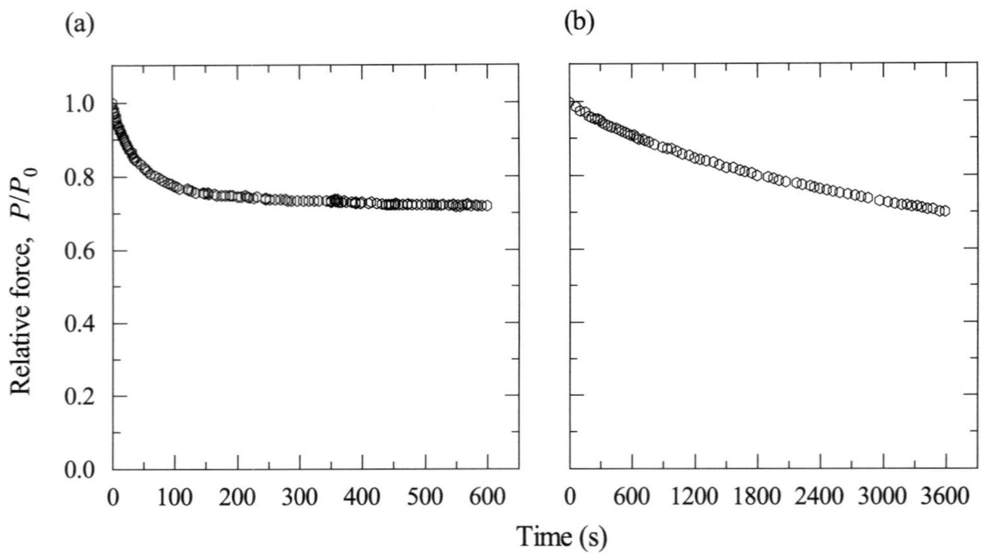

Figure 11.8 Plots of force-time behavior for two hydrogel particle systems illustrating stress relaxation. (a) 0.46 mm particle (modified from Andrei, D.C et al. 1996). (b) 17 mm particle (adapted from James, J.D et al. 2020).

force P_0 over time scales of tens of minutes. Similar behavior was reported for relaxation under uniform compression and cylindrical and conical indentation of flat surfaces (Lee et al. 2009; Cai et al. 2010) and spherical indentation of particles (Berry et al. 2020). The implications of Figure 11.7 and Figure 11.8 are that the measured forces in monotonic failure tests are likely to be slightly overestimated relative to the equilibrium values. The force and displacement variations with time are negligible relative to the overall variations and differences in behavior shown in Figure 11.6 but extremely detailed interpretation of such plots should take relative time scales of tests into account.

11.2.2 Transverse

As shown in Figure 11.4b, a diametrally deformed sphere can be characterized by two transverse radii: the contact radius, a and the equatorial radius, b. Figure 11.9 shows a logarithmic plot of these experimental radii for several of the sphere systems discussed, using data derived from the published works cited. For ease of comparison the data are plotted as relative contact

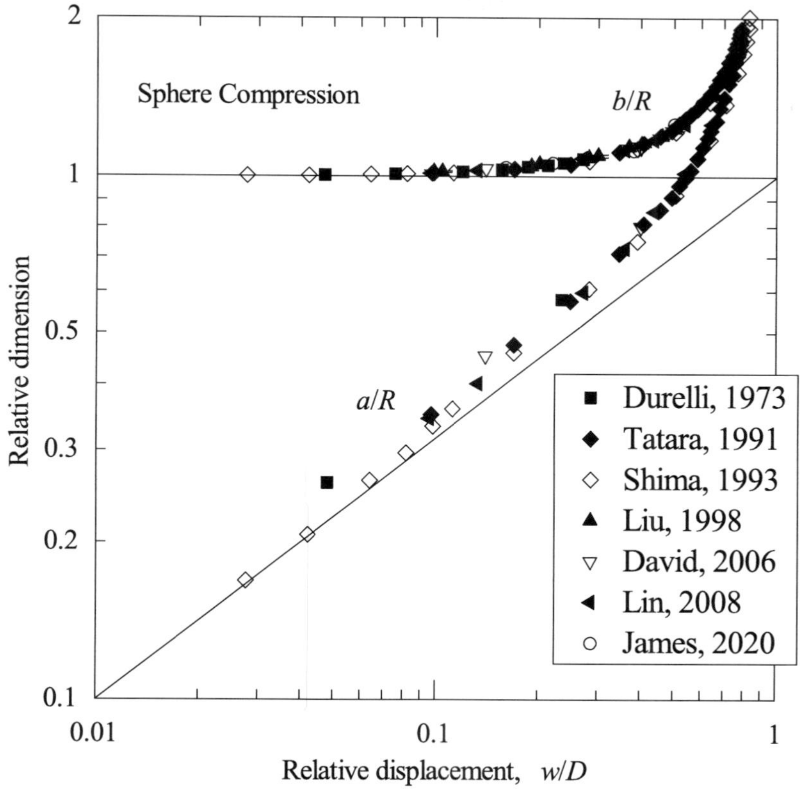

Figure 11.9 Logarithmic plot of relative contact radii (a/R) and relative equatorial radii (b/R) vs relative displacement (w/D) for a number of spherical elastomer and hydrogel particle systems. Note the universal behavior across a wide domain of particle types. Lines are guides to the eye representing small scale elastic behavior. First authors of works noted.

radius a/R or equatorial radius b/R vs relative displacement $w/D = \delta/R$. Symbols represent individual observations and lines are guides to the eye of value 1 or of slope 1/2, the Hertzian a/R variation. The data for all sphere systems in Figure 11.9, for both relative contact radius and relative equatorial radius, lie on universal responses. As in Figure 11.6, the contact radii a are described by Hertzian behavior, $(a/R) \sim (\delta/R)^{1/2} = (w/D)^{1/2}$, at small relative displacements, described by a predominant, slightly greater power-law response at moderate displacements, and exhibited a rapidly increasing, non-power-law dependence at large displacements. The equatorial radii b asymptotically approached the value 1 at small displacements, weakly increased from this value at moderate displacements, and exhibited a rapidly increasing, non-power-law dependence at large displacements. At very large axial displacements, $w/D \to 1$, the transverse equatorial and contact radii converge, $b/a \to 1$, characterizing the deformation of the sphere into a pancake (see Huang et al. 2007 for an example of reversible cylinder to pancake deformation).

The equatorial and contact radii mark key points on the full cross-sectional profiles of deformed spheres. In cylindrical polar coordinates (r, z) centered on the sphere with z aligned along the deformation axis, Figure 2.21, the profile of the sphere is given by a relation between z and r. Initially $z^2 + r^2 = R^2$. On deformation, in the same coordinate system, Figure 11.10, the profile within the contact radius is given by the position of the platen, and thus $z = R - \delta = h$ for $r \leq a$. Exterior to the contact radius, the sphere extends transversely in a curved profile. Observations suggest that the profile has a characteristic radius less than than that of the initial sphere and that the profile meets the platen at an angle greater than zero. A circular profile that meets these constraints is given by $(r - c)^2 + z^2 = e^2$ for $a \leq r \leq b$, where $(c, 0)$ are the center coordinates of an exterior circular arc of radius e, with $0 \leq c \leq R$ and $h \leq e \leq R$. The arc subtends an angle of 2θ, where $\tan\theta = h/(a - c)$, such that the profile meets the platen at an angle of $\theta - \pi/2$. The earlier mechanics-based approach to the sphere compression problem (Tatara et al. 1993) provides the exterior profile, but only after iterative numerical solution and then not in closed form. A previous empirical approach (Lin et al. 2008) simplified the above equation by setting $c = 0$ and $e > R$ but was not a good description of observations. A modification of this latter approach using an elliptical profile was more successful, for both compressible and incompressible materials, but was of greater complexity.

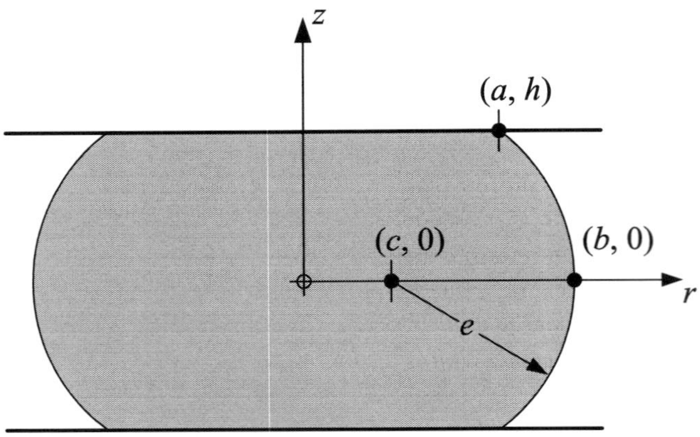

Figure 11.10 Schematic cross-section diagram illustrating the dimensions of large deformation of a sphere in diametral compression: a, contact radius; b, equatorial radius; e, radius of exterior circular profile; c, center of exterior circular profile. Deformed sphere shown shaded. Coordinate system (r, z) same as Figure 2.21.

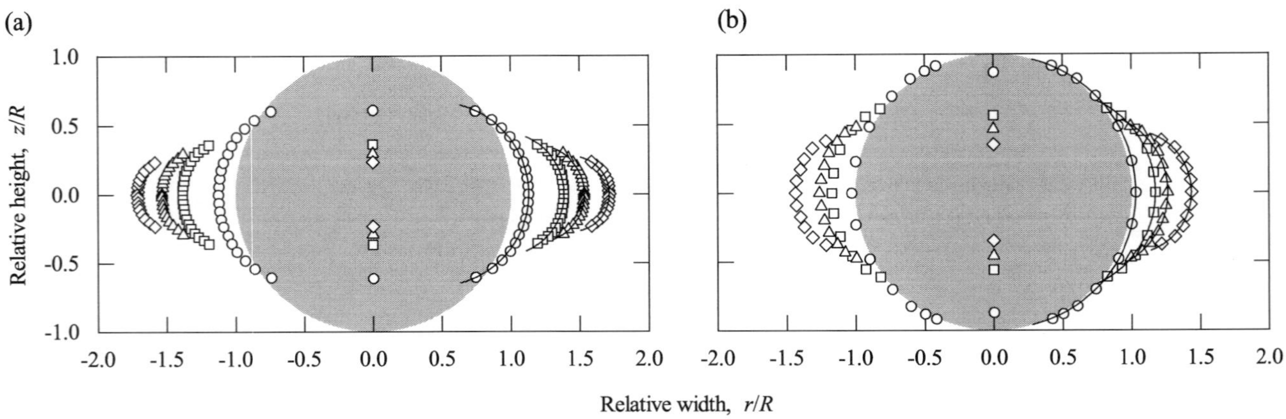

Figure 11.11 Deformation profiles for compressed elastomer spheres represented in relative radial (z/R) and axial (z/R) coordinates. Shaded gray discs represent profiles of undeformed spheres. Symbols represent individual measurements of deformed spheres; different symbols indicate different states of compression. Lines represent best fit circular arcs, see Figure 11.10. (a) Adapted from Shima, S et al. (1993). (b) Adapted from Lin, Y.-L et al. (2008).

Figure 11.11 shows experimental deformation profiles for two incompressible rubber sphere systems. The $r(z)$ results are from (a) Shima et al. (1993) and (b) Lin et al. (2008), using data derived from the published works cited. For ease of comparison, the data are shown as relative height and width values, z/R and r/R, respectively, and the profiles are centered for each state of deformation. Four profiles of four states of deformation from a single sphere experiment are shown in each plot and the symbols represent individual measurements of points on a profile. As guides to the eye, the initial sphere profiles are indicated as gray discs and the center symbols in each plot indicate the reported macroscopic positions of the platens for each deformation profile. It is clear that the equatorial profile radius e decreases significantly as deformation proceeds and that the measurements of system (a) represent greater states of deformation than those of system (b). It is also clear that measurements of the initial profile for system (b) are altered very little from those of the undeformed sphere, consistent with the initial Hertzian deformation. The lines shown on the right in each plot in Figure 11.11 represent best fits to each profile of the above shifted circles, treating c and e as unconstrained fitting parameters. The data in each case are clearly well described by such fits. For each system, c increased and e decreased as deformation $w = 2\delta$ proceeded. The angle subtended by the circular arc of each profile was approximately invariant at $2\theta = 120°$, implying invariance in the friction between the spheres and platens. Similar results were obtained for fits to the data for a sphere of a compressible foam material (Lin et al. 2008).

Figure 11.12 shows experimental deformation profiles at smaller levels of compression for an incompressible rubber sphere system and a water saturated hydrogel system. The $r(z)$ results are from (a) Durelli and Chen (1973) and (b) David et al. (2006), using data derived from the published works cited, and are plotted in the same manner as those in Figure 11.11. Profiles of two states of deformation from a single sphere experiment are shown in (a) and profiles of an unloaded initial state and two deformed states are shown in (b); the symbols represent individual measurements of points on the profiles.

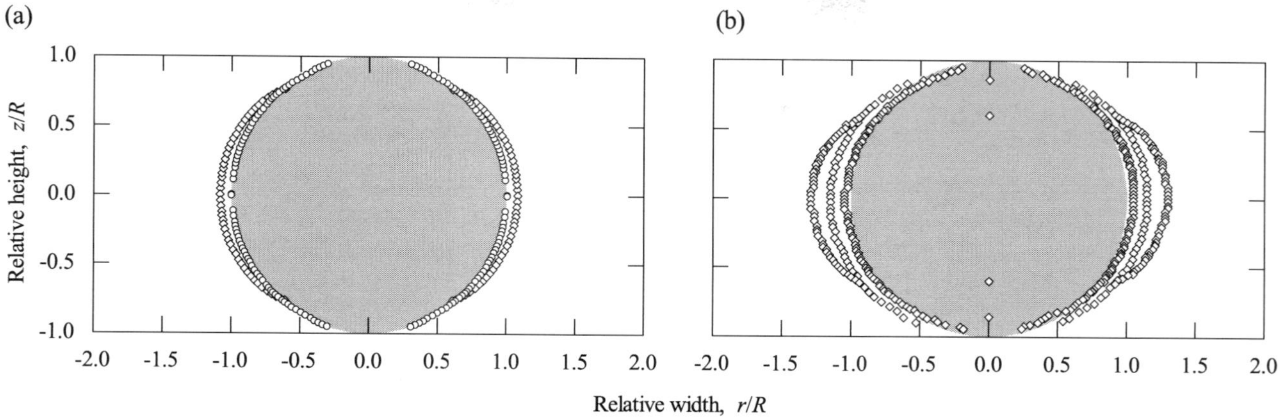

Figure 11.12 Deformation profiles for compressed (a) elastomer and (b) hydrogel spheres represented in relative radial (z/R) and axial (z/R) coordinates. Shaded gray discs represent profiles of undeformed spheres. (a) Adapted from Durelli, AJ et al. (1973). (b) Adapted from David, B et al. (2006). Note that these spheres are only weakly compressed relative to those in Figure 11.11.

The initial profile in (b) is clearly oblate. The deformed profiles in both systems are altered much less from that of the undeformed sphere than those in Figure 11.11. All profiles are fit by the shifted circles described above (for clarity, not shown). As above, c increased and e decreased as deformation $w = 2\delta$ proceeded and the angle subtended by the circular arc of each profile was approximately invariant at 120°.

The more restricted profile information in Figure 11.9 can also be used to determine the profile fits in Figure 11.11, as consideration of the circular form shows that

$$\begin{aligned} c &= (b^2 - a^2 - h^2)/2(b - a) \\ e &= b - c \\ \theta &= \arctan[h/(a - c)], \end{aligned} \tag{11.1}$$

or, alternatively,

$$\begin{aligned} b &= c + e \\ a &= c + e\cos\theta \\ h &= e\sin\theta. \end{aligned} \tag{11.2}$$

The parameters describing the exterior profile, (c, e, θ), are then given by the experimental dimensions (a, b, h). If the contact angle, and thus friction, θ is assumed invariant, description, of profiles are reduced to variation in two parameters (a, b).

The volume of the deformed particle, V, is given by the integral

$$V = 2\pi \int_0^h r^2(z)\mathrm{d}z, \tag{11.3}$$

which is evaluated easily by numerical methods once $r(z)$ is specified. In particular, experimental $r(z)$ data describing observed deformed particle profiles, as in Figures 11.11 and 11.12 can be integrated and compared with initial spherical particle volumes, $V_0 = 4\pi R^3/3$. Figure 11.13 shows the results of such calculations, for the particle systems considered above and some others, as a plot of relative volume V/V_0 vs relative displacement w/D. The symbols in Figure 11.13 represent the results of integration by Eq. (11.3) of a profile generated from Eq. (11.1) from data of James et al. (2020) in Figure 11.9, profiles from data of Shima et al. (1993), Lin et al. (2008), Durelli and Chen (1973), and David et al. (2006) in Figures 11.11 and 11.12, and a similar profile (not shown) of a foam sponge from Lin et al. (2008), in addition to direct determination of cylinder compression results of a gel-related material from Huang et al. (2007). The bars represent uncertainties estimated from profile determinations. The lines are guides to the eye of value 1 and slope −1. The open symbols represent the behavior of the rubber elastomer materials from Shima et al. (1993), Lin et al. (2008), and Durelli and Chen

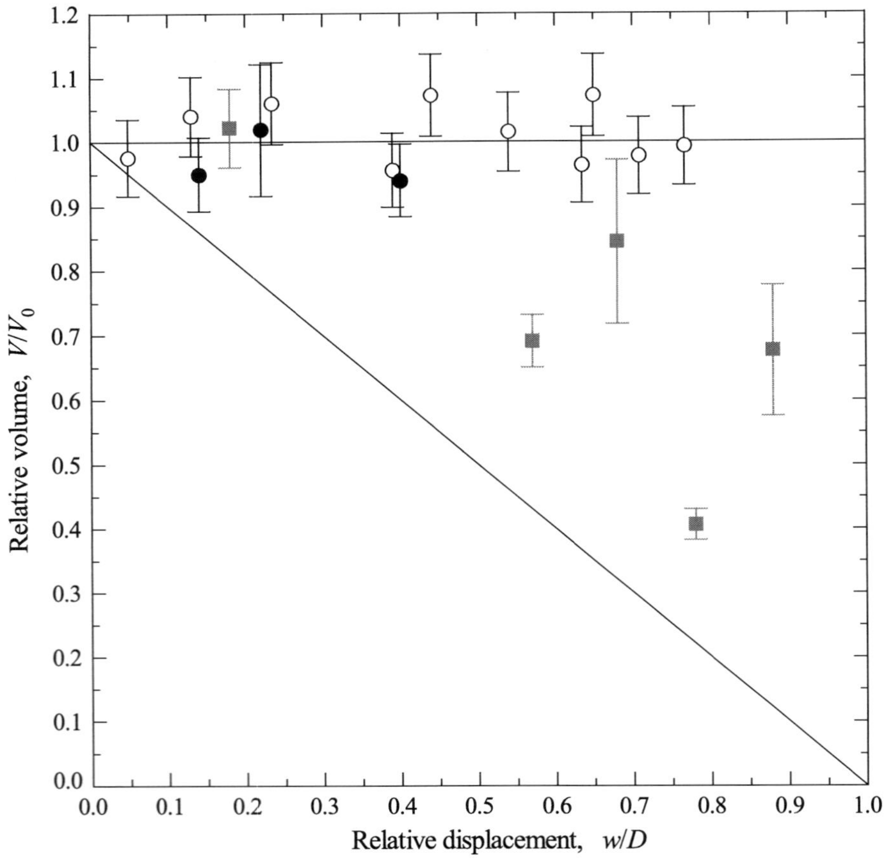

Figure 11.13 Plot of relative volume vs relative displacement for compressed spherical particles. Measurements from profile such as those in Figure 11.11. Open symbols, elastomers. Filled symbols, hydrogels. Gray symbols, foams. Note that elastomers and hydrogels deform similarly at constant volume.

(1973). The filled symbols represent the behavior of saturated hydrogel materials from James et al. (2020) and David et al. (2006) (taking the initial oblate shape into account). The lower grayed symbols represent the behavior of compressible rubber foam from Lin et al. (2008). The upper grayed symbols represent the behavior of a gel-derived foam in air from Huang et al. (2007). As anticipated (Treloar 1975), experimental results for the rubber materials display no significant change in volume on diametral compression. As might also have been anticipated, given the lack of water exudation on deformation (James et al. 2020), experimental results for the hydrogel materials also display no significant change in volume on diametral compression. Finally, also perhaps anticipated, the behavior of the sponge materials is completely different: the sponges decreased in volume significantly on diametral compression, and increased in volume on release—in fact, all materials in Figure 11.13 were elastic and recovered, albeit sometimes slowly, on load removal. Volume-conserving deformation behavior was assumed by Tatara (1991, 1993) in elastic analysis of the external profiles of deformed rubber particles, essentially reversing the logical order of Figures 11.13 and 11.11 and using $V/V_0 = 1$ as a constraint in predicting $r(z)$.

A major point of Figure 11.13 is that hydrogel particle deformation in diametral compression is much the same as another three-dimensionally interconnected polymer, rubber, in that volume is conserved. However, a clear difference between hydrogels and rubber is that for hydrogels the relative displacements are restricted to w/D values of less than about 0.4, whereas for rubber and sponges the values extend to about 0.8. The restriction in (normal, single network) hydrogels arises from the tendency to fracture—a tendency overcome in rubber and advanced hydrogels through increased crosslinking (see Cook and Oyen 2021 and reviews cited therein). Deformation under stress of normal hydrogel particles is truncated by fracture strengths, leading to fragmentation as shown in Figure 11.5d and in many other works (e.g. Gong et al. 2003; Topuz and Okay 2009; Cai et al. 2010; James et al. 2020), including under the action of the self-induced gravitational body force (for a cylinder example see Naficy et al. 2011; here, many of the 40 mm spherical particles fragmented after saturation, the smaller particles did not). The following section considers the effects of deformation rate on the fracture strengths of hydrogel particles loaded in diametral compression.

11.3 Strength

Figure 11.14 shows strength edf $Pr(\sigma)$ plots for samples of saturated hydrogel particles, $D = (17.5 \pm 0.5)$ mm, initially (3.10 ± 0.05) mm, from the work of James et al. (2020), using data derived from the published work. The work was motivated by the potential of hydrogels for tissue engineering in biomedical applications. Approximately 250 particles were tested in a conventional compression platen apparatus at different platen displacement rates. Symbols represent individual strength measurements and different symbols represent different rates, (i) 1 mm s^{-1}, (ii) 0.1 mm s^{-1}, and (iii) 0.01 mm s^{-1}. Failure displacements in all cases were approximately 10 mm, or characteristic compressive failure strains, w/D, of approximately 0.6. Failure stresses or strengths, σ, were determined from the peak forces, P_{max}, typically about 20 N, and the hydrated particle diameters using the HO equation, Eq. (2.41), $\sigma = 0.9P_{max}/D^2$. The characteristic strengths determined in this way were (10–200) kPa—about a factor of 10^3 smaller than many of the particles studied in previous chapters. The strengths are consistent with the much smaller estimated elastic modulus E, about 50 kPa, about a factor of 10^6 smaller than many of the particles studied previously. It is recognized that the strength expression is based on a small-scale elastic response of particles, but has the great advantage that it facilitates comparison with previous results. For the work here, the exact value of the 0.9 coefficient is not too important as the form and relationships of $Pr(\sigma)$ rather than the magnitude of the strengths is the focus. The strength values are slightly reduced from those reported in James et al. and a few (4/250) suspect small strength values censored. The data are well separated by particle size and exhibit narrow strength domains σ_N/σ_1 of

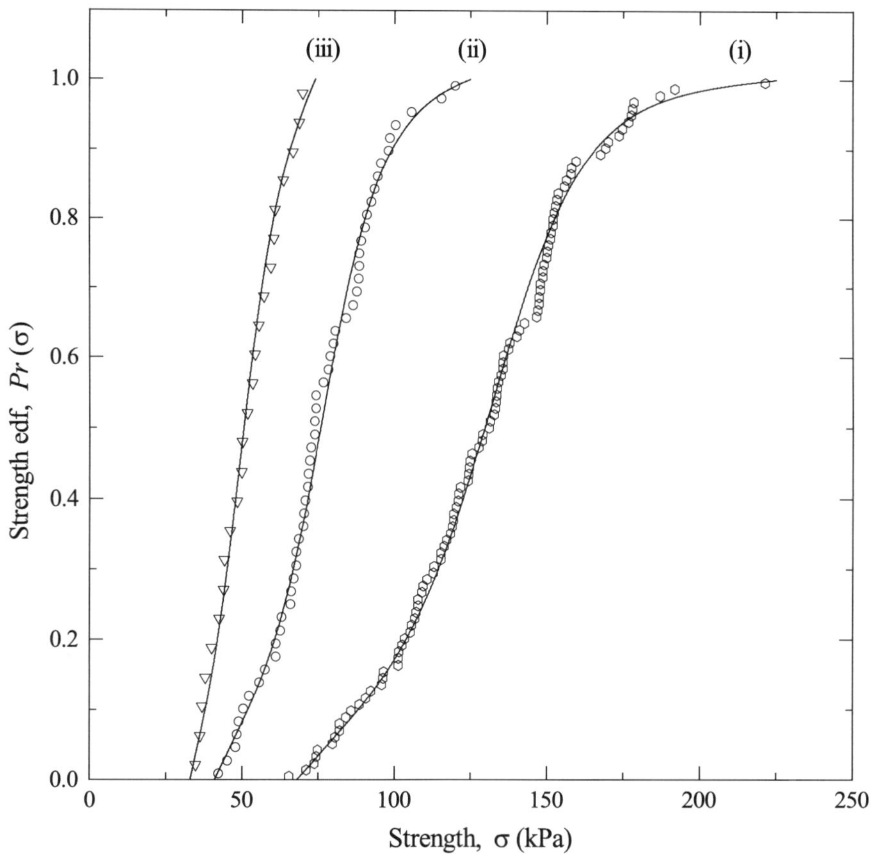

Figure 11.14 Plot of strength edf behavior, $Pr(\sigma)$, for hydrogel particles. Particle diameter, $D = 17$ mm, strength measurement displacement rates 1 mm s^{-1}, 0.1 mm s^{-1}, and 0.01 mm s^{-1}, labeled (i)–(iii), number of measurements, $N_{tot} = 254$ (Adapted from James, J.D et al. 2020). Solid lines show independent sigmoidal best fits. The separated lower bound threshold strengths are indicative of deterministic size effects.

approximately 3. A clear feature of Figure 11.14 is that the data exhibit threshold strengths that decrease with decreasing deformation rate, indicative of deterministic effects. The solid lines in Figure 11.14 represent unconstrained visual best fits to the data using the tri-linear sigmoidal function. Deterministic rate effects on strength are thus observed here as variations in both strength distribution domain and shape. The data are very similar to those observed for glass spheres, Figure 9.20 in particular, in which deterministic size effects were evident in sigmoidal strength distributions.

Figure 11.15 shows as solid lines the crack length pdf $h(c)$ variations underlying the hydrogel particle strength behaviors. The $h(c)$ variations were determined using $B = 0.7$ kPa m$^{1/2}$ and particle size and Eq. (7.2) as a constraint. The B value is about a factor of 10^3 smaller than that used to describe many of the particles studied in previous chapters. The value is consistent with the scaling $B \sim (ER)^{1/2}$, the elastic modulus E from above, and, similar to other single network hydrogel particles (Cook and Oyen 2021), $R = 2\gamma \approx 10$ J m^{-2} estimated from simple scission of polymer chains across a section of a saturated hydrogel (Zhao 2014). The largest crack length was about 0.4 mm, compared with the compressed dimension of the particles of about 7 mm, and the smallest crack length was about 0.01 mm. Decreasing displacement rate lead to crack length domains that remained approximately invariant in relative width, but which translated to increasing lengths with decreasing rate; the rate variations are labeled (i)–(iii). The crack distributions consisted of overlapping peaks that become less well formed as the displacement rate decreased. Consistent with deterministic effects, the crack length distributions varied in domain location and (slightly) in distribution shape as deformation rate changed. As might be anticipated, the data are very similar to those observed for glass spheres, Figure 9.21 particularly.

In interpreting Figure 11.15, it is important to remember several assumptions leading to the $h(c)$ variations. First, the strength analysis assumed that the stress acting on the strength controlling flaw was proportional to the applied force. Second, the fracture parameter B was assumed to be invariant with displacement rate. Both of these assumptions ignore potential effects of deformation kinetics. Such effects were demonstrated in Figures 11.7 and 11.8 and are most probably due to rate effects of poroelastic flow in the polymer network of the water saturated particles. Third, the crack lengths in Figure 11.15 reflect the cracks at particle failure. Fracture kinetic effects could lead to extension of cracks from initial,

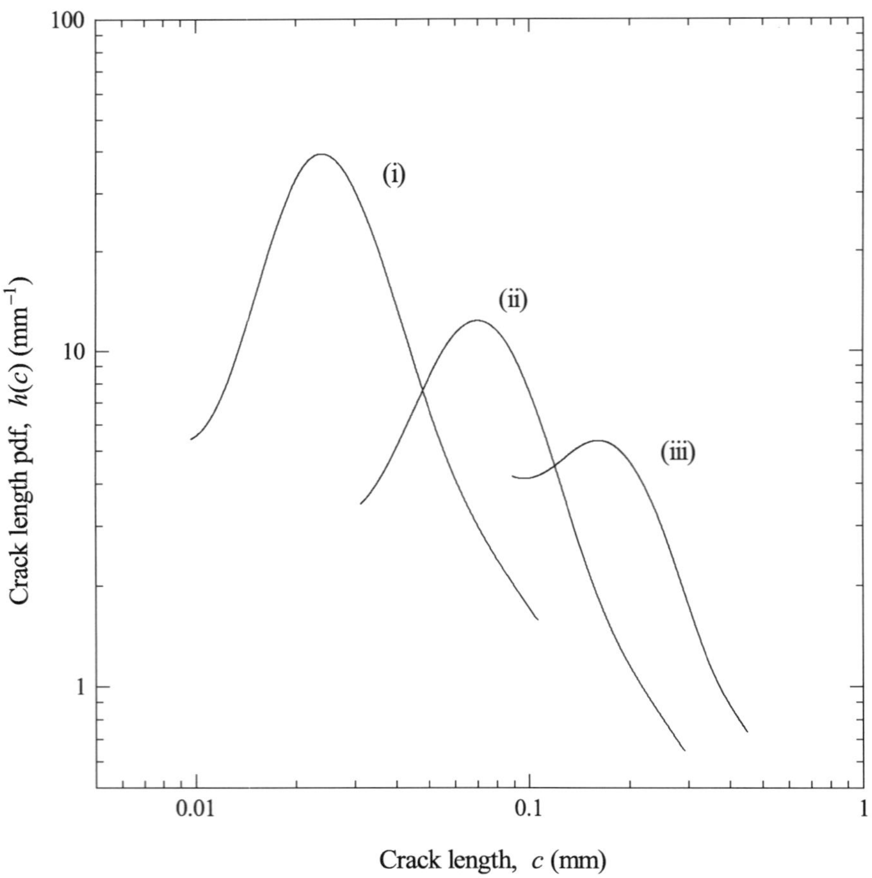

Figure 11.15 Plot in logarithmic coordinates of crack length pdf $h(c)$ variations for hydrogel particles. $D = 17$ mm, decreasing displacement rates labeled (i)–(iii); determined from Figure 11.14.

shorter, crack lengths at the start of loading. As all particles were identical prior to testing, it seems likely, in fact, that the different crack length populations at failure developed during loading. Distinguishing the poroelastic kinetics of stress generation from the fracture kinetics of crack extension is not possible from these experimental data alone. However, it is clear from the hysteresis and load relaxation observations, Figures 11.7 and 11.8, that stress acting on a flaw is probably less than linearly proportional to the applied force. Similarly, the increase in strength controlling flaw size with the loading period, Figure 11.15, suggests that final crack lengths are greater than initial crack lengths.

11.4 Summary and Discussion

This chapter has surveyed experimental observations of deformation and strength of compliant particles—particles that exhibit substantial deformation on loading in diametral compression to small forces. Two groups of materials forming compliant particles, both based on three-dimensional polymeric networks, were considered: the hydrogels, consisting of

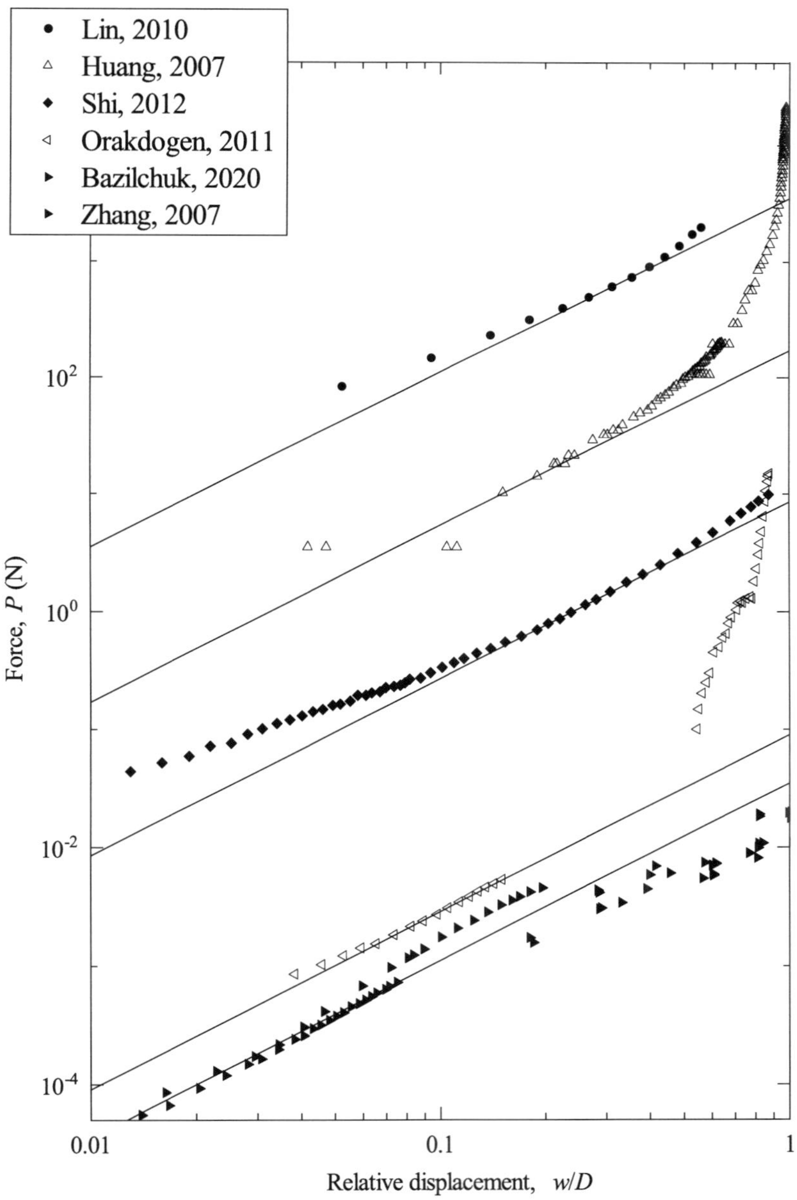

Figure 11.16 Logarithmic plot of force vs relative displacement for a number of particle systems, including blocks, a contact lens (a shell), and plastically deforming metal-coated spheres. Lines are guides to the eye representing small scale spherical elastic behavior. First authors of works noted.

sparse networks enclosing liquid water (Figure 11.1), and the elastomers or rubbers (Figure 11.3), consisting of dense networks, sometimes in the form of foams containing air. Elastomer particles very rarely fail in diametral loading, commonly sustaining relative displacements w/D of 0.9 or greater. Hydrogel particles, however, almost invariably fail in diametral loading by meridional fracture at relative displacements of approximately 0.5. Prior to a peak (non-failure) force, both elastomers and hydrogels exhibit force-displacement behavior that is non-linear but nearly reversible. Time-dependent effects controlled by viscoelastic (elastomers) and poroelastic (hydrogels) deformation perturb underlying elastic behavior—although such effects are usually not noticeable on monotonic loading. Hydrogels are brittle and fail extremely rapidly at peak force by the propagation of multiple cracks, usually with consequent fragmentation such that there is no force supported after failure. The deformation of both elastomer and hydrogel particles was considered here, providing background for consideration of strength of hydrogel particles.

A feature of hydrogel particle deformation on loading is the similarity to the deformation of elastomers. In particular, the diametral compression axial force-displacement responses for both sets of materials exhibit similar characteristics, Figure 11.6. Both tend to a Herztian $(w/D)^{3/2}$ response at small forces and displacements and a much greater dependence of force on displacement at large displacements. Other, particle-like systems display similar characteristics. Figure 11.16 shows a logarithmic plot of experimental axial force-displacement behavior for several large deformation systems loaded in compression, using data derived from the published works cited. Relative displacement w/D is used to enable comparison, although, in some systems, this is not a sphere diameter but the initial axial dimension of a block of material. The data included are (in order of decreasing supported force): poly(dimethylacrylamide) hydrogel cylinders, 8 mm diameter × 6 mm tall, saturated with water and loaded axially in uniform compression (Lin et al. 2010), styrene and butyl acrylate-based microsphere gel cylinders, (24–26) mm diameter × (12–18) mm tall, loaded axially in uniform compression in air (Huang et al. 2007), commercial silicone hydrdogel contact lenses, 14 mm diameter × 3.7 mm tall shells, with radius of 8.36 mm and thickness of 94 μm, saturated with isotonic solution and loaded axially between flat platens (Shi et al. 2012), gelatin spheres containing DNA (similar to David et al. 2006) formed with a cryogenic process, (10–14) mm diameter, saturated with water, and loaded in diametral compression (Orakdogen et al. 2011; see also Topuz and Okay 2009; Okay and Lozinsky 2014), metal-coated dense poly(methylmethacrylate) spheres, 10 μm diameter, loaded singly in diametral compression with a flat punch indenter (Bazilchuk et al. 2020), or in a group of approximately 100 between flat platens (Zhang et al. 2007). Symbols represent individual observations. The lines are of slope 3/2 and are guides to the eye *only* to enable comparison with Figure 11.6.

As in Figure 11.6, the diverse set of systems in Figure 11.16 leads to a large range of supported forces. The hydrogel sphere system of Orakdogen et al. follows the 3/2 Hertzian response at small displacements before exhibiting a significant increase in load at large displacements, similar to those in Figure 11.6. The polymer sphere systems of Bazilchuk et al. and Zhang et al., although clearly displaying predominant plastic deformation on loading, exhibit 3/2 behavior over the full displacement range. The measurements of Huang et al., Shi et al., and Orakdogen et al. are all anticipated to exhibit linear force-displacement behavior at small displacements and do so. Note that logarithmic coordinates are required to assess and compare the force-displacement behavior of the diverse systems of Figures 11.6 and 11.16—in linear coordinates all responses resemble the convex behavior of Figure 11.2.

There are also a few measurements of transverse deformation from other particle-like systems. These measurements display similar characteristics to those of the elastomer and hydrogel systems of Figure 11.9. Figure 11.17 shows a logarithmic plot of experimental contact radii for two spherical systems, using data derived from the published works cited. The form of the plot, including lines as guides to the eye is identical to Figure 11.9. The data are from the silicone hydrogel contact lens measurements of Shi et al. (2012) and the metal coated polymer particle measurements of Bazilchuk et al. (2020). Symbols represent individual observations. In both cases, the measurements follow the slope 1/2 Hertz-like line, consistent with the behavior observed in the axial measurements. The results of Shi et al. are consistent with elastic shell deformation models. The results of Bazilchuk et al. are consistent with a localized zone of plastic deformation in the particle, adjacent to the flat punch contact, as described in Chapter 2.

The similarity of the hydrogel particle strength and crack length behavior, Figures 11.14 and 11.15, to that of glass particle behavior, Figures 9.20 and 9.21, respectively, has been noted. In the case of glass, the data were interpreted as a deterministic variation in flaw population with particle size, leading to extreme value variation in the strength-controlling flaws. Implicit in this interpretation is that the resulting strength distributions were intrinsic to the particles and not significantly influenced by the diametral loading protocol. In Figures 11.14 and 11.15, the opposite is probably true—the particles were all the same and diametral loading protocol variation, here displacement rate, determined the failure condition and strength. Another system in which testing procedure rather than particle variation determined the failure condition and apparent

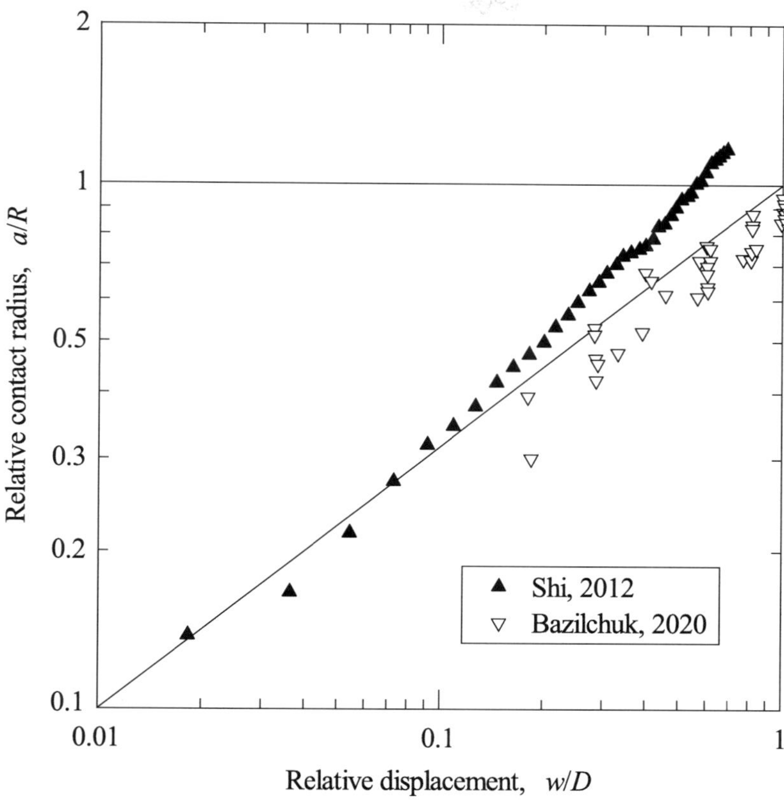

Figure 11.17 Logarithmic plot of relative contact radii (a/R) vs relative displacement (w/D) for two particle systems: a contact lens (a shell) and a plastically deforming metal-coated sphere. Lines are guides to the eye representing small scale elastic behavior. First authors of works noted.

strength is that of ceramic particles diametrally loaded between platens of different susceptibility to plastic deformation by Van Rooyen et al. (2010). Figure 11.18a shows a plot of strength edf responses for 800 μm diameter nested-sphere composite ZrO_2-C-SiC particles compressed between metal platens of varying hardness, using data derived from the published work. These were TRISO particles, considered in Chapter 5 and strength was determined from reported failure force using the HO equation. The strengths of the particles were limited by the strengths of the encapsulating 30 μm SiC shells. The metal platens varied from very soft annealed Al, hardness $H \approx 0.2$ GPa, to hardened steel, $H \approx 9.3$ GPa. Three strength distributions are shown in Figure 11.18, from measurements using the platens of bounding hardness and an intermediate Al set, $H \approx 0.78$ GPa. The results from increasing hardness platens are shown right to left, (i), (ii), and (iii). The data are well separated by particle size and exhibit narrow strength domains σ_N/σ_1 of approximately 2.5. A clear feature of Figure 11.18 is that the data exhibit threshold strengths that decrease with increasing platen hardness, indicative of deterministic effects. The solid lines in Figure 11.18a represent unconstrained visual best fits to the data using the tri-linear sigmoidal function. Deterministic platen effects on strength are thus observed here as variations in both strength distribution domain and shape.

Figure 11.18b shows as solid lines the crack length pdf $h(c)$ variations underlying the ceramic particle strength behaviors. The $h(c)$ variations were determined using $B = 0.1$ MPa m$^{1/2}$ and shell thickness and Eq. (7.2) as a constraint. The B value is similar to that used to describe many of the particles studied in previous chapters. The largest crack length was about 30 μm. Increasing platen hardness lead to crack length domains that remained approximately invariant in relative width, but which translated to increasing lengths with decreasing hardness; the hardness variations are labeled (i)–(iii). The crack distributions consisted of overlapping peaks that become slightly broader as the platen hardness increased. Consistent with deterministic effects, the crack length distributions varied in domain location and (slightly) in distribution shape as platen hardness changed. As might be anticipated, the data are very similar to those observed for hydrogel spheres, Figure 11.15.

There are two points to note in interpreting Figure 11.18. The first is that particle failure was controlled by flaws on the exterior of the particle, in or on the SiC shell, rather than by flaws from the particle interior. Tensile stress on

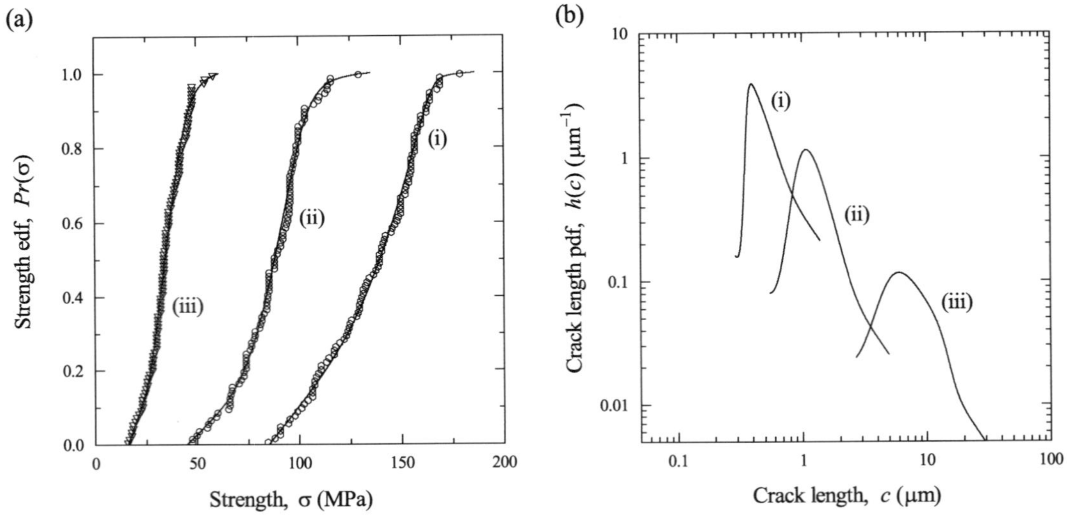

Figure 11.18 (a) Plot of strength edf behavior, $Pr(\sigma)$, for ceramic composite particles. Particle diameter, D = 800 μm, platen hardness values 0.2 GPa, 0.78 GPa, 9.3 GPa, labeled (i)–(iii), number of measurements, N_{tot} = 299 (adapted from Van Rooyen, G.T et al. 2010). Solid lines show independent sigmoidal best fits. The separated lower bound threshold strengths are indicative of deterministic size effects. (b) Plot in logarithmic coordinates of conjugate crack length pdf $h(c)$ variations for ceramic composite particles. Platen hardness values labeled (i)–(iii) as in (a).

the meridional plane at the exterior of spherical particles loaded in diametral compression was discussed in Chapter 2. For the extended contact area loading conditions considered, the exterior stress was about 0.7 of the interior stress, Figure 2.22. As discussed by Shipway and Hutchings (1993a, 1993b, 1993c), this ratio varies considerably with the particle-platen contact details, and is nearly always less than 1. Thus, specific combinations of platen and particle deformation properties and interior and exterior particle flaw populations are required to cause failure at the exterior of a diametrally loaded particle (and hence, attention in this book and elsewhere focuses on interior failure). The second point extends the first by detailing the dependence of the exterior stress on the platen hardness. Finite element analysis by Van Rooyen et al. showed that stress developed in the spherical SiC shell on particle loading depended significantly on the yield stress, and thus hardness, of the platen. For the failure forces and particle geometries considered, comparable to those in the strength measurements, stress in the shell increased by approximately a factor of 8 as platen yield stress increased by approximately a factor of 12. To first approximation, the HO equation can then be rewritten here as $\sigma = 0.6(H/H_{ref})P/D^2$. The first term on the right side represents the effect of the exterior stress location and the second term represents the effect of platen hardness variation, where H_{ref} is an empirical reference hardness. Although the details may vary, the consequence is that failure force P distributions are simply translated with varying platen hardness, with no change in particle flaw population. Or, alternatively, from the external frame of reference of experimental observation, apparent strengths and flaw distributions are translated with little change in shape, in agreement with experimental observations. In particular, particles tested with softer platens will support greater loads, and appear to contain smaller flaws.

A generalization of the ideas above is to rewrite P in the HO equation as βP, where β is a function that scales the force-stress relationship for the deterministic factor of interest. For the ceramic particle case, $\beta \sim H/H_{ref}$. A natural consideration for the hydrogel failure distributions of Figure 11.14 is $\beta \sim \dot{w}_{ref}/\dot{w}$ where $\dot{w}$ is the imposed displacement rate and $\dot{w}_{ref}$ is a reference rate. However, the hysteresis and relaxation behavior of Figures 11.7 and 11.8 suggest that displacement rate effects, although clearly an influence at the level of $0.3P$, are unlikely to account for the factor of 3 in strength variation with rate. Hence, it seems that the kinetic effects observed in strength behavior arise not from deformation kinetics linking stress development to displacement rate, but to fracture kinetics linking crack development to rate. In particular, non-equilibrium crack propagation in hydrogels, observed at the multi-millimeter scale by Baumberger and colleagues (see Naassaoui et al. 2018 for a list of works) and at the multi-μm scale in the related silsesquioxane materials (Cook and Liniger 1999) is probably responsible for the strength decreases with displacement rate in Figure 11.14. If this is the case, crack length in the Griffith equation is then effectively modified such that $c \sim \dot{w}_{ref}/\dot{w}$. Smaller displacement rates lead to larger, extended, cracks and thus smaller strengths. Such behavior has long been known in ceramics and glasses (Lawn 1993). Deciding between a kinetically limited stress development effect or a kinetically limited crack extension effect, or a combination, requires poroelastic and fracture modelling beyond the scope of this book. However, both procedures are straightforward (Ding et al. 2013; Cook 2015).

The results here illustrate application of compressive strength testing of spherical particles to provide qualitative and quantitative information regarding hydrogel mechanical properties. A key observation was the significant effect of deformation rate on strength, strength distributions in particular, reinforcing previous observations that deformation in these materials is controlled by poroelastic and viscoelastic kinetic effects. The results here showed that deformation rate effects also influence fracture. These results are placed in the context of hydrogel strength and the broader context of brittle fracture strength in Figure 11.19. Figure 11.19a shows in semi-logarithmic coordinates a ranked distribution of hydrogel strengths compiled in the perspective overview of Cook and Oyen (2021). A striking feature of the distribution is the large strength domain, a factor of nearly 10^4 has been reported for hydrogel strengths, although the median strength of approximately 100 kPa for "normal" single network hydrogels is clear. The shaded band in Figure 11.19a indicates the domain of "advanced hydrogels," (600–3000) kPa. As noted in Section 11.1, the increases in strength from normal to advanced hydrogels have been achieved by microstructure development. An example of strength increase by microstructure development is shown in Figure 11.19b from the work of Tonsomboon et al. (2017). The plot is a histogram of strengths of composite alginate hydrogels reinforced with gelatin fibers of various orientations. The semi-logarithmic strength scale and shaded bar are repeated from Figure 11.19a. The open bar (0) represents the strength of the unreinforced material. The hatched bars represent the strengths of composite materials with fibers oriented (1) perpendicular to the applied stress on testing, (2) randomly, (3) perpendicular and parallel in laminated plies, and (4) parallel to the applied stress. The significant increase in strength arising from microstructural manipulation, well into the domain of advance hydrogels is clear. The strength increase on displacement rate variation for a single network hydrogel is shown in Figure 11.19c from the work of James et al. (2020). The semi-logarithmic strength scale and shaded bar are repeated from Figure 11.19a. It is clear that kinetic effects associated with variations in applied displacement rate are no-where near as significant as microstructural effects in altering strength. It is also noted that microstructural manipulation appears to decrease relative strength variability, whereas kinetic effects appear to have very little effect on relative strength variability. The predominant effect of microstructure over kinetics in determining brittle fracture strength has been measured and analyzed in ceramic and glass materials for some time, including the differing effects on strength variability (Cook 2015).

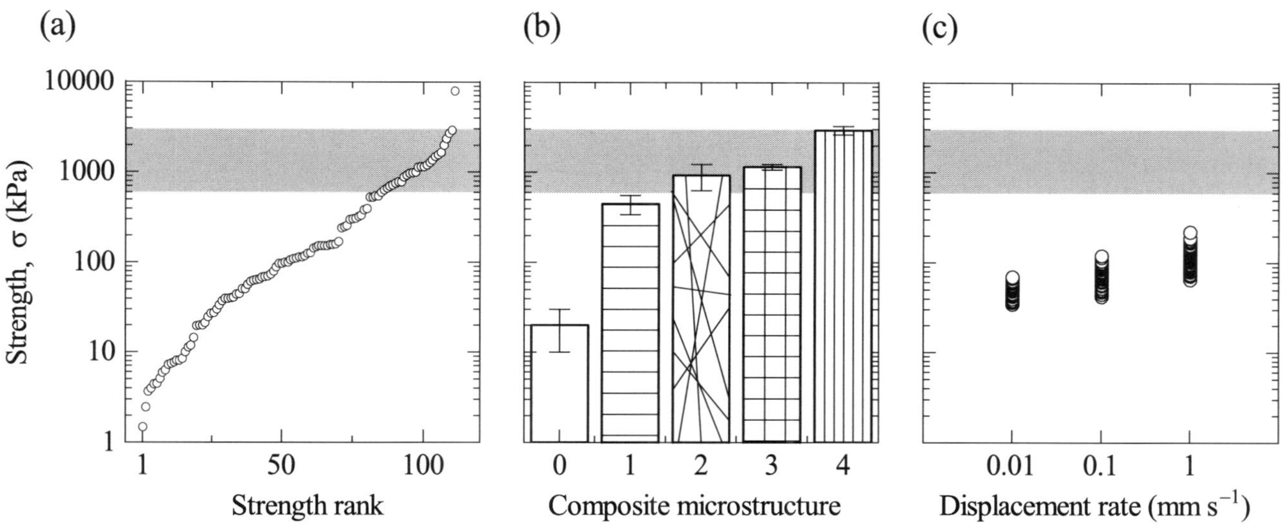

Figure 11.19 Aspects of strength behavior of hydrogel materials summarized in three panels. (a) Reported strengths of hydrogels ranked. The domain of strengths is broad. The strength of "advanced" hydrogels is indicated by the shaded band and repeated throughout. (b) Strengths of fiber composite hydrogels as a function of microstructure. Fiber alignment indicated in bars; parallel to the applied stress direction results in 100-fold increase in strength. (c) Strengths of single network hydrogels as a function of applied displacement rate. Increasing rate 100-fold increases median strength by about a factor of 3.

The analytical framework of fracture mechanics is well suited to describe the particle strength variations shown in Figure 11.19 (Lawn 1993; Cook 2015). In addition, a fracture mechanics framework enables a clear assessment of the physical meaning and quantitative estimates of the parameter B used throughout the book to couple strength to crack length through the Griffith equation, $\sigma = Bc^{-1/2}$. The following chapter, Chapter 12 develops and examines fracture mechanics analyses for application to particle strength measurements.

References

Aangenendt, F.J., Punter, M.T., Mulder, B.M., van der Schoot, P., and Wyss, H.M. (2020). Nonmonotonic swelling and compression dynamics of hydrogels in polymer solutions. *Physical Review E* 102: 062606.

Abd El-Rehim, H.A., Hegazy, E.S.A., and Abd El-Mohdy, H.L. (2006). Effect of various environmental conditions on the swelling property of PAAm/PAAcK superabsorbent hydrogel prepared by ionizing radiation. *Journal of Applied Polymer Science* 101: 3955–3962.

Andrei, D.C., Briscoe, B.J., Luckham, P.F., and Williams, D.R. (1996). The deformation of microscopic gel particles. *Journal de Chimie Physique* 93: 960–976.

Banerjee, H., Suhail, M., and Ren, H. (2018). Hydrogel actuators and sensors for biomedical soft robots: brief overview with impending challenges. *Biomimetics* 3: 15.

Bazilchuk, M., Sumigawa, T., Kitamura, T., Zhang, Z., Kristiansen, H., and He, J. (2020). Contact area measurement of micron-sized metal-coated polymer particles under compression. *International Journal of Mechanical Sciences* 165: 105214.

Bhattacharyya, A., O'Bryan, C., Ni, Y., Morley, C.D., Taylor, C.R., and Angelini, T.E. (2020). Hydrogel compression and polymer osmotic pressure. *Biotribology* 22: 100125.

Berry, J.D., Biviano, M., and Dagastine, R.R. (2020). Poroelastic properties of hydrogel microparticles. *Soft Matter* 16: 5314–5324.

Bertrand, T., Peixinho, J., Mukhopadhyay, S., and MacMinn, C.W. (2016). Dynamics of swelling and drying in a spherical gel. *Physical Review Applied* 6: 064010.

Cai, S., Hu, Y., Zhao, X., and Suo, Z. (2010). Poroelasticity of a covalently crosslinked alginate hydrogel under compression. *Journal of Applied Physics* 108: 113514.

Cook, R.F. (2015). Multi-scale eects in the strength of ceramics. *Journal of the American Ceramic Society* 98: 2933–2947.

Cook, R.F. and Liniger, E.G. (1999). Stress-corrosion cracking of low-dielectric-constant spin-on-glass thin films. *Journal of the Electrochemical Society* 146: 4439–4448.

Cook, R.F. and Oyen, M.L. (2007). Nanoindentation behavior and mechanical properties measurement of polymeric materials. *International Journal of Materials Research* 98: 370–378.

Cook, R.F. and Oyen, M.L. (2021). On the failure and fracture of hydrogels for cartilage replacement. *Journal of Physics: Materials* 4: 021001.

David, B., Barbe, L., Barthès-Biesel, D., and Legallais, C. (2006). Mechanical properties of alginate beads hosting hepatocytes in a fluidized bed bioreactor. *The International Journal of Artificial Organs* 29: 756–763.

Ding, Z., Liu, Z., Hu, J., Swaddiwudhipong, S., and Yang, Z. (2013). Inhomogeneous large deformation study of temperature-sensitive hydrogel. *International Journal of Solids and Structures* 50: 2610–2619.

Drury, J.L. and Mooney, D.J. (2003). Hydrogels for tissue engineering: Scaffold design variables and applications. *Biomaterials* 24: 4337–4351.

Durelli, A.J. and Chen, T.L. (1973). Displacement and finite-strain fields in a sphere subjected to large deformations. *International Journal of Non-linear Mechanics* 8: 17–18.

Fennell, E. and Huyghe, J.M. (2019). Chemically responsive hydrogel deformation mechanics: A review. *Molecules* 24: 3521.

Findley, W.N., Lai, J.S., and Onaran, K. (1976). *Creep and Relaxation of Nonlinear Viscoelastic Materials*. North-Holland.

Galli, M., Comley, K.S., Shean, T.A.V., and Oyen, M.L. (2009). Viscoelastic and poroelastic mechanical characterization of hydrated gels. *Journal of Materials Research* 24: 973–979.

Galli, M., Fornasiere, E., Cugnoni, J., and Oyen, M.L. (2011). Poroviscoelastic characterization of particle-reinforced gelatin gels using indentation and homogenization. *Journal of the Mechanical Behavior of Biomedical Materials* 4: 610–617.

Gibbs, D.M., Black, C.R., Dawson, J.I., and Oreffo, R.O. (2016). A review of hydrogel use in fracture healing and bone regeneration. *Journal of Tissue Engineering and Regenerative Medicine* 10: 187–198.

Gong, J.P. (2010). Why are double network hydrogels so tough? *Soft Matter* 6: 2583–2590.

Gong, J.P., Katsuyama, Y., Kurokawa, T., and Osada, Y. (2003). Double-network hydrogels with extremely high mechanical strength. *Advanced Materials* 15: 1155–1158.

Gu, W.Y., Yao, H., Huang, C.Y., and Cheung, H.S. (2003). New insight into deformation-dependent hydraulic permeability of gels and cartilage, and dynamic behavior of agarose gels in confined compression. *Journal of Biomechanics* 36: 593–598.

Henderson Jr, G.V.S., Campbell, D.O., Kuzmicz, V., and Sperling, L.H. (1985). Gelatin as a physically crosslinked elastomer. *Journal of Chemical Education* 62: 269–270.

Hua, J., Ng, P.F., and Fei, B. (2018). High-strength hydrogels: Microstructure design, characterization and applications. *Journal of Polymer Science Part B: Polymer Physics* 56: 1325–1335.

Huang, T., Xu, H.G., Jiao, K.X., Zhu, L.P., Brown, H.R., and Wang, H.L. (2007). A novel hydrogel with high mechanical strength: A macromolecular microsphere composite hydrogel. *Advanced Materials* 19: 1622–1626.

James, J.D., Ludwick, J.M., Wheeler, M.L., and Oyen, M.L. (2020). Compressive failure of hydrogel spheres. *Journal of Materials Research* 35: 1227–1235.

Ji, D. and Kim, J. (2021). Recent strategies for strengthening and stiffening tough hydrogels. *Advanced NanoBiomed Research* 1: 2100026.

Kilcast, D., Boyar, M.M., and Hudson, J.B. (1984). Gelatin photoelasticity: A new technique for measuring stress distributions in gels during penetration testing. *Journal of Food Science* 49: 654–655.

Kong, H.J., Lee, K.Y., and Mooney, D.J. (2002). Decoupling the dependence of rheological/mechanical properties of hydrogels from solids concentration. *Polymer* 43: 6239–6246.

Lawn, B.R. (1993). *Fracture of Brittle Solids, 2nd ed.* Cambridge.

Lee, S.Y., Pereira, B.P., Yusof, N., Selvaratnam, L., Yu, Z., Abbas, A.A., and Kamarul, T. (2009). Unconfined compression properties of a porous poly (vinyl alcohol)–chitosan-based hydrogel after hydration. *Acta Biomaterialia* 5: 1919–1925.

Lin, W.C., Fan, W., Marcellan, A., Hourdet, D., and Creton, C. (2010). Large strain and fracture properties of poly (dimethylacrylamide)/silica hybrid hydrogels. *Macromolecules* 43: 2554–2563.

Lin, Y.-L., Wang, D.-M., Lu, W-M., Lin, Y-S., and Tung, K.L. (2008). Compression and deformation of soft spherical particles. *Chemical Engineering Science* 63: 195–203.

Liu, K.K. (2006). Deformation behaviour of soft particles: A review. *Journal of Physics D: Applied Physics* 39: R189–R199.

Liu, Z., Toh, W., and Ng, T.Y. (2015). Advances in mechanics of soft materials: A review of large deformation behavior of hydrogels. *International Journal of Applied Mechanics* 7: 1530001.

Liu, K.K., Williams, D.R., and Briscoe, B.J. (1998). The large deformation of a single micro-elastomeric sphere. *Journal of Physics D: Applied Physics* 31: 294–303.

Long, R. and Hui, C.Y. (2016). Fracture toughness of hydrogels: Measurement and interpretation. *Soft Matter* 12: 8069–8086.

Naassaoui, I., Ronsin, O., and Baumberger, T. (2018). A poroelastic signature of the dry/wet state of a crack tip propagating steadily in a physical hydrogel. *Extreme Mechanics Letters* 22: 8–12.

Naficy, S., Brown, H.R., Razal, J.M., Spinks, G.M., and Whitten, P.G. (2011). Progress toward robust polymer hydrogels. *Australian Journal of Chemistry* 64: 1007–1025.

Normand, V., Lootens, D.L., Amici, E., Plucknett, K.P., and Aymard, P. (2000). New insight into agarose gel mechanical properties. *Biomacromolecules* 1: 730–738.

Okay, O. (2009). General properties of hydrogels. In: *Hydrogel Sensors and Actuators* (ed G. Gerlach and K.-F. Arndt), 1–14. Springer-Verlag, Berlin Heidelberg.

Okay, O. and Lozinsky, V.I. (2014). Synthesis and structure–property relationships of cryogels. In: *Polymeric Cryogels* (ed. O. Okay), 103–157. Springer.

Orakdogen, N., Karacan, P., and Okay, O. (2011). Macroporous, responsive DNA cryogel beads. *Reactive and Functional Polymers* 71: 782–790.

Oyen, M.L. (2014). Mechanical characterisation of hydrogel materials. *International Materials Reviews* 59: 44–59.

Pitt, K.G., Newton, J.M., and Stanley, P. (1989). Stress distributions in doubly convex cylindrical discs under diametral loading. *Journal of Physics D: Applied Physics* 22: 1114–1127.

Rivlin, R.S. and Thomas, A.G. (1953). Rupture of rubber. Part I. Characteristic energy for tearing. *Journal of Polymer Science* 10: 291–318.

Rosenthal, A.J. (2016). Contact stresses in gelatin spheres under compressive die loading. *Journal of Texture Studies* 47: 457–460.

Shi, J., Robitaille, M., Muftu, S., and Wan, K.T. (2012). Deformation of a convex hydrogel shell by parallel plate and central compression. *Experimental Mechanics* 52: 539–549.

Shima, S., Tatara, Y., Iio, M., Shu, C., and Lucero, C.J. (1993). Large deformations of a rubber sphere under diametral compression: Part 2: Experiments on many rubber materials and comparisons of theories with experiments. *JSME International Journal Series A* 36: 197–205.

Shipway, P.H. and Hutchings, I.M. (1993a). Fracture of brittle spheres under compression and impact loading. I. Elastic stress distributions. *Philosophical Magazine A* 67: 1389–1404.

Shipway, P.H. and Hutchings, I.M. (1993b). Fracture of brittle spheres under compression and impact loading II. Results for lead-glass and sapphire spheres. *Philosophical Magazine A* 67: 1405–1421.

Shipway, P.H. and Hutchings, I.M., (1993c). Attrition of brittle spheres by fracture under compression and impact loading. *Powder Technology*, 76: 23–30.

Tatara, Y. (1991). On compression of rubber elastic sphere over a large range of displacements–Part 1: Theoretical study. *Journal of Engineering Materials and Technology* 113: 285–291.

Tatara, Y. (1993). Large deformations of a rubber sphere under diametral compression: Part 1: Theoretical analysis of press approach, contact radius and lateral extension. *JSME International Journal Series A* 36: 190–196.

Tatara, Y., Shima, S., and Lucero, J.C. (1991). On compression of rubber elastic sphere over a large range of displacements–Part 2: Comparison of theory and experiment. *Journal of Engineering Materials and Technology* 113: 292–295.

Tonsomboon, K., Butcher, A.L., and Oyen, M.L. (2017). Strong and tough nanofibrous hydrogel composites based on biomimetic principles. *Materials Science and Engineering: C* 72: 220–227.

Tonsomboon, K. and Oyen, M.L. (2013). Composite electrospun gelatin fiber-alginate gel scaffolds for mechanically robust tissue engineered cornea. *Journal of the Mechanical Behavior of Biomedical Materials* 21: 185–194.

Topuz, F. and Okay, O. (2009). Macroporous hydrogel beads of high toughness and superfast responsivity. *Reactive and Functional Polymers* 69: 273–280.

Treloar, L.R.G. (1975). *The Physics of Rubber Elasticity, 3rd ed.* Oxford.

Van Rooyen, G.T., Du Preez, R., De Villiers, J., and Cromarty, R. (2010). The fracture strength of TRISO-coated particles determined by compression testing between soft aluminium anvils. *Journal of Nuclear Materials* 403: 126–134.

Wang, H.F. (2000). *Theory of Linear Poroelasticity*. Princeton.

Zhang, Z.L., Kristiansen, H., and Liu, J. (2007). A method for determining elastic properties of micron-sized polymer particles by using flat punch test. *Computational Materials Science* 39: 305–314.

Zhao, X. (2014). Multi-scale multi-mechanism design of tough hydrogels: Building dissipation into stretchy networks. *Soft Matter* 10: 672–687.

12

Fracture Mechanics of Particle Strengths

This chapter examines fracture mechanics of particle strengths and, in particular, provides interpretations of crack length c and the strength parameter B in detailed physical terms. Fracture is viewed in terms of both energy and stress frameworks and the equivalence of these analyzed. Fracture behavior of simple Griffith cracks, crack initiation and stability of cracks at inclusions, grains, and contacts, stable precursor extension of cracks under combined loading, sub-threshold flaws, ductile-brittle transitions, and the effects of positive and negative variations in fracture resistance are all considered. Clear relations between particle strength and externally controlled or measured experimental variables are given.

12.1 Introduction

This chapter develops and applies fracture mechanics analyses relevant to particle failure. One of the most important results derived is the Griffith equation, used throughout the book to relate measured strengths to underlying crack lengths. The derivation of this equation, from consideration of energy variations in extended components, provides a clear example of the relationship between the energy and stress viewpoints of fracture and insight into the ways in which fracture in particles might differ from that in other components. Although the energy viewpoint usually provides greater insight into physical fracture phenomena, the stress viewpoint is usually more amenable to fracture analyses, including investigation of variations in system geometry and application of multiple sources of loading. Hence, fracture analyses based on stress variations will be used to consider stable cracks, crack initiation, crack propagation, and the strengths of components containing residually stressed flaws. Although examples will be provided to illustrate specific points, the goal here is to develop and introduce a framework for analysis of particle fracture rather than apply fracture mechanics to specific particle fracture strength measurements. Fracture mechanics for metals, ceramics, polymers, and biomaterials is considered in detail elsewhere (Broek 1982; Williams 1984; Kanninen and Popelar 1985; Atkins and Mai 1985; Lawn 1993).

Fracture analysis begins with consideration of energy changes associated with cracks in loaded components. Figure 12.1 shows schematic diagrams of two prismatic tensile bars, both of length L, loaded at the ends by forces F indicated in dark gray. The bars differ in outline cross-section: (a) circular, radius R; (b) rectangular, dimensions b and h. Both bars are shown as including cracks, indicated in light gray, in the centers of the bars on planes perpendicular to the loading axes. The cracks differ in outline: (a) circular, radius c; (b) rectangular, $2c \times h$. The First Law of Thermodynamics for an adiabatic change between equilibrium states of the bars (heating is absent) is given by

$$\Delta U = w, \tag{12.1}$$

where ΔU is the change in internal energy of a bar and w is the work performed on a bar (Planck convention). This general expression is now rewritten to focus more specifically on fracture behavior.

In the bars of Figure 12.1 the internal energy change has two elements, the change in elastic energy ΔU_{E}, distributed throughout the bar, and the change in surface energy ΔU_{S}, localized to the crack surfaces, such that

$$\Delta U = \Delta U_{\mathrm{E}} + \Delta U_{\mathrm{S}}. \tag{12.2}$$

For an adiabatic change of state of the applied loading mechanism giving rise to the forces F, the First Law gives a similar equation to Eq. (12.1),

Particle Strengths: Extreme Value Distributions in Fracture, First Edition. Robert F. Cook.

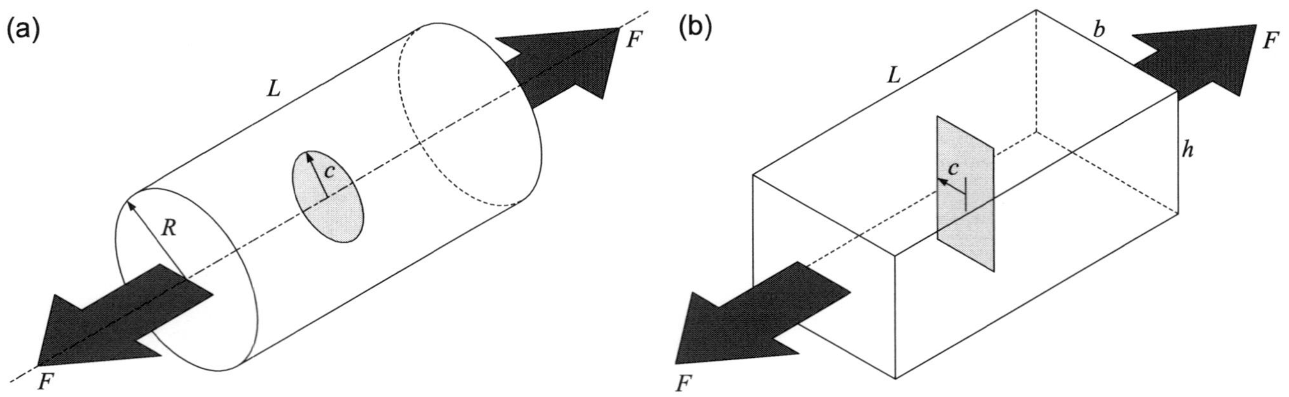

Figure 12.1 Schematic diagrams of loaded bars containing cracks under uniform tensile stress. (a) Circular cross-section bar containing crack with circular perimeter, radius *c*. (b) Rectangular cross-section bar containing through crack with straight perimeter, semi-width *c*.

$$\Delta U_A = w_A, \tag{12.3}$$

where ΔU_A is the change in internal energy of the applied loading mechanism and w_A is the work performed on the mechanism. If the applied loading and a bar are coupled to form a composite system such that the applied loading does work on a bar, $w = -w_A$, and thus

$$w = -\Delta U_A. \tag{12.4}$$

Combining Eqs. (12.2) and (12.4) into Eq. (12.1) gives a relation between the energy changes on transformations between equilibrium states of the composite bar-applied loading system:

$$\Delta U_E + \Delta U_A = -\Delta U_S. \tag{12.5}$$

The *mechanical energy* change ΔU_M is defined by

$$\Delta U_M = \Delta U_E + \Delta U_A, \tag{12.6}$$

such that the First Law equilibrium statement of Eq. (12.1) can be re-written for consideration of fracture as

$$-\Delta U_M = \Delta U_S. \tag{12.7}$$

Specifying these changes in terms of crack length c, applied force F, and component geometry L, R, b, and h is one of the major goals of fracture mechanics.

In particular, the application of fracture mechanics to particles is considered here. Figure 12.2 shows schematic diagrams of two loaded particles, illustrating the similarities and differences with the extended component tensile bars of Figure 12.1. The similarities are clear: The particles have defined geometry and loading, diameter D and loaded in diametral compression by forces F indicated in dark gray. The particles differ in shape: (a) spherical; (b) cylindrical, thickness h. Both particles are shown as including cracks, indicated in light gray, in the centers of the particles. The cracks differ in outline: (a) circular, radius c; (b) rectangular, $2c \times h$. Some of the differences are also clear: The cracks in Figure 12.1 are under direct tension, perpendicular to the loading axes. Those in Figure 12.2 lie on particle meridional planes parallel to the loading axes, under indirect tension, opposite in sign to the applied force. As discussed in Chapters 2, 5, and 6, the localized tensile stress in the particles restricts attention to cracks near particle centers, whereas the distributed uniform tensile stress fields in extended components implies that the behavior of all included cracks should be considered. Thus, a less obvious difference between cracks in particles and those in extended components is that stress fields in particles vary from tensile at the particle center to compressive adjacent to the loading contacts at particle poles. As a consequence, cracks in particles experience decreasing stress fields on extension from the center. Establishing fracture mechanics analyses to examine the consequences of such varying stress fields is a goal of this chapter. As a foundation, the following section considers cracks loaded in uniform tension.

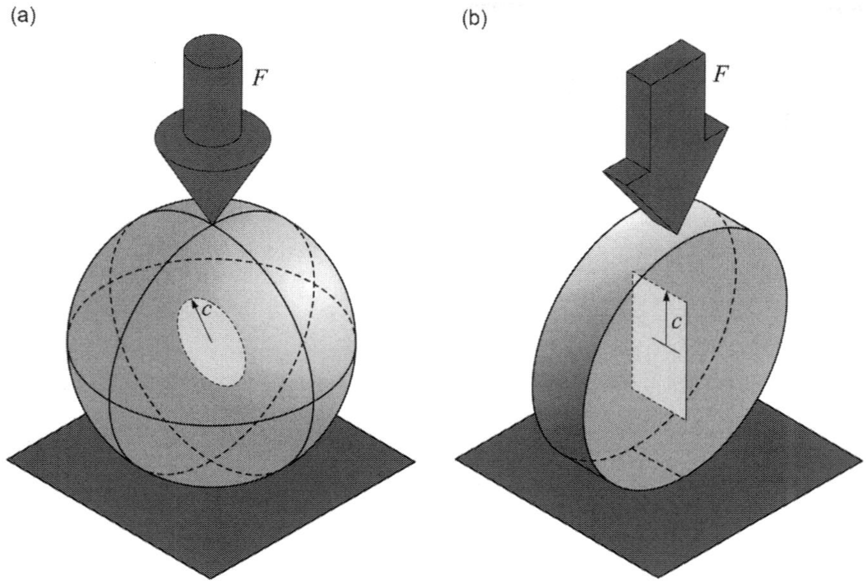

Figure 12.2 Schematic diagrams of particles in diametral compression containing cracks placed in tension. (a) Spherical particle containing crack with circular perimeter, radius *c*. (b) Cylindrical particle containing through crack with straight perimeter, semi-height *c*.

12.2 Uniform Loading

12.2.1 Work and Elastic Energy

Figure 12.1 shows circular and rectangular section bars loaded homogeneously by forces F applied uniformly over the cross sectional areas, πR^2 and bh, respectively (unless stated otherwise $bh = \pi R^2$ is assumed). The bars are composed of linear elastic material, Young's modulus E. As shown in the schematic diagram of Figure 12.3, loading the bars in the absence of cracks from zero force, state O, to a peak force F, state A, leads to a linearly increasing end displacement u to peak value u_0. The linear responses are characterized by $u = \lambda F$ or $F = ku$ where λ is the compliance and $k = 1/\lambda$ is the stiffness. For the uncracked bars the compliance values are $\lambda_0 = L/\pi R^2 E$ or $\lambda_0 = L/bhE$ and are the reciprocal of the slope of the line OA in Figure 12.3. The work w_0 performed by the applied loading in extension of the uncracked bars is the area enclosed by OAB in Figure 12.3, $w_0 = Fu_0/2$. As the system is linear elastic, the work generates identical elastic energy in the bars, $U_{\mathrm{E0}} = Fu_0/2$ (taking the initial condition $u = F = U_{\mathrm{E}} = 0$).

In the presence of cracks, many systems also exhibit linear force-displacement behavior, identical qualitatively but differing quantitatively, from uncracked systems. Such behavior is the basis of linear elastic fracture mechanics, and is assumed here. In this case, loading the cracked bars from zero force to peak value F, with no change in c, leads to an increasing linear end displacement to peak value u_c, state C, as shown in Figure 12.3. The linear responses in the cracked configuration are characterized by $u = \lambda_c F$ or $F = k_c u$ where λ_c and k_c are the compliance and stiffness of the cracked bars. The work w_c performed by the applied loading in extension of the cracked bars is the area enclosed by OCD in Figure 12.3, $w_c = Fu_c/2$, equal to the total elastic energy in the cracked bars, $U_{\mathrm{EC}} = Fu_c/2$. The difference in elastic energy for the cracked and uncracked bars loaded to identical force is $\Delta U_{\mathrm{E}} = U_{\mathrm{EC}} - U_{\mathrm{E0}} = F(u_c - u_0)/2 = F^2(\lambda_c - \lambda_0)/2$, and it is noted that $\Delta U_{\mathrm{E}} > 0$.

The work performed by the applied loading in the above cases, leading to state A or state C, is for static system configurations, loaded serially, either cracked (0) or uncracked (c). Work can also be performed by the applied loading in transformation from state A to state C for a dynamic system configuration in which the crack length changes from 0 to c in parallel with application of a force. Such a transformation is shown for a constant force path, line AC in Figure 12.3. The work performed by the applied loading on the cracking system in this case is the area enclosed by BACD, $w_F = F(u_c - u_0)$. This value illustrates Clapeyron's theorem, that the work performed on a system in a constant force transformation between two linear elastic states is twice the elastic energy difference between the states. Here, $w_F = 2\Delta U_{\mathrm{E}}$. From Eq. (12.3)

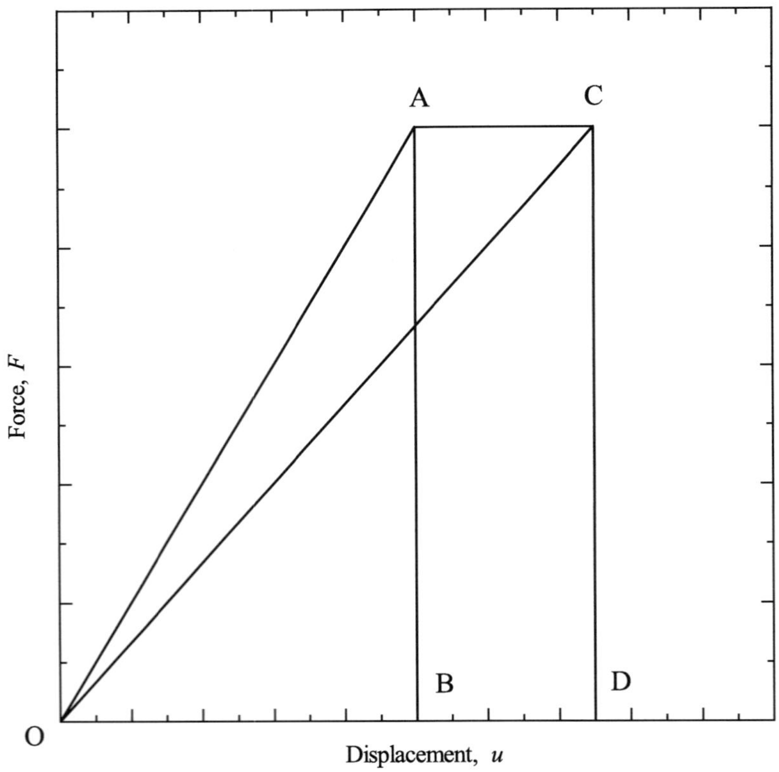

Figure 12.3 Linear elastic force-displacement responses of an uncracked component, path OA, and a cracked component, path OC.

$w_F = -w_A$, and thus from Eq. (12.4) $\Delta U_A = -2\Delta U_E$. From Eq. (12.7), $\Delta U_M = \Delta U_E - 2\Delta U_E = -\Delta U_E$. Description of the mechanical energy of the system in which the crack extends at constant force thus devolves to specification of the change in elastic energy on crack extension.

12.2.2 Mechanical Energy and Surface Energy

Following the discussion in Chapter 2 of intensive and extensive deformation quantities (with some notation changes), the peak tensile stress generated homogeneously in the uncracked cylindrical bar is $\sigma = F/\pi R^2$. Hence the elastic energy in the uncracked bar is given in extensive terms by $U_{E0} = \lambda_0 F^2/2$ and in intensive terms by $U_{E0} = (\sigma^2/2E)(\pi R^2 L)$. The first term in the second expression is recognized as the homogeneous elastic energy density of the uncracked bar, $\mathcal{U}_{E0}$, and the second term is the volume of the bar V, such that $U_{E0} = \mathcal{U}_{E0} V$. For the cracked bar, a simple scaling extension is then

$$U_{EC} = U_{E0}\,[1 + \upsilon(c)/V], \tag{12.8}$$

where $\upsilon(c)$ is the volume of material affected by the crack. Physically, elastic strain energy density is increased in material adjacent to the crack perimeter and decreased in material adjacent to the crack faces such that the increase dominates the decrease. For systems in which the cracks are small, $c \ll R, L$ in the cylindrical bar and $c \ll b, L$ in the rectangular bar, the volume $\upsilon(c)$ scales as the crack dimension alone,

$$\upsilon(c) = (4\pi\psi^2/3)c^3 \tag{12.9}$$

for the circular crack and

$$\upsilon(c) = 2\psi^2 hc^2 \tag{12.10}$$

for the straight crack. ψ^2 is a dimensionless geometrical parameter of order 1 that is different for the circular crack and the straight crack. In both cases,

$$\Delta U_E = U_{E0}(v/V) = \mathcal{U}_{E0}v. \tag{12.11}$$

Eq. (12.11) can be interpreted as representing a cracked bar containing a zone of material centered on the crack in which the strain energy density is greater than the surrounding material. In the case of the cylindrical bar, the zone of material is near spherical. The cross-sectional area of the crack A is given by $A = \pi c^2$, such that v for the circular crack may be expressed in terms of crack area as

$$v(A) = (4\psi^2/3\pi^{1/2})A^{3/2}. \tag{12.12}$$

For the rectangular bar, a near-cylindrical zone of material centered on the straight crack exists, in which the strain energy density is greater than the surrounding material. The cross-sectional area of the crack A is given by $A = 2ch$, such that v for the straight crack may be expressed in terms of crack area as

$$v(A) = (\psi^2/2h)A^2. \tag{12.13}$$

Combining the above expressions for ΔU_M, ΔU_E, $\mathcal{U}_{E0}$, and $v(A)$ gives the mechanical energy change in terms of crack area for cracks extending under constant stress. For a circular crack

$$\Delta U_M = -(2\psi^2/3E\pi^{1/2})\sigma^2 A^{3/2}, \tag{12.14}$$

and for a straight crack

$$\Delta U_M = -(\psi^2/4Eh)\sigma^2 A^2. \tag{12.15}$$

Eqs. (12.14) and (12.15) provide expressions for the term on the left side of fracture equilibrium expression, Eq. (12.7).

The term on the right side of the fracture equilibrium expression is the surface energy change. For both circular and straight crack systems this change can be expressed most simply as a linear proportionality in crack area:

$$\Delta U_S = 2\gamma A. \tag{12.16}$$

The proportionality constant γ is the excess areal density of energy associated with a cracked surface relative to the bulk material. The factor of 2 arises as there are 2 crack surfaces. Usually, the term "areal density" is omitted and implied and the factor of 2 used explicitly, such that in fracture contexts 2γ is usually referred to as the "surface energy." For most materials, $2\gamma \approx (1\text{–}10)\ \text{J m}^{-2}$.

12.2.3 The Griffith Equation

Setting $U_M = U_S = 0$ as initial conditions at $A = 0$, enables the total energy U_T of the loaded cracked system to be expressed in terms of the changes, Eqs. (12.14) or (12.15) and (12.16), as

$$U_T = 2\gamma A - (2\psi^2/3E\pi^{1/2})\sigma^2 A^{3/2} \tag{12.17}$$

or

$$U_T = 2\gamma A - (\psi^2/4Eh)\sigma^2 A^2. \tag{12.18}$$

Figure 12.4 shows a plot of $U_T(A)$ using Eq. (12.17) and the material properties $E = 100$ GPa and $2\gamma = 5\ \text{J m}^{-2}$, the crack geometry term $\psi^2 = 0.25$, and an applied stress $\sigma = 50$ MPa. The equivalent crack length c is shown on the upper axis, using $c = (A/\pi)^{1/2}$. The dashed lines show the asymptotes of the first and second terms, Eqs. (12.16) and (12.14), respectively, and the solid line shows the full response, Eq. (12.17). The total energy passes through a maximum. Setting $dU_T/dA = 0$, equivalent to imposing the equilibrium condition of Eq. (12.7) in the infinitesimal limit, gives the crack area at the maximum of

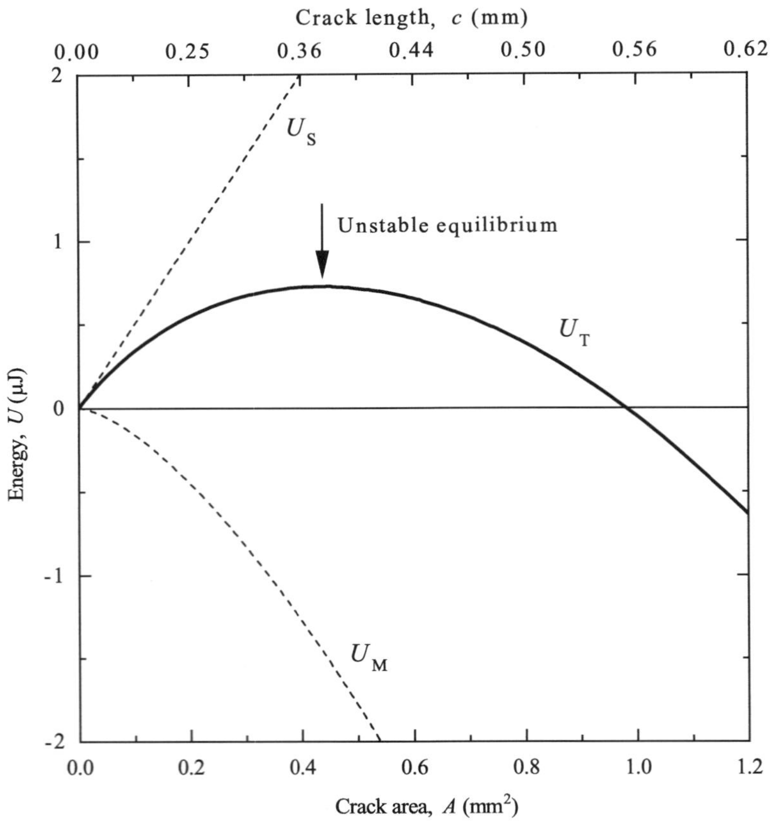

Figure 12.4 Energy vs crack area for a component containing a circular crack in uniform tension. The surface energy U_S and mechanical energy U_M are shown as the dashed lines. The total energy U_T is shown as the solid line and passes through a maximum of unstable equilibrium. Such behavior is the basis of the Griffith equation for brittle fracture strength.

$$A = \left[\frac{\pi^{1/2}}{\psi^2}\right]^2 \left[\frac{2\gamma E}{\sigma^2}\right]^2, \tag{12.19}$$

approximately 0.45 mm^2, and crack length, c, of approximately 0.38 mm. The variation of U_T gives $d^2U_T/dA^2 < 0$ at this point and hence the crack area and length at the maximum is a point of unstable equilibrium, indicated by the arrow in Figure 12.4. An interpretation of Figure 12.4 can be made in terms of progression along the path of fixed force, and thus fixed stress, AC in Figure 12.3. As this path is traversed, the crack grows in area until a point of unstable equilibrium is reached. Perturbations in F or u at this point that cause an increase in crack area lead to monotonic decrease in total energy, runaway crack extension, and component failure. A more intuitive interpretation of Figure 12.4 can be made in terms of progression along the path of fixed crack length, OC in Figure 12.3. As this path is traversed, the crack length is static and the force, and thus stress, increases until a point of unstable equilibrium is reached. Continued increase in F or u at this point lead to monotonic decrease in total energy, runaway crack extension, and component failure. The stress at the point of unstable equilibrium is the strength of the cracked component. Inverting Eq. (12.19) and replacing A with c gives this stress as

$$\begin{aligned}\sigma &= (2E\gamma)^{1/2}/\psi c^{1/2} \\ &= Bc^{-1/2},\end{aligned} \tag{12.20}$$

which is the Griffith equation, Eq. (3.1), used throughout the book (Griffith 1921). The expression for the straight crack is identical.

Equation (12.20) makes clear that $B = (2\gamma E)^{1/2}/\psi$ can be interpreted as the ratio of two terms—both recognized by Griffith. The first is a material properties term involving the resistance to elastic deformation and the excess energy associated with surface creation. The second is a dimensionless geometry term connecting crack length to the volume of stress-enhanced material adjacent to the crack (such that B is different for different shaped cracks in the same material). The ability of Eq. (12.20) to describe experimental observations has been has been demonstrated since Griffith, as noted in Figure 3.2. The value of B evaluated from the geometry and material parameters above and Eq. (12.20) is comparable to that used to describe glass, ceramic, rock, and mineral particle failure throughout this book, $B = 1.4$ MPa m$^{1/2}$. The analysis here differs from that of Griffith in emphasizing the physical scaling of the volume of material perturbed by a crack, $\upsilon(c)$, in favor of evaluation of the numerical value of ψ. Griffith considered straight cracks and used the description developed by Inglis (1913) for the stress field about an elliptical cavity in a plate to estimate ψ. It is recognized that application of the Griffith equation requires that rates of crack retraction in nonequilibrium states are much slower than those of crack extension.

12.2.4 Configurational Forces: $\mathscr{G}$ and R

Configurational forces acting on the crack area A may be defined from the mechanical energy U_M and the surface energy U_S. The *mechanical energy release rate* $\mathscr{G}$ is given by

$$\mathscr{G} = -\frac{dU_M}{dA}. \tag{12.21}$$

The *fracture resistance* R is given by

$$R = \frac{dU_S}{dA}. \tag{12.22}$$

The fracture equilibrium First Law statement of Eq. (12.7) can then be writteng generally as

$$\mathscr{G} = R. \tag{12.23}$$

Equilibrium of a crack system can then be interpreted as the equal and opposite action of two forces on the crack area, one acting to increase the crack area, $\mathscr{G}$, and the other acting to decrease the crack area, R. Similar configurational forces include the Peach-Koehler force and the osmotic force acting on the position of dislocations in plastically deforming materials and the osmotic pressure acting on separated solutions (Hull and Bacon 1984; Lubarda 2019).

For the specific case of the uniform applied stress, the mechanical energy release rate for both the circular and straight cracks can be written as

$$\mathscr{G} = \psi^2\sigma^2 c/E. \tag{12.24}$$

For different or more complicated applied stress or crack geometry configurations, the crack length dependence is more complicated than linear. Similarly, in general R is crack length dependent, $R = R(c)$, reflecting microstructural influences on the resistance to fracture. For the simple case here, R was invariant:

$$R = 2\gamma. \tag{12.25}$$

Stability of a crack system is determined by the relative signs of the derivatives of $\mathscr{G}$ and R. For $d\mathscr{G}/dA > dR/dA$ the system is unstable and for $d\mathscr{G}/dA < dR/dA$ the system is stable. For the case here of uniform stress acting on a crack in a homogeneous material, $d\mathscr{G}/dA > 0$ and $dR/dA = 0$, and the system is unstable.

Section 12.2.1 showed that the change in elastic energy of a system transformed at constant force F between equilibrium configurations was $\Delta U_E = F^2\Delta\lambda/2$, where $\Delta\lambda$ is the change in compliance between the configurations. Noting that for this condition $\Delta U_M = -\Delta U_E$, using Eq. (12.21) gives

$$\mathscr{G} = \frac{F^2}{2}\frac{d\lambda}{dA} \tag{12.26}$$

In the analysis of uniform stress considered above, the compliance of the cracked bar can be expressed similarly to the elastic energy as $\lambda_c = \lambda_0(1 + \upsilon/V)$. Applying Eq. (12.26) and the relevant equations above thus leads to $\mathscr{G}$ as in Eq. (12.24). Simple scaling of the compliance by the projected area of the fractured section leads to an increase that scales as $\Delta\lambda \sim A \sim c^2$ and assumes that the stress field in a bar remains homogeneous in the presence of the crack. The increased crack length dependence determined here, $\Delta\lambda \sim \upsilon \sim c^3$, is a consequence of the inhomogeneous stress field generated by the crack.

12.3 Localized Loading

12.3.1 Analysis

A schematic diagram of a loaded crack configuration at the opposite extreme of those in Figure 12.1 is shown in Figure 12.5. The crack is similar to that in Figure 12.1a as it has a circular outline, radius c, is embedded in the center of a large block of material, characteristic dimension L, and has tensile loading perpendicular to the crack plane. However, the crack in Figure 12.5 is loaded by a force F that is applied over a very small region at the center of the crack, radius a, and the geometry is known as a center-loaded penny crack. Similar to the cracks in Figure 12.1, the elastic strain energy in the block of material in Figure 12.5 increases as c increases at fixed F, but with decreasing crack length derivative rather than increasing. Hence, in contrast to the analysis above, focused on describing the unstable $F(c)$ configuration that defined component strength, attention here is focused on describing the *stable* $c(F)$ configuration that defines initial flaw size. In developing this description, the analysis uses energy scaling ideas first articulated by Roesler (1956). These ideas are incorporated into the framework and notation used earlier.

Analysis begins by recognizing that, as $c \ll L$, the characteristic stress σ in the fracture system of Figure 12.5 varies as $\sigma \sim F/c^2$. Similarly, the volume of material v surrounding the penny crack affected by the force F, and thus stress σ, varies as $v \sim c^3$. A characteristic strain energy density in the system is thus $\sigma^2/2E \sim (F/c^2)^2/E$ and a characteristic energy is $(F^2/Ec^4)(c^3) = F^2/Ec$. Noting that $A \sim c^2$, scaling of the mechanical energy change on crack extension can then be expressed as

$$\Delta U_{\mathrm{M}} \sim \frac{F^2}{EA^{1/2}} - U_{\mathrm{a}}, \tag{12.27}$$

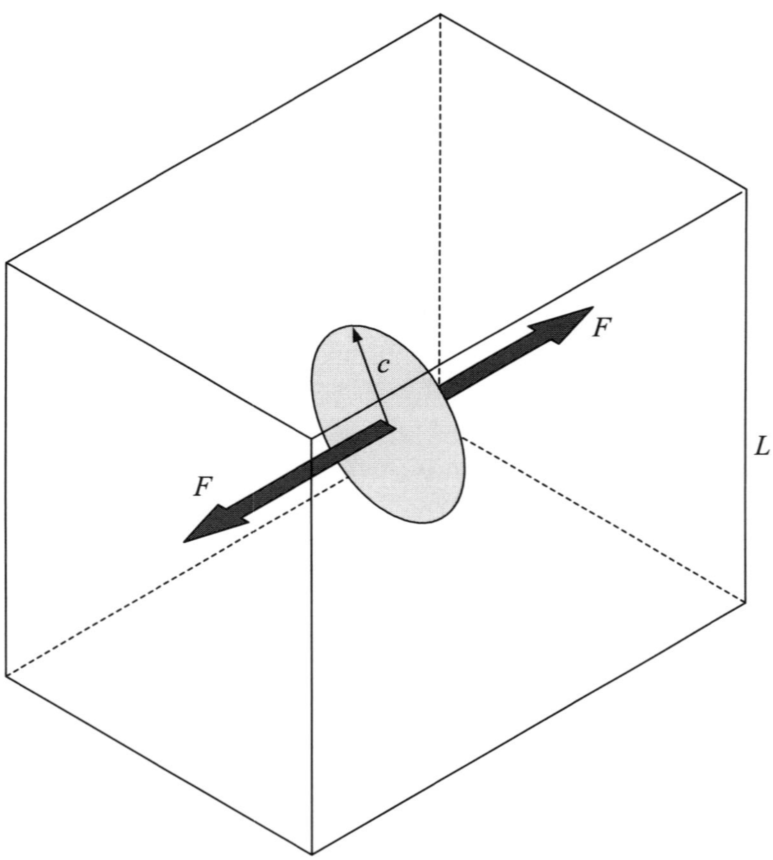

Figure 12.5 Schematic diagram of a block of material containing a crack loaded by a localized force. The crack has circular perimeter, radius c, and the fracture configuration is known as a center loaded penny crack.

where U_a is a maximum change in mechanical energy for large cracks and given by $U_a \approx F^2/Ea$. Note that the scaling of Eq. (12.27) leads to $\Delta U_E \sim U_a - F^2/EA^{1/2}$ and that the signs of all quantities remain unchanged from those above. Applying Eq. (12.21) in Eq. (12.27) gives

$$\mathcal{G} = -\frac{dU_M}{dA} \sim \frac{F^2}{EA^{3/2}} = \frac{\chi^2 F^2}{Ec^3}, \tag{12.28}$$

where the last term is expressed in terms of crack length c and a proportionality term χ^2 that includes all geometrical factors.

Setting $U_S = 0$ at $A = 0$ and $U_M = U_{max}$ at a bounding large crack area $A \approx L^2$ enables the total energy U_T of the loaded cracked system to be expressed in terms of the changes, Eqs. (12.27) and (12.16), as

$$U_T = 2\gamma A + (2\pi^{3/2}\chi^2/E)F^2 A^{1/2} - U_{max}. \tag{12.29}$$

Figure 12.6 shows a plot of $U_T(A)$ using Eq. (12.29) and the material properties $E = 100$ GPa and $2\gamma = 5$ J m^{-2}, and the χ and F terms adjusted as described below. The equivalent crack length c is shown on the upper axis, using $c = (A/\pi)^{1/2}$. The dashed lines show the asymptotes of the first and second terms, Eqs. (12.16) and (12.27), respectively, and the solid line shows the full response, Eq. (12.29). The total energy passes through a minimum. Setting $dU_T/dA = 0$, equivalent to imposing the equilibrium condition of Eq. (12.7) in the infinitesimal limit, gives the crack area at the minimum of

$$A = \left[\frac{\pi^{3/2}(\chi F)^2}{2\gamma E}\right]^{2/3}, \tag{12.30}$$

where the χF product was adjusted to give the minimum at a crack area of approximately 0.45 mm^2, and crack length, c, of approximately 0.38 mm, similar to that of the maximum in Figure 12.4. The variation of U_T has $d^2U_T/dA^2 > 0$ at this point

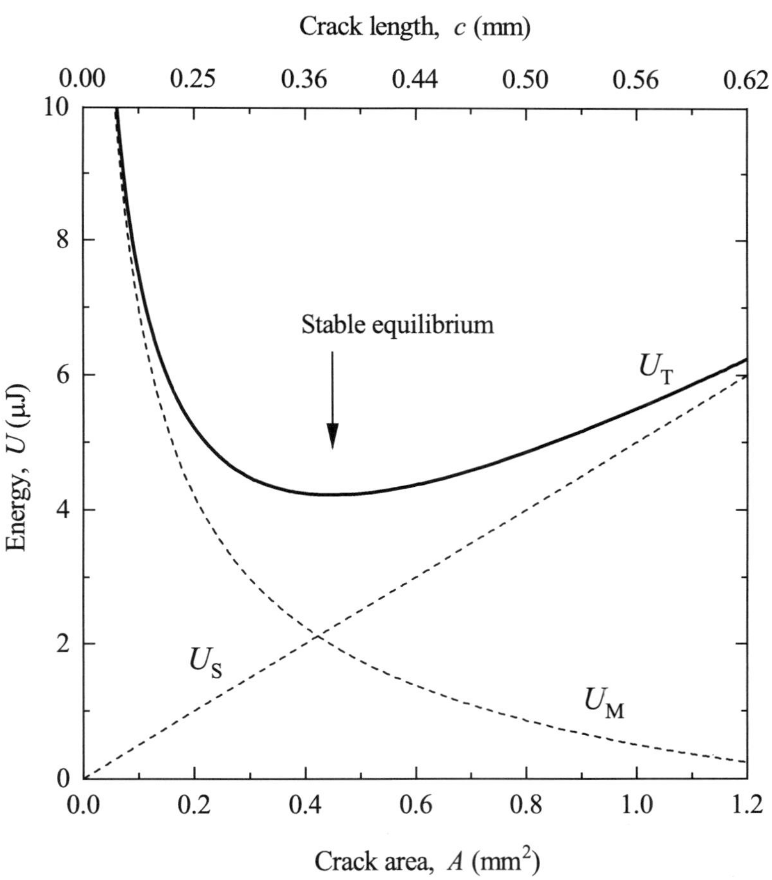

Figure 12.6 Energy vs crack area for a component containing a circular crack loaded by a localized force. The surface energy U_S and mechanical energy U_M are shown as the dashed lines. The total energy U_T is shown as the solid line and passes through a minimum of stable equilibrium. Such behavior is the basis of the Roesler equation for brittle crack lengths.

and hence the crack area and length at the minimum is a point of stable equilibrium, indicated by the arrow in Figure 12.6. An interpretation of Figure 12.6 can be made in terms of progression along the path of fixed force, AC in Figure 12.3. As this path is traversed, the crack grows in area until a point of stable equilibrium is reached. Perturbations in F or u at this point that cause an increase in crack area lead to increases in total energy and return of the crack to the equilibrium point.

The crack length as a function of the applied force at the point of stable equilibrium is a key experimental observation that provides insight into the likely size of strength controlling flaws in a material. Replacing A with c in Eq. (12.30) gives this crack length as

$$c = \left[\frac{\chi F}{(2E\gamma)^{1/2}} \right]^{2/3}, \tag{12.31}$$

a result found by Roesler (1956). This equation and the Griffith equation, Eq. (12.20), stand at opposite ends of fracture scaling. In the case here, the load is applied over an area small relative to the crack, leading to fracture stability. In the Griffith case, the load is applied over an area large relative to the crack, leading to fracture instability. In both cases, the product $(2\gamma E)^{1/2}$ appears as a proportionality constant linking crack length to either force or stress.

12.3.2 Examples

The most common method of assessing Eq. (12.31) is through indentation testing, in which the flat surface of a block of material is loaded by a probe to a specified force F. The probe may be spherical, leading to "blunt" indentation in which the contact is elastic and described by Hertzian stress fields (Johnson 1985; Lawn 1998), or conical or pyramidal, leading to "sharp" indentation in which the contact is elastic-plastic and described by a combination of the Boussinesq field and a blister field (Yoffe 1982; Johnson 1985; Cook and Pharr 1990). Figure 12.7 shows schematic cross-section diagrams of (a) blunt and (b) sharp indentations. In Figure 12.7a, a cone crack is generated, with truncated apex at the contact site. In Figure 12.7b, a set of perpendicular half-penny cracks is generated, emanating from a near hemi-spherical zone of compaction and plastic deformation. Both the cracks and the zone are centered at the contact site. In both geometries, the diameter of the cracks is taken as $2c$ as shown, noting that the dimension for the cone crack is sub-surface. Flat punches of hardened metal are often used in blunt indentation testing and Vickers diamond pyramids are often used in sharp indentation.

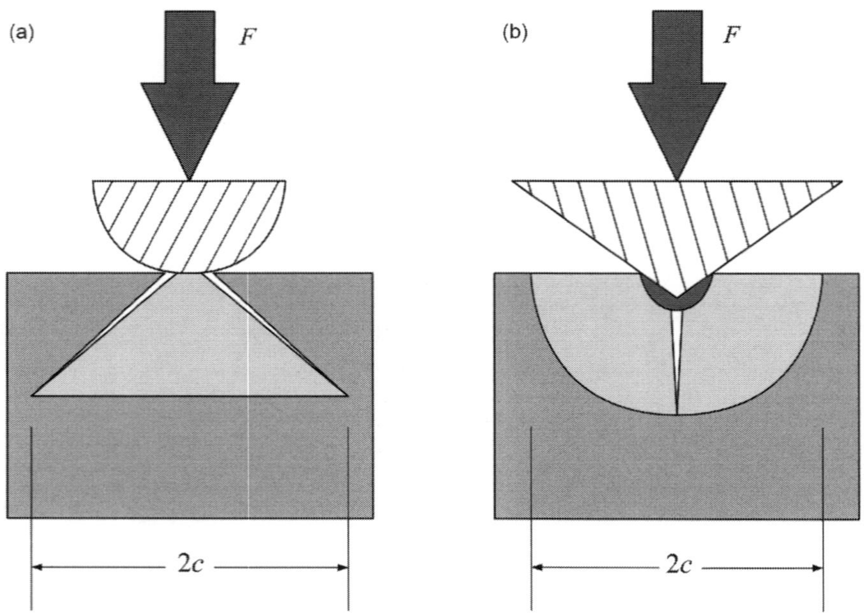

Figure 12.7 Schematic cross-section diagrams of indentation cracks generated by contact forces F. (a) Blunt, spherical indentation generating cone cracks of diameter $2c$. (b) Sharp, pyramidal or conical indentation generating plastic deformation zone (shaded dark) and half-penny cracks of diameter $2c$.

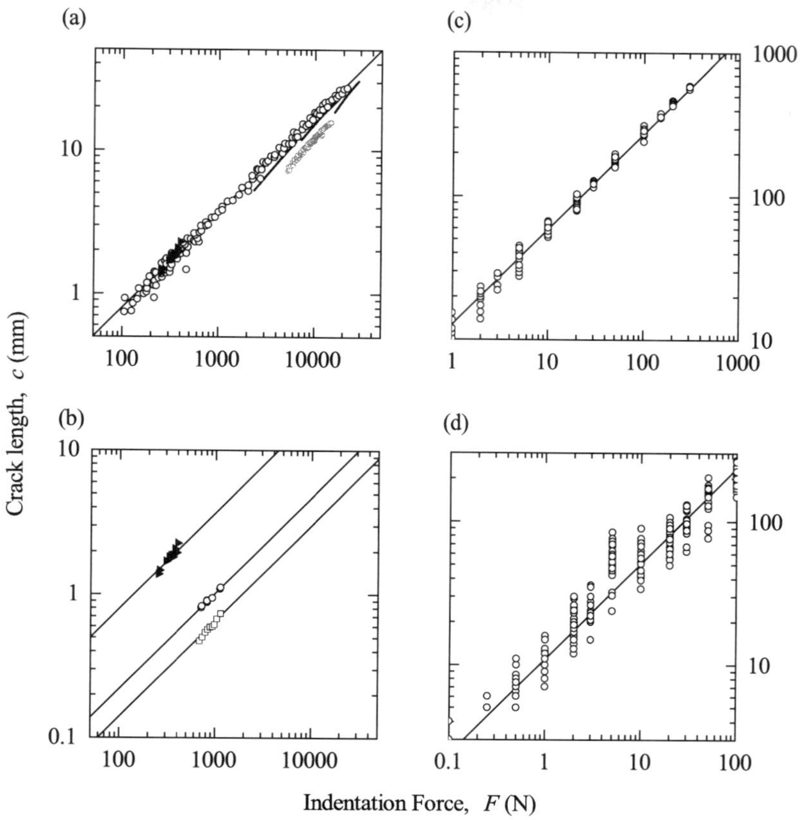

Figure 12.8 Crack length c as a function of contact force F for cone cracks and half-penny cracks. (a) Cone cracks in glass (Adapted from Roesler F. C 1956; Benbow, J. J 1960; Lawn, B. R et al 1975; Zeng, K et al. 1992a, 1992b). (b) Cone cracks in glass and ceramics (adapted from Zeng, K et al. 1992b). (c) Half-penny cracks in cordierite glass-ceramic. (d) Half-penny cracks in polycrystalline alumina.

Figure 12.8 shows logarithmic plots of measurements of stable crack length as functions of indentation force, $c(F)$, for a range of contact systems, using data derived from the published works cited. The plots are arranged with cone crack measurements on the left and half-penny crack measurements on the right. The plots are all at the same scale, vertical × horizontal of 100 × 1000. The fine solid lines are empirical guides to the eye of slope 2/3, consistent with Eq. (12.31). Figure 12.8a shows cone crack measurements in fused silica and glass indented with flat punches or WC spheres, (0.4–6.4) mm in diameter, from the works of Roesler (1956), Benbow (1960), Lawn and Fuller (1975) and Zeng et al. (1992a, 1992b). Open black symbols represent individual in situ measurements using a range of punches on fused silica (Benbow 1960). Bold solid lines represent in situ measurements using a single punch for each line shown on glass (probably soda-lime silicate, Roesler 1956). Open gray symbols represent individual in situ measurements in vacuum using a WC flat punch on soda-lime silicate glass (Lawn and Fuller 1975). Solid symbols represent ex situ post (spherical probe) contact measurements on polished sections of soda-lime silicate glass (Zeng et al. 1992b). The four sets of data are well described by the 2/3 trend lines and there is agreement betweeen the sets, with the exception that the vacuum observations of Lawn and Fuller exhibited shorter cracks at a given indentation force. The shortening reflects the environmental influence discussed in Chapter 2—crack lengths in these experiments were determined by inert conditions rather than a reactive moist air environment. A point to note is that cone crack systems are large: Forces of 100 N to over 20 kN are required to generate millimeter scale cracks. As the contacts are elastic, the cracks close on force removal, practically limiting extensive measurements to transparent materials under force. Figure 12.8b illustrates this effect, showing ex situ cone crack measurements in glass and two ceramics indented with WC spheres from the work of Zeng et al. (1992b). The solid symbols repeat the glass data from Figure 12.8a. The open symbols show the responses of a fine-grained Al_2O_3 (upper symbols) and an Al_2O_3-SiC whisker composite (lower)—measurements in these opaque ceramics are clearly restricted to ex situ techniques. For similar force domains the ranges of cone crack lengths in the ceramics are much smaller than that of glass. The 2/3 trend describes the limited ranges of data (from which Zeng et al. estimated material fracture properties, see Section 12.4).

Figure 12.8c shows half-penny crack measurements in a fine-grained glass ceramic indented with Vickers pyramids (Cook 2020). This is the same material represented in the strength measurements of Figure 3.2. Symbols represent individual measurements. The width of the force domain (a factor of 300), the relative dispersion of crack lengths at a given force (about a factor ± 20 %), and the description of the data by the 2/3 trend are all similar to those of Figure 12.8a. The obvious point of difference to note is that half-penny crack systems are small: Forces of 1 N to 300 N generate cracks tens of micrometers in scale. A characteristic of sharp indentations is that the contacts are elastic-plastic, generating localized zones of plastic deformation on loading that continue to force the surface traces of the half-penny cracks open on unloading. As a consequence, measurements are extended to opaque materials after force removal. Figure 12.8d shows half-penny crack measurements in a large-grained polycrystalline alumina indented with Vickers pyramids (Cook 2020). Symbols represent individual measurements. The data are described by the 2/3 trend, although there is a great deal of relative dispersion in the crack lengths at a given indentation force (about a factor ± 50 %) and scatter about the 2/3 trend is significant. An extensive survey (Cook 2020) shows that the results of Figure 12.8c and 12.8d are representative of sharp indentation cracks in a wide range of materials: the 2/3 trend is observed over wide domains of contact force and dispersion increases with microstructural scale. In addition to providing experimental support for the fracture mechanics considerations leading to Eq. (12.31), the observations of Figure 12.8 have several implications for particle strengths.

The mechanics of indentation of a flat surface by a spherical probe are similar to those of diametral compression of a sphere by a flat platen, as noted in Chapter 2. As a consequence, it is possible to generate cone crack fractures in particles on loading. Such cracks form at the poles of particles that are resistive to meridional fracture and thus failure from central or equatorial flaws. Figure 12.9a shows a schematic cross section diagram of cone crack generation in a spherical particle loaded by a flat. The morphology is similar to that of Figure 12.7a with the exception that the crack propagates nearly parallel to the particle surface leading to fine slivers of material attached to the bulk of the particle by thin sections. Figure 12.9b shows an image of a cone crack generated fracture surface at the pole of a glass particle, about 12 mm in diameter, loaded to a peak force of approximately 2 kN. On unloading, although the cone cracks were visible, the particle was intact. Slight vibration of the particle after loading lead to fracture of the thin sections and formation of the surface cone. The similarity of the fracture surface and the schematic diagram is clear, as is the millimeter scale of the cone to the measurements of Figure 12.8a. Similar images were shown by Gorham and Salman (1999) of a cone formed in a glass plate indented by a WC sphere and a cone formed in glass sphere by impact against a flat anvil Gorham and Salman (2005). During loading of the glass spheres to greater forces here, failure occurred by near meridional fracture from the contact pole to the exterior surface

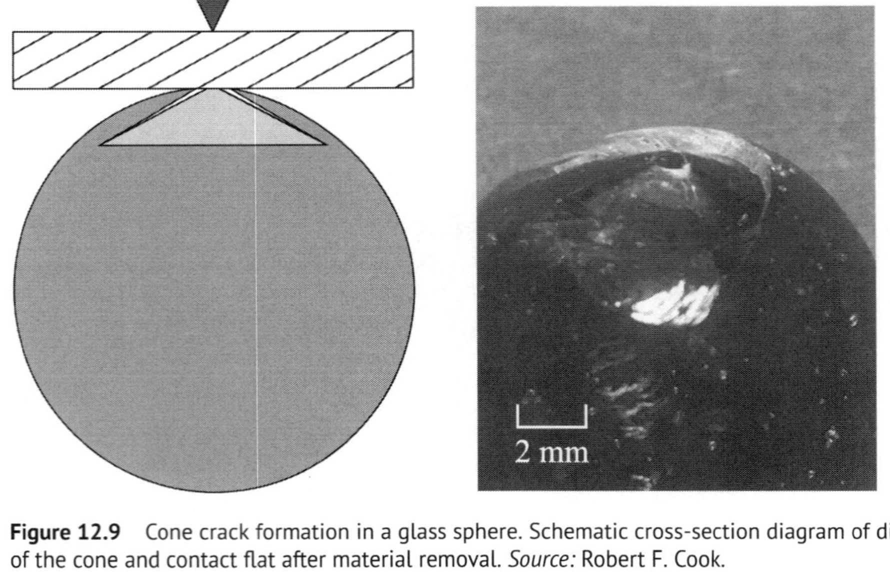

Figure 12.9 Cone crack formation in a glass sphere. Schematic cross-section diagram of diametral loading by a flat platen. (b) Image of the cone and contact flat after material removal. *Source:* Robert F. Cook.

below the particle equator. This was the failure mechanism investigated by Brzesowsky et al. (2011) for the submillimeter quartz sand particles shown in the strength plot of Figure 9.10. The characteristic stress for the cone-crack initiated failure process here can be estimated from the peak force and the contact area of the remaining contact flat in Figure 12.9b: $\sigma \approx 2\ \text{kN}/(0.7\ \text{mm})^2 \approx 4$ GPa, corresponding to $c = (B/\sigma)^2$ surface flaws of 30 nm using $B = 0.7\ \text{MPa m}^{1/2}$ for glass. The characteristic stress value is comparable to the strengths estimated by Brzesowsky et al. (2011) for surface failure of sand grains (Figure 9.10 calculates the center stress at failure). The implied surface flaw dimension is consistent with the particle size and the sub-wavelength-of-light flaw dimension is consistent with the shiny glass surface.

The stress developed in the center of the glass particle in Figure 12.9b at failure, using the HO equation, was about 12.5 MPa. This value is far smaller than the quartz particle strengths reported by Brzesowsky et al. or reported for the extensive sets of glass sphere tests considered here earlier, Figure 9.19 and following. In addition, an estimate of the strength-controlling flaw size, assuming this stress as the strength (using $B = 0.7\ \text{MPa m}^{1/2}$), is approximately 12.5 mm, which is clearly unphysical. The implication is that here surface flaws associated with cone crack initiation in the Hertzian contact field at the particle pole were strength controlling. The strength results of Brzesowsky et al., Kschinka et al. (1986), and Watkins and Prado (2015), all of which exhibit characteristic minimum central strengths of approximately 100 MPa, can be interpreted as cone crack failures in the same way. The glass sphere strength results of Huang et al. (2014) and Shan et al. (2018), which exhibit characteristic maximum central strengths of approximately 100 MPa, can not be interpreted so easily. An estimate of the strength controlling flaw size assuming this strength is $(0.7/100)^2 \approx 50\ \mu\text{m}$. This dimension is comparable to the pores visible in the glass interior in Figure 12.9b. As noted in Chapter 2, significant meridional tensile stresses are generated on the periphery of the equatorial plane of a diametrally loaded spherical particle ($\sigma_{\theta\theta}$ in Figure 2.22). Such tensile stresses are approximately 0.8 of the generated central tensile stress. Hence, an estimate of the strength controlling flaw size for equatorial surface failure, assuming a strength of 80 MPa and a B value for glass in moist air of $0.5\ \text{MPa m}^{1/2}$, is $(0.5/80)^2 \approx 40\ \mu\text{m}$. This dimension is comparable to that of the sharp contact flaws shown in Figure 12.8b, Figure 12.8c, and in many other materials (Cook 2020). Glass sphere failure in these cases could thus be determined by strength-controlling flaws either in the bulk or on the surface.

The above discussion highlights the importance of center loaded flaws in determining particle strengths. For particles that have small pre-existing flaws, e.g. pores, sharp contact induced half-penny cracks can lead to particle failure at small stresses and blunt contact induced cone cracks can lead to particle failure at large stresses. The following sections addresses particle strengths in greater detail by extending the above fracture analyses to address the superposed effects of uniform applied stress and local center loading (Figures 12.1 and 12.5 combined). This superposition is a very good model for strength testing of particles containing the commonly encountered strength-controlling flaws associated with sharp surface contacts or with misfitting bulk inclusions. Superposed loadings and variable stress fields (e.g. at pores and inclusions) are more easily analyzed from a fracture mechanics stress viewpoint. The definitions and terms used in this viewpoint and the relation to the energy viewpoint above are discussed first.

12.4 Spatially Varying Loading

12.4.1 Stress-Intensity Factor and Toughness

The stress distribution surrounding cracks in materials that behave as linear elastic continua is a universal function of relative crack tip coordinates. Figure 12.10 shows a schematic cross section of a planar crack in such a material, along with a crack tip cylindrical polar (r, θ, z) coordinate system and a representative material element in that system. The element is located at a distance r from the crack tip and at angle θ relative to an axis in the crack plane. The crack is loaded in the far field, remote from the crack tip, indicated by the representative dark arrow. The components of the stress tensor, σ_{ij}, for the element are given by

$$\sigma_{ij} = \frac{K}{(2\pi r)^{1/2}} f_{ij}(\theta), \tag{12.32}$$

where $f_{ij}(\theta)$ depends on ij. The derivation of Eq. (12.32), based on complex potential analysis, is given elsewhere (e.g. Broek 1982; Kanninen and Popelar 1985). An important aspect of Eq. (12.32) is that the amplitude, radial, and angular functional components are separable. The radial component decreases with increasing distance from the crack tip (and formally diverges at the tip, a consequence of the continuum assumption), reinforcing the energy-based ideas above that strain energy changes associated with fracture are localized to the crack. The tensor angular functions are listed in many

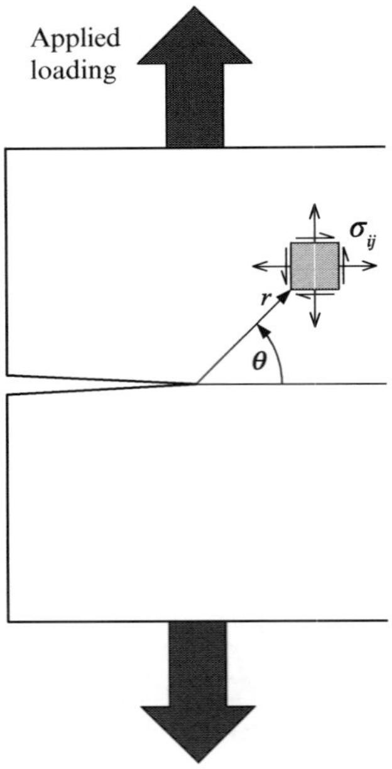

Figure 12.10 Schematic cross-section diagram of a loaded planar crack illustrating the cylindrical polar (r, θ) crack tip coordinates and stress field components σ_{ij}.

works (e.g. Broek 1982; Williams 1984; Kanninen and Popelar 1985; Atkins and Mai 1985; Lawn 1993) and in many cases $f_{ij}(0) = 1$.

The scalar amplitude component K in Eq. (12.32) is known as the *stress-intensity factor* (SIF). K includes all information regarding the boundary conditions of loading—the magnitude, orientation, and location of all applied forces and displacements, and, importantly, the crack length c. As $K(c)$ describes the crack tip stress field in a linear elastic material and is a linear stress amplitude term, $K(c)$ fields arising from different sources of applied loading on a crack are linearly additive. As might be anticipated, there is a close connection between K and $\mathcal{G}$. The connection between the two is known as the Irwin relation, derivable in a number of ways (see e.g. Lawn 1993), and is given by

$$\mathcal{G} = \frac{K^2}{E}. \tag{12.33}$$

Eq. (12.33) has the familiar structure of an energy term related to the ratio of a stress term squared and a modulus.

$K(c)$ functions for a wide variety of loading configurations are listed in several compendia (e.g. Sih 1973; Tada et al. 1973; Murakami 1987). Fracture mechanics analysis for a particular application typically involves selecting and adding the $K(c)$ SIF function or functions appropriate to the considered fracture configuration, imposing equilibrium and stability conditions, and solving for the quantity of interest (e.g. a force or a crack length). The equilibrium condition in the SIF framework is defined by direct correspondence with that in the energy framework, Eq. (12.23):

$$\mathcal{G} = R.$$

Implementing the Irwin relation on both sides of this equation gives

$$\frac{K^2}{E} = \frac{T^2}{E} \tag{12.34}$$

or

$$K = T, \tag{12.35}$$

where T is the *toughness* of the material, given by $T = (RE)^{1/2}$, and defines the value of the SIF in fracture equilibrium, noting that K is the net SIF. T may depend on crack length, $T = T(c)$, in microstructurally inhomogeneous materials that exhibit toughening, $\mathrm{d}T(c)/\mathrm{d}c > 0$. T may exhibit environmental dependence in materials that react with environmental species. In an inert environment, T takes a maximum value T_0, although the 0 is often implied and omitted. In a reactive environment, often ambient moisture, $T < T_0$. The surface energy is similarly notated, such that for homogenous materials in inert environments $T_0 = (2\gamma_0 E)^{1/2}$.

Using the Irwin relation, the SIF for uniform applied stress σ from Eq. (12.24) is

$$K = \psi\sigma c^{1/2}. \tag{12.36}$$

The SIF for a penny crack center loaded by localized force F from Eq. (12.28) is

$$K = \chi F/c^{3/2}. \tag{12.37}$$

Exact stress analysis gives $\psi = \pi^{1/2}$ for the straight crack and $\psi = 2/\pi^{1/2}$ and $\chi = \pi^{-3/2}$ for the circular crack (Sih 1973; Tada et al. 1973; Murakami 1987).

Imposing the equilibrium condition, $K = T$, in Eq. (12.36) returns the Griffith equation

$$\sigma = (T/\psi)c^{-1/2} \tag{12.38}$$

and the parameter B used extensively throughout the book is recognized as T/ψ. Similarly, imposing the equilibrium condition in Eq. (12.38) returns the Roesler equation

$$c = (\chi F/T)^{2/3}. \tag{12.39}$$

Inversion of this equation gives $T = \chi(F/c^{3/2})$, such that measuring $F/c^{3/2}$ from indentation crack lengths can provide an estimate of material toughness T (if χ is known). This was the approach above of Zeng et al. (1992b) using cone cracks and in many other studies using sharp indentation half-penny cracks in ceramics and glasses (see Cook 2020 for a review). Stability of a crack system is determined by the relative signs of the derivatives of K and T. For $\mathrm{d}K/\mathrm{d}c > \mathrm{d}T/\mathrm{d}c$ the system is unstable and for $\mathrm{d}K/\mathrm{d}c < \mathrm{d}T/\mathrm{d}c$ the system is stable. For the cases here, assuming that T is single valued and thus $\mathrm{d}T/\mathrm{d}c = 0$, the uniform stress system is unstable and the localized force system is stable. The following sections consider examples of the SIF framework that are relevant to strength of particles.

12.4.2 Crack at a Stressed Pore

Figure 12.11 shows a schematic diagram of a cylindrical particle, diameter $D = 2R$, containing a central cylindrical through pore, diameter $2a$. The particle is loaded in diametral compression by force F and the axis of the pore is perpendicular to the compression axis. Two straight cracks on the central meridional plane of the loaded particle extend symmetrically from the pore. With exception of the pore, the diagram is similar to Figure 12.2. The radius of the pore $a \ll R$ and the cracks emanating from the pore are each of length $c \ll R$, such the ensemble of pore + cracks can be considered as embedded in the uniform tensile stress field, $\sigma = 0.9F/D^2$, of the diametrally loaded particle (see Chapter 2). (The geometry is not that of a "ring" test, Chen and Hsu 2001; Zhu et al. 2004; or that of a notched disc, Shetty et al. 1985; Elghazel et al. 2015).

The SIF for the cracks in Figure 12.11 is given by (Murakami 1987)

$$K = \pi^{1/2}\sigma c^{1/2}\left[1 + 1/(2\xi^2 + 1.93\xi + 0.539) + 1/2(\xi + 1)\right],\ \xi = c/a. \tag{12.40}$$

The $K(c)$ response of Eq. (12.40) is shown as the solid line in the logarithmic plot of Figure 12.12. Eq. (12.40) has two crack length limits. For cracks long relative to the pore size, $\xi \gg 1$, $K \rightarrow \pi^{1/2}\sigma c^{1/2}$ consistent with the exact solution for a uniformly stressed crack in stress field σ. For cracks short relative to the pore size, $\xi \ll 1$, $K \rightarrow 3.35\pi^{1/2}\sigma c^{1/2}$ consistent with the exact solution for a uniformly stressed crack in stress field 3.35σ. The factor slightly greater than 3 arises from the familiar stress concentration of 3 at the periphery of a circular cavity in a uniaxially stressed sheet (Sadd 2009), plus the added effect of compliance of a crack on the opposing face of the pore. The two limits are indicated by dashed lines in Figure 12.12 and the full solution curves between them in a stabilizing branch highlighted by the shaded region.

Imposing the equilibrium condition, noting that the instability condition is satisfied throughout the crack length domain, although weakened in the shaded band, enables the strength to be expressed as

$$\sigma = (Ta^{-1/2})f(\xi), \tag{12.41}$$

where the first term is a characteristic strength formed by the product of the material toughness and the size of the strength controlling pore and the second term is a dimensionless function of the crack/pore aspect ratio, decreasing as the aspect ratio increases. Expressions similar to Eqs. (12.40) and (12.41) were developed and applied by Green (1980) and Chao and

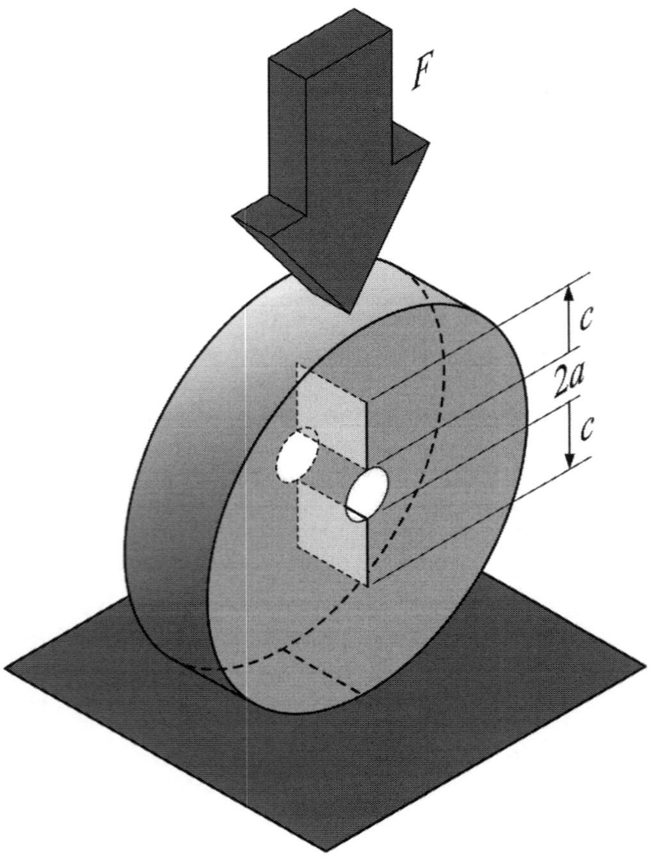

Figure 12.11 Schematic diagram of a through pore and meridional cracks in a diametrally loaded cylindrical particle.

Shetty (1992) to describe annular cracks surrounding spherical pores. Strength variability in brittle particles with strength-controlling flaws consisting of pore + crack combinations is interpreted from Eq. (12.41) as arising primarily from two separate effects: variability of the pore size, a, and variability of the crack/pore aspect ratio, ξ (variability in toughness T is usually minor; considered below). Usually, distributions in a are measurable and strength variability is interpreted in terms of likely fixed values of ξ and thus $f(\xi)$ (Chao and Shetty 1992). Here the opposite approach is taken, a is assumed fixed and and variations in c and thus ξ and $f(\xi)$ are considered. Figure 12.13 shows strength edf $Pr(\sigma)$ behavior in semi-logarithmic coordinates for three uniform distributions of crack lengths in domains described by Eq. (12.41) and Figure 12.12. The first domain considered is the shaded region in Figure 12.12 and the simulated edf resulting from stochastic selection of crack lengths from that domain is shown as the solid symbols. The solid line is a guide to the eye. The next two domains are the unshaded regions, left and right, in Figure 12.12 and the simulated edfs resulting from stochastic selection of crack lengths from these domains are shown as the open symbols, right and left. The solid lines are guides to the eye with the same slope as the central line. It is clear that strengths decrease as crack length increases. However, the major point to be taken from Figure 12.13 is that strength variability is decreased significantly if the behavior of strength controlling cracks is influenced by a stabilizing SIF field. Here the transition from a pore-affected SIF field to a pore-immune SIF field lead to a destabilizing influence and significant increase in strength variability (a factor of three). Comparison of the slope of the solid lines and the simulated edf responses in Figure 12.13 emphasize this point (noting that the logarithmic abscissa enables direct comparison of relative slopes). If the varying crack length/pore size aspect ratio considered here depended on particle size, the resulting strength behavior would be interpreted as a deterministic size effect (Chapter 9), consistent with the varying threshold strengths in Figure 12.13.

This section has illustrated the application of fracture mechanics to determination of particle strengths limited by a common flaw, the pore associated crack. The known SIF could be used directly and the behavior was completely destabilizing, leading to straightforward assessment of strength. The following section also considers the fracture mechanics of a common

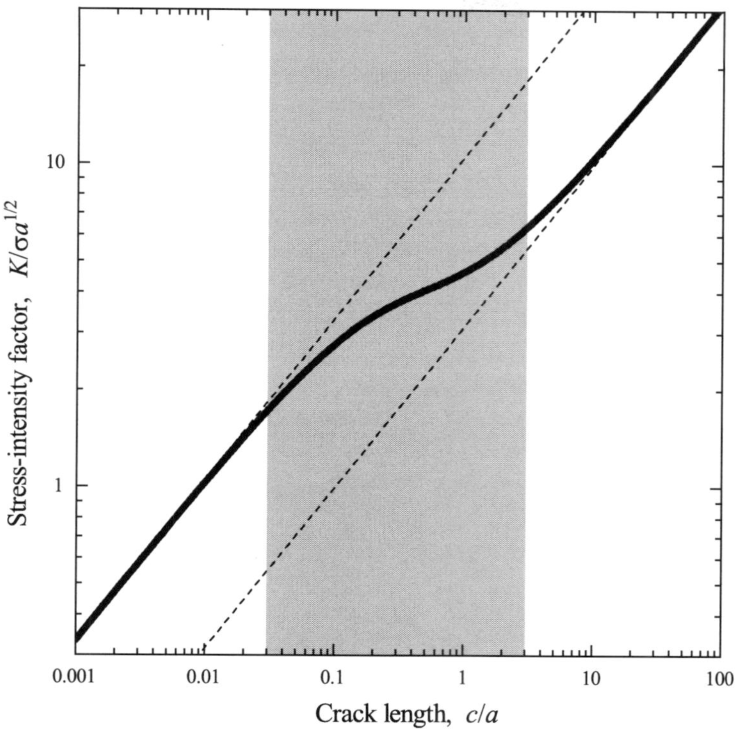

Figure 12.12 Plot of the stress-intensity factor K as a function of crack length c for the cracked pore system of Figure 12.11 (solid line). The asymptotic behavior at short and long crack lengths are shown as the dashed lines. The shaded band in indicates a transition region.

flaw in particles, that of a crack associated with a misfitting expanding inclusion. In this case, the SIF of the inclusion stress field is constructed through the common technique of integration using a weighting function derived from a simpler field. The resulting inclusion SIF field consists of a destabilizing-to-stabilizing transition that describes crack initiation.

12.4.3 Crack at a Misfitting Inclusion

Analysis of fracture at an expanding inclusion begins with consideration of a circular crack, radius c in an infinite body. Tensile forces F are applied on the opposing crack faces along circles of radius a centered on the crack. The forces are applied as uniform line forces, F_L (forces per line length), such that here $F_L = F/2\pi a$. A schematic diagram is shown in Figure 12.14a. The SIF for this crack is (Sih 1973; Tada et al. 1973; Murakami 1987)

$$K = \frac{2F_L}{(\pi c)^{1/2}} \frac{a}{(c^2 - a^2)^{1/2}}. \tag{12.42}$$

Cracks very large relative to the radius of the loaded circles are given by $c \gg a$. Taking this limit and replacing F_L with the expression in F gives Eq. (12.37) with the ideal value of χ, $K = F/(\pi c)^{3/2}$. For large cracks, the details of the loading configuration of F are unimportant and the fracture system behaves as a center-loaded penny crack.

A fracture configuration related to Figure 12.14a is that of a crack loaded by a uniform tensile stress σ_a over discs of radius a, as shown in the schematic diagram of Figure 12.14b. The SIF for this configuration is obtained by recognizing that the uniform stress can be re-cast as a series of line forces $dF_a = \sigma_a dr$ over the radial domain $0 \leq r \leq a$, where dr is an infinitesimal line width. An infinitesimal contribution dK to the disc loaded crack is then, from Eq. (12.42),

$$dK = \frac{2\sigma_a dr}{(\pi c)^{1/2}} \frac{r}{(c^2 - r^2)^{1/2}}. \tag{12.43}$$

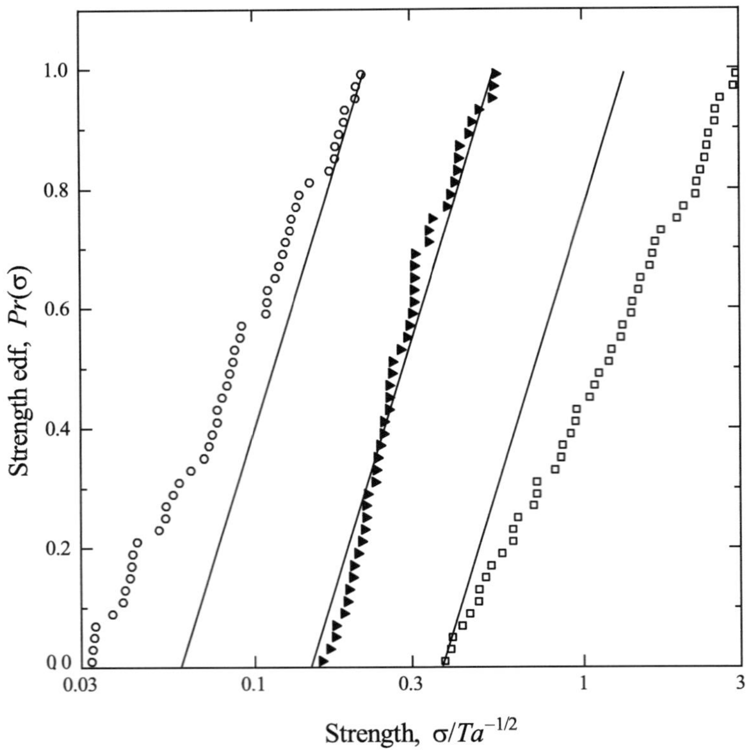

Figure 12.13 Simulated strength edf $Pr(\sigma)$ variations for components containing strength-controlling cracked pores. Filled symbols indicate cracks within the transition region of Figure 12.12. Left and right open symbols indicate cracks longer and shorter than the transition region—note the increased strength dispersion.

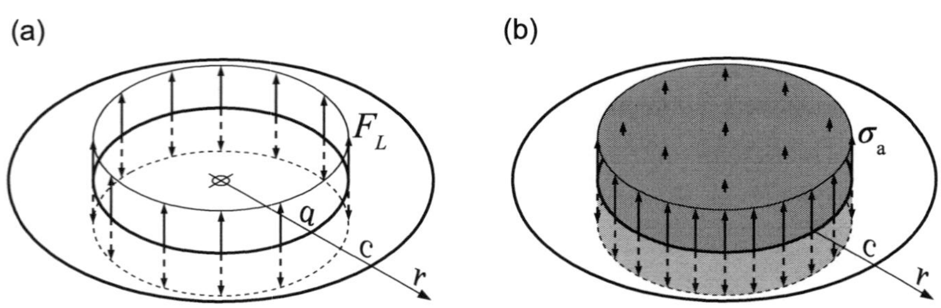

Figure 12.14 Schematic diagrams of loaded circular cracks, radius c. (a) Crack loaded by a line force F_L on a circle radius a. (b) Crack loaded by uniform tensile stress σ_a over a disc radius a.

The total SIF arising from the uniformly loaded discs is then given by a standard integral

$$K = \int_0^a \mathrm{d}K = \frac{2\sigma_a}{(\pi c)^{1/2}} \int_0^a \frac{r\mathrm{d}r}{(c^2 - r^2)^{1/2}}$$

$$= \frac{2\sigma_a}{(\pi c)^{1/2}} \left[c - (c^2 - a^2)^{1/2}\right], \tag{12.44}$$

in agreement with known SIF solutions (Sih 1973; Tada et al. 1973; Murakami 1987). For the condition $a = c$, the crack is completely uniformly stressed and Eq. (12.44) devolves to Eq. (12.36) with the ideal value of ψ for a circular crack, $K = (2/\pi^{1/2})\sigma_a c^{1/2}$. For cracks very large relative to the loaded disc, $c \gg a$. Taking this limit and replacing σ_a with $F = \sigma_a \pi a^2$ again gives Eq. (12.37) characterizing a center-loaded penny crack with the ideal value of χ, $K = F/(\pi c)^{3/2}$.

Figure 12.15a shows a schematic diagram of a spherical inclusion, radius a in an infinite body, surrounded by an annular crack of radial width c. The combined inclusion + crack system constitutes a flaw with an extent given by a radius $e = a + c$. In this case, the crack dimension c is not measured from the center of the flaw but from the edge of the inclusion, as

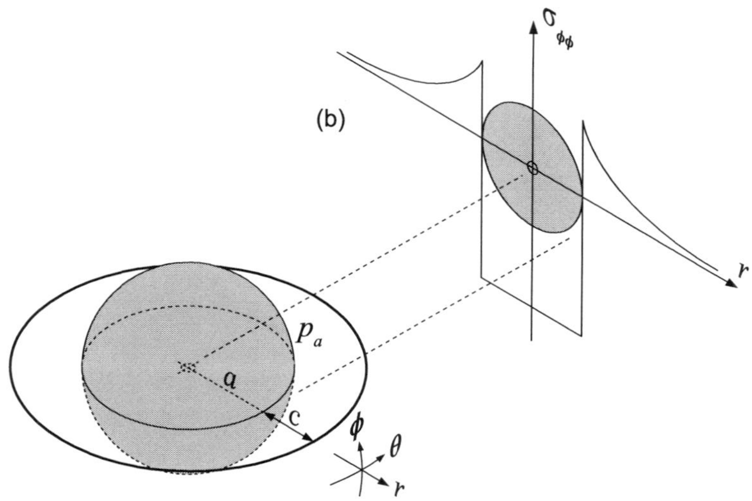

Figure 12.15 (a) Schematic diagram of an annular crack, radial width c, loaded by a misfitting spherical inclusion, radius a, with internal hydrostatic pressure p_a. Note the spherical polar (r, θ, ϕ) coordinate system. (b) The variation of the circumferential stress perpendicular to the crack plane, $\sigma_{\phi\phi}$, with radial coordinate, r, through the center of the inclusion and crack.

in experimental observations. The material within the inclusion has undergone a volume increase relative to the matrix material, from either the effect of phase transformation or thermal expansion difference. Such an increase is constrained by the elastic matrix, leading to a positive hydrostatic pressure, p_a, compressive stress, within the inclusion. The expression for the stress developed in the matrix, exterior to the inclusion, is well known, (e.g. Selsing 1961) and given in terms of radial coordinate r as

$$\sigma_{rr} = -2\sigma_{\theta\theta} = -2\sigma_{\phi\phi} = -p_a\left(\frac{a}{r}\right)^3, \tag{12.45}$$

using the spherical polar coordinate system of Figure 12.15a (note that this is different from the cylindrical system of Figure 2.21). The radial stress component $\sigma_{rr} < 0$ is compressive and the circumferential stress components $\sigma_{\theta\theta} = \sigma_{\phi\phi} > 0$ are half the magnitude and tensile. $\sigma_{\phi\phi}$ acts perpendicular to the crack plane and is the stress component of interest here, as illustrated in Figure 12.15b. The SIF acting on the crack at the misfitting inclusion is given here by a generalization of Eq. (12.44) for a radially varying stress field. Combining Eq. (12.45) into the integral of Eq. (12.44) and adjusting the limits gives

$$\begin{aligned} K &= \frac{2}{(\pi e)^{1/2}} \int_a^e \frac{p_a}{2} \frac{a^3}{r^3} \frac{r\mathrm{d}r}{(e^2 - r^2)^{1/2}} \\ &= \frac{p_a a^3}{(\pi e)^{1/2}} \int_a^e \frac{\mathrm{d}r}{r^2(e^2 - r^2)^{1/2}} \end{aligned} \tag{12.46}$$

It is noted that although the expression encompasses the entire flaw dimension e, the integral extends over only the stressed crack dimension c. A similar approximation has been used for cracks at indentation deformation zones (Lawn and Evans 1977), pores (Green 1980), and transforming particles (Green 1981; Swain 1981). This point is considered in the discussion. Eq. (12.46) is also a standard integral and the SIF $K(c)$ for a crack at the inclusion is given by

$$K(c) = \frac{p_a a^{1/2}}{\pi^{1/2}} \left[\frac{(1+\xi)^2 - 1}{(1+\xi)^5}\right]^{1/2}, \ \xi = c/a, \tag{12.47}$$

noting the definition of c for this geometry. A similar equation was obtained by similar means by Green (1981). In the limit of a crack large relative to the inclusion, $c \gg a$ and thus $\xi \gg 1$. In this case, Eq. (12.47) approaches the response of the SIF for a center loaded penny crack, $K(c) = F/(\pi c)^{3/2}$ with $F = p_a \pi a^2$. The stress field exterior to the inclusion, although distributed, is concentrated by the cubic radial dependence such that the distribution is not perceived by large cracks. In the limit of a crack small relative to the inclusion, $c \ll a$ and thus $\xi \ll 1$. In this case, Eq. (12.47) approaches the response of the SIF for a uniformly stressed crack, $K(c) = (2/\pi)^{1/2} p_a c^{1/2}$. The amplitude term reflects the geometry of the system and is not identical to that for a penny crack or a straight crack uniformly stressed by p_a. (The maximum stress experienced by the annular crack is $p_a/2$ and the crack does not extend to the center of the flaw.) Figure 12.16 is a logarithmic plot of the $K(c)$ response using the normalized coordinates of Eq. (12.47), shown as the solid line. The asymptotic behaviors for short $K \sim \xi^{1/2}$ (slope 1/2) and long $K \sim \xi^{-3/2}$ (slope −3/2) cracks discussed above are shown as the dashed lines. It is straightforward to show that the SIF passes through a maximum between these two asymptotes at $c^*/a = (5/3)^{1/2} - 1 \approx 0.29$, shown as the vertical gray line. It is clear from Figure 12.16 that the gray line separates destabilizing, $dK/dc > 0$, behavior at short crack lengths from stabilizing, $dK/dc < 0$, behavior at long crack lengths, and that the maximum represents a neutral transition point. This stability behavior in term of crack initiation at inclusions is now discussed.

Equilibrium of the fracture system is given by $K = T$. The $K(c)$ behavior of Figure 12.16 shows that the crack + inclusion system of Figure 12.15a can exist in states with either zero, one, or two fracture equilibria, depending on the composite amplitude term combining inclusion pressure and size, $p_a(a/\pi)^{1/2}$. These states are shown in Figure 12.17 for three $K(c)$ trajectories in a representative fracture system. The system consists of a material, toughness $T = 0.7$ MPa m$^{1/2}$, containing a flaw composed of an inclusion, $a = 100$ μm, and a small crack nucleus on the inclusion periphery characterized by a crack length $c = c_1 = 1$ μm. The inclusion undergoes a volume increase relative to the material matrix that generates a SIF field as described above, characterized by a pressure p_a that increases to ≈ 1 GPa. Note that such a stress in a material of elastic modulus $M \approx 100$ GPa corresponds to an inclusion volume strain of $p_a/M \approx 1$ %. This value is typical of phase transforming particle systems, e.g. NiS in glass (on cooling), 4 % (Swain 1981), ZrO_2 in ceramics (on cooling), 5 % (Muddle and Hannink 1986), and SiO_2 quartz in rocks (on heating), 5 % (Deer et al. 1966). Similarly, such a strain in a system with coefficient of thermal expansion mismatch between the (less expansive) inclusion and matrix of $\Delta\alpha \approx 7 \times 10^{-6}$ K^{-1} corresponds to a temperature decrease of $p_a/\Delta\alpha M \approx 1400$ K, both values typical of ceramic and rock systems (Kingery et al. 1975; Deer et al. 1966).

The horizontal line in Figure 12.17 indicates the $T = 0.7$ MPa m$^{1/2}$ matrix toughness. The left vertical solid line indicates the $c_1 = 1$ μm crack nucleus. The central dashed vertical line indicates the position of maximum K, $c^* \approx 29$ μm. The lower bold line indicates the $K(c)$ behavior for an inclusion pressure $p_a = 60$ MPa (a weakly transformed or cooled inclusion). For all values of c, the SIF is less than the toughness, $K(c) < T$. The crack nucleus is held open by kinetic or other constraints and, although mildly stressed, the system is metastably trapped in a nonequilibrium state; the number of equilibria is zero.

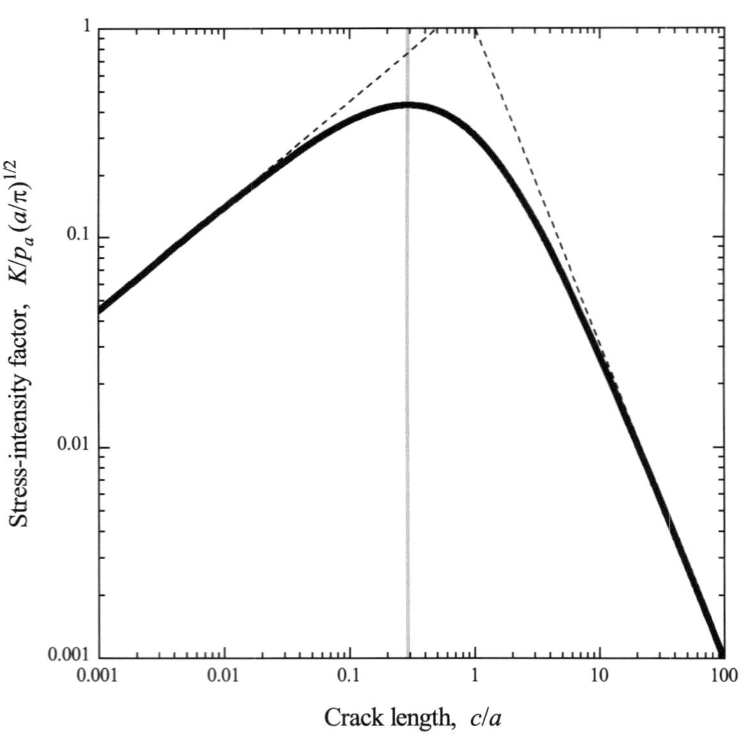

Figure 12.16 Plot of the stress-intensity factor K as a function of crack length c for the inclusion and crack system of Figure 12.15 (solid line). The asymptotic behavior at short and long crack lengths are shown as the dashed lines. The full solution passes through a maximum of neutral stability.

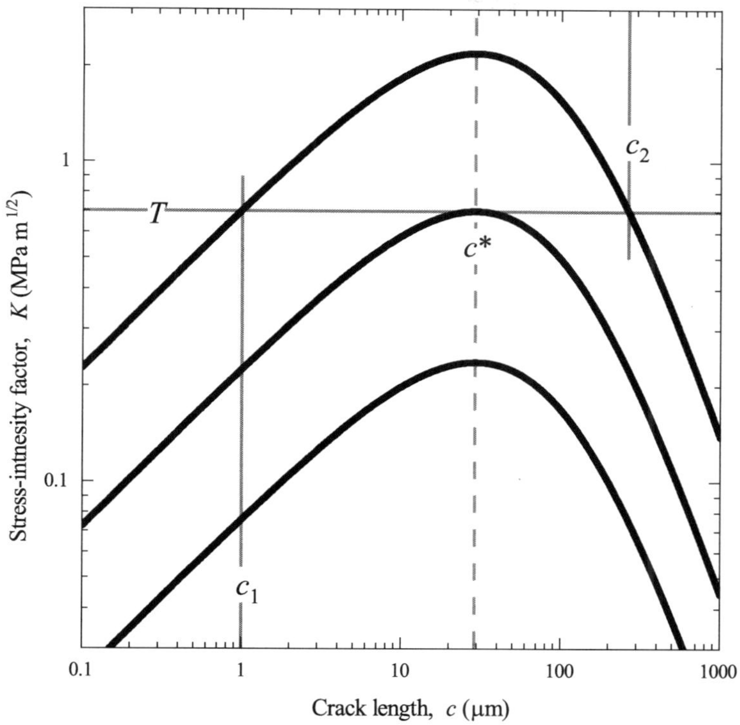

Figure 12.17 Plot of stress-intensity factor variations K as a function of crack length c for the inclusion and crack system of Figure 12.15 with three levels of increasing internal inclusion pressure, bottom to top. A crack nucleus size is indicated by the solid vertical line c_1 and a stable equilibrium crack length is indicated by the solid vertical line c_2. The dashed vertical line c^* indicates the transition crack initiation length.

The central bold line indicates the $K(c)$ behavior for an inclusion pressure $p_a = 177$ MPa (a moderately transformed or cooled inclusion). For this condition the maximum SIF equals the toughness at one crack length $c = c^* = 29$ μm such that $K(c^*) = T$. The system, although moderately stressed, remains metastably trapped in a nonequilibrium state; the number of equilibria is one, a neutral equilibrium at c^*. The upper bold line indicates the $K(c)$ behavior for an inclusion pressure $p_a = 555$ MPa (a nearly completely transformed or cooled inclusion). For this condition the SIF equals the toughness at two crack lengths, $c = c_1 = 1$ μm and $c = c_2 = 265$ μm such that $K(c_1) = K(c_2) = T$. The number of equilibria is two; the system exhibits an unstable equilibrium state at c_1 and a stable equilibrium state at c_2. The crack nucleus is in an unstable configuration in this state—any perturbation that increases c also increases K and thus leads to a nonequilibrium condition with positive fracture force that further increases the crack length. For such a perturbation from this state at $K = T$, the system follows the upper bold line with $c > c_1$ and $K > T$. The crack velocity dc/dt is limited by kinetics or dynamics (Lawn 1993) and passes through a maximum, until a new equilibrium state is reached at a crack length of c_2 and $K = T$. The crack is in a stable configuration in the final state such that any perturbation that increases c decreases K, leading to a nonequilibrium condition with negative fracture force that decreases the crack length and returns the system to the c_2 equilibrium configuration. Conversely, any perturbation that decreases c increases K and leads to a nonequilibrium condition with positive fracture force that increases the crack length and thus restores equilibrium.

The process described here, the formation of a crack from a crack nucleus, is termed *crack initiation*. If a distinction in size is made between nuclei, $c < c^*$, and cracks, $c > c^*$, at the above inclusions, it is apparent that all nuclei and cracks initiate or propagate to crack lengths of c_2 on increase of inclusion pressure to 555 MPa. If the increase in pressure from 177 MPa to 555 MPa is gradual, increasing numbers of cracks in the domain $c^* \leq c \leq c_2$ propagate through a sequence of stable equilibria as the pressure increases. Similarly, increasing numbers of nuclei in the domain $c_1 \leq c \leq c^*$ initiate from a sequence of unstable equilibria as the pressure increases. In both cases, c^* thus marks the minimum stable equilibrium crack length for the inclusion. If the increase in pressure from 177 MPa to 555 MPa is rapid, all nuclei and cracks in the domain $c_1 \leq c \leq c_2$ are placed in nonequilibrium states and a sequence of crack "pop-in" events occurs that is dependent on the perturbation kinetics. It is apparent from Eq. (12.47) that the condition for unstable equilibrium of a nucleus depends only on the product $p_a a^{1/2}$. Hence the discussion above could have been framed equally in terms of fixed pressure and varying inclusion size, variations in both, or fixed pressure and inclusion size and variations in nucleus size. Similarly, the condition for stable equilibrium of a large ($c \gg a$) initiated crack depends only on the product $p_a a^2$, and is thus not affected by the details of initiation or the separate contributions of p_a or a^2. If p_a is invariant then $c \sim a^{4/3}$ in the limit of large cracks.

Figure 12.18 shows a plot illustrating crack initiation behavior in soda-lime glass containing NiS inclusions from the work of Swain (1981), using data derived from the published work. The plot is in logarithmic coordinates of observed crack length c vs inclusion diameter $2a$. The hatched band indicates a domain of inclusion diameters $2a < 60$ μm for which no cracks were observed. The symbols represent individual inclusion and crack length measurements. The diagonal lines are guides to the eye: The lower line represents $c = c^* = 0.29a$, the position of the peak in the SIF variation of Figure 12.16, marking the lower bound for stable equilibrium crack lengths. The upper line represents $c = a$, the approximate lower bound for description as center loaded penny cracks. The stable crack length measurements lie between these bounds, at approximately 0.4–$0.5a$. Comparison with Figure 12.16 has two implications. First, that the cracks were initiated from nuclei in unstable states in the range 0.03–$0.1a$, 1–3 μm. Second, that the cracks in stable states were not sufficiently extended to be regarded as center loaded. Both implications are consistent with other parts of the experimental observations. The inferred nuclei sizes are consistent with the threshold observation, $a \approx 30$ μm. Further, the observed minimum crack length is consistent with the minimum length predicted from Figure 12.16, $0.29a \approx 10$ μm. In addition, the inferred nucleus size places a lower bound on the transformed inclusion pressure of $p_a > 400$ MPa, consistent with the estimate by Swain from thermoelastic constants of $p_a \approx 600$ MPa. Swain also noted that the crack lengths were short relative to the initiating inclusion size and that a description as center loaded penny cracks was not appropriate (in addition, a clear $c \sim a^{4/3}$ trend is not obvious).

Similar behavior was observed by Todd and Derby (2004) in ceramic composites of Al_2O_3 containing 20 % by volume of graded sizes a of SiC inclusions. SiC exhibits a smaller coefficient of thermal expansion than Al_2O_3 and hence a compressive stress developed in the SiC inclusions and compensating tensile stress developed in the polycrystalline Al_2O_3 matrix on cooling from the hot pressing temperature of 1700 °C. A threshold for cracking of $a = (10 \pm 3)$ μm at the SiC inclusions after cooling was observed. Direct measurements of stress in the hot-pressed composites were performed using neutron diffraction showing approximately −1.3 GPa in the inclusions and +0.4 GPa in the matrix for inclusion sizes smaller than the cracking threshold and −0.8 GPa and +0.3 GPa in the inclusions and matrix, respectively, for larger inclusions. As

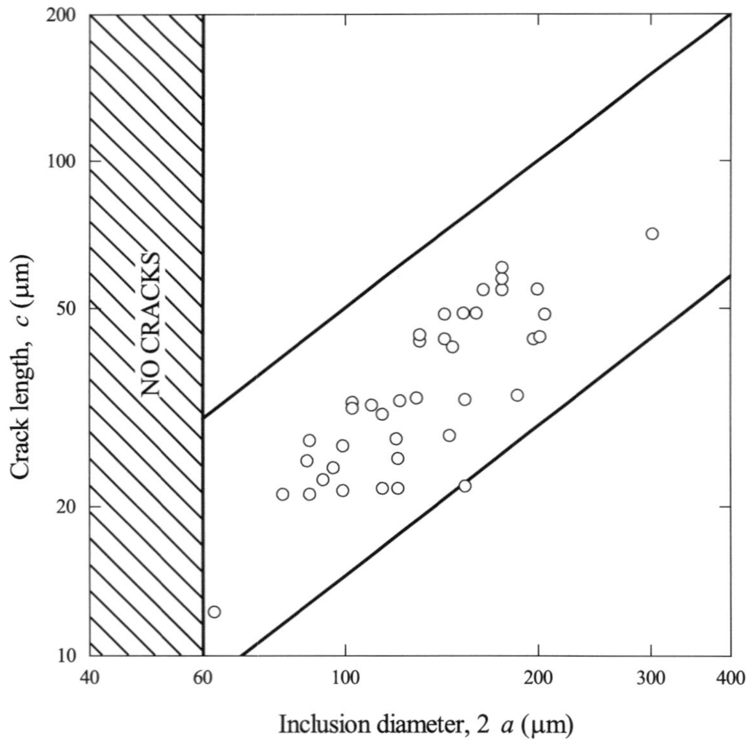

Figure 12.18 Plot of crack lengths in glass observed at NiS inclusions (Adapted from Swain, M. V 1981). The hatched band indicates small inclusions at which no cracks were observed. The solid lines indicate the bounds of crack lengths predicted from inclusion-based fracture mechanics considerations.

the inclusions were not dilute, post-threshold cracking of the matrix was pervasive throughout the composite, extending from inclusion to inclusion, probably explaining the significant stress relief. Using the measured threshold a value, the minimum crack length predicted from Figure 12.16 is $c^* = 0.29a \approx 3$ μm, about the inclusion spacing and consistent with the pervasive cracking. Using this c^* value and the measured inclusion stress p_a in the limiting short crack SIF equation provides a lower bound to the Al_2O_3 matrix toughness of 1.8 MPa $m^{1/2}$, consistent with many other Al_2O_3 measurements.

Cracking thresholds as a function of inclusion size were also observed by Padture and colleagues (Padture et al. 1993a, 1993b; Lawn et al. 1993) and Watts (2011) in ceramic composites of Al_2TiO_5-Al_2O_3 and SiC-ZrB_2, respectively. In the first system, Al_2TiO_5 has a smaller coefficient of thermal expansion than Al_2O_3 and in the second, SiC has a smaller coefficient of thermal expansion than ZrB_2. In both systems, the first named constituents were incorporated into the composites in the form of duplex microstructural grains. In the first system, the Al_2TiO_5 inclusions developed a compressive stress of $p_a \approx$ 7.7 GPa, estimated from thermoelastic constants, on cooling from the sintering temperature of approximately 1600 °C. The toughness of the Al_2O_3 matrix material was assumed from previous measurements as $T = 2.75$ MPa $m^{1/2}$. In the second system, the SiC inclusions developed a compressive stress of approximately 880 MPa, determined by X-ray diffraction and Raman spectroscopy, on cooling from the hot pressing temperature of 1800 °C. In both studies the threshold for matrix cracking associated with a critical minor phase inclusion size was identified by direct observation and by the clear onset of precipitous decreases in strengths of the composites. The Al_2TiO_5-Al_2O_3 materials exhibited inclusion size dependent cracking thresholds that decreased with increasing volume fraction of Al_2TiO_5: (7–12) μm (volume fraction = 0.1), (4–6.5) μm (0.2), (3–5) μm (0.3), and (2.5–4) μm (0.4). Using the matrix toughness and the inclusion compressive stress in the short crack limit here provides an estimate of the nucleus size of ≈ 0.2 μm, consistent with the threshold observations. The SiC-ZrB_2 materials exhibited a SiC inclusion size cracking threshold of approximately 11.5 μm (SiC volume fraction = 0.3). Estimating the minimum stable crack length as $0.29a \approx 3.4$ μm and using the inclusion compressive stress gives a lower bound to the matrix toughness of 1.3 MPa $m^{1/2}$, consistent with polycrystalline ZrB_2 toughness of 3.5 MPa $m^{1/2}$. Detailed analyses of inclusion stress, including the effects of inclusion shape, have been made by Li and Bradt (1989a, 1989b) consistent with these estimates.

The importance of considering cracking at inclusions is two fold: (i) inclusion + crack flaw systems are often strength-controlling in extended components and particles and (ii) the inclusion + crack fracture system describes many other strength-controlling flaws. The following section extends the above analysis to consider two other flaw systems that are pervasive and strength-controlling in materials: ansisotropic grains and sharp contacts.

12.4.4 Crack at an Anisotropic Grain or Sharp Contact

The fracture mechanics analyses of crack initiation and stabilization considered above are extended easily to other fracture systems that act as strength-controlling flaws in particles. These systems include cracks in polycrystalline materials composed of single phases with ansiotropic thermal expansion coefficients as in Figure 12.19, and cracks at sharp contacts on surfaces as in Figure 12.20. The physical similarities of Figure 12.19 and Figures 12.20 to 12.15 are clear and this section outlines the different analytical descriptions and interpretations (note the slightly different coordinate systems).

In many polycrystalline materials, a characteristic stress σ_λ is developed at grain boundaries on cooling from the processing temperature. This stress is given by $\sigma_\lambda \sim \Delta\alpha\Delta T M$. Here $\Delta\alpha$ is the maximum difference in the coefficients of thermal expansion as a function of orientation in the constituent phase. ΔT is a characteristic change in temperature between the processing and (cooled) application steps, and M is a characteristic elastic modulus. In the diagram of Figure 12.19a, a grain in a polycrystal is represented with the direction of least coefficient of thermal expansion in the z direction of the (r, θ, z) cylindrical polar coordinate system shown. The grain dimension λ is in the perpendicular r direction. Crack propagation c in the r direction is on grain boundary facets primarily in the $(r\theta)$ plane. The component of thermal expansion mismatch stress σ_λ controlling fracture is $\sigma_{zz}(r)$, shown in Figure 12.19b. Typically, the $\Delta\alpha$ values in single phases generated by anisotropy are less than those generated in multi-phase composites by inhomogeneity, $(2–3) \times 10^{-6}$ K^{-1} rather than $(5–7) \times 10^{-6}$ K^{-1}, and as a consequence the stress values σ_λ are less, ≈ 100s of MPa rather than ≈ 1 GPa, as discussed earlier. Nevertheless, the amplitude term in the SIF describing crack initiation, Eq. (12.47), can be significant and in this case varies as $\sigma_\lambda \lambda^{1/2}$, where λ is a characteristic grain size. The amplitude term thus increases with increasing grain size λ, and λ is often interpreted as the size of the largest grains in a polycrystal. For a polycrystalline material with grain boundary toughness T_λ the crack initiation condition can thus be written fully as $\Delta\alpha\Delta T M\lambda^{1/2} \sim T_\lambda$, a scaling relation shown by Bradt and colleagues in early consideration of cracking in ceramic polycrystals (Kuszyk and Bradt 1973; Cleveland and Bradt 1978) and later reiterated by Lawn (1991).

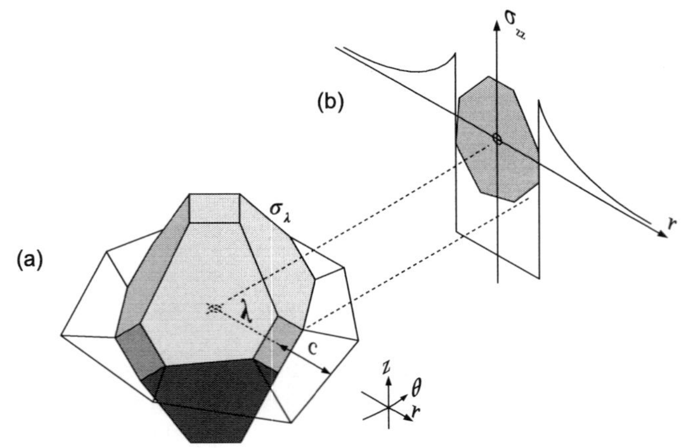

Figure 12.19 (a) Schematic diagram of a faceted intergranular crack, characteristic width c, loaded by a thermal expansion mismatched grain, characteristic size λ, with internal stress σ_λ. Note the cylindrical polar (r, θ, z) coordinate system. (b) The variation of the circumferential stress perpendicular to the crack plane, σ_{zz}, with radial coordinate, r, through the center of the grain and crack.

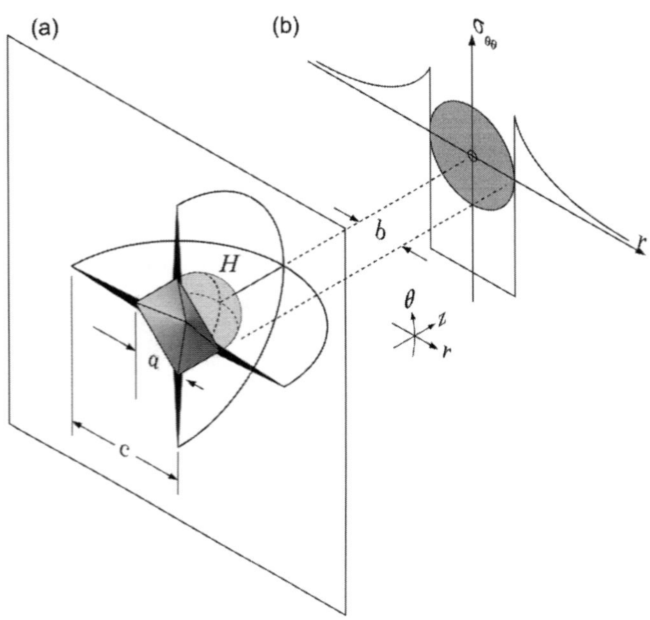

Figure 12.20 Schematic diagram of half-penny cracks, radius c, loaded by a localized plastic deformation zone, characteristic size b, with internal pressure H. Note the cylindrical polar (r, θ, z) coordinate system in indentation orientation. (b) The variation of the circumferential stress perpendicular to the crack plane, $\sigma_{\theta\theta}$, with radial coordinate, r, through the center of the zone and cracks.

This relation can be interpreted in several ways with regard to crack initiation in a polycrystalline material on cooling from a stress-free maximum processing temperature. The simplest interpretation is if $\Delta\alpha = 0$, as in polycrystals of cubic materials, and no cracking is possible. The next simplest is if λ is small, as in fine-grained polycrystals, and the equilibrium fracture condition is not met for any grain-grain boundary combination. The system is described by the lower curve in Figure 12.17 and no cracking is possible. However, if λ is large enough, the equilibrium condition for crack initiation can be met and the system is then described by the upper curve in Figure 12.17 and cracking is expected. Experiments on Al_2O_3 in which λ was varied were performed by Blendell and Coble (1982), in which extensive cracking was observed to develop as grain size was increased from an apparent crack-free $\lambda \approx 50$ μm, to the point that the samples had no integrity and could not be handled for $\lambda > 150$ μm. More detailed interpretations are possible. The grain size required to observe thermal expansion anisotropy induced cracking effects, such as decrease in elastic modulus or decrease in strength, scales as $\lambda \sim (\Delta\alpha)^{-2}$ for different materials, and was demonstrated by Cleveland and Bradt (1978). The temperature difference required to observe such effects scales as $\Delta T \sim \lambda^{-1/2}$ for a single material, and was demonstrated by Case et al. (1980) and Ohya et al. (1987) (the latter verified cracking through in situ acoustic emission). The equilibrium fracture condition can be written as $\Delta\alpha\Delta T M\lambda^2/c^{3/2} \sim T_\lambda$, noting that this equation applies to cracks long relative to the large grained precursor, $c \gg \lambda$ and thus provides a lower bound for cracks in the surrounding small-grained matrix. The scaling relation for grain boundary crack length as a function of grain size is thus $c \sim \lambda^{4/3}$. The non-linearity explains why "spontaneous microcracking" in polycrystalline anisotropic materials appears so grain size dependent. Assuming that $c = \lambda = 150$ µm in the completely fragmenting Al_2O_3 materials above (Blendell and Coble 1982), this scaling gives $c = 34$ µm for the $\lambda = 50$ µm material, about a grain facet size and consistent with the lack of observed cracks. Lawn (1991) extends this idea to show that "activated microcracking" in systems with a superposed uniform applied stress are also unlikely to exhbit many grain boundary cracks for small grain size materials.

Sharp contacts on material surfaces generate permanent deformation impressions that remain after the load-unload contact cycle (as discussed in detail in Chapter 2). Such irreversible deformation can be termed plastic, generally, describing predominantly dislocation-mediated effects in metals, and predominantly discrete shear crack (or shear fault) and compaction effects in other materials. In particular, brittle materials such as dense ceramics, glasses, minerals, and semiconductors plastically deform under sharp surface contacts by a combination of shear fracture and compaction; porous materials predominantly compact. The characteristic surface dimension of the residual contact impression at a sharp contact is a. For conical contacts, a is the radius of the contact disc, such that the projected area of contact is πa^2. For square pyramidal (Vickers) contacts, a is the semi-diagonal of the contact diamond, as shown in Figure 12.20a, such that the projected area of contact is $2a^2$. Noting the cylindrical contact oriented (r, z, θ) coordinate system in Figure 12.20a, the contact surface is the $(r\theta)$ plane and the contact axis is the z direction. For many materials the mean supported or projected contact stress, determined by the peak contact force P, applied along z (commonly referred to as the "load"), and the projected area, in $(r\theta)$, is invariant with contact force and impression dimension and termed the hardness H (Cook 2020). H is a characteristic stress in the contact event and scales as $H \sim P/a^2$ (e.g. for Vickers indentations, $H = P/2a^2$). In brittle materials with limited plasticity, a permanent deformation zone is formed beneath the contact area in a localized volume that is semi-ellipsoidal in shape (shown shaded). The zone contains sheared and compacted material and is embedded in the surrounding plastically undeformed matrix. The characteristic dimension of the plastic deformation zone is b, and simple interpretation is that b is a characteristic zone radius, Figure 12.20a. In reaction to the imposed local plastic deformation at the contact, the matrix deforms elastically in a distributed manner. Modeling the zone as a pressurized spherical cavity in an elastic-plastic material gives scaling of $b/a \sim (M/H)^{1/2}$, where M is a characteristic elastic modulus (Lawn et al. 1980).

The localized plastic deformation zone generates a circumferential stress field component, $\sigma_{\theta\theta}(r)$, Figure 12.20b. The stress field is compressive within the zone (significant shear stress is also present) and tensile exterior to the zone and decreases with distance from the contact. These characteristics are identical to those discussed earlier for inclusions and grains and lead to identical fracture behavior of crack initiation and stabilization. Two half-penny cracks are initiated at sharp contacts from shear fault intersections at the plastic zone periphery, Figure 12.20a (cracks shown in black). The cracks lie on the (rz) and (θz) planes and have surface trace lengths of c. The cracks can be approximated by the SIF for center-loaded penny cracks. The relationship between the center force F and the contact load P can be derived as follows. The characteristic volume strain in the plastic deformation zone is $-a^3/b^3$. The corresponding hydrostatic pressure within the zone is $p_a \sim Ma^3/b^3$. This pressure applies an outward force $F \sim p_a b^2$ on the zone-matrix boundary and thus crack surfaces, such that $F \sim Ma^3/b$. Using the definition of hardness and the elastic-plastic cavity scaling for the deformation zone gives $F \sim (M/H)^{1/2}P$. The SIF for a half-penny crack generated at a sharp contact P in a brittle material is thus $K \sim (M/H)^{1/2}P/c^{3/2}$. It is common to describe cracks generated by Vickers indention as $K = \chi P/c^{3/2}$, expressed in term of the experimental value of P and in which χ contains the M/H term and all geometrical factors. The stress field driving the cracks is $\sigma_{\theta\theta} = (p_a/2)(b/r)^3$; that for the (r, z) crack is shown. The half-penny crack approximation strictly applies for $c \gg b$ although experimental observations show that the approximation is widely obeyed (Cook 2020). Applying the equilibrium condition $K = T$ gives an expression for indentation crack length as a function of indentation force, $c \sim (F/T)^{2/3}$, examples of which are shown in Figures 12.8c and 12.8d and in an extensive related survey (Cook 2020).

The previous two sections have developed and applied fracture mechanics analyses to describe cracking at misfitting inclusions, anisotropic grains, and sharp contacts. Such analyses show that the decreasing stress fields surrounding these features, residual from thermal or deformation effects, lead to the common phenomena of crack initiation and formation of stable equilibrium cracks. Quantitative and qualitative experimental observations in all three systems support the analyses. A conclusion is that the most common strength controlling flaws in particles and other components—inclusions, grains, and contacts—are well described by the equilibrium state defined by a stabilizing SIF field and material toughness. An implication is that the action of such SIF fields persist on subsequent application of a uniform applied stress field. The strengths of particles and other components controlled by such flaws are thus determined by the superposition of two SIF fields. The following section further illustrates the power of the additivity of SIFs within the linear elastic fracture mechanics framework and develops a strength formulation for residually stressed flaws.

12.5 Combined Loading

12.5.1 Strength of Post-Threshold Flaws

The strength of a brittle component, including a particle, can be viewed as the outcome of a two-step sequential fracture process. In the first step, a flaw is generated in the component. In the second step, the component is stressed to failure. The maximum stress—the strength—is an inverse measure of the potency of the flaw. Here, the strengths of spherical particles are considered within the framework of this sequence, building on the flaw generation analysis above and the stress analysis of Chapter 2. In the first step here a strength-controlling flaw is introduced into a particle, either by a thermal effect in the bulk of the particle (a localized transformation or expansion) or by a mechanical effect on the surface of the particle (a localized sharp contact). Figures 12.21a and 12.21b illustrate these first steps for thermal and mechanical effects, respectively. In these examples, the flaws are characterized as equilibrium center-loaded penny-like cracks that are large relative to the central regions of force application, but small relative to the particles. These flaws are termed "post-threshold," as the threshold criteria for crack initiation have been met or exceeded, as described earlier. (In describing strength-limiting flaws, the post-threshold term is usually omitted and implied, unless distinction is required with uncracked flaws. Such "sub-threshold flaws" are considered in the following section.) In the second step the particles are loaded in diametral compression such that the flaws are placed in uniform tensile stress fields. For the thermally induced flaw associated with a bulk inclusion or grain, the flaw is located near the center of the particle with the crack plane parallel to a particle meridional plane. For the mechanically induced flaw associated with a surface contact, the flaw is located on the surface of the particle near the particle equator with the crack plane also oriented parallel to a particle meridional plane. Figures 12.21c and 12.21d illustrate these second steps for thermal and mechanical effects, respectively. Using the notation of the cylindrical polar coordinates of Figure 2.21, the center bulk crack is under the influence of a localized force in the r direction and the

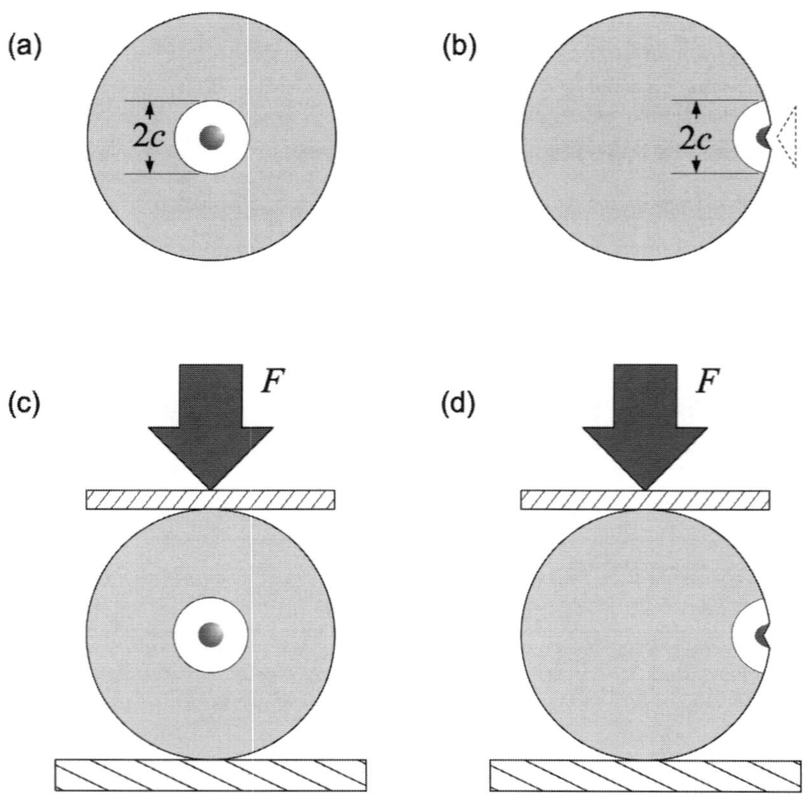

Figure 12.21 Schematic cross-section diagrams of the two step strength process in spherical particles. Step 1: (a) Flaw generation at the center of a particle by an expanding inclusion that initiates a circular crack, radius c; or (b) Flaw generation on the surface of a particle by sharp contact that initiates a half-penny crack, radius c. Step 2: (c) Loading of the particle in diametral compression, such that the central flaw is placed in tension; or (d) Loading of the particle in diametral compression, such that the equatorial flaw is placed in tension. In all cases, the crack is on a meridional plane.

uniform stress component σ_{rr} and the equatorial surface crack is under the influence of a localized force in the θ direction and the uniform stress component $\sigma_{\theta\theta}$.

The two flaw systems are characterized by identical SIF based fracture mechanics descriptions. The SIF for the localized central force is given by

$$K_F = \chi F/c^{3/2}, \tag{12.48}$$

where c is the crack radius. The force F represents the effects of a constrained volume at the center of the crack, characterized by a generalized pressure p and area a^2, such that $F \sim pa^2$. Examples of p and a in specific systems were given previously. The dimensionless parameter χ is a geometrical factor. The SIF for the uniform applied stress is given by

$$K_A = \psi\sigma_A c^{1/2}. \tag{12.49}$$

The stress σ_A represents the effects of diametral loading of the particle. The dimensionless parameter ψ is a geometrical factor. Details are given in Chapter 2. The total SIF for the flaw is given by

$$K = K_F + K_A = \chi F/c^{3/2} + \psi\sigma_A c^{1/2} \tag{12.50}$$

Equilibrium of the fracture system is given by $K = T$, where T is the toughness of the material, here assumed invariant and uninfluenced by environmental or microstructural effects. Two equilibrium limits to Eq. (12.50) are apparent. First, a lower bound to the stable equilibrium crack length c_0, is given by setting $K_A = 0$, such that $K_F = T$ and thus

$$c_0 = (\chi F/T)^{2/3}. \tag{12.51}$$

as shown earlier, Eq. (12.39). Second, an upper bound to the unstable equilibrium strength, σ_0, is given by setting $K_F = 0$, such that $K_A = T$ and thus, using $c = c_0$,

$$\sigma_0 = (T/\psi)c_0^{-1/2} = (T^{4/3}/\psi)(\chi F)^{-1/3}. \tag{12.52}$$

as shown earlier for the first equality, Eq. (12.38). The second equality reflects the addition of an intermediate step in the sequence, in which K_F is set to 0 subsequent to crack formation but prior to stress application. Experimentally, this modified sequence usually represents the addition of an intermediate thermal annealing step leading to reduction of the localized field and $\chi \approx 0$ during subsequent stressing. Note that the expression for strength in terms of the localized crack stabilizing force, $\sigma(F)$ in Eq. (12.52), exhibits an increased dependence on toughness relative to the Griffith equation as a crack length dependence is incorporated through F.

Between the limits of Eqs. (12.51) and (12.52), the full SIF Eq. (12.50) describes longer stable equilibrium crack lengths, $c \geq c_0$, and lesser unstable equilibrium stresses or strengths, $\sigma \leq \sigma_0$. Setting $K = T$ in Eq. (12.50) and inverting gives an equilibrium $\sigma_A(c)$ trajectory:

$$\sigma_A(c) = (1/\psi c^{1/2})(T - \chi F/c^{3/2}). \tag{12.53}$$

This trajectory is shown as the fine line in Figure 12.22 using $\psi = 1$, $T = 1$ MPa m$^{1/2}$, and $\chi F = 1$ N, leading to $c_0 = 100$ μm, and $\sigma_0 = 100$ MPa, all typical for a brittle material. The trajectory is shown in the common format in which the independent variable σ_A is shown on the ordinate. For $\sigma_A = 0$, $c = c_0$ and as σ_A increases, c increases, until a maximum value of σ_A is attained. The maximum is a point of instability as $\mathrm{d}K/\mathrm{d}c = 0$ here implies $\mathrm{d}\sigma_A/\mathrm{d}c = 0$. The instability is indicated by a dashed gray arrow in Figure 12.22. Imposing the instability condition in Eq. (12.53) gives the crack length and stress coordinates (c_m, σ_m) at the maximum:

$$c_m = (4\chi F/T)^{2/3} = 4^{2/3}c_0 \approx 2.52c_0, \tag{12.54}$$

and

$$\sigma_m = 3T^{4/3}/4^{4/3}\psi(\chi F)^{-1/3} = (3/4^{4/3})\sigma_0 \approx 0.47\sigma_0. \tag{12.55}$$

The coordinates (c_m, σ_m) give the configuration of the system at failure of the particle—σ_m is the particle strength. The localized force adds a stabilizing factor to the failure process, leading to precursor stable crack extension from c_0 to c_m and decreased strength from σ_0 to σ_m. The failure trajectory is shown as the bold gray line in Figure 12.22.

Eq. (12.54), Eq. (12.55), and Figure 12.22 have several important implications for interpretation of strength behavior of particles. The first implication is that failure from the most common strength-controlling flaws in particles—cracks at inclusions, grains, and contacts—is not described by the mechanics of failure pertaining to Griffith flaws. Griffith flaws

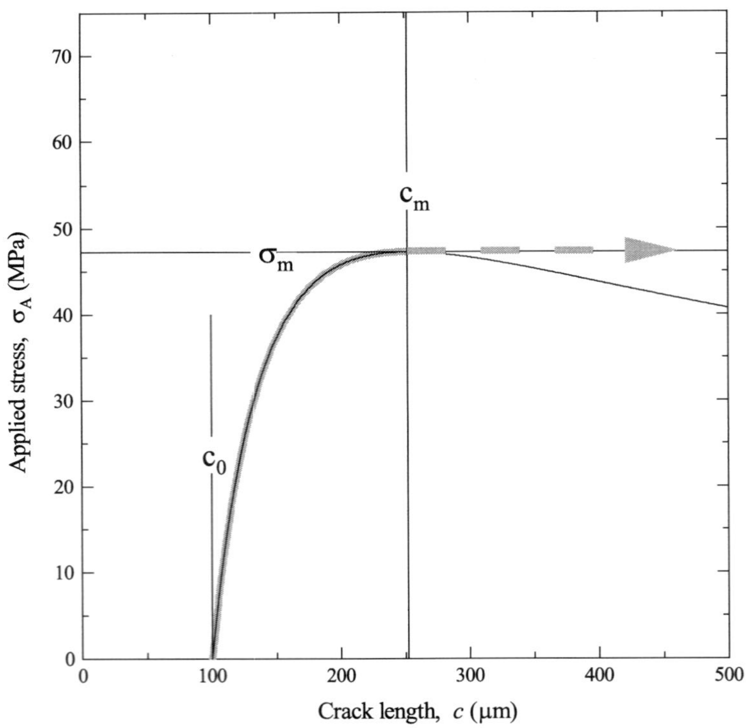

Figure 12.22 Equilibrium stress-crack length trajectory for a residually stressed inclusion, grain, or contact. Note the stable precursor crack extension from c_0 to c_m as the stress increases from 0 to the strength and instability at σ_m.

are simple cracks that are trapped at length c_0 by kinetic effects, with no attendant stabilizing field. The localized stress field that initiated and stabilized the cracks has passed or been eliminated. Examples of such transient fields include those to due to mechanical or thermal shock, those annealed away by thermal effects, or those removed by corrosion or other environmental degradation or fine abrasion. Griffith (1921) removed the residual stress field associated with sharp scratch contact flaws by annealing glass test components prior to strength measurement. Application of a uniform stress to a Griffith flaw simply increases the SIF K_A with no change in crack length, until the equilibrium condition $K_A = T$ is reached at which point the component fails. The instability condition is automatically fulfilled by the sole, destabilizing K_A field. The implication is important for two reasons: (1) The stress required for failure from stabilized flaws can be greatly overpredicted if Griffith mechanics are assumed to act on observed initial crack lengths (neglects the magnitude of K_F). (2) The effects of environment and microstructure on failure from stabilized flaws can be be greatly underestimated if precursor extension prior to failure is ignored (neglects the functional dependence on crack length of K_F).

A second implication of the above equations is that they are testable and useful. In particular, the recognition that the localized stabilizing force at an indentation flaw F is proportional to the indentation contact load P (see earlier paragraph) enables direct experimental assessment of Eq. (12.55). Tests of $\sigma_m \sim P^{-1/3}$ behavior, and deviations from that behavior, enable direct comparisons of materials and elucidation of environmental and microstructural effects (on T) and stress field and flaw geometry effects (on χ). The sophistication of these tests and analyses and application in interpreting and predicting the behavior of "natural" flaws is shown in Lawn et al. (1993), Padture (1993a, 1993b) and Cook (2015). Such "controlled flaw" testing is certainly applicable to particle materials and may even be applicable directly to particles.

A third implication is that the strength equation, Eq. (12.55), is extendable to other controllable or measurable flaw scale variables beyond the localized stabilizing force F. Two examples include the combination of Eqs. (12.51) and (12.55), leading to $\sigma_m \sim c_0^{-1/2}$, or use of the related indentation load, leading to $\sigma_m \sim P^{-1/3}$. A simple extension of this last relation is to use the connection between contact load and impression dimension, $P \sim Ha^2$, to give $\sigma_m \sim a^{-2/3}$. Impact by a sharp object is often characterized by the kinetic energy U_K of the object. If the contact is elastic-plastic, the force-displacement behavior during contact can usually be approximated with the quasi-static relation $P(\delta) \sim \delta^2$ (see Chapter 2). If the kinetic energy is completely converted into work of deformation, the peak indentation load is given by $P \sim U_K^{2/3}$ (Wiederhorn and Lawn 1979) and thus $\sigma_m \sim U_K^{-2/9}$. Such considerations can be used to diagnose particle failures, or to predict potential particle failures or the forces and energies required. This last point is considered in Chapter 13.

A fourth implication of Eq. (12.54), Eq. (12.55), and Figure 12.22 is that they can all accommodate, without modification, variations in T by environmental and microstructural effects. The earlier analysis has used T implicitly as the toughness in an inert environment, T_0, as this circumvents concerns regarding loading rate effects. However, identical fracture mechanics to that above applies if the toughness T_{env} characterizing fracture equilibrium in a reactive environment pertains. In such environments loading rate effects are often clear, although attaining reactive equilibrium at very slow rates is an experimental issue. Nevertheless, the analysis above describes behavior in both environments and the loading rate dependent transition between the two (Cook 2015). The analysis above can be extended to include microstructural effects, characterized by a crack length dependent toughness $T = T(c)$, in weakly toughened systems. Such systems are characterized by variations in $T(c)$ such that $\mathrm{d}T/\mathrm{d}c \leq \mathrm{d}K_{\mathrm{A}}/\mathrm{d}c$, as this constraint does not alter the instability condition used above to determine strength. Functionally, if toughness is described by a power law in such systems, $T(c) \sim c^{\tau}$, the constraint can be written as $\tau \leq 0.5$. Long cracks, extending over many grain diameters, approaching "steady-state" toughness, are often well approximated by such behavior. Consideration of environmental effects is required in failure of porous particles or particles with strengths controlled by surface flaws. Consideration of microstructural effects in the weak toughening limit is unlikely to apply to many particle systems, as most strength-controlling cracks in particles are short and remain so during failure. Nevertheless, the limit provides a lower bound to microstructural effects. "Short" crack microstructural effects are considered in Chapter 13.

12.5.2 Strength of Sub-Threshold Flaws

The sequential two step strength determination process outlined above, flaw generation followed by component loading, also applies to particle systems in which the flaw does not contain a well-formed crack. In these systems, the flaw consists of a structural feature that is a center of expansion (inclusion, grain, or deformation zone) and fracture nuclei on or near the interface of the feature and the surrounding tensile matrix. The nuclei are small, crack-like discontinuities in the material (e.g. grain boundary pores, shear fault intersections) and are much smaller than the structural feature of interest. As earlier, the flaw is considered small relative to the particle; the configurations are similar to those in Figures 12.21a and 12.21b without the surrounding cracks. Such flaws are termed "sub-threshold" as the criteria for crack initiation have not been met.

The SIF for a putative crack initiated from a sub-threshold flaw is shown in Figure 12.23 for the inclusion system considered in Figure 12.17. The SIF variation, notated here as $K_{\text{inc}}(c)$, was determined using Eq. (12.47) and the maximum value of the internal pressure considered in Figure 12.17, $p_a = 555$ MPa. Such a pressure generates an unstable equilibrium state for a nucleus with a characteristic size of 1 μm. The $K_{\text{inc}}(c)$ variation extends in a nonequilibrium trajectory from c_1 and is indicated by the dashed line labeled 1 in Figure 12.23. Diametral compression of the particle results in a uniform stress σ_{A} superposed on the flaw, as in crack-free versions of Figure 12.21c and 12.21d, and a resulting SIF $K_{\mathrm{A}}(c)$, Eq. (12.49). The total SIF for the system in this state is given by $K = K_{\text{inc}} + K_{\mathrm{A}}$. The $K_{\mathrm{A}}(c)$ variation for $\sigma_{\mathrm{A}} = 16$ MPa is shown as the dashed line labeled 2 in Figure 12.23. The bold solid gray line in Figure 12.23 is the sum of $K_{\mathrm{A}}(c)$ and $K_{\text{inc}}(c)$ (1 + 2) and describes the behavior of the stressed sub-threshold flaw. The most noticeable change to the SIF for the stressed flaw (gray line) relative to the unstressed flaw (dashed line 1) is the appearance of a minimum at large crack lengths. The SIF for the stressed flaw intersects the toughness (T) line at two crack lengths either side of the minimum. The shorter of these crack lengths is a position of stable equilibrium that is slightly longer than that in the unstressed state. Although not readily visible, the SIF for the stressed flaw is slightly increased at the much shorter crack length of the nucleus size, c_1. The imposition of the applied stress in this case thus perturbs the sub-threshold system so that the crack initiates, $K(c_1) \gtrsim T$, and generates a slightly longer stable equilibrium post-threshold crack. The unstable equilibrium crack length, longer than the position of the minimum in the $K(c)$ trajectory, is not physically realizable as it can only be accessed if the system passes through a sub-equilibrium region of $K < T$.

The $K_{\mathrm{A}}(c)$ variation for $\sigma_{\mathrm{A}} = 47$ MPa is shown as the dashed line labelled 3 in Figure 12.23. The solid black line in Figure 12.23 is the sum of $K_{\mathrm{A}}(c)$ and $K_{\text{inc}}(c)$ (1 + 3) and describes the behavior of the sub-threshold flaw at this greater stress level. Two notable changes are apparent in the SIF for this stressed flaw (black line) relative to the unstressed flaw (dashed line 1). Again, a minimum has formed at large crack lengths. More noticeable however, is that the SIF for the greater stress does not intersect the toughness line; there are no equilibrium configurations. Now more visible, the SIF for the greater stressed flaw is also increased at the much shorter crack length of the nucleus size, c_1. The imposition of the greater applied stress in this case thus perturbs the sub-threshold system so that the crack initiates, $K(c_1) > T$, and remains in a nonequilibrium state on extension. The unchecked nonequilibrium extension indicates particle failure and hence the strength of the particle in this case lies in the range $16 < \sigma < 47$ MPa. This strength range is to be compared with the strength of a particle containing a crack of the minimum stable length at this inclusion ($c^* = 29$ μm, see previous lines) acting as

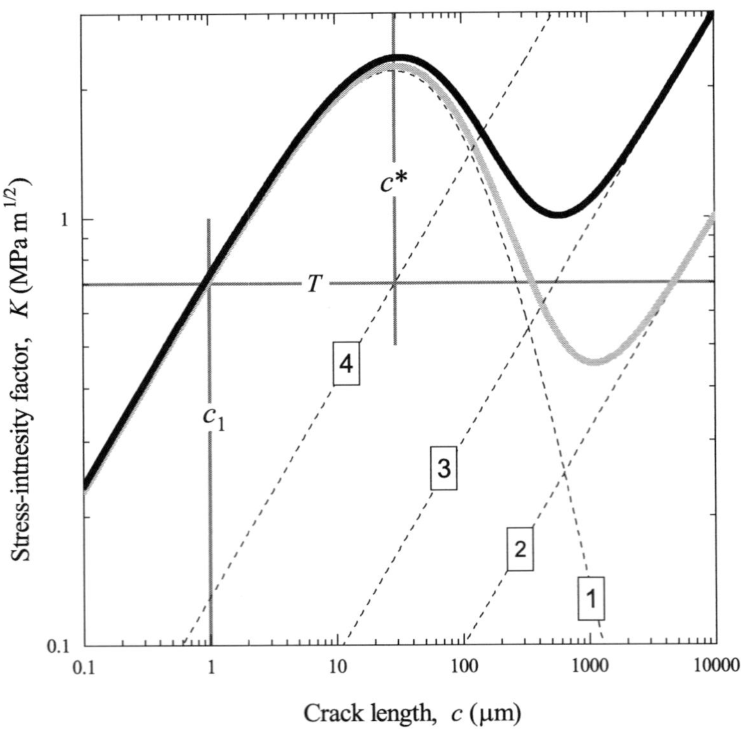

Figure 12.23 Plot of stress-intensity factor variations K as a function of crack length c for the inclusion and crack system of Figure 12.15 with two levels of superposed uniform stress. A crack nucleus size is indicated by the solid vertical line c_1. The vertical line c^* indicates the crack initiation length. For small levels of superposed stress, indicated by the gray line, the stress initiates a crack. For larger levels of superposed stress, indicated by the black line, the stress causes component failure. Dashed lines indicate limiting K fields.

a Griffith flaw, $\sigma \approx 115$ MPa. The SIF for such a system is indicated by dashed line 4 in Figure 12.23. Hence, the applied stress range over which the fracture behavior of the particle + inclusion is altered, from sub-threshold to post-threshold to failed, is small (0–47) MPa relative the maximum tensile stress in the matrix adjacent to the inclusion (278 MPa) and the compressive stress within the inclusion (−555 MPa). However, this stress range is a substantial fraction of the representative Griffith strength of the particle (115 MPa). The conclusion from these comparisons is that small changes in stress relative to the stresses acting within a particle can induce large changes in fracture initiation and the related failure behavior.

This section, Section 12.5 has considered the fracture systems of loaded particles that contain flaws and their associated cracks that are small relative to particle. In these systems, measurement focus is on the conditions defining unstable fracture equilibrium and the instability conditions leading to strength of the particle and large excursions from equilibrium. The equilibrium fracture condition itself is usually studied using extended components containing large cracks that are a substantial fraction of the component. In these studies, the measurement focus is on crack propagation and fracture conditions that are very close to equilibrium, such that the material equilibrium fracture property, toughness, can be determined. The following section considers large (long) crack toughness determination experiments that are relevant to particles and which demonstrate further application of the SIF fracture framework.

12.6 Long Cracks in Particles

12.6.1 Polymer Discs

Figure 12.24 shows schematic cross sections of two-dimensional cracked disc components loaded in tension. The disc diameters are D and the thicknesses (not shown) are b. Both components contain through cracks of length c and approximately $D/2$. The component in Figure 12.24a is loaded by forces F applied at the edge of the component at the crack mouth. The component in Figure 12.24b is loaded by forces F applied at pins through the body of the component, separated from the crack plane by distance a. The crack planes are perpendicular to the applied forces. Such geometries are known in the

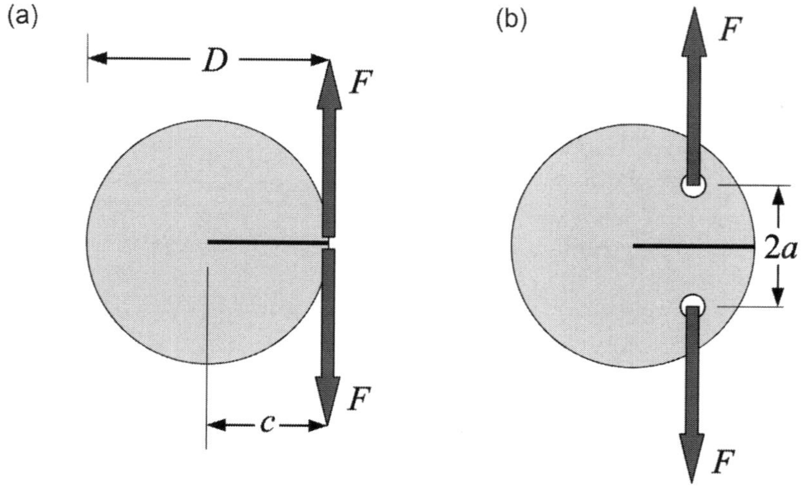

Figure 12.24 Schematic cross-section diagrams of cylindrical particles containing equatorial cracks c comparable to the particle size D. (a) Edge loaded by forces F. (b) Body loaded by forces F through pins separated by $2a$ across the crack plane.

fracture community as "compact tension specimens" (Sih 1973; Tada et al. 1973; Murakami 1987). The SIF for the crack in the edge loaded system, Figure 12.24a, is (Kendall and Gregory 1987)

$$K = \frac{F}{bD^{1/2}}\xi^{-1/2}\left[\frac{e_1\xi}{(1-\xi)^{3/2}} + \frac{e_2}{(1-\xi)^{1/2}}\right], \ \xi = c/D. \tag{12.56}$$

The definition of ξ here is noted and $e_1 = 2.81$ and $e_2 = 2.07$ are dimensionless geometrical factors. Similarly, the SIF for the crack in the pin loaded system, Figure 12.24b, is

$$K = \frac{F}{bD^{1/2}}\xi^{-1/2}\left[\frac{p_1\xi}{(1-\xi)^{3/2}} + \frac{p_2}{(1-\xi)^{1/2}}\right], \ \xi = c/D. \tag{12.57}$$

Here $p_1 = e_1(1-a/D)$ and $p_2 = (e_2/2)(1-2a/D)$ are dimensionless geometrical factors taking into account pin placement. (There is a typographical error in Kendall and Gregory 1987; the equations for force and K contain a factor of 2 that is inconsistent with the presented force data.) Inspection of Eqs. (12.56) and (12.57) shows that the SIFs consist of three elements. The first element is an amplitude term that includes the force F. The second element is a stabilizing $\xi^{-1/2}$ term that is the long straight crack analog of the large circular crack $\xi^{-3/2}$ dependence noted earlier (section 12.4.3). In the systems here, the straight crack is long relative to the line of force application at the crack mouth or at the pins. The third element is a destabilizing term that takes into account the lengths of the cracks in the finite components. For small cracks in the edge loaded system, $\xi \to 0$ and this term $\to e_2$. For long cracks in this system, $\xi \to 1$ and this term diverges as $\to e_1\xi^{1/2}/(1-\xi)^{3/2}$. The behavior of the pin loaded system is similar. Figure 12.25 shows a logarithmic plot of the $K(c)$ behavior from Eqs. (12.56) and (12.57) using geometrical parameters from the experiments of Kendall and Gregory (1987) on poly(methyl methacrylate) (PMMA) discs, $D = 50$ mm, $b = 3$ mm, $a = 14$ mm, and a typical force value, $F = 50$ N. The similar short crack stabilizing asymptotes and long crack destabilizing divergences are clear in Figure 12.25, as is the reduced K value for the pin system relative to the edge loaded system (the crack tip is closer to the pins than to the component edge). The shaded band indicates the crack length domain of most experiments, $0.2 \leq c/D \leq 0.8$, and suggests that most measurements with these geometries take place in destabilizing fracture fields. Maintaining the system near fracture equilibrium with controlled crack propagation requires ever greater decreases in the applied force as the crack extends. The difficulties are described by Kendall and Gregory.

Figure 12.26a shows a schematic cross section of a two-dimensional cracked disc component loaded in diametral compression. The disc dimensions are as shown earlier. The disc contains a through crack of length c, approximately $D/2$. The disc is loaded by force F applied at the pole and thus the crack mouth. The crack plane is parallel to the applied force. The SIF for this crack is (Kendall and Gregory 1987)

$$K = \frac{F}{bD^{1/2}}\xi^{1/2}\left[\frac{d_1}{(1-\xi)^{3/2}}\right], \ \xi = c/D. \tag{12.58}$$

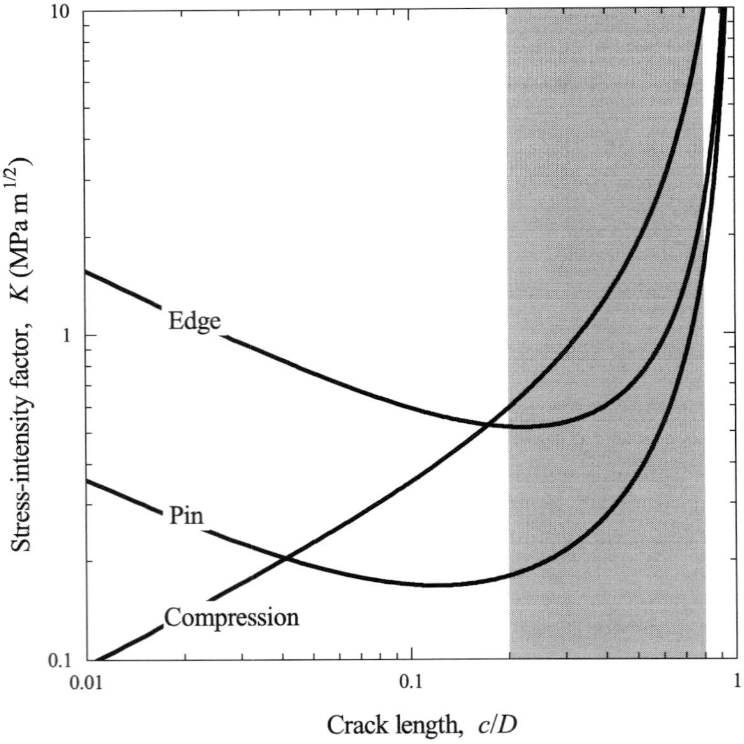

Figure 12.25 Plot of stress-intensity factor variations K as a function of crack length c for the edge, pin, and compression loaded equatorial crack system of Figures 12.24 and 12.26. The shaded gray band indicates the usual experimental domain of crack lengths, in which all three K fields are destabilizing.

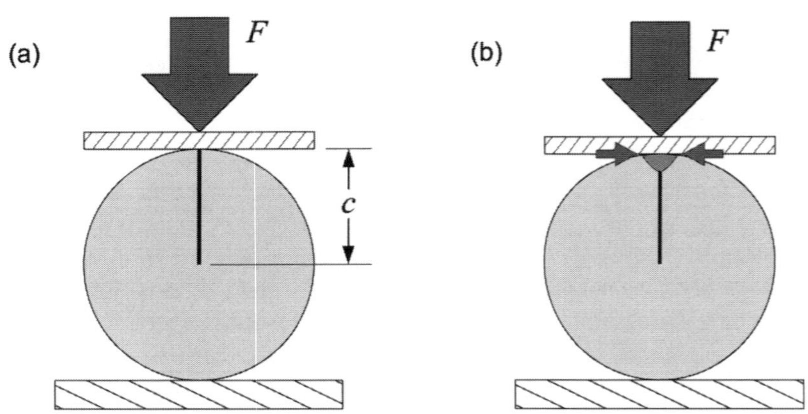

Figure 12.26 Schematic cross-section diagrams of cylindrical particles loaded in diametral compression by force F and containing meridional cracks c comparable to the particle size. (a) Elastic loading. (b) Elastic-plastic loading such that a plastic deformation zone develops at the contact and transverse tractions develop due to frictional effects.

The dimensionless geometrical constant $d_1 = 1.27$. Inspection of Eq. (12.58) shows that this SIF also consists of three elements. The first and third elements are as those shown previously, an amplitude term that includes the force F, and a destabilizing term that takes into account the length of the crack in the finite component. Distinct from the previous, the second element is a destabilizing $\xi^{1/2}$ term characteristic of uniform applied stress (section 12.4). For small cracks in the diametral compression system, $\xi \to 0$ and third term $\to d_1$. For long cracks in this system, $\xi \to 1$ and the entire SIF diverges as $\to d_1\xi^{1/2}/(1-\xi)^{3/2}$, similar to the edge and pin loaded systems seen earlier. Figure 12.25 shows the $K(c)$ behavior from Eq. (12.58) using the geometrical parameters seen here and a typical force value for this geometry, $F = 500$ N. The short crack destabilizing asymptote and long crack destabilizing divergence are clear. The behavior of K in the shaded band suggests that this geometry is more difficult to control and maintain near fracture equilibrium than those seen earlier, as noted by Kendall and Gregory. Kendall and Gregory also noted two further points regarding this geometry. The first point was that at large forces, a zone of plastic deformation, approximately 3 mm in diameter, formed at the polar contact and for short cracks completely precluded fracture. The second point was that Eq. (12.58) assumes that lateral constraint at the contact, leading to transverse compression, is negligible. At the large forces used (≈ 1 kN) this may not be so. Supporting this idea, a split probe that allowed lateral motion across the crack plane at the contact was observed to improve stability for long cracks. Figure 12.26b shows schematic representations of these points in the cracked disc diametral compression geometry. The points are considered below in the context of toughness estimation.

Figure 12.27 shows a semi-logarithmic plot of toughness estimates of PMMA using the disc geometries of Figures 12.24 and 12.26, from the work of Kendall and Gregory (1987) using data derived from the published work. The horizontal gray line indicates the toughness previously measured for this material, approximately 1.1 MPa $m^{1/2}$. The filled symbols show measurements from the tensile disc geometries using Eqs. (12.56) and (12.57). Despite a systematic minimum in the data as crack length was varied, the tensile disc geometry estimates agree with the previous measurement, although the range of values, (1.0 ± 0.3) MPa $m^{1/2}$, is large. As noted by Kendall and Gregory, the forces observed in the compression disc geometry measurements were significantly greater than those expected from the known toughness. Direct application of Eq. (12.58) to the measured force and crack length values gives toughness values too great by a factor of 2–3. However, two alterations to Eq. (12.58) according to the points above can "correct" the values. The first alteration is to assume that the crack lengths in the diametral compression experiments were shortened by the plastic deformation zone at the contact, such that $c \rightarrow c' = c - 3$ μm. The second alteration takes advantage of the additivity of SIFs. The divergent element of the SIF for the crack in the edge-loaded disc, the large ξ limit of Eq. (12.56) discussed earlier, is identical in form to the SIF for the crack in the diametral compression loaded disc, Eq. (12.58). At large crack lengths in the diametral geometry the transverse compression of the lateral constraint thus acts as a superposed *negative* edge loaded SIF asymptote. As the two SIFs have the same functional form, the effect of transverse compression is to simply reduce the amplitude of the SIF. Here, the alteration was $F \rightarrow F' = 0.5F$. The open symbols in Figure 12.27 show the altered measurements from the compressive

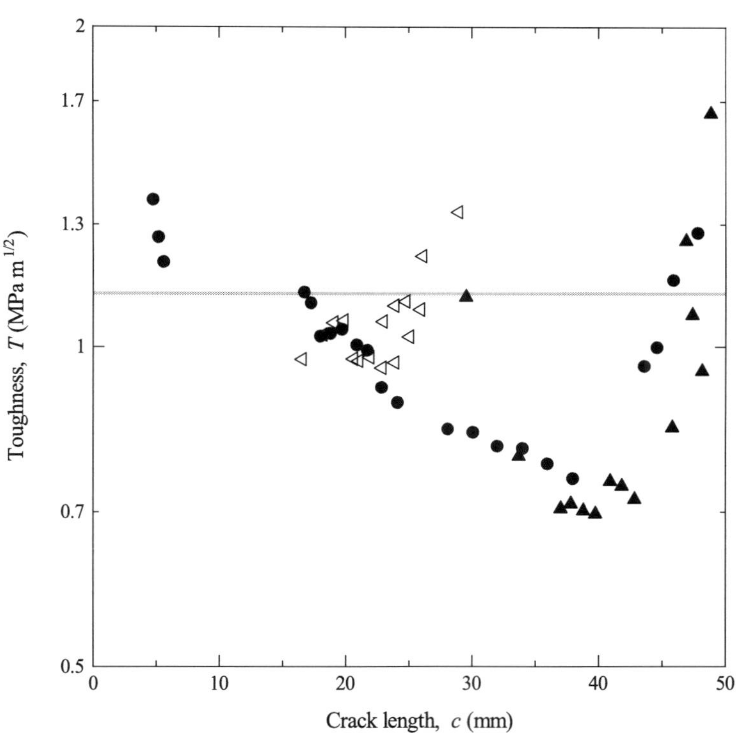

Figure 12.27 Toughness estimates for PMMA using cracked cylindrical disc geometries. Solid symbols indicate results from edge and pin loaded components, Figure 12.24. Open symbols indicate results from compression loaded components, Figure 12.26, corrected for local deformation zone and transverse frictional force effects.

disc geometry, using Eq. (12.58). The measurements now agree with the known toughness value, although the agreement is somewhat empirical.

12.6.2 Microcellulose Tablets

In an application of Eqs. (12.56) and (12.57) by Roberts and Rowe (1989), the toughness of compacted microcellulose tablets was evaluated using cracked diametral compression and edge-loaded tensile geometries. The tablets were 15 mm in diameter and compacted with various peak forces to generate porosities in the range 0.1–0.4, similar to those considered in Chapter 10. Cracks, approximately 5–13 mm long were formed in the tablets by a driven blade and measured by optical microscopy prior to loading. Both Roberts and Rowe (1989) and Jarosz and Parrott (1982) (see Chapter 10) used both compressive and tensile loading geometries to measure the fracture properties of tablets. However, the toughness measurements of Roberts and Rowe are distinguished from the strength measurements of Jarosz and Parrot by the use of deliberately included cracks.

Figure 12.28 shows a plot of toughness as a function of porosity from the work of Roberts and Rowe (1989), using data derived from the published work. Filled symbols represent edge loaded tensile measurements (Figure 12.24a), open symbols represent diametral compression measurements (Figure 12.26a). The data, reflecting the use of Eqs. (12.56) and (12.58) by Roberts and Rowe, were altered as discussed earlier by reductions of factors of $2^{1/2}$ and 2, respectively. The straight line is a guide to the eye. Both testing geometries exhibit a decrease in toughness with increasing porosity. The decrease is linear, $T \sim (1-P)$, probably reflecting linear decreases with porosity in both elastic modulus, $M \sim (1-P)$, and fracture resistance, $R \sim (1-P)$, as discussed in Chapter 10 and as $T = (ER)^{1/2}$.

Although the data from the two testing geometries appear separated in Figure 12.28, the diametral compression data represent a lower bound as a major effect that lead to the factor of ≈2 alteration in Figure 12.27, transverse constraint at large force contacts, was probably reduced considerably. Compaction at the contacts, equivalent to contact plasticity in the PMMA discs, was observed in the tablet measurements and was likely a significant factor in crack length reduction. Hence the data separation in Figure 12.28, although exaggerated relative to that published, is probably a consequence of unknown scalar amplitude factors. It is noted that the values here, although reduced from those reported are still very large

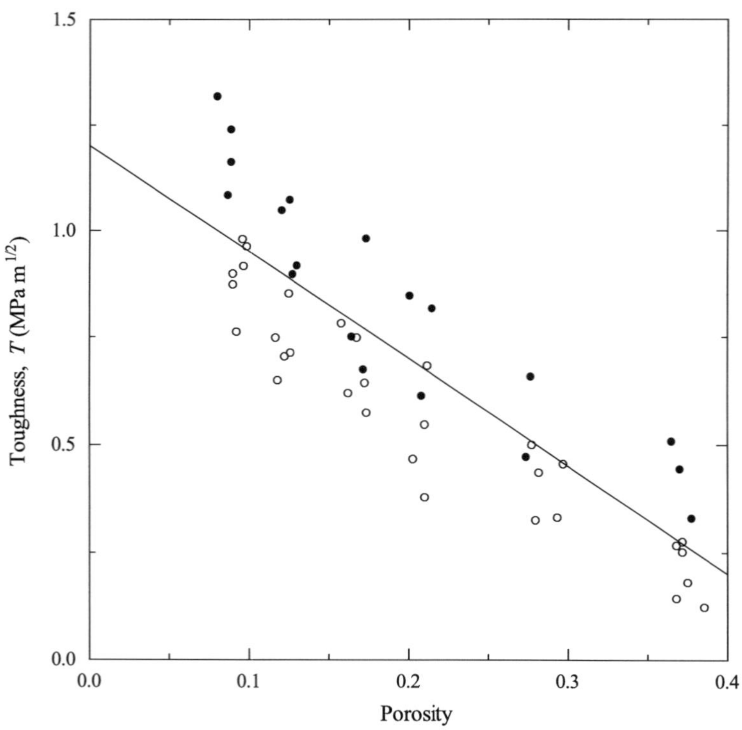

Figure 12.28 Toughness estimates for microcrystalline cellulose using cracked cylindrical disc geometries. Solid symbols indicate results from edge loaded components, Figure 12.24. Open symbols indicate results from compression loaded components, Figure 12.26, corrected for local deformation zone and transverse frictional force effects. The diagonal line is a guide to the eye.

for compacted porous tablets, ≈ 0.2–$1\ \mathrm{MPa\ m^{1/2}}$—the toughness of dense glass is $0.7\ \mathrm{MPa\ m^{1/2}}$. (The reported upper bounds of $\approx 2\ \mathrm{MPa\ m^{1/2}}$ are comparable to those of dense polycrystalline Al_2O_3, which seems even more unlikely.) An implication is that the cracks introduced into the specimens here were notch like and that the tests were closer to strength tests than toughness tests. A broader implication is that this observation may be true of all porous agglomerate fracture in which the crack tips are not well approximated as sharp in a continuum sense, but are rounded at the scale of the sub-particle porous microstructure.

12.6.3 Ductile-Brittle Transitions

In materials that can exhibit both plasticity and fracture, a characteristic length scale exists for deformation that divides predominantly ductile behavior, plasticity, from predominantly brittle behavior, fracture. The length scale is determined easily by simple scaling of material properties. The characteristic elastic energy density, energy per volume, required to deform a material at a longitudinal yield stress Y is Y^2/E, where E is the Young's modulus. The characteristic surface energy density, energy per area, required to fracture a material is R. The ratio of these two densities gives a characteristic length scale $a^* = (RE)/Y^2$. Deformation events larger than a^* are likely to lead to fracture and events smaller than a^* are likely to lead to plasticity. Increasing the system scale considered, a, from smaller than a^* to larger than a^* thus passes through a *ductile-brittle transition*. A classic example is sharp particle contact, for which the proportionality of hardness to yield stress is recognized, $H \approx 3Y$, and toughness T is recognized as $T = (RE)^{1/2}$. Substituting these terms into the ratio gives $a^* \sim T^2/H^2$, and, using $P \sim Ha^2$, $P^* \sim T^4/H^3$. Contact forces greater than P^* or contact impressions larger than a^* are likely to lead to cracking and are thus termed (cracking) threshold conditions, noting that the two parameters are not independent but connected by a material parameter, in this case the hardness or yield stress. Such scaling was recognized in the early studies of sharp indentation crack initiation (Lawn and Evans 1977) and shown more recently for blunt indentation (Rhee et al. 2001). A ductile-brittle transition in particle behavior was recognized by Kendall at about the same time as the earlier work (Kendall 1978). Kendall noted that although large particles could usually be fractured in comminution processes, small particles often could not and exhibited plastic deformation under comminution conditions. In experiments on single particles loaded in diametral compression, Kendall demonstrated that small particles plastically deformed and large particles fractured. Kendall developed a ductile-brittle analysis that described this finding. The analysis of Kendall is reinterpreted here within the fracture mechanics framework used earlier for consideration of long cracks in particles.

The analysis of Kendall makes three key physical assumptions that can be encapsulated mathematically by modification to Eq. (12.58). The schematic diagram of Figure 12.29 illustrates the relevant dimensions. The first assumption is that all particles under consideration are planar (thickness b, not shown) and contain cracks of length c that are in fixed proportion relative to the particle size D, such that $\xi = c/D$ in Eq. (12.58) is invariant. It is noted that the value of ξ selected does not alter the destabilizing nature of K: Kendall and colleagues appear to have $\xi \approx 0.5$ in mind (Kendall 1978; Roberts et al. 1997). The second assumption made by Kendall is that the geometry term of the SIF can be modified by an expression $(1 - w/D)$, where w is a characteristic contact width of the loading force at the pole of the particle. Using these two assumptions, the SIF from Eq. (12.58) can be written as

$$K = \frac{F}{bD^{1/2}} d_2 \left(1 - \frac{w}{D}\right), \tag{12.59}$$

where $d_2 = d_2(\xi)$ is a dimensionless parameter that depends on the value of ξ used and is invariant throughout this analysis. The third assumption made by Kendall is that the contact generates plasticity in the particle such that the contact width w developed is always related to the contact force F by a characteristic yield stress Y of the material, $F = Ybw$. It is recognized that bw is the contact area in plane problems and thus Y connects a force and a dimension similarly to hardness discussed above for indentations. Combining this assumption into Eq. (12.59) gives

$$K = \frac{F}{bD^{1/2}} d_2 \left(1 - \frac{F}{YbD}\right), \tag{12.60}$$

which is a negative quadratic in F.

The physical interpretation of Eq. (12.60) is that the SIF passes through a maximum when the mean stress on the equatorial plane of the particle, F/bD is some fraction of the yield stress. A equivalent negative quadratic in w can also be formed

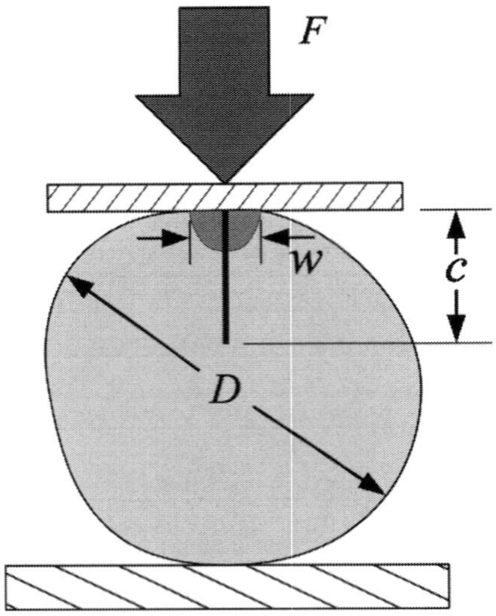

Figure 12.29 Schematic cross-section diagram of a particle loaded in diametral compression exhibiting both contact induced plasticity and meridional fracture. As particle size increases the deformation tendency changes from plasticity to fracture and the particle undergoes a ductile to brittle transition.

$$K = \frac{Yw}{D^{1/2}} d_2 \left(1 - \frac{w}{D}\right). \tag{12.61}$$

Equation (12.60) is related to the form originally proposed by Kendall with $d_2 = (3/2)^{1/2}$. Setting $\mathrm{d}K/\mathrm{d}F = 0$ or $\mathrm{d}K/\mathrm{d}w = 0$ gives the parameters for the maximum SIF:

$$\begin{aligned} w^* &= D/2 \\ F^* &= YbD/2 \\ K^* &= YD^{1/2} d_2/4 \end{aligned} \tag{12.62}$$

The conditions for the SIF to reach a maximum for the crack in the particle are that the contact width is half the particle size or that the mean stress on the equatorial plane is half the yield stress.

Setting the fracture equilibrium for the maximum in the SIF, $K^* = T$ gives the essential ductile to brittle result,

$$D^* = \left(\frac{4T}{Yd_2}\right)^2 = \frac{16ER}{d_2^2 Y^2}. \tag{12.63}$$

This particle size is the minimum size that will fracture on loading. The result is identical to that obtained by Kendall through the equivalent of setting $K = T$ in Eqs. (12.60) or (12.61) and solving the quadratic in F or w (Kendall 1978; Roberts et al. 1997). (Dimensional analysis will of course give all these results, as seen earlier, but without the numerical terms.) For $D \geq D^*$, the variation of fracture force F with particle size is the quadratic lower bound (−) solution of

$$F = \frac{YbD}{2} \left\{1 \pm \left[1 - (D^*/D)^{1/2}\right]^{1/2}\right\}. \tag{12.64}$$

For $D \gg D^*$, the asymptotic variation of the above is $F \rightarrow TbD^{1/2}/d_2$. For $D \leq D^*$, the maximum force is not given by the above but by the condition for complete yielding of the particle, $F = YbD$. Note that these limiting behaviors are determined by a combination of fracture (T) or plasticity (Y) parameters, and that at the transition the supported force decreases by a factor of 2. Comparison of Eqs. (12.61) and (12.58) using the value of d_2 used by Kendall and $w/D = 0.5$ suggests that a lower bound value of $\xi \approx 0.146$ enables the two SIF formulations to be matched (for these values, $(3/2)^{1/2}(1-w/d) \approx 0.61$).

Figure 12.30 shows a plot of experimental observations of a ductile-brittle transition in polystyrene components from the work of Kendall (1978), using data derived from the published work. The components were elongated pentagonal shapes, pre-cracked at a vertex at which diametral loading was performed. Note that the exact component shape does not matter except for (i) a single dimension D and (ii) the invariance of ξ. Symbols represent peak force values for individual measurements (using the cited $b = 3$ mm). The filled symbols indicate ductile deformation at the force indicated and the open symbols represent brittle failure. The solid line is a best fit to the data using Eq. (12.64) and the coupled complete yielding condition. The fit parameters used were $D^* = 4.5$ mm and the cited yield stress $Y = 80$ MPa. The D^* value is very close

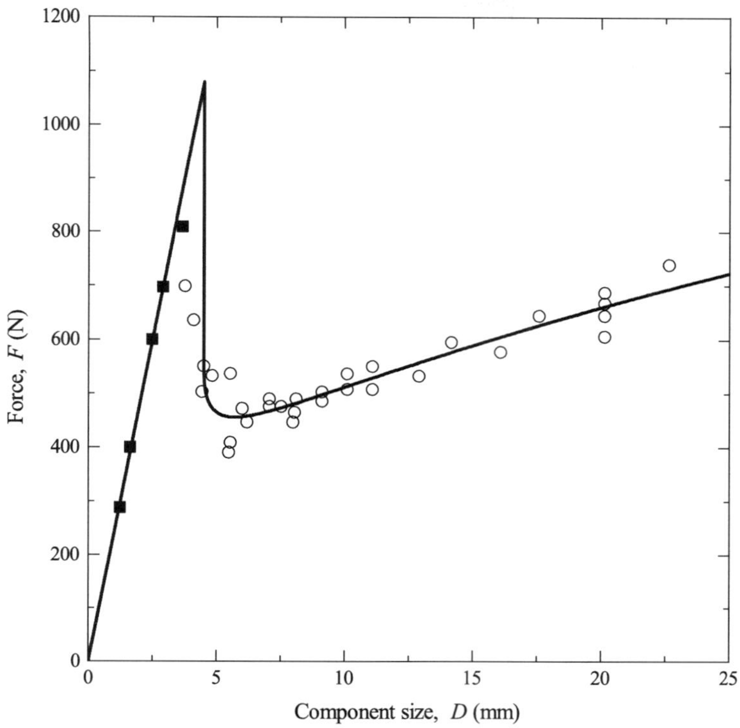

Figure 12.30 Plot of maximum supported force as a function of size for cracked polystyrene components loaded in diametral compression. Filled symbols indicate a ductile response and open symbols indicate a brittle response (Adapted from Kendall, K 1978). The solid line is a best fit to the data assuming a ductile-brittle transition model and independently determined yield and fracture parameters.

to 4.48 mm calculated by Kendall using Eq. (12.63) and the cited $E = 2.8$ GPa, $R = 960\ \mathrm{J\ m^{-2}}$, and $d_2 = 1.22$, noting that $T = (RE)^{1/2}$. The fit to the observations is clear and confirms the essential elements of the ductile-brittle model.

12.6.4 Agglomerate Compaction

An SIF formulation related to the ductile-brittle transition is that for fracture during compaction of multi-dimensional arrays or agglomerates of particles. This phenomenon was considered by Roberts et al. (1997) in experiments using cuboidal NaCl particles and in analysis that built on the ductile-brittle formulation of Kendall (1978). The analysis of Roberts et al. is re-expressed here in SIF terms, extending Eq. (12.59). The particle system considered by Roberts et al. is shown in the schematic cross section diagram of Figure 12.31. The particles are equiaxed and thus vary from those of Kendall in shape such that all dimensions are $\approx D$, but otherwise contact width and associated plastic zone dimension w and crack length c are identically defined. The contacts, from other particles in the array, are on all four sides and the axis of the particle lies in the plane of the included single crack. The major distinction in this system from that of Kendall is that in addition to the axial compressive force F, there is a transverse compressive force of magnitude X that characterizes the influences of neighboring particles in the array. (Plastic deformation zones at these contacts not shown.) The SIF describing this system is given by

$$K = K_F + K_X = \frac{F}{D^{3/2}}\left(1 - \frac{w}{D}\right) d_2(\xi) - \frac{X}{D^{3/2}} d_2(\xi)\xi, \tag{12.65}$$

where on the right side K_F is identified with the first term and K_X is identified with the second. The crack restraining action of the transverse compressive forces is apparent in the negative K_X term (the factor $2X$ used by Roberts et al. is not necessary in the SIF formulation).

In the ductile-brittle analysis developed earlier, attention was focused on the w/D loading geometry term and d_2 was taken as invariant. Here the opposite is the case, w/D is taken as invariant and attention is focused on the crack length dependence of the K terms. For simplicity, and without loss of generality in the phenomena described, $d_2(\xi)$ may be taken

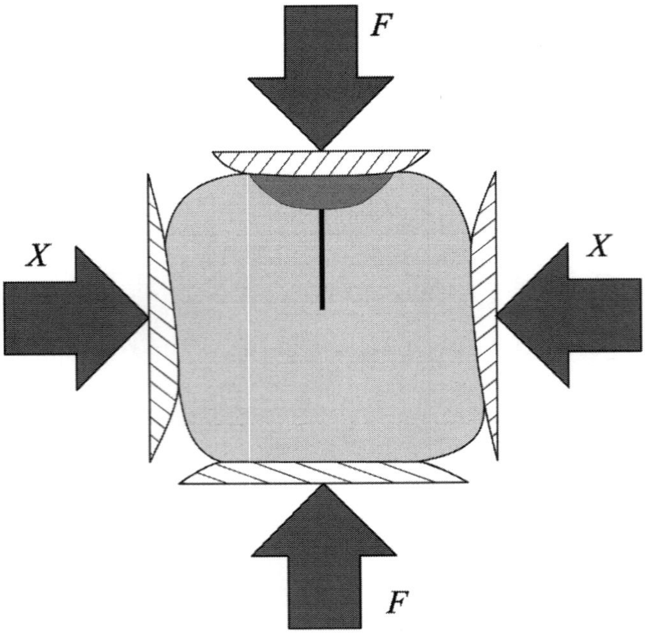

Figure 12.31 Schematic cross-section diagram of a particle loaded in diametral compression in both axial, *F*, and transverse, *X*, directions and exhibiting both contact induced plasticity and meridional fracture.

to leading order as $\xi^{1/2}$, such that Eq. (12.65) can be written as

$$K = \frac{\psi_F F}{D^2} c^{1/2} - \frac{\psi_X X}{D^3} c^{3/2}, \tag{12.66}$$

where ψ_F and ψ_X are dimensionless geometrical terms. A major similarity of Eq. (12.66) and Eq. (12.50) is that both consist of the sum of two independent terms, a destabilizing ($\mathrm{d}K_F/\mathrm{d}c > 0$) term and a stabilizing ($\mathrm{d}K_X/\mathrm{d}c < 0$) term. However, several major differences exist: (i) Total SIF $K > 0$ is required to ensure crack opening. As $K_X < 0$, the transverse force is thus constrained to a maximum value of $X \leq \psi_F FD/\psi_X c$. (ii) Both SIF terms are particle size dependent and have different dependencies. The relative importance of the K_X term diminishes at large particle sizes. (iii) Most importantly, the destabilizing element K_F dominates at small crack lengths and the stabilizing element K_X dominates at large crack lengths—the opposite of Eq. (12.50). The behavior of Eq. (12.66) is thus more akin to the crack initiation and stabilization SIF field of Eq. (12.47) and Figure 12.16. Hence, the position of maximum in the SIF of (12.66) gives the minimum stable equilibrium crack length in a particle, c^*. Setting $\mathrm{d}K/\mathrm{d}c = 0$ gives

$$c^* = \left(\frac{\psi_F F}{\psi_X X}\right)\frac{D}{3}. \tag{12.67}$$

If the compacted agglomerate is isotropic, the first term in parentheses is expected to be close to unity. Hence, to leading order the lower bound to the stable crack lengths in a compacted array of multi-sized particles is $c^* \approx D/3$. Inserting the crack length of Eq. (12.67) into the SIF expression of Eq. (12.66) gives the value of the maximum SIF,

$$K^* = \frac{\psi_F F}{D^{3/2}}\left[\left(\frac{\psi_F F}{\psi_X X}\right)^{1/2} \frac{4}{3^{3/2}}\right]. \tag{12.68}$$

Imposing the equilibrium conditions for fracture, $K^* = T$, and deformation, $F = YwD$, and recognizing that the term in brackets is near unity gives the minimum particle size in which cracks can form,

$$D^* = \frac{T^2}{Y^2}\left[\frac{1}{\psi_F(w/D)}\right]^2. \tag{12.69}$$

If the term in brackets is neglected, the expected ductile-brittle scaling as a function of particle size results, $D^* \sim T^2/Y^2$ and $F^* \sim T^4/Y^3$.

In more detail, required for prediction in a specific system, the SIF considerations show that $\psi_F \approx d_1 \approx 1.3$. The post compaction NaCl particle microstructural observations of Roberts et al. show that $w/D \approx 0.2$, such that $D^* \approx 15T^2/Y^2$.

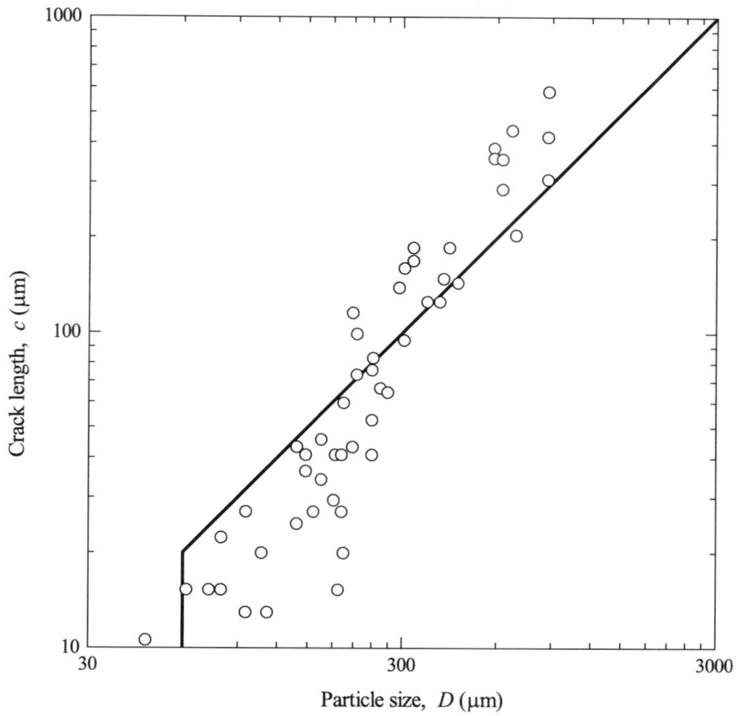

Figure 12.32 Plot of induced crack length as a function of size for cracked NaCl particles compacted in an array. Symbols indicate individual measurements (Adapted from Roberts, R. J et al. 1997). Solid line is a prediction assuming a combined loading stress-intensity factor model and independently determined yield and fracture parameters.

Using the cited values for the system of toughness $T = 0.18$ MPa m$^{1/2}$ and yield stress $Y = 90$ MPa, Eq. (12.69) gives $D^* \approx$ 60 μm. Figure 12.32 shows a logarithmic plot of observed crack length as a function of particle size from the measurements of Roberts et al., using data derived from the published work. The solid line represents the prediction of the threshold particle size for cracking and the ensuing crack length variation, using Eqs. (12.67) and (12.69). The symbols represent individual measurements. The agreement between the lower bound prediction and observations is within the factor of two arising from assumptions of values for geometry constants and material properties, although the threshold appears to be considerably underestimated. Nevertheless, the fit to the observations is clear and confirms the essential elements of the multi-axial fracture model.

12.7 Discussion and Summary

This chapter has outlined two fracture mechanics frameworks that describe brittle particle cracking and strength. The frameworks are from the viewpoints of either energy or stress changes during fracture. In the energy viewpoint, fracture was shown to be the conversion of mechanical work on a component into elastic energy distributed within the bulk of the component and then into surface energy localized to cracks created by fracture of the component. In the mathematically equivalent stress viewpoint, fracture was shown to be the result of force and displacement conditions imposed at the boundary of a component leading to intensification of stress surrounding a crack tip in the component. Fracture equilibrium was established by equality of quantities derived from boundary conditions, mechanical energy release rate or stress-intensity factor, and a material property, fracture resistance or toughness, in the energy or stress frameworks, respectively. The simplest loading geometries, uniform applied stress and localized applied force, were used as examples to illustrate the equivalency of, and relationship between, the energy and stress frameworks, and these geometries were shown to encompass the domain of stability for most fracture systems. A particularly useful formulation in engineering applications described the fracture strength of a component containing a simple crack under uniform applied stress, the Griffith equation, used throughout the book.

The stress framework was demonstrated as extremely effective in describing the diverse fracture behavior of brittle components—primarily through the use of the stress-intensity factor (SIF). The SIF, K, characterizes the enhanced stress field about crack tips and derives from the linear elastic behavior of fractured components. K is the major analysis tool in linear elastic fracture mechanics as it has the great advantage of scalar linear additivity for superposed sources of normal loading on cracked components (superposed shear loadings must be added vectorially—not considered here). Some very simply expressed and interpreted K functions were shown to describe the range of fracture phenomena observed in particles, especially strength phenomena: fracture at pores; crack initiation at misfitting inclusions, grains, and sharp contacts; crack propagation at stressed flaws; ductile-brittle transitions in small particles; and stable crack formation in particle compacts. Of particular interest for the subject of this book is the use of the SIF framework to interpret strength measurements in terms of underlying flaw characteristics. Analyses throughout have used the Griffith equation, as $\sigma = Bc^{-1/2}$, to relate strength σ to a strength-controlling flaw size characterized by a crack length c. The proportionality parameter B was recognized as a composite term that characterized both the fracture properties of the material and the nature of the flaw. The flaws described in the Griffith equation result from transient force effects, e.g. from shock, that leave cracks in a component kinetically trapped in metastable states, as the transient no longer exerts any significant stress in the fracture system. The cracks are thus characterized simply by a crack length c_0 and a dimensionless scalar ψ that characterizes crack geometry (e.g. circular, straight). The strength of components that contain such cracks is given by $\sigma_0 = Bc_0^{-1/2}$. The proportionality term is given by $B = T/\psi$, where T is the toughness of the material, the equilibrium value of K in the, usually inert, fracture environment.

Crack initiation forces in many cases are not limited to transient effects and significant residual stress fields often result from crack initiation events. The residual stress fields stabilize cracks at equilibrium lengths and significantly affect subsequent fracture behavior. A frequent cause of crack initiation in brittle materials is surface sharp contacts that generate localized plastic deformation zones and surrounding, stabilizing, residual stress fields. The superposed fields decrease the strengths of components containing such strength-controlling flaws from the Griffith strength to $\sigma_m = B'c_0^{-1/2}$, where $B' = (3/4^{4/3})B \approx 0.47B$. Note that σ_m reflects the two-step strength sequence of Figure 12.21: (i) flaw generation, followed by (ii) component loading. Quantitative specification of the flaw can be extended to variables beyond crack size, to features or parameters more descriptive of the flaw initiation event. In the case of contact flaws, sharp or blunt, flaws are more conveniently specified by the peak contact load P. For misfit inclusions or large grains, the parameter is the inclusion or grain size a. For impact flaws, the parameter is the impacting element kinetic energy U_K. The analyses of this chapter shows that the resulting strengths can all be expressed in a common format:

$$
\begin{aligned}
\sigma_0 &= Bc_0^{-1/2} \qquad \text{(Griffith)} \\
\sigma_m &= B'P^{-1/3} \\
\sigma_m &= B'a^{-2/3} \\
\sigma_m &= B'U_K^{-2/9}
\end{aligned}
\tag{12.70}
$$

In each case B' is appropriately defined in terms of material and geometry variables, e.g. for contact flaws the pre-factor to $P^{-1/3}$ is $B' = 3T^{4/3}/4^{4/3}\psi\chi^{1/3}$, Eq. (12.55). If flaw generation is extended to include relaxation of the initiating stress field, the Griffith crack is seen to be a specific example of the two-step strength sequence that describes shocked, annealed, corroded, or abraded flaws. In terms of the analyses in this book, Eq. (12.70) makes clear that all the interpretations of strength in terms of underlying crack lengths, $c = (B/\sigma)^2$, used extensively in Chapters 4–11, can be recast as $x = (B'/\sigma_m)^y$, where x is a characteristic flaw scale parameter and y is the conjugate exponent relating strengths and flaws (e.g. P and 3). Equation (12.70) makes clear that the analyses of Chapters 3 and 4 are thus completely general in terms of strength interpretation. The well-known Griffith equation provides a means of comparing behavior of a wide range of systems, but other equations, used within the same strength analyses, may provide greater insight for specific systems.

In the SIF framework, the parameters P, a, U_K appearing in Eq. (12.70) are amplitude terms in functions describing destabilizing to stabilizing transitions in residual $K(c)$ fields describing crack initiation. The far field crack length dependencies, describing stable equilibrium cracks and applicable in strength analyses, are usually of the form $K \sim c^{-r/2}$. The exponent r takes limiting values of $r = 1$ and $r = 3$ depending on the symmetry of the system. For straight cracks with residual fields that act in the far field as localized line forces, r takes the value 1. For circular cracks with residual fields that act in the far field as localized point forces, r takes the value 3. The near field crack length dependencies, describing unstable equilibrium crack nuclei and applicable in initiation analyses, are nearly always of the form $K \sim c^{1/2}$ and associated with limiting peak values of stress at crack initiating features. The $K(c)$ behavior between these two limits is always characterized by a maximum, but the detailed shape depends on the assumed symmetries of the system and the assumed functional form of the stress field surrounding a feature. In early applications of $K(c)$ determination, Frank and Lawn (1967) and Lawn (1968) used linearly varying or Hertzian stress fields and straight crack symmetry to (successfully) describe circular cone cracks at blunt indentations. Lawn and Evans (1977) used a linearly varying stress field approximation and circular crack symmetry to describe crack initiation at sharp indentations. The same formulation was used by Swain (1981) to describe crack initiation at inclusions in glass. Green (1980, 1981) used the spherical void stress field and circular symmetry to describe cracking at pores and a decreasing cubic field and circular symmetry to describe cracking at inclusions. Cook and Thurn (2002) used an empirical formulation matched to the limiting behaviors of short and long straight cracks to describe fracture of dielectrics by metallic lines in microelectronic circuits. The analyses here most closely matched those of Green (1980, 1981). Examination of all these cases suggest that the choices of symmetry and stress field do not affect the description of crack initiation. The scaling of the relevant parameters through a standard integral resembling Eq. (12.46) always results: linear dependence on the amplitude term, inverse dependence on the crack length, and an integration limit that describes the system size.

The analyses in Sections 12.5 and 12.6 have been performed in the crack tip frame of reference. In these analyses all influences on a crack have been evaluated as $K(c)$ terms. Fracture equilibrium and stability were considered through the behavior of a sum of the $K(c)$ terms relative to the value of an invariant, intensive material property, toughness T. Toughness is related to other intensive material properties, surface energy and elastic moduli. This frame of reference and mode of analysis is unambiguous if there is a single influence on a crack (e.g. a residual field alone), and natural, if multiple influences are active and all represent positive, crack driving, forces (e.g. a residually stressed flaw under the influence of a uniform applied stress). However, different frames of reference may be appropriate if multiple influences act on a crack. These frames of reference lead to different modes of analysis, particularly if some of the influences are negative, crack resistive forces. In such systems, the frame of reference of the predominant cracking driving force, usually perceived as that of the "applied loading," may provide greater insight and simpler interpretation of the extensive behavior of the cracked component. In this frame, the fracture behavior of the material is extensive and depends on crack length. Such dependence is quantified in the concept of fracture resistance curves, commonly known as R-curve behavior.

Generation of stable cracks in compacted arrays of particles provides an example of material R-curve behavior perceived in the applied loading frame of reference. The equilibrium fracture condition for fracture of particles in the array, described in the crack-tip frame by Eq. (12.66) is

$$K = K_F + K_X = \frac{\psi_F F}{D^2} c^{1/2} - \frac{\psi_X X}{D^3} c^{3/2} = T_0, \tag{12.71}$$

where the 0 in T_0 emphasizes the minimum, invariant nature of this toughness value. This equation can be re-arranged so that equilibrium is given by

$$K_F = \frac{\psi_F F}{D^2} c^{1/2} = T_0 - K_X = T_0 + T_X = T_0 + \frac{\psi_X X}{D^3} c^{3/2}. \tag{12.72}$$

The change in the frame of reference leads to the identification of a toughening element with a negative SIF, $T_X = -K_X$. This identification leads to a perception in the reference frame of the axial applied loading F that the material becomes tougher as cracks extend, $T = T(c)$. The existence of the transverse force X is not perceived in the axial frame. In the axial frame, the stability of the meridional cracks is a consequence of the material toughness derivative exceeding the loading SIF derivative $\mathrm{d}T/\mathrm{d}c > \mathrm{d}K_F/\mathrm{d}c$.

From an energy perspective, the Irwin relation, Eq. (12.33), shows that the fracture resistance of the material in the applied loading frame of reference is given by $R = T^2/E = [T_0 + T_X(c)]^2/E = R(c)$ and the fracture resistance is given by a function of c, an R-curve. If attention is restricted to the SIF description, the varying toughness is given by a T-curve, $T(c)$. The T-curve above was associated with transverse forces acting on a particle to close the crack. A frequently encountered T-curve is is that associated with microstructural tractions acting behind a crack tip that act to close a crack. Such tractions are represented by $K_\mu(c) < 0$ and have been most intensively studied in ceramic materials using controlled indentation flaws (Lawn 1993; Lawn et al. 1993; Padture et al. 1993a; Cook 2015). In this case the revised equilibrium condition is given by an extension of Eq. (12.50):

$$K = K_A + K_F + K_\mu = T_0. \tag{12.73}$$

Re-arranging as above gives

$$K = K_A + K_F = T_0 - K_\mu = T_0 + T_\mu. \tag{12.74}$$

Note that in this case the applied loading frame of reference includes both the uniform applied stress and the localized indentation field. Judicious choice of the functional form of $T_\mu(c)$ ($\sim 1 - c^{-3/2}$) enables Eq. (12.74) to be solved similarly to Eq. (12.50) for the applied stress at equilibrium instability—the strength—of toughened materials (Cook 2015). The small flaw region of such strength calculations is often of great interest. Intrinsic flaws in materials are usually small and the behavior of such flaws in toughened materials thus determines the intrinsic strength. Such intrinsic flaws do not typically act in the weak toughening limit (see Section 12.5) and fracture stability as a function of crack length is altered by material toughening. Most particles are expected to exhibit short-crack toughened fracture behavior described by Eq. (12.74). This point is addressed in detail in Chapter 13.

In addition to microstructural affects remote from the crack tip, and therefore often best studied in the applied loading frame of reference, kinetic effects are active at the crack tip and are best studied in the crack tip frame. Kinetic effects are important in determining the strengths of particles if reactive species can gain access to a tensile region of the particle during loading. This is particularly important for the pervasive case of inorganic oxide or related particles, in which water or moisture is extremely surface reactive leading to stress-corrosion cracking. Such cracking generates loading rate effects in brittle material strength tests as the kinetics of crack-tip bond reactions affect crack propagation velocities and thence fracture stability. At slow loading rates, the strength is invariant, reflecting near complete reaction, for water, hydrolysis, of the cracked surface. As the rate increases the strength increases, reflecting incomplete reaction and greater surface energy. The phenomena are discussed in detail elsewhere (Lawn 1993; Cook 2015), with the essential point that inorganic particle strength should increase with increasing loading rate. Such effects have been observed (Yashima, 1989, as reported by Ryu and Saito 1991) for borosilicate glass, quartz, and limestone particles, in which the strength doubled as the loading rate was increased by about a factor of 10^7, consistent with other brittle material oxide observations (Cook 2015). Similar strength doubling changes with changes in loading time of about a factor of 10^6 were observed in compressed sand arrays and attributed to fracture kinetic effects (Karimpour and Lade 2010) and increases in strength were observed for individual sand particles with increased static loads (Liu and Wang 2018). In all the particle works, the particle materials were dense and crack propagation effects were likely restricted to particle surfaces. If the particles contained open porosity, e.g., the bauxites of Bertrand et al. (1988), moisture assisted crack propagation effects could affect the tensile central volume of the particle. Such effects leading to strength increases with stressing rate tests were observed in coral (Ma et al. 2019).

This chapter has focused on energy and stress analyses applicable to the fracture behavior of single particles. Insight was provided into the nature of the strength-controlling flaws underlying the strength distributions considered in earlier chapters (6–9 in particular). The following chapter, Chapter 13 considers the consequences of fracture behavior of multiple particles and the energies, stresses, and forces required to fracture entire distributions. Such considerations require application of the linkages between the intensive properties of materials and extensive properties of particle components.

References

Atkins, A.G. and Mai, Y-W. (1985). *Elastic and Plastic Fracture.* Ellis Horwood.

Benbow, J.J. (1960). Cone cracks in fused silica. *Proceedings of the Physical Society* 75: 697–699.

Bertrand, P.T., Laurich-McIntyre, S.E., and Bradt, R.C. (1988). Strengths of fused and tabular alumina refractory grains. *American Ceramic Society Bulletin* 67: 1217–1221.

Broek, D. (1982). *Elementary Engineering Fracture Mechanics.* Martinus Nijhoff.

Brzesowsky, R.H., Spiers, C.J., Peach, C.J., and Hangx, S.J.T. (2011). Failure behavior of single sand grains: Theory versus experiment. *Journal of Geophysical Research* 116: B06205.

Case, E.D., Smyth, J.R., and Hunter, O. (1980). Grain-size dependence of microcrack initiation in brittle materials. *Journal of Materials Science* 15: 149–153.

Chao, L.-K. and Shetty, D.K. (1992). Extreme-value statistics analysis of fracture strengths of a sintered silicon nitride failing from pores. *Journal of the American Ceramic Society* 75: 2116–2124.

Chen, C.S. and Hsu, S.C. (2001). Measurement of indirect tensile strength of anisotropic rocks by the ring test. *Rock Mechanics and Rock Engineering* 34: 293–321.

Cleveland, J.J. and Bradt, R.C. (1978). Grain size/microcracking relations for pseudobrookite oxides. *Journal of the American Ceramic Society* 61: 478–481.

Cook, R.F. (2015). Multi-scale effects in the strength of ceramics. *Journal of the American Ceramic Society* 98: 2933–2947.

Cook, R.F. (2020). A critical evaluation of indentation crack lengths in air. *Journal of the American Ceramic Society* 103: 2278–2295.

Cook, R.F. and Pharr, G.M. (1990). Direct observation and analysis of indentation cracking in glasses and ceramics. *Journal of the American Ceramic Society* 73: 787–817.

Cook, R.F. and Thurn, J. (2002). Stable dielectric fracture at interconnects from electromigration stresses. *Acta Materialia* 50: 2627–2637.

Deer, W.A., Howie, R.A., and Zussman, J. (1966). *An Introduction to the Rock Forming Minerals*. Longman.

Elghazel, A., Taktak, R., and Bouaziz, J. (2015). Determination of elastic modulus, tensile strength and fracture toughness of bioceramics using the flattened Brazilian disc specimen: Analytical and numerical results. *Ceramics International* 41: 12340–12348.

Frank, F.C. and Lawn, B. (1967). On the theory of Hertzian fracture. *Proceedings of the Royal Society of London. Series A. Mathematical and Physical Sciences* 299: 291–306.

Gorham, D.A. and Salman, A.D. (1999). Indentation fracture of glass and mechanisms of material removal. *Wear* 233: 151–156.

Gorham, D.A. and Salman, A.D. (2005). The failure of spherical particles under impact. *Wear* 258: 580–587.

Green, D.J. (1980). Stress intensity factor estimates for annular cracks at spherical voids. *Journal of the American Ceramic Society* 63: 342–344.

Green, D.J. (1981). Stress-induced microcracking at second-phase inclusions. *Journal of the American Ceramic Society* 64: 138–141.

Griffith, A.A. (1921). The phenomena of rupture and flow in solids. *Philosophical Transactions of the Royal Society London A* 221: 163–198.

Huang, J., Xu, S., Yi, H., and Hu, S. (2014). Size effect on the compression breakage strengths of glass particles. *Powder Technology* 268: 86–94.

Hull, D. and Bacon, D.J. (1984). *Introduction to Dislocations, 3rd ed.* Butterworth-Heinemann.

Inglis, C.E. (1913). Stress in a plate due to the presence of cracks and sharp corners. *Transactions of the Institution of Naval Architects* 55: 219–241.

Jarosz, P.J. and Parrott, E.L. (1982). Factors influencing axial and radial tensile strengths of tablets. *Journal of Pharmaceutical Sciences* 71: 607–614.

Johnson, K.L. (1985). *Contact Mechanics.* Cambridge.

Kanninen, M.F. and Popelar, C.H. (1985). *Advanced Fracture Mechanics.* Oxford.

Karimpour, H. and Lade, P.V. (2010). Time effects relate to crushing in sand. *Journal of Geotechnical and Geoenvironmental Engineering* 136: 1209–1219.

Kendall, K. (1978). The impossibility of comminuting small particles by compression. *Nature* 272: 710–711.

Kendall, K. and Gregory, R.D. (1987). Fracture of radially edge-cracked discs. Journal of Materials Science 22: 4514–4517.

Kingery, W.D., Bowen, H.K., and Uhlmann, D.R. (1975). *Introduction to Ceramics.* Wiley.

Kschinka, B.A., Perrella, S., Nguyen, H., and Bradt, R.C. (1986). Strengths of glass spheres in compression. *Journal of the American Ceramic Society* 69: 467–472.

Kuszyk, J.A. and Bradt, R.C. (1973). Influence of grain size on effects of thermal expansion anisotropy in $MgTi_2O_5$. *Journal of the American Ceramic Society* 56: 420–423.

Lawn, B.R. (1968). Hertzian fracture in single crystals with the diamond structure. *Journal of Applied Physics* 39: 4828–4836.

Lawn, B.R. (1991). Fundamental condition for the existence of microcrack clouds in monophase ceramics. *Journal of the European Ceramic Society* 7: 17–20.

Lawn, B.R. (1993). *Fracture of Brittle Solids, 2nd ed.* Cambridge.

Lawn, B.R. (1998). Indentation of ceramics with spheres: A century after Hertz. *Journal of the American Ceramic Society* 81: 1977–1994.

Lawn, B.R. and Evans, A.G. (1977). A model for crack initiation in elastic/plastic indentation fields. *Journal of Materials Science* 12: 2195–2199.

Lawn, B.R., Evans, A.G., and Marshall, D.B. (1980). Elastic/plastic indentation damage in ceramics: The median/radial crack system. *Journal of the American Ceramic Society* 63: 574–581.

Lawn, B.R. and Fuller, E.R. (1975). Equilibrium penny-like cracks in indentation fracture. *Journal of Materials Science* 10: 2016–2024.

Lawn, B.R., Padture, N.P., Braun, L.M., and Bennison, S.J. (1993). Model for toughness curves in two-phase ceramics: I, Basic fracture mechanics. *Journal of the American Ceramic Society* 76: 2235–2240.

Li, Z. and Bradt, R.C. (1989a). Micromechanical stresses in SiC-reinforced Al_2O_3 composites. *Journal of the American Ceramic Society* 72: 70–77.

Li, Z. and Bradt, R.C. (1989b). Thermoelastic micromechanical stresses associated with a large α-silicon carbide grain in a polycrystalline β-silicon carbide matrix. *Journal of the American Ceramic Society* 72: 459–466.

Liu, S. and Wang, J. (2018). Static fatigue of sand particles. *Canadian Geotechnical Journal* 55: 1682–1687.

Lubarda, V.A. (2019). Dislocation Burgers vector and the Peach–Koehler force: A review. *Journal of Materials Research and Technology* 8: 1550–1565.

Ma, L., Li, Z., Wang, M., Wei, H., and Fan, P. (2019). Effects of size and loading rate on the mechanical properties of single coral particles. *Powder Technology* 342: 961–971.

Muddle, B.C. and Hannink, R.H. (1986). Crystallography of the tetragonal to monoclinic transformation in MgO-partially-stabilized zirconia. *Journal of the American Ceramic Society* 69: 547–555.

Murakami, Y. (Ed.) (1987). *Stress Intensity Factors Handbook.* Pergamon.

Ohya, Y., Nakagawa, Z.E., and Hamano, K. (1987). Grain-boundary microcracking due to thermal expansion anisotropy in aluminum titanate ceramics. *Journal of the American Ceramic Society* 70: C-184–C-186.

Padture, N.P., Bennison, S.J., and Chan, H.M. (1993a). Flaw-tolerance and crack-resistance properties of alumina-aluminum titanate composites with tailored microstructures. *Journal of the American Ceramic Society* 76: 2312–2320.

Padture, N.P., Runyan, J.L., Bennison, S.J., Braun, L.M., and Lawn, B.R. (1993b). Model for toughness curves in two-phase ceramics: II, Microstructural variables. *Journal of the American Ceramic Society* 76: 2241-2247.

Rhee, Y.W., Kim, H.W., Deng, Y., and Lawn, B.R. (2001). Brittle fracture versus quasi plasticity in ceramics: a simple predictive index. *Journal of the American Ceramic Society* 84: 561–565.

Roberts, R.J. and Rowe, R.C. (1989). Determination of the critical stress intensity factor (K_{IC}) of microcrystalline cellulose using radially edge-cracked tablets. *International Journal of Pharmaceutics* 52: 213–219.

Roberts, R.J., Rowe, R.C., and Kendall, K. (1997). The influence of lateral stresses on brittle–ductile transitions in the die-compaction of sodium chloride. *Journal of Materials Science* 32: 4183–4187.

Roesler, F.C. (1956). Brittle fractures near equilibrium. *Proceedings of the Physical Society. Section B* 69: 981–992.

Ryu, H.J. and Saito, F. (1991) Single particle crushing of nonmetallic inorganic brittle materials. *Solid State Ionics* 47: 35–50.

Sadd, M.H. (2009) *Elasticity.* Elsevier.

Shan, J., Xu, S., Liu, Y., Zhou, L., and Wang, P. (2018). Dynamic breakage of glass sphere subjected to impact loading. *Powder Technology* 330: 317–329.

Selsing, J. (1961). Internal stresses in ceramics. *Journal of the American Ceramic Society* 44: 419.

Shetty, D.K., Rosenfield, A.R., and Duckworth, W.H. (1985). Fracture toughness of ceramics measured by a chevron-notch diametral-compression test. *Journal of the American Ceramic Society* 68: C-325–C-327.

Sih, G.C. (1973). *Handbook of Stress-Intensity Factors.* Lehigh University.

Swain, M.V. (1981). Nickel sulphide inclusions in glass: an example of microcracking induced by a volumetric expanding phase change. *Journal of Materials Science* 16: 151–158.

Tada, H., Paris, P.C., and Irwin, G.C. (1973). *The Stress Analysis of Cracks Handbook.* Del Research.

Todd, R.I. and Derby, B. (2004). Thermal stress induced microcracking in alumina–20% SiCp composites. *Acta materialia* 52: 1621–1629.

Watkins, I.G. and Prado, M. (2015). Mechanical properties of glass microspheres. *Procedia Materials Science* 8: 1057–1065.

Watts, J.L. (2011). Stress measurement and development of zirconium diboride-silicon carbide ceramics. Doctoral Dissertation. 2010. https://scholarsmine.mst.edu/doctoral dissertations/2010

Wiederhorn, S.M. and Lawn, B.R. (1979). Strength degradation of glass impacted with sharp particles: I, Annealed surfaces. *Journal of the American Ceramic Society* 62: 66–70.

Williams J.G. (1984). *Fracture Mechanics of Polymers.* Ellis Horwood.

Yoffe, E.H. (1982). Elastic stress fields caused by indenting brittle materials. *Philosophical Magazine A* 46: 617–628.

Zeng, K., Breder, K., and Rowcliffe, D.J. (1992a). The Hertzian stress field and formation of cone cracks–I. Theoretical approach. *Acta metallurgica et materialia* 40: 2595–2600.

Zeng, K., Breder, K., and Rowcliffe, D.J. (1992b). The Hertzian stress field and formation of cone cracks–II. Determination of fracture toughness. *Acta metallurgica et materialia* 40: 2601–2605.

Zhu, W.C., Chau, K.T. and Tang, C.A. (2004). Numerical simulation on failure patterns of rock discs and rings subject to diametral line loads. *Key Engineering Materials* 261: 1517–1522. Trans Tech.

13

Applications and Scaling of Particle Strengths

This chapter examines application of particle strengths and strength distributions in four examples: energy consumption in particle comminution, reliability of grinding particles, mass effects in particle handling, and microstructural effects on particle failure. Scaling effects in particle behavior and strength data, enabling transformations of strength information from intensive characterization parameters into extensive design parameters, are demonstrated in each example. The examples build on experimental observations presented in earlier chapters and the fracture mechanics analyses of Chapter 12. The importance of unbiased coordinates in strength distribution reporting and presentation is emphasized.

13.1 Introduction

To be useful in application, particle strength distributions must be converted from the intensive terms of materials science (e.g. density, stress, strain) to the extensive terms of materials engineering (e.g. mass, force, dimension). Such conversions rely on simple power-law scaling relating intensive and extensive properties. Material density ρ is converted into particle mass m by the particle diameter D cubed, $m \sim \rho D^3$. Stress σ is converted into force F by the particle diameter squared, $F \sim \sigma D^2$. Strain ε is converted into a change in dimension ΔL by the particle diameter, $\Delta L \sim \varepsilon D$. Associated with stress and strain in a material is the elastic energy density $\mathcal{U}_{\mathrm{E}}$ that scales as stress squared and inversely with an elastic modulus M of the material, $\mathcal{U}_{\mathrm{E}} \sim \sigma^2/M$. Combining this expression with that for stress shows that the total elastic energy in a particle loaded by force F scales inversely with particle size, $U_{\mathrm{E}} \sim F^2/MD$. A variation in such scaling was used in Chapter 12 in analysis of contact forces.

In this chapter, four applications of particle strength distributions are analyzed using such scaling. The first two applications reflect the divergent approaches to particle strength noted in Chapter 1. In some cases particle fracture and fragmentation is a desired outcome, as in ore and cement crushing in mining and construction. The first section here estimates the energy required to crush particles based on particle size-dependent strength distributions. In some cases fracture and fragmentation are to be avoided and particle longevity is desired, as in surface grinding in manufacturing. The second section estimates particle reliability based on particle size-dependent complementary strength distributions. The second two applications reflect decisions often faced in particle fabrication involving two forms of strength-controlling flaws, surface flaws generated by contacts and bulk flaws influenced by material microstructure. The third section estimates particle failure force distributions based on particle size and surface flaws generated by particle impact. The fourth section estimates particle failure force distributions based on particle material grain size and bulk flaws controlled by material toughening effects. The third and fourth sections extend the fracture mechanics analyses of Chapter 12.

13.2 Particle Crushing Energy

Figure 13.1a shows plots of strength cdf $H(\sigma)$ variations repeated from Figure 9.31. The variations are fits by a deterministic extreme value model to the strength edf $Pr(\sigma)$ behavior for quartz particles shown in Figure 9.7 from the work of Tavares and King (1998). The particle sizes were $D = 4, 2, 1, 0.5$, and 0.25 mm, shown left to right in Figure 13.1a; the largest 4 mm particles are indicated by the black solid line, the smallest particles are indicated by the black dashed line, and intermediate sizes by the gray lines. This convention will be followed throughout this section. The fits were based on an invariant concave

Particle Strengths: Extreme Value Distributions in Fracture, First Edition. Robert F. Cook.

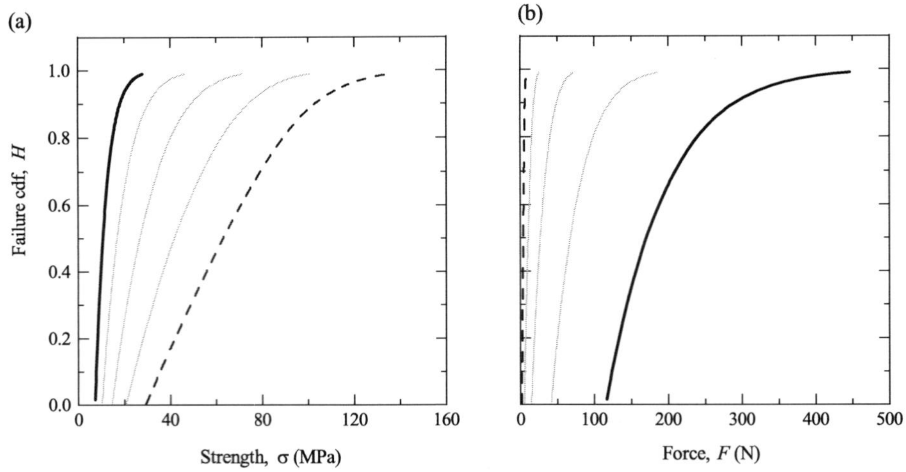

Figure 13.1 (a) Strength cdf $H(\sigma)$ variations for quartz particles, sizes $D = 4$ mm (dashed line), 2 mm, 1 mm, 0.5 mm, and 0.25 mm (bold line), left to right, from Figure 9.7. (b) Failure force cdf $H(F)$ variations (same notation) determined from (a) using $F = \sigma D^2$. Note reversal, size effect now right to left.

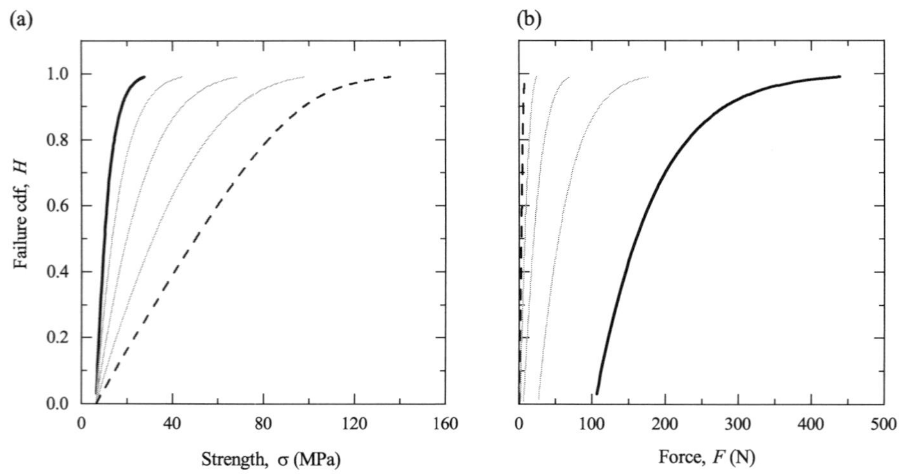

Figure 13.2 (a) Strength cdf $H(\sigma)$ variations for quartz particles, sizes $D = 4$ mm (dashed line), 2 mm, 1 mm, 0.5 mm, and 0.25 mm (bold line), left to right, predicted from $D = 4$ mm response in Figure 13.1a and assuming a stochastic size effect. (b) Failure force cdf $H(F)$ variations (same notation) determined from (a) using $F = \sigma D^2$. Note reversal, size effect now right to left, and similarity to Figure 13.1b.

population of fundamental element strengths (not shown) with bounds of 6.5 MPa and 165 MPa and conjugate crack lengths determined using $B = 0.4$ MPa m$^{1/2}$. The number of cracks in each particle was given by $k = 20(D/4)$ and the upper bound crack length in each particle was given by $c_U = 3(D/4)$ μm. The data display the familiar patterns of decreasing strength with increasing particle size and separated thresholds indicative of a deterministic system. The fits describe the data well; details are given in Chapter 9. Figure 13.1b shows plots of failure force cdf $H(F_{max})$ variations determined from Figure 13.1a using $F_{max} \sim \sigma D^2$. The most obvious feature of Figure 13.1b is that the failure distributions are reversed relative to Figure 13.1a—the failure force distributions increase with increasing particle size. In addition to the force scaling, crack length scaling gives $\sigma \sim c^{-1/2}$ (Griffith equation) and $c \sim D$ (deterministic model used earlier). The overall expected scaling of failure force is thus $F_{max} \sim D^{3/2}$, as observed.

A variation of the above is to assume stochastic extreme value behavior. Predicted strength distributions as a function of particle size are then generated using the invariant strength population and varying only the number k of flaws in each particle. Figure 13.2a shows such predictions using $k = 20(D/4)$. The distributions, with the exception of the invariant threshold characteristic of stochastic systems, are very similar to those of Figure 13.1a and display decreasing "fanned out" strength distributions with increasing particle size. Figure 13.2b shows plots of failure force cdf $H(F_{max})$ variations determined from Figure 13.2a using $F_{max} \sim \sigma D^2$. The most obvious feature here is that Figure 13.2b closely resembles Figure 13.1b—the

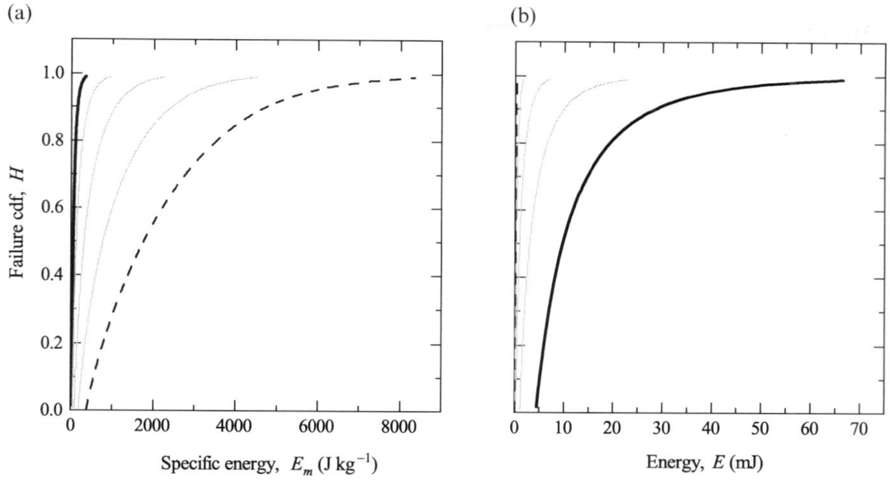

Figure 13.3 (a) Specific failure energy cdf $H(E_m)$ variations for quartz particles, sizes $D = 4$ mm (dashed line), 2 mm, 1 mm, 0.5 mm, and 0.25 mm (bold line), left to right, determined from Figure 13.1. (b) Failure energy cdf $H(E)$ variations (same notation) determined from (a) using $E = E_m D^3$. Note reversal, size effect now right to left.

failure force distributions are reversed relative to those of strength and the distributions are separated. The stochastic distribution feature of an invariant strength threshold is removed in considerations of failure force. Comparison of Figures 13.1 and 13.2 emphasizes that assessment of extreme value scaling effects (e.g. deterministic vs stochastic) must be performed with intensive rather than extensive quantities.

Figure 13.3a shows plots of specific failure energy (energy/mass) cdf $H(E_m)$ variations determined from Figure 13.1a using $E_m \sim (\sigma/1.5)^2$, as detailed in Chapter 2 (the relation averages over specific elastic modulus variations for common brittle particles). These variations closely replicate and are in the same units (J kg^{-1}) as those published by Tavares and King (1998). As earlier, when plotted as an intensive quantity, here E_m, the failure distribution of the largest particles exhibits the smallest values. Figure 13.3b shows plots of failure energy (energy/particle) cdf $H(E)$ variations determined from Figure 13.3a using $E \sim E_m \rho D^3$, where ρ is the particle density (here assumed 3 Mg m^{-3}). Once again, the conversion from an intensive to an extensive variable reverses the trend of failure distributions such that the largest particles exhibit the largest failure energies E. The scale of Figure 13.3b, tens of mJ/particle, should be noted. Very little energy is required to fracture a single particle. Figure 13.3b, based on single particle strength tests, is the basis for prediction of the energy required to crush particles. As noted by comparison of Figures 13.2 and 13.1, assuming stochastic as opposed to deterministic effects makes little difference to energy considerations.

The means and standard deviations of the failure energy distributions of Figure 13.3b are shown as a function of particle size by the filled symbols and bars in the logarithmic plot of Figure 13.4. Note the reversed abscissa with large particles on the left of the plot. The lower solid line is of slope 2, indicating that mean failure energy/particle $\bar{E}$ varies as $\bar{E} \sim D^2$. A sequential fragmentation model can use data such as these to estimate the *cumulative* energy required to reduce the size of particles in a crushing operation. Such an operation is represented by proceeding from large to small particles, left to right in Figure 13.4. Figure 13.5 shows a schematic diagram of such sequential fragmentation, using a cubic particle as an illustration. The left to right sequence in Figure 13.5a shows an initially intact particle, dimension D_0, first divided into two fragments, then the two fragments divided to generate four fragments, and finally the four fragments divided into eight. In each step $i = 0, 1, 2, 3$ in this sequence ($i = 0$ is the initial condition), the number of fragments increased as 2^i, such that the volume of the fragments halved at each step. The mean linear dimension, the fragment size D_i decreased as $D_i = D_0 2^{-i/3}$. After $i = 3$ steps, there were 8 fragments, the mean fragment size was $D_3 = D_0/2$, and mean fragment volume was $D_0^3/8$. The energy E_i required to fracture a single fragment of size D_{i-1} at step i, following the experimental trend exhibited in Figure 13.4, is $E_i = E_0(D_{i-1}/D_0)^2$, where E_0 is the energy required to fragment the initial particle. Fragmentation steps 3, 4, 5, 6 are shown in Figure 13.5b, and fragmentation steps 6, 7, 8, 9 are shown in Figure 13.5c. The filled symbol data in Figure 13.4 represent steps $i = 0, 3, 6, 9, 12$. Combining the scaling relations of fragment size, energy, and number gives the energy required at each step as $E_i = E_0 2^{(i-1)/3}$ and thus the cumulative energy E_{tot} for N steps as

$$E_{tot} = E_0 \sum_{i=1}^{N} 2^{(i-1)/3}, \tag{13.1}$$

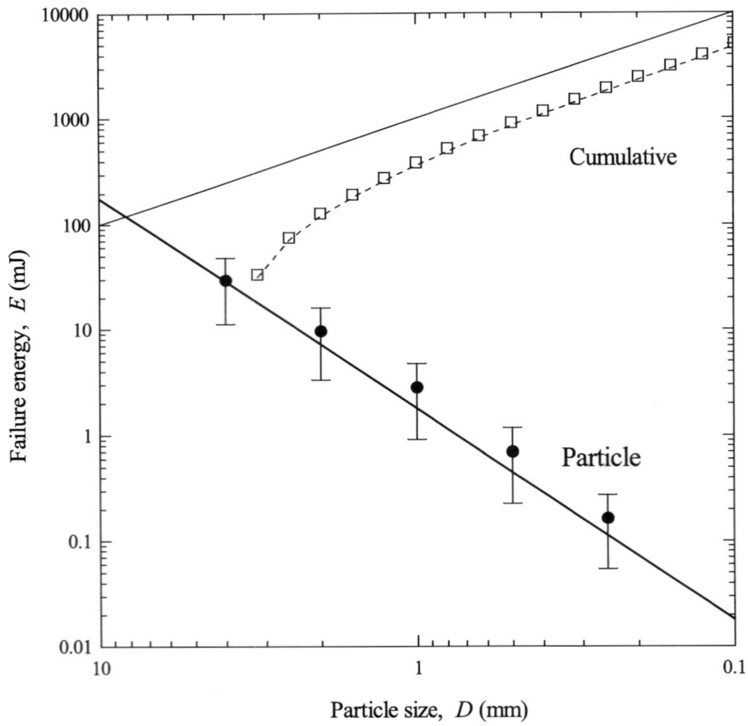

Figure 13.4 Failure energy E as a function of particle size D for quartz particles. Note reversed abscissa, largest to smallest left to right. Filled symbols, failure energy/particle, means and standard deviations, from Figure 13.3b. Open symbols, cumulative energy/particle determined from filled symbol trend. The lines are guides to the eye.

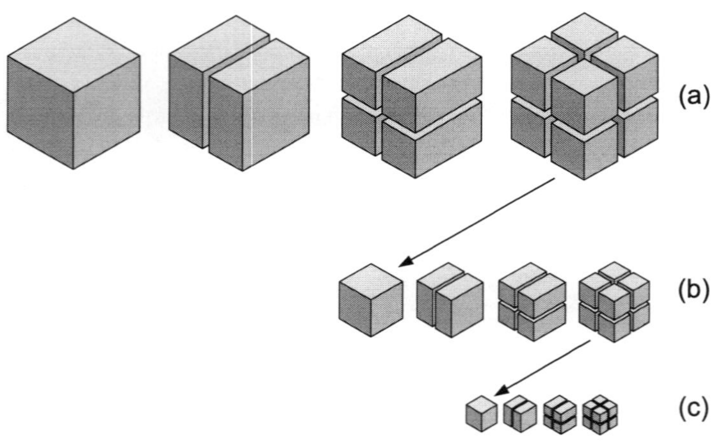

Figure 13.5 Schematic diagram of sequential fragmentation of a particle. (a) Initial cubic particle separated into 2, 4, and 8 fragments, left to right. (b) Smallest fragment from (a) separated into 2, 4, and 8 fragments, left to right. (c) Smallest fragment from (b) separated into 2, 4, and 8 fragments, left to right. The initial particle is thus separated into 2^9 fragments.

where the fragment or particle size at each step is given by

$$D_i = D_0 2^{(-i/3)}. \tag{13.2}$$

Eqs. (13.1) and (13.2) form a simple parametric set to describe the total energy required to crush particles to a given size. The open symbols in Figure 13.4 show such energy for the quartz particles described by the filled symbols. The total energy required increases markedly with decreasing particle size—although the energy requirement for fragmenting a particle decreases significantly with decreasing size there are many more particles. The upper solid line in Figure 13.4 is a guide to the eye of slope 1, consistent with scaling expectations: although the fragmentation energy/particle decreases with each particle refinement step as D^{-2}, the number of particles increases as D^3. The dashed line in Figure 13.4 is a quasi-empirical fit to the cumulative response of the form $E_{tot} \sim (D_0/D_i - 1)$, consistent with some of the earliest considerations of this topic (Tanaka 1966; Oka and Majima 1970) and more recent consideration of ultrasonic fragmentation (Kusters et al. 1994).

It is useful to place Figure 13.4 in the context of industrial operations. The total energy required to crush a single 4 mm quartz particle to 0.1 mm (= 100 μm) powder according to Figure 13.4 is approximately 5 J. The mass of a single particle is approximately 0.2 g, such that crushing 10^6 g (10^3 kg, a tonne) of such particles would require approximately 26 MJ of energy, equivalent to 7.2 kW h or running a 100 W light bulb continuously for three days. However, industrial comminution operations are extremely inefficient, at best with efficiencies of 0.05 (Tromans 2008). The energy required to crush the brittle particles here in an industrial context is thus approximately 500 MJ/t. This value is a substantial fraction of the energy required to run an entire modern-day cement clinker production facility, approximately 3000 MJ/t (Schneider et al. 2011). Although these values are approximate, they point to the usefulness of analysis of particle strength tests in improving industrial comminution processes.

13.3 Grinding Particle Reliability

In contrast to the previous section in which particle fragmentation and energy minimization were the goals, selection of particles as grinding media often have the longevity of intact particles as a goal. Hence, a major application of strength distribution measurements of grinding particles is particle reliability—an assessment of particle performance over *time*. Here, reliability is taken as the proportion of a population of equal sized grinding particles that survive an applied stress sequence. (As the particles are of identical size, such a sequence is equivalent to a similar sequence of applied forces.) Reliability estimates overall can be divided broadly into those that focus on flaw behavior, introducing time dependence through the kinetics of flaw development, or those that focus on driving force behavior, introducing time dependence through a component loading history. For the brittle fracture limited reliability considered here, this division falls into a focus on crack growth or a focus on the applied stress time course. The first of these involves the non-equilibrium fracture properties of the material: see Cook (2018) for an overview. The second approach involves the near-equilibrium strength properties of the considered components and was used by Gaither et al. (2013) to estimate reliability of MEMS structures. Both deterministic and stochastic methods of reliability estimation were outlined and demonstrated by Gaither et al. The stochastic method of Gaither et al. will be implemented here in estimating the reliability of grinding particles, using similar analytical elements: an assumed applied stress sequence, interpreted as a time course, and an experimental strength distribution of components, here the $D = 1.29$ mm alumina grinding media from the work of Huang et al. (1995), Figure 9.5.

The starting point of analysis is specification of the applied stress sequence. Here, as assumed by Gaither et al. (2013), an exponential distribution for the probability density $f(\sigma)$ of applied stress values σ is used:

$$f(\sigma) = (1/\mu)\exp(-\sigma/\mu), \tag{13.3}$$

where μ is the mean of the distribution (Forbes et al. 2011). Various values of μ were used here, μ = (10, 20, 30, and 50) MPa and 3600 values of σ were selected stochastically from the resulting distributions. The 3600 values represent a sequence of 3600 applied stresses exerted on particles during a 1 h grinding operation, such that each value represents the instantaneous stress at an average time interval of 1 s. Figure 13.6a shows a plot of stress as a function of time for μ = 20 MPa. The white line indicates the mean of the distribution. It is clear that many stress events greatly exceed the mean value (and the distribution would be regarded as marginally "heavy tailed"; see Chapter 3). (The spacing of stress values decreases as time proceeds from 0 to 3600 to accommodate strength distribution variation, see following text.) As noted throughout the book, it is not the mean stress that determines the fracture behavior of an ensemble of components, or the instantaneous stress, but the maximum stress applied in the period considered. All components with strengths less than the moving maximum will have broken in that period, or, of more use here, only components with strengths greater than the moving maximum will remain intact in that period. Figure 13.6b shows a plot of the instantaneous stress, repeated from Figure 13.6a in gray, and the moving maximum of stress as a black line resembling a staircase. The number of "stairs" is far fewer than the number of instantaneous stress values and many local stress maxima are bypassed. Maximum stress profiles such as these are utilized in reliability analyses.

Figure 13.7a shows a plot of the best-fit strength cdf $H(\sigma)$ for the D = 1.29 mm Al_2O_3 grinding particles from Figure 9.5 (from the work of Huang et al. 1995). The strength distribution extends from a lower bound threshold of approximately 50 MPa to an upper bound of approximately 400 MPa and is concave. In order to combine this strength information with applied stress maxima profiles such as Figure 13.6b, simple transformation of the $H(\sigma)$ response is required: The complementary cumulative distribution function (ccdf) $\bar{H}(\sigma) = 1 - H(\sigma)$ is calculated and plotted in axes rotated by $\pi/2$

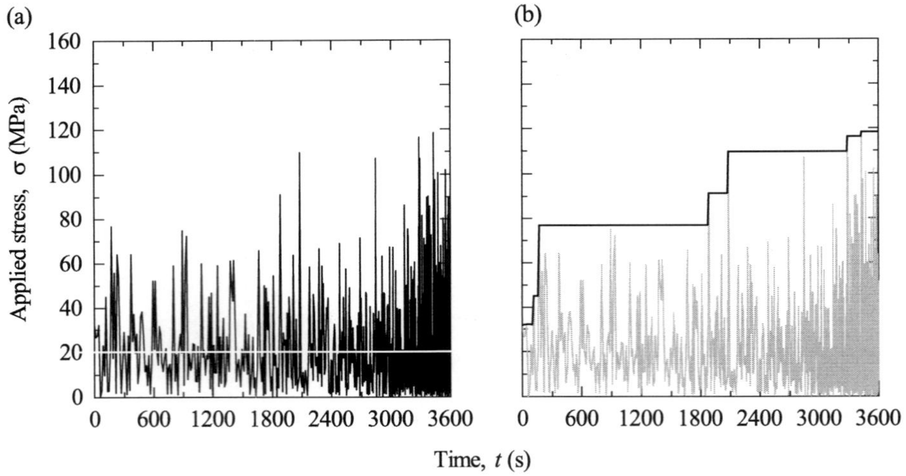

Figure 13.6 (a) Plot of applied stress sequence as a function of time for an exponentially decreasing stress distribution. Mean of stress of 20 MPa indicated by white line. Note that there are many large stress excursions. (b) Sequence repeated from (a) in gray. Moving maximum indicated in black appears as much more slowly evolving staircase and determines component reliability.

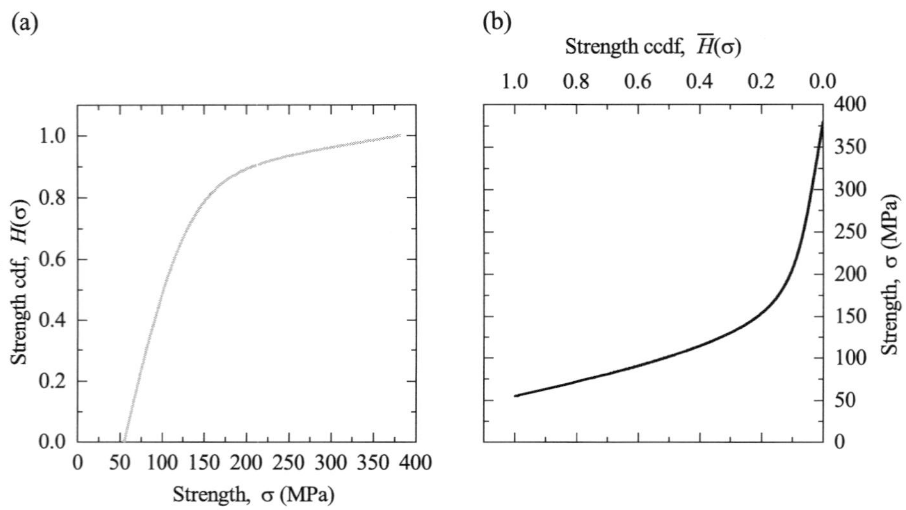

Figure 13.7 (a) Strength cdf $H(\sigma)$ variations for Al_2O_3 particles, size $D = 1.29$ mm, from Figure 9.5. (b) Rotated strength ccdf $\bar{H}(\sigma)$ response determined from (a), $\bar{H} = 1 - H$.

relative to Figure 13.7a. This transformation is shown in Figure 13.7b. The transformed response is convex and can also be obtained by a reflection of the original response in the plot diagonal. Importantly, the transformed component response can be compared with the applied stress profile as both have stress on the ordinate and both proceed from left to right as the number of potential unbroken components diminishes. In reliability terms, the reliability decreases from left to right (Walpole and Myers 1972). Strength ccdf variations such as these are utilized in reliability analyses.

Figure 13.8 combines Figures 13.6b and 13.7b in a common plot that provides a visual display of reliability. The bold line indicates the particle strength ccdf. The fine line indicates the applied stress maximum profile. The gray shaded regions indicate the condition in which the maximum stress was less than the strength of the particles. The hatched region indicates the condition in which the maximum stress was greater than the strength of the particles. The relative size of the gray and hatched areas provides an approximate assessment of the particle reliability. For approximately the first 300 s of the applied loading, the maximum stress did not exceed the threshold of the particle strength distribution and no particles failed. At 300 s, a stress value exceeded the threshold (left scale) (approximately 50 MPa) and about 0.25 of the particles failed (top scale). This maximum stress was exceeded at 1800 s when peaks of approximately 100 MPa occurred and an additional 0.25 of the particles failed. There were a few minor increases in the maxima toward the end of the 3600 s cycle, such that just over half the particles were predicted to fail in this sequence.

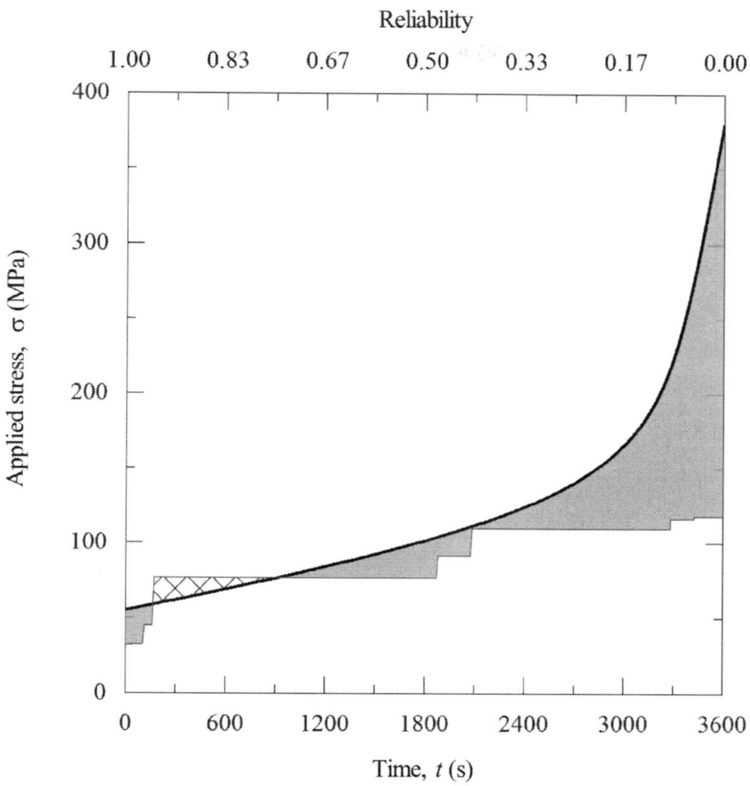

Figure 13.8 Overlaid applied stress-particle strength distribution plots from Figure 13.6 and Figure 13.7. Reliability decrease indicated by gradual contraction of shaded area with time. Failed particles indicated by hatched area.

The simulation technique illustrated in Figure 13.8 can be used in two ways to assess the reliability of grinding media. The first is to simply repeat the stochastic simulation of Figure 13.8 using a fixed mean stress. Such simulations enable assessment of the effects of the assumed form of the applied stress probability density, in this case Eq. (13.3). The second is to perform the simulation with different values of mean stress. Such simulations assess the effects of the relative scales of varying applied stress and fixed component strength distribution. Figure 13.9 shows repeated reliability plots using a fixed mean stress of $\mu = 20$ MPa. Figure 13.9a repeats Figure 13.8. The remaining plots, Figure 13.9b, 13.9c, 13.9d are very similar, leading to a range of 0.5–0.75 particles failed at the conclusion of the simulated stress sequences. Figures 13.9a and 13.9b are similar in that there is one major stress event that leads to a considerable proportion of failed particles. Figures 13.9c and 13.9d are similar in that there are several minor stress events leading to smaller proportions of failed components. Conventional reliability measures, such as the time to first failure and mean time between failures (Bazovsky 2004) are easily determined from plots such as Figure 13.9.

Figure 13.10 shows reliability plots using mean stress values of $\mu =$ (a) 10 MPa, (b) 20 MPa, (c) 30 MPa, and (d) 50 MPa. Figure 13.10b repeats Figure 13.8. In this case, the plots are all very different, leading to a range of 0 to all particles failed at the conclusion of the simulated stress sequences. In Figure 13.10a, the maximum applied stress barely exceeds the threshold strength and almost zero particles have failed at the completion of the applied stress sequence; there are no hatched regions. As discussed, in Figure 13.10b, the maximum applied stress exceeds the threshold strength and approximately 0.5 of the particles have failed at the completion of the applied stress sequence; there is one small hatched region. In Figure 13.10c, the maximum applied stress greatly exceeds the threshold strength and approximately 0.9 of the particles have failed at the completion of the applied stress sequence; there is one large hatched region. In Figure 13.10d, the maximum applied stress exceeds the upper bound of the strength distribution and all particles have failed at the completion of the applied stress sequence; there is only a large hatched region and the shaded grey region has been completely eliminated. The domain of mean stress levels, (10–50) MPa, of the exponential stress density, Eq. (13.3), encompasses the full range of particle reliability, from zero fails to complete failure of the entire sample. The domain thus places bound on the stresses and forces

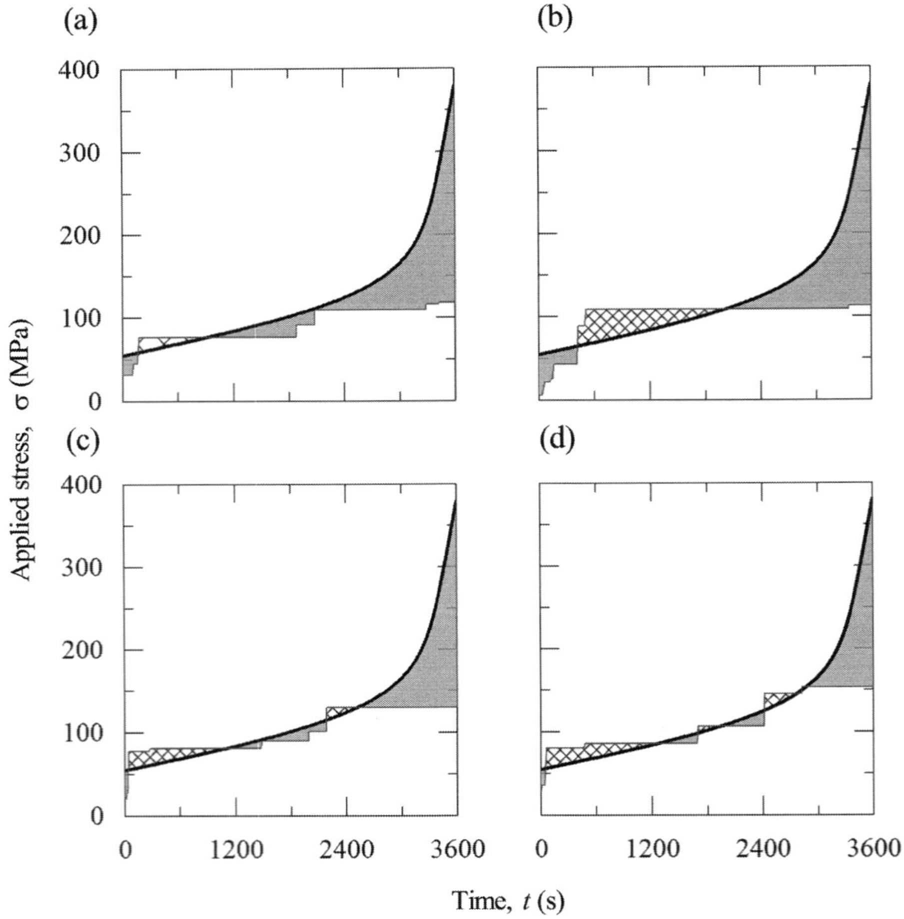

Figure 13.9 Repeated stochastic variations in reliability behavior for the system of Figure 13.8 (same notation) using a mean applied stress of 20 MPa. End of stress sequence particle survival varies from (a) ≈ 0.5 to (d) ≈ 0.75.

used in grinding if particle lifetime is an engineering concern. Conventional reliability measures such as wear out time and replacement time (Bazovsky 2004) are easily determined from plots such as Figure 13.10. In addition, it is clear that stress sequences with fewer large events, i.e. not long tailed, will lead to greater reliability and are easily simulated.

13.4 Mass Effects on Particle Strength

The two previous sections considered applications of particle strength distributions in predictions of engineering performance outcomes of energy consumption and reliability. In this section and the next the effects of particle fabrication choices are considered in determining such strength distributions, building on the fracture mechanics foundations of Chapter 12. This section 13.4 considers particle size, characterized by particle diameter D, and the related particle mass $m \sim \rho D^3$, where ρ is the particle material density (for a spherical particle, $m = \pi \rho D^3/6$). The effects of particle size will be considered in the two-step strength scenario outlined in Chapter 12: flaw generation by a local event in the first step, followed by global particle loading through diametral compression in the second step. The local event considered here is impact of a moving particle against a stationary object, resulting in sharp contact damage on the particle surface. A fundamental scaling relation for moving particles is that the particle kinetic energy $U_K = mv^2/2$, where v is the particle velocity, varies with particle size as $U_K \sim D^3$ for a given particle velocity and density.

Figure 13.11a shows a schematic diagram of several moving particles impacting a fixed surface. The impacts lead to rapid deceleration of the particles and thus forces exerted on the particles by the surface. The surface is rough, composed of small, sharp asperities that are very stiff and unyielding relative to the particle material. The particles are all composed of the same material and therefore have the same density and mechanical properties. All particles are traveling toward the surface at the

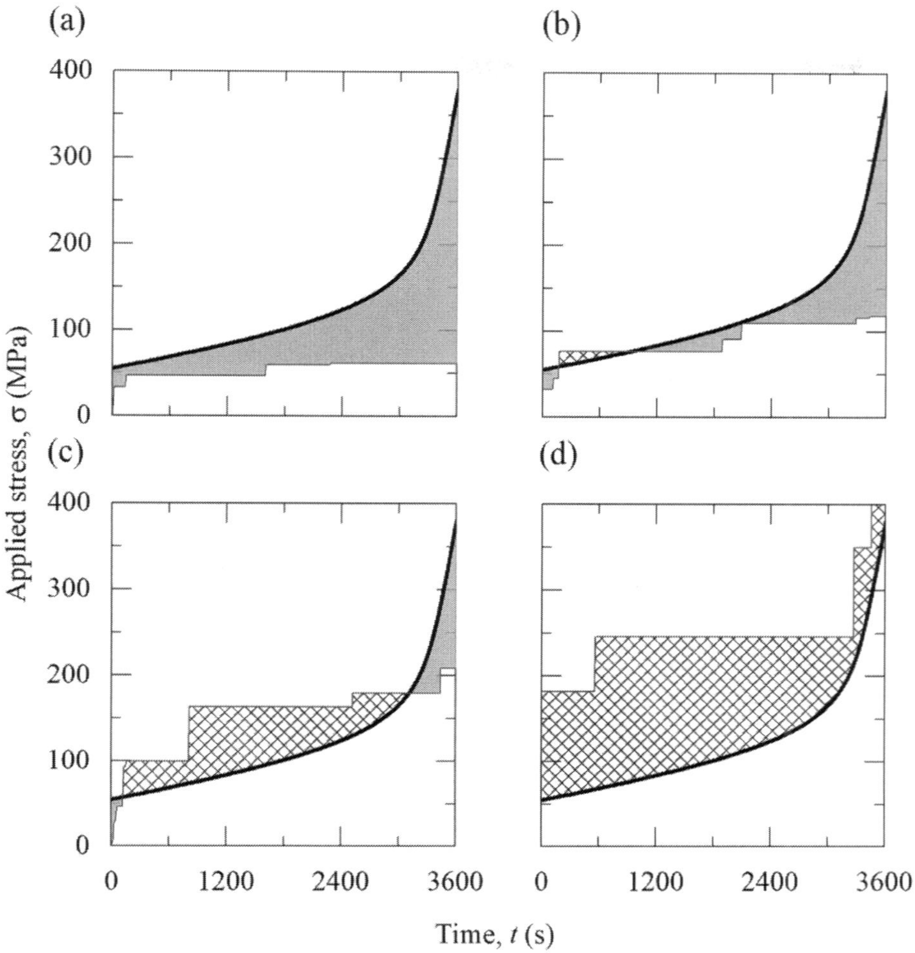

Figure 13.10 Reliability behavior for the system of Figure 13.8 (same notation) using mean applied stress values of (a) 10 MPa, (b) 20 MPa, (c) 30 MPa, and (d) 50 MPa. End of stress sequence particle survival varies as (a) 1 (all survive), (b) 0.5, (c) 0.1, and (d) 0 (all fail).

same velocity v. The impact of a particle on a surface asperity gives rise to damage in the particle typical of elastic-plastic sharp contacts: A residual contact impression is generated on the surface of the particle and an associated sub-surface localized compaction and plastic deformation zone is formed. As a result, elastic deformation, characterized by a residual stress field, is generated in the surrounding particle matrix material. (If there are multiple contacts the largest is considered here.) The peak axial force generated by the asperity on the particle is F. The transverse radius of the residual contact impression is a, and is related to the contact peak axial force by an invariant material property, the hardness $H \sim F/a^2$. During the contact deceleration, the instantaneous axial displacement of the asperity into the particle δ_i is related to the instantaneous contact force F_i by

$$F_i(\delta_i) = \frac{H\delta_i^2 \tan^2 \phi}{1 + \gamma H/M} = H'\delta_i^2, \tag{13.4}$$

where the terms 2ϕ, the included angle of the asperity, M, the elastic modulus of the particle, and γ, a dimensionless geometrical factor (Marshall et al. 1983) are collected into H'. Integrating Eq. (13.4) gives the work w performed in generating the contact deformation

$$w = \int_0^{\delta} F_i \mathrm{d}(\delta_i) = H'\delta^3/3 = F^{3/2}/3H'^{1/2}, \tag{13.5}$$

where δ is the peak axial contact displacement (for plastic contacts, $H \ll M$, $\delta = a/\tan \phi$). If the contact is adiabatic and there are no other losses, conservation of energy requires $w = U_K$ and thus the peak contact force is related to the particle

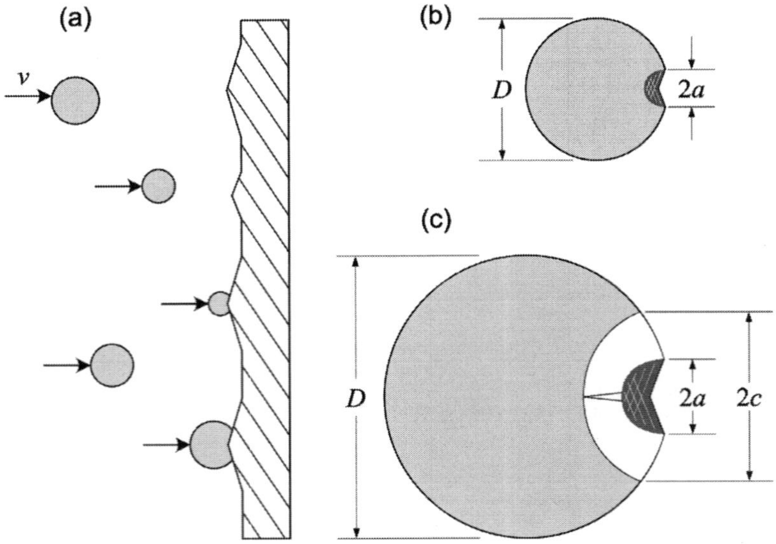

Figure 13.11 (a) Schematic diagram of multiple particle impacts on a rough surface. Particles are different sizes, but all have initial velocity *v*. (b) Schematic cross section diagram of a small particle after sub-threshold impact. There is residual plastic deformation in the form of shear faults and compaction at the contact site but no cracking. (c) Schematic cross section diagram of a large particle after post-threshold impact. There is residual plastic deformation, compaction, and cracking at the contact site. The contact cross sections vary linearly with particle size in (a), shown as *a* and *D* in (b) and (c).

kinetic energy by $F \sim U_K^{2/3}$. This relationship and the invariance of material hardness $H \sim F/a^2$ implies the invariance of the quantity $(U_K^{1/3}/a)^2$. Using the scaling of particle kinetic energy with particle size, $U_K \sim D^3$, gives $(D/a)^2$ also as an invariant quantity. Thus for the sharp impact conditions considered, the residual contact impression scales linearly with the particle size, $a \sim D$, as indicated in Figure 13.11a.

Deformation within the sub-surface plastic deformation zone is not homogeneous, consisting of discrete crack-like shear faults. In the initiation of half-penny cracks at the contact impression the intersections of these shear faults at the zone and impression periphery act as crack nuclei that increase in size with the zone and contact sizes. In Section 12.4 crack initiation at an inclusion was analyzed in terms of an invariant crack nucleus in an increasing stress field. Crack initiation at a sharp contact impression is the converse of this and is analyzed as a crack nucleus that increases in size in an invariant stress field. A simple expression of these effects is the stress-intensity factor (SIF) K acting on a half-penny crack nucleus of size c_1 at the periphery of a contact impression:

$$K = \Omega_1 H c_1^{1/2} = \Omega_1 H(\Omega_2 a)^{1/2}. \tag{13.6}$$

Ω_1 and Ω_2 are dimensionless geometry constants characterizing the coupling of the crack nucleus to the contact stress field magnitude H and the proportionality of the crack nucleus to the contact impression size a, respectively. More detailed analyses of multiple crack initiation events in the inhomogeneous stress fields at sharp contacts, greatly extending Eq. (13.6), are given elsewhere (Cook 2019). Using the invariance of the D/a ratio enables Eq. (13.6) to be re-written as the SIF K_{D1} acting on a crack nucleus at an impact by particle size D:

$$K_{D1} = \psi_D H D^{1/2}, \tag{13.7}$$

where ψ_D dimensionless geometry constant. Setting equilibrium, $K_{D1} = T$, where T is the toughness of the particle, gives a threshold particle size D^*

$$D^* = (1/\psi_D)^2 (T/H)^2. \tag{13.8}$$

Small particles, $D < D^*$, that impact the rough surface will not meet the fracture equilibrium criterion for crack initation and the impact damage will thus remain as sub-threshold contact impressions, Figure 13.11b. Large particles, $D > D^*$, that impact the rough surface will exceed the fracture equilibrium criterion for crack initation, noting that Eq. (13.6) is destabilizing, and the impact damage thus transforms into post-threshold flaws consisting of contact impressions + half-penny cracks, Figure 13.11c. Note that the threshold particle size criterion has the familiar $D^* \sim T^2/H^2$ scaling.

Following the analyses of Chapter 12, for large particles the SIF K acting on post-threshold half-penny cracks formed on impact is

$$K = \chi F/c^{3/2} \sim \chi H a^2/c^{3/2}, \tag{13.9}$$

where c is the crack length, F and a are the peak contact force and impression dimension generated, respectively, and χ is a dimensionless geometry constant. Once again, the invariance of the D/a ratio enables Eq. (13.9) to be re-written as the SIF K_{D2} acting on a half-penny crack at an impact as a function of particle size D:

$$K_{D2} = \chi_D D^2/c^{3/2}, \tag{13.10}$$

where χ_D dimensionless geometry constant. The detailed analysis of Section 12.5.1 considering combined loading of flaws, now follows directly. The effects of a uniform stress superposed on post-threshold flaws described by Eq. (13.10) (step 2 shown earlier) is given by a total SIF

$$K = K_{D2} + K_{\mathrm{A}} = \chi_D D^2/c^{3/2} + \psi\sigma_{\mathrm{A}} c^{1/2}. \tag{13.11}$$

Particle strength is thus given by

$$\sigma_{\mathrm{m}} = 3T^{4/3}/4^{4/3}\psi\chi_D^{1/3}D^{2/3}. \tag{13.12}$$

Eq. (13.12) represents another variation of Eq. (12.70), $\sigma_{\mathrm{m}} = B'D^{-2/3}$, and, as discussed, all the analyses of the previous chapters are thus applicable. Such analyses are mostly intensive. From an extensive perspective, the failure force F_{m} of particles described by Eq. (13.12) scales as $F_{\mathrm{m}} \sim \sigma_{\mathrm{m}} D^2 \sim D^{4/3}$. Failure force is thus predicted to increase slightly greater than linearly with particle size, consistent with Figure 13.1b. Similarly, the failure energy E of particles described by Eq. (13.12) scales as $E \sim \sigma_{\mathrm{m}}^2 D^3 \sim D^{5/3}$. Failure energy is thus predicted to increase slightly less than quadratically with particle size, consistent with Figure 13.4. Post-threshold fracture mechanics analysis and size and mass scaling of particles provides a clear framework, supported by experimental observations, for considering energy consumption in comminution of particles.

A semi-quantitative discussion of the effects of superposed uniform applied stress on sub-threshold flaws was given in Chapter 12. The discussion highlighted the effects of applied stress on the complete destabilizing to stabilizing transition described by the initiating SIF. Here that discussion is extended and made more quantitative by restricting attention to the destabilizing limit of the contact induced SIF, K_D, Eq. (13.7). Combining Eq. (13.7) with the SIF applicable to a uniform applied stress σ_{A} acting on a crack nucleus gives

$$K = K_{D1} + K_{\mathrm{A}} = \psi_D H D^{1/2} + \psi\sigma_{\mathrm{A}} D^{1/2}. \tag{13.13}$$

Note that the nucleus size c_1 is defined by the contact impression size and therefore the particle size D ($c_1 \sim a \sim D$), and that the values of ψ_D and ψ will thus not approach the ideal values. A similar equation for sub-threshold sharp indentation flaws in loaded extended components was given by Jung et al. (2004a, 2004b). Setting equilibrium $K = T$ in Eq. (13.13), inverting, and using Eq. (13.8) gives an expression for strength, noting that both terms in Eq. (13.13) are destabilizing,

$$\sigma_{\mathrm{sub}} = \left(\frac{\psi_D H}{\psi}\right)\left[\left(\frac{D^*}{D}\right)^{1/2} - 1\right]. \tag{13.14}$$

This strength is the value of the applied stress required to initiate a crack from a contact impression nucleus generated by particle impact against a sharp asperity. The value is thus expressed in terms of particle size, D. Similar to the considerations in Chapter 12 regarding ductile-brittle transitions, the plastic deformation at the contact impression was assumed to be in equilibrium with the contact force applied to the particle on impact. Hence the proportionality to particle hardness H. An upper limit to the applicability of this expression is $D \leq D^*$, as this is the threshold particle size at which a half-penny crack initiates on impact. An interpretation of Eq. (13.14) is that the applied stress "assists" in the initiation of a crack and thence failure of a particle and that the requirement for assistance diminishes as the particle size increases.

Figure 13.12 shows a logarithmic plot of strength as a function of particle size using Eq. (13.12) and Eq. (13.14). A threshold particle size of $D^* = 5$ mm was used, indicated by the fine vertical line. The straight bold line indicates the variation of post-threshold strength σ_{m} using $B' = 15$ MPa mm$^{2/3}$. The curved bold line indicates the variation of the sub-threshold strength σ_{sub} using $(\psi_D H/\psi) = 50$ MPa. The fine diagonal line is a guide to the eye of the asymptotic $D^{-1/2}$ small D response of this strength behavior. The asymptotic large D response is shown by the fine vertical line. In the post-threshold region, $D \geq D^*$, only σ_{m} strengths can be observed. In the sub-threshold region, $D \leq D^*$, both σ_{m} strengths and σ_{sub} can be

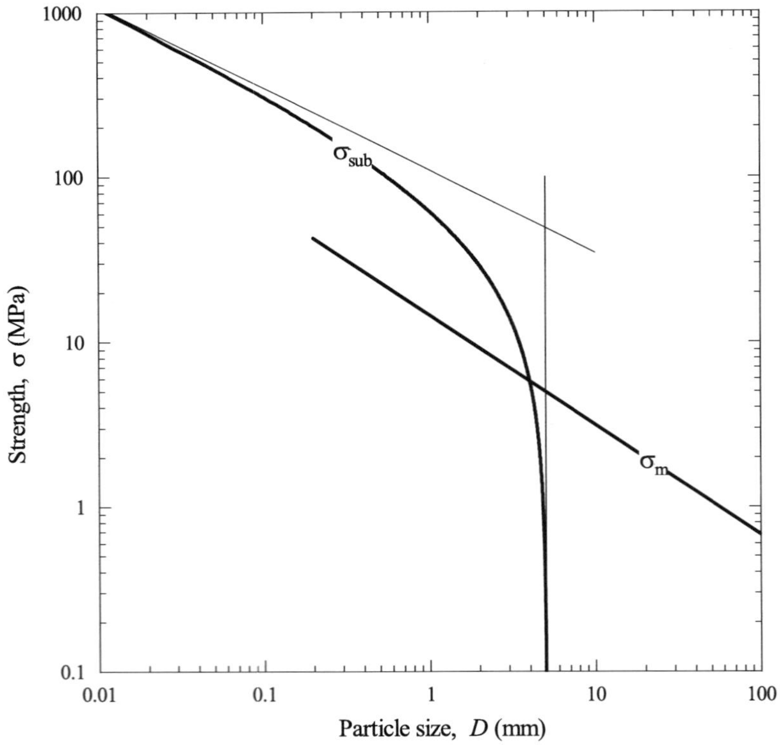

Figure 13.12 Plot of strength as a function of particle size for impacted particles. Upper bold line indicates behavior of particles containing uncracked sub-threshold flaws. Strength decreases precipitously with particle size until the upper limit of spontaneous crack pop-in is reached. Asymptotic behavior indicated by fine lines. Lower bold line indicates weaker variation of particles containing cracked post-threshold flaws.

observed, although the probability of observing a post-threshold σ_m value decreases as D decreases. Hence, there is some possibility of observing bi-modal strength behavior for particle sizes and masses smaller than the threshold value, in this case smaller than approximately 5 mm or 0.25 g (this is a form of brittle-ductile transition). Note that the abscissa of Figure 13.12 extends over a factor of 10^4 in particle size and therefore a factor of 10^{12} in mass. The fabrication and experimental difficulties associated with such large domains suggest that generating clear demarcations of post- and sub-threshold strength behavior is unlikely, although possible (Jung et al. 2004a, 2004b). Nevertheless, Figure 13.12 provides a framework for particle manufacturing or handling applications in selection of particle sizes, masses, and impact conditions to optimize strength. Figure 13.12 also provides a framework for interpreting particle behavior, e.g. why some nominally identical small particles exhibit large strengths and extensive fragmentation on failure and some exhibit small strengths and little fragmentation, and why it is very difficult to eradicate a long weak tail from some strength distributions.

13.5 Microstructural Effects on Particle Strength

The effects of microstructure on the fracture of brittle components has long been a concern to materials processors and designers. This is particularly so in ceramics (Kingery et al. 1975), and microstructural effects in fracture of both polycrystalline and multi-phase ceramics have been studied extensively. In both microstructural forms preferred fracture paths exist: In polycrystalline materials these are weak grain boundaries; in multi-phase materials these are weak interfaces between phases. In addition, in both forms of microstructure thermal expansion plays a large role: In polycrystalline materials thermal expansion anisotropy effects often generate large stresses within microstructures on cooling from a processing temperature; similarly, in multi-phase materials thermal expansion inhomogeneity effects often generate large stresses. The two ceramic materials that typify these behaviors are polycrystalline alumina (Al_2O_3) and glass + crystalline oxide multi-phase porcelain, respectively. Both Al_2O_3 and porcelain particles and extended components have been considered in this book (e.g. see Chapters 5 and 9). The effects of weak grain boundaries and anisotropic coefficient of thermal expansion

(CTE) effects in crack initiation were considered in Chapter 12 and the effects of weak interfaces and CTE inhomogeneity on strength were considered briefly in Chapter 5. Here the effects of weak grain boundaries and CTE stress effects on crack propagation and the resultant strengths in polycrystalline ceramic materials are considered. Such materials are often used in particle form and hence the analysis here is applicable to designing and processing microstructures of particle materials with optimized strengths.

The mechanical basis of microstructural effects on fracture of polycrystalline materials is shown in Figures 13.13 and 13.14. In Figure 13.13a a schematic cross section of a polycrystalline microstructure is shown. The grain boundaries are indicated by fine black lines and all the grains have similar shapes and sizes. The microstructure is part of a component that is placed in tension, indicated by the dark arrows, such that a crack propagates through the microstructure along the weak grain boundaries, left to right, indicated by the bold black line. (Similar images of such behavior in polycrystalline ceramics are shown in many works, see e.g. Kingery et al. 1975.) In this material, the angular and spatial deviations in the crack path from the overall crack propagation direction and extension are small. As a consequence, the material parts easily on the crack surface to generate two fracture surfaces and two fragments, Figure 13.13b. Apart from the slight undulation of the fracture plane and the obvious decrease in fracture resistance of the grain boundaries relative to the crystalline grains, fracture such as this would not be regarded as exhibiting significant microstructural effects.

In Figure 13.14a a schematic cross section of a polycrystalline microstructure is also shown. In this microstructure, however, some of the grains are larger and have elongated sections relative to the surrounding matrix of smaller equi-axed grains. As before, the microstructure is part of a component that is placed in tension, shown by the dark arrows, and a crack propagates through the microstructure along the weak grain boundaries, left to right, indicated by the bold black line. In this material, the angular and spatial deviations in the crack path are not small relative to the overall propagation direction and extension. In fact, the large grains in the crack path lead to crack propagation locally perpendicular to the overall propagation direction and deviations in the crack path 3–4 times the usual grain facet distance. As a consequence, this material does not part easily on the crack surface. The large grains act as "ligamentary bridges" that exert localized compressive forces across the crack surface and oppose crack opening and propagation. These forces originate from frictional shear tractions acting across contacting fracture surfaces that are perpendicular to the overall propagation direction. Such contacting surfaces and tractions occur between the large grains and the small grain matrix. The frictional tractions exerted by the large grains on the surrounding matrix are indicated by dark arrows in Figure 13.14b. (Again, similar images of such behavior in polycrystalline ceramics are shown in many works, see e.g. Kingery et al. 1975, including the apparently discontinuous crack propagation segments.)

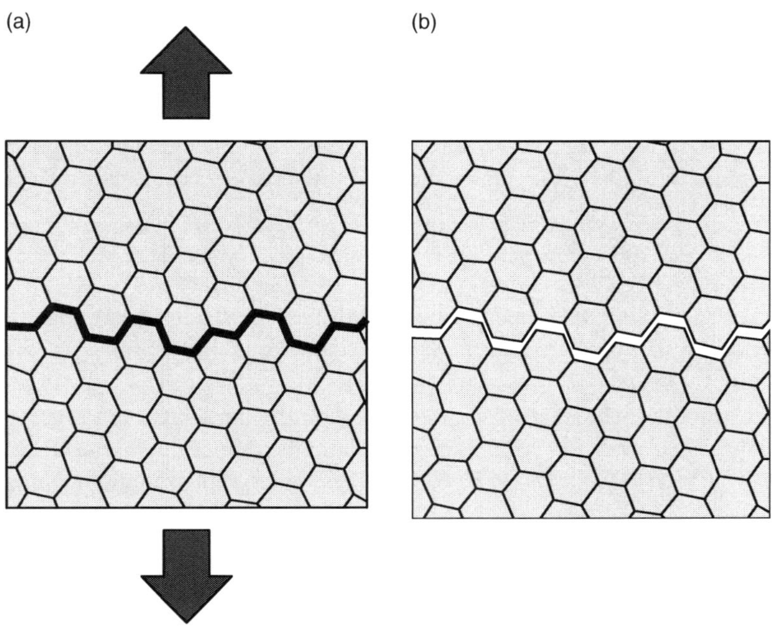

Figure 13.13 (a) Schematic diagram of a polycrystalline microstructure consisting of equiaxed grains; grain boundaries are indicated as fine lines. The microstructure is loaded vertically by the dark arrows and a potential crack path with mean horizontal orientation is indicated as the bold line. (b) Cracked and separated microstructure from (a).

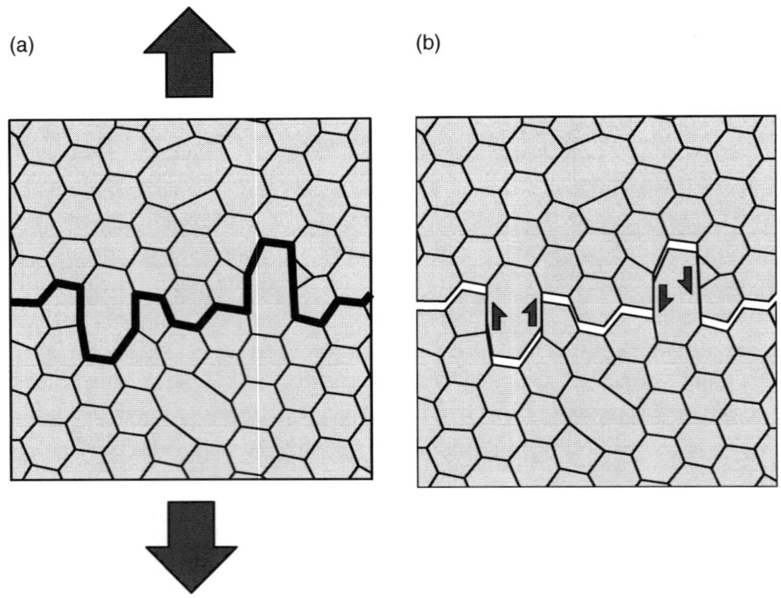

Figure 13.14 Schematic diagram of a polycrystalline microstructure consisting of mostly of equiaxed grains but containing some larger tabular grains; grain boundaries are indicated as fine lines. The microstructure is loaded vertically by the dark arrows and a potential crack path with mean horizontal orientation is indicated as the bold line. Deviations in crack path are significant. (b) Cracked and partially separated microstructure from (a). The large grains are sites of frictional ligamentary bridges that act to close the crack and impede separation of the fragments. Frictional shear tractions indicated by dark arrows.

Most ceramic microstructures contain a much greater diversity of grain sizes and shapes than shown in the schematic diagram of Figure 13.14. Hence, effects on crack propagation by microstructurally induced ligamentary bridging are pervasive in polycrystalline ceramics. Such bridges commonly hold globally fractured components together (weakly) through the collective action of the localized forces. Anisotropic CTE- and anisotropic elastic modulus-induced local stress effects often greatly influence crack paths and bridge formation and subsequent bridging action. In particular, "clamping" of large grains by compressive CTE stresses perpendicular to the grain elongation axis is thought to contribute significantly to bridging effectiveness. As microstructural bridging effects act to oppose crack propagation they are frequently analyzed as toughening agents that act to increase the toughness of a material in the frame of reference of the applied loading. Analyses of bridging derived toughening and the consequent effects on strength have been brought to a considerable level of sophistication by Lawn and colleagues for both single phase materials (Bennison and Lawn 1989; Chantikul et al. 1990) and multi-phase materials (Lawn et al. 1993; Padture et al. 1993). In particular, the analyses of Lawn and colleagues incorporate the effects of tensile CTE stresses on initiation and propagation of short cracks (similar to the initiation considerations of Chapter 12) along with the effects of compressive CTE stresses on clamping at bridges of long cracks.

A simplified model of bridging that is amenable to strength and microstructural analysis is described here. The model omits explicit consideration of local tensile stresses and omits explicit consideration of the force-displacement behavior of bridges (both features of the Lawn analyses), although both are incorporated implicitly. The model has been applied extensively in descriptions of the strength behavior of a wide range of ceramic materials containing a wide range of flaw sizes, in both equilibrium and nonequilibrium fracture conditions (see Cook 2015 for an overview). The model is based on the recognition that ligamentary bridges act as microstructurally induced compressive tractions perpendicular to the crack surface in a zone adjacent to the crack tip. The compressive tractions have a microstructurally controlled characteristic force scale set by the bridge spatial density, size, and average frictional traction. The zone has a microstructurally controlled characteristic length scale set by the bridge spatial density and bridge failure strain (failed bridges, defined by complete large grain pullout at large crack opening, delimit the boundary of the compressive traction zone). Hence, the microstructural traction zone can be represented by a compressive line force (force/length), magnitude f^*, acting a distance δ behind the crack tip. For a circular crack of radius c, this is a circular line force acting over radius $(c - \delta)$, as shown in the schematic diagram of Figure 13.15. This diagram is clearly similar to that of Figure 12.14a and the resulting SIF acting on the crack is described by similar analysis. The distinction for this case is that the dimension of the circle describing the line force location is not fixed, a in Figure 12.14a, but maintains a constant separation δ from the crack perimeter.

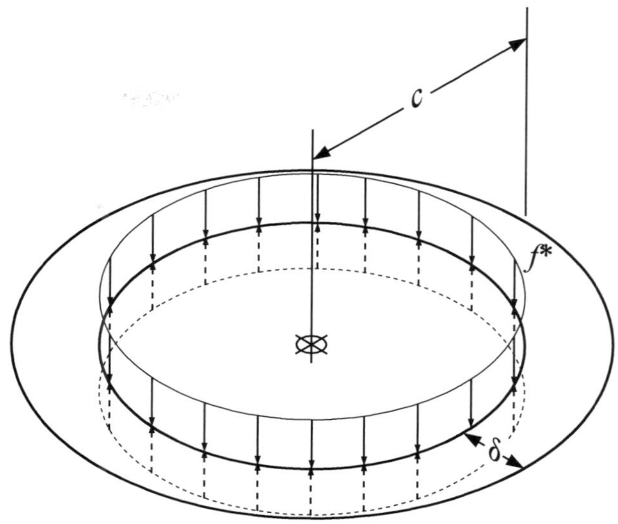

Figure 13.15 Schematic diagram of circular crack, radius c, loaded in compression by a line force f^* acting over a circle radius $(c - \delta)$. This geometry models microstructural effects in bridged polycrystalline materials.

The microstructural traction field of Figure 13.15 is described simply by a SIF given by Eq. (12.42) but the resulting expression is not incorporated simply into strength analyses. An algebraic form that very closely approximates the SIF arising from Figure 13.15, that is amenable to strength analyses, is given by

$$K_\mu(c) = -\mu f^* \delta^{-1/2}[1 - (\delta/c)^{3/2}],\ c \geq \delta. \tag{13.15}$$

The negative sign indicates that K_μ acts to oppose crack propagation as f^* is the magnitude of a compressive line force. μ is a dimensionless microstructural geometry factor. The flaws considered here will be simple Griffith cracks loaded by a uniform applied stress, such that the applied loading SIF is given by the familiar form $K_A = \psi\sigma_A c^{1/2}$ (see Chapter 12). The total SIF K acting on the crack is the sum of the microstructural and applied terms,

$$K = K_\mu + K_A = -\mu f^* \delta^{-1/2}[1 - (\delta/c)^{3/2}] + \psi\sigma_A c^{1/2}. \tag{13.16}$$

Equation (13.16) can be interpreted in two ways depending on whether the emphasis is on toughness or strength. In the toughness interpretation, the SIF expression of Eq. (13.16) is set to fracture equilibrium, $K = T_0$ and the K_μ term interpreted as a toughening contribution $T_\mu = -K_\mu$, as for the transverse loading geometry discussed in Chapter 12. Thus the equilibrium expression transforms as

$$\begin{aligned} K = K_\mu + K_A &= T_0 \\ K_A = T_0 + T_\mu &= T \end{aligned} \tag{13.17}$$

and the total toughness $T = T_0 + T_\mu$ in this case is given by

$$T = T_0 + \mu f^* \delta^{-1/2}[1 - (\delta/c)^{3/2}]. \tag{13.18}$$

The subscript 0 on T_0 serves to emphasize the crack length invariance of the base (grain boundary) toughness. As discussed in Chapter 12, the switch in Eq. (13.17) is equivalent to a switch from the crack tip frame of reference to the applied loading frame of reference. In interpreting Eq. (13.18) it is convenient to recognize that material toughness in the applied loading frame of reference is crack length dependent, $T = T(c)$, increasing with crack length to a maximum value. The maxiumum increment of toughening ΔT, for $c \gg \delta$, is

$$\Delta T = \mu f^* \delta^{-1/2}, \tag{13.19}$$

such that the maximum toughness of the material, T_∞, appropriate for very long cracks, is

$$T_\infty = T_0 + \Delta T. \tag{13.20}$$

In the strength interpretation, the toughness expression of Eq. (13.17) is rewritten in intermediate form using the above notation and gathering crack-length dependent and independent terms separately as

$$\psi\sigma_A c^{1/2} + \mu f^* \delta / c^{3/2} = T_\infty. \tag{13.21}$$

As earlier, in consideration of particle mass effects in Section 13.4, this equation is recognized as identical to that in Section 12.5.1 describing combined loading. Rewriting Eq. (13.21) provides the equilibrium $\sigma_A(c)$ trajectory with microstructural toughening influence,

$$\sigma_A(c) = (1/\psi c^{1/2})(T_\infty - \mu f^* \delta / c^{3/2}). \tag{13.22}$$

This trajectory passes through a maximum at crack length c_m^μ,

$$c_m^\mu = \left(\frac{4\mu f^* \delta}{T_\infty}\right)^{2/3} \tag{13.23}$$

and stress σ_m^μ,

$$\sigma_m^\mu = 3T_\infty / 4\psi (c_m^\mu)^{1/2} = 3T_\infty^{4/3} / 4^{4/3} \psi (\mu f^* \delta)^{1/3}. \tag{13.24}$$

This stress can be interpreted as the strength of toughnened components containing strength controlling flaws with initial crack lengths $c < c_m^\mu$. The equilibrium trajectory for an untoughened Griffith flaw is

$$\sigma_0(c) = (1/\psi c^{1/2}) T_0. \tag{13.25}$$

Equating σ_0 and σ_m^μ provides a lower bound crack length c_0^μ for applicability of Eq. (13.24) in interpretation as a strength:

$$c_0^\mu = c_m^\mu (4T_0 / 3T_\infty)^2. \tag{13.26}$$

Note that

$$c_m^\mu = \delta (4\Delta T / T_\infty)^{2/3} \tag{13.27}$$

such that

$$c_0^\mu = \delta (4^{8/3} / 3^2)(\Delta T^{2/3} T_0^2 / T_\infty^{8/3}). \tag{13.28}$$

Although not readily apparent from Eqs. (13.27) and (13.28), the lower and upper bound crack lengths for the applicability of the strength expression Eq. (13.24) bracket the microstructural length scale δ, $c_0^\mu < \delta < c_m^\mu$. As an example, for $\Delta T = T_0$ such that $T_\infty = 2T_0$, $c_0^\mu \approx 0.71\ \delta$ and $c_m^\mu \approx 1.59\ \delta$.

The strengths of components containing strength controlling flaws with cracks c_w longer than the upper bound, $c_w > c_m^\mu$, are given by Eq. (13.22) directly, which is a generalization Eq. (13.25) using $T(c)$ rather than T_0. Hence, the strength σ_w in the long crack domain is

$$\begin{aligned} \sigma_w &= (1/\psi c_w^{1/2})(T_\infty - \mu f^* \delta / c_w^{3/2}) \\ &= \sigma_m^\mu (c_m^\mu / c_w)^{1/2} \left[4 - (c_m^\mu / c_w)^{3/2}\right] / 3. \end{aligned} \tag{13.29}$$

This expression has a strength upper bound of $\sigma_w = \sigma_m^\mu$ for $c_w = c_m^\mu$ and an asymptotic lower bound of $\sigma_w \rightarrow T_\infty / \psi c_w^{1/2}$ for $c_w \gg c_m^\mu$. The subscript w indicates that behavior in this long crack domain is in the weak toughening limit, defined by $dT/dc \leq dK_A/dc$ (note: not by the value of $T(c)$, which is approaching a maximum in this domain). Conversely, behavior in the microstructural domain, indicated by superscript μ, is determined by $dT/dc \geq dK_A/dc$ at the upper bound of the short crack domain.

The strength behavior discussed here can be classified into three contiguous regions on a plot similar to Figure 12.22. The regions are centered on the microstructural bridging and toughening length scale δ, indicated by the vertical dashed line in Figure 13.16. The three regions are identified in Figure 13.16 as those in which strength is determined by Griffith behavior (region G), by microstructural toughening effects (region M, shown shaded), or by weak toughening (region W). The regions are demarcated by initial crack length c and region M is further sub-divided into two sub-regions. The regions and crack length behaviors in Figure 13.16 are constructed using the above analysis and $\delta = 100\ \mu m$, $T_0 = 1$ MPa m$^{1/2}$, $T_\infty = 2$ MPa m$^{1/2}$, and $\psi = 1.5$, resulting in $c_0^\mu = 71\ \mu m$, $c_m^\mu = 159\ \mu m$, and $\sigma_m^\mu = 79$ MPa. In all regions the strength-controlling flaws are simple, metastably trapped, Griffith cracks. The regions and sub-regions are now discussed.

Region G: $c < c_0^\mu$. In this small crack region there are no microstructural effects. As the applied stress increases, the crack remains fixed at an initial length until the equilibrium instability condition of Eq. (13.25) is reached and the crack propagates unhindered until component failure. The strength in this region is σ_0 and is indicated by the upper dashed line in region G of Figure 13.16.

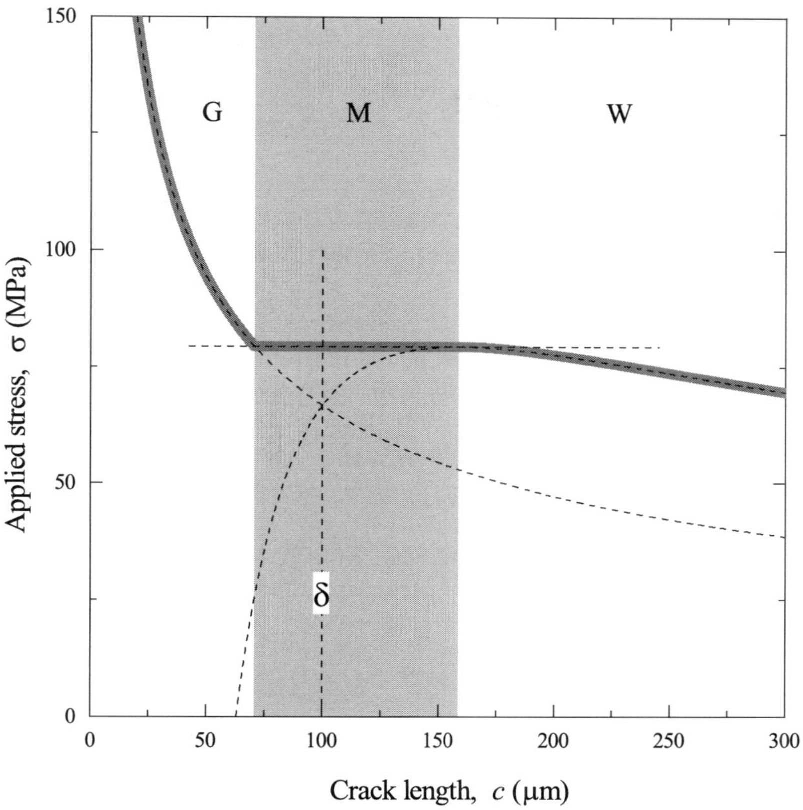

Figure 13.16 Equilibrium stress-crack length trajectories for Griffith cracks and cracks in microstructures toughened by ligamentary bridging. Three regions of behavior are indicated: Griffith (G), microstructure (M), and weakly toughened (W). The characteristic microstructural length scale δ is indicated by the vertical dashed line. The characteristic microstructural strength $\sigma_{\mathrm{m}}^{\mu}$ is indicated by the vertical dashed line; strength is invariant in region M. The full response is shown as the bold line.

Region M: sub-region $c_0^{\mu} \leq c \leq \delta$. In this intermediate crack length region, microstructural effects, although initially absent, eventually lead to crack arrest. As the applied stress increases, the crack remains fixed at an initial length until the equilibrium instability condition of Eq. (13.25) is reached. The crack then propagates at constant stress until the stable equilibrium branch of Eq. (13.22) is encountered and the crack arrests in the microstructural field. (The behavior is identical to that considered in Chapter 12 for crack initiation adjacent to a misfitting inclusion or crack initiation within a transversely compressed particle. In the earlier systems, the SIF passed through a maximum and the toughness was invariant. Here, the SIF increases monotonically and the toughness is initially invariant and then increases steeply.) Further increases in stress lead to stable extension of the crack as described by Eq. (13.22) and shown by the lower dashed line that originates in region G and extends into region M of Figure 13.16. Extension leads to eventual instability at the crack length and stress conditions of $(c_{\mathrm{m}}^{\mu}, \sigma_{\mathrm{m}}^{\mu})$, Eqs. (13.23) and (13.24), and the crack thence propagates unhindered until component failure. The strength in this region, independent of the specific initial crack length, is $\sigma_{\mathrm{m}}^{\mu}$, as indicated by the horizontal dashed line in region M of Figure 13.16.

Region M: sub-region $\delta \leq c \leq c_{\mathrm{m}}^{\mu}$. In this intermediate crack length region, microstructural effects lead to immediate toughening and stable crack extension. As the applied stress increases, the crack remains fixed at an initial length until the stable equilibrium branch of Eq. (13.22) is encountered. Further increases in stress lead to stable extension of the crack as described by Eq. (13.22) and shown by the lower dashed line that originates in region G and extends into region M of Figure 13.16. Extension leads to eventual instability at the crack length and stress conditions of $(c_{\mathrm{m}}^{\mu}, \sigma_{\mathrm{m}}^{\mu})$, Eqs. (13.23) and (13.24), and the crack thence propagates unhindered until component failure. The strength in this region, independent of the specific initial crack length, is $\sigma_{\mathrm{m}}^{\mu}$, as indicated by the horizontal dashed line in region M of Figure 13.16.

Region W: $c > c_{\mathrm{m}}^{\mu}$. In this long crack region, microstructural effects lead to immediate toughening. As the applied stress increases, the crack remains fixed at an initial length until the instability condition of Eq. (13.22) is reached, indicated by the upper dashed line in region W of Figure 13.16, and the crack propagates unhindered until component failure. The toughening in this region is substantial, as indicated by the separation of the upper and lower dashed lines in this region, but increases only weakly with crack length. The strength in this region is dependent on initial crack length and is less than that in the prior region ($\sigma < \sigma_{\mathrm{m}}^{\mu}$), as indicated by the upper dashed line in region W of Figure 13.16.

Overall strength behavior as a function of crack length in the system of Figure 13.16 is indicated by the bold dark line. The predominant feature is strength invariance in the intermediate crack length region. Such invaraince is generated by stabilizing microstructural effects that increase significantly with crack length. Either side of this region are regions of decreasing strength. At short crack lengths, there are no microstructural effects and at long crack lengths microstructural effects approach a steady-state maximum. For both short and long cracks the stabilizing influences of microstructure are absent and component failure is marked by a lack of stable precursor crack extension.

Figure 13.17 replots the overall strength vs crack length behavior of Figure 13.16 in expanded logarithmic coordinates, along with the behaviors of two other systems. The initial system is shown as the bold line and the bridging zone lengths δ are indicated. Consistent with experimental observations (Chantikul et al. 1990; Cook 2015), the behavior of the system with the longer zone length was determined using a slightly expanded toughness range $(T_0, T_\infty) = (0.9, 3.5)$ MPa m$^{1/2}$, and the the system with the shorter zone length was determined using a slightly contracted range, (1.1, 1.5) MPa m$^{1/2}$. For all three systems, the overall strength behavior decreases as $c^{-1/2}$ with perturbation by a central invariant strength plateau. The coupled δ and T effects change the strength behavior from a response weakly perturbed by microstructural effects ($\delta = 50$ μm) to one significantly perturbed by microstructural effects and exhibiting a prominent strength plateau ($\delta = 400$ μm). Such flaw tolerance at intermediate crack lengths was a focus of the earlier studies (Bennison and Lawn 1989; Chantikul et al. 1990; Lawn et al. 1993; Padture et al. 1993; Cook 2015).

In more detail, the effects of a microstructural length scale λ on such strength behavior can be considered through the effects on the steady-state toughness T_∞,

$$T_\infty = T_0 + \mu f^* \delta^{-1/2} \tag{13.30}$$

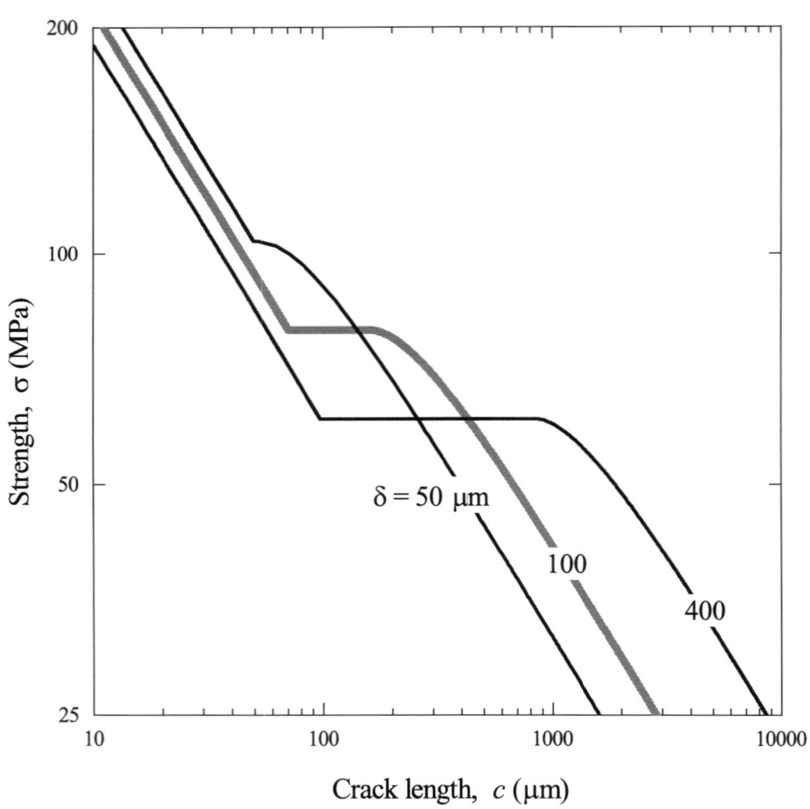

Figure 13.17 Plot of strength as a function of crack lengths for three microstructural parameter sets, with microstructural length scale δ indicated. Larger zones give rise to strength invariance.

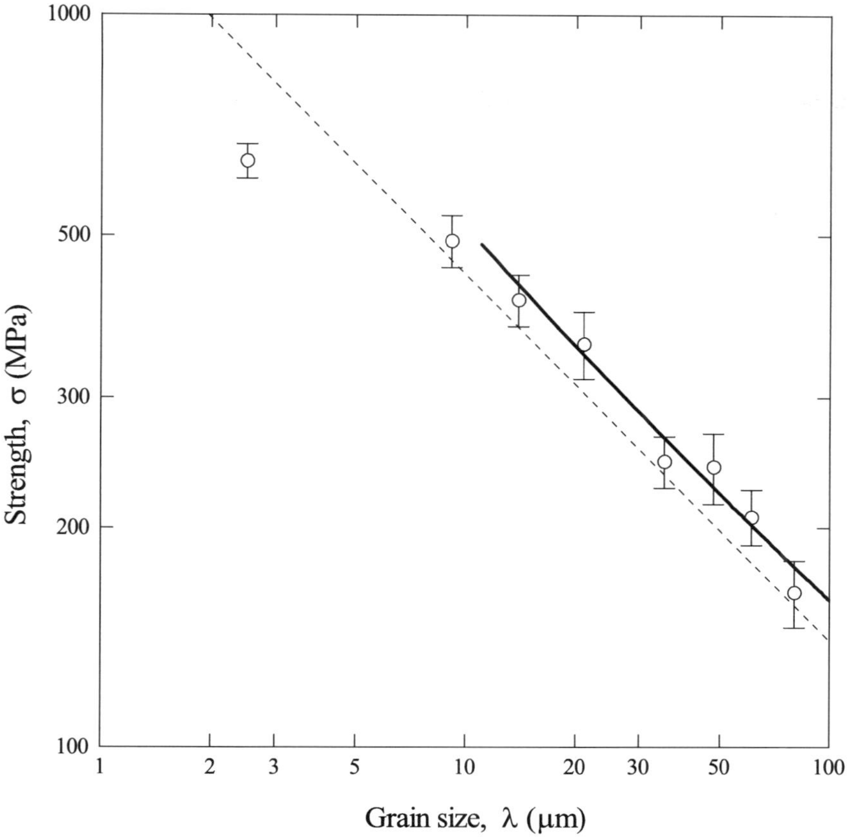

Figure 13.18 Plot of strength σ as a function of grain size λ for polycrystalline alumina. Symbols indicate experimental observations (Adapted from Chantikul, P et al. 1990). Dashed line is a guide to the eye of form $\sigma \sim \lambda^{-0.5}$. Solid line is an empirical fit from a simplified bridging model.

and the microstructurally controlled strength $\sigma_{\mathrm{m}}^{\mu}$,

$$\sigma_{\mathrm{m}}^{\mu} = \frac{3(T_0 + \mu f^* \delta^{-1/2})^{4/3}}{4^{4/3}\psi(\mu f^* \delta)^{1/3}}, \tag{13.31}$$

where Eqs. (13.19), (13.20), and (13.24) have been used. The length scale λ is a grain size dimension in polycrystals (e.g. mean grain size, largest grain, separation of large grains) or a phase size dimension in composites (e.g. second phase size, phase separation distance). Attention here will focus on the grain pullout-based polycrystal toughening depicted in Figure 13.14 and λ will be interpreted as mean grain size. Intuition suggests that the most significant scaling effect is that of zone length and thus $\delta \sim \lambda$ is taken as a basis. Consideration of microstructural effects in the context of the model parameters suggests that (i) f^* should increase with λ, taking into account that larger grains exhibit greater pullout distances and therefore exert greater effective line forces; and (ii) T_0 should decrease with λ, taking into account that strength controlling flaws will be under the influence of crack initiating tensile stresses. Simple empirical dependencies are $f^* \sim \lambda^x$ and $T_0 \sim \lambda^{-y}$ and examination of Eq. (13.30) and Eq. (13.31) shows that the exponents are constrained by $0.5 < x < 1$ and $0 < y < 0.5$. Figure 13.18 is a logarithmic plot of the intrinsic strength of polycrystalline Al_2O_3 from Chantikul et al. (1990), using data derived from the published work. The symbols represent the means and standard deviations of at least four biaxial strength tests at the grain sizes indicated. The dashed line is a guide to the eye of slope $-1/2$. The solid line is an empirical fit to the data using Eq. (13.31), the above scaling, and $x = 0.6$ and $y = 0.3$. The fit describes the data well and the scaling terms are consistent with the above expectations (the scaling is only applicable for increasing T and thus for $\lambda > 10$ μm).

The considerations above of grain-based microstructural effects on strength imply that the effects are well understood and simply modeled within the SIF fracture framework. From the practical viewpoints of design, fabrication, and selection of polycrystalline microstructures for optimized strengths of particles, the guidance is clear. Large-grained microstructures with weak grain boundaries will convey significant flaw tolerance to particles (extended strength plateaus) but at the expense of reduced strengths. The enhanced strengths associated with enhanced long-crack toughness are unlikely to affect common particle mechanical performance. Large rock particles composed of many grains are a possible exception. Small-grained microstructures with greater grain boundary integrity will generate particles with greater maximum strengths but at the expense of reduced flaw tolerance (contracted strength plateaus). The enhanced strengths associated with enhanced short crack toughness are also unlikely to to affect common particle mechanical performance. Small ceramic particles composed of very few grains are a possible exception. The effects of particle size D on microstructurally controlled strength can be approached in several ways. At the simplest level of scaling is the expectation that large particles contain large grains and vice versa (e.g. mineral ore vs grinding media) and thus $\lambda \sim D$. The resulting strength behavior follows Figure 13.18 and thus $\sigma \sim D^{-1/2}$. The next level of scaling is if the parameters f^* and δ are known as a function of D, for example as power laws as used earlier. The resulting strength behavior in this case follows Eq. (13.31) and deviations from Figure 13.18 occur, and perhaps initial crack length dependent behavior as in Figure 13.17. The most comprehensive level of scaling, within the analyses presented here, is to combine Eqs. (13.10) and (13.21) such that an additional crack driving force and the initial crack length depend on particle size. The full SIF based equilibrium expression for such a system is given by

$$K = \mu f^* \delta / c^{3/2} + \chi_D D^2 / c^{3/2} + \psi \sigma_A c^{1/2} = T_\infty. \tag{13.32}$$

It is convenient to define a characteristic particle size D_μ that scales the microstructural effect by

$$D_\mu^2 = \mu f^* \delta / \chi_D. \tag{13.33}$$

Particle strength is then given by simple modification of Eq. (13.24)

$$\sigma_m = 3T_\infty^{4/3} / 4^{4/3} \psi \chi_D^{1/3} (D^2 + D_\mu^2)^{1/3}. \tag{13.34}$$

This is the same analysis used to describe microstructural effects in the strengths of indented extended components (Cook 2015), with particle size parameter D_μ^2 replacing indentation load P. (Similarly, the same analysis was used to describe local tensile stress effects in Chapter 12 by $\chi F / c^{3/2}$.) For an identical strength response of impacted particles, Eq. (13.34) shows a shift to smaller particle sizes in the presence of microstructural effects. As the particle size decreases, initial crack length decreases, toughness decreases, and thus strength decreases. Equation (13.34) represents another variation of Eq. (12.70), $\sigma_m = B'(D^2 + D_\mu^2)^{-1/3}$, and, as discussed, all the analyses of the previous chapters are applicable.

13.6 Discussion

There are two clear features of the analyses of particle strength distributions described in this chapter. The first feature is that the analyses are all readily applied in industrial contexts to many of the particle systems discussed in this book. Energy consumption in particle crushing, fragmentation, and comminution is of great concern in rock mining, ore refinement, and cement and concrete production. Particle reliability is of concern in any application in which time is a design factor, including the useful lives of grinding media, environmental remediation aggregates, railway ballast, and catalyst beds. Size and mass effects are of concern in applications involving handling and transport of particles, such as iron ore pellets and proppant particles. Concerns regarding effects of microstructure on particle strength are pervasive but probably of greatest concern in applications that are almost wholly structural, such as particle reinforcement in refractories, aggregates in concrete, and dispersions in composites.

The second feature is that all the applications and interpretations here are directly testable using experimental strength distributions, as either input parameters (energy consumption and reliability) or output parameters (size and mass and microstructural effects). For example, the energy required to totally fragment a particle relies on experimental observations of strength distributions that show that the mean single fracture energy of a particle varies with particle size D as D^2. In another example, the stress required to fracture particles after constant velocity impact events is predicted to vary as $D^{-2/3}$. From this, if the particle size distribution is known, the strength distribution can be predicted.

The two features reinforce a point made throughout the book: Applications and interpretations of particle strengths would benefit greatly from presentation of particle strength distributions in unbiased formats. Such formats would facilitate analyses of particle strength information as shown in the examples in this chapter. In addition, such formats would enable size scaling, including extreme value, deterministic, and stochastic effects, and material behavior, including heavy tailed flaw distributions, concave strength distributions, and effects of porosity, to be assessed easily. Almost none of the findings on these topics detailed in this book could have been made from the published data. The dominant tendency to publish strength edf $Pr(\sigma)$ data in transformed coordinates—over three quarters of the works cited—significantly obscured strength distribution shapes, magnitudes, and relative behavior, rendering physical insight almost impossible. The historical motivation for this tendency is that transformed coordinates lead to easily fit straight-line behavior of the data if an assumed linearized analytical form for $Pr(\sigma)$ is followed. In many works however, the transformation was implemented to follow historical precedent, regardless of any fit and little use, if any, was made of the parameters describing the assumed form of $Pr(\sigma)$.

The most common transformation was based on the assumption of a two-parameter powered exponential function for the extreme value strength distribution. This function is popularized in the Weibull description as $H(\mu) = 1 - \exp(-\mu^m)$, using the notation of Chapter 3 and setting $\sigma_{\mathrm{th}} = 0$, $\sigma_{\mathrm{u}} = \sigma_0$, and $\mu = \sigma/\sigma_0$, to leave two degrees of freedom, m and σ_0. The analogous assumed form for the strength edf is $Pr(\sigma) = 1 - \exp[-(\sigma/\sigma_0)^m]$. In this form, the assumed edf is sigmoidal and has zero threshold. Inspection of the particle strength data in unbiased form in this book show that these assumptions very rarely describe particle strength observations. Hence, the two-parameter powered exponential description is usually a very poor fit to most particle data. Linearization into the form $y(x) = ax + b$ is achieved by the transformations $y = \ln[\ln(1/(1 - Pr(\sigma)))]$ and $x = \ln\sigma$, such that $m = a$ and $\sigma_0 = \exp(b/a)$, and these are the parameters commonly cited. However, the data in transformed coordinates are invariably still clearly concave, as noted in the earlier review of large specimen measurements by Shih (1980), and as may be seen in the historical progression of original particle works cited here (e.g. Brecker 1974; Wong et al. 1987; Huang et al. 1995; McDowell 2002; Wang et al. 2015). Comparison between studies is often impeded by the use of base 10 rather than natural logarithms and various units for σ included or not in the logarithm. Of greatest impediment, however, is that linearizing this assumed form requires multiple, contracting, logarithmic transformations that completely obscure the real shape of edfs, obscure the relative widths of edfs, and most crucially, obscure the relative threshold positions of edfs. Interpretation is often made more difficult by the tendency to implement contracted graph axes, rendering the data almost vertical and falsely enhancing the linearity, e.g. in the strength measurements of ceramic tubes by Shelleman et al. (1991) and of cellulose compacts by Keleş et al. (2015). Less common transformations, largely applied to specific failure energy data of particles, include that based on the normal distribution and error function, $H(\mu) = [1 + \mathrm{erf}(\mu)]/2$, (King and Bourgeois 1993; Tavares and King 1998; Tavares and das Neves 2008) and a range of other transformations similarly based on assumed sigmoidal behavior, such as the Gumbel, $H(\mu) = \exp[-\exp(-\mu)]$, logistic, $H(\mu) = 1/[1 + \exp(-\mu)]$, and Gamma distributions (Gustafsson et al. 2013; Cavalcanti and Tavares 2018; Tavares et al. 2018). Non-linear perturbations include $\mu \to \ln\mu$ and $\mu \to \mu^p$, noting that the last form was used to perturb the incomplete Beta function in recent works (Cook and DelRio 2019a, 2019b; Cook et al. 2019; Cook 2020).

There are three major reasons to believe that emphasis on such functions in describing strength distributions is misplaced. First, empirical. As demonstrated here and elsewhere (Cook and DelRio 2019a, 2019b), the use of unbiased strength edf coordinates provides insights regarding strength distribution shapes and thresholds that are obscured in transformed coordinates. Second, numerical. The application of modern-day desktop computers has rendered the need for linearized $Pr(\sigma)$ forms obsolete; smoothing of experimental strength data is easily accomplished. Third, the fitting of common probability distributions, e.g. powered exponential, Gumbel, or normal, to strength measurements suggests a fundamental physical interpretation that is unlikely to be true. As noted throughout this book, the fundamental physical characteristic controlling failure of particles is the flaw population. Stochastic and deterministic size effects lead to ensembles of flaws in individual particles. Extreme value effects lead to distributions of maximally potent flaws in groups of particles. Strength measurements are simply probes of these extreme flaws. Given the non-linear relationship between strength and flaw size (usually the Griffith equation) and the non-linear nature of extreme-value size effects, it is extremely unlikely that strength distributions are simple or easily interpretable. The complexities are outlined in Chapter 3 and in consideration of a simple quadratic flaw population in Cook and DelRio (2019a).

As noted in Chapter 1, distribution parameters for strength and strength-related properties (e.g. specific failure energy, King and Bourgeois 1993, failure velocity, Salman et al. 2002) are of great importance for particle applications in engineering, e.g. estimation of energy needs in mining or concrete manufacture (Tavares 2004; Tromans 2008; Schneider et al. 2011). Hence, there is a practical need for succinct description of behavior, e.g. by the normal distribution. However, the observations reviewed here suggest that even strength distributions of particles, which might be physically small enough and tested in large-enough numbers, very rarely provide direct insight into flaw populations by evading extreme value considerations. Hence, interpreting a particular functional form for a strength distribution in physical terms (e.g. Fischer et al. 2002; Lu et al. 2002; R'Mili et al. 2012; Weiss et al. 2014; Taloni et al. 2018) is likely to be inconclusive. (Especially if data are described by experimental truncation of an extended function rather than by complete application of a bounded function.) Of potentially greater importance is that interpretation of the powered exponential form is likely to be misleading.

As emphasized here and earlier, in works considering ceramics (Shih 1980), hazard (Todinov 2010), fibers (Zok 2017), and MEMS (Cook and DelRio 2019a), the concept of flaw independence, equivalent to weakest-link behavior, is completely separate from any consideration of any form of the flaw population. Historically, however, the two concepts have become conflated, as can be observed in many of the works cited here, such that weakest-link behavior is now often regarded as the basis of Weibull "theory", e.g. Brzesowsky et al. (2011). As a consequence, observations of apparent Weibull curves using transformed coordinates are taken as implying underlying weakest link behavior, when in fact no such implication can be drawn. The engineering and physics consequences are that stochastic size effect scaling based on a light-tailed flaw population are erroneously assumed. This book shows that both assumptions for particles are likely to be in error. Reservations regarding the applicability of the Weibull description are not new: as noted by Gorski (1968) regarding reliability, "nobody has been able to present a sufficient argument for the Weibull case."

References

Bazovsky, I. (2004) *Reliability Theory and Practice.* Dover.

Bennison, S.J. and Lawn, B.R. (1989). Flaw tolerance in ceramics with rising crack resistance characteristics. *Journal of Materials Science* 24: 3169–3175.

Brecker, J.N. (1974). The fracture strength of abrasive grains. *Journal of Engineering for Industry* 96: 1253–1257.

Brzesowsky, R.H., Spiers, C.J., Peach, C.J., and Hangx, S.J.T. (2011). Failure behavior of single sand grains: Theory versus experiment. *Journal of Geophysical Research* 116: B06205.

Cavalcanti, P.P. and Tavares, L.M. (2018). Statistical analysis of fracture characteristics of industrial iron ore pellets. *Powder Technology* 325: 659–668.

Chantikul, P., Bennison, S.J., and Lawn, B.R. (1990). Role of grain size in the strength and r-curve properties of alumina. *Journal of the American Ceramic Society* 73: 2419–2427.

Cook, R.F. (2015). Multi-scale effects in the strength of ceramics. *Journal of the American Ceramic Society* 98: 2933–2947.

Cook, R.F. (2018). Long-term ceramic reliability analysis including the crack-velocity threshold and the "bathtub" curve. *Journal of the American Ceramic Society* 101: 5732–5744.

Cook, R.F. (2019). Fracture sequences during elastic-plastic indentation of brittle materials. *Journal of Materials Research* 34: 1633–1644.

Cook, R.F. (2020). Single particle strength distributions: Heavy tails and extreme values. http://doi.org/10.5281/zenodo.4024618 (accessed July 4, 2021).

Cook, R.F. and DelRio, F.W. (2019a). Material flaw populations and component strength distributions in the context of the Weibull function. *Experimental Mechanics* 59: 279–293.

Cook, R.F. and DelRio, F.W. (2019b). Determination of ceramic flaw populations from component strengths. *Journal of the American Ceramic Society* 102: 4794–4808. (typographical error in Eq. (9)).

Cook, R.F., DelRio, F.W., and Boyce, B. L. (2019). Predicting strength distributions of MEMS structures using flaw size and spatial density. *Microsystems & Nanoengineering* 5: 1–12.

Forbes, C., Evans, M., Hastings, N., and Peacock, B. (2011). *Statistical Distributions, 4th ed.* Wiley.

Gaither, M.S., Gates, R.S. Kirkpatrick, R., Cook, R.F., and DelRio, F.W. (2013). Etching process effects on surface structure, fracture strength, and reliability of single-crystal silicon theta-like specimens. *Journal of Microelectromechanical Systems* 22: 589–602.

Gorski, A.C. (1968). Beware of the Weibull euphoria. *IEEE Transactions on Reliability* 17: 202–203.

Gustafsson, G., Häggblad, H.-Å., and Jonsén, P. (2013). Characterization modelling and validation of a two-point loaded iron ore pellet. *Powder Technology* 235: 126–135.

Huang, H., Zhu, X.H., Huang, Q.K., and Hu, X.Z. (1995). Weibull strength distributions and fracture characteristics of abrasive materials. *Engineering Fracture Mechanics* 52: 15–24.

Jung, Y.G., Pajares, A., and Lawn, B.R. (2004a). Effect of oxide and nitride films on strength of silicon: A study using controlled small-scale flaws. *Journal of Materials Research* 19: 3569–3575.

Jung, Y.G., Pajares, A., Banerjee, R., and Lawn, B.R. (2004b). Strength of silicon, sapphire and glass in the subthreshold flaw region. *Acta Materialia* 52: 3459–3466.

Keleş, Ö., Barcenas, N.P., Sprys, D.H., and Bowman, K.J. (2015). Effect of porosity on strength distribution of microcrystalline cellulose. *Pharmaceutical Science and Technology* 16: 1455–1464.

King, R.P. and Bourgeois, F. (1993). Measurement of fracture energy during single-particle fracture. *Minerals Engineering* 6: 353–367.

Kingery, W.D., Bowen, H.K., and Uhlmann, D.R. (1975). *Introduction to Ceramics*. Wiley.

Kusters, K.A., Pratsinis, S.E., Thoma, S.G., and Smith, D.M. (1994). Energy–size reduction laws for ultrasonic fragmentation. *Powder Technology* 80: 253–263.

Lawn, B.R. (1993). *Fracture of Brittle Solids, 2nd ed.* Cambridge.

Lawn, B.R., Padture, N.P., Braun, L.M., and Bennison, S.J. (1993). Model for toughness curves in two-phase ceramics: I, Basic fracture mechanics. *Journal of the American Ceramic Society* 76: 2235–2240.

Lu, C., Danzer, R., and Fischer, F.D. (2002). Fracture statistics of brittle materials: Weibull or normal distribution. *Physical Review E* 65: 067102.

Marshall, D.B., Evans, A.G., and Nisenholz, Z., (1983). Measurement of dynamic hardness by controlled sharp-projectile impact. *Journal of the American Ceramic Society* 66: 580–585.

McDowell, G.R. (2002). On the yielding and plastic compression of sand. *Soils and Foundations* 42: 139–145.

Oka, Y. and Majima, H. (1970). A theory of size reduction involving fracture mechanics. *Canadian Metallurgical Quarterly* 9: 429–439.

Padture, N.P., Runyan, J.L., Bennison, S.J., Braun, L.M., and Lawn, B.R. (1993). Model for toughness curves in two-phase ceramics: II, Microstructural variables. *Journal of the American Ceramic Society* 76: 2241–2247.

R'Mili, M., Godin, N., and Lamon, J. (2012). Flaw strength distributions and statistical parameters for ceramic fibers: The normal distribution. *Physical Review E* 85: 051106.

Salman, A.D., Biggs, C.A., Fu, J., Angyal, I., Szabó, M., and Hounslow, M.J. (2002). An experimental investigation of particle fragmentation using single particle impact studies. *Powder Technology* 128: 36–46.

Schneider, M., Romer, M., Tschudin, M., and Bolio, H. (2011). Sustainable cement production–present and future. *Cement and Concrete Research* 41: 642–650.

Shelleman, D.L., Jadaan, O.M., Conway Jr, J.C., and Mecholsky Jr, J.J. (1991). Prediction of the strength of ceramic tubular components: Part II—Experimental verification. *Journal of Testing and Evaluation* 19: 192–200.

Shih, T.T. (1980). An evaluation of the probabilistic approach to brittle design. *Engineering Fracture Mechanics* 13: 257–271.

Taloni, A., Vodret. M., Costantini, G. and Zapperi, S. (2018). Size effects on the fracture of microscale and nanoscale materials. *Nature Reviews | Materials* 3: 211–224.

Tanaka, T. (1966). Comminution laws. Several probabilities. *Industrial & Engineering Chemistry Process Design and Development* 5: 353–358.

Tavares, L.M. (2004). Optimum routes for particle breakage by impact. *Powder Technology* 142: 81–91.

Tavares, L.M., Cavalcanti, P.P., de Carvalho, R.M., da Silveira, M.W., Bianchi, M., and Otaviano, M. (2018). Fracture probability and fragment size distribution of fired iron ore pellets by impact. *Powder Technology* 336: 546–554.

Tavares, L.M. and das Neves, P.B. (2008). Microstructure of quarry rocks and relationships to particle breakage and crushing. *International Journal of Mineral Processing* 87: 28–41.

Tavares, L.M. and King, R.P. (1998). Single-particle fracture under impact loading. *International Journal of Mineral Processing* 54: 1–28.

Todinov, M.T. (2010). The cumulative stress hazard density as an alternative to the Weibull model. *International Journal of Solids and Structures* 47: 3286–3296.

Tromans, D. (2008). Mineral comminution: Energy efficiency considerations. *Minerals Engineering* 21: 613–620.

Wang, Y., Dan, W., Xu, Y., and Xi, Y. (2015). Fractal and morphological characteristics of single marble particle crushing in uniaxial compression tests. *Advances in Materials Science and Engineering* 2015: 537692.

Walpole, R.E. and Myers, R.H. (1972). *Probability and Statistics for Engineers and Scientists.* Macmillan.

Weiss, J., Girard, L., Gimbert, F., Amitranod, D., and Vandembroucq, D. (2014). (Finite) statistical size effects on compressive strength. *Proceedings of the National Academy of Sciences* 111: 6231–6236.

Wong, J.Y., Laurich-McIntyre, S.E., Khaund, A.K., and Bradt, R.C. (1987). Strengths of green and fired spherical aluminosilicate aggregates. *Journal of the American Ceramic Society* 70: 785–791.

Zok, F.W. (2017). On weakest link theory and Weibull statistics. *Journal of the American Ceramic Society* 100: 1265–1268.

Index

Page numbers referring to figures are indicated in italics.

Particle Strengths: Extreme Value Distributions in Fracture, First Edition. Robert F. Cook.